Electric Power Distribution

Electric Power Distribution

A S Pabla
Consulting Engineer
and
Former Chief Engineer
Punjab State Electricity Board

McGraw-Hill
New York Chicago San Francisco Lisbon London
Madrid Mexico City Milan New Delhi San Juan
Seoul Singapore Sydney Toronto

The McGraw-Hill Companies

Cataloging-in-Publication Data is on file with the Library of Congress

4 5 6 7 8 9 10 QVR/QVR 1 9 8 7 6 5 4 3 2

ISBN 0-07-144783-0

The sponsoring editor for this book was Stephen S. Chapman and the production supervisor was Sherri Souffrance. The art director for the cover was Anthony Landi.

Printed and bound by IBT Global.

This book was previously published by Tata McGraw-Hill Publishing Company Limited, New Delhi, India, copyright © 2004, 1997, 1992, 1989, 1981.

To
My wife Surinder,
My daughter Kirandeep,
and
The memory of my parents

Author's Profile

A S Pabla is presentely working as a consultant. He was till recently Chief Engineer with the Punjab State Electricity Board. He began his career with Punjab State Electricity Board in 1964 and has a rich experience in diverse areas including distribution systems, research and development, rural electrification, substation and transmission designs, power planning and development, tarrifs and billing, thermal power plant operation and maintenance, hydro-power plant design, project and monitoring, and material management.

Mr. Pabla obtained his BSc Engineering (Electrical) degree from Panjab University and M Eng Sc (Electrical Engineering) from University of New South Wales, Australia. He has contributed 41 technical papers in various journals and has received many awards for some of his contributions. He was honoured by the Institution of Engineers (India) for his outstanding professional contribution in the field of power systems. A Fellow of Institution of Engineers (India), he has also authored a book titled *Electrical Power Systems Planning.*

Contents

Preface

This book will serve as a valuable desk reference for engineers, regulatory commissions, and consultants and enable them to discover better solutions to current electrical power challenges. The book will also be helpful to students. To promote active learning, practical problems appear at the end of each chapter. As a teaching aid, a summary of each chapter, brief solutions to the numerical problems and hints to other problems, and the data of the Indian power distribution system can be downloaded from the website

www.tatamcgraw.com/digital_solutions/aspabla

In the preparation of this book I gratefully acknowledge the assistance given by A. B. Chance Company, USA, which authorized the use of the photograph of the product line 'fault detector'; Mr. Patrick McMullan from Electricity Supply Association of Australia, who supplied me a CD-ROM containing the proceedings of the Distribution 2000 Conference for reference; and my daughter Rujbir Kaur Pabla, a System Analyst with Lucent Technologies, Sydney, who provided valuable assistance in preparing the material on electronic billing. I also acknowledge, with deep gratitude, the help received from my friend Mr. I. S. Anand, New Delhi.

I welcome comments from readers, who can contact me at aspabla02@hotmail.com.

A S Pabla

Chapter 1

Power System

1.1 General Concept

Electricity is modern society's most convenient and useful form of energy. Without it, the present social infra-structure would not at all be feasible. The increasing per capita consumption of electricity throughout the world reflects a growing standard of living of the people. The optimum utilization by society of this form of energy can be ensured by an effective distribution system.

Figure 1.1 shows the typical parts of the electrical power system in a schematic single line diagram. The typical working voltage in the various sections of the system are as follows [1, 8]:

(a) Generation of ac (hydro, thermal, nuclear, diesel, non-conventional sources etc.) 0.4/0.44, 6.6, 10.5, 11, 13.8, 15.75, 21 and 33 kV.

(b) (i) Tieline (interconnecting line between two power systems/areas) and transmission: 220, 400, 500, 750, 765, 800 kV, etc.
(ii) High voltage dc transmission: 500, 800, 1400 kV, etc. for stable power flows over long distances beyond 800–1600 km.

(c) High voltage subtransmission: 33, 66, 110, 132, 220 kV, etc.

(d) High voltage primary distribution: 3.3, 6.6, 11, 22, 33, 66, 132 or 220 kV. 11 and 22 kV is the main primary distribution voltage used in India. The higher voltages are used for large consumers.

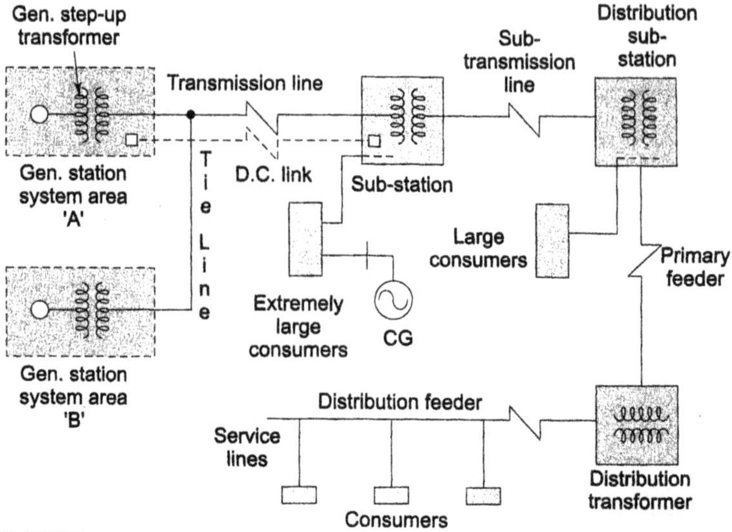

Fig. 1.1 *Schematic diagram of power system; CG = Captive generation*

(e) Low-voltage (secondary) distribution, ac, 415/240 V and 433/250 V, three-phase four-wire; 240/250 V single phase to neutral; 440/220 V (from HT SWER).

(f) Declared voltage for utilisation is 400/230 V +/–6%.

(g) The consumer load may be of the following devices (individual or any combination):
- Motors
- Heating
- Chemical equipment, e.g. electroplating, battery charging
- Electronic equipment
- Lighting

In practice, a power system is complicated. It has a number of power stations of different types, interconnected by a system of tielines, transmission lines, subtransmission lines and distribution networks to supply different types of loads to various consumers. The distribution system is that part of an electric power system which is dedicated to delivering electric energy to the end user. In December 2001, there were 105 million consumers—70 million domestic, 12.5 million commercial, 13.1 million agricultural with industry and others making up the rest. Annual (2000–01) per capita electricity consumption is about 380 kWh. The installed capacity of the power system in the country is given in Table 1.1.

Table 1.1

Power system in India

	MW (Ending July 2001)	Energy generated TWh (2000-01)
1 *Generation*		
Coal thermal	61025.88	
Gas thermal	10737.40	
Nuclear	2860	
Hydro	25289.53	
Diesel	870.89	
Non-conventional: (wind; micro-hydel; bio-mass; photovoltaic (PV))	(1339.96; 1341; 308; 58)	
Total	103830.7	510
2 Sub-stations (numbers)		
400 kV	315	
220 kV	1285	
110/132 kV	~ 4500	
33/66 kV	~ 9800	
22/11 kV	~ 2.20 millions	
3 Lines (Ckm)		
HVDC (+/-500 kV)	3420	
800 kV	1155	
400 kV	48313	
220 kV	95204	
110/132 kV	~115000	
33/66 kV	~265000	
22/11 kV	~1980000	
LT (240/415 V)	~4550000	
4 Average load density	1.97 kW/km^2	
5 Villages electrified (numbers)	508159 (86.53%)	

There is great diversity in the country in terms of topography, resources, daily peak time due to differences in patterns of annual peak load timings (winter or summer). Hence, the country has been divided into five regions for economic planning of power resources and economical exchange of energy between regions. The five regions are:

- Northern Region
- Western Region
- Southern Region
- Eastern Region
- North-Eastern Region

Islands (Andaman, Nicobar etc.) are separate from the above regions.

The distribution system immediately affects the consumer. About 75% of the total system losses occur during distribution. More than 90% of system faults are in the distribution area. Official recorded T & D losses are 22–23%. There is manipulation in estimating the unmetered agriculture consumption on higher side in the official statistics which makes actual losses higher by about 15%. Realistically, distribution losses should be 9%. Higher distribution losses are due to inadequate investment in the distribution system. The system is overloaded and there is lack of reactive support. Nearly 95% of all system network length lies in distribution. The efficiency of the distribution system including the consumer end is around 20%, extremely poor by any standard. Latest data on the Indian power system may be assessed from the website: indianelectricity.com/.

1.2 Distribution of Power

The distribution system is part of the system between transmission and the consumer service point. It contains:

- Sub-transmission circuits in voltage ratings, usually between 33 kV and 220 kV, which deliver energy to distribution substations;
- The distribution substation which converts the energy to lower primary system voltage for local distribution and usually improves facilities for voltage regulation of the primary voltage;
- Primary circuits of feeders usually operating in the range of 11 kV to 33 kV supplying the load in a well-defined geographical area;
- Distribution transformers usually installed on poles or on pads or near the consumers' sites which transform the primary voltage to the secondary voltage, usually 240/415 V;
- Secondary circuits at service voltage which carry energy from the distribution transformers along the streets etc.
- Service lines which deliver the energy from secondary circuits to the consumer premises at declared voltage of 400/230 V ± 6%.

For ac working, the standard frequency in India and many other countries is 50 Hz. Some special voltage such as 25 kV is used for different purpose, e.g. traction. Standard voltages, frequencies and their range of variation in distribution systems in India have been specified in IS: 12360-1988. The frequencies and distribution supply voltages prevalent in various countries of the world are given in Appendix I.

High voltage distribution may be a *three-phase*, *single-phase* or *single wire earth return* (SWER) system. In areas which are reticulated by three-phase and single-phase systems, there is little attraction in using a SWER at the expense of increase in cost in providing spares and training to the staff in operation and maintenance of an unfamiliar system. In large rural areas the SWER system has been found to be successful and does not pose any serious problem. It is more reliable and the expenditure incurred is also considerably reduced. This system may be applied as a means to meet basic energy needs and promoting productivity in sparsely populated and isolated villages where the loading is scattered and small and not likely to increase substantially in 5–10 years time.

Low voltage distribution can be single-phase, three-phase three-wire, three-phase four-wire, or single-phase three-wire from HT SWER system. Typical distribution systems are shown in Fig. 1.2.

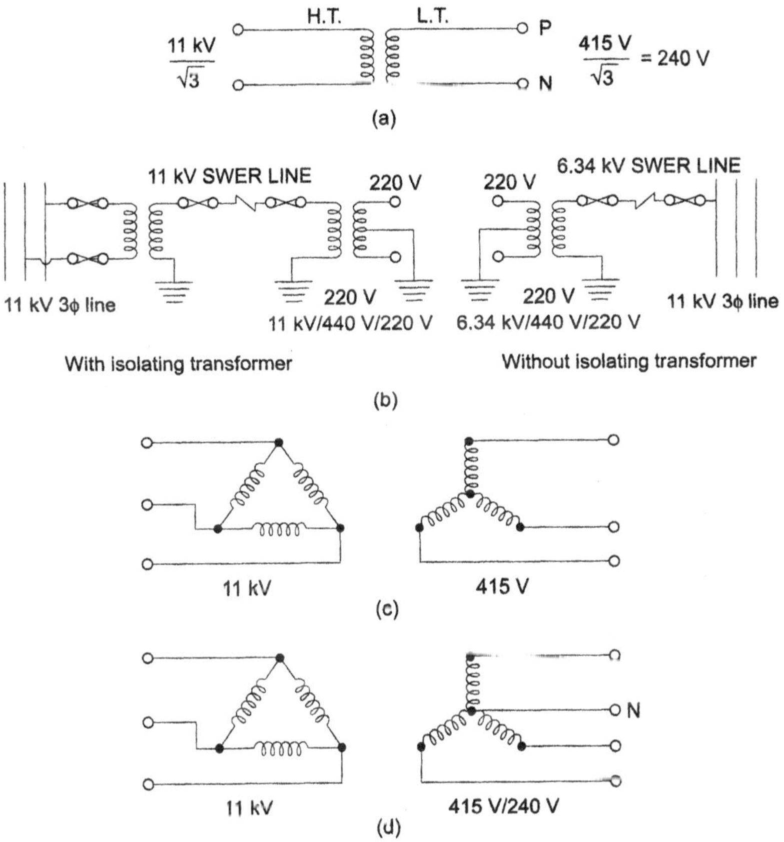

Fig. 1.2 *Distribution systems: (a) Single-phase; (b) SWER; (c) Three-phase three-wire (d) Three-phase four-wire*

1.2.1 Basic Distribution Systems

The six basic distribution systems used by utilities (shown in Fig. 1.3) are as given below:

(i) Radial

A radial system is connected to only one source of supply (see Fig. 1.3a). It is exposed to many interruption possibilities—the most important of which are those due to overhead line or underground cable failure or transformer failure. Each event may be accomplished by a long interruption. It has lower reliability. Both components (feeder and transformer) have finite failure rates and such interruptions are expected and statistically predictable. Feeder breaker reclosing or temporary faults are likely to affect sensitive loads. This system is suitable for small loads.

(ii) Primary Loop

A great improvement over a radial system is obtained by arranging a primary loop, which provides power from two feeders (see Fig. 1.3b). This is also called **open ring system**. Power flow to the consumer is by way of a single path at any one time from either side of the loop, depending upon the open/close status of sectionalizers and reclosers. The loop is normally operated with the sectionalizer switch open. Any section of the feeder can be isolated without interruption, and primary faults are reduced in duration to the time required to locate a fault and do the necessary switching to restore service. Each line of the loop must have sufficient capacity of carry all the load. The additional line exposure tends to increase the frequency of faults, but not necessarily the faults per consumer. Sensitive loads are affected by reclosing under temporary fault conditions.

(iii) Primary Selective

It uses the same basic components as in the primary loop. Each transformer can have supply from two sources (see Fig. 1.3c). High voltage automatic switching is provided ahead of the consumer's transformer. In the event of loss of feeder, transfer to the second feeder is automatic and the interruption duration can be limited to two or three seconds. Each service now represents a potential two feeder outage if the open switch

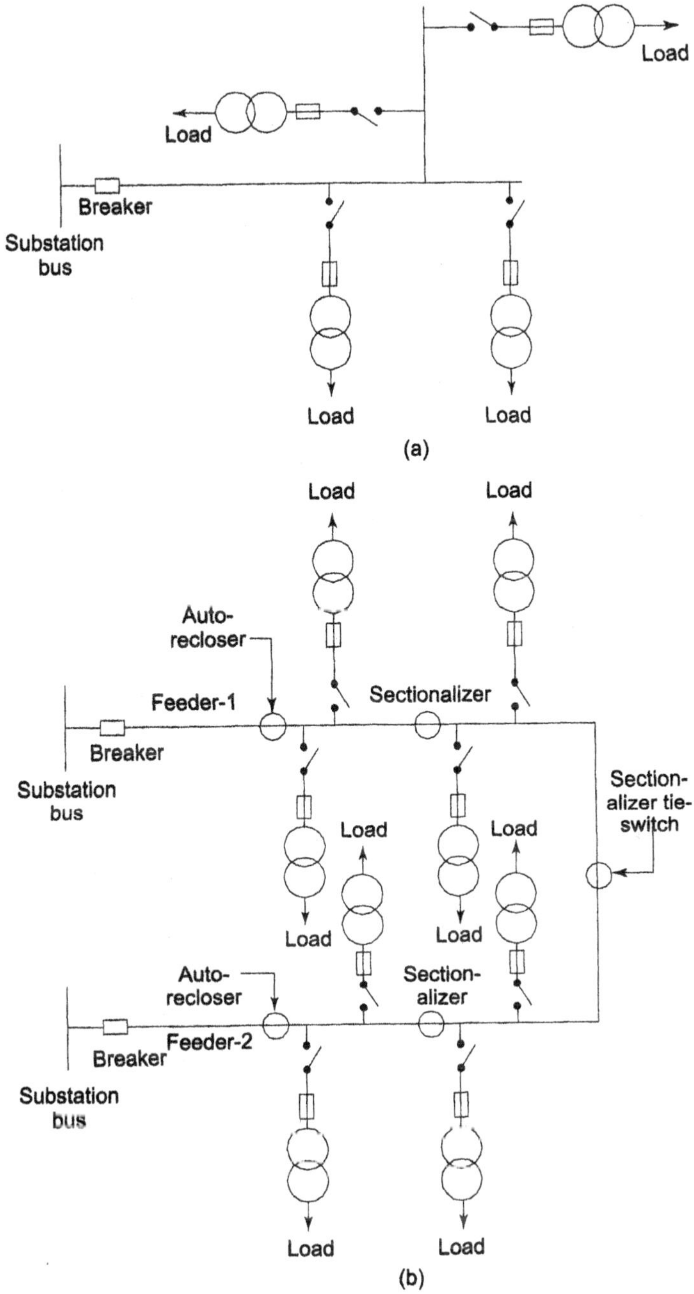

Fig. 1.3 *Basic distribution systems: (a) Radial (b) Primary loop*

fails, but under normal contingencies, service restoration is rapid and there is no need to locate the fault, as with the loop switching system, reliability is high. It also offers little advantage to sensitive load-like computer problems caused by temporary faults. This scheme is normally used for large, essential or continuous process industrial consumers.

(iv) Secondary Selective

This system uses the two transformers, each from a separate primary feeder and with low voltage switching (see Fig. 1.3d). The load is generally divided between two LT buses, with both transformers continuously energized. The tie switch on the secondary tie bus is normally open and is interlocked with the secondary feeder's switches. This system is commonly used for industrial plants and institutions like hospitals. Primary operational switching is eliminated. Duplicate transformers virtually eliminate the possibility of a long interruption due to failure. Each transformer and feeder must have sufficient capacity to supply the entire load. Transfer is automatic upon loss of voltage in either feeder with static switching equipment. Sensitive equipment can be effectively served. Reliability is better than in the primary selective system because of the additional redundancy of transformers. This is also called Open Ring Main System.

(v) Spot Network

Maximum service reliability and operating flexibility for most loads are obtained by use of the network, using two or more transformer units in parallel. It is similar to the closed ring system (see Fig. 6.3). The low voltage bus (spot network bus) is continuously energized by all units operating in parallel. Automatic disconnection of any unit is obtained by sensitive reverse (directional) power relays in the protection unit (see Fig. 1.3e). If one feeder develops a fault, it is isolated by the protection on that feeder.

Switching for the maintenance of the primary feeder can be done without consumer interruption. Spot networks are generally used in metropolitan or high load density areas for large continuous process industries and essential services load such as water works etc. A spot network, because of its multiplicity of feeders and transformers, is highly reliable. Momentary and long duration outages are almost non-existent.

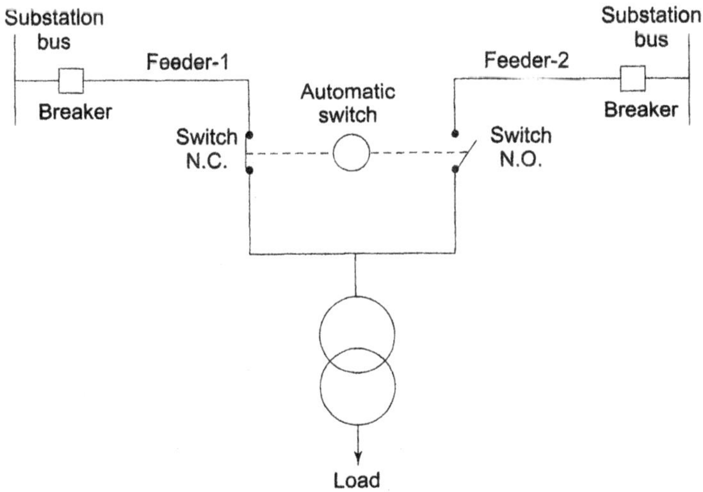

(c)

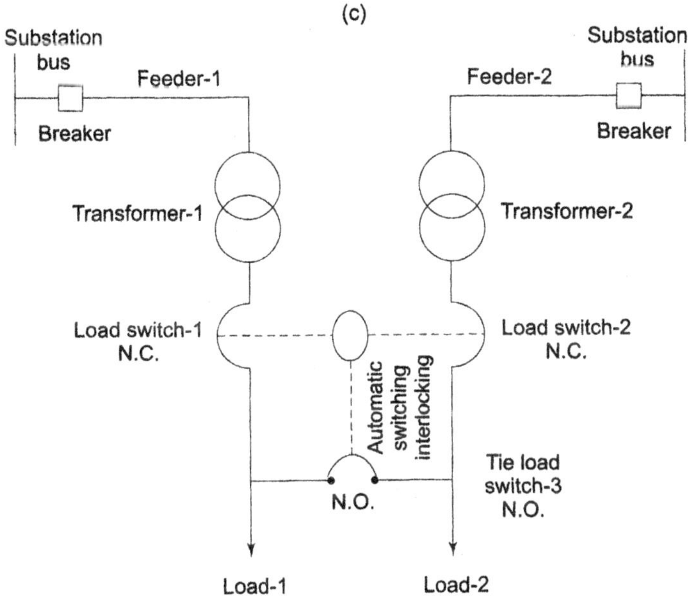

(d)

Fig. 1.3 *Basic distribution systems: (c) Primary selective (d) Secondary selective*

(vi) Grid Network

Grid networks provide maximum reliability and operating flexibility. These networks are the most economical and effective method in serving

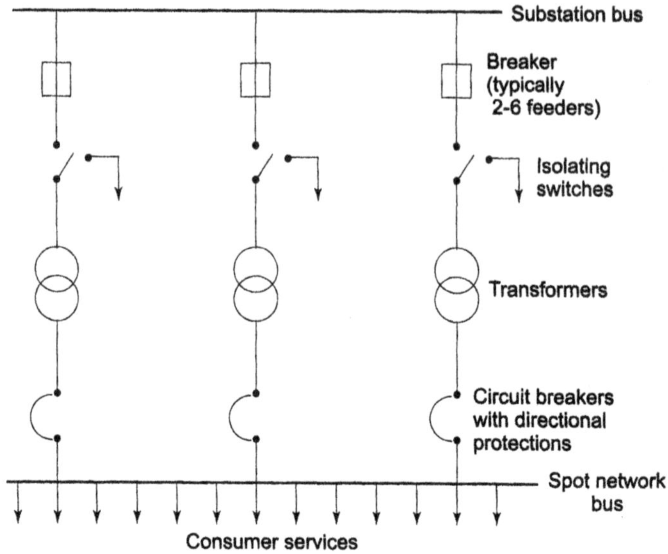

(e)

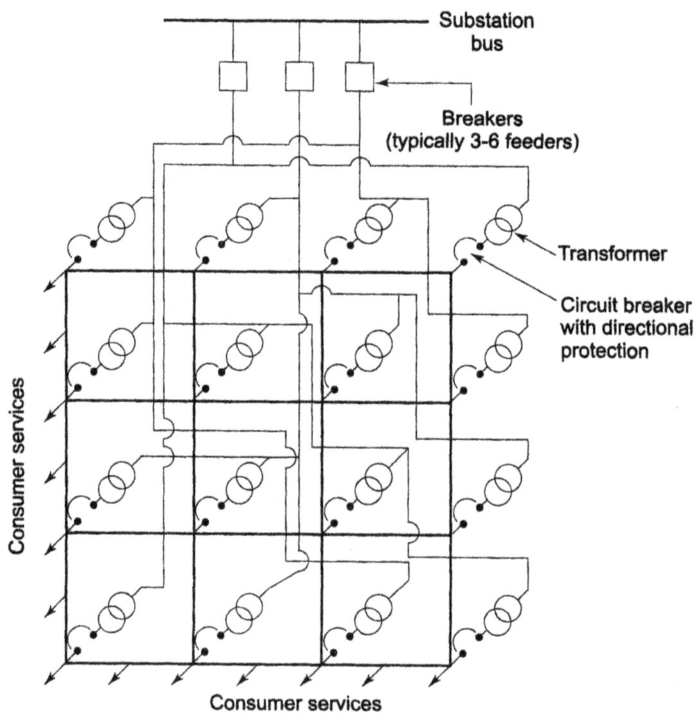

(f)

Fig. 1.3 *Basic distribution systems: (e) Spot network (f) Grid network*

the high density loads in metropolitan cities. The grid is simultaneously supplied from several feeders. In the grid network, no consumer outage is caused by switching off the primary feeder for scheduled maintenance. Voltage regulation is improved since power flow to the consumer is through several transformers operating in parallel (see Fig. 1.3f). The grid can handle abrupt load changes and disturbances associated with large motor starting, without severe voltage dips or surges. A strong grid network is sufficiently stiff and a fault in one unit does not disrupt voltage outside the sensitive load tolerance limits.

1.2.2 Systems Comparison

Up to 50% of the capital investment in supply industry is made in distribution system up to subtransmission level. The effectiveness with which a distribution system fulfills its function is measured in terms of voltage regulation, service continuity, reliability and cost. Other characteristics such as frequency and wave shape regulation is important due to increasingly wide use of high technology in various fields. Table 1.2 gives a general comparison in respect of Urban and Rural systems.

A comparative cost, voltage drop, losses of the various systems [3] as shown in Fig. 1.4 (a to h) is given in Table 1.3 on the basis of single-phase two-wire circuit as unit and for the same balanced loads, the same voltage between conductors, and the same conductor size.

Table 1.2

Distribution system—urban and rural

Item	*Urban*	*Rural*
Load density	Higher	Lower
System type	Underground or overhead	Generally overhead
Reliability criteria	Stringent	Moderate
Network:		
(a) Secondary (LT)	(a) 5 or 6 wire	(a) 2 to 4 wire
(b) Primary (HT)	(b) Feeder inter-connected or parallel or loop or ring main or radial	(b) Generally radial

Contd.

Item	Urban	Rural
HT/LT line ratio	2 to 1	0.4 to 1
HT line/transformer	0.5–0.9 km	1–1.5 km
Line cost/km	Higher	Lower
System capital cost per kW	Low	High
Losses %	Lower	Higher
Power factor	Better	Poor
Average distribution transformer size	Higher	Lower
Fault level	Higher	Lower
Renewable energy development scope	Municipal waste based small thermal power station	(i) Agro waste small thermal power station (ii) Micro hydel power station (iii) Wind power (iv) Solar power photovoltaic

Predominantly, the single-phase system is used for small loads and scattered load in rural area. Two-phase system is now obsolete. Three-phase system is widely used for power distribution. Six-phase (6-wire) system is used for supply to rectifiers or synchronous rectifiers and six-phase (7-wire) can be used for power distribution.

1.3 Quality of Supply

System should operate at normal frequency and voltage to maintain quality of supply. In the case of sensitive loads like computers, clocks, VCRs, etc. (see Sec. 2.2.5), power quality can be defined as the relative absence of utility-related voltage variations—particularly, the absence of outages, sags, surges and harmonics as measured at the point of service (see Table 7.3). Generally, in adverse cases, the system tends to operate on low voltage and low frequency.

On low voltage, different types of load respond differently to voltage variation. For every one per cent drop in voltage, approximate wattage reduction will be 1.6% in filament lamps, 1.4% in fluorescent lamps and 2% in resistance loads. Induction motors draw more current, result-

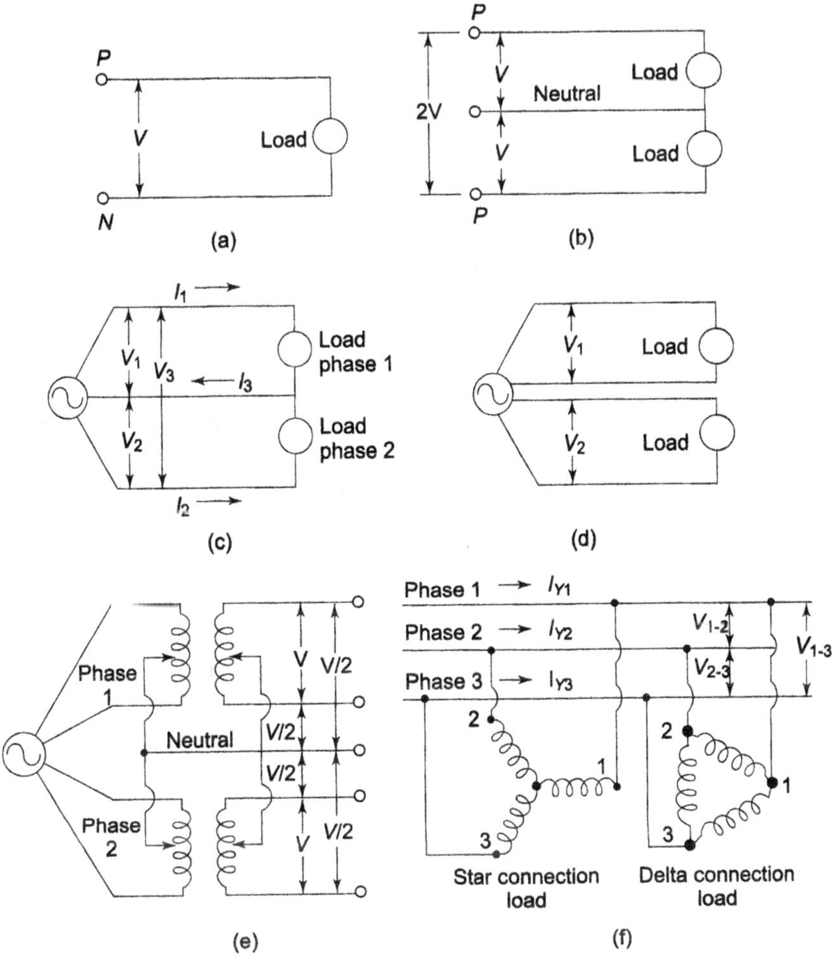

Fig.1.4 *AC supply systems: (a) 1 – ϕ, 2 wire (b) 1 – ϕ, 3 wire (c) 2 – ϕ, 3 wire (d) 2 – ϕ, 4 wire (e) 2 – ϕ, 5 wire (f) 3 – ϕ, 3 wire*

ing in overheating and reduction in the life of the motors. Insulation of motor winding may be weakened due to overheating, which may cause short circuit and burning of motors in due course. The line losses will be high due to increase in current. This will further increase the voltage drop in the line.

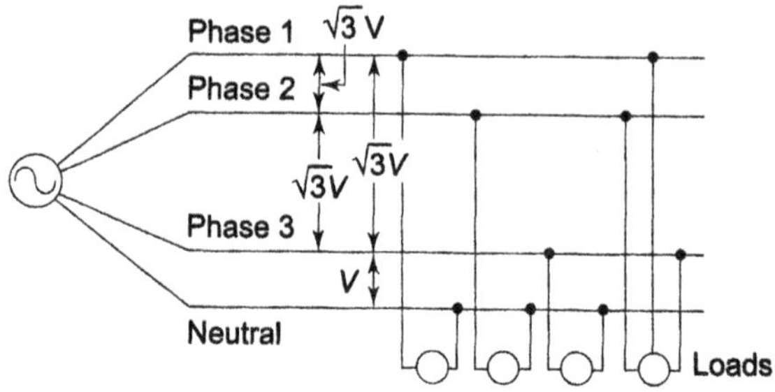

(g)

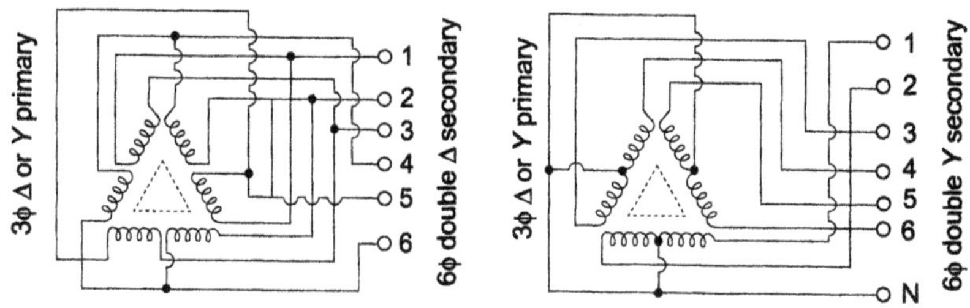

(h)

Fig. 1.4 *AC supply systems: (g) 3 – ϕ, 4 wire (h) 6 – ϕ, wire; 6 – ϕ, 7 wire*

Table 1.3

System comparison

System type		*Amount of conductor*	*Power loss voltage drop (approximate) each*
Single-phase	2-wire	1.00	1.00
	3-wire	1.5	0.25
Two-phase	3-wire	1.5	0.50
	4-wire	2.0	0.25
	5-wire	2.5	0.25
Three-phase	3-wire (Star)	1.5	0.167
	3-wire (Delta)	1.5	0.50
	4-wire (Star)	2.0	0.167
Six-phase	6-wire	3.0	0.042
	7-wire	3.5	0.042

Low frequency adversely effects the performance of equipment. Lower frequency results in lower speed, lower output and lower efficiency of industrial/agriculture drives. The motors are over-excited and overheated. According to a study carried by Bhakra Beas Management Board, it has been observed that more excitation is required by the alternators at reduced system frequency, to meet the increased reactive power requirement. The flux density, excitation current and losses are thus increased during under-frequency operation. The decreased speed of the generating machine results in poor ventilation and hence overheating. The line current and active power requirement in transmission and distribution system increases with the decrease of supply frequency for loads of induction motors and lighting. This leads to overloading of the system and increase in line losses. Better load management (see Sec. 8.12) will help maintaining the frequency within the permissible limit.

1.4 System Study

1.4.1 Requirements

In studying the aspects of any power system, we have to first consider the consumers requirements. A number of questions arise:

- What type of loads are present or will be present?
- How are they to be supplied by the sub-transmission?
- What are the capacities of the feeders and transformers?
- How much power is to be obtained?
- How much space is required and how much of it is available for a particular installation?
- How can power be distributed economically?
- How reliable should the system be?

The questions are numerous and their answers complicated. In this book, we will try to answer the various questions.

1.4.2 Systems Engineering

Technology offers solutions to many problems confronting society such as those involving energy, natural resources, pollution, inflation, etc. Yet there is sound evidence that the application of technology to social problems has often led to less than the desired results. What is needed is a symbiotic blending of technology with social, political and economic needs. This almost constitutes the subject matter of systems engineering.

Problems in the process of systems engineering may be: problem definition, goal setting, system synthesis, system analysis and choosing among various alternative systems. The successful systems engineer must not only be a competent engineer, but also a competent economist and manager. As an engineer, he must know the theory, application, reliability, testing techniques, state-of-art, concepts and their limitations. As an economist, he must know how to plan and control the fiscal and schedule aspects of the programme. And finally, as a manager, he must be able to communicate effectively to his staff or public the problems they are to deal with.

1.5 Benchmarking

There is a natural law: *There is always room for improvement.* Rethinking is necessary to bring about changes in an enterprise. Ways to improve practices is a continuous process in order to reduce costs, improve efficiency, add consumer value, improve environment, increase reliability, increase maintenance intervals etc. Benchmarking is a process:

Learn⟶Benchmark⟶Change

This involves measurement, analysis and comparison of a utility's operations with other better managed power utilities at the national and international levels and will help to identify areas for improvement. Benchmarking of the best practices teaches efficiency and introduces innovative ways of eliminating unnecessary processes. It implies use of timely, detailed information [7]. It is a process of working to stay ahead (see Fig. 1.5). To strive to be the number one power utility in the country or in the world.

To identify benchmarks, the following steps are desirable:

(a) Interview experts in the key functions. Such people understand the efficiencies and possibly how to streamline the system. The discussion with Central Electricity Authority (CEA), Rural Electrification Corporation (REC), Power Finance Corporation (PFC) may be useful.

(b) Review market and consumer surveys. For example, various surveys show that consumers are ready and willing to pay more for consistent supply of power, if they are assured of it meeting the

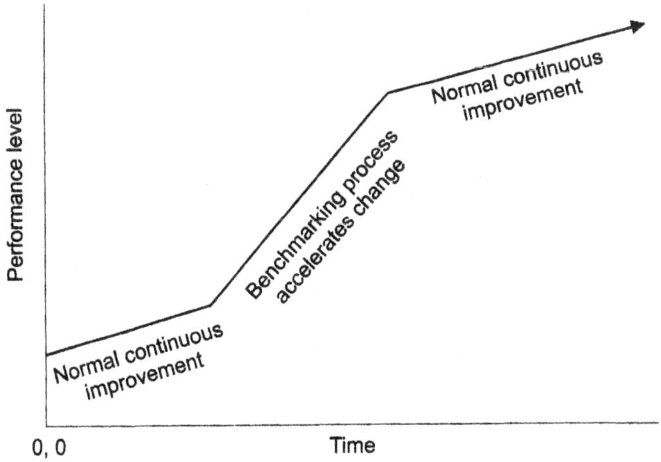

Fig. 1.5 *Benchmarking accelerates the performance level*

benchmarks of reliability and quality.

(c) CEA Annual General Review.
(d) Annual Reports of SEBs, power utility companies.
(e) Annual reports of the Ministry of Power.
(f) Utilities' websites etc.
(g) Planning Commission Reports.
(h) Asian Development Bank Report.
(i) World Bank Report.
(j) Publications on Global Benchmarks [10].
(k) External Audit: Third party external audit will measure the performance against past achievement.

Some of the benchmarks for Indian power utilities are:

Activity	*Target*
I Metering and billing:	
Manual meter-reading	
Urban consumers/day	600
Rural consumers/day	200
Handheld meter-reading instruments	
Urban consumers/day	500
Rural consumers/day	150
Meter-reading and billing cost per consumer	< Rs 5/-
Bill-collection ratio	> 95%

2 Distribution Network	
Availability	> 95%
(a) Mean time between failures (MTBF)-specific make-wise	
Transformers	> 20 years
Switchgear	> 15 years
Cables	> 15 years
Cable joints	> 15 years
Meters	> 15 years
(b) Annual damage rate	
%age maximum annual transformer damage	3
%age maximum annual energy meter damage	1
(c) Power interruptions:	
Number of 11 kV trippings /100 km of line /month	<1
Number of consumer no-current complaints/1000 consumers/day	<0.5
(d) Distribution system's maximum losses of primary feeder	<5%
(e) Agriculture pumpset efficiency	>85%
(f) Reliability (SAIDI, SAIFI)	
Five worst feeders (11 kV) every year	20% improvement
Other feeders (11 kV) every year	5% improvement
3 Labour productivity [12]:	
Number of employees/million kWh/year delivered	1.5–1
Number of employees/1000 consumers	< 5
Operation and maintenance cost (% of capital)	< 0.5%
Consumers/employee	150–250
Ratio of skilled to unskilled labour	6
4 Financial	
Annual revenue requirement	Fix optimum operative expenditure plus depreciation plus taxes plus rate of return
Revenue realised lead time	<one month
Distribution cost/kW (Rs)	<10000
Distribution cost/kWh (Rs)	<0.50

1.6 Electricity Reforms

1.6.1 Need

According to the Asian Development Report (2000), ineffective governments in Asia are a leading cause of poverty. The state electricity boards formed as per the Electricity Act, 1948, contributed greatly in extending electricity in the country particularly to rural areas and the weaker sections of society. They have done a good job of serving political and social obligations when sufficient state funding has been available. The state funding structure was favourable for the development of electricity. Loans were granted to state electricity boards. They had to pay back a nominal interest and principal in a set priority only if there was cash surplus as per Sections 59 and 67. Most of the boards did not even pay back any interest and the principal amount.

As per the direction of the Central Government, 50% of the state loans were converted into equity from time to time to reduce the interest plus principal liability. With time, state funding has dwindled. Inherent inefficiency and lack of accountability in the state electricity boards have reduced the quantity, quality and reliability of power. This resulted in increased cost of power supply, leading to huge financial losses. At the same time, the consumer's expectation for better supply has increased. More and more industrial consumers are now installing captive power as a backup.

Unbundling is one urgent alternative as smaller organisations find it easier to develop team working. The other alternative is to create smaller business/accountable units within existing SEBs. However, in both cases support of integrated information technology (e.g. ERP) is important to bring about efficiency and transparency in working. This is why the need for reforms arises.

A well functional National Grid is essential for the exchange of power between power-surplus and power-shortage regions. Open access to a grid/distribution system is a must for competitive marketing by power developers anywhere in the country. Connectivity of captive generation to the grid is necessary. The major indicators of the results of reforms are cost cutting and better service to the consumer.

Privatisation can contribute to the improvement of the distribution system. EDF in France and TVA in the U.S.A. are very competent public power utilities in the world. Reliance Energy, BEST and Tata

Power in Mumbai are successful power utilities in private sector in India. The power utility may be public or private, but for effective working, good governance is required. There is a need to create a good corporate culture of consciousness, strategic awareness, transparency, openness, accountability, professional and ethical approach. In a nutshell, the recipe for reforms is professional management and effective application of information technology. The utility may be public or private, vertically integrated (SEBs) or unbundled. The issues are well understood but political strength, courage and strong will are important to push for implementation of the solutions.

1.6.2 Model

The state/central regulatory commission should focus on the consumer interest of getting reliable power supply of acceptable quality at a reasonable cost and bring about an improvement in the performance of the power utility. For the purpose of accountability, generation, transmission and distribution may be separated as independent, competitive companies. Alternately, existing SEBs may be reformed to set up accountable business units. The structure may change from state to state. Each state must set milestones in the reforms process for timely implementation.

Retailing supply and a free electricity market will work only if there is a surplus of available power and transmission and distribution capacity is adequate [9]. Efficient operation of the electricity market needs to be guided for the prioritization of society's needs if sustainable energy systems are to be achieved.

Distribution has been privatised in Orissa. It is in the transition stage and the results are not positive yet. Similarly, Haryana, Andhra Pradesh, Karnataka and Maharashtra and Delhi are in the process of privatisation. People participation at the utility and regulatory levels is necessary for the success of any privatisation model.

1.7 Future Distribution Systems

This is the age of competition and fast change in technologies and social innovations. Future power distribution systems need to be highly service-oriented and caring for consumer values. Every power utility should have a well-drawn vision. Compelling vision, values and mission give a sense of direction to serve the needs of the consumers through a strategic plan. This will be possible through SWOT (strengths, weaknesses, opportunities and threats) analysis and benchmarking exercises within the power utility.

Leading the transformation process is painful. It is essential to expose the truths. Value-oriented, shared vision leadership can only achieve continual renewal. Inspiring vision creates values for consumers, employees and the community. Vision and action together bring change. A vision document must be prepared to spell out the strategy for future growth.

1.7.1 Vision

A clear vision for the future communicates a sense of the kind of organisation, power utility needs to become, and outlines the kind of results it must achieve. The vision provides a continuing focus and constantly reminds the employees and consumers of what the company is trying to change. Otherwise, employees easily become side-tracked and diverted. Every power utility should have a vision statement, which could be as stated in the following examples:

- 'We want to be the number one power utility in the country.'
- 'We are committed to give continuous power supply to our villages.'

A powerful vision contains three elements—first, it focuses on operations; second, it includes measurable objectives and metrics; and third, if it is really powerful, it changes the basis for competition in the industry.

1.7.2 Mission

For a particular task there can be a mission statement. In India, above 70% of the households in rural areas and 35% of the households in urban areas, are without electricity. The important mission statements could be:

- 'Electricity for all in India by the year 2012 AD.'
- 'We ensure that every consumer service complaint will be attended within 30 minutes in urban areas and within two hours in rural areas.'

1.7.3 Values

The power utility leadership must create values such as consumer satisfaction, environmental responsibility, service culture, technical excellence, commercial orientation etc., for the consumers.

1.7.4 Vision 2010

Every utility should prepare scenarios (see Sec. 2.13.2) to anticipate the changes [6] and to proact. The key expected changes in India by the year 2010 AD can be:

Consumers will look for the best service at the lowest price. This will be possible if power utilities manage their assets effectively. Limited investment needs to be directed on a priority at those networks with the poorest performance and highest operational costs.

Electricity will be sold as a 'brand', stamped with reliability and quality.

The role of demand-side management will be important as an alternative to the expansion of generation capacity.

Electric power supply content constitutes about 13% of India's end energy use [8] but in terms of fuel consumed or pollution emitted, it is about 40% of the Indian energy economy. Non-conventional renewable energy resources like solar photovoltaic, wind power, micro/mini hydel, fuel-cell, bio-mass, municipal waste, tidal and wave will be the most cost-effective schemes. Their contribution by the year 2010 will rise to about 10% of the total installed capacity in the country [6].

Internet penetration will be around 50% in the country. Meter reading, billing and collection will be made speedily through the power utility's website for consumers having a connected load of more than 50 kW.

There will be increased demand for consultancy to generate innovative solutions. The consultant will move from the job of 'advising' to 'doing'. Resolving the sourcing of new ideas and converting them into commercially viable products will be the main role of consulting firms. Consultants will add value to the power utility.

Skills will be highly valued. Power utilities will recruit good hands, retain the better workers and allow the best to advance.

Due to all-round environmental degradation, predominantly by power plants, there will be increased severity and frequency of monsoon storms and flooding in the Himalayas and rivers downwards. This phenomena will require review of the design of certain line structures and/or concrete disaster planning.

Due to an electricity market coming up in the near future, wholesale trade and third-party transactions will overload the system. This tendency will tend to reduce system reliability. Multi-circuit line design is the need of the future.

Electric energy storage system using superconducting materials will be developed.

Information Technologies will play an important role. Power utilities will become integrated for engineering and construction, as well as for utility-wide operations. The **Power Utility Information Technology Model** will be as shown in Fig. 1.6. The world power utilities will be inter-connected through the Internet or Super Highway, for better exchange of information on productivity, effectiveness and resources. The Internet is a computerized network of communication and information systems. To get on the Internet, one needs a personal computer with a modem and an Internet software and service package.

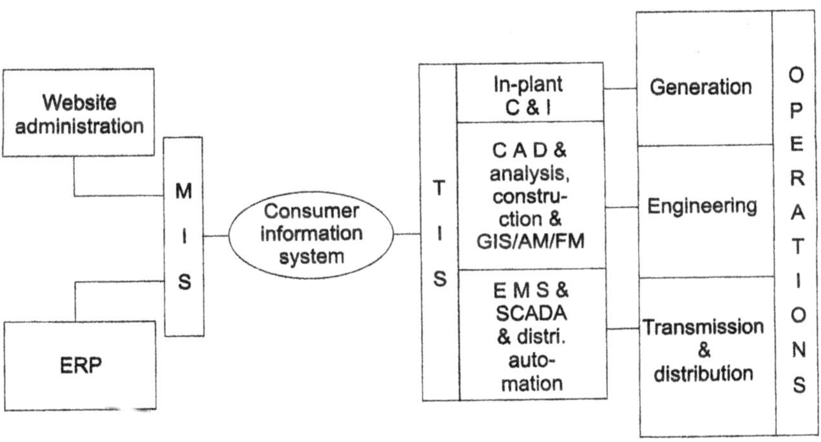

Fig. 1.6 *Information Technology Model*
MIS: Management Information System
TIS: Technical Information System
ERP: Enterprise Resource Planning

Power utility business will be 'process-oriented'. The process is a collection of activities, that takes one or more kinds of input and creates an output that is of value to the consumer. A process is '*a number of interrelated activities needed to accomplish a specific task*'. The right process in this context is '*a planned set of activities which produce desirable and predictable electricity delivery*'.

Power utilities would need to reinvent themselves constantly, since there would be constant change in markets and technologies. They will be transformed from mere distributors of electric energy, to producers of electric energy solutions. Competitive benchmarking will be the continuous process of measuring products, services and practices against the recognised leading power utilities.

To conserve energy, energy services will be evolved.

PROBLEMS

1. What is the maximum voltage that is allowed fornominal system voltages of 11, 33 and 66 kV as per the Indian Standard?
2. A power system having coal thermal power generation supplies power to an agriculture consumer with 5 H.P. motor-pumpset running at 0.8 power factor lagging. If motor and pump efficiencies are 82% and 35% respectively, estimate the overall energy efficiency from the first step of energy input at the generating station to real work done and the degradation in environment when the consumer runs the pumpset for four hours.
3. A concentrated group of 30 consumers each having 230 V rated load of impedance of 30+j40 ohms can be given a supply from a 500 kVA, three-phase, 11/0.4 kV transformer, three-phase, four-wire system or 230 V, single-phase, two-wire system. Calculate: (i) the line current; (ii) the total real power required; and (iii) which system is economical and by how much.
4. A star-connected three-phase load consumes a power of 70 kW at a power factor of 0.85 lagging when supplied from an 11 kV line. Calculate the phase voltage and line current.
5. How will you frame a strategy? State a few strategic goals for the power distribution vision for 2020 AD in India.

BIBLIOGRAPHY

1. (i) IS: 12360–1988 (revised).
   (ii) International Electrotechnical Commission Publication: 38. Voltage and frequency for ac transmission and distribution systems.
2. "Feasibility studies of SWER system", *CBI&P Report* No. 11, Central Board of Irrigation and Power, New Delhi, June, 1979.

3. Pansini, J. Anthony, *Electrical Distribution Engineering*, McGraw-Hill Book Company, New York, 1983, p. 19.
4. *New Renewable Energy Resources—A Guide to Future*, World Energy Council, Kogan Page, London,1994, p. 59
5. *Transmission and Distribution*, Vol. 82, No. 2, February 2000, p. 4
6. Cole John, *Global 2050—A Basis for Speculation*, Nottingham University Press, U.K., 1999.
7. MacDonald John and Steve Tannes, *Understanding Benchmarking*, Hodder & Stoughton, 1998.
8. Pabla AS, *Electrical Power Systems Planning*, Macmillan India, New Delhi, 1998, pp. 25–26.
9. Faruqui Ahmed and Clark W. Gellings, *California Energy Crisis 2000*, ELECTRA, No. 194, February 2001, pp. 16–21
10. Duke Energy Corporation: *Labor Productivity Benchmarks and International Gap Analysis*, ICON Group International, USA, October, 2000
11. Berrie T W, '*Electricity Economics and Planniing*', IEE Power Series 16, 1992, p. 96
12. Khalema-Redeby Lucy and others, Planning & Management of African Power Sector, Zed Books Limited, London, 1998, pp. 278–79.

Chapter 2

Electricity Forecasting

2.1 Power Loads

2.1.1 Load Demand

The load requirement of an area depends upon its terrain, its population and their living standards, its present and future development plans, cost of power, etc. In India, Annual Power Survey Committees under the auspices of the Central Electricity Authority are entrusted with the task of preparing the national prospective demand for power [4]. Based on the category-wise forecast, the 16th Electric Power Survey of India [4] has estimated the pattern of utilisation of electrical energy in the country as given in Table 2.1.

Table 2.1

Pattern of utilisation of electrical energy (per cent)

Category	*1997-98 (Actual)*	*2004-05 (Estimated)*
Domestic	20.69	22.99
Commercial	6.91	6.92
Agriculture	30.54	26.98
Industry	35.22	37.22
Others	6.65	5.89
Total	100	100

The Planning Commission and State Electricity Boards also undertake certain exercises. The techniques used by these agencies are mainly restricted to trend, end use and sheer's formula only. There is a continuing need, however, to improve the methodology for forecasting power demand more accurately.

2.1.2 Load Characteristics

(a) Nature of Loads

It is necessary to know the general nature of load which is characterized by the demand factor, load factor, power factor, diversity factor, utilisation factor and power factor. These are discussed in Chapter 8 (refer also Sec. 19.10).

(b) Types of Loads

In general, the types of load can be divided into the following categories.

Domestic: This consists mainly of lights, fans, domestic appliances, such as heaters, refrigerators, airconditioners, mixers, ovens, heating ranges and small motors for pumping, various other small household appliances, etc. The various factors are: demand factor 70–100%, diversity factor 1.2–1.3 and load factor 10–15%.

Commercial: This consists mainly of lighting for shops and advertisements, etc., fans, airconditioning, heating and other electrical appliances used in commercial establishments, such as shops, restaurants, market places, etc. The demand factor is usually 90–100%, diversity factor is 1.1–1.2 and load factor is 25–30%.

Industrial: These loads may be of the following typical power range:

- Cottage industries — < 5 kW
- Small-scale industries — 5-25 kW
- Medium-scale industries — 25-100 kW
- Large-scale industries — 100-500 kW
- Heavy industries — above 500 kW

The last two types of loads need power over a longer period and which remains fairly uniform throughout the day. For large-scale industrial loads the demand factor may be taken as 70–80% and the load factor 60–65% and for heavy industries the demand factor may be taken as 85–90% with a load factor of 70–80%.

Municipal: This load is for street lighting and remains practically constant throughout the night. For this the demand factor is 100% while the diversity factor can be taken as 1. Street lights are required mainly at night but there may be the small load of traffic signals throughout the day also. The load factor for street light is usually taken as 25–30%. Another type of municipal load is for water supply and drainage.

Agriculture: This type of load is required for supplying water for irrigation by means of suitable pumps driven by electric motors. The load factor is generally taken as 20–15%, the diversity factor as 1–1.5 and the demand factor as 90–100%.

Other Loads: Apart from the above mentioned loads, there are other loads such as bulk supplies, special industries such as paper, textile, etc., and traction and government loads which have their own peculiar characteristics.

(c) System Power Factor

All electrical equipment, except synchronous motors, resistance heaters and incandescent lamps, consume power at lagging power factors. The approximate average lagging power factor of various equipments is given below and in more detail in Appendix IX.

Fluorescent lamps	0.6-0.8
Neon signs	0.4-0.5
Arc lamps used in auditoriums	0.3-0.7
Fans	0.5-0.8
Induction heaters	~0.85
Resistance furnaces	0.6-0.9
Arc furnaces	0.8-0.9
Induction furnaces	0.75-0.85
Arc welders	0.3-0.5
Resistance welders	~0.65
Induction motors	0.55-0.85
Fractional HP motors	0.4-0.75

Induction motors are widely used in industry and agriculture and constitute a major part of the load in a power system. The overall power factor of a system is likely below 0.7 lagging, unless corrective measures are taken to improve it.

(d) System Load Diversity

For better load management, it is important to analyse statistically the summer and winter load diversity in an area. Also, it is desirable to evaluate the diversity of the load in different states or regions to plan for the peak power demand [7]. For example, in the Northern region during summer, Himachal Pradesh and J&K (hill-states) have lean power demand and they sell power to Punjab and other adjoining states as these states (plains) have large air-conditioning units, and agriculture loads. The situation is reversed during winter when Punjab has surplus power and the same is sold to Himachal Pradesh and J&K to cater to their heating demands. These power exchanges are made on the basis of bilateral agreements based on forecasts of available energy. Overall, India is a hot country having an agriculture base and where summer demand is greater than that of winter. It is important to determine system load sensitivity with respect to the variation of system voltage and frequency in the utility's area in terms of MW/kV, MVAr/kV and MW/Hz, MVAr/Hz. This is an important area of research in India.

(e) Feeder Load Characteristics

The load characteristics of a distribution feeder at the sub-station bus is important in terms of its load composition and its electrical characteristics. The load composition approximately is known by the type of feeder. Categories like domestic, commercial (non-residential), industrial, agriculture, etc. are expressed as percentage of total load at a time. The percentage load of each category in the total demand is dependent on the time of the year, time of the day, geographical location, socio-economic conditions and diversity factors of the different categories of loads. Typical load curve of an urban feeder is shown in Fig. 2.1. The yearly peak in the plain areas is usually in the summer season and in the hilly area, it is during severe winter season. Each category load may be function of voltage, frequency and time derivatives.

The segregation of category demand on the feeder can be done by measure such as:

- Knowing the connected load of each category of consumer. The product of connected load and suitable demand factor of each load will give the demand for each category. Industrial demand could be found when comparing the industrial off-day demand with

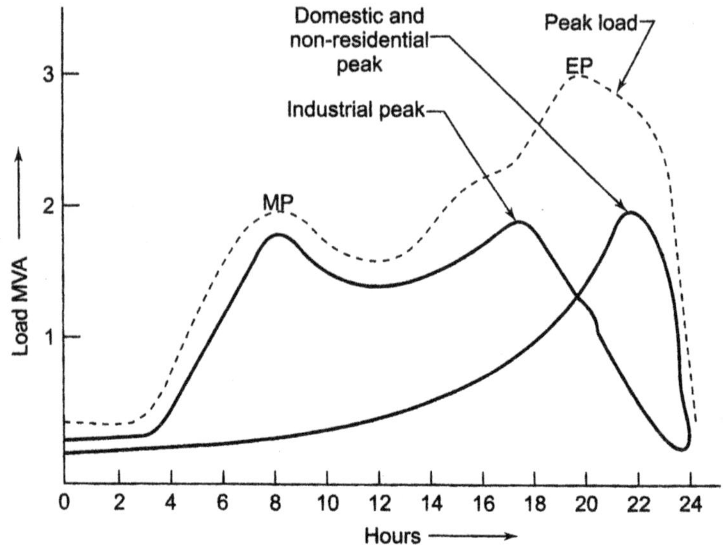

Fig. 2.1 *Typical urban feeder load*
MP = Morning Peak
EP = Evening Peak

other normal-day demand of the week. The difference of two demands will give the industrial demand.

- On the rainy days, the agricultural demand of the feeder will be almost nil. Comparing it with the pre-rain normal day of the week will give the agricultural demand.
- The major industrial load can be determined by checking the actual demand or contract demand.

(f) System Load

The character of electric load itself changes as a function of the number of consumers. As the number of consumers in a group or a system is increased, the contribution to the group or the system peak of each consumer (coincidence) is decreased and the high rate of fluctuations decrease.

The whole picture of load is based on the statistical laws, whereas the individual load may be entirely random in character. The certain average pattern is recognizable at the distribution transformers. At the sub-transmission level, this average effect is still more pronounced.

Finally, at the transmission level, load profile is an almost predictable situation.

The system load may be divided into motive load, heating load, lighting load and system losses. The general load characteristics of the system load are shown in Fig. 2.2. With the characteristics linearized for the values of voltage (Fig. 2.2a) and for the values of frequency (Fig. 2.2b) close to the normal one, the respective voltage and frequency dependence can be summarized as follows:

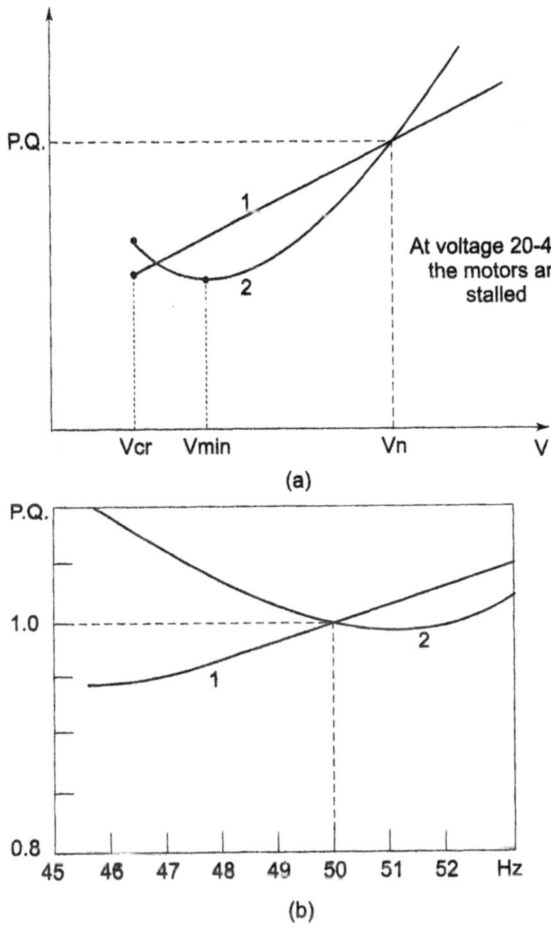

Fig. 2.2 *System load characteristics:*
Curve 1—active power (P), Curve 2—reactive power (Q)

In a region of 80-120% of rated (nominal) voltage, the voltage dependence of composite system loads can be expressed [1] by the relations:

$$P = \left[\frac{V}{V_n}\right]^a \cdot P_n$$

$$Q = \left[\frac{V}{V_n}\right]^b \cdot Q_n$$

Where the suffix n stands for nominal values and the values of the exponent a lies in the range 0.6 to 1.4 and b in the range 1.5 to 3.2. However, the average values of exponents $a = 1$ and $b = 2$ can be assumed for general purpose, that is, the real power consumption varies directly with the voltage, while the reactive power requirement varies directly with the square of the voltage. Similarly, the frequency dependence of composite system load in the region of ± 5% of nominal frequency can be expressed by

$$P = \left[\frac{H_z}{H_{zn}}\right]^{a'} \cdot P_n$$

The exponent a' ranges from 1.5 to 3 for active load component. For general purpose, the average values of $a' = 2$ can be assumed.

Generally, distribution networks operate above their surge impedance loading and hence absorb more VAr than they generate. The regional grids have been operating at a frequency varying from 47.5 to 53 Hz. Both high and low power frequencies create additional losses in the distribution system and at the consumer end. In case of higher frequency, the motors speeds up, and they would all meet higher load torques, thus require the motors to pull more power from the network. Low frequency causes a drop in active load, but considerably increases the reactive load due to magnetic circuit saturation in transformers and the motors, and also increases loading in the distribution system, thereby incurring additional active and reactive loss.

The above system load behaviour is based on individual load working in the system. As an example of actual test studies, the individual load's steady state active power (P) and reactive power (Q) variation with change of voltage from 50% to 120% of nominal rating is given by voltage relation given in Table 2.2.

Table 2.2

Variation of P and Q for excursion of voltage 50% to 120% of nominal

Individual load	*Variation of*	
	P	*Q*
Filament lamps	$V^{1.4-1.5}$	0
Discharge lamps	Non-linear function +	Probability function of lighting
Resistive load (heater, dryer, water heater, oven range, etc.)	V^2	0
Stallable induction motor (compressor, high head pump, etc.)	Almost constant in stable region + stalling phenomena	Saturation effect + stalling phenomena
Non-stallable induction motor (fan, low-head pump, etc.)	Almost constant	Saturation effect
Transformer	Almost exponential iron loss + linear copper loss	Almost exponential magnetizing loss
Power capacitor	0	V^2

EXAMPLES

(i) On the British system as a whole [2], a reduction of voltage of 1% produces a 5% reduction in reactive power demand. It also produces a 1.4% drop in real power demand.

(ii) In India, a reduction of 1 Hz frequency in the Northern Power Grid, produces 800 MW load drop.

2.2 Connected Load

Installed load is the sum of the rated inputs of electrical apparatus installed at the consumer's premises. Connected load is that part of the consumer's installed load as computed by the power supply utility. The definition of connected load varies from utility to utility. The major difference is in computing the load of spare plugs, loads connected in parallel through change-over switch. Generally, 1/3rd plugs are counted for load but some utilities count as low as 1/10th. In case of two parallel loads through a change-over switch, some utilities count only one load whereas others count both. For example, in Punjab, 1/3rd plugs and in Delhi 1/10th plugs are counted towards the connected load. The average connected load per consumer for various categories in India [12] as on 31.3.98 is given in Table 2.3.

Table 2.3

Connected load in India as on 31.3.1998

Category	*Average connected load/consumer (kW)*
Domestic	0.85
Commercial	1.34
Industrial (LV and MV)	11.16
Industrial (HV)	696.95
Agriculture	4.04
Traction	9977.11
Public lighting	3.11
Public water works	13.04
Miscellaneous	5.54
All categories	1.95

2.2.1 Industrial Loads

These types of loads are discussed below:

(a) Cottage Industries

These are very small electric motors for driving looms, small lathes, etc. The power required is approximately as:

Load	*Power required kW*
Silk throwing (reeling)	1/2
Single unit power loom	1/2
Knitting machine	1/2
Wooden toy-making	1/2
Small machine shops	1/4

(b) Small Scale Industries

For this the average power requirements are given below:

Load	Average power required kW
Flour mill vertical type	7
Flour mill, horizontal type	12
Expeller, baby type	10
Expeller, standard type	12.5-15
Oil filter	3
Groundnut decorticator	25
Ginning	4
Rice huller	5
Dal (pulse) roller	12
Saw mill (small)	5
Cinema	15
Ice factory	2.5 per tonne of ice
Small workshop	7.5
Foundry	2.5
Ice candy	2.5
Welding sets with welding current (A), 100 Vac	
150	6
200	7.5
250	9.5
300	11.5
400	20
500	28
Arc furnace	350 per tonne of steel
Rice shellar	2 per tonne of rice
Sugar mill	0.5 per tonne of sugarcane

2.2.2 Street Lighting

In small towns, the average size of the street light lamp is generally 40 W at 230 or 250 V with average support span of about 45-50 m and mounting height 7.5-9 m. The spacing and height depends upon the traffic speed and safety criteria. For larger towns, the size of the street lamp may be 40-100 W. These may be incandescent, fluorescent or mercury/sodium vapour lamps. Special street lighting fixtures, arrangements and the consequent load development on this account should be determined separately for each large town. IS-1944 (revised) gives the code of practice for the design of public lighting (see Sec. 19.16). Lighting intensity depends upon the traffic speed and safety criteria.

Lighting technology has now developed tremendously. Lamps with better efficiency, longer life and good colour-rendering index are becoming available. Their use for public lighting provides much saving in energy. It is much cheaper to use 18 W low pressure sodium vapour lamp (SOX) for village street lighting which has its illumination level equivalent to 300 W incandescent lamp (GLS). Another advantage is that the SOX lamp is comparatively less sensitive to voltage fluctuations. These lamps, are more efficient and should have wider application on account of energy shortages. The 160 W Mercury vapour lamp (MLL) requires no special fitting but an ordinary holder and is being largely used for rural public lighting in many states. The use of these lamps is quite economical on the basis of life cycle costs. High pressure sodium lamps (SON) are five times more efficient in lumens output as compared to GLS and are quite suitable for city roads. Solar PV street lanterns are most suitable for remote villages.

2.2.3 Water Supply

This load may approximately be estimated from the water requirement. The water requirement for rural areas may be taken as 50 litres per day per person; for semirural areas as 70 litres per day per person; for urban areas as 90 litres per day per person and in metropolitan cities as 120 litres per day per person. These are the minimum requirements based on individual house connection and sewerage programme. If *panchayats* or municipal bodies can provide better facilities, water requirement can be surveyed and estimated separately. World Health Organization (WHO) literature on water supply to villages places the minimum drinking water need level as low as 23 litres per person per day. This standard of per capita is modest and realistic for small villages in India in which case the water supply programme is not accompanied by individual house connection of water and sewerage but based on public standposts alone. For example, in Tanzania, the rural water supply programme focuses almost entirely on provisions of low cost public standpost installations.

2.2.4 Irrigation

The power requirement for this type of load depends on the number of waterings required for each kind of crop, area under the crop, nature of soil and depth of the ground water level. The horsepower requirement of electric motors on the basis of average discharge is shown in Table 2.4.

Table 2.4

Motor horsepower for average discharge

Ground water level (m)	*HP for 20 L/s discharge*	*HP for 40 L/s discharge*
0–2	3	5
2–5	3.5	5–7.5
5–10	5	7.5
10–15	7.5	10
15–20	10–15	15
20–30	15–20	25

This horsepower is indicated assuming reasonable efficiency (60-80%) of the centrifugal hydraulic pump. Efficiency depends on factors such as specific speed (hydraulic design), size of pump, inlet head, discharge head, internal running clearances, surface roughness (casing and impeller material), stuffing box friction (stuffing box pressure), etc. Field observations show certain wrong designs selections of pumps system with efficiencies as low as 30% while in operation.

In subsoil strata, where the ground water is inadequate, a higher horse-power may be required.

For efficient centrifugal pumps, it is necessary that the water table level should not be more than 5-6 m below the pump level. Where the water table is low, it becomes necessary to install the centrifugal pump in the well pits. Studies show that when the depth of pit exceeds 15 metres, it is advantageous to go for a submersible pump, rather than use the centrifugal pump. A submersible pump requires a minimum of 150 mm *bore* size housing pipe. The performance of a submersible pump of various stages against total head is given in Table 2.5.

Table 2.5

Performance of submersible pumps

Submersible pump stages	Motor HP	Discharge in litres/minute						
		700	800	900	1000	1100	1200	1300
1	3	9	8.5	8	7.6	7	6	5
2	6	18	17	16	15	14	12	10
3	9	27	25	24	23	21	18	15
4	12	36	34	32	30	28	24	20
5	15	45	42	40	38	35	30	25
6	17.5	55	51	48	46	42	36	30
7	20	63	60	54	53	49	42	35
		(Total head in meters)						

2.2.5 Sensitive Loads

The proliferation of computer-based products, which are most often non-linear loads, are called sensitive loads. Figure 2.3 illustrates the typical computer tolerance limits for various electrical power disturbances such as per IEEE Std 446. The voltage envelope between

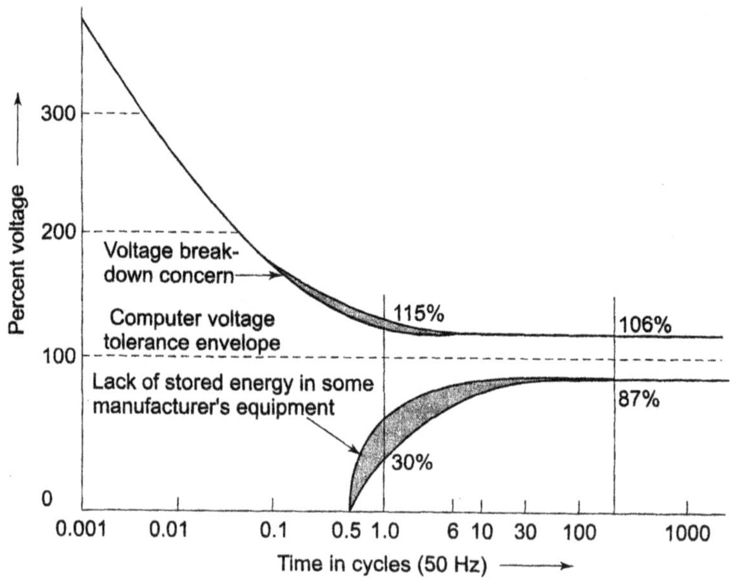

Fig. 2.3 *Minimum surge withstand capability of computer*

the two curves, represents the limits in which a typical computer can withstand these disturbances without malfunction or damage. The shaded area demonstrates the ranges in which various computers become susceptible to problems from corrupted voltage. In addition to voltage, such sensitive loads typically require the frequency to be within ± 0.5 Hz, the rate of change of frequency less than 1 Hz/Sec. voltage waveform distortion under 5% and voltage unbalance less than 3%. For specific applications, the power quality requirements should be obtained from the manufacturer of the sensitive equipment.

2.2.6 Dirty Loads

Dirty loads are known for producing harmonics, voltage fluctuations and light flickering, and large variations in the demand for reactive power. This phenomenon creates extra system losses, thermal over-loading, dielectric stressing, produce metering, control and protection relaying errors and disruption. Proportions of such loads are increasing in the modern power systems. These loads are computers, TVs, discharge lamps, arc furnace, induction furnace, welding sets, re-rolling mills, adjustable speed drives, large motors, railway traction and other heavy loads etc. It is desirable to safeguard the interests of other consumers to reduce or avoid the spread of the bad effects of these loads on them. Remedial measures are:

(i) to provide suitable filters, delta star connected transformers or underground feeder circuits in case of harmonics;

(ii) to provide such industrial consumers supply on high voltage or extra high voltage or/and to provide parallel transformers, or to provide power conditioning equipment in case of voltage fluctuations or dips or surges;

(iii) to provide thyristor controlled capacitors at such industrial loads, to provide reactive power at point of time to reduce voltage fluctuations.

EXAMPLE

The computer draws current in gulps and an arc-furnace draws current in swings which result in a heating effect (I^2t) which is considerably higher than the average load.

Consider the four sets of load currents varying in time on a per unit (pu) basis as shown in Fig. 2.4.

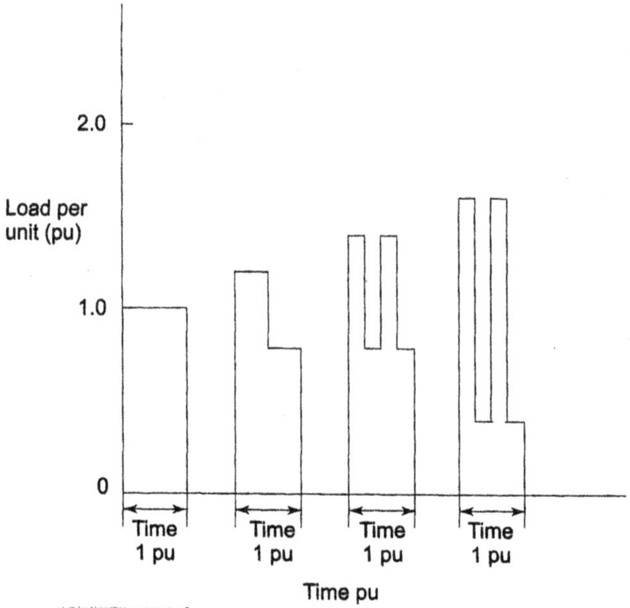

Fig. 2.4 *Heating effect of transient loads*

(i) The heating for load current of 1 pu for 1 pu time will be $= I^2 \times 1 = 1$ pu

(ii) The current variation is ± 20% pu over 1 pu time.
I^2t heating effect will be:

$$= 1/2\ (1.2)^2 + 1/2\ (0.8)^2 = 1.04 \text{ pu}$$

Here, the average load is 1 pu but the heating effect has increased by 4%.

(iii) Two equal current swings to 1.4 pu and down to 0.6 pu each of 1/4 pu time over 1 pu time. Net heating effect is

$$= 2\,(1.4)^2 \times 1/4 + 2\,(0.6)^2 \times 1/4$$
$$= 1.16 \text{ pu.}$$

It can be seen that in this case, heating effect increased by 16% although the average load over 1 pu is 1 pu.

(iv) This case has two equal swings upto 1.6 pu and down to 0.4 pu. The average load is still 1 pu but the heating effect is:

$$= 2\,(1.6)^2 \times 1/4 + 2\,(0.4)^2 \times 1/4$$
$$= 1.36 \text{ pu}$$

2.3 Load Forecasting

2.3.1 Introduction

There is a certain difference between the meanings of the words 'forecasting' and 'prediction' although these are often loosely used for the same purpose. Prediction is generally a statement or a guess of something which will happen regardless of whether the people want it or not and is something beyond one's control, e.g. one might predict rain or bad weather. Forecasting, on the other hand, is a statement of what will happen if certain conditions or trends continue and assumes that the causes of the events are under human control. Thus, if the forecast is undesirable it is within one's ability to change it.

Over the years forecasting has been refined considerably and has now reached a stage where it is more precise and unbiased. It has been extensively used in areas as diverse as electrical load forecasting, economic trends, market surveys, etc. In power systems, there is a great need for accurately forecasting the load and energy requirements because a great amount of money is spent in the electricity distribution industry. Medium-range forecasting of annual peak power, and energy requirements for each month of the coming year are necessary to plan for fuel e.g. coal, oil etc., for thermal plants; reservoir water releases for hydro projects, imposing of any restrictions on loads like air-conditioning during summer and heating in winter. Real time load forecasting for short durations varying from a few minutes to 2-24 hours has been in vogue by power utilities for system operation. If the energy forecast is too conservative, there is every likelihood of the generating capacity falling short of the actual demand, resulting in restrictions being imposed on the power supply which will be detrimental to the economic well-being of the country. If the forecast is too optimistic, it may lead to the creation of an excess generating capacity, resulting in part of the investment not giving immediate returns.

A developing country such as India, with considerable pressure on her limited financial resources, can ill afford either of these two situations. It is therefore, only appropriate that considerable importance is given to the forecasting of power demand. In order that a system plant and apparatus of the most economic size can be constructed at the correct place and right time to achieve maximum utilization of plant, it is important to estimate both (i) the magnitude of system load and (ii) the location of these loads.

The forecast of energy requirement is also important to power generation authorities so that the water or fuel requirements can be calculated and generator allocation schedules worked out. Forecasting methods as applied to the electrical industry fall into two broad categories: (i) estimates based on existing trends and (ii) econometric models.

2.3.2 Statistical and Econometric Models

A statistical model is a mathematical model which relates one variable with one or more variables in the form of assumptions and hypotheses. All variables have well-defined statistical properties.

In general, an economic model is a model which relates the behaviour of one economic function in terms of other economic functions. Economic models which qualify as statistical models, i.e. where all the variables are definable and mathematically measurable, are called econometric models. Often, as in the case of power system load forecasting, the model will consist of just one equation, in which case the model is called *regression model.* Most load forecasting models do not include economic factors directly, though they certainly do play a part. Thus, a complete load forecasting model involving economic indicators would be a form of an econometric model.

2.4 Definitions of Some Basic Concepts in Statistics

Time Series

A sequence of numbers or observations which relates a variable with respect to time. In forecasting, the process, whether continuous or not, is modelled as a time series. Time series analysis uses only the time series history of the variable being forecasted, in order to develop a model for predicting future values.

Markov Process

A process in which the probability of its changing from one state to another is independent of its past history.

Mean Value

The mean value (μ) of a series of numbers such as the time series x_i $i = 1$ to n is given by

$$\mu = \frac{\sum_{i=1}^{i=n} x_i}{n}$$

Weighted Mean

The weighted mean (μ') of a series of numbers is given by:

$$\mu' = \frac{\sum_{i=1}^{n} w_i x_i}{\sum_{i=1}^{n} w_i}$$

where w_i are *weighting factors* to be given to each element in the series and x_i is a series of numbers $i = 1$ to n.

Variance

The square of the standard deviation:

$$\sigma^2 = \frac{\sum_{i=1}^{n} (x_i - \mu)^2}{n}$$

where, σ is the standard deviation.
The expected value for a discrete series x_i where the probability that the process can have a value x_i is p_i is given by:

$$E(x) = \sum_{i=1}^{n} p_i x_i$$

The process cannot have values $\neq x_i$.

Normal Distribution

For a continuous variable x, mean μ and standard deviation σ, the normal distribution is given by:

$$f(x) = \frac{\exp\left(-\frac{1}{2}(x-\mu)^2/\sigma^2\right)}{\sigma\sqrt{2\pi}}$$

The probability that x lies between a and b i.e., $P_r(a < x < b)$ is given by the area under the curve of $f(x)$

$$P_r(a < x < b) = \int_a^b f(x)\, dx$$

For values of x about mean μ, we get 68.27% within $\pm$ 1 σ of the mean, 95.45% of 2σ and 99.73% between 3σ. Typically, a random continuous process is represented by a normal distribution and "95% confidence limits" based on 2σ are used to give an indication of the upper and lower bounds.

Moving Average

A series of average (mean) values for subsets of variables form the complete series. The subsets are so chosen that the nth subset consists of the $(n-1)$th subset less the lowest element and the next higher element from the time series added. For example, consider the series of numbers x_i, $i = 1, \cdots, n$. We can define a moving average of order m to be given by the following sequence of arithmetic means $(m < n)$:

$$\frac{\sum_{i=1}^{m} x_i}{m}, \frac{\sum_{i=2}^{m+1} x_i}{m}, \frac{\sum_{i=3}^{m+2} x_i}{m}, \cdots, \frac{\sum_{i=n-m}^{n} x_i}{m}$$

Sampling Frequency

In order to obtain meaningful information from a series which is cyclic in nature (or contains cyclic terms), there is a certain sampling frequency below which information is lost. This is called *Nyquist frequency* and is equal to the frequency at which a minimum of two observations per cycle is made.

Sampling Theory

It is practically impossible to measure all possible points in a time-series or population. Instead, samples are taken from which certain statistical properties, e.g. mean and variance, are calculated. A series of samples from which a series of means and variances are calculated themselves constitute *sampling distribution.*

The distribution of a sample refers to a mean centered on the actual population mean, but the mean of the distribution of sample variances has to be multiplied by $(n - 1)/n$ to obtain an unbiased estimate of the actual population variance.

The confidence limits to a sample statistic calculated in this way are given by standard sampling distribution curves, such as 'students *t*' distribution (for sampling means) and 'Chi square' distribution (for sample variances).

2.5 Regression Analysis

2.5.1 Trend

Regression or trend analysis is the study of the behaviour of a time series or a process in the past and its mathematical modelling so that future behaviour can be extrapolated from it. Two general approaches followed for trend analysis are:

(a) The fitting of continuous mathematical functions through actual data to achieve the least overall error, known as *regression analysis.*

(b) The fitting of a sequence on discontinuous lines or curves to the data.

The latter is more prevalent in short-term forecasting. A time varying event such as power system load can be broken down into the following four major components [10].

(a) Basic trend.

(b) Seasonal variation (Monthly or yearly variation of load).

(c) Cyclic variation which includes influences of periods longer than the above and causes the load pattern to be repeated for two or three years (or even longer cycles).

(d) Random variations which occur on account of the day-to-day changes and in the case of power systems, are usually dependent on the time of the week, e.g. weekend, week day, weather, etc.

The last three variations have a long-term mean of zero as shown in Fig. 2.5. In forecasting, either a continual variation process model can be used, or only certain points at regular intervals of the process. For example, either a complete load curve for a power system could be predicted, or, alternatively, only the yearly system peak demands. In the latter case the process is modelled as *time series* as shown in Fig. 2.6. The reason for this is that often, in power system planning, the network plant capacity is designed around the maximum system peak for the year, although the cyclic overloading factors are embodied in the design. Thus, interest is always centred on the yearly system peaks and not on the whole load curve.

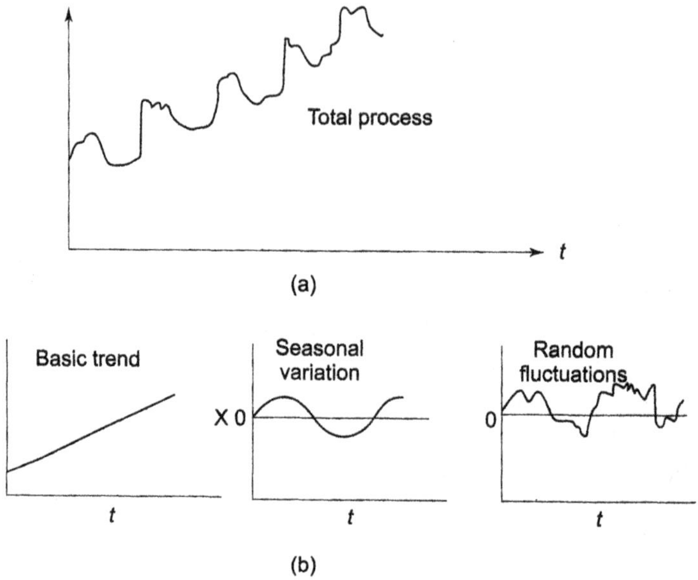

Fig. 2.5 *Decomposition of typical load growth curve: (a) Total process; (b) Decomposition*

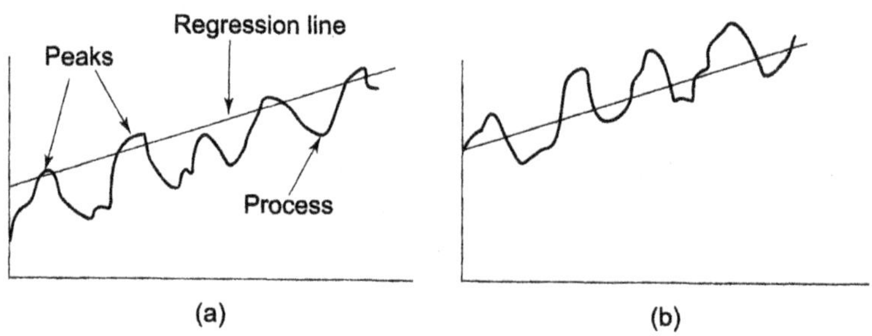

Fig. 2.6 *Regression curves: (a) Regression curve through process peaks; (b) Regression curve representing process mean*

EXAMPLES

(i) *Linear Trend:* A past trend where the increase in consumption from year to year is more or less constant. Tabulate the past consumption data and plot it on an arithmetical graph which will give a straight line. The projection of this line can give a forecast of future demands. But in real life, such a growth trend is unlikely in the power supply industry. The such growth trend in the power industry can be mathematically expressed as follows:

$$C_t = a + bt$$

where,

C_t = electricity consumption in any year t
a = consumption for base year $t = 0$
b = constant annual increase in energy consumption
t = cardinal number of year t with reference to the base year, i.e. equal to $T - 1 + n$ where T is the number of years for which the statistical trend is studied and n is the number of years for which the forecast is required.

let
$a = 2$ GWh
$b = 0.18$ GWh
$n = 5$
$t = T - 1 + n$
$t = 11 - 1 + 5 = 15$

then

$$C_{15} = 2 + (0.18 \times 15)$$
$$= 2 + 2.7$$
$$= 4.7 \text{ GWh}$$

The above forecast can also be read directly from the straight line trend graph.

(ii) *Exponential Trend:* In this case past data are drawn on a logarithmetic graph to give straight line projection for forecasting. Its mathematical expression is:

$$C_t = C_0\,(1 + m)^t$$

where,

m = mean annual rate of growth observed during T years. Other notations remain the same as given in the linear method relationship.

By converting the data to logarithms to base 10, the graph becomes a straight line as shown below:

$$\log C_t = a + bt$$

where,

$a = \log C_0$

$b = \log (1 + m)$

say

$T = 21$ is the reference period on which the calculation is based

$t_0 = 0$

$t_{20} = T - 1 = 20$

Consumption in reference year $t = 0$ is 4862 GWh (say), then

$$a = \log C_0 = \log 4862$$
$$= 3.686789$$

Mean rate of annual growth = 9.65% (say)

= 0.0965

$(1 + m) = 1 + 0.0965 = 1.0965$

$b = \log (1 + m) = 0.0401866$

Hence the equation for representation time is:

$$\log C_t = 3.686789 + 0.0401866\ t$$

Now suppose we wish to forecast consumption at the end of 25 years, i.e. $n = 5$. Then

$$t = T - 1 + n$$
$$= 21 - 1 + 5 = 25 \text{ years}$$

The forecast will be

$$\log C_{25} = 3.686789 + 0.0401866 \times 25$$
$$= 4.691455$$

Taking antilog,

$$C_{25} = 49143 \text{ GWh}$$

The forecast for energy consumption with the help of this method is considered to be the best. Other methods should be used for cross-checking and comparing the results to arrive at a reliable forecast.

2.5.2 Regression Functions

The principle of regression theory is that any function $y = f(x)$ can be fitted to a set of points (x_1, y_1), (x_2, y_2) so as to minimize the sum of errors squared at each point i.e.

$$\sum_{i=1}^{n} \{y_i - f(x_i)\}^2 = \text{minimum}$$

Sum of squared errors is used as it gives a significant indication of "goodness of fit".

Typical regression curves used in power system forecasting are:

(a) Linear $y = A + Bx$

(b) Exponential $y = A(1 + B)^x$

(c) Power $y = Ax^B$

(d) Polynomial $y = A + Bx + Cx^2$

(e) Gempertz $y = Ae^{Be^{Cx}}$

The coefficients used in these equations are called "regression coefficients".

EXAMPLES

(i) *Least Square Line:* The line $y = a_0 + ax$ is fitted to the sets of points $(x_1, y_1), (x_2, y_2) \cdots (x_n, y_n)$, i.e.

$$\varepsilon^2 = \sum_{i=1}^{n} [y_i - (a_0 + a_1 x_i)]^2 = \text{minimum}$$

Partial differentiation with respect to the regressions coefficients (a_0 and a_1) is made and the equations set to zero to obtain the minimum error criterion. This gives us a set of simultaneous equations in a_0 and a_1:

For a_0: $2\,\Sigma\,[y_i - (a_0 + a_0\, x_i)] = 0$, we get,

$$\Sigma a_0 + a_1\, \Sigma\, x_i = \Sigma\, y_i$$

or

$$Na_0 + a_1\, \Sigma\, x_i = \Sigma\, y_i$$

For a_1: $2\,\Sigma\,[y_i - (a_0 + a_1\, x_i)]\, x_i = 0$

$$a_0\, \Sigma\, x_i + a_1\, \Sigma\, (x_i)^2 = \Sigma\, x_i\, y_i \tag{2.1}$$

which yield

$$a_0 = \frac{(\Sigma y)(\Sigma x^2) - [\Sigma(x)]\cdot \Sigma xy}{N\Sigma x^2 - (\Sigma x)^2}$$

$$a_1 = \frac{N\Sigma xy - (\Sigma x)\cdot(\Sigma y)}{N\Sigma x^2 - (\Sigma x)^2} \tag{2.2}$$

(ii) *Least Square Parabola:* The curve $y = a_0 + a_1\, x + a_2\, x^2$ is fitted to minimize ε^2

$$\sum_{i=1}^{n} [y_i - (a_0 + a_1\, x + a_2\, x^2)]^2 = \text{minimum.}$$

Partial differentiation with respect to the three regression coefficients and setting the equation to zero, give us following three simultaneous equations:

$$\begin{aligned} a_0 N + a_1 \Sigma x_i + a_2 \Sigma x_i^2 &= \Sigma y_i \\ a_0 \Sigma x_i + a_1 \Sigma x_i^2 + a_2 \Sigma x_i^3 &= \Sigma x_i y_i \\ a_0 \Sigma x_i^2 + a_1 \Sigma x_i^3 + a_2 \Sigma x_i^4 &= \Sigma x_i^2 y_i \end{aligned} \tag{2.3}$$

which can be solved for a_0, a_1 and a_2.

(iii) *Least Square Exponential:* This is obtained by performing a transformation of the variables to obtain a linear equation, e.g. if the regression equation is $y = e^{Bx}$, this can be transformed into $V = A' + B'U$ when $V = \ln y$, $U = x$, $B' = B$, and $A' = \ln A$. In the solution of equations (2.1), therefore $\Sigma\, y$ is replaced by $\Sigma \ln y$ and the regression coefficients found; the coefficients are then transformed back using $A = e^{A'}$.

(iv) *Multiple Regression:* Two or more variables can be treated by an extension of the same principle, e.g. if we wish to fit an equation $z = a_0 + a_1\, x + a_2\, y$ to a series of points (x_1, y_1, z_1), (x_2, y_2, z_2),$\cdots$ it is a multiple linear regression; multinonlinear regressions are also used. As earlier, we minimize

$$\varepsilon^2, \text{ i.e. } \sum_{i=1}^{n} [z_i - (a_0 + a_1\, x + a_2\, y)]^2 = \text{minimum.}$$

Differentiation with respect to a_0, a_1 and a_2, in turn gives us the following simultaneous equations:

$$\begin{aligned} a_0 N + a_1 \Sigma x_i + a_2 \Sigma y_i &= \Sigma z_i \\ a_0 \Sigma x_i + a_1 \Sigma x_i^2 + a_2 \Sigma x_i y_i &= \Sigma x_i z_i \\ a_0 \Sigma y_i + a_1 \Sigma x_i y_i + a_2 \Sigma y_i^2 &= \Sigma y_i z_i \end{aligned}$$

which can be solved in a similar fashion as Eq. (2.3).

2.6 Correlation Theory

Correlation is a measure of how well a line or curve fits a given set of data. In general, it is a measure of how one variable is related to another. Typical correlations are shown in Fig. 2.7.

The standard error of estimate is defined by

$$S_{y \cdot x} = \sqrt{\frac{\Sigma (y_i - y_{\text{est}})^2}{N}} \quad \text{for } y \text{ on } x$$

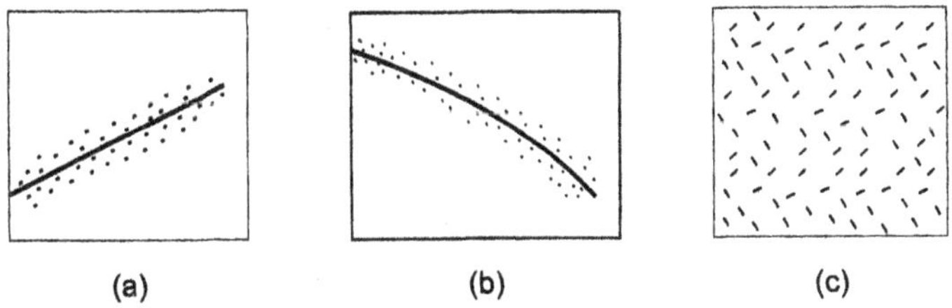

Fig. 2.7 *Typical correlations: (a) Positive linear; (b) Negative nonlinear; (c) No correlation*

where y_{est} is the estimated value of the dependent variable y for each value of the independent variable(s), here x. This is analogous to standard deviation.

The coefficient of correlation, often called "goodness of fit", is given by:

$$\gamma = \pm\sqrt{\frac{\Sigma(y_{est} - \bar{y}^2)}{\Sigma(y_i - \bar{y})^2}} \quad \text{where } \bar{y} \text{ is sample mean}$$

This varies from – 1 to + 1 and is ± 1 for perfect + ve/– ve correlation (or fit) and 0 for no correlation. As the standard deviation of y is

$$S_y = \pm\sqrt{\frac{\Sigma(y_i - \bar{y})^2}{N}}$$

we can define γ in terms of S_y and S_{yx}

$$\gamma = \sqrt{1 - \frac{S_{yx}^2}{S_y^2}}$$

2.7 Analysis of Time Series

Typical power system load curves can be represented by the equation:

$$Y = T \times C \times S \times I$$

where,

T = long-term trend
C = cyclical trend (often over several years)
S = seasonal trend (1 year cycle)
I = irregular movements ("noise")

The "noise" component is due, in part, to temperature effects. A reasonable correlation between demand and temperature has been found in most power systems. Some time series can be represented by a sum of these factors, i.e.

$$Y = T + C + S + I$$

2.7.1 Estimation of Trend

The least square regression method is most suitable for long-term forecasts. The moving average method can also be used, as this eliminates cyclic, seasonal and irregular patterns. The method is simple and often used for short-term trend estimations. Its disadvantages are that the beginning and ending data are lost and the results can be affected by extreme values. A weighted moving average can overcome this disadvantage, as can exponential smoothing.

2.7.2 Estimation of Seasonal Variations

It is sometimes required to determine how the data in a series vary throughout the year. For this seasonal indexes are helpful. These show the relative values of a variable (related to the average or trend) throughout the cycle. Various methods of calculation include:

(a) Average Percentage Method

The data for each month of the yearly cycle are expressed as percentages of the whole year. The process is repeated for different years and the average of the results obtained. As the base over the whole year is taken as 100%, the average value of the indexes should also be 100% over the whole year, i.e. the sum of the indexes should be 1200%. If this is not so, they should be adjusted by a suitable multiplier.

(b) Ratio to Trend Method

The data for each month are expressed as ratios to the trend values. Again, the values for the corresponding months are averaged and these are adjusted, if necessary, to give an average of 100% to give the required indexes. The seasonal index will include cyclical and irregular variations in the seasonal index as $Y/T = CSI$.

(c) Link Relative Method

The data for each month are expressed as percentages of those of the preceding month. These percentages are called *link relatives.* From these twelve average link relatives, one can obtain the relative percentages of each month with respect to the month of January for each year, which could be 100%. The relative percentages for the following January would not be 100% because of the trend. The various percentages can be adjusted to allow for this trend, producing the required index.

2.7.3 Estimation of Cyclical Variations

The data can be "deseasonalized" by dividing by relevant monthly seasonal index numbers. The data will still include trend, cyclical and irregular movements. We can then adjust for trends by dividing by the corresponding trend values. Irregular variations can then be removed by calculating moving averages over a period of a few months. The cyclical variations would remain. If these are periodic, cyclical indexes can be computed in the same manner as the seasonal ones.

2.7.4 Estimation of Irregular (Random) Variations

The removal of seasonal and cyclic variations leaves out the irregular component. In practice it is found that these are small in magnitude and follow a normal distribution. Thus, the confidence limits in the range of these fluctuations can be established using student's distribution.

2.7.5 Use of Auto-Correlation Coefficients

An auto-correlation coefficient plot for a time series is often useful for identifying cyclic variations, especially when these are not obvious from visual inspection of the time series plot itself. In an analogous fashion to the correlation coefficient which relates similar variations in two or more variables, auto-correlation coefficients indicate the dependence of a process on itself at some previous point in time.

The auto-covariance of a time series is defined as

$$R_{xx}(\tau) = \frac{\sum_{t=\tau+1}^{T} x_t x_{t-\tau}}{T-1-\tau}$$

i.e. the average value of the lagged products of the series, the maximum lag should not exceed $\tau_{max} = T/20$. The auto-correlation coefficient is then defined by

$$\rho(\tau) = \frac{R_{xx}(\tau)}{R_{xx}(0)}$$

The coefficient for $\tau = 0$ is clearly unity and generally drops off as τ increases. Cyclic factors are indicated by peaks in this coefficient. Purely random signals ("noise") have zero auto-correlation except between samples which are identically equal. Figure 2.8 shows a typical auto-correlation of the average demand of a power system showing a four-year cycle.

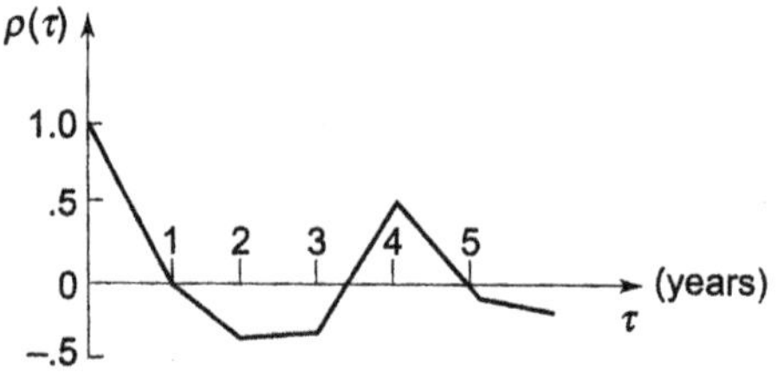

Fig. 2.8 *Typical auto-correlation of average power demand*

2.7.6 Summary of Fundamental Steps in Time Series Analysis

A brief procedure for analysis of time series due to M.R. Spiegel, consists of the following steps:

(a) Collect all data for the time series, ensuring that the data are reliable. Related information should also be collected. If necessary, adjust the data for comparability, e.g. adjust for leap years in energy consumption studies etc.

(b) Draw a graph of the time series. Visual inspection often shows the presence of long term, cyclical and seasonal variations.

(c) Construct the long-term trend, e.g. by the least squares method. If the trend is suspected to be nonlinear, the use of log paper may be advantageous.

(d) Obtain a seasonal index if seasonal variations exist and adjust the data for these variations, i.e. "deseasonalize" the data.

(e) Adjust the data for trend. The resulting data will contain cyclic and random fluctuations.

(f) Graph the cyclic variations of (e). A similar procedure to (d) can be used, if necessary, to obtain cyclic indices.

(g) Forecasts and ranges of errors can be made using adjusted data.

2.7.7 Correlation of Demand with Temperature

There is a fair amount of correlation of the power system demand with temperature. The random variations left in demand after deseasonalizing and removal of the trend effect are largely due to temperature variations. There are two portions of the power system load which are temperature dependent; domestic and commercial loads which increase with cold on account of the use of heating devices, and with heat which necessitates the use of fans, coolers, airconditioners and refrigerators resulting in loads increase.

Some people feel that the yearly seasonal variation in load is due purely to temperature and the correlation between seasonal demand and temperature variations is in fact high. The removal of temperature affects from load readings, however, still leaves cyclic and random effects. This is because similar weather conditions at different times of the year do not cause similar human responses. Other factors, such as wind and rain seem important, but are hard to account for, as the repetition of a certain set of exact weather conditions (e.g. cold night, rain) is unlikely. Thus, a correlation study based on 10–15 years may suffer from the Nyquist criterion, i.e. insufficient sampling data. A typical temperature-demand correlation is shown in Fig. 2.9.

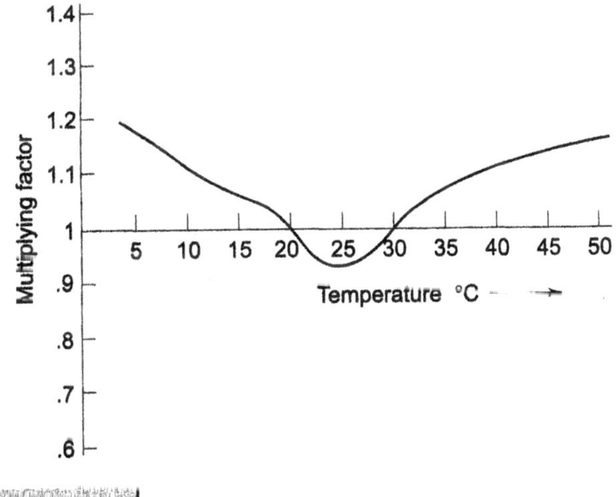

Fig. 2.9 *Typical temperature-demand correlation*

It is not necessary to desensitize load figures to temperature to check the validity of forecast models. Confidence limits can be established from raw data using the analysis techniques described earlier since temperature variations and their effects on demand are already included in the seasonal and random load variations.

2.8 Factors in Power System Loading

2.8.1 Econometric

Certain economic factors which influence the system load growth are as follows:

(a) Business and economic cycle (cyclic variations).
(b) Growth of gross national product (GNP) (long term variations).
(c) Growth in population (long term trend).

Most of these factors only affect the long term trend which will not be picked up in a normal model based on, say, a past history of 10 years (i.e., Nyquist condition). Of course, changes in government policy in, say, population, railway electric traction and integrated social-economic development of rural areas, will result in a change in the long term trend.

EXAMPLE

An examination of various electrical energy forecasts in India reveals that the energy demand with regard to population and GNP leads to a satisfactory linear regression model. The regression model is of the form [4][3]:

$$\ln y = 20.74773 + 2.8815 \ln x_1 + 1.3695 \ln x_2$$

where,

y = electrical energy demand in GWh
x_1 = population in millions
x_2 = GNP in crores of rupees

x_1 and x_2 are graphical projections based on data available from the Planning Commission or the concerned ministry, such as Finance or Social Welfare.

2.8.2 Single Factor Modelling

Single factor model indication may be defective because of the following reasons.

1. It is too general.
2. A sector-like industry occupies a far larger share in the consumption of electricity, say ~40%, as compared to its contribution to its GNP which is only ~23%. On the other hand, agriculture may have a larger share in the GNP, i.e. ~40% but a lower share in electricity consumption ~29%.
3. It is known that the rate of growth of various sectors of the economy is not the same. It is, therefore, preferable to have separate single factor models for electricity consumption for domestic, commercial, industrial, agricultural and other uses.

Domestic

For example, the forecast of domestic consumption by use of population forecast and other variables such as number of domestic consumers and per capital consumption can be a good model of trend line. In the UK, GDP and average temperature gave a reliable model for energy forecasts. The model fitted is of the form:

$$C_t = K + 0.7167 \log \text{GDP}$$
$$= 0.7082 \log C_t - 1 - 0.4957 \log T_t$$

where,

K = constant (can be calculated by the regression method)
T_t = average temperature over a period t (in °F)
GDP = Gross domestic product

Industrial

For industrial projection, the following two trends are important:

(a) The growth in industry as represented by the index of production and growth in the sale of electricity by the utility per unit growth in industry: Two separate graphs can be drawn. For any point of time, if the two quantities are multiplied then the total electricity consumption for the industrial sector can be arrived at.

EXAMPLE: Forecast, say, for 2007

Index of production as projected from graph = 180, (say) Electricity sold per unit of industrial index (from the graph)
= 37 GWh, say (projected). Hence, total electricity sales
= 180 × 37 GWh = 6660 GWh

(b) The growth in workers employed in industry and electrical energy consumed per worker: It should be possible to obtain data regarding industrial workers from either the Central Statistical Organisation or the Ministry of Labour. A trend graph can be established to show a forecast of industrial workers employed in industry at any point of time. A second trend graph has to be plotted for the electrical energy consumed per worker. From these two trend graphs, a forecast can be made for the requirement of electrical energy for the industrial sector.

EXAMPLE: The forecast for, say, 2007:

(i) Number of industrial workers projected = 0.86 million, say
(ii) Industrial electrical energy sales per worker projected = 7750 kWh, say
(iii) Forecast of electrical energy sales
for industrial sector = 7750 × 0.86 GWh = 6665 GWh

For industrial units having a demand of 1 MW and above, trend projections are made on the basis of electricity consumption in the past as per information furnished by such industrial units in the states. The projection of captive power/co-generation is deducted to arrive at the demand on the utility system.

Agriculture

The electricity demand for agriculture can be processed in the same manner as industrial consumption, the independent variable being the agriculture output or added value.

Alternatively, the trend of installation of agriculture irrigation pumps can be established keeping in view the targets fixed by the Planning Commission and perspective plan envisaged by the Ministry of Agriculture and Irrigation or states. A second trend graph can be established for the consumption of electricity per pump. From these two graphs, a trend a graph can be established for electricity consumption for the agriculture sector. A model similar to the following has been adopted by CEA [4]:

$$Y = NSH$$

where,

Y = Electricity consumption in kWh

N = Number of pumpsets as per the programme of energisation in the middle of the target year

S = Average capacity of a pumpset (connected load) in kW in the middle of the year determined on trend basis.

H = Average consumption per year per kW of connected load determined on the basis of the trend.

Other Uses

Forecast of street lighting, water works, sewerage, railways, auxiliary consumption, transmission and distribution losses, etc. can be made by establishing trend graphs based on time series study. Alternately, these projections may be made on the basis of plan targets, wherever deemed feasible.

2.9 Unloading the System

Unloading the system with ***supply-side and demand-side management*** will be a priority for meeting energy and power demands in the 21st century. It is cheaper to save a kW than to generate a kW. In environmental terms, saving energy is saving the earth. The more you know about conserving energy, the less energy you will use. Looking into the Indian power system progressive load curve, the annual growth rate in peak demand is higher than the annual average load growth rate. Economic growth in the country has led to rapid increase in the living standards of the people, resulting in the use of air-conditioners, more home appliances and computers. Also, the rate of increase in the peak demand is higher compared to the connected load. Thus, the system annual load factor is progressively reducing. In Punjab, annual connected load growth is about 90 MW/year and peak demand is growing at the rate of about 150 MW/year. There is an urgent need to take measures for shaving peak and flattening the load curve.

2.9.1 Supply-side Management

Supply-side management (SSM) is planning, implementing and evaluating a utility's improvement programme of the distribution system.

The system losses are proportional to the square of the load. Fig. 2.10 depicts a typical annual load duration curve of a power utility and a corresponding losses duration curve at a 220 kV substation. The load factor and loss factor can be calculated respectively from the annual load duration curve and annual losses duration curve. The respective

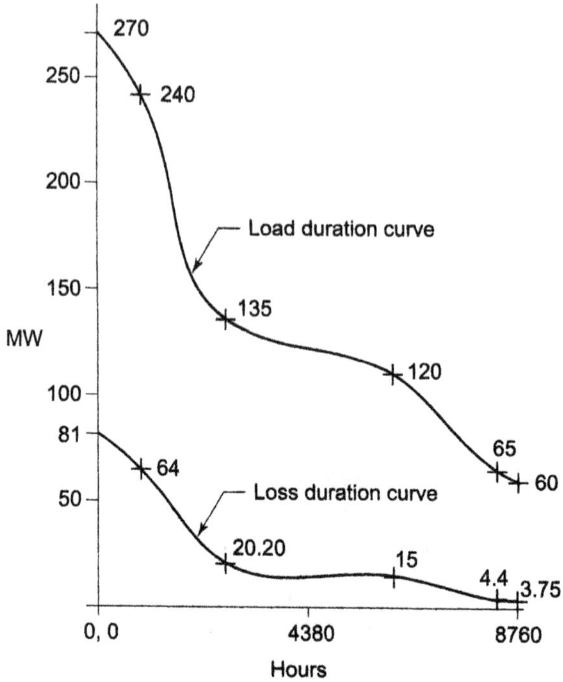

Fig. 2.10 *Typical annual load and the respective losses duration curve for power supply from a 220 kV substation*

factor shall be the area under the curve divided by the area of corresponding rectangle formed by a maximum *x-y* co-ordinates of the respective curve. Average losses in the Indian power system are estimated at 21%. However, the losses are higher during the peak hour of the year or during the peak month of the year. A 6% reduction in average losses is shown in Fig 2.11. Reducing the average losses to 15% (i.e. 6% reduction) can reduce peak month losses in the year to 24% (i.e. 8% reduction) and yearly peak hour losses to 28% (i.e. 12% reduction). This shows that flattening the load curve and consequently improving the load factor leads to significant reduction of losses. Reducing the peak of the load curve will significantly reduce system average losses. The following methods can help to reduce costs:

- Introduce summer season tariff for agriculture. During this time, agricultural load is heavy. The rate may be the cost plus rate of return (say 3%) for four months (May to August). For the remaining months, the tariff could be subsidized.

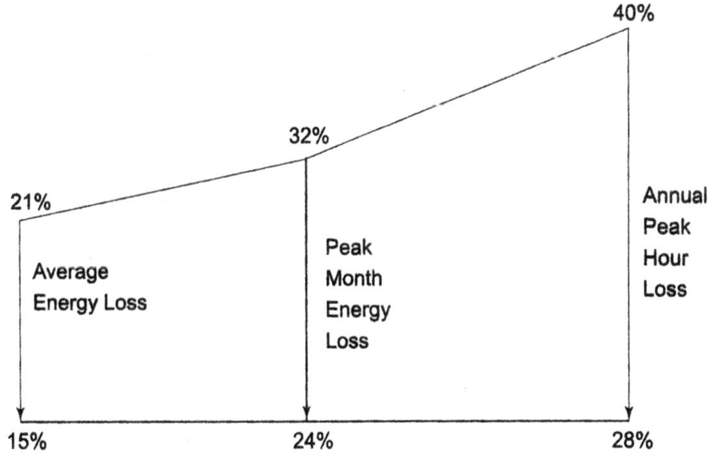

Fig. 2.11 *Effect of a 6% average energy loss reduction*

- Introduce Time-of-Day tariff for industry (large and medium), commercial and bulk power supplies.
- Power connection for the industry during night shift (10 pm to 6 am) should be given at a concessional tariff. This tariff may be based on energy-related cost only. This can divert significant consumption from day time to the night.
- Shop timings should be shifted/staggered during evening peak hours by 1/2 hour between the five different zones in the country. Such a step during the critical months shaves peak significantly. For example, advancing the closing hours of shops from 8 pm to 7 pm in Punjab shaved the summer peak by about 80 MW during the year 2001.
- Stagger weekly holidays of industrial complexes and shopping centres on different days of the week in different zones. This arrangement in Punjab during summer (2001) shaved peak by 150 MW.
- Levy peak load restrictions to reduce demand and to reduce the relatively higher system losses during peak hours.

2.9.2 Demand-side Management

Demand-side management (DSM) is planning (see Section 3.1), implementing and evaluating utility-sponsored programmes to effectively utilise consumer energy and power usage. The utility activates designs to help the consumer use electricity more efficiently. It is energy management across the meter. Among the various options studied (1999) by the World Bank in Thailand, DSM was considered cheapest as given below:

Option	Cost (US Cents/kWh)
Demand-side management (DSM)	2.1
Hydro power import from Laos	2.6
Combined cycle gas plant	4.0
Lignite with high-efficiency FGD	5.1
Low-sulphur imported coal without FGD	5.2
Imported LNG	5.4

DSM projects may be prepared for implementation on a cost-benefit basis. These can be:

- No cost.
- Low cost.
- Requiring investment in energy-efficient equipment and appliances.
- Energy conservation reduces demand and energy consumption by efficient use. The utilities' need for a new generation and distribution system and consequently investment is reduced. Energy conservation measures at the consumer's end may be taken as explained in Section 8.12.
- Reducing pilferage of energy (see Section 8.14).
- Keep tabs on consumer capacitors at the consumer's premises for healthy working. A system of penalty may be introduced for faulty units.
- It is seen that 1 MW saved at consumption point is equivalent to saving an installed generation capacity of 2.3 MW [9]. This also helps the environment: Each MW saved leads to an approximate saving of:
  (a) 17 tonnes of coal/day
  (b) 3.4 tonnes of coal dust/day
  (c) 0.13 tonnes of SO_2/day
  (d) 0.18 tonnes nitrogen oxides

In India, there is scope for a 25–30% reduction in energy consumption and about 40% demand reduction when energy audit and conservation programmes are implemented. The basic requirements for conserving energy in any system/plant are:

- Reduction in specific energy consumption
- Full capacity utilisation
- Proper maintenance
- Technology upgradation

Equipment procurement should be made on the basis of life cycle costs so that the cost of energy consumed by the equipment is capitalised. Proper energy conservation services in the country need to be developed. Energy Saving Performance Contracting (ESPC) is one method that can be used to reduce consumption/demand at no capital cost to the consumer. Under ESPC, an Energy Service Company (ESCo) pays all costs involved in identifying, installing, operating and maintaining energy-efficient equipment and improvements. ESCo is compensated by receiving a share of the savings resulting from these improvements. At the end of the contract period (10 to 25 years, depending upon the share), the consumer owns all the improvements and receives the continuing savings. For example, Tata Honeywell Ltd (THL) is executing 'Performance Contracts' where it guarantees 20–40% power savings to clients who are traditionally burdened with high-energy bills. Without any investment on the part of the consumer, THL does an energy audit and demarcates the area where energy consumption can be reduced using automated power equipment. After the contract period is over, the new power-saving equipment is reverted to the client (consumer) at no cost. In the Philippines and Thailand, feeder-wise energy audit is being done by ESCo and benefits are shared.

2.10 Forecast of System Peak

With the major categories of energy sales (domestic, commercial, industrial, agricultural and some very heavy individual consumers, if any, such as a steel plant) forecast, remote points established, intermediate values fixed and trends determined, the total requirement of energy is obtained by adding certain other items of consumption, such as public lighting, water works, sewerage pumping, electric traction (railway/tram/trolley bus), auxiliary consumption in generating stations and a reasonable allowance for losses. The total losses are expressed as a percentage of the total energy accounted for. After all adjustments and additions have been made, the final estimate of the energy requirement can again be checked for accuracy by a direct projection of the energy requirement which goes back to several years.

(i) Only a forecast of the load factor is now required to arrive at an estimate of the system peak.

$$\text{Annual System Peak} = \frac{\text{Energy requirement}}{8760 \times \text{Load factor}}$$

To project the load factors, these are plotted against time series to establish the trend graph. Also, some correlation between these load factors and economic activity can be made. The highest load factors tend to accompany great economic activity, conversely the lowest load factors are often associated with low economic conditions. However, it can be safely said that the relationship is not linear, but still some weightage may be given to the expected improvement in economic activity, as anticipated by economists, for forecasting the load factor.

(ii) For any given year, the system peak P can be represented as:

$$P_1 = K + dR_1 + c\,C_1 + fI_1$$

and for any other year:

$$P_2 = K + dR_2 + c\,C_2 + fI_2$$

where,

P = yearly peak
R = yearly domestic sales
C = yearly commercial sales
I = yearly industrial sales
K = an additional constant
d, c, f = multiplication constants.

With the use of data P, R, C and I for an adequate number of past years say 10, values for constants k, d, c, f which statistically must nearly fit the conditions, can be determined by the method of least squares. Thus, the equation is solved with the help of historical data.

Two such equations which have proved useful in the USA for forecasting the December peak in terms of December sales are:

$$P = 110 + 2.408\,R + 3.2\,C + 2.157\,I \quad \text{(I)}$$

$$P = 98.5 - 5.249\,T + 2.559\,R + 3.14\,C + 2.525\,I \quad \text{(II)}$$

Here P is in MW and all sales in GWh adjusted for average December metering interval with the addition in second equation of a factor T representing time, i.e. the number of years which have elapsed since the base year when the time series on which the equation is based began. These equations gave an excellent forecast of system peaks, the average error being less than one per cent. One relationship of GNP (Gross

National Production) to electricity consumption in USA has been found to hold is:

$$E = -766 + 2.15\ (\text{GNP})$$

where E is in TWh and GNP is in billions of US dollars.

The relation of GDP (Gross Domestic Production) to energy consumption is one which has been studied throughout the world and is regarded as an important economic indicator. Clearly, changes in technology, pressure on energy resources or redirection of emphasis in the types of energy source will not only affect the electric energy consumption, but the energy consumption generally on a national and world-wide scale and, consequently, on the world's living standard.

The elasticity of consumption, with respect to GDP in India during 1980–81 to 1998–99 works out to 1.41. This implies that an increase in GDP by 1% resulted in a 1.41% increase in consumption. Plan-wise, the elasticity has declined from over 3 in the First and the Second Five Year Plans, to 0.97 during the Eighth Plan, as can be seen from Table 2.5. [13]

Table 2.5

Elasticity of electricity consumption with respect to GDP in India

National plan	*Elasticity*
First Plan	3.14
Second Plan	3.38
Third Plan	5.04
Fourth Plan	1.85
Fifth Plan	1.88
Sixth Plan	1.39
Seventh Plan	1.50
Eighth Plan	0.97

2.10.1 Capacity Forecast Model

In the residential, industrial and commercial sectors, electric energy forecasts are made on a national basis, as discussed earlier in Section 2.8.2. National projection of electrical energy is converted into regional peak demand expectations. From the peak demand, regional capacity requirements are developed over the study period. Current regional generating capacities and planned capacity additions, using the specific plans of the utilities, are removed from these gross capacity needs. Added to this, are the planned retirement or decommission of the units. The net results are the regional needs for new capacity, as shown in Fig. 2.12.

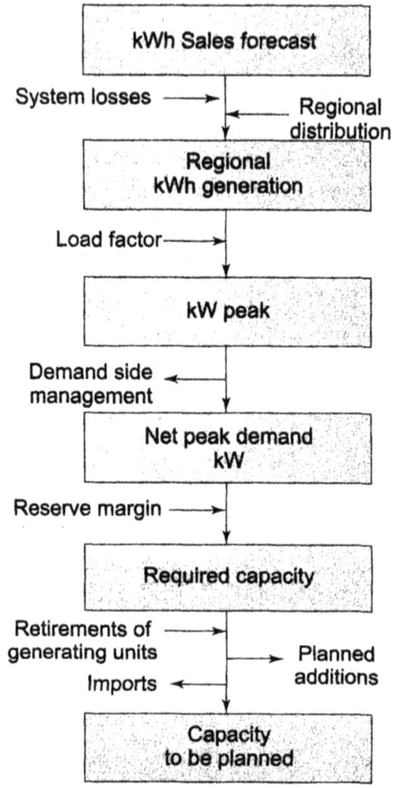

Fig. 2.12 *Capacity forecast process*

2.11 Strategic Forecasting (10)

Strategic forecasting is becoming increasingly important nowadays, and involves the explicit examination of the factors and issues affecting future growth. It recognises the impact that policy decisions can have on future loads. This requires details of consumer's operations, their current and potential demand for electricity, their competitiveness in the market place and their options with respect to production processes, switching alternatives, etc.

In the industrial sector, this implies combining elements of the econometric approach with the technology detail found in end-use/process models. Strategic models must be capable of doing more than merely forecasting future requirements. They must be able to provide planners with additional information to help shape the future demand.

2.12 Spatial Load Forecasting

The load forecast used in distribution planning must also indicate their locations in sufficient graphic detail, to allow the planner to locate and size distribution equipment additions, such as substations, feeders and their spurs. In the first place, a different consumer mix on each feeder will lead to differential growth rates. In the second place, electricity demand growth is a combination of increase in specific consumption and increase in the number of connections. The first is required mainly for reinforcement of existing feeders; the second is more likely to be associated with expansion of the system due to peri-urban or ribbon development.

Details of specific consumption and the number of consumers should be collected for each tariff group. These can be extrapolated using simple methods of time series analysis. Large consumers supplied at transmission or subtransmission voltage should be excluded. Feeder metering exercises, as well as consumer billing data for identified feeders, should be carried out. These can be used as sample basis for determining *Diversity Load Factor* for a consumer group and distribution losses can also be determined. On a feeder by feeder basis, likely growth should be estimated. For inner city areas, it is likely that growth will be purely due to increase in specific consumption by the associated consumer base. Development plans for suburban and peri-urban and rural areas will probably be determined from the authorities, over the period under consideration, as planned targets for the development of respective areas.

Further change in land-use-projections of where, how much and what type of new residential, commercial and industrial consumers will occur on small-area basis (grid of small area) and a 5 year or more time-frame can be used for spatial forecast. Every new small area growth pattern generally follows what is known as an 'S' curve (see Fig. 2.13). At low spatial resolution (i.e. when using a large number of small areas), the

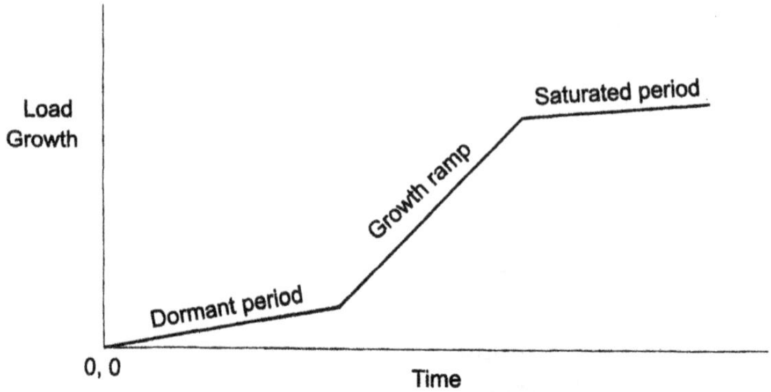

Fig. 2.13 *New small area load forecast*

sharpness of the growth ramp of the 'S' curve is less, compared to high spatial resolution (i.e. a smaller number of small areas) where most of the growth appears as a sharp burst. However, for big-area forecasting (e.g. a whole 11 kV feeder), the growth is smooth, because the timing of the growth ramp of the 'S' curve differs in each small area of the feeder (see Fig. 2.14). A small area is usually 1 km^2 in urban areas and constituting each town or village in sub-urban or rural areas. The land-use map is converted to electric load, by using kW per acre of load on land-use class basis. Agriculture load can also be projected on the basis of projections of land irrigation, based on pending applications of prospective agriculture consumers and the requisite availability of ground water in the area.

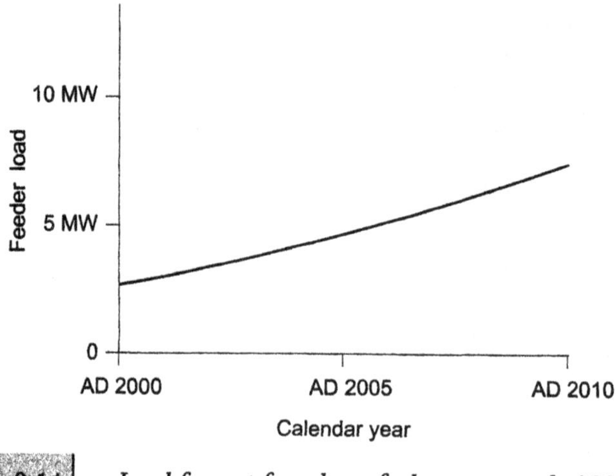

Fig. 2.14 *Load forecast for a large feeder area, nearly 200 km^2*

The judicious use of this data can lead to reliable and consistent spatial load forecasts. It is not expected that the methodology be analytically rigorous. The expectation is that a fair amount of judgement and local knowledge will be used.

The demand on the subtransmission system, will be based on the coincident demands of the distribution feeders and consumers, supplied at the subtransmission voltage. All consumers at this level have maximum demand metering. The data is generally available to determine coincidence factors from the substations.

EXAMPLE

The 16th Power Survey Committee [4] has estimated energy requirement and load factor for the year 2004-05 as 635065 GWh and 70.96% respectively. Therefore, annual peak demand for the year 2004–05 is

$$= \frac{635065}{8760 \times 0.7096} = 102161 \text{ MW}$$

2.13 Technological Forecasting

Adoption of new technologies affect load demand. Distribution automation, use of electric cars, providing roof-top solar photovoltaic panels or inverters or portable generating sets by various categories of consumers will curtail the power utility's maximum demand.

2.13.1 The Delphi Technique (15)

The Delphi technique was a product of the Rand Corporation think tank and utilizes a group approach to provide generalized forecasts. It works on the principle that "many minds are better than one", but adopts a procedure which avoids the pitfalls into which committees can usually fall.

There are advantages in the use of panels or committees of experts in a field such as forecasting. But often, a number of dangers occur in the proper running of such a committee, e.g. a "personality clash" is often found among various experts who have strong views about the correctness of their own opinions. This may result in the opinions of a strong personality or majority overpowering the opinions of the less domineering or minority, though these may be of equal importance. The outcome of such a committee, therefore, is often not as objective as it could have been.

The use of experts, i.e. persons with a particular knowledge in a field or related fields in which the forecast is required, can be of great benefit especially where the event has considerable implications in the social, ecological and technical areas. The Delphi technique circumvents the problems faced by earlier committees by using three main features. Anonymity is achieved by physically separating each panel member so that they never meet face to face or are identified to each other. A panel director is appointed to pose questions to each panel member in writing to which responses are also made in writing. They are then circulated to the other members without giving any indication of the source. This controlled feedback is completely channelled through the panel director who not only poses the original questions and receives back responses, but also edits them to ensure that extraneous material is removed. It can be seen that the role of the panel director is a rather demanding one and, as in a normal committee, the calibre of the chairman will make or break the results of the group.

The Delphi technique is most useful in areas which are too new to have accumulated enough historical data to permit the use of other mathematical methods, or interdisciplinary areas where a number of separate fields of knowledge have a bearing on the matter. It can be seen that the forecasts derived under the Delphi technique would only be in general terms and in the electric power industry, would probably be more applicable to the determination of long-term trends.

An example of how a Delphi panel operates is given below.

Suppose that each member of the panel is asked to predict the probability of certain events occurring by certain dates. The panel is presented with this initial questionnaire as the first 'round'. The questionnaire is framed in an open ended manner to permit the panel members to insert any items which they feel are pertinent, but which were not named in the questionnaire. The panel director receives the responses, tabulates the dates of named events and also includes any extra material listed by the panelists. He then compiles a list of the events and dates along the following lines.

Event No. 1	2007	50% chance
	2008	30% chance
Event No. 2	2008	40% chance
	2010	30% chance

This list with the dates then becomes the list for the second 'round' when the panel is instructed again to give a 50% chance date for all events. (The reasons for the estimates given could also be included.) The second set of responses are then tabulated by the panel director who produces a new list with further reasons. This in turn becomes the basis of a third questionnaire for the panel. The process is repeated until either some degree of convergence has been obtained or some definite opinions are formed by the panel members. Usually, three 'rounds' are necessary to achieve satisfactory results; occasionally, four or five may become necessary. The final result gives the most favoured opinion, along with the degree of divergence. At times, convergence will not occur. This generally indicates that the problem has not been adequately defined, in which case it is best to restate the problem and start afresh with another panel.

Some areas where a Delphi panel could possibly be used in the electrical, power industry, might be to predict the effects on demand, say, of a major change in technology, such as electric cars, widespread use of solar water heaters or natural gas, etc.

2.13.2 The Scenario Technique (5)

The scenario technique is basically a method for viewing the future in a quantitative fashion. A narrative describing the probable or possible sequence of events is developed. The events are either sequentially or simultaneously recorded and built up into a story or case history. At branch or decision points, the narrative is often directed towards a set year. All possible outcomes are investigated. The sort of decisions or assumptions which might be made by a firm developing such a scenario might be: "What should be our staff size?" "Should we in the future retire people at the age of fifty?" Or perhaps questions more relevant to the power industry: "Should we make all the city lines underground after a certain date?", "When switching over to computer based distribution automation?" and so on.

The major differences of the scenario technique with normal planning processes are:

(a) As with the Delphi technique, this is a quantitative forecasting method, generally applicable for a very long term forecast.
(b) It is not confined to the historical past for checking alternatives.
(c) It allows longer time spans to be considered in detail.
(d) It demands increased attention to detail.

2.14 Sources of Error: Regulating the Model

Forecasting, like all fields of statistics, can never be expected to yield *exact* answers. As outlined in Sec. 2.8, many economic and other factors influence the load and a model cannot hope to identify or use these factors. However, used intelligently, mathematical trend curves are a useful forecasting tool, which for reasonable span of time, can produce unbiased forecasts within a defined band of error.

Generally speaking, errors in forecasting arise as a consequence of one or more of the following factors:

(a) The trend in the future may not follow the kind of curve chosen on the basis of past data. There may be fundamental changes in certain factors which were not embodied in the model. These factors are usually economic.

(b) The normal statistical errors which were discussed in Sec. 2.6 are due to the derivation of parameters for trend curves based on limited data and scatter of points about any curve.

(c) Certain exact conditions may not repeat themselves during the sample (e.g. in demand/weather correlation studies) and thus the model may be based on insufficient data in case of certain variables.

A rule of thumb approach to these errors particularly in the case of (a), is that the uncertainty of the forecast will tend to increase as the square of the time interval between the forecast date and the last sample date. Therefore, it is clear that caution should always be taken in preparing long-range forecasts. Power authorities may attempt to produce quantitative forecasts for periods exceeding 10–12 years ahead and to be continuously updated every five years. The percentage of annual load growth can be calculated as below:

$$\%\text{age annual load growth} = [(\text{Load ending base year/Load forecast ending } n\text{-years horizon})^{1/n} - 1] \times 100$$

PROBLEMS

1. A domestic consumer installs the following points/equipment at his premises:

(a) Incandescent lamp points	10 nos.
(b) Fluorescent tubes 1200 mm	15 nos.
(c) 6A socket outlets	14 nos.
(d) 16A socket outlets	9 nos.

(e) 2.5 kW air conditioner installed separately on 32A socket outlet
(f) Two electric bells

(g) Ceiling fans 8 nos.

Compute the connected load.

2. Which of the following are the constant impedance loads: computer system, fan, cooler, room heater, pure resistance.
3. If a system is at an operating frequency of 47 Hz, how does it affect the domestic refrigerator?
4. What do you understand by the System Regulation Coefficient? What is its value for various regional operating grids in the country?
5. A 50 MW hydro-generator delivers 320 million kWh during the year. Calculate the plant load factor.

BIBLIOGRAPHY

1. Murthy, P.S.R., "*Power System Operation and Control*," Tata McGraw-Hill, New Delhi, 1984, p. 47.
2. Harrison, J.A., '*The Essence of Electric Power Systems*', Prentice-Hall of India, New Delhi, 1997, p. 62.
3. Satsangi, P.S. and A.K. Garhwal, "Electrical energy demand forecast for India". *Proceedings*, Fifth National Systems Conference, Punjab Agricultural University, Ludhiana, Part 3, 4–6 September 1978, pp. 56–59.
4. 16th Electric Power Survey of India, Central Electricity Authority, Govt. of India, New Delhi, September 2000, p. 24.
5. Ringland Gill, *Scenario Planning–Managing For Future*, John Wiley & Sons, Chichester, 1998.
6. Swift Louise, *Mathematics and Statistics for Business, Management and Finance*, Macmillan, 1977, pp. 864–896.
7. Ready, T.A., "Statistical Analyses of Electricity Use", *Energy*, Vol. 15, No. 1, Pergamon Press, U.K., January 1990, pp. 45–61.
8. '*Public Electricity Supply*', All India Statistics 1997–98, General Review, Central Electricity Authority, New Delhi, pp. 203–207.
9. Pabla, A.S., '*Electrical Power Systems Planning*, Macmillan India Ltd., New Delhi, 1998, p. 283.
10. Ahmad Faruqui et al., "Demand Forecasting Technologies", *Energy*, Vol. 15, Nos. 3, 4, Pergamon Press, U.K., March–April 1990, p. 287.
11. *Directory of Computer Software for Power System*, Central Board of Irrigation and Power, New Delhi, April 1988, pp. 72, 74.
12. '*National Power Plan 1985–2000*', Central Electricity Authority, New Delhi, June 1987, pp. 49–55, 372.
13. '*Annual Report on the working of State Electricity Boards and Electricity departments*', Planning Commission, Government of India, June 2001, p. 13.
14. 'Progressive Farming', *Journal of Punjab Agricultural University*, August 1993, p. 20.
15. Shim Jae K. '*Strategic Business Forecasting*', St. Lucia Press, New York, 2000, p. 254.

Chapter 3

System Planning

3.1 Planning Process

Planning is the process of taking a careful decision. The main input in planning is quality systematic thought. It involves selecting the vision, values, mission and objectives, and deciding what should be done to attain them. Basically, the objective of distribution planning is to provide satisfactory service at the lowest possible cost. The components of the planning process are shown in Fig. 3.1. In a power utility, this process seeks to identify the best schedule of future resources and actions to achieve the utility's goals.

Planning is driven by two inputs: Future needs and time to fulfil these needs with defined priorites in Master Plan. Long-term planning (5- to 10-year horizon) determines the power energy forecasts and optimum network arrangements. What investment would be required and the timing of these to obtain maximum benefits. Network planning covers individual investments in one or two years as a medium-term planning within the period of a long-term plans. A short-term plan covers the current or next year ie annual plans for each year on the horizon. This covers engineering design of each network component, smaller and more common type of investments for system improvements. The following steps are involved in the planning process:

(i) Feasibility studies are carried out to identify, evaluate and finalise the best plan.
   Define the problem ⟶ Find the alternatives ⟶ Evaluate the alternatives ⟶ Select the best one.

(ii) A project report for long, medium and short-term works along with the action plan/pert chart/bar chart for each activity/work is prepared. Dates are set for milestones i.e. important events measurable along the path to the fulfilment of the plan.

(iii) Final approval is accorded after financial and economic appraisal.

(iv) Once the best plan has been selected, the next process of implementation begins.

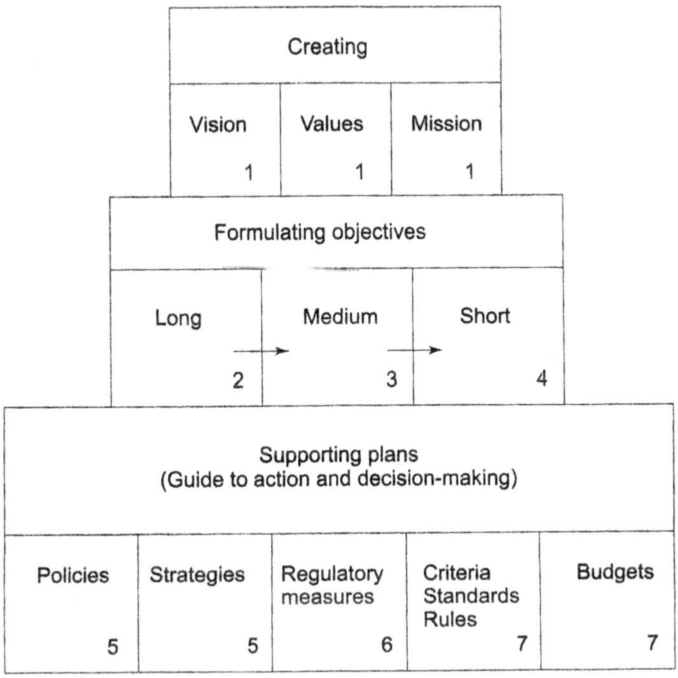

Fig. 3.1 *Components of the planning process*

In executing the plan, monitoring is important. Develop a detailed task list that supports the milestones as '*the devil is in the detail*'. Without detailing, who will do what and when, the plan may not be successfully implemented. The planning action is shown in Fig. 3.2.

3.1.1 Basic Principles in Distribution System Planning

Any distribution system planned is used to transport a certain amount of power to maximum capability from the source point at one location to another location with certain consequences. In essence, sub-

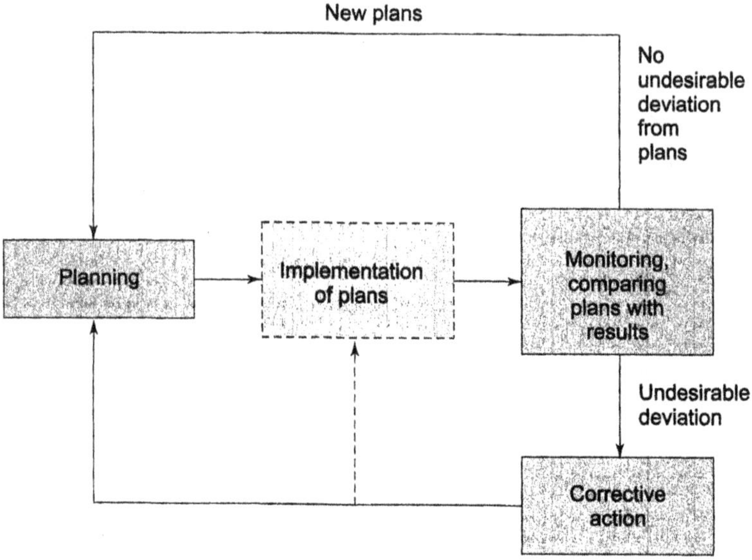

Fig. 3.2 *Planning actions*

transmission and distribution lines transport power from one bulk power location to the consumer site, and transformers change the voltage level of the power, considering the following basic principles:

- It is more economical to transport power at a high voltage. The higher the voltage, the lower the cost/kW to transport power to a distant point.
- Electricity travels as per Kirchhoff's current and voltage laws. It follows the least resistance path.
- Power must be delivered in relatively small quantities at a low voltage (e.g. 400/230 V) level.
- Voltage drop occurs from the source point to the end location.
- Losses in power are incurred, creating a cost.
- Equipment and labour come at a cost.
- Operation and maintenance add to service cost.
- Future growth accounting is survival.
- When power is used for any purpose by the consumer, the responsibility lies on the consumer to share the degradation of environment on this account.
- Nominal rated system voltage is the most efficient voltage for equipment operation. A rise above this voltage tends to reduce the power factor of equipment (see also Sections 2.1.2 (f) and 15.1.1).

- Segregation and rostering of agriculture supply on a feeder is a step in the direction of economy and to supply power at low cost to agriculture consumers.
- *Electricity market:* Wholesale, select retail, bi-lateral contracts will cut-down the cost of supply if adequate power surplus and grid links are available.

EXAMPLE

At a 66 kV substation (namely Almgir in Punjab) having 66/11 kV, 20 MVA power transformer, 11 kV bus incoming load was 465 amperes with bus voltage 11 kV and power factor 0.96 lagging. The load was mixed feeders, rural lighting, agriculture and predominantly industrial. The shunt capacitors installed at the substation were out of circuit and not connected to the 11 kV bus. When the bus voltage was raised to 11.2 kV with the help of the transformer's OLTC, the power factor was 0.95 lagging. On increase in voltage to 11.3 kV, the power factor further reduced to 0.94 lagging.

The main steps in planning power distribution are shown in Fig. 3.3. The goal should be to provide electricity at the lowest possible economic and social cost.

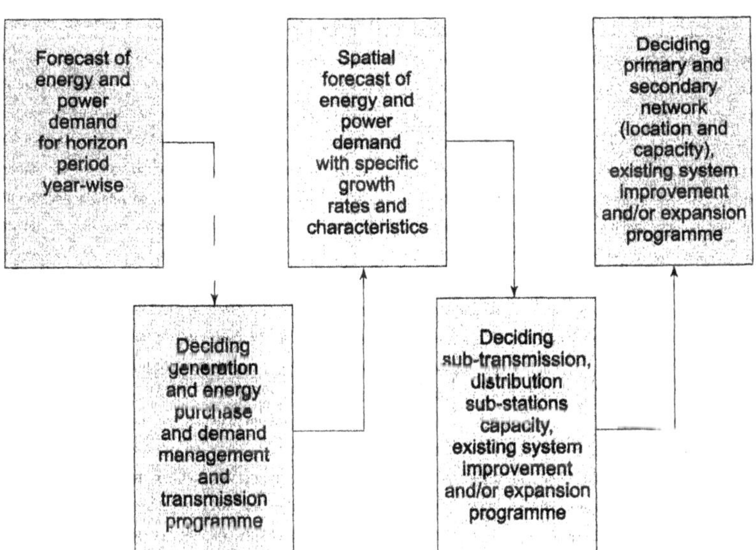

Fig. 3.3 *Steps for Power System Planning*

3.1.2 Planning Methods

Planning of integrated resources requires the following two methods:

(i) Traditional Least Cost Planning

It is a process by which utilities minimise the cost of supplying a given amount of electricity. It is the method of acquiring resources at the lowest cost, taking into account all possible means of meeting electricity needs and all resource costs including construction, operation, sub-transmission, distribution, consumer and environment costs.

Firstly the existing system inadequacies are identified as:

- Poor voltage regulation;
- Higher system losses;
- Higher equipment failures/breakdowns and/or higher line break-downs/trippings;
- Bad quality of power supply; and
- No scope for future load growth.

The initial system improvements can be very cost-effective in removing the above inadequacies as compared to the other alternative of laying a new extended system. Thus, there are two options:

(a) *System improvement:* Augmentation and strengthening of the existing system; improving the reliability and quality of supply; reduction of commercial and technical losses, and/or

(b) *Expansion of the existing network:* The least cost optimal solution from various alternative schemes may be worked out considering the capital cost of the proposed works and present values of the kW and energy losses over the expected life of equipment in case of expansion of network. When the augmentation and strengthening of existing system are involved, the benefits of saving in losses (kW and energy), net revenue increase due to additional sale of power and energy after adjusting the expenditure incurred on generation of the additional energy. The net present values of alternate plans are compared to choose the least-cost solution. Also, financial analysis of the chosen scheme is done to satisfy the funding organization. The economical and financial analyses are discussed in Sections 3.5, 6.2.1 and 12.18.1.

(ii) Demand-side Planning (DSP)

It is the process by which power utilities quantify and assess programmes to alter the pattern and level of their consumers' demand for electricity. This is planning at the consumer level and often has a long planning period, much longer than distribution planning and often as long as that of generation. It can take years of slow progress to obtain meaningful levels of participation. According to a study, the low-cost demand side options can be:

Item	Approximate cost (2000)
Implementing energy conservation programmes	Rs 1500/kW
Providing vigilance and detection of theft	Rs 50/kW
Providing metering	Rs 300/kW

As explained in Section 2.9.2, demand-side planning (DSM) measures require special programmes that try to mobilize cost-effective savings in electricity and peak demand. Numerous studies in India, China and other countries have found that cost-effective DSM programmes can reduce electricity use and peak demand by approximately 20 to 40 per cent. DSM benefits households, industry, agriculture, utilities and society in the following ways:

- Reduces consumer energy bills.
- Reduces the need for power plant, transmission and distribution construction.
- Stimulates economic development.
- Creates long-term jobs that benefit the economy.
- Increases the competitiveness of local enterprises.
- Can reduce maintenance and equipment replacement costs.
- Reduces local air pollution.
- Reduces emissions that contribute to national and international environmental problems such as acid rain and global warming.
- Enhances national security by easing dependence on foreign energy sources.
- Can increase the comfort and quality of work spaces, which in turn can increase worker productivity.

Utility DSM programmes generally fall into three main categories:

1. Conservation programmes

Reduce energy use with programmes to improve the efficiency of equipment (like lighting and motors), buildings and industrial processes as per the Energy Conservation Act, 2001.

2. Load-management programmes

Redistribute energy demands to spread it more evenly throughout the day. Some of the ways of doing this are: Load-shifting programmes (reducing air-conditioning loads during periods of peak demand and shifting these loads to less critical periods), time-of-use rates (charging more for electricity during peak demand) and interruptible rates (providing rate discounts in exchange for the right to reduce consumers' electricity allocation each year during a few hours when electricity demand is the highest).

3. Strategic load growth programmes

Increase energy use during some periods, e.g. encourage cost-effective electrical technologies that operate primarily during periods of low electricity demand.

Within these categories, the following approaches can be used:

1. General information programmes to inform consumers about generic energy-efficient options.
2. Site-specific information programmes that provide information about specific DSM measures appropriate for a particular industry, agriculture or home.
3. Financing programmes to assist consumers to pay for DSM measures including loans, rebates and shared-savings programmes.
4. Direct installation programmes that provide complete services to design, finance and install a package of efficiency measures.
5. Alternative tariff programmes including time-of-use tariff, interruptible tariff and load-shifting tariff. These programmes generally do not save energy but they are effective ways to shift loads to off-peak periods.
6. Bidding programmes in which a utility solicits bids from consumers and energy service companies to promote energy savings in the utility's service area.
7. Market transformation programmes that seek to change the market for a particular technology or service so that the efficient technology is in widespread use without continued utility intervention.
8. Load Limiters are effective in demand-side management, as they limit the maximum power that the consumer draws from the supply. Widespread use of Load Limiters for low-consumption

rural/urban slum consumers instead of meters can result in substantial savings on transmission, distribution and generating equipment (see Section 14.7.1).

The process of designing and implementing DSM programmes generally consists of the following steps:

- Identifying sectors, end-uses and efficiency measures to target;
- Developing programme designs;
- Conducting cost-effective screening;
- Preparing an implementation plan;
- Implementing programmes and
- Evaluating programmes.

3.2 Planning Criteria and Standards

Criteria and standards together form a set of requirements against which the planning process can compare alternatives in the evaluation and final choice. A distribution plan must provide good economy and also satisfy various criteria and standards. Criteria are rules or procedures. Standards are the specifications to ensure that the system is built with compatible equipment that will fit and function together when installed and maintained in an economical manner. Standards and criteria and their applications to the planning process depends upon vision, mission and the value system of the utility. Criteria and standards convey direction in the Master Plan. The following can be the typical criteria for planning:

- A Perspective Plan for the next 15 years to meet the anticipated load growth and forecast load centres. The plan will be reviewed yearly on the basis of annual plans with respect to targets achieved.
- Detailed Project Reports be framed to identify the system strengthening works on long-term and short-term bases:
  - (i) Feeders having poor performance: re-configuration (bifurcation or trifurcation, etc.) of feeder or augmentation of line conductors and distribution transformers;
  - (ii) New technology deployment for system improvement; and
  - (iii) Loss minimization plan.
- Demand-side Management Projects Reports be undertaken on pay-back period financial analysis to achieve tangible reduction in demand and energy consumption in the planned horizon year.

- *Security:* (i) In industrial cities, alternative source of supply be provided by using sub-transmission open ring circuit of 33 or 66 or 132 or 220 kV.
  - (ii) 11 kV open ring main system be provided in all urban estates.
  - (iii) In case of important or essential low voltage consumers, alternate supply arrangement from adjacent distribution transformer be provided.
  - (iv) Separate independent feeders be laid down for major industrial consumers.
  - (v) In case of rural area, separate feeders be provided.
- The following voltage levels be used for release of power connections to consumers:
  - (i) Connected load upto 10 kW to be supplied at 230 V, single-phase, two-wire.
  - (ii) Connected load between 10 kW and 50 kW to be supplied at 415/230 V, three-phase, four-wire.
  - (iii) Load demand between 50 kW and 5 MW to be supplied at 11 kV.
  - (iv) Load demand between 5 MW and 30 MW to be supplied at 33 or 66 kV.
  - (v) Load demand between 30 MW and 50 MW to be supplied at 132 kV.
  - (vi) Load demand above 50 MW to be supplied at 220 kV.
- Economic appraisal of alternate plans be done on least net present values.
- Power utility would create and use load research facilities in order to identify consumer load profiles in the respective geographical area of the system to forecast changes in the load.
- The distribution system for historical buildings of national importance be underground.
- The number of 11 kV outgoing feeders at the distribution sub-station should not exceed 15. The length of 11 kV outgoing feeder emanating from the sub-station upto tail end not to be more than 12 km.
- Loss minimization could be achieved by the following measures:
  - (a) LT line not exceeding 0.8 km. In city area, the aerial bunched cables be used for LT.
  - (b) Improved metering, i.e. electronic meters be provided for all types of consumers.

- Three-phase fault levels should not exceed 2000 MVA and 750 MVA respectively at 66 and 33 kV level. Fault level at 11 kV not to be more than 350 and 250 MVA in urban and rural areas respectively.
- Total harmonic distortion at any voltage level should be within 5 per cent.

Planning standards exist for reasons of efficiency, to achieve the greatest economy or utility convenience. For example, code of practice for equipment design, layout, loading, performance, voltage and service quality standards, location of substations and methods of economic evaluation assure the electrical needs of its consumers. The system may conform to various Indian Standards, Indian Electricity Rules, Rural Electrification Corporation standards, IEC, ISO and the Electricity Act, 2003. Better, improved or additional criteria and standards may be required by a power utility to achieve their objectives. The Indian Electricity Rules can be amended by a state government to serve its ends in developing the power distribution network. The planning standards can be of the following types:

- Development of a standard cost structure for material and labour rates for different voltage systems to be used in the estimation.
- Standards for system voltages as per Indian Standards and Voltage regulation as per IE Rules.
- The load growth of at least 10 years will be taken into account to prepare new or system improvement schemes.
- Shunt capacitor fixed/switched type shall be installed in the distribution system at substations to improve the power factor and voltage profile and reduce transmission and distribution losses. The size and location will be determined by load flow studies for maximum and minimum load conditions.
- Fixed LT capacitors on the distribution transformers shall be installed.

3.3 System Development

Large amounts of power are generated at power plants and sent to a network of high-voltage (220, 132, 110, 66 or 33 kV) sub-transmission lines. This system deals with longer distances and increased power requirements. These lines supply power to distribution centres (distribution substations) feeding distribution primary system, which supply power to the still lower voltage (0.415 kV) distribution secondary system. Thus, the total network is a complex grid of interconnected lines. This network has the function of transmitting power from the points of generation to the points of consumption. Power utilities should plan their investment programme 5 to 10 years in advance through annual plans with a detailed list of investment. The route and locations of sub-transmission lines and distribution substations is made after carrying the computer-based load flow studies of various alternatives. The distribution system is particularly important to an electrical utility for two reasons: Its proximity to the ultimate consumer and its high investment cost. The objective of ***distribution system planning*** is to ensure that the growing demand for electricity, with growing rates, can be satisfied in an optimum way, mainly to achieve minimum total cost of the distribution system expansion. Therefore, the distribution system planner partitions the problem of planning the total distribution system into a set of sub-problems that can be handled by using available, usually heuristic methods and techniques [18].

3.3.1 Sub-transmission

The sub-transmission designates the circuits which deliver energy from the transmission system to the primary distribution system. Usually the sub-transmission system is supplied by the transmission sub-stations, and is still referred to as the sub-transmission. Many sub-transmission systems were previously the transmission lines. Load growth and demand for more power resulted in the transmission voltage being too low. As a result, voltages from 220 kV down to 33 kV are found in sub-transmission systems. Distribution is considered as consisting of four elements: Its sub-transmission, the sub-station itself, the feeder system and the consumers. A sub-station contains all equipment involved in the switching or regulating of electricity. Sub-stations can be large or small. Their control can be automatic or manual.

Power transformers constitute an important part of the sub-station. The transformer is a static device which transfers electrical energy from one circuit magnetically coupled with another and transforms voltage levels. On-load tap changing transformers are used to regulate voltages. Switchgear constitutes an essential component of the sub-station. Under normal operating conditions, it provides the means to perform routine switching operations, e.g. disconnecting and isolating various equipment for maintenance, inspection or replacement, transferring load, isolating regulators, etc. Under abnormal conditions, switchgear provides the means for automatically isolating parts of the system in trouble to prevent damage and to localize the problem. The main components of the switchgear include circuit breakers, disconnecting switches, fuse, instrument transformers, buses and connections, supporting insulators, protective and control relays, and control switches.

For the development of the system, the spatial load demands are worked out and imposed on the existing system in order to assess the inadequacy of the system to meet the demand in the horizon year.

The system augmentation/strengthening is then worked out to meet the proposed demand as well as to identify the constraints in the back-up system. The options are:

- Augmentation of power transformer at capacity at the existing distribution sub-station (66 or 33 kV) and/or;
- Re-arranging or re-configuration the sub-transmission (66 or 33 kV) feeders from the new transmission sub-station (e.g. 220/66 kV) nearby, augmenting the line conductors and/or;
- Establishing new 66/11 or 33/11 kV sub-stations nearer to the load centres and re-distributing the load between the existing and new sub-stations as also strengthening the existing 11 kV feeders and adding new 11 kV feeders.

Load flow studies enable the computation of losses for various alternatives. The least-cost optimal solution is worked out by considering the capital cost of the proposed works and the present values of peak demand kW losses and energy losses over the expected life of the equipment.

Power utilities should prepare a *Code of Practice for Network Expansion and Demand Side Management* to meet the increasing demand, and to improve the reliability and quality of supply. This includes determining planning standards, criteria and strategies. System expansion for the purpose of releasing of power connections is done as per the Distribution Code (see Section 8.5).

The other considerations can be: (i) designs as per the Design Code (see Section 4.1.1), and (ii) construction standards (see Section 8.9.3).

3.3.2 Distribution Substations—Siting

Planning of the substation is best done by considering the impact of any siting or sizing decision on all four levels. The main criteria for selecting a substation site is:

(i) *Proximity of load:* Some sites are close to the existing transmission line or can be reached at a low cost. Other sites require lengthy or underground access, thus adding to costs.

(ii) *Out-going feeder space:* Getting a feeder out of a substation requires right-of-way.

(iii) *Geographic:* Nearby terrain or public facilities may constrain feeder routing and raise costs.

(iv) *Site preparation:* The slope, drainage and underlying soil and rock determine the cost of preparing the site for a substation and building the foundation etc.

(v) *Cost of land:* Some sites cost more than others.

(vi) *Weather exposure:* Sites on hilltops are more exposed to lightning and adverse weather, increasing some operation and maintenance costs.

Size

As a thumb-rule [10], the minimum economical capacity (MVA) for a substation is approximately equal to one-fourth of high-side voltage (kV). A 66 kV can serve about 16 MVA.

Service Area Location

The service area for a substation should be, as far as is practical, circular. The consumer should be served from the nearest substation. This will make the supply line distance as short as possible to reduce losses, costs and service interruption exposure. To apply this concept, the best approximation is made by the perpendicular bisector rule. It consists of the following steps [10]:

(i) Draw a straight line between a proposed substation site and each of the substations surrounding it.

(ii) Then draw a perpendicular bisect of each of these lines.

(iii) The area enclosed by the perpendicular bisectors around the proposed substation will be the service area (see Fig. 3.4).

(iv) The shifting of the load of nearby substations can be determined from the area falling within the polygon. Let's say the purpose is to shift the specific load of a particular by substation, 'C'. If this is not accomplished, then the proposed site should be moved closer to that substation and repeat the above three steps.

(v) So the '*optimal site*' for a new substation is determined by an iterative process.

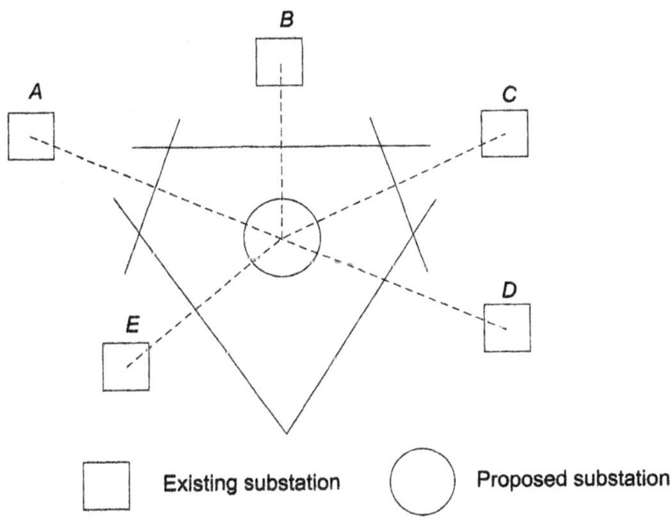

Fig. 3.4 *Process of substation siting*

3.3.3 Feeder System

Feeder is part of the distribution system tailored to load locations and needs. Voltage drop, power flow, power quality and cost are important points of consideration. More than 80% of the distribution worldwide is accomplished using a radial feeder in which there is only one path between any consumer and the substation (see Fig. 1.3a). In most cases, the feeder system is physically interlinked with normally open switches (see Figures 1.3b, 1.3c, 1.3d) at suitable points, which are operated as radial. The various types of feeders are:

- *Radial Feeder:* Radial circuits are low-cost and easy to analyse and operate though reliability is low. Any equipment failure will interrupt service to at least all the consumers downstream from it.
- *Loop Feeder:* Two feeders can be constructed and operated as loop feeder circuits and are tapped for consumers (see Figs. 1.3b and 3.5) in which the power flows into each end of a feeder. There is a '*null point*' somewhere on the loop where no power flows. This is basically a dynamic radial circuit with an open point (null point) shifting as the load changes. When constructed and protected properly, it provides a high level of reliability for the consumer.
- *Feeder Network:* This consists of a group of feeders which are interconnected so that there is always more than one route between any two points in the feeder network (see Fig. 1.3). It is designed with sufficient capacity protection throughout. This system gives a very high level of reliable power to the consumer. The cost is very high compared to the radial system. Voltage drop, fault behaviour and load flow studies are somewhat complicated. Computer programs are now available to carry out such studies.

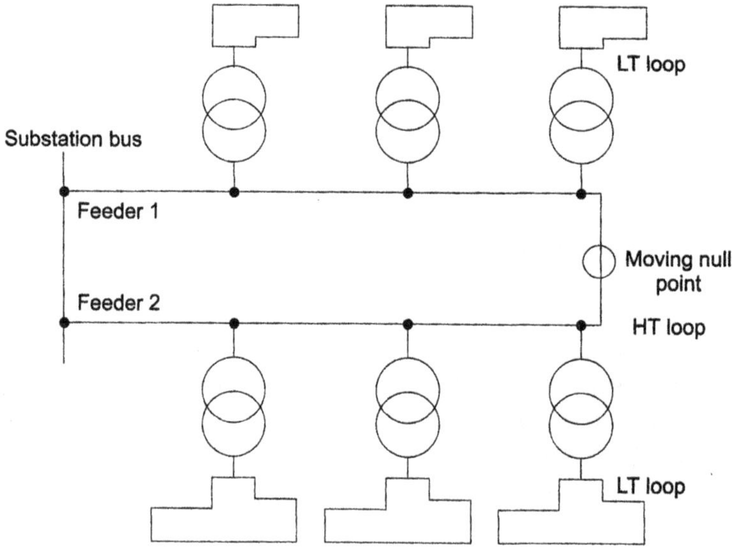

Fig. 3.5 *HT and LT loop feeders*

3.4 Dispersed Generation

This is when the generating plant is connected to a distribution network capable of supplying the consumer load directly. This enhances reliability and the quality of power supply. Now expectations for reliability especially for data centres, call centres, computer-controlled critical processes/businesses and continuous process industry are four nines (99.99 per cent). Also, line losses will be nominal since no transmission system is required. Dispersed generation with local radial distribution network will play an important role in the 21st century. The industry will go for captive power/co-generation. Dispersed generation is generally from local renewable resources. This saves environment degradation compared to conventional thermal generation. The most promising technologies are [16]:

- Fuel cells
- Solar photovoltaics
- Wind power
- Tidal and wave power from ocean
- Small-hydro (up to 25 MW capacity)
- Geo-thermal
- Bio-mass
- Municipal and industrial waste

The following issues concerning connection to grid need careful analysis:

- Voltage rise
- Island operation
- Increase in short circuit level
- Impact on power quality

A typical grid connection of dispersed generation is shown in Fig. 3.6. On the basis of life cycle costs, including the social cost of environmental degradation, dispersed generation has been found to be cost-effective.

Net Metering: The kilowatt-hour meter (import-export meter) can be used to accurately register the flow of electricity in either direction. This means the 'netting' process associated with net metering happens automatically—the meter spins forward (in the normal direction) when the consumer needs more electricity than is being produced and spins backward when the consumer is producing more electricity than is needed in the house or building. Net metering simplifies this arrangement by allowing the consumer to use any excess electricity to offset electricity used at other times during the billing period.

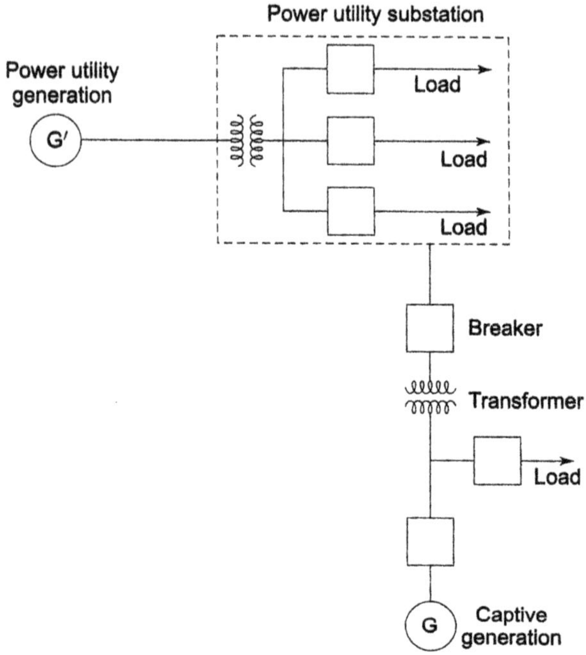

Fig. 3.6 *Captive power connection to grid supply*

Studies and estimates (2001) undertaken by the Ministry of Non-conventional Sources (MNES) reveal that India has the potential for generating 45,000 MW from wind, 15,000 MW from small hydro, 19,500 MW from biomass, 1,700 MW from waste, 10,000 MW from geo-thermal and 15,000 MW from ocean tides. The power generation as in December, 2001 was: wind: 1507 MW; small hydro: 1423 MW; biomass: 400 MW; municipal and industrial wastes: 400 MW; solar PV photovoltaics: 1.8 MW (grid connected), more than 6,00,000 solar lighting systems in use, more than 4200 SPV pumps in use for agriculture. Under the New Renewable Energy Policy (NREP), it is envisaged that by 2012, 10 per cent of the total addition to generation capacity will be from renewable resources. About 80,000 villages still remain unelectrified. Of this, 60,000 villages can still be connected to the grid. The above non-conventional energy resources are being explored for electrifying the remaining 20,000 remote villages which cannot be provided with viable power supply through the power grid.

3.4.1 Power Generation from Solar Energy

Solar energy has great potential in India and could be of great help in mitigating the power crisis.

The daily peak clear sky intensity available at a horizontal surface is about 5 kWh/m^2. Though the seasonal variation around the equator is minimal, the temperate regions experience large variations. For example, in Trivandrum the ratio of variations between the highest and lowest daily value is nearly 1.4. It is as high as 3.6 at Srinagar. At Hyderabad, the daily average energy is about 4.5 kWh/m^2. For the whole of India, the total annual incidence is nearly 6×10^7 kWh per hectare of land surface.

Of course, the solar energy also has its drawback. It is intermittent and has a low intensity; it is also dispersed and unreliable. But there are its advantages too. Solar energy is limitless, universally available and, more important in the present context, pollution free.

Direct energy conversion methods are now receiving increasing attention and are under intensive development. With the advent of semiconductor technology, direct generation of electricity from solar energy has become possible by using the photovoltaic effect in a *p-n* junction. Photovoltaic generators are simple to fabricate and easy to operate. They have a higher power output per weight ratio and have a life of 20–25 years. Further, their present efficiency of conversion is far higher than of any other solar energy converter. The solar cell (silicon solar cell) power has been reliably utilized in space, street lighting, irrigation pumping. Solar panels can be installed on roof taps. The solar photovoltaic (PV) capacity installed in the country is 55 MW ending 2001.

3.4.2 Wind Power

The department of power has fixed up the target for the development of power from wind of around 5,000 MWs by the turn of the century. This would mean harnessing only 25% of the known potential of at least 20,000 MW in the country for wind-generated electricity. The development has been of 1300 MW capacity until the end of 2001. The wind power varies as the cube of its velocity. In the regions where the wind velocity is high, power can be generated economically. With the indigenous technology now developed, the cost of setting up future units has come down.

The wind power has great potential in coastal area of Gujarat, Maharashtra, Tamil Nadu, Orissa, Andhra Pradesh. Hilly terrain of Eastern Frontier Area States, foothills of entire Himalayan ranges from Pathankot in Punjab to Terai in U.P., desert area of Rajasthan. On a rough estimate, significant potential is available for power generation at 22 km per hour wind on the coastal stretch in Gujarat (Kandla/Mandvi) and 17 to 20 km/h on the Saurashtra (Veraval, Okha, Dwarka, Rajkot and Bhavnagar), Orissa (Puri), Tamil Nadu (Tuticorin) and Maharashtra (Deograh, Dahanu) coast lines, in some inland locations in Tamil Nadu (Coimbatore), Western Rajasthan (Phalodi) and Madhya Pradesh (Indore), and in some peninsular locations.

In the North-East Frontier Area, where the terrain is so difficult and setting up power plants is almost unthinkable, small windfarms set up even at a higher cost is worthwhile. Decentralized windfarm projects starting with 0.5–1 MW can be installed in farflung areas and can easily fit into the existing grid network, without expensive transmission and development investment.

3.5 Distribution System Economics and Finance

Economic analysis is carried out to determine the low cost plan among various alternatives. Financial analysis determines the rate of return and risk involved on the investment to be made on the plan.

The investment to carry out a planned distribution is either:

- Annual expenses–obtained from operating revenue; or
- Capital expenditure–obtained from financing, reinvested reserve, reinvested earnings, consumers' contribution for service connections; or
- Both annual expenses and capital expenditure.

Investment decision is made on alternative proposals with the following methods.

3.5.1 Economic Analysis (14)

Minimum Revenue Requirement

A choice is made on the basis of the present value of all future annual costs. That is, the economic choice is the one with the lowest present value of all future costs. The economic comparison between alternatives involves two steps:

- For each alternative, estimate the annual cost for each year;
- If the annual costs are not uniform, calculate their present value.

Time Value of Money: Money has time value and interest on its use has to be paid. The rate of interest is determined by the Reserve Bank of India according to economic conditions at the time. There is a similar mechanism to determine the interest rate at the international level also. An alternative which requires the least expenditure immediately would be the best, other things being equal. If the prevailing interest is 10%, then Rs 100 today is equivalent to Rs 100 (1 + 0.10) or Rs 110 a year from now. From today, 'n' years in future, this amount is compounded by the factor $(1 + i)^n$ and the future value of Rs 100 after 'n' years will be $100\,(1 + i)^n$.

The process of taking money and finding its equivalent value at some future date is called *future value* calculation. The process of finding the equivalent value at some earlier time is called *present value* calculation. Present value is the reverse of future value calculation. We can say that Rs 110 a year from now has a present value of Rs 100 today. Since the future value factor for n years is $(1 + i)^n$, the present worth factor is $1/(1 + i)^n$.

Revenue Requirement of Investment: The total revenue requirements of investment is the sum of the annual charges extending over the service life. It includes:

- Return on investment
- Depreciation
- Insurance expenses
- Operating and maintenance expenses
- Interest on loan capital and working capital
- Taxes etc.

The above charges can be conveniently estimated as a percentage of the original investment.

Three formats are given below to calculate the present value:

Cumulative Present-worth Method: As stated above, when the annual revenue requirement are non-uniform, the sum of the present value of the revenue requirements for each alternative is calculated. The most economical alternative will have the minimum present value of the revenue requirement. In case the annual revenue requirements are uniform throughout the service life or study period, the alternative with the minimum annual revenue requirements will be the most economical.

Levelised Annual Cost Method: The level annual carrying charge is the percentage by which the capital investment (P) can be multiplied to determine its annual cost of capital on a uniform basis. It is desirable to calculate the annual costs of capital investment made in each alternative scheme. The value of the annual carrying charge depends upon the expected life of the equipment because depreciation varies with life expectancy. The capital investment (P) made today is converted to an equivalent annual annuity (R) using the uniform capital recovery factor $i(1 + i)^n/(1 + i)^n - 1$ as shown in Fig. 3.7a.

Thus $$R = \frac{P \cdot i(1+i)^n}{(1+i)^n - 1}$$

Uniform Series Present Worth Method: In economic analysis, there often exist a uniform series of annual costs or payment (R') (an annuity) that extend from today through n years. In order to compute the present worth (P') for the uniform annual series of payments, the formula is (see Fig. 3.7b):

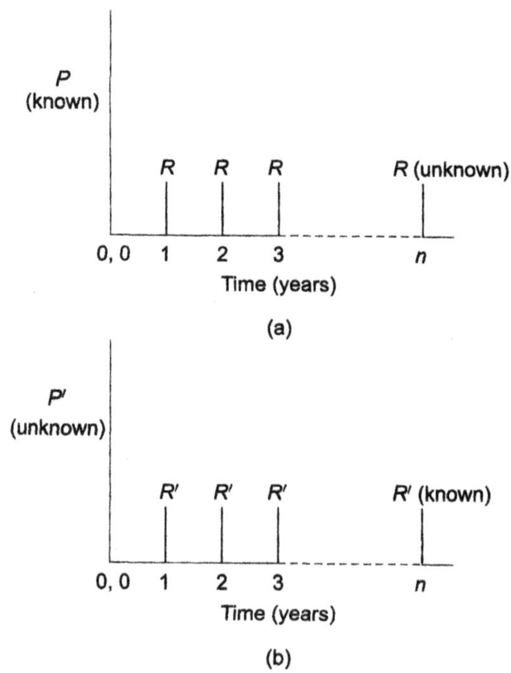

Fig. 3.7 *(a) Uniform Annual Charge (R), (b) Uniform Series Present Worth*

Thus $$P' = \frac{R' \cdot (1+i)^n - 1}{i(1+i)^n}$$

For example, the cost of cable conductor losses in a project of cable laying is equal to *uniform annual costs* over the life of the cable laid.

These three revenue requirement methods are all equivalent. The minimum cost selection is, therefore, the same regardless of the method because they differ only in the way in which the results are presented.

3.5.2 Financial Analysis

Benefit/Cost Ratio: This method ranks a project by the ratio of the present value of revenue and/or benefits earned to the present value of the costs incurred. System improvement works are generally evaluated on a benefit-cost basis by the Rural Electrification Corporation of India (see Section 14.4.2). The Central Electricity Authority (CEA) also uses this method for such schemes, based on lifecycle (25–30 year period) cost-benefits. The works proposed for strengthening, upgradation and improvement of the sub-transmission and distribution system help in reducing technical and commercial losses by the horizon year. The system would also be able to meet the load demand in the horizon year with a resultant increase in sale of energy. The financial analysis is usually done for the lifetime of the project considering the year-wise costs including interest, depreciation, operation and maintenance charges, and the expenditure on additional sale of energy, and the year-wise gross benefits including the revenue from additional energy, and cost saving due to saving in losses. The year-wise net present value of the net benefits is calculated and the cumulative net present value of the net benefits is evaluated.

Internal Rate of Return (IRR): The design and sustainability of the project should take into account the level of incentive for undertaking and maintaining a project investment. The financial incentive takes the form of increased income that the investment generates with the execution of the project. This method calculates the rate of interest needed for the present value of the returns to be the same as the present value of the investment needed, i.e. net present value to be zero. This is the point at which the project breaks even. REC specifies the IRR as 15 per cent while sanctioning short-term (seven-year repayment) loan schemes.

Payback Period

The length of time required to recover the initial investment is computed for each alternative. This method does not consider the time value of money and the life of investment after the payback period. But this measure is used as an indication of the amount of the investment's risk (see Section 15.6.2). The alternative having the lowest payback period is selected. Acceptable payback period for utility projects range from 1 to 5 years.

3.6 Mapping

3.6.1 Global Positioning System (GPS)

GPS is a system in which earth-orbiting satellites provide precise information on time and position enabling GPS receiving devices to compute position on earth. Signals must be received from at least three satellites in order to establish the latitude and longitude of the receiver position. A fourth satellite is required to calculate the altitude. There are rules for sharing the use of GPS satellites for a Geographical Information System (GIS) job [7, 17]. GIS deals with spatial information and requires the following:

(i) 24 US Naval GPS satellites in different orbits about 17,700 km above the earth. The orbits are such that at any time, at least four satellites are above the horizon for an observer.
(ii) Portable mobile GPS receivers (antenna plus console).
(iii) Base receiver station positioned at a known geographical position in order to perform differential correction of the satellite signals.

GPS technology is used to locate tap-off points, transformers and other facilities of power distribution networks and consequently to map the system with an accuracy of up to one metre. GPS can be used to capture network data for 11 kV and above voltage level distribution/sub-transmission lines and substations/distribution transformers for mapping. However, for LT network for each distribution transformer, spatial or geographical maps on paper can be drawn based on judgment and eye approximation. The length of the feeder is generally fixed, based on estimates and pole-to-pole standard distances. These maps can be prepared on *Survey of India sheets.*

3.6.2 Background Maps

Background paper maps with the latest geographical information are necessary for drawing a correct digital network diagram. Geographical information of various states is usually maintained by agencies like the Survey of India, the aerial survey and satellite imagery departments of National Remote Sensing Agency (NRSA). The topographical sheets covering each state are available from the Survey of India at Rs 20–25 per sheet. Also, digitized geo-coded maps of a state are available on CD-ROMS at a cost of about Rs 7,000.

3.6.3 Digital Mapping

Digital mapping (see also Section 5.8) software can be used to create an integrated and automated facilities model in which paper maps are first digitized and then suitably linked. A digitizer board is used for digitizing nodes with respect to a reference point pre-fixed on the map. One person can read co-ordinates on the digitized board and the other can feed it into a computer. Once digitization is complete, the software numbers the nodes, draws the network diagram and calculates the lengths. Networks drawn with a computer then undergo field verification. A computer processing centre will then issue a prescribed format to obtain electrical loading data of feeders, transformer capacity, size/type of the feeder for the concerned area. The following steps should be taken to prepare the maps:

(a) The complete area should be covered and duplication avoided.
(b) The 132-33 kV system be digitized first followed by 11 kV and finally the 400 kV network.
(c) Differential survey with GPS instruments and VHF sets may be required for dense urban areas to capture various attributes of the system and ensure high accuracy in digitized maps since for every reading of the location of a facility, the location is checked with the base station and corrected.
(d) Oracle RDBMS (latest version 9i) is suitable for data storage. The process of digitally mapping the network is shown in the block diagram in Fig. 3.8.

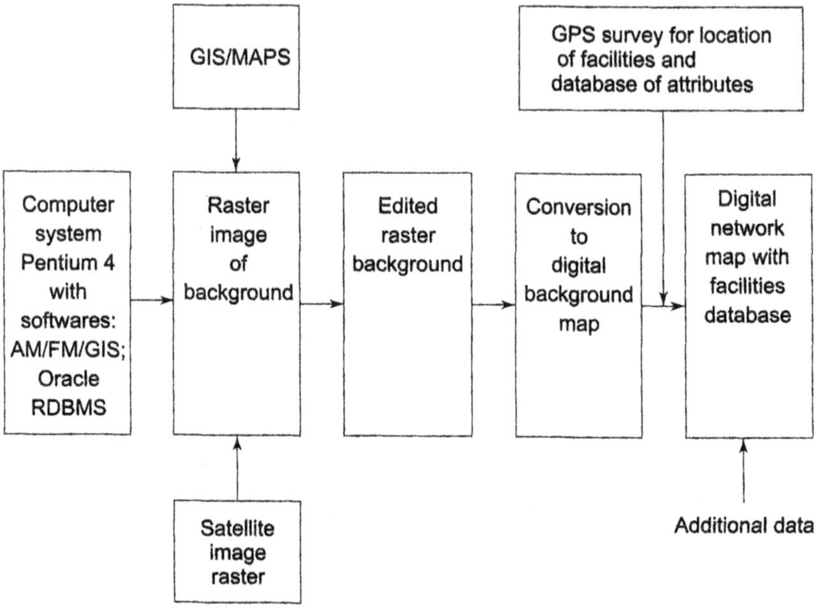

Fig. 3.8 *Process of digitally mapping a network*

3.6.4 Automated Mapping (AM)/Facilities Management (FM)/Geographical Information System (GIS)

Every transaction, whether it is a work order, billing, operation, system improvement or maintenance is tied to a location. AM and FM provide an integrated tool to create an automated facilities model and help to convert paper maps to a digital environment where the system details and its changes are fed and stored in a graphic and tabular database. The graphic data is a spatial representation of the network and is the backbone of an AM/FM/GIS system and the driving force for all the applications. FM software links the graphic network map to tabular and non-graphic data associated with the databases, work orders, drawings etc. GIS establishes the connection to these databases through a record server and provides the best characteristics of AM and FM systems to provide an integrated information system.

The Role of GIS

GIS is a system of mapping of complete electrical network including low voltage system and consumer meter. Database plays a central role in the operation of planning, where analysis programme form a part of the system supported by a *database management system* which stores, retrieves and modifies various data on the distribution system. Electrical utility companies need two types of geographical information: Details on the location of facilities, and information on the spatial interrelations between them. The integration of geographically referenced database, analytical tools and in-house software tools will allow the system to be designed more economically and operated closer to its limits resulting in more efficient, low-cost power distribution systems.

Additional benefits such as improved material management, inventory control, preventive maintenance and system performance can be accomplished in a systematic and cost-effective manner. All information systems are built around Relational Database Management System (RDBMS) and are constantly updated. Establishing links between these information systems and GIS is only in defining a relationship between objects in the two systems. The number of different information systems in the same utility (technical information system, consumer information system etc.) or several overlapping technical information systems can be updated.

The ultimate aim of AM/FM/GIS is to integrate the dynamic side of the operation with the relatively static side of the utility facility record for effective utilisation by all functional departments and personnel. The distinguishing features of the AM/FM/GIS system are:

- Explicit spatial representation of the facilities network linked to related data.
- Integrated facility database and its management with multi-user access.
- Management of different stages of the facility's life cycle.
- Optimum data for facility applications.

The software platform (e.g. Oracle) with database and document warehouse may include the following:

- Installed facility records
- Change orders: Pending, implemented
- Standard equipment costs
- Standards, practices and criteria for design, operation and maintenance

- Land use
- Maintenance and inspection records
- Equipment: Operations, trouble history and records
- Facility operating data (real-time data link)

Applications:

(i) The planning department requires field data of land records, load growth based on population growth and demand, existing network layout and general land base data. The data is needed for feeder layout/routing and network planning, load flow analysis, protective devices co-ordination, location and layout and design of substation, manpower requirement.

(ii) *Estimation and costing:* For a planning engineer to develop estimates and prepare costing reports, bill of materials etc. based on standard costs.

(iii) *Management reports:* Reports like ongoing works; transformers in operation, damaged, replaced or newly-commissioned during a certain period; new consumers connected; pending change order data.

(iv) *Design/operational logistics:* The operational requirement may need re-configuration of the electrical system during abnormal conditions and also electrical system analysis.

(v) *Maintenance:* To carry out maintenance, testing and fault root-cause analysis.

(vi) *Troubleshooting:* To support help desks and improve consumer services.

3.7 Enterprise Resource Planning (ERP)

ERP system [12, 15] strengthns the planning and execution system. It integrates all the processes of the organisation in the most synergistic way possible. It is basically a cost reduction programme with the best practices incorporated. A successful ERP system is one which everyone uses because it makes their work easier. Within this construct, authority and responsibility are shifted from the management level to the operational level, often erasing job categories along with boundaries between various departments within the enterprise. This structure facilitates the flow of information among all the functions within the organisation including power delivery, logistics, finance and human resources.

An ERP system is as good as the data it receives. The enterprise-wide database attempts to consolidate all business activities into a single computing environment. The system forces co-operation among departments as individuals across the enterprise log trouble calls, create work-order entries, acquire material procurement and schedule work that is required to keep the business running smoothly. Enhanced work-management processes encompass those activities involving installation, repair and retirement of outside facilities that are geographically dispersed and that are serviced by a variety of work gangs and service vehicles. The most efficient process is supported by work management principles that will use standard references to set up task lists. ERP enables power utilities to take better care of their consumers in releasing service connections, repair work and routine maintenance and consumer information system in an effective manner. AM/FM/GIS systems provide comprehensive infrastructure for building an ERP. The typical functional diagram of an ERP system's human resources, commercial and financial modules for a power utility is shown in Fig. 3.9.

PriceWaterHouse Associates, India, have developed an ERP solution with software capabilities to run on platforms like Oracle and Microsoft Windows particularly for the power sector in India. It includes modules on financial, commercial, billing and collection, energy accounting, consumer enquiry, material management, power plant management, coal accounting, work order monitoring and personnel information. The client/server enables ERP to run across various database back-end through Open Base Connectivity (OBC). Most of the utilities have gone in for a standard package. For example, a three-tier-server open system architecture SAP/R3 is mostly used by power utilities in European countries. SAP modular packages (see mySAP.com) are available. Leading ERP software vendors are:

- SAP AG (Germany)
- QAD Inc (USA)
- Peoplesoft (USA)
- Oracle Corporation (USA)
- Bann Invensys (Dutch)
- J.D. Edwards (USA)
- IBM (USA)

Fig. 3.9 *Typical enterprise resource planning (HR, commercial and financial modules)*

- JBA Holdings (USA)
- Wipro (India)
- Micro ERP 2000 (India)
- Radix Infotech (India)

3.8 Modelling

Models are descriptions of behaviour and response of some elements of the distribution system. It is used as a proxy for that element during steady state analysis and evaluation of planning options. In engineering, models can be of four types:

Level	*Language*	*Modelling function*
Technical	Schematic diagrams	Component function
Physical	Graphs, networks, 3D figures	Phenomenon modelling
Mathematics	Block diagrams	Equation modelling
Algorithmic	Programming language	Assignment modelling

Often, the models used involve equations whose solutions are best computed by some numerical method: An algorithm which is usually implemented by a computer program. For example, a planning engineer in a power utility may decide to model load flow on a feeder:

$$\text{Load current} = \frac{\text{Sending end voltage}}{\text{Line impedance + load impedance}}$$

$$= \frac{V_s}{Z_{\text{line}} + Z_{\text{load}}}$$

This can be called a basic model of power flow for electrical behaviour. In order to apply this model to the network, apply this equation to various segments and loads in the system. This means to solve a set of simultaneous equations to compute voltages and currents. The particular way selected is the algorithm. In simple cases, the line, transformers, capacitors and loads are represented by phase impedance. The power (P) transmitted through the line is given by equation:

$$P = \frac{V_s V_r \sin\delta}{X}$$

Where V_s, V_r, δ, X are line sending-end voltage, receiving-end voltage, line angle and line reactance respectively.

The power (P) and reactive power (Q) generated by a power plant is arrived at by the following equations:

$$P = \frac{EV \sin \delta_t}{X_s}$$

$$Q = \frac{EV \cos \delta_t}{X_s} - \frac{V^2}{X_s}$$

Where E, V, δ_t, X_s are generator no load e.m.f., generator terminal voltage, power or torque angle of generator and generator synchronous reactance respectively.

The circuit model contains representation of lines, loads and other equipment along with connectivity information to link them together so that load flow, fault current, reliability analysis and costing studies can be made. Generally, the distribution system is measured in node resolution. Nodes are required to represent explicitly every consumer metre, every unit of equipment and every pole-to-pole segment as a separate entity. Several identical line spans can be lumped into one segment for load flow studies. But for a reliable analysis, actual pole/span count is used because failures are related to actual pole count, not just the line length etc. Most algorithms (e.g. load flow, fault current studies, reliability prediction) compute results only at nodes. More nodes mean more locations where information is available about performance. Large node count provide more detail in representation of circuit topology and equipment (refer Section 3.6 AM/FM/GIS).

It is a planner's first responsibility to make certain that the models used in planning are appropriate to their task of understanding their limitations, shortcomings and range of applications. Many approaches to this numerical problem are applicable including Gauss-Seidel, Newton-Raphson etc. These are discussed in Section 3.13.1.

The mathematical model for sizing a substation in urban areas is [11]:

Optimum rating of substation (MVA) = 50 (load density in MVA/km^2)$^{1/3}$.

3.9 System Calculations

The planning, design and operation of power distribution systems require continuous and comprehensive analysis in order to evaluate the system performance and to determine the effectiveness of alternative plans for system expansion. These studies play an important role in providing a high standard of power system reliability, security and quality and in ensuring the maximum utilization of capital investment.

The analysis usually implies the computation of network voltage and currents under a given set of conditions. In many cases, the computation is organized to give a particular kind of information for a special purpose. For example, it may be desirable to determine the current flowing through a line at a particular spot to enable the setting up of a relay or determining the effect of a load producing harmonics on a network. The location and requirements of capacitor banks for the system can also be determined through such analysis.

The steady state analysis of a network is carried out for both normal and faulty networks. The unbalanced system operation is, however, usually ignored. Often, when dealing with normal or near normal operation, balanced operation is assumed and the network solved on a per phase basis. The results are then extrapolated to obtain information for the remaining two phases for a three-phase network. This results in great saving of time and effort and usually gives solutions of reasonable accuracy. When the system is unbalanced, other methods must be used. A method generally preferred is that of "symmetrical components". This method allows the use of per phase analysis of system having unbalanced loads or terminations, such as short circuits or faults.

Today, the computer [6] is an indispensable tool in all phases of power system planning, design and operation. The distribution engineer should be capable of bridging the gap between theory and computer utilization. With the remarkable decline in the cost of computing equipment and development of efficient computational techniques, computer technology provides more efficient means of utilizing engineering talent by relieving the need for tedious hand calculations. This allows for more effective taking of engineering decisions since alternative solutions can be more easily calculated. It also enables the undertaking of studies which were previously prohibitive due to the volume of calculations involved. Typical system software packages are now easily available with various computer service institutions [4, 9].

3.9.1 Computer Computations Procedure

Problem-solving using computers involves the following five distinct steps (refer also subsection 4.7.5).

- Choice of a method
- Designing the algorithm
- Flow-charting
- Programming
- Computer execution

(i) *Method* It is a mathematical formula for finding the solution of a given problem. There may be more than one method available to solve the same problem. The method which best suits the given problem should be chosen. The inherent assumptions and limitations of the method must be studied carefully.

(ii) *Algorithm* It is a complete and unambiguous set of computational steps to be followed in a particular sequence to obtain the solution. The algorithm tells to the computer where to start, what information to use, what are the operations to be carried out and in which order, what information is to be printed, and when to stop. An algorithm has the following essential features:

- Finiteness : It must terminate after a finite number of steps.
- Definiteness : Each step of an algorithm must be clearly defined or the action to be taken must be unambigously specified.
- Inputs : It must specify the quantities which must be read before the algorithm can begin.
- Outputs : It must specify the quantities which are to be outputted and their proper place.
- Effectiveness : It must be effective, i.e. all operations specified must be executable.

(iii) *Flow-chart* It is a graphical representation of a specific sequence of steps (algorithm) to be followed by a computer to produce the solution of a given problem. It makes use of the flow-chart symbols to represent the basic operations to be carried out. The various symbols are connected by arrows to indicate the flow of information and processing. While drawing a flow-chart, any logical error in applying the algorithm can easily be detected and corrected.

(iv) *Program and execution* The flow-chart can be translated into any suitable high-level language, such as FORTRAN, PASCAL, C, BASIC, etc. after which is called a program or software. This can be executed on the computer [5].

The writing of software requires a good understanding of numerical analysis and the art of programming. A good software must satisfy the certain criteria–selfstarting, accuracy, reliability, minimum number of levels, good documentation, ease of use and portability (i.e. software should be usable on any other computer with no or minimal modification).

3.10 Introductory Methods

3.10.1 Per Unit Calculation

The calculations of network performance are generally carried out using a per unit representation of voltage, current, power, reactive power, apparent power and impedance. The numerical per unit value of any quantity is its ratio to a chosen base quantity of the same dimensions. Thus a per unit quantity is a 'normalized' quantity with respect to a chosen base value.

The numerical values of the above quantities are so related that if we specify the base values of voltage and current, the base values of the other quantities will be defined. Apparent power S and voltage V are generally taken to be the base values. The same voltampere base S is used in all parts of the system. A given base voltage is arbitrarily selected. All other base voltages must be related to an arbitrarily selected one by the turns ratio of the connecting transformer.

For a single-phase or three-phase system where current refers to line current, voltage to phase-to-earth voltage and apparent power to apparent power per phase, we can write

$$\text{Base power VA} = S_b = I_b\, V_b$$

$$\text{Base current A} = I_b = \frac{S_b}{V_b}$$

$$\text{Base impedance } \Omega = Z_b = \frac{V_b}{I_b} = \frac{V_b^2}{S_b}$$

For a three-phase system where base power refers to the total power for all the three phases and base voltage as the line-to-line voltage, we have the following expressions.

$$\text{Base power } S_b = \sqrt{3}\ I_b\, V_{\text{line } b}$$

$$\text{Base current } I_b = \frac{S_b}{\sqrt{3}\, V_{\text{line}^b}}$$

$$\text{Base impedance } Z_b = \frac{V_{\text{line}^b}}{\sqrt{3}\, I_b} = \frac{V^2_{\text{line}^b}}{S_b}$$

$$\text{pu impedance } Z_{\text{pu}} = Z/Z_b$$

$$\text{pu admittance } Y_{\text{pu}} = \frac{1}{Z_{\text{pu}}}$$

To convert impedance from one base system (S_b and V_b) to another one (S'_b and V'_b), we write

$$Z'_{\text{pu}} = Z_{\text{pu}} \frac{S'_b}{S_b} \frac{V_b^2}{V_b'^2}$$

The ohmic value of the resistance or leakage reactance of a transformer depends on whether it is measured from the high or low voltage side,

$$Z_{\text{LV}} = \left(\frac{V_{\text{LV}}}{V_{\text{HV}}}\right)^2 Z_{\text{HV}}$$

It can be shown that

$$(Z_{\text{pu}})_{\text{LV}} = (Z_{\text{pu}})_{\text{HV}}$$

Therefore, if the same power is selected on both sides of the transformer and the voltage base on its either side is selected to have the same ratio as the turns ratio of the transformer winding, the per unit impedance will be the same when expressed on the voltage base of either side of the transformer.

3.10.2 Matrix Algebra

With the advent of digitial computers, the use of matrix algebra for the formulation and solution of complex engineering problems has become increasingly important. Matrix operations provide a concise method for expressing problems in a manner which is easily adapted to computer solution [11].

Matrix notation is a shorthand means of writing simultaneous equations in a concise form. A matrix is defined as a rectangular array of numbers called elements, arranged systematically in m rows and n columns. The nomenclature a_{ij} designates the element of the matrix in the *i*th row and *j*th column. Matrix representation **A** of $m \times n$ elements is given as follows:

$$\mathbf{A} = \begin{bmatrix} a_{11} & a_{12} & \cdots & a_{1j} & \cdots & a_{1n} \\ a_{21} & a_{22} & \cdots & a_{2j} & \cdots & a_{2n} \\ \vdots \\ a_{i1} & a_{i2} & \cdots & a_{ij} & \cdots & a_{in} \\ \vdots \\ a_{m1} & a_{m2} & \cdots & a_{mj} & \cdots & a_{mn} \end{bmatrix}$$

The following elementary operations of matrices are important.

(a) Equality.
(b) Addition and subtraction.
(c) Scalar multiplication.
(d) Matrix multiplication.
(e) Transpose of a matrix.
(f) Determinant.
(g) Cofactor.
(h) Inverse of a matrix.
(i) Diagonal matrix.
(j) Adjoint of a matrix.

3.10.3 Symmetrical Components

Fortesque* has shown that three unbalanced phasors of a three-phase system can be resolved into three balanced systems of phasors known as positive, negative and zero sequence components (shown in Fig. 3.10).

Defining the operator a as $1 \angle 120°$, the three line currents and voltages are given by:

$$\mathbf{I}_{abc} = \begin{bmatrix} I_a \\ I_b \\ I_c \end{bmatrix} \qquad \mathbf{V}_{abc} = \begin{bmatrix} V_a \\ V_b \\ V_c \end{bmatrix}$$

Defined in terms of three symmetrical components of currents and voltages, we get:

$$\mathbf{I}_{012} = \begin{bmatrix} I_{a0} \\ I_{a1} \\ I_{a2} \end{bmatrix} \qquad \mathbf{V}_{012} = \begin{bmatrix} V_{a0} \\ V_{a1} \\ V_{a2} \end{bmatrix}$$

*Fortesque C.L., "Method of symmetrical co ordinates applied to the solution of poly-phase networks", *AIEE, Trans.*, Vol. 37, Part II, 1918, pp. 1027–1140.

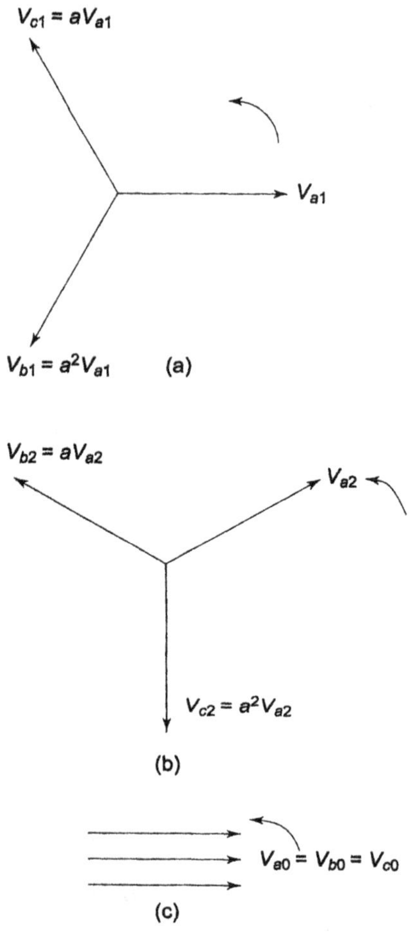

Fig. 3.10 *Symmetrical components: (a) Positive phase sequence; (b) Negative phase sequence; (c) Zero phase sequence*

such that

$$\mathbf{I}_{abc} = \mathbf{A}\mathbf{I}_{012}$$

and

$$\mathbf{V}_{abc} = \mathbf{A}\mathbf{V}_{012}$$

where

$$\mathbf{A} = \begin{bmatrix} 1 & 1 & 1 \\ 1 & a^2 & a \\ 1 & a & a^2 \end{bmatrix}$$

The same relationship applies to the voltage vector. Conversely, the relationship between the symmetrical components and phase currents and voltage are:

$$\mathbf{I}_{0\ 2} = \mathbf{A}^{-1}\,\mathbf{I}_{abc} \text{ and } \mathbf{V}_{012} = \mathbf{A}^{-1}\,\mathbf{V}_{abc}$$

where

$$\mathbf{A}^{-1} = \frac{1}{3}\begin{bmatrix} 1 & 1 & 1 \\ 1 & a & a^2 \\ 1 & a^2 & a \end{bmatrix}$$

The impedance is then given by

$$\mathbf{Z}_{012} = \mathbf{A}^{-1}\,\mathbf{Z}_{abc}\,\mathbf{A}$$

Power from Symmetrical Components

The power at any point is the sum of the individual powers computed in the individual phases, i.e.

$$P_{3\phi} = Re\ V^T_{abc}\,\mathbf{I}^*_{abc}$$

where V^T_{abc} is the transpose matrix of V_{abc} and $\mathbf{I}^*_{abc}$ is the conjugate of $\mathbf{I}_{abc}$. In terms of symmetrical components one can write for power at any point

$$\begin{aligned} P_{3\phi} &= 3Re\ V^T_{012}\ \mathrm{u}\ \mathbf{I}^*_{012}, \text{ u being a diagonal matrix} \\ &= 3Re\ V^T_{012}\,\mathbf{I}^*_{012} \\ &= 3Re[\mathbf{V}_{a0}\,\mathbf{I}^*_{a0} + \mathbf{V}\mathbf{I}^*_{a1} + \mathbf{V}_{a2}\,\mathbf{I}^*_{a2}] \end{aligned}$$

The total three-phase power is a function of the symmetrical components of voltage and current of the same phase sequence and there is no coupling of power from the negative sequence current reacting with positive or zero sequence voltage.

Sequence Networks

In many problems the unbalanced portion of the physical system, such as an unbalanced fault or unbalanced load supplied by a balanced network, can be isolated for separate study, with the rest of the system considered as balanced.

A sequence network is a copy of the balanced system to which the fault point is connected and contains the impedance of the network for each sequence. Once the fault point is determined, each sequence network can be analysed by Thevenin's Theorem, i.e. by obtaining the open circuit voltage and equivalent impedance for that sequence looking into the network from the point of unbalance. Only the positive sequence network has a voltage source, since, by definition, the only voltages generated in the three-phase system have a positive sequence (Fig. 3.11).

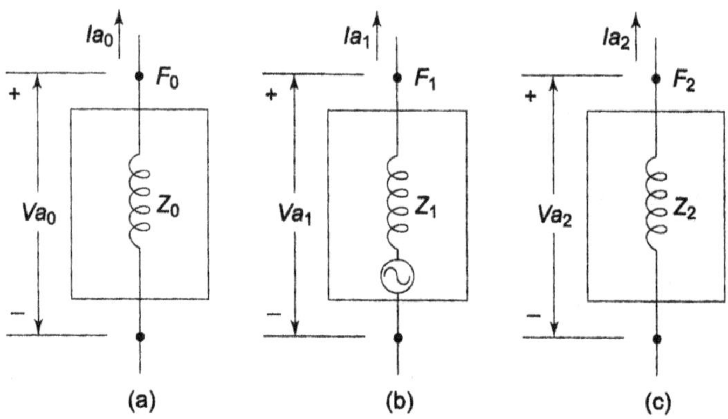

Fig. 3.11 *Sequence networks: (a) Zero sequence; (b) Positive sequence; (c) Negative sequence*

3.11 Network Elements

In analysing any electrical network, the first major requirement is that the voltage current characteristics of the elements in the circuit are known. A thorough discussion of the characteristics of circuit elements is beyond the scope of this chapter. However, some of the more pertinent points are described below.

For most calculations, the phases of the network elements are assumed to be symmetrical. Thus, there is no variation of characteristics between the phases.

In determining the network characteristics, we need to distinguish between positive, negative and zero sequence impedances. For positive and negative sequence impedances, a balanced three-phase voltage is applied to the three phases of the respective sequences. For zero sequence impedances, a single-phase voltage is applied to the three phases of the network element.

The negative sequence network is identical to the positive sequence network except that the negative sequence reactances of synchronous machines are usually less. The zero sequence network differs from the positive and negative sequence impedances.

3.11.1 Overhead Lines (2)

Resistance

The resistance of a conductor is obtained from its resistivity and cross-sectional area. The ac resistance is increased because of skin and proximity effects. This adds 0–2% to the resistance (see Appendix II).

Positive and Negative Sequence Inductance (H/km)

For single-phase circuit each conductor

$$L = 2 \times 10^{-4} \ln \frac{d_{12}}{0.779\,r} = 2 \times 10^{-4} \left(\ln \frac{d_{12}}{r} + \frac{1}{4} \right)$$

For three-phase symmetrically transposed circuit

$$L = 2 \times 10^{-4} \ln \frac{d_m}{0.779r} \text{ H/km/phase}$$

where, d_{12}, d_{23}, d_{31} = distance separating the conductors (mm)

$d_m = (d_{12}\, d_{23}\, d_{31})^{1/3}$ i.e. geometric mean spacing

r = radius of conductor (mm)

The inductance is approximately 4% greater for flat spacing.

Inductive Reactance $(X_l) = 2\pi fL$

Capacitance (Farads 1km)

$$C = \frac{1}{2} \times 1.05 \times 10^{-8} \cdot \frac{1}{\ln \frac{d_m}{r}} \text{ for single phase circuit}$$

$$C = 1.05 \times 10^{-8} \cdot \frac{1}{\ln \frac{d_m}{r}} \text{ per phase for 3-phase circuit}$$

$$\text{Capacitive Reactance } (X_c) = \frac{1}{2\pi fC}$$

Zero Sequence Impedance

For the purpose of fault calculation, the average value of zero sequence reactance (X_0) can be obtained from Table 3.1. This does not include the tower/pole footing resistance.

Table 3.1

Type of line	X_0/X_l Single circuit	X_0/X_l Double circuit
No ground wire	3.5	5.5
Steel ground wire	5.5	5
Nonmagnetic ground wire	2.0	3

The impedance of a overhead line depends largely on the construction of the line and the nature of earth conductors (Fig. 3.12).

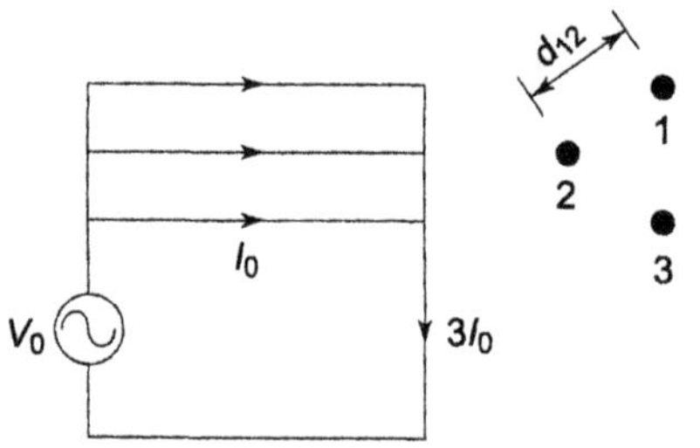

Fig. 3.12 *Equivalent zero sequence circuit*

$R_0 = R_c + \dfrac{3\pi \times 10^{-4}\omega}{2}$ Ω/km/phase, R_c is conductor resistance, R_0 is zero sequence resistance per km/per phase.

$$X_0 = 6\omega \times 10^{-4} \ln \frac{\delta}{\left(0.779r\, d_m^2\right)^{1/3}} \ \Omega/\text{km/phase}$$

where,

δ = effective depth of earth return current (m)

$= 660 \sqrt{\rho/f}$

$d_m = (d_{12}\, d_{23}\, d_{31})^{1/3}$ (m)

ρ = earth resistivity (Ω m)

f = frequency (Hz)

With an earth wire present, zero sequence current will flow in this earthwire. It varies depending on the arrangement of any overhead earth wires.

3.11.2 Transformer

Positive Sequence

The impedance of a transformer is represented by the sum of the resistance and leakage reactance of the winding. The reactance is generally much greater than the resistance. The impedance is generally expressed as a percentage of the rated output of the transformer. It is independent of the connections of the transformer.

Zero Sequence Impedance

The zero sequence impedance varies greatly depending on the type of connection and construction of the transformer. Capacitance currents are generally disregarded in the calculation. However, for higher frequencies, this may not always be satisfactory. Conductors connected to transformer windings with a delta connection or with star zig zag connection with an insulated neutral point cannot carry zero sequence currents. The equivalent circuits used for various transformer connections are shown in Fig. 3.13.

Standard Connections

Groups

Delta/Y-11 and Y/Y-0 are two winding connection groups commonly used in core type, distribution transformers. Star/Star(Y/Y-0) winding is most economical as the number of turns per phase and the amount of insulation is minimum. This group is widely used in China. Delta/Y-11 winding group is largely used in India and most of the other countries. The operation of unbalance loading (single-phase loads) has a different effect with these two type of windings [13]. In the LT network, most of the consumer appliances are single-phase. The degree of asymmetry transferred to the primary of distribution transformer depends upon the transformer winding arrangement chosen. As shown in Fig. 3.14a, in Y/Y-0 (both neutrals earthed) winding, the load on each primary phase is proportional to that of the corresponding secondary phase. In Fig. 3.14b, in delta/Y-11 (secondary neutral earthed) winding, the delta-connected primary will permit single-phase unbalance secondary loading. The secondary load current will cause a primary current from phase *R* to phase *Y*.

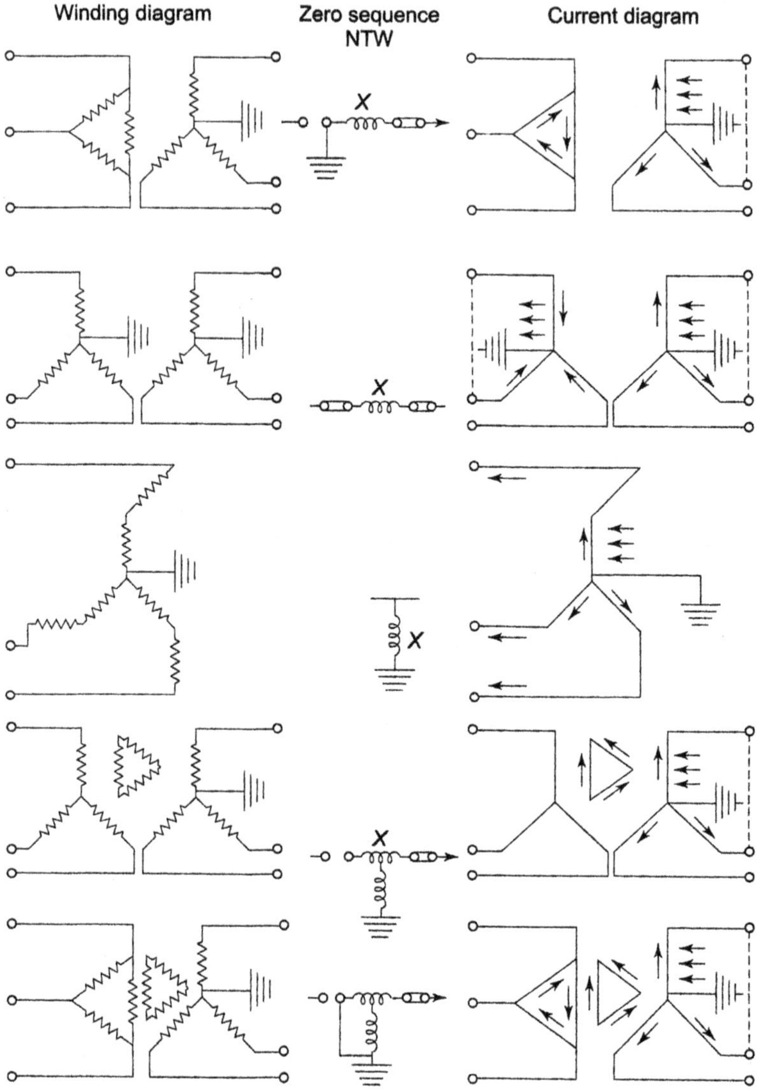

Fig. 3.13 *Transformer equivalent zero sequence circuit*

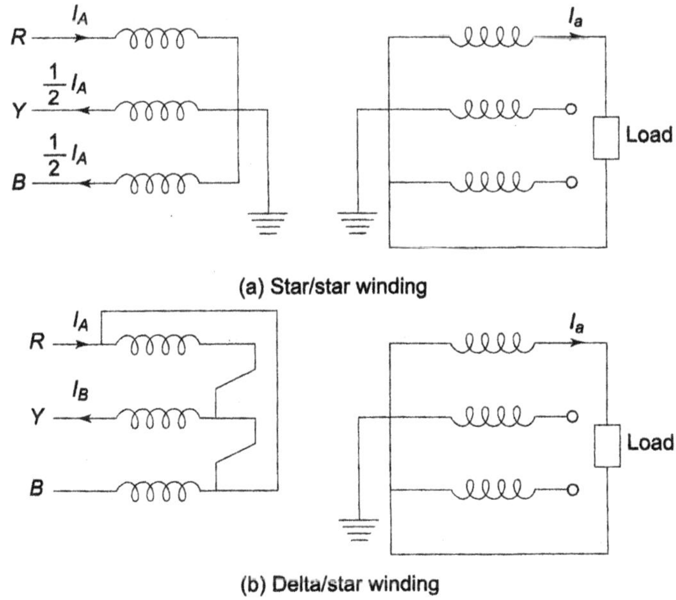

Fig. 3.14 *Typical 11/0.433-0.240 kV transformer windings single-phase load unbalance*

3.11.3 Cables

3 Core cables: $Z_0 = 3Z_1$ to $5Z_1$
Single core cables: $Z_0 = 1.25\ Z_1$.
Z_1 is positive sequence impedance; Z_0 is zero sequence impedance.

3.12 Load Flow

In order to evaluate the performance of a power distribution network and to examine the effectiveness of proposed alterations to a system in the planning stage, it is essential that a load flow analysis of the network is carried out. The load flow studies are normally carried out to determine:

(a) The flow of active and reactive power in network branches.
(b) No circuits are overloaded, and the busbar voltages are within acceptable limits.
(c) Effect of additions or alterations on a system.

(d) Effect of loss of circuits under emergency conditions.
(e) Optimum system loading conditions.
(f) Optimum system losses.

The load-flow in distribution networks is carried out in essentially two forms. In the case of simple radial networks, the load-flow is largely self-evident. However, in interconnected networks having a number of sources or parallel paths, the problem is more complex.

The use of the digital computer is now almost universally accepted in studying load flow. Ready access to computers allows for a more continuous and thorough review of system performance.

EXAMPLE:

A city indoor substation has two 11/0.433 kV transformer, 750 kVA and 320 kVA and are operating in parallel to cater load of 800 kW at power factor of 0.8. The percentage impedance of each is 5.5% and load losses are 7 and 3.5 kW, respectively. Find load sharing.

Solution: 320 kVA Transformer:

$$\text{Resistance} = \frac{\text{Load losses} \times \text{kV}^2}{(\text{kVA})^2} = \frac{I^2R \times \text{kV}^2}{(\text{kVA})^2}$$

$$= \frac{3.5 \times 11^2 \times 10^3}{320^2} = 4.13\ \Omega$$

$$\text{Reactance} = \frac{\%\text{impedance} \times \text{kV}^2 \times 10^3}{\text{kVA}}$$

$$= \frac{5.5 \times 11^2 \times 10^3}{100 \times 320} = 20\ \Omega$$

$$\text{Impedance} = 4.13 + j\,20 = 20.42\ \angle 78.3°\ \Omega$$

750 kVA Transformer:

$$\text{Resistance} = \frac{7 \times 11^2 \times 10^3}{750} = 1.5\ \Omega$$

$$\text{Reactance} = \frac{5.5 \times 11^2 \times 10^3}{100 \times 750} = 8.87\ \Omega$$

$$\text{Impedance} = 1.5 + j\,8.87 = 9\ \angle 80.4°\ \Omega$$

$$\text{The total load on transformers} = 800\ \text{kW at 0.8 pf}$$
$$= 800 + j\,600\ \text{kVA}$$
$$= 1000\ \angle 36.9°\ \text{kVA}$$

On parallel running, the load on each transfer will be divided in inverse ratio of impedances, i.e.

$$\text{Load on 320 kVA transformer} = \frac{1000\,\angle 36.9^\circ \times 9\,\angle 80.4^\circ}{(4.13 + j20) + (1.5 + j8.87)}$$

$$= \frac{1000\,\angle 36.9^\circ \times 9\,\angle 80.4^\circ}{29.59\,\angle 79.1^\circ}$$

$$= 306\,\angle 38.2^\circ = 306 \text{ kVA}$$

$$\text{Load of 750 kVA transformer} = \frac{1000\,\angle 36.9^\circ \times 20\,\angle 78.3^\circ}{29.59\,\angle 79.1^\circ}$$

$$= 694\,\angle 36.1^\circ = 694 \text{ kVA}$$

By quick look, it can be said that load is shared in proportion to rated capacity of the transformers.

3.12.1 Formulation of the Problem

The network matrix equation provides an extremely suitable mathematical model of an electrical network. With the help of this equation it is possible to describe the characteristics of the network as well as the interconnection of these elements.

The selection of independent variables determines the elements of the networks matrix. The independent variable can be voltage or current and correspondingly the elements of the network matrix are impedance or admittance.

The form of network matrices used in the equation depends on the frame of reference, namely, bus, loop or branch. In the bus (or nodal) frame of reference, the variables are nodal voltages and currents. In the loop frame of reference the variables are loop voltages and loop currents.

Since the loop frame of reference does not yield sets of voltages or currents, it is rarely used in the study of load-flow problems. The complexity of data preparation further restricts its use.

In the bus frame of reference, the performance of the interconnected network is described by $(n - 1)$ independent nodal equations, where n is the number of nodes. The matrix notation in impedance form is

$$\mathbf{E}_{\text{bus}} = \mathbf{Z}_{\text{bus}}\,\mathbf{I}_{\text{bus}} = [\mathbf{Y}_{\text{bus}}]^{-1}\,\mathbf{I}_{\text{bus}}$$

In admittance form

$$\mathbf{I}_{\text{bus}} = \mathbf{Y}_{\text{bus}}\,\mathbf{E}_{\text{bus}}$$

where $\mathbf{E}_{\text{bus}}$ – vector of node voltages measured with respect to the reference node

$\mathbf{I}_{\text{bus}}$ = vector of impressed node currents
$\mathbf{Z}_{\text{bus}}$ = bus impedance matrix
$\mathbf{Y}_{\text{bus}}$ = bus admittance matrix

In the formulation of the load-flow problem, fixed data concerning network and its interconnections are first assembled. The power generation or loading at any particular busbar is then specified. The power specified at any node can take the following forms:

(a) At any busbar k, the total assumed complex power ($S_k = P_k + jQ_k$) injected into or moved from the network is specified.

(b) At any busbar k, usually a generator bus or a busbar connecting two sub-networks referred to as voltage regulated bus, the power P and voltage V are specified.

(c) With transmission losses of the network unknown at least one generator output S_j is unknown. However, the voltage can still be specified both in magnitude and phase. This is called the *floating bus* or *swing bus.*

Due to nonlinearity of algebraic equation describing the power system, their solution is based on iterative technique. The solution must satisfy Kirchhoff's laws, i.e. the algebraic sum of currents at a node or voltages around a loop must be zero. One of these laws is used as a test for convergence of the solution in the iterative computational method.

The voltage and current variables at any node k are related to the network by the power requirements. These are usually given by:

$$S_k = V_k^* I_k = P_k + jQ_k$$

where S_k is the complex conjugate of the power.

$$\text{Node current } I_k = \frac{P_k + jQ_k}{V_k^*} = \frac{S_k}{V_k^*}$$

If the shunt elements are not included in the parameter matrix, the total current at bus k is

$$I_k = \frac{P_k + jQ_k}{V_k^*} - Y_k V_k$$

The iterative solution involves setting the busbar voltages at an initial starting point and then, through successive iterations, adjusting the estimated busbar voltages such that the specified power requirements of the network are satisfied.

3.12.2 Formulation of Network Matrix

The electrical characteristics of individual network components can be conveniently presented in the form of a primitive network matrix which represents the characteristics of each element. It does not, however, provide any information about the connections of these network elements. In order to describe the geometrical structure of the network, the network components are replaced by single line segments (elements), irrespective of their characteristics.

The following methods are used for constructing a network matrix.

Connection Table

A connection matrix describes the connection of the elements of the network. The type of transformation used depends on the frame of reference used, i.e. bus, loop or branch. The bus incidence matrix shows the incidence of elements of nodes of a connected graph.

In the bus incidence matrix A, any node is selected as the point of reference (usually taken to be ground). The elements in the matrix of the other nodes are determined with respect to this reference. The elements of the matrix are:

$a_{ij} = 1$, if the *i*th element is incident to and oriented away from the *j*th node.

$a_{ij} = -1$, if this element is oriented towards the *j*th node.

$a_{ij} = 0$, if the *i*th element is not incident to the *j*th node.

For example, for the network shown in Fig. 3.15(a) the bus incidence matrix is shown next to it.

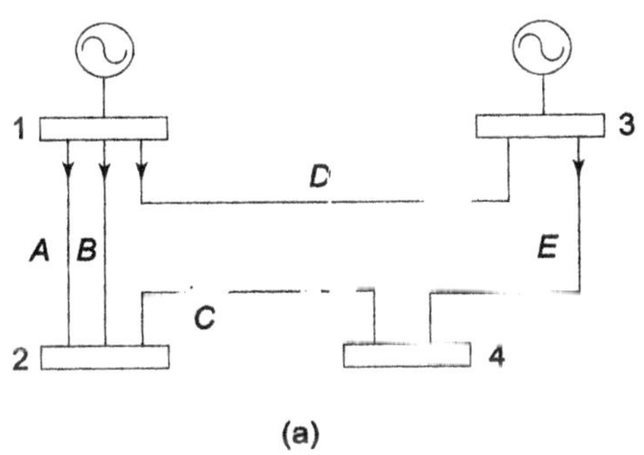

(a)

	1	2	3	4
A	1	−1		
B	1	−1		
C		1		−1
D	1		−1	
E			1	−1

(b)

Fig. 3.15 *Network and bus matrix: (a) Network: (b) Bus incidence matrix*

Therefore,

$$\mathbf{A} = \begin{bmatrix} 1 & -1 & 0 & 0 \\ 1 & -1 & 0 & 0 \\ 0 & 1 & 0 & -1 \\ 1 & 0 & -1 & 0 \\ 0 & 0 & 1 & -1 \end{bmatrix}$$

From this connection matrix, the bus admittance matrix is formed using the following matrix multiplication:

$$\mathbf{Y}_{\text{bus}} = \mathbf{A}^T y\,\mathbf{A} = \mathbf{A}^T g\,\mathbf{A} + j\,\mathbf{A}^T b\,\mathbf{A}$$

where y is the primitive admittance matrix such that $\mathbf{y} = \mathbf{g} + j\,\mathbf{b}$.

To form the bus impedance matrix, the bus admittance matrix is inverted.

The loop impedance matrix $\mathbf{Z}_{\text{loop}}$ is formed from the primitive matrix $\mathbf{z}$ and the loop incidence matrix $\mathbf{C}$, i.e.

$$\mathbf{Z}_{\text{loop}} = \mathbf{C}^T \mathbf{z} \mathbf{C}$$

and

$$\mathbf{Y}_{\text{loop}} = \mathbf{Z}_{\text{loop}}^{-1}$$

Algorithm Method

This is an alternative method of formulating a network matrix directly from system parameters and code-node numbers. This method is generally used for large interconnected networks. The algorithm simulates the construction of the network, one element at a time.

Two types of elements in terms of network topology can be added to a network—a branch or a link. If a branch is added, a new node is added, increasing the dimension of the matrix but not changing the values of the elements of the former matrix. If a link is added, no new node is added but the elements of the bus impedance matrix must be again calculated.

The algorithm allows for the determination of the added elements to the matrix and for the change of the matrix elements when a link is added. In cases where there are no mutual coupling between the element to be added and the rest of the network, equations for the formation of the bus impedance matrix are shown in Table 3.2.

Table 3.2

Equations for formation of bus impedance matrix (no mutual coupling)

Added element $p-q$	p is not a reference bus	p is a reference bus
Branch	$Z_{qi} = Z_{pi}$ $i = 1, 2, \cdots, m$ $i \neq q$ $Z_{qq} = Z_{pq} + Z_{pq,pq}$	$Z_{qi} = 0$ $i = 1, 2, \cdots, m$ $i \neq q$ $Z_{qq} = Z_{pq,pq}$
Link	$Z_{li} = Z_{pi} - Z_{qi}$ $i = 1, 2, \cdots, m$ $i \neq l$ $Z_{ll} = Z_{pl} - Z_{ql} + Z_{pq,pq}$	$Z_{li} = -Z_{qi}$ $i = 1, 2, \cdots, m$ $i \neq l$ $Z_{ll} = -Z_{ql}$

3.13 Automated Planning

The computer is only a tool, an electronic device that is completely subject to human will. It is essentially a tool for problem solving. It can provide decision information that was too costly before and took too long to process. It enables solving complex problems containing too many variables for the human mind to handle. The computer forces us to think through our problem logically, by system analysis. Tangible benefits are choice of solution with higher quality and lower cost. It is a learning tool. Optimization-based methods contribute to an engineer's understanding of the interplay of costs, standards and performance and their trade-offs.

Software Needs

A few large power utilities have MIS/IT teams and can afford in-house software development. Most utilities purchase commercially available package software due to the difficulty of producing quality software. It is risky to undertake software development without proper information technology infrastructure, both in terms of hardware and software. Software development is very human resource intensive. Unless project implementation is guided and monitored very well, it tends to get out of hand. Also, the issue of maintenance is often overlooked. If technical documentation is not done satisfactorily, maintenance becomes difficult.

Outsourcing is another alternative. Find a cost-effective reliable software company to deliver the right quality software in time and keep up maintenance contract upfront. The success of a software development project will ultimately depend upon the involvement of the power utility with the developer. They must have a partnership approach.

Before purchasing any package, its strength and weaknesses must be ascertained to find out its suitability for the job. Load un-balancing, low voltage conditions, two-phase supply or single-phase 11 kV supply for agriculture in some states need special consideration in the choice of software (see Section 4.1.1.)

A very impressive computer programme, implementing a clever algorithm, will produce nothing of value if it is built upon a model that is inadequate when measured against the planning requirements. Generally software is selected on the basis of the *least lifecycle cost or least ownership costs (hiring or purchase + training + maintenance costs)*. The other points given below can be checked out depending upon the importance attached to each.

- Best overall performance;
- Best solution; and
- Best open architecture system.

Some of the vendors for Distribution System Studies software (AM/FM/GIS/Network system analysis) are:

- Electricite de France, Paris (PRAO);
- Scott & Scott, New Delhi-110 017 (SYNERGEE);
- Swed Power Sydkraft International, Malmo, Sweden (SWENET);
- Milsoft Integrated Solutions Inc, Abilene, Texas;
- Power Technologies South-Asia Pvt. Ltd., New York 12301-1058 (PSS/U, NE PLAN);
- ABB, Bangalore-560 058;
- Trident Technologies Pvt., New Delhi-110 088 (CYME);
- Global Energy Consulting Engineers Private Ltd., Hyderabad-500482 (POWERNET);
- KLG System Ltd., Gurgaon (SPARD);
- Datagen Power Systems (P) Ltd., New Delhi (POWERGEN, NETWORKX);
- International Computers, New Delhi (DINIS);
- Indicos Information Technology (P.) Ltd., Mumbai 400 604 (GIS); and
- Autodesk®, Bangalore, (GenMap)-AM/FM/GIS.

Analysis of the System

The computer-aided power flow studies of the existing sub-transmission and distribution system and proposed expansion of network upto the meter point should be carried out to assess the following:

- Active and reactive power flows, location of capacitors;
- Voltage variation (at each node or bus, per cent variation);
- Technical peak power loss and energy loss;
- Computation of commercial loss from feeder or sub-station energy balance sheets;
- Inadequacy or over-loading of transformers and lines;
- Adequacy of new expansion; and
- Security and reliability of the power supply.

System Study Software Features Required

- Network creation with detailed modelling of all the network elements/equipment, geographical mapping of cable routes, overhead lines, sub-station and other physical details;
- Creation of network connectivity from database;
- Geographical Information Systems;
- Calculation of network, load and operational parameters;
- Global and spatial load forecast;
- Load flow analysis for voltage regulation, active and reactive power flows and loss calculations (power and energy);
- Short circuit analysis;
- Sub-station sizing and locations;
- Capacitor placement and sizing;
- Network re-configuration and reinforcement;
- Voltage control through series capacitors and voltage regulators, etc.;
- Load modelling;
- Load balancing;
- Mechanical design of distribution system elements;
- System optimization studies; and
- Cost estimates and financial and economic analysis.

Data

The success of any information technology system is highly dependent on the accurate, up-to-date and relevant field data for improvement in the network and the system. The data is collected on appropriate prescribed formats as per the requirement of a particular software applied for planning studies. For data formation, the electrical network at the distribution level is composed of hundreds to thousands of nodes (locations) consisting of power delivery points (buses) at the sub-stations, tee-offs or spurs, switching places, fuse points, feeder sectionalizing points, distribution transformers, service tap points, consumer meters etc. (see Section 3.8). The details in representing the distribution system can be measured as a node resolution, i.e. the ratio of network database nodes to the number of consumers fed by the network. A greater number of nodes provides greater potential for accuracy for two reasons. Firstly, large nodes in a circuit provide greater detail in representation of the circuit topology and equipment. Secondly, most programmes (load flow, reliability studies, etc.), compute results only at nodes. The radial network operates as a tree, i.e., a hierarchy of network nodes showing the topology and connections between networks and sub-stations. A GIS system (see Section 3.6) offers an effective environment for preparing the common database. The local detailed data may be collected in respect of:

- *Nodes:* Sub-stations, busbars, tee-offs points, witching places, fuse points, feeder sectionalizing points, distribution transformers, service tap points, consumer meters, etc.;
- *Single line diagrams:* Distribution network single line diagrams of distribution sub-stations, primary feeders and each distribution transformer along with LT lines showing line conductor sizes and lengths (see Section 18.1.1), equipment (transformers, capacitors, circuit breaker, isolator, CTs, PTs, distribution transformers, etc.) parameters; sub-transmission feeder: from S/S to, name, circuit km, conductor size, average and peak demand;
- *Distribution sub-stations:* Name, voltage ratio, power transformer capacity, maximum demand (MVA, MVAr), typical daily load profile for each month of the past year, load growth rate, number and names of primary feeders;
- *Primary (11 kV) feeder:* From distribution S/S, name, total length, segment-wise conductor size, peak demand, distribution transformers' number rating-wise and their total kVA;
- *Distribution transformer and secondary (LT) feeders:* Name of 11 kV feeder, name of distribution transformer, size, installation date, cable size, number of LT feeders and their lengths, total services connected (category-wise) on each LT feeder, with connected load, per cent loading;

- *Consumer data:* Name, address, account number, area, location (feeder and distribution transformer) codes etc., type of connection, contract demand, energy consumption profile, supply voltage, connected load, bill and payment record, meter make, service cable size and length, size of the capacitors size; and
- *Load data:* Peak load and load growth rate of each distribution transformer and feeder; power factor, diversity factor, load factor and loss factor at various voltage levels.

Planning studies should be based on actual feeder load conditions and measurements at various points in the system. For example, the annual load duration curve may be derived by aggregating the typical hourly readings of the system power demand at grid sub-station for typical working days and holidays in each month during the year so as to take into account the seasonal variation as well. An annual loss load duration curve may be accordingly derived from the load duration curve by squaring the ordinates. The *load factor* and *loss load factor* for the system may be computed from the load-duration curve and loss load-duration curve respectively.

Database

The collection of data is called a *database.* A common database is required to ensure that it is available to all sections of the power utility. In a computer, it is possible to store vast quantities of data economically and retrieve the desired data with access times of the order of seconds. Operations on the database are performed by the *Database Management System* a software which acts as an intermediary between applications and data files.

3.13.1 Solution Techniques

The techniques used for solving load-flow problems are:

1. Gauss iterative method.
2. Gauss-Seidel iterative method.
3. Newton-Raphson iterative method (also referred to as Ward Hale method).
4. Improved Newton-Raphson method.

Many modifications and variations to these methods are also available. Most of these variations are introduced in terms of numerical techniques to either speed up the computation or reduce the required computer resources [11].

Gauss Iterative Method

The solution is initiated by assuming voltages for all buses except the slack bus where the voltage is specified. Currents are calculated for all buses except the slack bus from the equation:

$$I_p = \frac{P_p + jQ_p}{V_p^*}$$

The performance of the network is then obtained from the equation:

$$I_{bus} = Y_{bus} E_{bus}$$

This then allows the bus voltages to be calculated from the expression:

$$E_p = \frac{1}{Y_{pp}} \left[I_p - \sum_{\substack{q=1 \\ p \neq q}}^{n} Y_{pq} E_q \right], \quad \underset{p \neq q}{p} = 1, 2, \cdots, n$$

This new voltage is then substituted back into the expression to recalculate bus currents for a subsequent solution of the bus voltages. This is continued until changes in the bus are negligible.

Gauss-Seidel Iterative Method

This method is almost identical to the Gauss iterative method except that in substituting the bus voltages into the expression to obtain the new value of the bus voltage, the most recent value of bus voltage is substituted into the expression. In other words, while in the Gauss method only the bus voltages from the previous iterations are used, in the Gauss-Seidel method, for evaluating the nth bus voltage, the voltages for the iteration are used for the 1 to $(n - 1)$ buses while the bus voltage for the previous iterations are used for the rest of the bus voltages.

Newton-Raphson Iterative Method

This method of solving nonlinear equations was adopted by Ward and Hale. It is by far the most sophisticated of the basic methods. Not only it will work in most cases without risk of divergence, but it will also as a rule converge faster than the proceeding methods described.

The method is based on the following theory. Consider the simultaneous equations:

$$f_1(x_1, x_2, \cdots, x_n) = 0$$
$$f_2(x_1, x_2, \cdots, x_n) = 0$$
$$\vdots$$
$$f_n(x_1, x_2, \cdots, x_n) = 0$$

Let us initially guess that $x_1^{(0)}, x_2^{(0)}, \cdots, x_n^{(0)}$ are solutions of these n equations and the error in each guess $\Delta x_1^{(0)}, \Delta x_2^{(0)}, \cdots, \Delta x_n^{(0)}$.
We can then write that

$$f_1 = (x_1^{(0)} + \Delta x_1^{(0)}, x_2^{(0)} + \Delta x_2^{(0)}, \cdots, x_n^{(0)} + \Delta x_n^{(0)}) = 0$$
$$f_2 = (x_1^{(0)} + \Delta x_1^{(0)}, x_2^{(0)} + \Delta x_2^{(0)}, \cdots, x_n + \Delta x_n^{(0)}) = 0$$
$$\vdots$$
$$f_n = (x_1^{(0)} + \Delta x_1^{(0)}, x_2^{(0)} + \Delta x_2^{(0)}, \cdots, x_n^{(0)} + \Delta x_n^{(0)}) = 0$$

These equations can then be expanded using the Taylor's series. Neglecting the second order terms, since the errors Δx_1 etc. are small, we get

$$f_1 = (x_1^{(0)}, x_2^{(0)} \cdots) + \Delta x_1^{(0)} \left(\frac{\partial f_1}{\partial x_1}\right)^{(0)} + \Delta x_2^{(0)} \left(\frac{\partial f_1}{\partial x_2}\right)^{(0)} + \cdots = 0$$

$$f_2 = (x_1^{(0)}, x_2^{(0)} \cdots) + \Delta x_1^{(0)} \left(\frac{\partial f_2}{\partial x_1}\right)^{(0)} + \Delta x_2^{(0)} \left(\frac{\partial f_2}{\partial x_2}\right)^{(0)} + \cdots = 0$$

Writing this in a matrix form, we have

$$\mathbf{f}^{(0)} + \mathbf{J}^{(0)} \Delta x^{(0)} \approx \mathbf{U}$$

where $\mathbf{J}$ is referred to as the Jacobian and is equal to:

$$\mathbf{J}^{(0)} = \begin{bmatrix} (\partial f_1/\partial x_1)^{(0)} & \cdots & (\partial f_1/\partial x_n)^{(0)} \\ (\partial f_2/\partial x_2)^{(0)} & \cdots & (\partial f_2/\partial x_n)^{(0)} \\ \cdots\cdots\cdots & \cdots & \cdots\cdots\cdots \\ (\partial f_n/\partial x_1)^{(0)} & \cdots & (\partial f_n/\partial x_n)^{(0)} \end{bmatrix}$$

The corrections to the initial guess $\Delta x^{(0)}$ are equal to

$$\Delta x^{(0)} \simeq -[\mathbf{J}^{(0)}]^{-1} \mathbf{f}^{(0)}$$

In terms of the load-flow problem, the power at bus p is

$$S_p = P_p - jQ_p = E_p^* I_p = E_p^* \sum_{q=1}^{n} Y_{pq} E_q$$

Expressing E_p and Y_{pq} in cartesian form:

$$E_p = e_p + jf_p \qquad Y_{pq} = G_{pq} - jB_{pq}$$

The real and imaginary components of power become

$$P_p = \sum_{q=1}^{n} \{e_p(e_q G_{pq} + f_q B_{pq}) + f_p(f_q G_{pq} - e_q B_{pq})\}$$

$$Q_p = \sum_{q=1}^{n} \{f_p (e_q G_{pq} + f_q B_{pq}) - e_p(f_p G_{pq} - eq B_{pq})\}$$

This equation is used to obtain the partial differentials with respect to e and f to form the coefficients of the Jacobian giving the equation:

$$\begin{bmatrix} \Delta P \\ \Delta Q \end{bmatrix} = \begin{bmatrix} J_1 & J_2 \\ J_3 & J_4 \end{bmatrix} \begin{bmatrix} \Delta_e \\ \Delta_f \end{bmatrix}$$

Here ΔP and ΔQ are equal to the difference between the scheduled power and the power derived from the previous iteration. This is then used to obtain values for e and f such that the new values of the bus voltages can be obtained as follows

$$e_p^{k+1} = e_p^k + \Delta e_p^k, \qquad f_p^{k+1} = f_p^k + \Delta f_p^k$$

and the new values of power are evaluated.

When the values of ΔP_p and ΔQ_p are sufficiently small, the solution is found and the line currents etc. are evaluated.

Improved Newton-Raphson Methods

There are many improved methods based on the Newton-Raphson method which have been developed and which aim to improve the speed of the computations or to decrease the computer requirements. These are:

(a) *Acceleration:* The correction to the bus voltage is changed by a multiple of the calculated correction.

(b) *Decoupling:* The elements J_2 and J_3 of the Jacobian are set to zero.

(c) *E-coupling:* The decoupling method is modified to accelerate the iteration and to ensure convergence.

The relative speeds of convergence of various methods of load flow analysis are difficult to ascertain since this will vary from computer to computer. In addition, the effectiveness of a technique is a function of the speed of convergence, the time taken per iteration and its sensitivity to the network parameters.

The Newton-Raphson method is preferred since it converges very much faster than the Gauss-Seidel method. It is claimed that the E-coupling technique offers convergence times of approximately 40% that of the Newton-Raphson method.

3.14 Fault Studies

When a fault occurs on a power distribution network, abnormal currents may flow through various elements of the network. These must be interrupted by protective equipment. Firstly, it is important that values of the currents which flow are known so that the protective devices may be placed at the correct settings. Secondly, it is important to ascertain that the fault level throughout the network can be determined and to check that it does not exceed the fault rating of any of the equipment.

In general, fault calculations fall into two categories: (i) Three-phase symmetrical short circuits and (ii) unsymmetrical faults.

3.14.1 Three-phase Symmetrical Faults

Balanced voltage and currents of positive phase order or sequence are called *positive sequence* voltages and currents. Three-phase generators generate positive sequence voltages only. A three-phase balanced system must have the same impedance to positive-sequence currents in each phase. Thus, the instantaneous symmetrical fault current for a symmetrical three-phase fault may be calculated by drawing a one-line diagram of the impedances of one phase to the fault and applying the phase to neutral voltages to this equivalent circuit. A neutral conductor having zero impedance is used as the return path for the current. Generators feeding into the network at various points are assumed connected to a common busbar.

Fault level analysis is carried out by any one of the two methods given below.

Network Reduction

As distinct from the load-flow problem, where an iterative solution has to be found in order to reach the required boundary conditions, the fault level analysis is readily carried out using the network reduction technique.

In this method the impedances (or reactances where resistance is small) are represented in per unit form on a convenient MVA base. For example, 100 MVA base is commonly used. Alternatively, the elements are represented in ohms on a common voltage base throughout the network. The network is then progressively reduced into a simple impedance which enables the determination of current distribution factors.

EXAMPLE:

Consider the network shown in Fig. 3.16. Determine the three-phase fault current at the point of fault *F*.

Assume: Base power $S_b = 100$ MVA

$$\text{Base impedance at 33 kV, } Z_b = \frac{(\text{kV}_{\text{line}})^2 \times (10^3)^2}{100 \times 10^6}$$

$$= \frac{(33)^2 \times (10^3)^2}{100 \times 10^6} = 10.8952\ \Omega$$

$$\text{Base impedance at 11 kV, } Z_b = \frac{(11)^2 \times (10^3)^2}{100 \times 10^6} = 1.21\ \Omega$$

$$\text{Source impedance} = \frac{\text{Base}_{\text{MVA}}}{\text{Fault level}} = \frac{100 \times 10^6}{1500 \times 10^6}\ \text{pu}$$

$$= j0.067$$

$$\text{Transformer impedance, } Z'_{\text{pu}} = Z_{\text{pu (rating)}} \frac{S_b}{S_{\text{rating}}}$$

$$\text{15 MVA transformer, } Z'_{\text{pu}} = 0.10 \times \frac{100}{15} = 0.667\ \text{pu}$$

$$\text{500 kVA transformer, } Z'_{\text{pu}} = 0.04 \times \frac{100}{0.5} = 8.0\ \text{pu}$$

$$\text{11 kV line impedance, } Z_{\text{pu}} = \frac{Z}{Z_b} = \frac{3(0.15 + j0.25)}{1.21}$$

$$= 0.37 + j0.62\ \text{pu}$$

Therefore, the fault level at low voltage busbar

$$= 100 \times \frac{1.0}{0.41 + j9.14}$$

$$= 10.9\ \text{MVA}$$

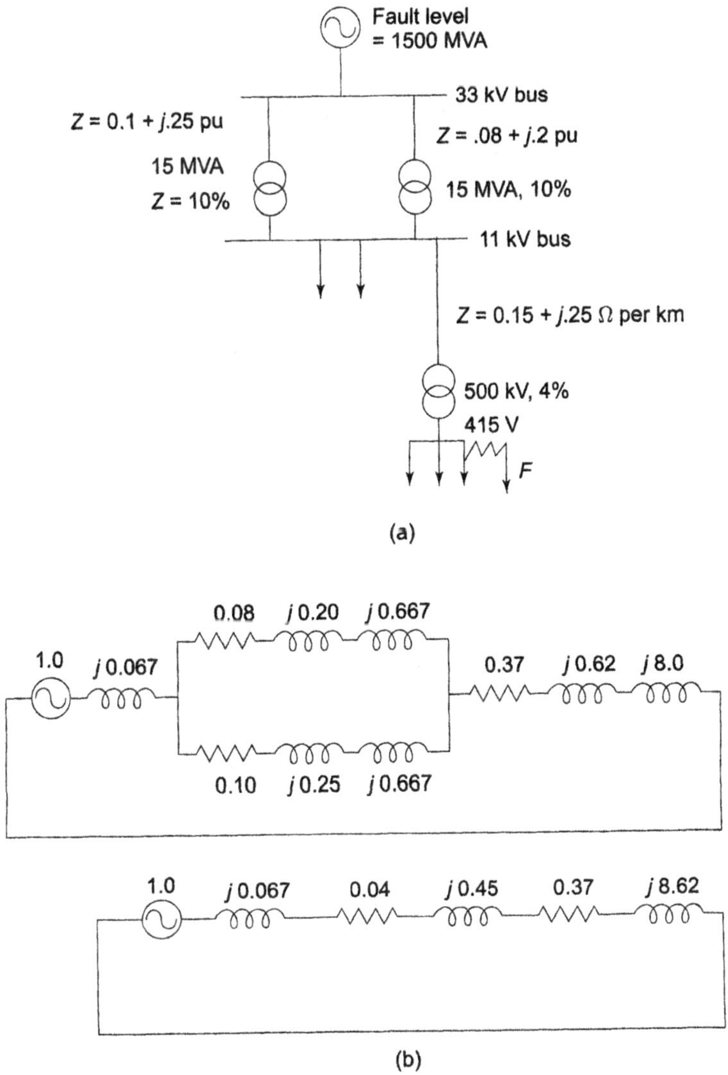

Fig. 3.16 *Equivalent circuit under fault conditions: (a) Network; (b) Equivalent circuit*

Computer Solution

Digital computer techniques can be readily applied to the balanced fault study problem. For a simple network, the network reduction techniques can be applied. For larger networks, however, this becomes impractical and tedious. Accordingly, a more systematic method based on Thevenin's theorem, is used. The bus impedance matrix provides a convenient means of calculating short circuit currents and voltages when the ground is selected as reference (Fig. 3.17).

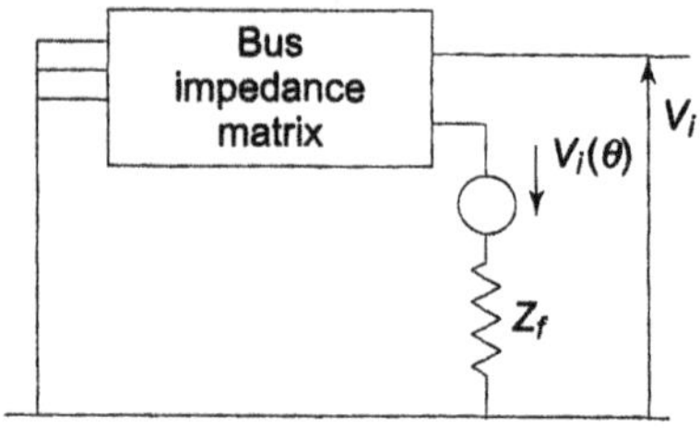

Fig. 3.17 *Thevenin theorem: Equivalent circuit*

The effect of a short circuit is equivalent to that of connecting the branch of impedance between the faulted bus and ground. Thevenin's theorem states that: ***The changes in the network currents and voltages caused by the added branch are equivalent to those caused by the added voltage equal to the bus voltage prior to the fault, and all other sources short circuited.***
Mathematically

$$\mathbf{V}_T = \mathbf{Z}_{\text{bus}} \mathbf{I}_f, \text{ where } \mathbf{V}_T = \text{Thevenin's bus voltage}$$

But

$$\mathbf{V}^f_{\text{bus}} = \mathbf{V}^{(0)}_{\text{bus}} + \mathbf{V}_T = \mathbf{V}^{(0)}_{\text{bus}} + \mathbf{Z}_{\text{bus}} \mathbf{I}_f$$

Now given that with a fault at bus q

$$\mathbf{V}^f_q = \mathbf{Z}^f \mathbf{I}^f$$

where $\mathbf{Z}^f$ is the fault impedance. Therefore,

$$\mathbf{I}_f = \mathbf{V}^{(0)}_q / (\mathbf{Z}_f + \mathbf{Z}_{qq})$$

$$\mathbf{V}^f_i = \mathbf{V}_i^{(0)} - \mathbf{Z}_{iq} / (\mathbf{Z}^f + \mathbf{Z}_{qq})\, \mathbf{V}^{(0)}_q,\ i \neq q$$

$$\mathbf{V}^f_q = \mathbf{Z}^f / (\mathbf{Z}^f + \mathbf{Z}_{qq})\, \mathbf{V}^{(0)}_q$$

where $\mathbf{Z}_{iq}$ can be obtained from the bus impedance matrix.

3.14.2 Unbalanced Faults

The impedance matrices of symmetrically operated three-phase generators, transformers and transposed transmission lines are diagonal with all diagonal elements equal. This fact results in decoupling between phases.

In an unsymmetrically faulted or loaded system, neither the currents nor voltages will possess three-phase symmetry. The impedance matrices of generators, transformers and lines will all be nondiagonal. It is no longer possible to limit the analysis to one-phase because coupling exists between all three phases. It is necessary to treat the different phases individually. This added complexity is considerably offset by the systematizing features of the symmetrical components described previously.

Calculation of System Currents and Voltages

The application of symmetrical components method may be summarized as follows:

(a) Set up network impedances in the form of a single line diagram for each of the three sequence components (positive, negative and zero).

(b) Determine the single equivalent impedance of each of the individual networks and distribution factors in the individual branches.

(c) Connect the three single impedances as appropriate to the type of fault and determine the sequence components of current at the point of fault.

(d) Determine the distribution of sequence components of current throughout the network by the application of distribution factors.

(e) Combine sequence components of current at any point to obtain phase currents.

(f) Determine sequence components of voltage throughout the network from the current distribution and sequence impedances to obtain phase voltages.

(g) Combine the sequence components of voltages at any particular point to obtain the phase voltages.

(h) Pay particular attention to delta-star transformers.

Connection of Sequence Networks

The connections of the sequence networks for the various types of faults are determined by establishing the boundary conditions at the point of fault (refer Figs. 3.18 and 3.19).

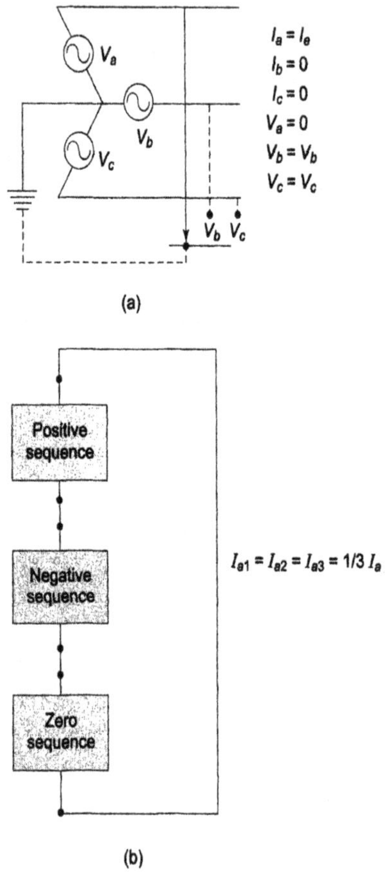

Fig. 3.18 *Single-phase fault through zero fault impedance: (a) Single-phase to earth fault; (b) Equivalent circuit*

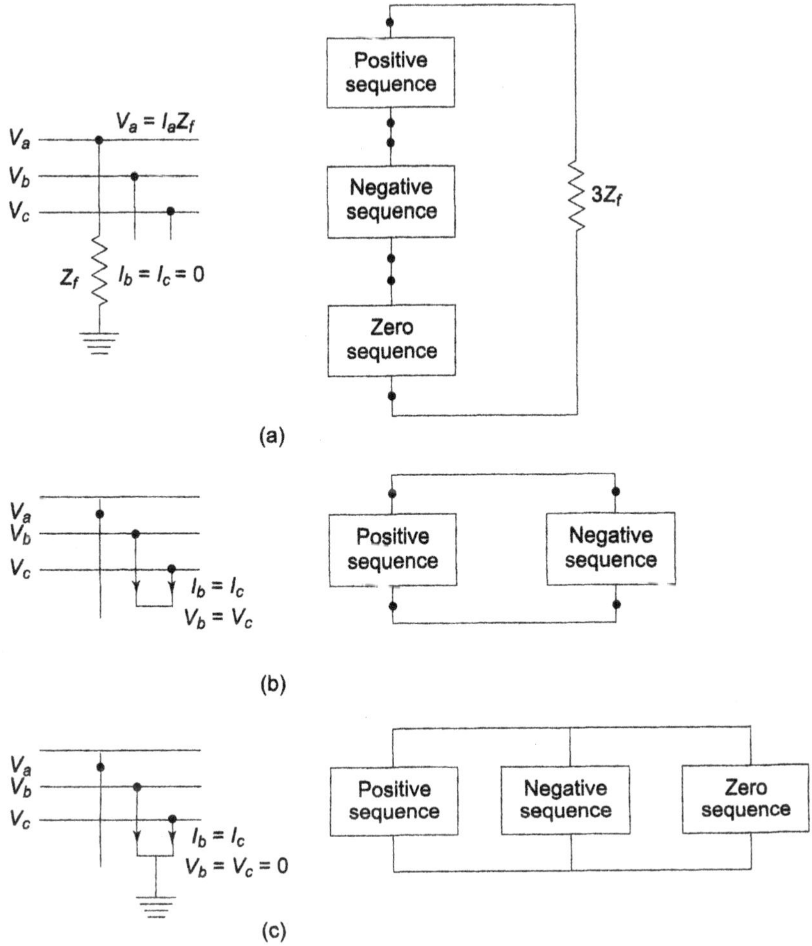

Fig. 3.19 *System faults: (a) Single phase to earth through fault impedance; (b) Phase to phase fault; (c) Phase to phase to earth fault*

3.14.3 Types of Faults

There are four types of faults which can occur in distribution system:

- Three-phase grounded or ungrounded fault (3ϕ)
- Phase-to-phase (or line-to-line) ungrounded fault (L-L)
- Phase-to-phase (or double line-to-ground) grounded fault (2 LG)
- Phase-to-ground (or single line-to-ground) fault (SLG)

SLG faults take place more due to the multigrounded construction. The relative numbers of the occurrence of different fault types depend upon various factors, e.g. circuit configuration, the height of ground/ neutral wires, voltage class, method of grounding, relative insulation levels to ground and between phases, speed of fault clearing, quality of maintenance, number of stormy days per year, and atmospheric conditions. The average probabilities of occurrence of various types of faults are:

$$\text{SLG faults} = 70\%$$
$$\text{L-L faults} = 15\%$$
$$2\,\text{LG faults} = 10\%$$
$$3\ \phi\ \text{faults} = 5\%$$

The actual fault current is usually less than the dead short-circuit three-phase value. However, the SLG fault often produces a greater fault current than the 3ϕ fault especially where the associated generators have solidly grounded or low-impedance neutral with star-grounded of delta-star transformers. According to Anderson, SLG fault is the most severe, with the other faults severity following in decreasing order as 3ϕ, 2 LG and L-L.

The maximum and minimum fault currents are calculated for a system by load flow studies. The maximum fault current is calculated on the following assumptions:

- All generators are in service
- The fault is a dead short-circuit
- The load is maximum, i.e., on-peak load

The minimum fault current is calculated on the assumptions:

- The number of generators connected is minimum
- The fault is not dead short-circuit, i.e., the fault impedance is not zero but has a value somewhere between 30 and 40Ω
- The load is minimum, i.e., off-peak load.

Overload distribution systems are subjected to transient faults and permanent faults. Approximately 75 to 90% of the total number of faults are transient in nature and faults occur when phase conductors electrically contact other phase conductors or ground momentarily, due to trees, birds or animals, public accidents, high winds, lightning flashovers, etc. Transient faults are cleared by circuit breaker or fuse cut-outs. Permanent faults are those which require repairs in terms of (i) replacing burned-down conductors, or any other damaged apparatus,

(ii) removing tree branches from the line, etc. The number of consumers affected by a fault is minimized by properly selecting and locating the protective apparatus on the feeder main, at the tee-off point of each branch, and at critical locations on branch circuits. The number of faults occurring on an underground system is relatively much less than that on the overhead system.

3.15 Effect of Abnormal Loads

In a distribution network, the loads which are actually connected to the system are almost invariably other than the balanced three-phase constant loads. Usually some characteristic of the device results in the load being either unbalanced, intermittent or irregular in current waveform. For example, in domestic areas, individual houses are usually supplied with a single-phase supply from the three-phase system which results in an unbalanced load. In addition, in homes various appliances are switched on and off irregularly both by the consumers and by the thermostatic devices. Further, in industrial areas there are many devices which are of an unbalanced nature or intermittent in waveform, such as welding, thermostatically controlled heaters, extrusion presses, lathes, induction furnaces, variable speed drives, etc.

Due to the random nature of these individual loads, however, the sum of many individual loads either within the installation itself or on the network itself, usually results in a reasonably well-balanced load on the system. Accordingly, load-flow analysis as discussed earlier on the transmission and subtransmission networks and voltage drop calculations in distribution networks, usually assumes that the load is of a constant balanced form.

It is interesting, however, to consider in detail the effect of some abnormal loads on the distribution system. Large fluctuations of individual loads, although in a balanced form, can result in sudden voltage drops appearing at other points in the network. The quality of the voltage supply to adjoining consumers is accordingly affected and it may result in what is known as *flicker* problem. Many loads on the system have nonlinear characteristics, such as, variable speed drives, mercury arc rectifiers, computers, electronic loads etc. These devices result in harmonic currents flowing in the system which produce voltage distortions in the network.

3.16 Line Circuits

The performance of the lines is known by the efficiency and voltage regulation. The efficiency of the power line is determined as below:

$$\% \text{ efficiency} = \frac{\text{Power delivered at the receiving end}}{\text{Power delivered at the receiving end + line losses}} \times 100$$

Voltage regulation of a line is defined as the change in the receiving end voltage, expressed in per cent of full load voltage, from no load to full load, keeping the sending voltage constant.

3.16.1 Short Lines

Power lines, less than approximately 3% of the wavelength of the line for a particular frequency, are considered to be electrically short. Accordingly, the equivalent circuit of the line is formed by 'lumping' the line parameters together.

$$\text{Wavelength } (\lambda) = \frac{3 \times 10^8}{\text{Frequency (Hz)}} \qquad \text{(m)}$$

The length of the line considered short for various frequencies is as follows:

Frequency (Hz)	*Short line length (km)*
50	180
150	60
250	36
350	26
550	16

For a very short line at fundamental frequency, the line capacitance, which is a shunt element, may be neglected and so the line is represented by the series impedance of the line generally up to 22 kV level.

Figure 3.20 shows the equivalent circuit and phasor diagram with the sending end voltage $\mathbf{E}_s$, receiving end voltage $\mathbf{E}_R$ and line current $\mathbf{I}$.

Here $\mathbf{E}_R$ = reference voltage

$$\mathbf{I} = I \cos \phi_R + jI \sin \phi_R, \; Z = R + jX$$

$$\mathbf{E}_S = (E_R + IR \cos \phi_R - IX \sin \phi_R) + j(IX \cos \phi_R + IR \sin \phi_R)$$

For a leading power factor, ϕ_R is positive and for a lagging power factor ϕ_R is negative.

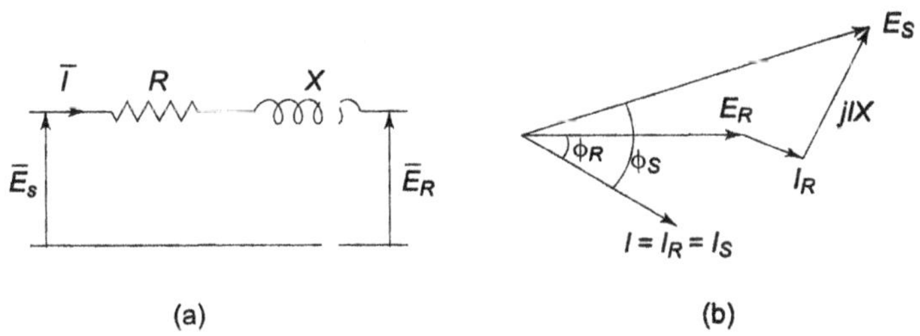

Fig. 3.20 *Short line: (a) Equivalent circuit; (b) Phasor diagram*

If the *IR* and *IX* drop are not over 10% of $\mathbf{E}_R$, then $\mathbf{E}_S$ can be determined within an error of 0.5% if the quadrature component is ignored, i.e.

$$\mathbf{E}_S = \mathbf{E}_R + IR\cos\phi_R - IX\sin\phi_R$$

$$\text{Voltage regulation (\%)} = \frac{(|E_s| - |E_R|)}{|E_R|} \times 100$$

3.16.2 Moderately Long Line

For electrically medium lines in which the above assumption cannot be made, the effect of the shunt capacitance must be considered. The capacitance of the line increases with line length. In addition, for higher harmonics the reactance of the capacitance is reduced and must therefore be taken into account. Such lines are generally in voltage levels of 33 to 132 kV.

The equivalent circuit can be taken as either an equivalent '*T*' or π section.

For the equivalent '*T*' circuit, where $\mathbf{E}_S$ and $\mathbf{E}_R$ are defined as before, $\mathbf{I}_S$ and $\mathbf{I}_R$ are the sending and receiving end currents respectively (Fig. 3.21) such that

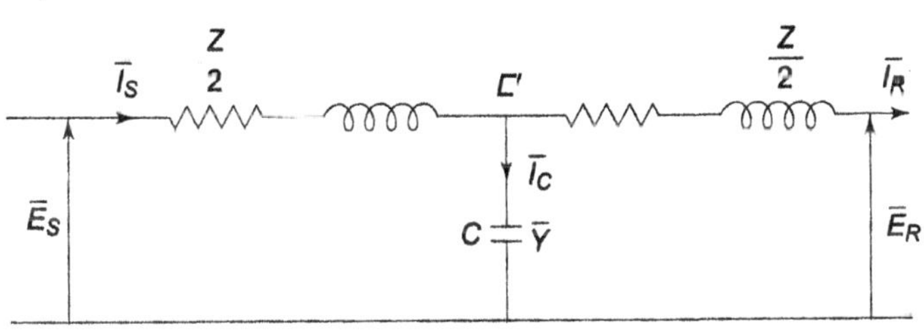

Fig. 3.21 *Moderately long line–equivalent 'T' circuit*

Load admittance $\mathbf{Y}_R = \mathbf{I}_R/\mathbf{E}_R$

$$\mathbf{I}_S = \mathbf{I}_R + \mathbf{I}_C = \mathbf{Y}_R \mathbf{E}_R + \left(\mathbf{E}_R + \mathbf{I}_R \frac{\mathbf{Z}}{2}\right)\mathbf{Y}$$

$$= \mathbf{E}_R\left[\mathbf{Y}_R + \mathbf{Y}\left(1 + \mathbf{Y}_R \frac{\mathbf{Z}}{2}\right)\right]$$

$$\mathbf{E}_S = \mathbf{E}' + \mathbf{I}_S \frac{\mathbf{Z}}{2} = \mathbf{E}_R\left(1 + \mathbf{Y}_R \frac{\mathbf{Z}}{2}\right) + \mathbf{E}_R\left[\mathbf{Y}_R + \mathbf{Y}\left(1 + \mathbf{Y}_R \frac{\mathbf{Z}}{2}\right)\right]\frac{\mathbf{Z}}{2}$$

$$= \mathbf{E}_R\left[1 + \mathbf{Y}_R\mathbf{Z} + \mathbf{Y}\frac{\mathbf{Z}}{2}\left(1 + \frac{\mathbf{YZ}}{2}\right)\right]$$

3.17 Urban Distribution

The general comparison between urban and rural distribution systems is given in Table 1.1. The reliability requirement of power distribution in urban areas, is stringent in respect of continuity and quality of power supply. Load density is high, the feeder network is generally looped or parallel or ring main or grid network etc. An important feature of the urban distribution system is that each load point can be supplied by an alternative route during an emergency. It is not necessary to wait for the faulty element to be repaired before the supply is restored. Rural networks are generally both designed and operated in a radial configuration. Urban networks are generally designed in configuration which permits emergency supply and operated in a radial configuration. The high voltage (11 or 22 kV) network consisting of outgoing feeders from grid/distribution substations can be suitably inter-connected to give enough tie capacity, so that the load can be transferred during an emergency. Cities should have UHV or EHV ring main network which have a high degree of reliability. Fault currents are high and therefore the protection should be fast. SCADA and EMS are desirable for better management and load control. Underground cables, packaged substations and/or Gas Insulated Substations (GIS) are preferred in metropolitan areas.

3.17.1 Gas Insulated Substations (GIS)

GIS are installed outdoors in city areas, industrial areas, along the sea shore, and in snowy climates. However, most GIS are installed indoors in buildings, in urban or metropolitan areas. Due to the compact size of the switchgears, relatively inexpensive buildings are needed. The indoor arrangement allows easier installation and maintenance, especially for those countries with severe climatic conditions. The packaged SF_6 substation is available in all types of bus arrangements including double busbar, single breaker, breaker and a half, ring-bus and single-bus schemes. The main elements built into the substation are completely enclosed in the SF_6 installation and include transformers, circuit breakers, load break switches, disconnecting switches, ground switches, current and potential transformers, busbars, coupling capacitors, and SF_6 loadout busing insulators, for connection to overhead lines, transformer or other external equipment. Equipment such as shunt capacitors, power-line carrier line traps, etc. are not manufactured as part of the SF_6 insulation systems. However, long runs of bus using SF_6 insulation may extend some distance from the substation, in order to make the connection to the other conventional equipment.

3.18 Outsourcing

In India, the power industry was built on an engineering culture. The task of electrification was a pure construction and engineering exercise. However, the industry is now moving to a commercial and competitive environment. Outsourcing is an initiative for taking up various tasks in this environment. It can deliver distinct improvements in:

- Productivity
- Power utility employee satisfaction
- Consumer service
- Quality
- Safety
- Technical performance
- Resource management
- Efficiencies
- Cost cutting up to 50%

The success of outsourcing depends on contract strategy. Contracts are used to procure people, plant, equipment, materials and services. It can be an Engineering, Procurement and Construction (EPC) contract. The contract should satisfy the Indian Contract Act, 1872, and meet the following basic requirements:

- Contract work objectives (time, cost, quality and risk)
- Organisation for design, construction and commissioning
- Risk allocation
- Terms of payment
- Conditions of contract
- Tendering procedure: Competitive or negotiated
- Training and smooth transfer of project/work

How will the power utility maintain control of contracted performance? Performance measures and standards must be part of the process. These must be written into the contract to establish clear expectations. Of course, the utility will carry out strong monitoring and inspections. Select a contractor who is commercially sound and can provide expertise, experience and innovation with vision.

Outsourcing should be done for various tasks like preparation and issue of electricity bills, bill distribution, maintenance of the distribution system, capacitor installation, energy conservation projects etc. Outsourcing is a viable solution for most of the information technology jobs. Consumer satisfaction survey and performance improvement strategies can be given to 'Call Centre' services available in the country.

PROBLEMS

1. A feasibility study for the release of a 30 MW power connection for a continuous process chemical industry has to be made. What checks and balances will you carry out to complete the study?
2. What is the difference between the economic and financial analysis of a project? A 15 km, 11 kV overhead feeder having a maximum demand of 200 A with a loss factor of 0.4 has a 50 mm^2 ACSR conductor. The feeder was augmented to 80 mm^2 ACSR at the cost of Rs. 30 lakh. Carry out its financial analysis assuming the interest rate to be 12 per cent.
3. A 20 MVA transformer with 66 kV primary and 11 kV secondary sides has a reactance of 8.7 ohms referred to the primary side. What is the per-unit reactance on the primary side? What is the per-unit reactance referred to the secondary side?
4. A symmetrical three-phase short circuit occurs on the 11 kV busbar of the circuit shown below in Fig. 3.22. Calculate the fault apparent power and fault current.
5. Perform a real-power flow in branches (P_{ab}, P_{bc}, P_{dc}, P_{ad}) of the network loop shown in Fig. 3.23.

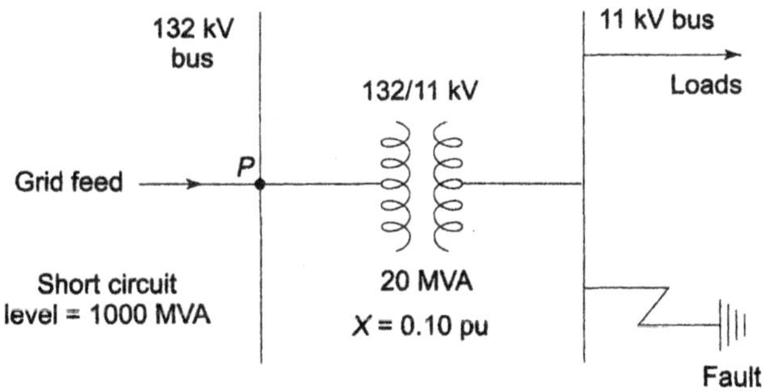

Fig. 3.22 *Single-line Diagram*

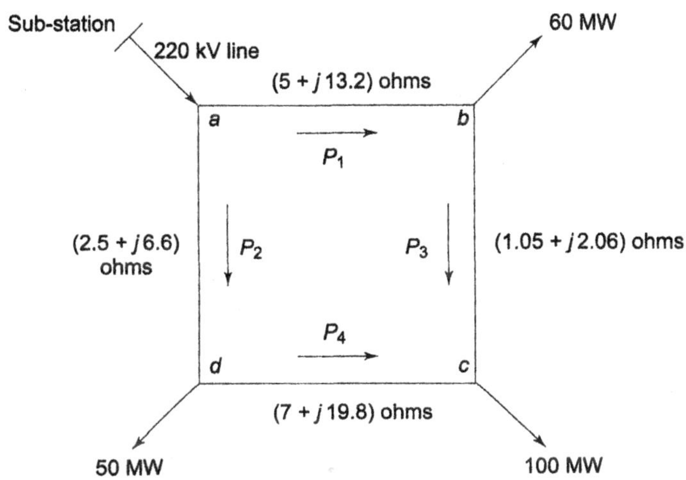

Fig. 3.23 *Single-line Diagram of Loop Network*

BIBLIOGRAPHY

1. Jennings, Bunch B., Richard D. Miller, and James E. Wheeler, 'Distribution System Integrated Voltage and Reactive Power Control', *IEEE Transac.*, PAS-101 Feb. 1982, pp. 284–286.
2. Wagner, C.F. and R.D. Evans, *Symmetrical Components as Applied to the Analysis of Unbalanced Electrical Circuits*, McGraw-Hill Book Company, 1933, pp 137–181.
3. Jain, M.K., S.R.K. Iyenger and R.K. Jain, *Numerical Methods for Scientific and Engineering Competition*, Wiley Eastern Ltd., New Delhi, 1987.

4. *Directory of Computer Software for Power System*, Central Board of Irrigation and Power, New Delhi, April, 1988.
5. Wallach, Y., *Calculations and Programs for Power Systems Networks*, Prentice-Hall, New Jersey, 1986.
6. Arrillaga, J. and C.P. Arnold, *Computer Analysis of Power System*, John Wiley and Sons, 1990, pp. 256–291
7. Rees Gareth, *The Remote Sensing Data Book*, Cambridge University Press, 1999, pp. 119–120.
8. Pabla A.S., *Electrical Power Systems Planning*, Macmillan India Limited, 1998, pp. 212–220.
9. *1997 U.S. Electric Utility Industry Software Directory*, PennWell Publications, U.S.A.
10. Willis H. Lee, '*Power Distribution Planning Reference Book*' Marcel Dekker, Inc. NY 1997, p. 209.
11. Stevenson, D. William, '*Elements of Power System Analysis*', McGraw-Hill Kogakusha, Tokyo, 3rd edition, 1975.
12. Hazan Earl 'ERP Applications Gain Popularity', *Transmission and Distribution World*, April 1999, pp. 36–38.
13. Heathcote J. Martin '*The J&P Transformer Book*', 12th Edition, Newness, 1998, p. 630.
14. Stoll Harry G., '*Least-Cost Electric Utility*', John Wiley & Sons, New York, 1989, p. 74.
15. Mensah S., 'Distribution Management Platform Architecture for Enterprise Integration', Fifth International Distribution Conference (Distribution 2000) November 1999, Brisbane (Australia) *REFOCUS*-International Renewable Energy Magazine, September 2001.
16. *REFOCUS-International Renewable Energy Magazine*, ISES, November-December 2001.
17. '*Geographic Information System Planning for Power System Planning*', United Nations Publications, June 1997.
18. *Guidelines For Development of Sub-transmission and Distribution Systems, Volume I*, Central Electricity Authority, Ministry of Power, New Delhi, November 2001.

Chapter 4

Design and Operation

4.1 Engineering Design

Engineering in electric power distribution is the application of scientific and technological knowledge to planning, design, construction, operation and maintenance of various electric supply schemes for the benefit of society. Any engineering design starts with an engineering problem. A typical flow diagram for a design problem solution is given in Fig. 4.1. It involves the following steps:

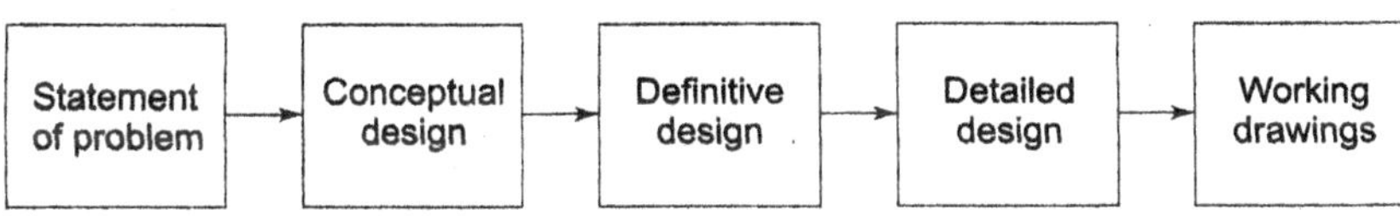

Fig. 4.1 *Flow diagram of the design process*

- Definition of the problem: A clear, simple definition of the problem.
- Conceptual design: Basic principles, design ideas and alternatives.
- Definitive design: This is selection/evaluation of a scheme from alternative design schemes. There is no such thing as a perfect design solution to an engineering problem. The design that is ultimately adopted is the '*best solution*' to the problem. It is a compromise that the engineers have to make to solve the problem in the most effective and economically feasible manner. Each design alternative will have its particular strengths and weaknesses that will need to be compared and evaluated to determine the best solution.

- Detailed designs: This process is called '*engineering*'. It involves determining equipment size, specifications, quality, reliability and costs.
- Layouts: These are working drawings.

In this chapter, basic design and operation aspects have been discussed but the major subject matter is dealt within the respective chapters.

Firstly existing system must always be considered for improving voltage profile, reducing losses and improving the quality and reliability of powers:

- Strengthening re-configuration of feeders (bifurcation or trifurcation and/or part of feeder shifting to other lightly loaded feeder nearby, adding additional sectionalizing points on the feeder, adding loop/inter-connection points between adjacent feeders etc.); renovation of sub-stations, distribution transformers and lines (joints, jumpers etc.); capacitors' installation at proper locations;
- Augmentation of the system-power transformers, distribution transformers, line conductors and reduction of losses to a reasonable level in order to ensure the delivery of power to consumers at a proper voltage;
- De-augmentation of distribution transformers and reduction of the HT/LT line ratio;
- Shifting of the load from overloaded distribution transformers to lightly loaded distribution transformers;
- Introduction of technical innovations or new technologies into the system such as:
  - (i) Use of modern and effective equipments like electronic meters, Automatic Meter Reading (AMR), all aluminium alloy conductors, aerial/insulated overhead conductors in theft-prone and forest areas, covered conductors for H.V. lines, low loss amorphous core distribution transformers, SF_6 or vacuum type circuit-breakers/load switches, XLPE cables, ABC system, ring main units and automatic sectionalizers;
  - (ii) Use of microprocessor devices in the protection and adoption of distribution automation techniques and SCADA; and
  - (iii) Consumer billing through the Internet.

The new innovations or technologies are generally determined on the basis of lifecycle costs (see Section 7.2.1).

4.1.1 Design Tools

The tools of electric power distribution engineering design are well developed, powerful and readily available in both analytical and computer expressions to those who wish to apply them. Care should be taken to choose a programme that takes into account all the parameters. Programmers should show full calculations carried out so that the engineer can check the results. Potentially, the computer method will be more accurate and fast compared to the manual method. This, of course, is fundamentally dependant upon the quality of the software. Various software programme available should be carefully assessed in handling the system equations/formulae correctly. The advice of other users of particular packages can be a source of information. The electrical design engineer still requires full understanding of the fundamentals of electrical engineering principles. With the availability of a database system, different task programmes/modules, various planning, design and analysis studies can be carried out. The exception report for each study can be printed showing the weak points. Some weak points are given below:

- Feeders where voltage drops is above the permissible limit
- Lines where losses are above the prescribed limit
- Lines and transformers operating closer to or exceeding the thermal limit
- Lines and other equipment where the fault level is closer to or exceeds the fault rating.

4.1.2 Design Criteria and Standards

The rules of distribution system design are:

- Total generation at any moment is equal to total electricity consumption and distribution losses in the system. Losses should be minimised.
- Electricity is allowed to flow through the grid network in accordance with physical laws.
- The distribution system must be designed for some contingency enabling continuity of supply.

As per the planning criteria, distribution sub-stations are located, primary feeders routed and the distribution system needs to be designed in detail i.e. switching, protection, automation etc. All these design decisions affect reliability and cost. Automated design tools can be made to minimise total costs. Engineering design criteria are framed to specify precisely how equipment such as primary feeder lines are to be assembled with suitable conductor size and spacing with support structures to handle the weight, wind, ice loading, load, fault current to be expected, the changes in the specifications of the discrete equipment and layout standards, service area assignment, coincident of peaks and other specific requirements. To a certain extent, the sub-transmission and distribution system must be designed as a whole.

Design standards specify a set of specific equipment types and size and how the pieces are put together (layouts):

Standardization of sizes and ratings of equipment for each voltage viz. 230/440, 11000, 33000, 66000, 132000 V

- Conductors and cables
- Supports and structures
- Insulators and lightning arresters
- Transformers
- Isolator and switchgear etc.

Standard layout for sub-stations and distribution transformers to fulfil safety and operational requirements. For example, the following designs may be developed and brought out as part of the *design code:*

- Sub-station layout may be standardised to provide individual circuit breakers to control the primary and secondary side of the power transformer, single 11 kV bus for transformers up to 5 MVA capacity, independent control of individual 11 kV feeder.
- Layout for distribution transformers up to 200 kVA as pole-mounted and the higher capacity units with plinth-mounted arrangement may be standardized. For LV protection of distribution transformers, 100 kVA capacity and above shall be provided with suitable MCCBs and transformers below 100 kVA shall have suitable wire-fuse.
- Loading in any current-carrying capacity components of the distribution system e.g. conductors, switchgear, cables, other equipment etc. shall not exceed 80 per cent of the thermal limit.
- Earthing system type for sub-stations and distribution transformers shall be solid earthing.
- km-kVA tables shall be prepared for reference for various standard sizes of conductors for LT, 11 kV and 33/66 kV feeders.
- The power transformer capacity for distribution sub-station (66/11 kV) will be 6.3/8, 10/12, 16/20 MVA.

- The standard rating of three-phase distribution transformers will be 25, 63, 100, 200, 500 and 630 kVA. The voltage ratio for rural supply will be 11/0.433 kV as per the REC Standard. For urban areas, this ratio will be 11/0.415 kV. For rural areas, wherein the transformer size will not be of more than 100 kVA.
- The voltage drop at 66 or 33 or 11 kV feeders shall not exceed 5 per cent at the farthest end under peak load conditions and the voltage regulation at the consumer's end shall be as per I.E. Rules.
- The voltage drop at 415/240 V feeders shall not exceed 4 per cent at the farthest end under peak load conditions.
- The voltage drop at service mains shall not exceed 0.5 per cent.
- The tap change on power transformers will be +/– 10 per cent.
- The 11 kV cable for distribution sub-station will be XLPE, 300 mm^2, Al, and for LT, PVC cables shall be used.
- The mechanical design of the lines shall have safety factors as per the relevant I.E. Rules.

4.2 Operation Criteria and Standards

The criteria and standards cover the procedure and practices for safe and efficient operation of the distribution system. The design engineer must work within the criteria that specify how equipment can be used. The power utility shall prepare an Operational Code to incorporate all the criteria and standards of the distribution system operation. The criteria defines the principles of operation. For example, the following items need to be covered:

- Notification in advance to consumers about the outage programme.
- Contingency planning: The power utility will stipulate the steps to restore and maintain power supply to the consumers in a quick and efficient manner when there is system failure. Sectionalizing the distribution system and scheduling, in order of priority, the essential and non-essential loads to be connected during the restoration process is necessary. Supply to water works, telephone services, continuous process industries and urban residential consumers can be prioritized.
- Issue of public notification before the start of the year for month-wise power and energy shortage expected in the power utility area.

- Any peak load restrictions will be notified in advance.
- Restricted supply arrangement for agriculture will be notified in advance.
- Metering arrangement for energy audit at the substations to prepare the feeder-wise energy balance-sheets and the whole sub-station.
- Electronic meters for consumer metering shall be adopted. For these meters, the CTs of accuracy class 0.2S and 0.5S as per IS: 2705 (part 2)-1992 shall be used. Energy, demand and power factor metering shall be adopted for consumers having a connected load above 70 kW.
- The power utility's business unit (e.g. the circle area in SEBs) will prepare the monthly energy balance-sheet annually along with efficiency benchmarks to be achieved for each 11 kV feeder:
  (a) Energy received at each distribution sub-station (e.g. 66/11 kV);
  (b) Energy consumption at the sub-station;
  (c) Energy sent out from each 11 kV feeder of the sub-station;
  (d) Energy billed for each feeder (also detailing consumers category-wise);
  (e) Total kWh loss for each feeder;
  (f) Technical loss computed in each feeder;
  (g) Non-technical loss in each feeder; and
  (h) Un-metered energy.

Operation standards cover operating costs, load monitoring, load management and voltage flicker limits (see Section 4.11).

The flicker limits depend upon the fault level. For example, the flicker limits can be:

Service area	*Voltage variation*
Metropolitan area	3%
Urban/suburban	3%
Rural	4%

Operating standards may comprise some of the following items:

(i) Consumer meters shall be periodically tested/calibrated for correct working as per Section 26 of the Indian Electricity Act, 1910. Complete schedule for different categories of consumer meters shall be prepared.

(ii) Any breakdown in system 33 kV and above will be analysed and an investigation report prepared for remedial measures.

(iii) A schedule for maintenance of lines and other equipment shall be prepared.

(iv) Data logging will be done for all feeders.

(v) Load management: Policy in the use of frequency trend relays or discrete frequency relays for automatic load shedding in an emergency on selected 132 and 66 kV feeders.

(vi) Energy audit exercises on a monthly basis shall be carried out for each 11 kV feeder for energy balance to prepare the energy balance-sheet.

(vii) Safety standards: Safety code prepared in accordance with I.E. Rules. Safety Code Test pass will be compulsory for the operating staff.

(viii) The power utility shall issue a service-level time schedule for the following consumer services:

- For restoring power supply after fault in case of different categories of consumers.
- For providing supply and meter for a consumer of various categories.
- Response time for consumer's meter billing complaints.
- Time taken to investigate the consumer's voltage complaints.

(ix) Parallel operation of power transformers at the substations shall be carried out in city areas but in rural area, power transformer will be operated in isolation.

(x) Negative unbalance voltage not exceeding 1.5 per cent shall be allowed in LT lines.

(xi) The consumer power factor for industrial consumers will not be less than 0.9 lagging failing which a penalty will be levied.

(xii) At distribution sub-station on 11 kV bus, the power factor of the incoming load (incoming feeder) would not be less than 0.95 lagging and not beyond 0.95 leading. For any outgoing load (outgoing feeder), the power factor will not be less than 0.90 lagging.

4.3 Sub-transmission

The sub-transmission line supplies power from the bulk power source to a distribution sub-station at a voltage ranging from 33 to 220 kV. The distribution sub-station containing power transformers with controlling circuit breakers, buses, protection equipment etc. reduces subtransmission voltage to lower primary voltage in the range of 22 to 11 kV for local distribution. Distribution transformers in rating from 10 to 630 kVA are connected to a primary feeder. They reduce the voltage to 433/250 or 415/240 V to delivery utilization or service voltage of 400/230 V +/–6 per cent at the consumer end. The main consideration for sub-transmission design is economics and reliability. The system design options are: Radial system, ring main and grid or network type as given in Fig. 4.2. The ring main can be closed or open. An open ring main is more common and is cheaper. It provides the isolating switch or circuit breaker as a link (normally open) for alternative supply. But supply restoration takes a comparatively longer time. The higher reliability and higher quality of power is obtained when the sub-transmission system is operated in a closed ring but this requires an expensive protection system and the operation becomes more complicated. Exhaustive computer-based system studies are necessary to determine load flow, fault current and protection requirements in the closed ring main system.

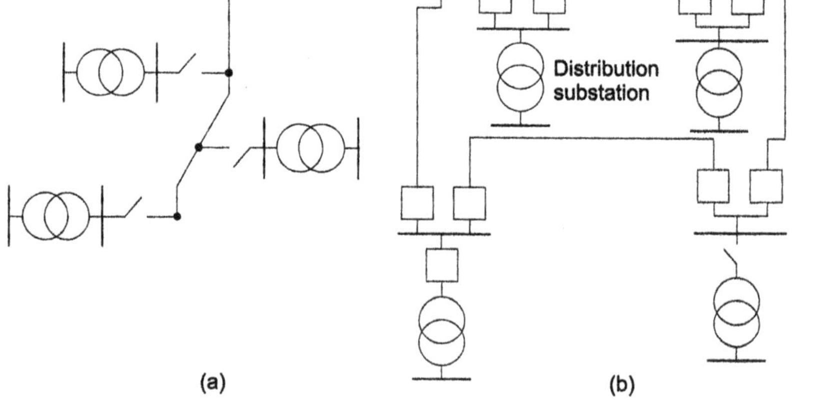

Fig. 4.2 *Typical sub-transmission system designs: (a) Radial type; (b) Ring main*

Unnecessary circulating currents (which create losses and block the system capacity) need to be avoided. For main industrial cities or important security areas, a 33 or 66 or 132 or 220 kV closed ring main system is desirable.

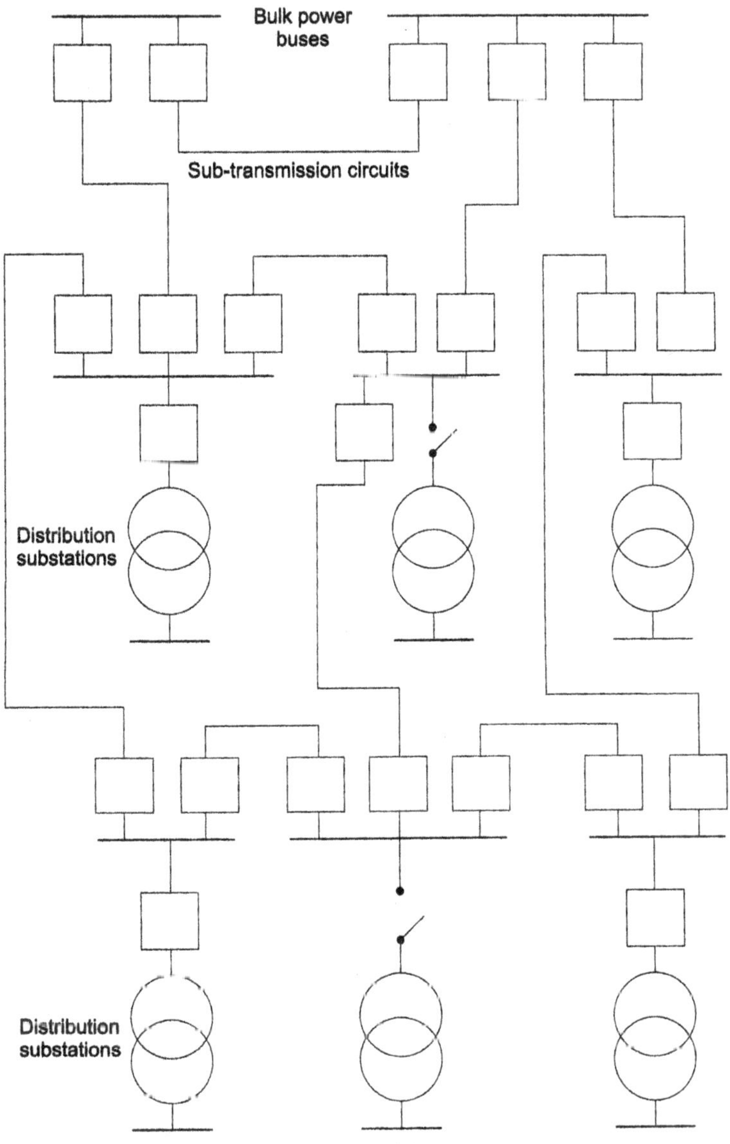

Fig. 4.2 *Typical subtransmission system designs: (c) Grid or network type*

Beyond sub-transmission, the distribution network design process can be divided into three independent parts:

Load forecasting

- Load growth of the geographical area served by the substation;
- Determination of load magnitude and its geographic location;
- Consumer load characteristics.

Design of secondary system

- Optimal sub-station allocation and transformer sizing;
- Secondary circuitry routing and sizing.

Design of primary system

- Optimal distribution sub-station allocation;
- Primary circuitry routing and sizing.

To reduce the complexity of a problem each part of the design process is divided into functional sub-problems. Each of these sub-problems is then much easier to manage. Although only some independent parts of the design process interact, i.e. placement of the sub-station will influence secondary routing which in turn will influence primary routing. The number of possible design solutions that might satisfy a given set of spatial, technical and economic constraints is numerous.

4.4 Sub-station and Feeder

Sub-stations transform the electrical energy to a different voltage and transfer electrical energy from one line to another. Usually, planners try to locate a sub-station near the centre of the load. A feeder is an electrical distribution circuit fed from a single source point: Breaker or fuse at the sub-station. It operates at the primary distribution voltage and delivers power to an assigned area. Together, the feeders emanating from a sub-station serve all the loads and cover all the areas assigned to that particular sub-station. This area should be an approximate circle, polygon or hexagon and the sub-station should be located approximately at the central point in cases where the load is more or less uniform. Sometimes, there are constraints of geography–river, canal, forest etc. or just poor planning in the past which can cause an exception to this rule. A feeder consists of a single route, leaving the sub-station which branches with spurs. This splits the power flow into smaller-capacity tee-offs for delivery as power flows from the sub-station to the consumer. Feeders are planned by starting from the substation with the main trunk portion of the largest economical conductor and generally follow streets, roads, highways and property boundaries.

4.4.1 Distribution Substation

The distribution substation is the convenient point for the control and protection of the distribution network. A typical substation may have the following equipment: Substation bus, power transformer, circuit breaker, isolating switch, CTs, PTs, shunt capacitors, protection relays, lightning arresters, station batteries, earthing, structures etc. Substations are located at the load centre as far as possible (see Section 6.6). The different bus schemes are shown in Fig. 4.3.

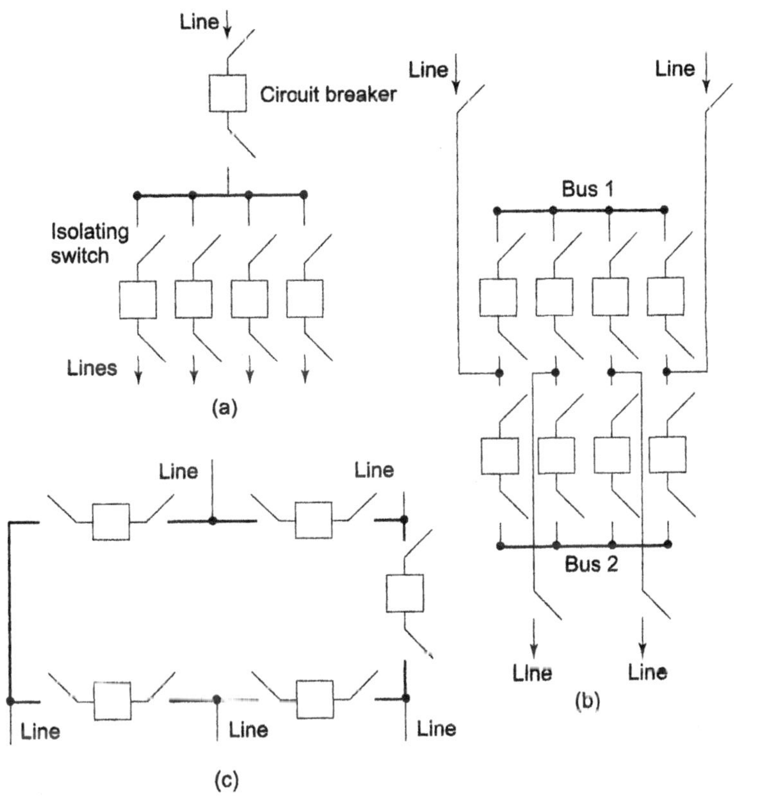

Fig. 4.3 *Typical bus schemes: (a) Single bus; (b) Double bus double breaker, (c) ring or mesh bus*

- Single bus
- Double bus
- Main and transfer bus
- Mesh or ring bus
- Breaker and a half

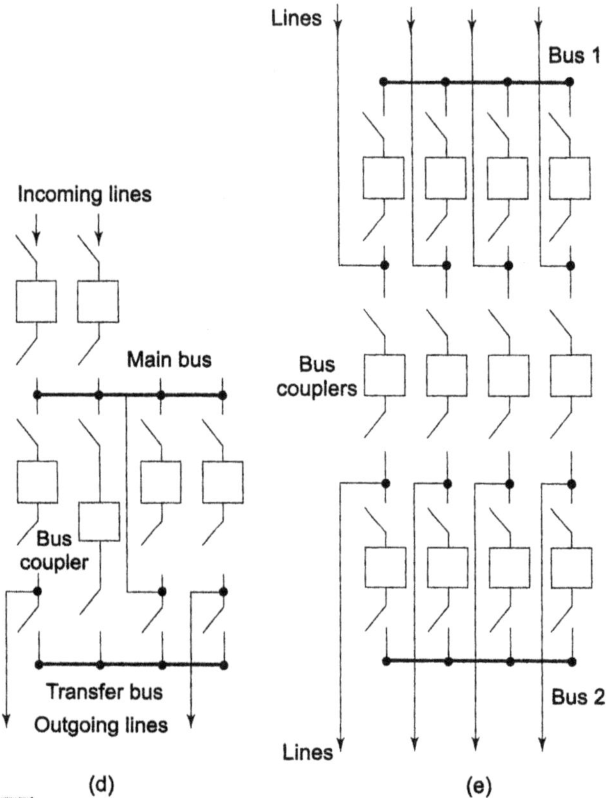

Fig. 4.3 *Typical bus schemes: (d) Main and transfer bus; (e) Breaker and a half*

An important factor for selecting the type of bus-bar is the degree of reliability of supply during maintenance and fault. The amount of duplication and sectionalizing determines reliability. Future expansion is another consideration. The additional substation's capacity to cater to increasing loads can be met by:

- Either keeping the service area of a given substation constant and increasing its capacity; or
- Developing new distribution substations and thereby keeping the rating of the given substation as standard.

The first option is helpful for short-term distribution planning and the second for long-term planning.

4.4.2 Feeder System

Most distribution systems are designed as radial distribution systems. This system is characterized as having only one power flow path between each consumer and sub-station. Any interruption in the path results in a complete loss of power to the consumer. It is cheaper and simpler in planning, design and operation. An alternative to the purely radial feeder design in a *loop or ring main system* consisting of a distribution design with two paths between the sub-station and the consumer. Most radial feeder systems in urban areas are constructed as loops with other feeders (see Fig. 1.3b), but operated radially by opening the switch. The major disadvantages of loop systems are capacity and cost. At loop must be able to meet all power and voltage regulation requirements when fed from only one end and not both ends. The loop system is more reliable and costly. The *network or grid system* is much more complicated, and most reliable and is used for a secondary underground system in high load density area (see Fig. 1.3f). The loading and power flow, fault currents and protection must be determined through computer studies. Typically, a metropolitan city may have a 66 kV closed loop or a closed ring main system feeding 66/11 kV distribution sub-stations with a primary feeder system (11 kV) as an open loop or open ring main.

Primary Systems

The primary feeder is usually three-phase with spurs: Three-phase, single-phase or two-phase. The feeder is controlled by a circuit breaker or auto-recloser at the distribution sub-station. The main feeder and spurs are usually protected and sectionalized with fuse-GO switches. The voltage condition of the feeder can be improved by installing shunt or series capacitors. The feeder rating depends upon the nature of the load, area load density, load growth of the area, quality and continuity of service required. Feeder voltage level depends upon the given load, permissible voltage drop and given feeder length. The various forms of primary feeders are (see Fig. 1.3):

- Radial
- Loop or Ring Main

(a) Open ring main or open loop (see Fig. 1.3b). In the open ring main, the load switch or circuit breaker (normally open) is used as a sectionalizer. For example, in Mumbai, the Bombay Reliance Energy Limited has an 11 kV Open Ring System at the primary distribution level.
(b) Closed ring main system: In this case the sectionalizing load switch or circuit breaker is normally closed.

Economic Conductor Size

Feeder design starts with determining the conductor size from the following data:

(a) Primary voltage
(b) Conductor km-kVA (thermal limit)
(c) Conductor initial installation cost/km
(d) Annual loss factor
(e) P_V factor
(f) Annual O & M cost
(g) Annual losses cost
(h) Present value of above (f) + (g)
(i) Total present value cost (P_V) = (h) + (c)
(j) Maximum load current including growth for the next 5–10 years.

Knowing the maximum load and its total present value (see Section 3.5) for declared service life in design code (say 30 years), the curves for different conductor sizes can be drawn as standards (see Fig. 4.4). The

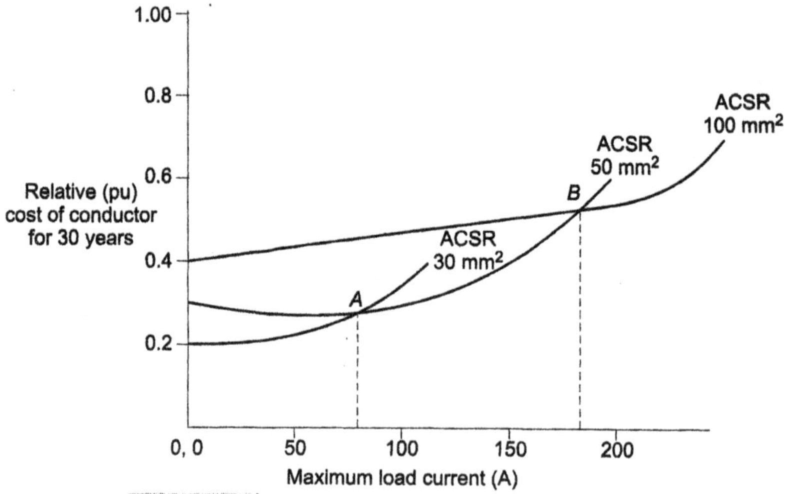

Fig. 4.4 *Standardisation of conductor sizes*

conductor which gives the lowest P_v should be selected. You may see that at the point *A*, 30 mm^2 conductor is more economical for loads lower than indicated at this point and for loads beyond this point, 50 mm^2 is economical. Similarly, at point *B*, 50 mm^2 conductor is economical at loads less than that indicated at the point *B* and beyond this point, 100 mm^2 is economical.

The larger conductor size has a shorter economic reach (electrical distance) because the voltage does not drop in inverse proportion to the increasing conductor size as do the losses. Losses are the function of resistance which varies roughly in inverse proportion to the conductor ampere capacity. But voltage drop is a function of impedance (resistance and reactance). Reactance mostly depends on line phase spacing. As a thumb rule, doubling a line's ampere capacity will cut resistance by half but will reduce reactance by only 10 per cent. Lines having high X/R ratios have higher voltage drops for the corresponding economic loading and thus have a shorter load reach than the smaller conductor size. It is thus advisable to always make an economic selection.

Secondary System

The secondary distribution system includes step-down distribution transformers, secondary circuits, consumer services and consumer energy metre. To minimize the cost and the secondary line length, distribution transformers should be located close to the load centres (see Section 6.6). The main considerations for the design of the secondary system are: Load losses in transformer and secondary circuit; permissible voltage drop, voltage flicker etc. The various types of secondary system used are shown in Fig. 4.5:

- Radial;
- Parallel distribution transformers: Giving advantage over the radial system with improved voltage regulation, reduced voltage flicker, improved reliability and improved flexibility in accommodating load growth at a low cost.
- Secondary network or grid (see Fig. 1.3f). This is the most suitable system for high-load density areas in metropolitan cities.
- Open ring main: It gives advantage over radial of improved reliability. For example, in Mumbai, BSES has open ring main system at LT secondary distribution level.

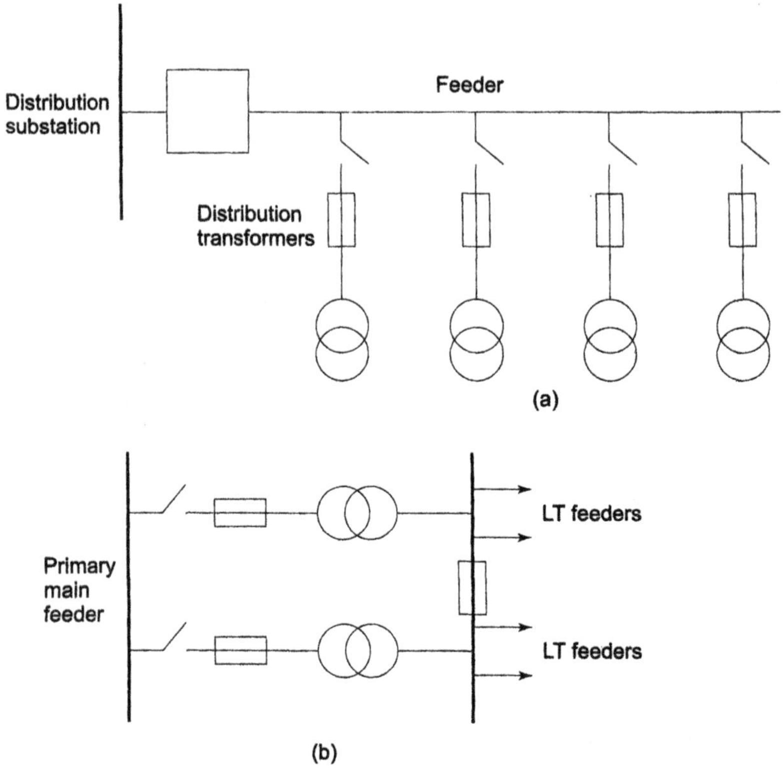

Fig. 4.5 *(a) Radial secondary system (b) Secondary banking*

Service Lines

The service line is the connection between the LV secondary network and the consumer end. In case of an underground network, feeder pillars are used to take out service lines (see Fig. 19.4). For an overhead system, services are taken from the nearest pole or support. Not more than four service lines are taken from a support.

4.5 Low Voltage—Three-phase or Single-phase

In India, the three-phase/single-phase nominal consumer standard voltage of 400/230 V has been adopted with a distribution transformer standard voltage ratio of 11/0.415 kV and 11/0.433 kV (REC Standard). A single-phase system is used for smaller loads, generally up to about 10 kW. For higher loads, a three-phase system is used. Advantages of the three-phase system over a single-phase system are:

- For a given amount of power to be transmitted through, the three-phase system requires conductors with a smaller cross-sectional area. The installation is thus cost-effective for higher loads.
- Two voltages, i.e. phase voltage and line voltage, are available at the consumer distribution board.
- Three-phase motors are very robust, relatively cheap, generally smaller, provide a steadier output and require little maintenance compared to single-phase motors. Torque in a three-phase motor is balanced and runs smoothly compared to a single-phase motors. For a given amount of copper and steel, a three-phase machine has greater output compared to a single-phase one.
- Low-line losses.

Delta and Star Configuration

Most three-phase power systems are a combination of delta and star connected equipment (see Figs 4.6 and 4.7). Predominantly, high-voltage transmission and sub-transmission as delta and distribution as star are connected. Delta configuration is usually cheaper for high voltage and star configuration is cheaper for low voltage.

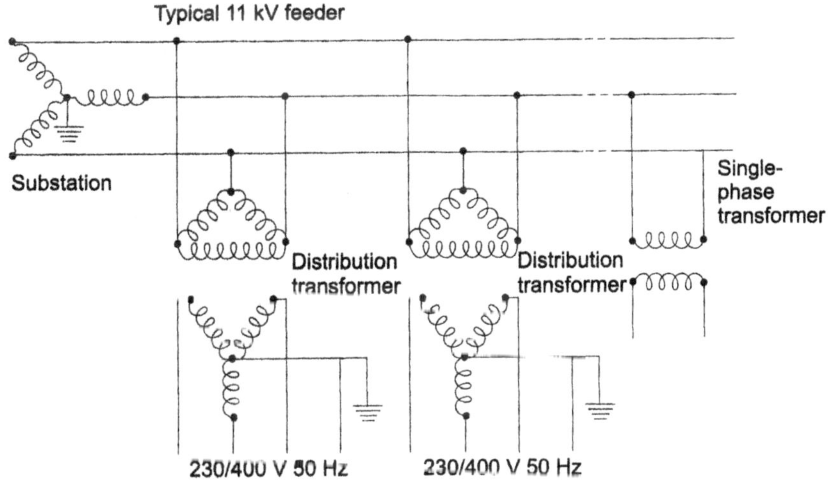

Fig. 4.6 *European distribution system*

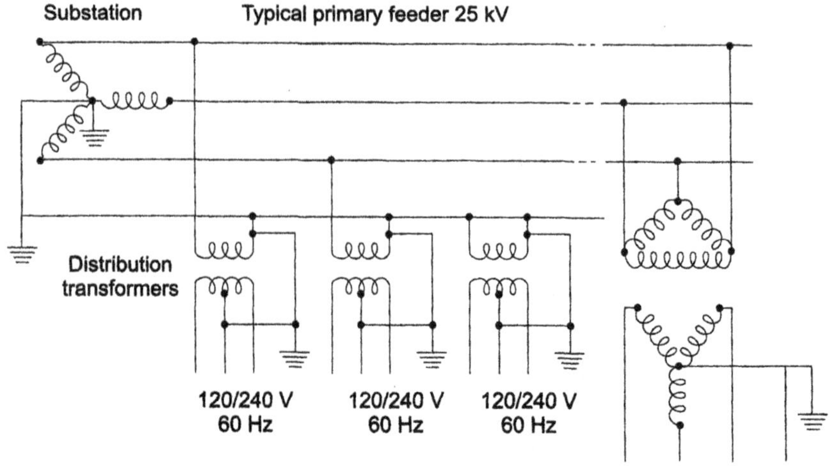

Fig. 4.7 *North American distribution system*

4.6 Practices

System

All distribution systems share a number of common features. To begin with, they all have primary circuits that originate a sub-stations supplied from the sub-transmission system. Each of the primary (HT) feeder leaves the sub-station as a three-phase circuit and supplies a number of distribution transformers to which on the secondary side are connected low tension (LT) feeders. The distribution transformers and LT feeders are rated to maintain the voltage received by consumers within a prescribed tolerance over the full range of loading conditions. While utility distribution systems can vary substantially, for the most part, they can be grouped according to whether their roots trace back to Europe or North America (U.S.A., Canada, Mexico etc) which have fundamental differences of approach. In India, we follow the European practice based on cost benefits. Recently, some pilot projects have followed a modified North American practice—using HT single-phase transformers for supply to agricultural loads in Andhra Pradesh and Kerala. The performance has not yet been concluded.

Comparison of the North American System and European System

For a given loading and design objectives, the European system allows 230 V LT circuit (see Fig. 4.6) to be longer than that of the 120 V of the North American practice. In addition, the European one uses three-phase configuration on LT supply circuits extensively, since the net current in the neutral conductor is lower than the single-phase 230 V supply. This system carries all the phases into most of the service points. The three-phase primary feeder supplies mostly large three-phase distribution transformers or in some cases large single-phase transformers (see Fig. 4.6) in rural areas.

The North American system uses a good deal of the single-phase. However, large factory loads are given primary high-voltage three-phase supply with three-phase distribution transformers. The dual low-voltage 120/240 V arrangement is standard supply in North America (see Fig. 4.7). The general practice is to minimise low-voltage circuits by using smaller single-phase transformers.

The greater average length of LT circuit allows a larger number of consumers to be grouped together on each distribution transformer in the European system. This in turn leads to the use of large three-phase transformers being fed from three-phase primary circuits for most, if not all, of their length. This is a marked contrast to North American systems where single-phase branches and spurs constitute the greater part of a typical primary feeder (see Fig. 4.7).

With the changeover from the three-phase to single-phase usually occur on the secondary LT system instead of primary system, primary feeder typically do not carry a neutral conductor on European style system. Whereas on most North American style systems, single-phasing (phase-neutral) is done on primary feeders. We can see the advantages and disadvantages of each system as given below:

(i) In the European system, there is considerable flexibility in selecting both their location and routing of primary feeders. The transformer sizes are high and LT network is extensive. There is a usually an H-pole construction for transformer mounting. For the American system, there is a single-pole construction for transformer mounting as the transformers are smaller in size and there are shorter LT circuits. The North American approach results in primary feeders being routed past for all consumers which gives the greatest feeder approach in most congested areas of the highest load density.

(ii) The European system can more readily accommodate load growth resulting from new consumers located in newly-developed areas that require extension to the distribution system. By contrast, the North American system has the advantage of meeting load growth resulting from both increased use by existing consumers and new consumers who are located in areas that are already serviced.

(iii) In terms of physical configuration, a higher proportion of LT circuits in the total feeder (HT) length dominate the European style, whereas in the North American system, primary HT circuit length is proportionately high compared to the LT circuit length.

(iv) For a given load, the North American system allows an inherent margin of additional load-carrying capacity. That is, system km-kVA capacity is high by virtue of a higher proportion of circuit length of the primary feeder.

(v) In the North American system, to serve a growing load within an existing service area, it is necessary only to change out individual transformers or add an additional transformer and a new service connection. In the European system, increasing load is more likely to require construction of primary feeder extension for supply to a new transformer location and laying an LT line to the site of the consumer.

(vi) *Earthing:* The North American practice uses a multiple-earth neutral conductor that is often common to both the primary and secondary circuits. It ensures a low earthing resistance. The neutral current is shared between the metallic neutral conductor and the earth. Current-sharing also occurs under fault conditions. In the European system, only the secondary circuit (LT) is a multiple-earthed neutral conductor.

(vii) *Theft:* Pilferage of energy due to illegal tappings are more prone in the European style due to extensive LT circuits which are easy to tap than primary circuits, predominant in the American system. Power theft is approximately 2 per cent in the U.S.A.

(viii) In the European system, particularly in the UK, the practice is to connect the primary winding of single-phase distribution transformers across two phases of the primary feeder (11 kV system) in rural areas. In the U.S.A., the practice is to connect single-phase distribution transformers between phase and neutral.

(ix) *Vee-phase distribution:* In North America, vee-phase distribution system configuration is also used. In this system, supply is provided by the neutral and only two of the three-phase conductors, increasing the line capacity when compared to single-phase distribution and permitting low-cost access to three-phase power.

Other Practices

United Kingdom

- *Transformers:* In addition to three-phase distribution transformers, single-phase distribution transformers of ratings 25, 50, 100 kVA are used in the rural areas. The transformer life is considered to be 40 years.
- *Conductors:* Aluminium alloy conductor is for 11 kV and LT lines. ACSR conductor is used for 132 kV lines.
- *Street light:* In Greater London city areas, each street light point has a solar-photovoltaic control switch for on/off control of the lamp point.

Australia

- *Alternate supply:* In cities, the 11 kV locked G.O. switch (normally open) is used to provide loop for alternate supply. Also, LT alternate is available with Normally Open contacts (NO) with HRC cut-out provided and taken out at loop point.
- *Distribution transformer protection:* LT side of the distribution transformer is protected by HRC fuses.

U.S.A.

- *Losses:* System losses in Philadelphia are standardised as 4 per cent in transmission and 4 per cent in distribution.

4.7 Location of Sectionalizer

A sectionalizer is placed at the line to sectionalize it for the purpose of increasing line reliability. Under fault conditions, the faulty line section is isolated. The time to detect fault and to carry out repairs is curtailed. Manual gang-operating-air-switches or automatic vacuum sectionalizer are generally used as a thumb rule at every 2–3 km on the trunk route. However, the optimum location of the pole-mounted sectionalizer/isolator-switch at the line is determined by comparing its benefits in terms of reduction in outage costs at various possible locations. Risk of circuit failure is proportional to the length of the line beyond that point where the sectionalizing switch is to be installed. Therefore, risk is proportional to km-kVA beyond that point. The cost of outage is proportional to energy not supplied.

Supply can be restored from alternative sources (feeders) via the interconnections using switching operations. During re-configuration, a Normally Open (NO) switch is closed and forming a loop.

4.8 Voltage Control

All equipment connected to the power utility system is designed to be used within a certain voltage range. Because voltage drop exists in each part of the system, the consumer who is electrically farthest from the sub-station receives the lowest voltage. Voltage on a primary distribution feeder varies from a maximum value for consumers on the first distribution transformer nearest to the sub-station to a minimum value at the tail end of the circuit i.e. the last consumer on the last distribution transformer (see Fig. 4.8). Since all consumers have appliances/ equipment of the same standard of service voltage, it is necessary to provide them almost the same service voltages. A compromise is needed, however, between what voltage range equipment will operate satisfactorily. If the limits of voltage provided by the power utility are too broad, the cost of equipment, especially computers, will be high because they will have to be designed to operate satisfactorily within those limits. On the other hand, if the voltage limits required for satisfactory operation of the equipment are too narrow, then the cost of providing the power without exceeding these limits will also be high. The maximum and minimum values of voltage at the consumer end are prescribed in I.E. Rules, 1956. Both voltage drop and losses depend on the impedance of the line as well as its loading. Generally, lower-line voltage drop and line losses are desirable and a larger conductor size can be used for that purpose. However, use of a conductor above a certain size yields diminishing returns in terms of voltage drop and losses. Impedance ($Z = R + jX$) does not drop much where R is already small and further reduction in reactance X is a function of conductor spacing, which does not change with conductor size. Therefore, for any loading an optimum conductor size is desirable.

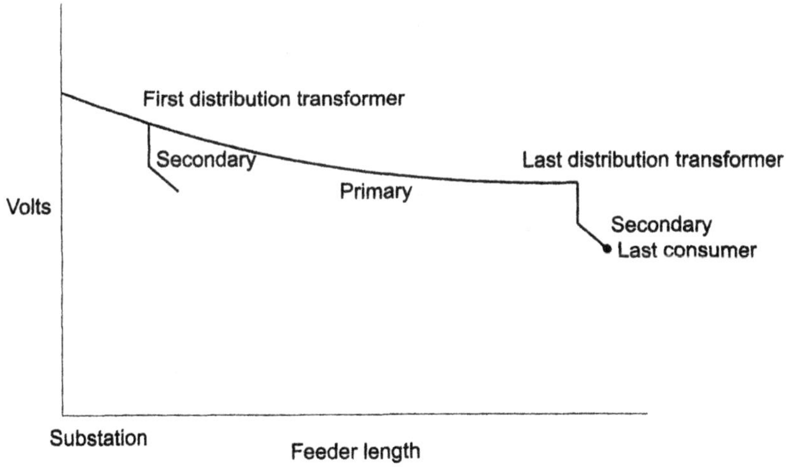

Fig. 4.8 *Typical feeder voltage profile*

4.8.1 Regulation

Before we go further it is necessary to establish what we understand by *system voltage regulation.* It describes the concept of maintaining a relatively stable voltage at the consumer's terminals, the fluctuations being at an acceptable level and within the limits of permissible spread.

For example, regulation values given in Appendix II are for an undiversified maximum regulation. Since, not all the loads reach a maximum at precisely the same time, it is necessary to modify these values by applying a *coincidence factor* defined as

$$\text{Coincidence factor (CF)} = \frac{1}{\text{Diversity factor (DF)}}$$

However, regulation values for individual loads with various power factors may be obtained by using the regulation formula given below.

$$\text{Percentage regulation} = \left(\frac{V_s - V_R}{V_R}\right) \times 100$$

$$= \left[\frac{IR\cos\phi + IX\sin\phi}{V_R}\right] \times 100$$

where I = line current, R = line resistance, ϕ = pf angle, and X = line reactance. There are three main considerations for voltage regulation in general:

(a) The voltage must not be subjected to serious fluctuation or flicker.
(b) The voltage must approximate to some optimum level.
(c) The voltage spread must not be greater than certain established limits.

The problems of voltage fluctuation or flicker are important but are usually associated with abnormal loads on power systems. Technically, the two latter requirements can be met but problems arise when economic considerations are applied and we discover that if the selected level is incorrect or limits are too narrow, costs become prohibitive. Further, in some cases it is not feasible to meet all responsibilities fully at all times. The best that can be done is to provide the majority of consumers with acceptable supply practically all the time; and in the case of the less fortunate consumers, concentrate on improving this situation when resources permit.

Our task is then to investigate and decide on the measures to be adopted in order that the system voltage to consumers may be optimized, bearing in mind that the distribution voltages must be controlled within acceptable limits. IE Rules, 1956 (Rule No. 54) prescribe ± 6 per cent and + 6, – 9 per cent voltage limits for LT and HT supply respectively to consumers. For extra high tension voltage (66 kV and above) supply, the limits are – 12.5 to 10 per cent.

4.8.2 kVA-km Conductor-loading

We know that voltage regulation is calculated thus:

$$\text{\%age voltage regulation} = \frac{IR \cdot \cos\phi + IX \cdot \sin\phi}{V_R} \times 100$$

where V_R = Nominal voltage to neutral
I = Phase current in amperes
R = Resistance per phase in ohms at conductor temperature (say 60°C)
X = Reactance per phase in ohms
$\cos\phi$ = Power factor (say 0.8)

Let us find out the kVA-km loading for the voltage regulation for a 11 kV-line, conductor size = 6/3.35 mm + 1/3.35 mm = 50 mm^2/A1

Resistance in ohms/km at 20°C = 0.5524 ohms

[see IS : 398 (part 2)-1996]

Resistance at 60°C = 0.7136 ohms

Inductive reactance at 50 Hz in ohms/km (for equivalent spacing of 1000 mm) for 11 kV line = 0.421 ohms

Assuming,

Length of line = 1 km

kVA loading = 1000 kVA

$$\text{Percentage regulation} = \frac{52(0.7136)\times 0.8 + 0.421\times 0.6}{6351}\times 100$$

$$= 0.674\%$$

For a 1% voltage regulation, the kVA-km loading will be

$$= \frac{1000}{0.674} = 1484$$

For 5% voltage regulation, the kVA-km = 1484 × 5 = 7420

Similarly, we can find the kVA-km for any %age regulation, power factor, line voltage and conductor size and temperature.

4.8.3 Voltage Drop Calculations

There are two methods for voltage drop calculations. These are:

(i) When data of the maximum demand on the primary feeder (11 kV) is not available, the sum of kVA ratings of all distribution transformers installed on the feeder and diversity factors is used to find out the maximum demand on the feeder and the voltage drop.

Maximum demand (A)

$$= \frac{\text{Sum of kVA ratings of distribution transformers}}{\text{Diversity factor}}$$

Per cent voltage drop

$$= \frac{\text{Voltage drop per km.kVA}\times(\text{total km.kVA})}{\text{Diversity factor}}$$

In case of secondary (LT) feeders, the voltage drop is calculated on the basis of consumer connected loads and the consumer loads diversity factor.

Also since the voltage drop depends on the type of load distribution on the feeder, a relationship between the Distribution Factor (*df*) and km-kVA or km-kW needs to be developed. The distribution factor (*df*) is a combination of the type of load, the demand factor and the diversity factor. In place of the *diversity factor* used above, then the *distribution factor* (*df*) is used.

(ii) When the maximum demand (A) on the primary feeder (11 kV) is available from the sub-station log records, the demand factor is calculated as follows:

$$\text{Demand factor} = \frac{1.732 \times 11 \times \text{maximum demand}}{\text{Sum of kVA ratings of distribution transformers}}$$

$$\text{Per cent voltage drop} = \text{Voltage drop per km.kVA} \times (\text{total km.kVA}) \times \text{demand factor}$$

EXAMPLE: A 11 kV feeder (Fig. 4.9) from the main substation has 11/0.433 kV transformers with total capacity of 1503 kVA with spur lines at points *A, B, H, C, I, D, R, E, F.* Determine the feeder conductor size and voltage drop at tail end *M*, considering Diversity Factor (DF) as 2.5 and load power factor 0.85.

Solution: The total transformer capacity from the Fig. 4.9 is

$$\begin{aligned} 10 \times 100\text{ kVA} &= 1000\text{ kVA} \\ 6 \times 63\text{ kVA} &= 378\text{ kVA} \\ 5 \times 25\text{ kVA} &= 125\text{ kVA} \\ \hline \text{Total } &1503\text{ kVA} \end{aligned}$$

$$\text{The maximum demand} = \frac{1503}{2.5} = 601\text{ kVA}$$

$$\text{Maximum feeder current} = \frac{601}{\sqrt{3} \times 11} = 31.5\text{ A.}$$

From Table A.2.2., Appendix II, the minimum ACSR conductor size is 6/1/2.11 mm. Its current carrying capacity is 107 A. Although feeder current is only 31.5 A, a conductor size of below 6/1/2.11 mm is not permitted because of the requirement of tensile strength. We must also select size for voltage drop within limits. For calculating voltage drop, taking moments of kVA about the section points of the main feeder, we have:

$$\begin{aligned} OA &= 1503 \times 1.0 = 1503.00 \\ AB &= 1403 \times 1.5 = 3104.50 \\ BH &= 1215 \times 1.5 = 1822.50 \\ HC &= 1115 \times 0.5 = 557.50 \\ CI &= 927 \times 1.0 = 927.00 \\ ID &= 764 \times 1.0 = 764.00 \\ DR &= 601 \times 2.0 = 1202.00 \\ RE &= 413 \times 2.0 = 826.00 \\ EF &= 288 \times 2.5 = 720.00 \\ FG &= 100 \times 1.0 = 100.00 \\ \hline \text{Total kVA} &\times \text{km} = 10526.60 \end{aligned}$$

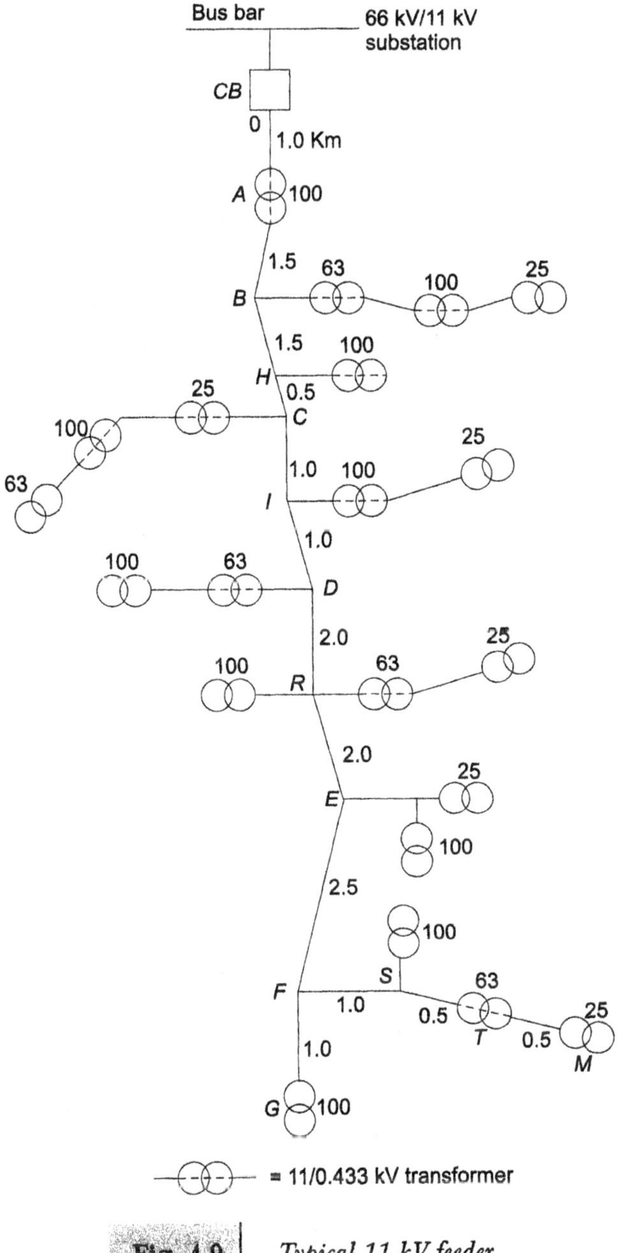

Fig. 4.9 *Typical 11 kV feeder*

$$\% \text{ voltage drop} = \frac{\text{Voltage drop per km.kVA} \times (\text{total km.kVA})}{DF}$$

Referring to Table A-2.5 Appendix II for 6/1/2.11 mm ACSR conductor (20 mm^2), the

$$\% \text{ voltage drop} = \frac{6/4.41 \times 10526.5}{2.5 \times 1000} = 5.73\%$$

which is within line voltage regulation limits of 6% for 11 kV feeders.

Therefore, the conductor size is all right. The moments up to the point *F* are 10,426.50 km kVA. Moments for the spur line *FM* using the same conductor size are:

$$FS = 188 \times 1 = 188.00$$
$$ST = 88 \times 0.5 = 44.00$$
$$TM = 25 \times 0.5 = 12.50$$

Total	244.50 km.kVA

% Voltage drop up to point *F*

$$= \frac{10426.50 \times 6/4.41}{(1000 \times 2.5)} = 5.67\%$$

% Voltage drop between point *F* and *M*

$$= \frac{244.5 \times 6/4.41}{(1000 \times 2.5)} = 0.13\%$$

Therefore, % voltage drop at the tail point *M*

$$= 5.67 + 0.13 = 5.8\%$$

4.8.4 Correction of System Voltage Problems

The Methods available to achieve voltage control are summarized below.

(a) Adjust sub-transmission voltage levels at supply points.
(b) Utilize grid or main sub-station on load/off load tap changing arrangement of transformers.
(c) Add additional feeders.
(d) Increase conductor size of the existing feeders.
(e) Rearrange the system, transfer loads.
(f) Balance loads between phases.
(g) Convert SWER or single-phase system to three-phase system.
(h) Close loops where advantageous.

(i) Add distribution transformer capacity.
(j) Alter tap settings on distribution transformers.
(k) Add line drop compensation.
(l) Install voltage regulators (see REC specification 37/1993).
(m) Increase distribution voltage levels.
(n) Install shunt capacitors, switched/nonswitched.
(o) Install series capacitors (refer Section 15.2.1).

The voltage control on the primary distribution feeder through the use of distribution transformer tap-changer, voltage regulator and shunt capacitors has been shown in Figures 4.10 and 4.11 respectively. For heavily loaded long feeders, series capacitors are the most suitable (see Fig. 15.6c). The task is complicated by the fact that in most cases power engineers are generally concerned with increasing and adding to their existing system capacity, rather than devising entirely new systems. Further (unknown) requirements add still further complications especially if the development is irregular. An interactive computer programme can optimize the use of various voltage controls or correction devices, with the engineer having the facility to fix parameters at individual locations to meet specific network requirements [refer to sub-section 4.12.2].

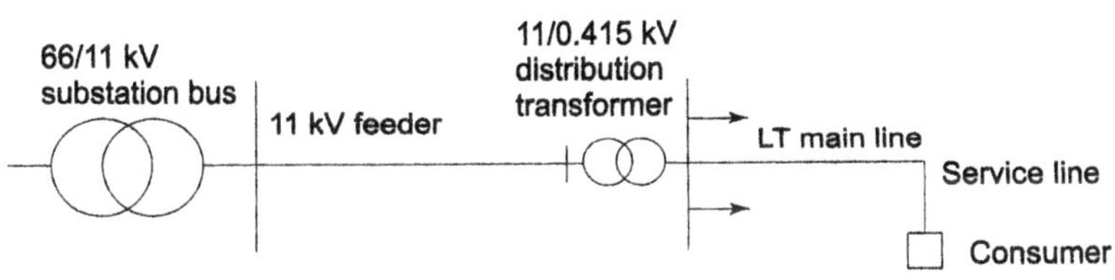

(a) Single-line circuit diagram

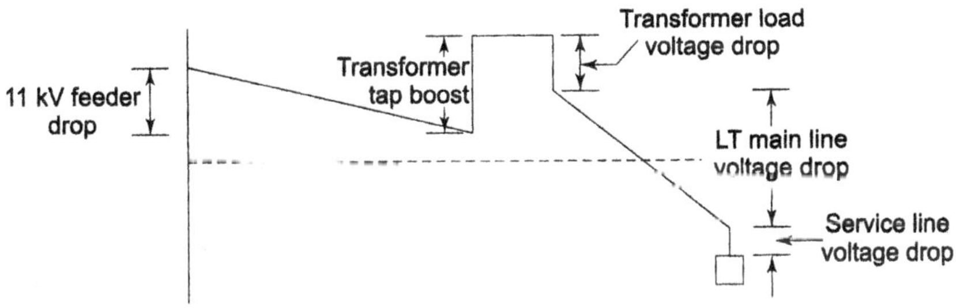

(b) Voltage drop

Fig. 4.10 *Typical voltage regulation of 11 and 0.415 kV feeders*

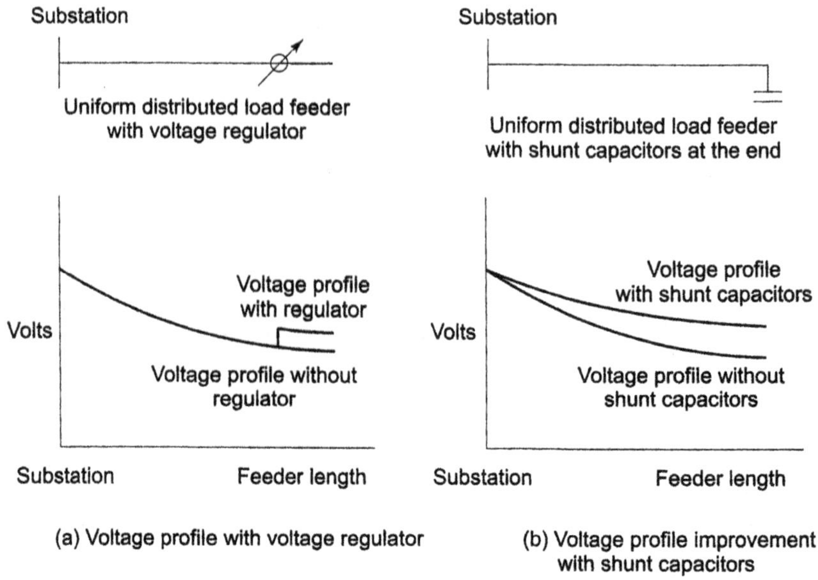

Fig. 4.11 *Feeder voltage control with regulator or capacitors*

For better voltage regulation, the distribution system are rated in the following order of merit (see Fig. 1.3).

1. Grid network
2. Spot network
3. Primary selective
4. Secondary selective
5. Primary loop
6. Radial

4.8.5 Computation

The formulation of a problem in a form suitable for solving on a digital computer is called *programming*. The procedure adopted for this is as follows:

(a) The problem is expressed in terms of mathematical equations, e.g. $V = IZ$; $P = IV$.

(b) An appropriate numerical method must now be found for solving these equations, such as the iterative process for finding the voltage drop around a loop.

(c) The numerical equation must now be broken down into separate arithmetical operations, and specific logical orders which determine the nature of operation of the computer must be introduced. These orders comprise to computer program.

(d) Finally, the problem must be translated into the language of the computer (called coding).

A voltage regulation program can be developed to calculate diversified voltage drops along 11 kV feeders. The program can cope with closed loops and spurs and some detailed comments on the iterative technique as applied to the loop can be applied to technical/economic investigation or optimization studies. The procedure to develop an 11 kV feeder voltage regulation calculation involves (i) compiling and (ii) processing of information.

In a manner similar to that adopted for 'manual' calculation, a standard sheet is prepared for recording data required for computer application. The data entry operator obtains all pertinent information relating to distribution transformers, loadings, length of line sections, conductor sizes and the position of loops and spurs and network mapping.

This can be summarized as follows:

1. To ensure that the voltage supplied to any consumer conforms to the accepted standards, it is essential that the characteristics and capabilities of the distribution system are known. The acquisition of this information depends on the selection, accuracy and flexibility of essential records and is strongly influenced by the procedures and routines devised to process available data into usable form.
2. In areas where system voltage problem exists, the task of correction or control can be achieved on an incremental basis, provided the approach is sound. The most rewarding area for initial investigation is considered to be the actual HV feeder system, exclusive of all other equipment. Other important aspects which can be considered as gains accruing from this initial effort are, for example, that the information collected when establishing voltage regulation values can be readily applied to system forecasting and development problems and also utilized for operational purposes.
3. A precise account of system characteristics can result in major capital savings, for example, in the form of postponed subtransmission and grid sub-station work.
4. A computer can be readily utilised for voltage control calculations, particularly if efficient 'Manual' methods already exist.

4.9 Harmonics

4.9.1 Harmonic Distortion

With increasing use of the nonlinear devices, harmonic distortion of the voltage waveform is a problem which is receiving the considerable attention. Harmonic currents are generated with the use of devices, such as rectifiers, inverters, thyristor-controlled variable speed drives, induction furnaces, arc furnaces, fluorescent lamps, TVs, UPS, computers and saturable reactors. These currents result due to the fact that the device either has an impedance which varies during each half cycle of applied emf or it generates a back-emf of nonsinusoidal shape. It is necessary to determine the distortion of the voltage waveshape due to the flow of harmonic current throughout the network and the effect of such distortion. IS : 325-1996 limits it up to 5 per cent at any instant in supply voltage up to 11 kV.

The magnitude and order of harmonics generated by a load is governed by the nature of the device. For example, with rectifiers mercury arc or semiconductor devices), the harmonic currents can be determined analytically. It is found that the larger the number of rectifier phases on the device, the fewer the number of harmonics generated and, correspondingly, the smaller their magnitude. Where the rectifiers are controlled (e.g. thyristors), the harmonics generated are a function of the delay angle of the rectifier firing.

Figure 4.12 shows the waveforms of the DC current in the load, together with the waveforms of the phase voltages and currents for a fully controlled thyristor circuit. It also shows a frequency spectrum of the harmonic currents flowing in the phase current for a delay angle of 0°.

Any periodic waveform of nonsinusoidal form can be synthesized by expressing it as the sum of a series of harmonics of the fundamental frequency using Fourier analysis.

$$f(t) = \frac{a_0}{2} + a_1 \cos \omega t + a_2 \cos 2\,\omega t + a_3 \cos 3\,\omega t + \cdots$$
$$+ b_1 \sin \omega t + b_2 \sin 2\,\omega t + b_3 \sin 3\,\omega t + \cdots$$

where, $a_0 = \frac{1}{T}\int_0^T f(t)\,dt,\ a_n = \frac{2}{T}\int_0^T f(t)\cos n\,\omega t\,dt$

$$b_n = \frac{2}{T}\int_0^T f(t)\sin n\,\omega t\,dt$$

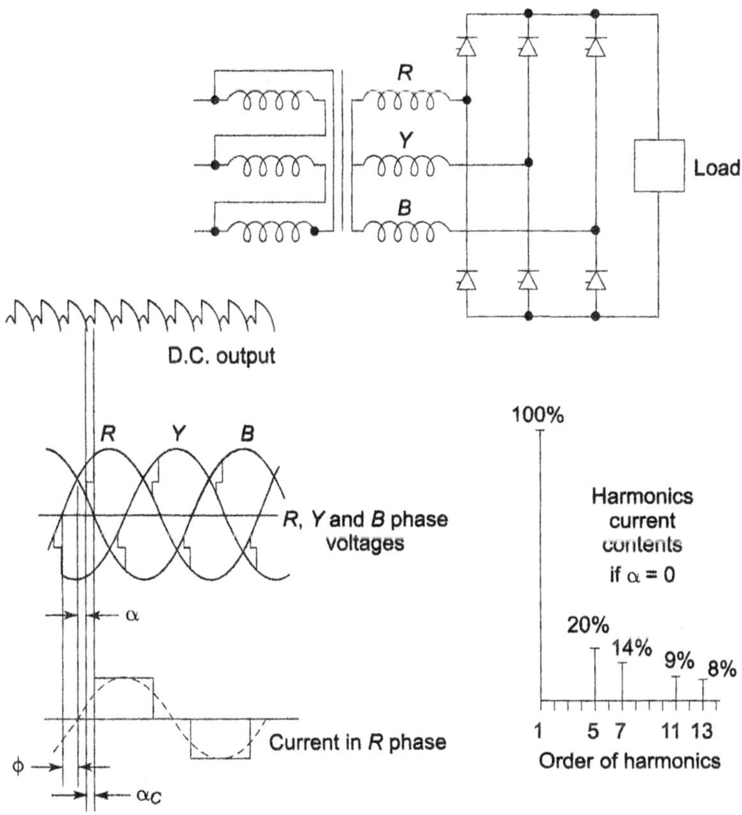

Fig. 4.12 *Harmonics in thyristor controlled rectifier*

Harmonic analysis can also be carried out on the network itself with the aid of a harmonic analyzer. The waveform is sampled and the analyzer sweeps through the range of frequencies and tunes out various frequencies. Harmonic penetration studies software packages are commercially available.

Total Harmonic Distortion (THD) is usually expressed as a percentage of fundamental voltage by the expression:

$$\sqrt{\sum_{n=2}^{n=n_{max}} \left(\frac{V_n}{V_1}\right)^2} \times 100\%$$

where V_1 = Fundamental frequency voltage component.
V_n = nth harmonic voltage component.

In balanced 3-phase system, the triple harmonics (3rd, 6th, 9th, etc.) have the same instantaneous magnitude in each phase. They can be considered as equivalent zero sequence values. When triple harmonics appear in a low-voltage system, they can overload the neutral conductor when three-phase components add arithmetically. These also circulate in the delta windings of distribution transformers. The 4th, 7th, 10th, 13th, etc. $(3n + 1)$ harmonics have the same phase sequence as the fundamental (positive sequence) while 2nd, 5th, 8th, 11th, etc. $(3n - 1)$ harmonics have the reverse phase sequence and are therefore of negative sequence.

The 5th, 7th, 11th and 13th harmonics can cause resonance in capacitor banks and are a source of noise in communication circuits. Underground feeder circuits, because of the high capacitance of the cables, also help to dampen these harmonics. The cables work as a natural sink for some harmonics having frequencies near their tuning circuits. The third harmonics is the most prevalent and is generally of the highest magnitude. Delta/star connected transformers are excellent natural filters for 3rd harmonics. The applications of special filters are not straightforward, since it may increase other harmonics for which these are not tuned. The filter resonates near the system harmonic frequency. The filter may sink harmonic currents from distant loads, as a low impedance short circuit and current carrying capacity of the line conductors may need to be substantially increased. Increasing the short circuit level of the system by using lines and/or transformers in parallel, reduces the effect of harmonics. Based on the ability to prevent the spread of harmonics, the various distribution systems are rated in the following order of merit (see Fig. 1.3).

Best:	•	Spot network
Moderate:	•	Grid network
	•	Primary selective
	•	Secondary selective
	•	All underground systems
Worst:	•	Overhead radial
	•	Overhead primary loop

4.9.2 Effect of Harmonics on Networks

Harmonics in power distribution networks can cause the following effects:

(a) *Overloading* of power factor correction capacitors due to tuning for a particular frequency.
(b) *Resonance* between capacitances and transformer reactances resulting in excessive voltages or currents.
(c) *Interference* with telephone circuits and broadcasting due to zero-sequence harmonic currents.
(d) *Malfunctioning* of control equipment as a result of distorted waveform affecting firing position of thyristor circuits; incorrect operation of some ripple control systems and protection relays.
(e) *Metering errors* in energy rotating disc meters.
(f) *Overheating* of rotating machinery due to increased iron losses due to eddy currents as well as loss of torque; and overheating of conductors as the skin effect increases with frequency increase.
(g) *Overloading of delta* connected windings of transformers on account of either excessive third harmonics or excessive exciting currents due to the flow of dc currents through the winding.
(h) *Overloading of neutral* conductor of low voltage distribution system where large single phase electronic loads are connected.
(i) Corrupting of data in computers.

4.9.3 Flow of Harmonic Currents

All the non-linear loads draw nonsinusoidal currents which cause distortion in the voltage waveform, not only within the individual plant or equipment, but also in the network supplying power to it, i.e. in transformers, lines, circuit breakers, motors etc. As the fraction of non-linear loads is increasing in the distribution system, so the effect of these loads needs to be limited. The harmonics flow through the distribution systems and may be transmitted from one consumer to another.

Individual harmonic frequency magnitudes can be represented as a percentage of the fundamental component. The crest factor measures one type of distortion. The effect of harmonics in different situations can vary, and hence different methods of characterising them are required. The optimum solution to reduce harmonics should involve equipment designs that do not result in significant harmonic generation.

The supply generator sends out a perfectly sinusoidal 50 Hz voltage into the network. The nonsinusoidal current into the nonlinear device can be represented by the fundamental current plus the series of harmonics. The fundamental wave of the current is alone responsible for the transport of energy into the load. The harmonic currents do not contribute to the transport of energy and they represent an undesirable phenomenon as they reduce the power factor. The load, in fact, operates as if it were a generator of harmonic currents. Accordingly, the analysis is carried out by assuming that each harmonic is injected into the system by a current source at the point of nonlinear load on the system, as shown in Fig. 4.13 (a).

Unless known otherwise, it is generally assumed that the reactive inductance at the harmonic frequency is n times the fundamental frequency value and the capacitive reactance is $1/n$ times the fundamental value, the resistance remaining constant.

The equivalent circuit of this network with respect to the harmonic generator is shown in Fig. 4.13 (b). The fundamental wave frequency content g is the ratio of the fundamental frequency current I_1 to the total AC line current I_L and the harmonic content k is given as follows:

$$g = \frac{I_1}{I_L}, \; k = \frac{\sqrt{\sum_{n=2}^{\infty} I_n^2}}{I_L}$$

such that

$$g^2 + k^2 = 1$$

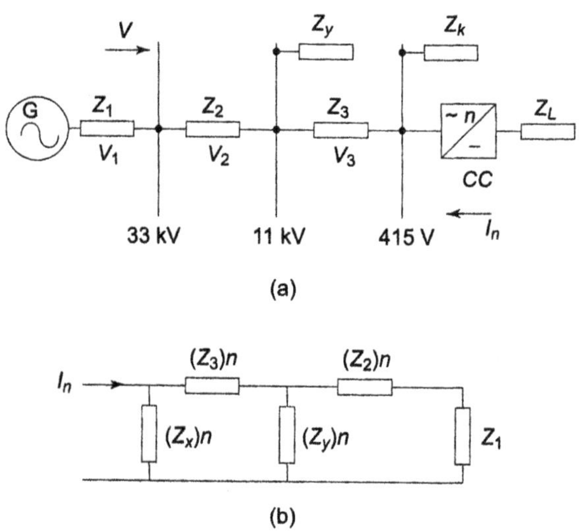

Fig. 4.13 *Harmonics in system: (a) Harmonic current flow; (b) Equivalent circuit*

4.8.4 Limits of Harmonics

At present, there is discussion going on about the maximum allowable generation of harmonics by various load [12]. This is a result of the increasing use of thyristor controlled equipment, computers etc.

Various international and national bodies have either issued standards or are considering them. CENELEC, a standards' body in Europe, was one of the first to issue a standard dealing with harmonics in domestic appliances. Since then standards have been published by Dutch, Swedish, German, French, IEEE and British authorities [17].

In the U.K. 'Engineering Recommendation G 5/2' was issued by the Chief Engineers' Conference in1967. It sets limits in terms of the actual current for each harmonic at the point of common coupling, i.e., the point where the consumer load first couples with the load of any other consumer on the system. For example, for 11 kV busbar, it limits the harmonic currents to 9.7 A to 5th Harmonic, 6.3 A to 7th and 1.0 A to 11th.

In the light of previous experience, an amended recommendation G 5/3 has been issued in 1976. It is somewhat less conservative than G 5/2. One item of considerable difficulty has been the addition of independent loads each producing harmonics. Arithmetic sums were previously used. It is now accepted that probabilistic techniques should be used and the sum is not likely to be more than the rms value of the sum of the harmonics.

The German Standard DIN 57160 Part 2-1981 places quite specific limits on various loads. For three-phase bridge rectifiers, for example, the power rating of the converter is limited to 1 per cent of the systems fault level; any single harmonic must not exceed 5 per cent of the fundamental up to the 15th harmonic and then decrease logarithmically; the total harmonic voltage must not exceed 10 per cent; depth and the duration of 'commutation notches' is specified; the system voltage may drop to 80 per cent for periods up to 3.89 per cent of a cycle (14°).

The limits on harmonics in various countries are given in Table 4.1 [9, 16, 17]. However, good recommendations or standards for harmonic voltage or current limits should consider the stochastic nature of the harmonics and focus on limit values of the probability of exceeding of 0.5 per cent, 1 per cent or 5 per cent for each harmonic separately and also for each harmonic group (i.e. $3n$, $3n - 1$, $3n + 1$).

Table 4.1

Limits on harmonics at common coupling in various countries

Country	*System voltage level (kV)*	*Total harmonic voltage distortion % (THD)*	*Individual harmonic voltage distortion % (THD)*	
			Odd	Even
Australia	33	5.0	4.0	2.0
	110	1.5	1.0	0.5
Sweden	0.43	4	—	—
	3.3–24	3.0	—	—
	36-72	2.0	—	—
	84	1.0	—	—
Finland	1.0	5.0	4.0	4.0
	110	1.5	1.0	1.0
U.K.	0.415	5.0	4.0	2.0
	6.6–11	4.0	3.0	1.75
	33–66	3.0	2.0	1.0
	132	1.5	1.0	0.5
U.S.A.	2.4–69	5.0	—	—
	115	1.0	—	—

In India, the following maximum harmonics were observed in the case of some consumers [17].

Consumer	Maximum % THD
Railway traction	22.43
Cement industry	6.50
Casting plants	7.79
Chemical plants	4.5

Microprocessor-based electronic meters which can measure harmonics (individual as well as total % THD) should be installed at feeders from the sub-station, and for industrial, commercial or other consumers with predominant non-linear load. The consumer should be levied a penalty for any harmonics above the stipulated limits. Modern electronic equipment have a low power factor (average 0.65 lagging) mainly due to harmonics. For example, a typically well-designed APC make, online, 620 VA UPS output produces THD of 15% and a maximum individual harmonic distortion of 14% as per the name plate rating.

4.9.5 Filters

A series-tuned harmonic filter consists of a capacitor bank with a reactor (inductor) in series with it. The series combination provides a low impedance path for a specific harmonic component. A typical configuration for low voltage applications is shown in Fig. 4.14. The filter provides a low impedance path for harmonic currents, thereby minimizing harmonic voltage distortion problems. The filter is usually tuned slightly below the harmonic frequency of concern. This method allows for tolerance in the filter components, and prevents the filter from acting as a short circuit for the offending harmonic current. This allows the filter to perform its function of providing a low impedance at the harmonic frequency, while helping to reduce the duty on the filter components:

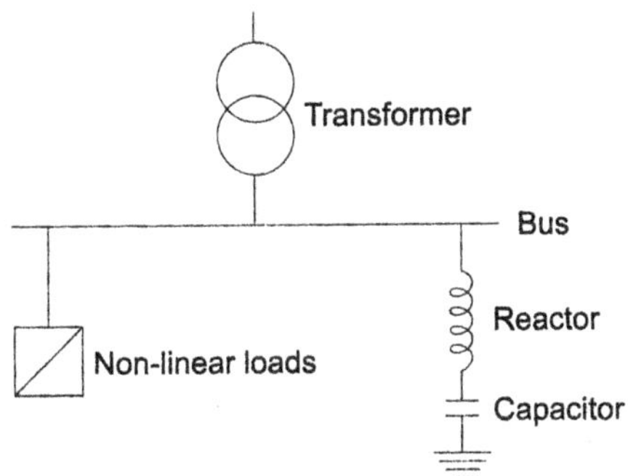

Fig. 4.14 *Single line diagram of harmonic filter*

The general guidelines for applying filters are:

- Apply on single-tuned shunt filter first and design it for the lowest generated frequency.
- Determine the voltage distortion level at the low voltage bus. The commonly applied limit is 5%.
- Vary the filter elements according to the specified tolerances and check the filter effectiveness.
- Check the frequency response characteristics, to verify that the newly created parallel resonance is not close to a harmonic frequency.
- If necessary, investigate the need for several filters such as 5th and 7th.

- Because consumer loads and the overall system characteristics are continually changing, the requirements for filtering can change with time. This often makes design of a standard, passive filter difficult.

Therefore concerned individual consumers should install Harmonic Filters at their premises.

4.9.6 Higher Level of Harmonics

In India, harmonics are increasing in the system due to the following reasons:

- Increasingly weak supply system, on account of the generation falling short of demand, due to the rapid growth of demand.
- Low value of ratio of system fault level (MVA) to the capacity (MW) of the harmonic generating loads. Normally, if the value of the above ratio is less than 20, the effects of harmonics will be more.
- The massive electrification programme undertaken by the Indian Railways at the rate of 800–1000 route kms per year. The modern thyristor controlled locomotives add to the harmonics.
- Increased thrust on energy conservation, leading to the rapid proliferation of several electronic energy conservation devices such as electronic fan regulators, electronic chokes for tube lights, electronic energy controllers for motors etc.

For example, harmonics generated by some of the energy conservation typical devices are given in Table 4.2 [18].

Table 4.2

Typical harmonics in electronic devices

Device (electronic)		Harmonic content %			
		3rd	5th	7th	9th and above
Fan regulator	Voltage	65	26	21	12
	Current	70	21	9	8
Chokes	Voltage	1	0.9	1.2	0.15
	Current	80	65	44	30
Motor energy controllers	Voltage	1.8	15.7	7.3	7
	Current	90	50	18	8

- Single-phase electronic loads are increasing in three phase distribution systems, having large single-phase electronic loads i.e. commercial buildings systems are rich in third harmonic circuits. These currents add in the neutral conductor, and can be up to 1.7 times the phase current for converter loads.
- The large use of shunt capacitors to improve power system power factor and voltage, has significant influence on harmonic levels. Capacitors do not generate harmonics, but provide a circuit for possible resonant conditions. If the addition of the capacitors in the system are tuned to resonate near a harmonic frequency present in the load current or system voltage, large currents or voltage at that frequency will be produced. The resonant frequency of a low voltage system with a capacitor bank can be found from:

$$n = \left[\frac{Q_s}{Q_c}\right]^{1/2}$$

where,

n is order of the harmonic at which resonance may occur.
Q_s is the available short circuit kVA.
Q_c is the kVAr rating of the capacitor bank.

4.9.7 Standards

Existing harmonic standards can be broadly classified into two types—the system (connection) standards and the equipment standards. The system standards deal with the connection of large harmonic-producing loads to supply systems, while the equipment standards deal with the harmonic performance of a piece of equipment. Since many harmonic-related problems result from interaction between the utility system and the consumer load, the system standards are of more concern to utilities at present. The equipment standards can be further divided into the harmonic emission standards and the harmonic susceptibility standards. In many cases, these standards are applied to the equipment of utilization voltage level (e.g. 230 V). The equipment standards are, therefore, of more interest to manufacturers. From this perspective, several existing international harmonic limits can be classified as follows:

System Standards

IEEE 519–1992	Low, medium and high voltage system
IEC–1000	Low voltage systems
IEC–77B	Medium voltage systems
CIGRE, WG 36.05	Medium voltage systems

Equipment Standards

IEEE 446–1987	Disturbance susceptible standards
IEC 555	Harmonic emission standards

The IEEE Standard 519, as shown in Table 4.3, gives the harmonic current limits based on the size of the consumer, with respect to the size of the power system to which the consumer is connected. The ratio of I_{sc}/I_{load} is the ratio of short circuit current, available at the Point of Common Coupling (PCC) to the nominal fundamental frequency load current. As the size of the consumer load decreases in comparison to the size of system, the percentage of harmonic current the consumer is allowed to inject into the utility system, becomes larger. This protects other consumers on the same feeder, as well as the utility, which is required to furnish a certain quality of power to its consumers.

Table 4.4 indicates the recommended values from IEEE Std. 519 which are low enough to ensure that properly designed equipment will operate correctly, provided no additional harmonics are generated by the sensitive load itself.

Table 4.3

Harmonic current limits as per IEEE Std. 519

I_{sc}/I_{load}	Harmonic order					Total harmonic distortion (%)
	11	11-16	17-22	23-24	35	
			HD%			
20	4.0	2.0	1.5	0.6	0.3	5.0
20–50	7.0	3.5	2.5	1.0	0.5	8.0
50–100	10.0	4.5	4.0	1.5	0.7	12.0
100–1000	12.0	5.5	5.0	2.0	1.0	15.0
1000	15.0	7.0	6.0	2.5	1.4	20.0

Table 4.4

Maximum voltage distortion as per IEEE Std. 519

Maximum distortion		System voltage		
		2.3–69 kV	69–138 kV	138 kV
Individual harmonic	%	3.0	1.5	1.0
Total harmonic	%	5.0	2.5	1.5

4.10 Load Variations

4.10.1 Voltage Fluctuations

When the load supplied by a network changes, the voltage drop along the line also changes correspondingly. Accordingly, the voltage as seen by other consumers along the length of the line will have fluctuations. If these are very frequent, the continuous variations may become annoying or irritating to the consumers. At the same time, voltage fluctuations may adversely affect the operation of many commonly used devices and appliances. For example, computer installations and their power supplies are sensitive to voltage fluctuations and 'spikes'. In addition, the continuous process plant can be seriously disrupted in its production if there is an inadvertent operation of an undervoltage relay.

The most sensitive monitor of voltage fluctuation is, however, the incandescent light bulb. The light output of tungsten filament lamp is extremely sensitive to variations of voltage about the rated voltage of the lamp, and any variations in the system voltage become quite noticeable. The tolerance of consumers to flicker has been established to be a function of the frequency of the voltage fluctuation. Many different curves have been published which attempt to establish this relationship. A set of commonly used curves which give levels of perception and toleration of lamp flicker is shown in Fig. 4.15. In case of fluorescent lighting, the light output is quite insensitive to voltage variations. The extensive use of these light fittings has resulted in reducing the awareness of the community to flicker problems.

The types of loads which are likely to cause flicker problems are welders, rolling mills, colliery winders, motors, compressors, induction furnaces and arc furnaces. The method of analysis of the problem for motor starting and for continuously varying loads is described below.

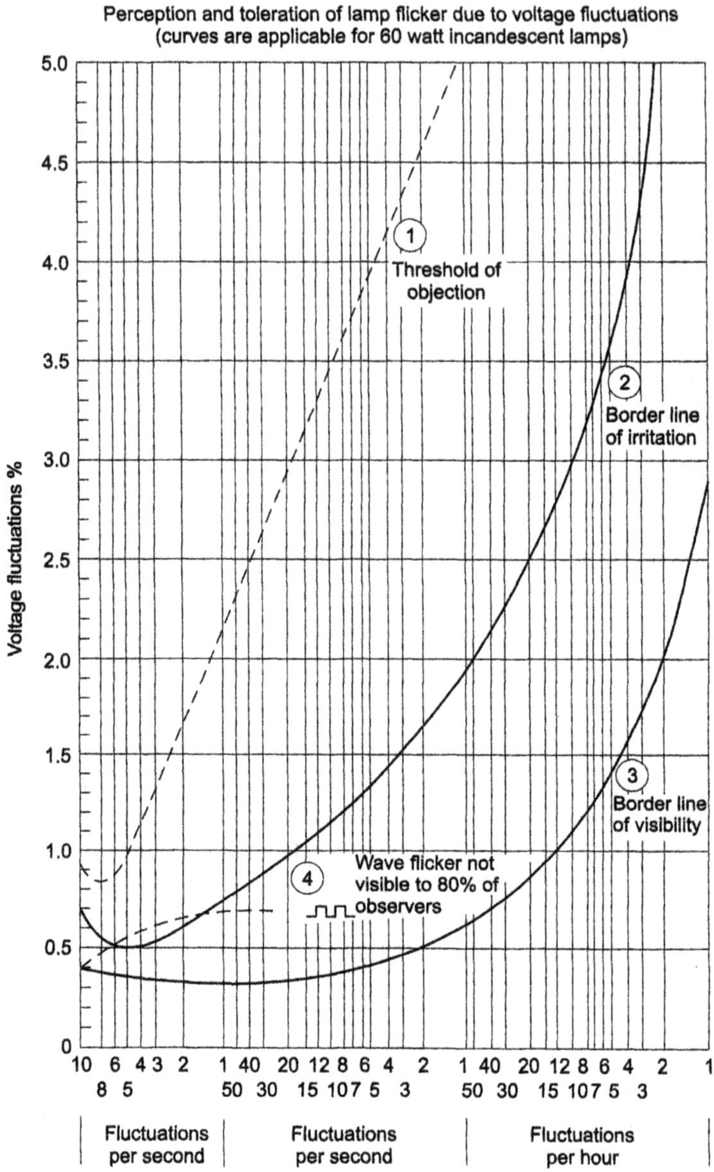

Fig. 4.15 *Lamp flicker perception and tolerance for 60 watt incandescent lamps*
Sources: Curve 1-'Electrical World', November 1958 (L. Brieger); Curve 2, 3-'Electrical Engineering', July 1956 (A. Kroneberg); Curve 4-'ERA Report' V/T 146-1962 (R. Thomas & P. Kendall)

4.10.2 Motor Starting

The voltage drop due to the starting current of an induction motor will depend on the size of the motor, power factor when starting, impedance of the system and method of starting. For a three-phase motor, the voltage drop is extrapolated after analyzing the equivalent single phase case.

From Section 3.16.1 we get the voltage drop equation as

$$E_s = E_R + I_R \cos \phi_R - I_X \sin \phi_R$$

where,

E_s = phase-to-neutral supply voltage
E_R = phase-to-neutral voltage at load
I = motor starting current
R = system resistance
X = system reactance
$\cos \phi_R$ = power factor of load on starting

$$\text{Voltage drop (\%)} = \frac{\sqrt{3}}{E_1} (IR \cos \phi_R - IX \sin \phi_R) \times 100$$

where, E_1 is the line-to-line supply voltage and ϕ_R is negative for lagging power factor loads and positive for leading power factor loads.

It can easily be taken that the power factor of the motor will be approximately 0.3 on starting, and the starting current for 'direct on line' (DOL) motor starting will be in the range from four to seven times the rated full load current (usually taken to be 6 if not known).

The starting current of motors can be reduced with the use of starting equipment such as Auto-Transformer starting and Star-Delta starting. The starting current from the power system is given by:

Auto-Transformer starting $I = \left[\frac{V_m}{V_1}\right]^2 I_{dol}$

V_m = auto-transformer secondary volts
V_1 = supply volts
I_{dol} = direct on line starting current

Star-Delta starting:

$$I = \frac{I_{dol}}{3}$$

4.10.3 Simultaneous Operation

Where there are many similar devices in an installation which may operate independently of one another and each device is on for a certain fixed period, there may perhaps be no flicker problem. However, when the devices operate simultaneously, flicker may become unacceptable. Devices such as thermostat controlled heaters, welding and multi-motor installations may fall into this category.

The frequency of occurrence of simultaneous operations may be calculated using the probability theory.

Consider an installation with n units labelled 1, 2, $\cdots$, j, $\cdots$, n and in a period T, the unit j is operated f_j times and is on for a period a_j and off for a period $(T - a_j)$.

The probability that the unit j is on is, therefore,

$$P_j = \frac{a_j}{T}$$

The probability that the unit 1, 2, $\cdots$, k are on simultaneously and the rest of the units off is given by

$$P(k) = p_1 \, p_2 \, p_3 \cdots p_k (1 - p_1)(1 - p_2) \cdots (1 - p_k)$$

The average duration of simultaneous operation of these k units is

$$t(k) = \frac{T}{f_1/p_1 + f_2/p_2 + \cdots + f_k/p_k + \dfrac{f_{k+1}}{1 - p_{k+1}} + \cdots + f_n/1 - p_n}$$

The frequency of simultaneous operation of k units is given by

$$F(k) = \frac{P(k)}{t(k)}$$

4.10.4 Continuously Varying Loads

As distinct from motor loads which have a definite voltage variation on starting, there are many loads which have a very irregular current pattern. This makes it difficult to apply the procedures previously described to such loads.

The most notable types of loads which in recent years have caused considerable interest are those of the arc furnace, induction furnace and rolling mills. In the arc furnace, the arc voltage is subject to stochastic changes due to the melting process. There are frequent short-circuits between electrodes and scrap-metal charge. Often, when the molten scrap metal drops away from an electrode, the arc is extinguished and no current flows. Since fluctuating load currents to the furnace passes through the supply network, corresponding fluctuations are impressed at the bus-bar linking the other consumers on the same network i.e. the Point of Common Coupling (PCC). Arc movements which result in current fluctuations in the frequency band 0.05 to 30 Hz can affect consumer lighting, TV sets, electronic apparatus etc. The most sensitive frequency is around 10 Hz where modulation of only 0.2% on supply voltage causes visible flicker effects. In the induction furnace the load varies continuously as the coupling of the load in the furnace with the induction winding varying during the process of the melt.

These fluctuations are very rapid. In Fig. 4.16 typical voltage fluctuations are shown for a sub-station busbar which is supplying a 12 MW induction furnace installation via a set of cables. To measure the various voltage fluctuations, a meter with an appropriate response characteristics is necessary to give a realistic assessment of the level of fluctuations (flickermeter).

The method of dealing with high speed voltage variations is under considerable discussion. The concept of fluctuation voltage V_f is used which is the modulation envelope of the fundamental frequency waveform as a measure of the flicker severity on the human eye. Extensive tests by the ERA have enabled the determination of the critical values of fluctuation voltage as a percentage of the supply voltage as shown in Table 4.5.

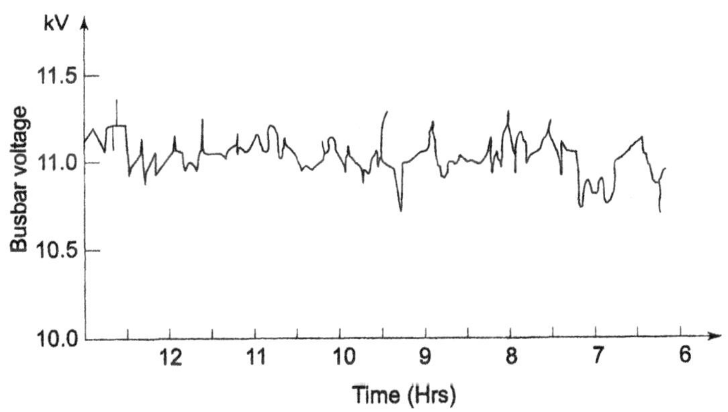

Fig. 4.16 *Load voltage fluctuations*

Table 4.5

Critical values of fluctuation voltage

V_f(%)	*Reaction*
0.20	Just perceptible but not annoying
0.25	Obvious, but not annoying
0.30	Uncomfortable but tolerable
> 0.30	Intolerable

Arc furnace acceptability on the basis of the shape of the critical voltage cumulative probability functions, i.e., the curve showing the percentage of total time that a particular value of V_f will be exceeded can be made. A gauge point fluctuation voltage V_f is used which corresponds to the value of fluctuation voltage which will be exceeded for only 1% of the time.

Also, the percentage voltage dip due to continuous irregular load changes during the melting stage of arc furnaces can be calculated with the following empirical formula [2].

Percentage voltage dip

$$= \frac{0.45 \times P_r \times n \times D \times 100}{P_s}$$

where,

P_r = MVA rating of one furnace transformer of the highest rating
P_s = short circuit level in MVA of supply bus
n = number of furnaces in simultaneous melting
D = diversity factor of furnaces during simultaneous melting operation

4.10.5 Measures to Reduce Flickering

Practical countermeasures for flickering on the power supply side include:

(a) An increase of power supply transformer capacity.
(b) Separation at supply voltage to consumers of varying load equipment. This would amount to segregating the varying load and steady load consumers.
(c) An increase of equivalent three-phase short circuit capacity of the system through insertion of series capacitors etc.

The countermeasures on the load side are:

(a) The lowering of absolute value of current variation through insertion of a series reactor in varying load circuit.

(b) Installation of thyristor controlled shunt capacitors.

(c) Suppression of overcurrent by insertion of a series reactor.

EXAMPLE

A prospective flour mill consumer, having 50 HP, 400 V, induction motor has applied for power connection. Before releasing connection, how will you check the voltage dip tolerance for other consumers fed from the same distribution sub-station?

Solution: Assuming the power factor of motor 0.85 at maximum efficiency of 90%, LT cable length for this consumer 100 metres, having resistance of 0.0926 ohm and reactance of 0.0084 ohm, we have

$$\text{Full load current} = \frac{50 \times 746 \times 100}{\sqrt{3} \times 400 \times 0.85 \times 90}$$

$$= 70.3 \text{ A}$$

$$\text{Current on direct on line start (DOL)} = 6 \times 70.3 = 421.8 \text{ A}$$

$$\text{Current on star-delta starting} = \frac{421.8}{3} = 140.6 \text{ A}$$

Percentage voltage drop:

(i) Direct on line start

$$= \frac{\sqrt{3}}{400} (421.8 \times 0.0926 \times 0.3 + 421.8 \times 0.0084 \times 0.954) \times 100$$

$$= 6.5\%$$

(ii) Star-delta start:

$$\text{Percentage voltage dip} = \frac{6.5}{3} = 2.83\%$$

Observing the curve 2 (extrapolated) in Fig. 4.15, to avoid flicker irritation, only 3 starts of the motor for every 2 hours are allowed if DOL is permitted. In case of Star-Delta starting, 12 starts of motor per hour are allowed which usually does not happen in actual practice. Hence to avoid irritable voltage flicker, we may release the power connection on Star-Delta starting 50 HP induction motor.

EXAMPLE

For power supply to an arc furnace consumer, having two furnaces each of 10 tonnes (5 MVA) capacity and auxiliary load. Power utility has earmarked a 16/20 MVA, 66/11 kV power transformer at its 66 kV sub-station where fault level is 1090 MVA on 66 kV bus and 266 MVA on 11 kV separate bus for the proposed transformer. Find the percentage voltage dip during melting operation of the furnaces when operating in sequence and simultaneously. How do you advise power supply arrangement for auxiliary load and to other consumers from the proposed transformers?

Solution: The voltage fluctuation is given by the formula:

$$\text{Percentage voltage dip} = \frac{0.45 \times P_r \times n \times D \times 100}{P_s}$$

When one furnace is working at a time, $D = 1$,

When two furnaces are operating simultaneously $D = 0.8$

(i) Now for supply given at 11 kV,

(a) Voltage variation for one furnace is

$$\frac{0.45 \times 5 \times 1 \times 1 \times 100}{266} = 0.846\%$$

(b) Voltage variation for simultaneous operation is

$$\frac{0.45 \times 5 \times 2 \times 0.8 \times 100}{266} = 1.351\%$$

Variations in both cases are more than the permissible limit of 0.3%. Beyond this limit light flicker becomes intolerable and voltage variations are injurious to the working of sensitive electronic equipment. Hence, auxiliary load and other consumers should not be supplied from 11 kV supply from this transformer.

(ii) Checking the supply to auxiliary and other consumers from this 66 kV sub-station.

(a) Percentage dip on 66 kV station bus for one furnace is

$$\frac{0.45 \times 5 \times 1 \times 1 \times 100}{1090} = 0.206\%$$

(b) For two furnaces

$$\frac{0.45 \times 5 \times 2 \times 0.8 \times 100}{1090} = 0.330\%$$

We see that for one furnace, 66 kV bus voltage variations are much below the permissible limits and in case of two furnaces, the variations are just permissible. Hence, we conclude that auxiliary load and other consumers should be given proper supply from a separate 66/11 kV transformer at this 66 kV sub-station.

4.11 Impact Loading of Transformer

Empirical guidelines on varying or impulse or impact loading can be used to approximate the accelerated use of transformer life, if the over-current magnitude, duration and frequency of occurrence are known. Instruments such as overcurrent monitor [1] are available to collect impact load data: the monitor provides information regarding fault magnitude, duration and frequency. The threshold limits on magnitudes and durations are variable. The units may be placed in parallel with varying threshold limits to obtain specific quantitative levels of overcurrent magnitudes and durations.

Impact loading data relating to normal life expectancy have accumulated as a result of experience in industrial applications such as arc furnaces and rectifier applications as well as full scale short circuit testing. Restricting the data to disc type, core form construction of transformer, maximum rms value of a current pulse in per unit rated current may be expressed as the number of overcurrent pulses per hour as given below for 55°C rise (average winding temperature rise by resistance) rated, self-cooled transformer [3]:

$$n = (4.76/I_{pu})^4$$

where,

n = number of current pulses per hour for normal life

I_{pu} = maximum rms value of current pulse in pu based on 55°C winding rise rated current (as per IS: 2026–1977 revised for power transformers).

4.12 Ferroresonance

A ferroresonant circuit is characterized by a series circuit consisting of an iron-cored and, therefore, non-linear inductance in series with a capacitance and the circuit excited by an ac voltage. As a result of the non-linear nature of inductance the circuit can exhibit a number of separate forms of behaviour all of which are referred to as ferroresonance:

(a) Fundamental frequency stable subresonance.
(b) Fundamental frequency stable ferroresonance.
(c) Stable subharmonic frequency oscillations.
(d) Self-excited unstable but continuous oscillations.
(e) Transient oscillations reverting to subresonance.

The fundamental frequency stable ferroresonant state is the one which is potentially most dangerous in electrical distribution networks. This circuit can be formed by the single-phase switching of an unloaded or lightly loaded delta connected distribution transformer in conjunction with a length of cable. In a distribution system, generally Delta (HV)-star (LV) connected transformers are used. An underground cable at distribution voltage has a capacitance to ground per unit length (average figure of about 0.31 μf/km) of about fifty times that of an overhead line. This factor means that the underground cable connections are much more likely to cause the ferroresonant overvoltage problem than overhead conductors.

A single-line diagram of such an arrangement of a three-phase system is shown in Fig. 4.17. Under these conditions, voltages

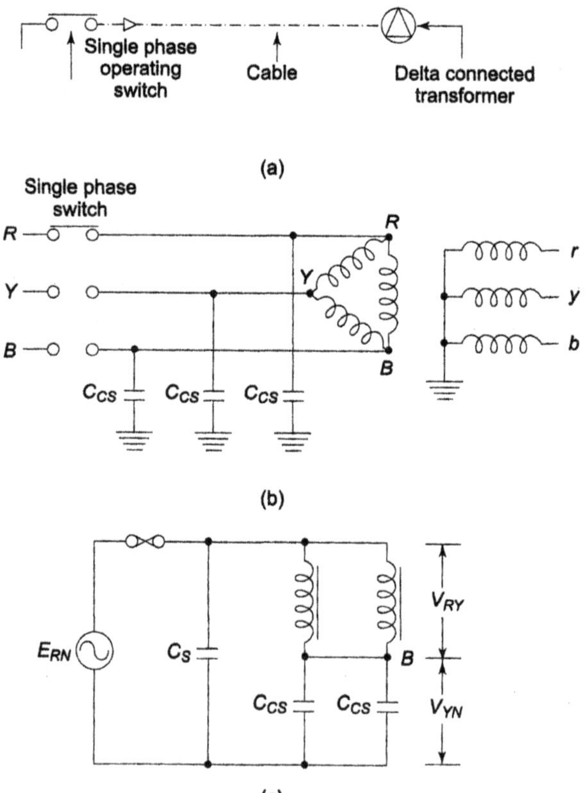

Fig. 4.17 *Ferroresonance: (a) Single-line diagram of ferroresonance circuit arrangement; (b) Three-phase equivalent circuit; (c) Approximate single-phase circuit*

considerably greater than the applied voltage (typically three to four times) will appear across the series capacitance in the circuit. The full length of the cable, a transformer winding, cable terminations and open phases of the switchgear are subjected to this overvoltage.

There are several possible switching arrangements which result in the same equivalent circuits that will result in ferroresonant conditions. This switching may be:

(a) Any type of switch used in overhead network (air-break switches, etc.)

(b) Drop out fuses or horn gap fuses replaced by hot-stick.

(c) Circuit breakers.

A single-phase switching condition occurs due to either the blowing of one or two HV fuses or unequal closing time of air-break switch blades or circuit breaker poles. In most practical situations, ferroresonant overvoltages are in the range of 1.5 to 2 pu.

It may be noted that a ferroresonant circuit occurs when both the transformer and a length of cable are being single-phase switched at the same time. Accordingly, ferroresonant conditions could be avoided by energizing the cable first and then the transformer and conversely, de-energizing the transformer first and then the cable.

4.13 System Losses

System losses in different states vary up to 46.5%. Actual losses in most of the state power utilities are even higher, as the losses have been shown as agriculture consumption which is un-metered supply in most of the states. According to a study, the losses in some of the cities in the country (2000–01) are given below:

City	Per cent losses
Ambala	33
Bangalore	16.5
Chennai	13
Delhi	55
Hyderabad	50
Kolkata	11
Ludhiana	23
Mumbai	11.5

Losses vary in each state because of varying system characteristics such as load density; urban/sub-urban/rural network; predominant sub-transmission voltage 33, 66 or 132 kV; voltage of supply—LT, or EHT. Other factors such as inadequate designs and bad maintenance contribute to higher losses. There may be significant proportion of unaccounted losses due to inaccuracy in meters and metering equipment, flat rate tariffs, inaccurate estimation of non-metered supplies, error of consumer billings, pilferage of energy and un-authorized use of electricity. The complete analysis is desirable in this respect. The losses in different utilities or areas should be compared with due care to point out the better managed or mismanaged power utilities. For example, comparing the Tata Power system with its 300 or so large consumers at HT and EHT with the Maharashtra State Electricity Board having a far flung mixed urban and rural LT, HT hybrid system is not fair. However, comparison with the Ahmedabad Electricity Company, in so far as its HT system of supply to mills is concerned would be in order. Comparison of losses in different similar countries such as China having about 13% T & D loss with India having about 22% will be meaningful. The main reason attributed to low losses in China is installation of generating stations near load centres such as numerous mini/micro-hydel plants, nuclear plants, etc. The losses in other countries like Pakistan, Thailand, Philippines, Korea, Taiwan, U.S.A., Japan are 28%, 18%, 18%, 12.5%, 9%, 8%, 7% respectively (1999).

4.13.1 Losses in Components

Power system losses constitute the loss in step up transformers, transmission lines, step down transformers, sub-transmission lines, primary distribution feeders, distribution transformers, LT lines and metres. The general idea of comparative losses in various sections of power systems can be made from a typical values given below from a study of a system losses in a power utility.

Section	*% losses*
Total system	100
Generation step up transformers and transmission sub-station	11.58
Transmission and sub-transmission lines	20.66

Sub-transmission or grid sub-stations	12.85
Primary feeders and line equipment	25.27
Distribution transformers	17.22
Secondary and services lines and grounds	11.82
Metres	0.6

The range of average losses at full load, at rated voltage in the various equipment are [21]:

Watt Losses

Switchgear (33 to 66 kV)	0.005–0.02%
Transformers	0.40–1.90%
Load-break switches	0.003–0.025%
Busway (440 V and below)	0.05–0.50%
Low-Voltage switchgear	0.13–0.34%
Motors 1–15 HP	14–35%
20-200 HP	6–12%
200–1500 HP	4–6%
1500 HP and above	2.3-4%
Rectifiers (large)	3–9%
Static variable-speed drives	6–15%
Capacitors	0.5–2 Watts/kVAr

kVAr Losses

Primary and secondary lines:	
Rural Area	0.8–1%
Urban Area	0.4–0.6%
Distribution Transformers:	
Up to 100 kVA	7–9%
200–1000 kVA	6–8%

4.13.2 Reduction

Computer studies could be helpful in reducing system losses. Alternative studies of the system with different input data can be made. Energy balance-sheet for each feeder should be prepared and analysed for the billing period. The reduction in system losses can result in substantial saving in energy as well as an increase in the power capacity. Various means for reducing the system losses are given below.

1. Optimizing line capacity (see Section 6.7.1 and Appendix II)

(a) Selecting appropriate kVA-km capacity based on the required voltage regulation and normal power factor for the conductor used for LT distribution circuits.
(b) Select appropriate MW-km capacity for the standard conductors used in primary distribution, thereby limiting lengths of primary feeders.

2. Optimizing transformer capacity, location and use (see Section 6.7)

EXAMPLE

A city 132 kV sub-station has two transformers each of 10 MVA, 132/11 kV are run in parallel. Each having no-load losses = 18 kW and load losses = 60 kW, percentage impedance = 10.5%, no=load current = 0.9%. Calculate the losses, power factor at HV Bus if load is 12 + *j* 7.2 MVA.

Solution: The parallel transformers with equal impedance will share load equally.

(i) Losses

(a) $$\text{Total losses} = 2 \times \text{load loss}\left[\frac{\text{MVA actual}}{\text{MVA rated}}\right]^2 + 2 \times \text{no load losses}$$

$$= 1/2 \times 60\, \frac{(12^2 + 7.2^2)}{10^2 \times 10^3} + 2 \times 0.018$$

$$= 0.0948 \text{ MW}$$

$$\text{Percentage kW loss} = \frac{0.0948 \times 100}{2 \times 10} = 0.474\%$$

(b) Reactive Losses: (See Section 15.1)

Reactive losses = 2 × % impedance voltage

$$\times \frac{(\text{MVA actual})^2}{\text{MVA rated}} + 2 \times \text{magnetizing loss}$$

% No load current = % Magnetizing current = % Magnetizing loss

$$\text{Reactive losses} = 1/2 \times \frac{10.5}{100} \times \frac{(12^2 + 7.2^2)}{10} + \frac{2 \times 0.9 \times 10}{100}$$

$$= 1.03 + 0.18 = 1.21 \text{ MVAr}$$

(ii) Power factor (pf)

$$\text{Low voltage Bus pf} = \frac{12}{\sqrt{12^2 + 7.2^2}} = 0.858 \text{ Lagging}$$

Since kW loss is negligible

$$\text{High voltage Bus pf} = \frac{12}{\sqrt{12^2 + (7.2 + 1.21)^2}} = 0.82 \text{ Lagging}$$

The power factor variation:

$$= 0.858 - 0.82 = 0.038$$

3. Maintaining appropriate voltage levels in distribution systems.
4. *Installation of shunt capacitors:* Apply shunt capacitors across individual inductive load, at distribution transformer, lines and at distribution sub-stations (see Chapter 15 and Section 14.4).
5. Selecting SWER system for sparsely populated rural areas having lesser potential for future development (see Section 14.5).
6. *Limit Unbalanced in LT Supply:* As a result of unequal load on individual phases; negative and zero phase sequence components cause overheating of transformers, cables, conductors and motors, thus increasing the losses and the motors malfunction under unbalance voltage conditions. Keeping the system negative phase sequence voltage within limits amounts to savings in capital (as otherwise equipment is derated) as well as energy losses.

IS: 325-1996 (clause 4.2) limits Negative Phase Sequence (NPS) of 1.5% voltage. A spot check on the unbalance can be obtained by measuring all three-phase voltages.

EXAMPLE

Assume the following.

Sets of measured line voltage (V)			*Approximate NPS %*
410	415	420	1.5
405	415	425	3
405	415	415	1.5
395	415	435	6

The percentage unbalance is calculated from the general expression:

$$\text{Percentage NPS} = \frac{71}{V_{\text{average}}}\left(V_{\text{high}} - \frac{1}{4}V_{\text{middle}} - \frac{3}{4}V_{\text{low}}\right)$$

The percentage NPS voltage can also be recorded, say over a week, with a recorder.

EXAMPLE

A 11 kV feeder has unbalance loads on *R*, *Y*, *B*, phases as 100 A, 200 A, 150 A respectively. Find saving in losses if load is balanced. The line conductor resistance is 4 Ω.

Solution: Losses = $I_R^2\,R + I_Y^2\,R + I_B^2\,R$ where R is line phase resistance = 4 Ω

Unbalance load losses $= 100^2 \times 4 + 200^2 \times 4 + 150^2 \times 4$
$= 290{,}000$ W

On balancing, load on each phase will be 150 A

Balance load losses $= 150^2 \times 4 + 150^2 \times 4 + 150^2 \times 4$
$= 270{,}000$ W

The loss savings $= 290{,}000 - 270{,}000$
$= 20{,}000$ W

Under unbalanced conditions, additional losses occur in neutral wire/earth of secondary system due to flow of unbalanced current or due to circulating current in delta winding of the distribution transformer. On balancing, these losses are also saved.

7. *Miscellaneous Measures:* There is large scope for energy saving in areas such as use of computerized control for network monitoring and management; use of solid-state or digital energy meters instead of conventional induction type; laying efficient public lighting with MLL, SOX, SON lamps; installation of low loss amorphous metal alloy core distribution transformers; staggering of agricultural loads in rural areas; and using of proper minor irrigation motor pumps for agriculture; silicon diode and thyristor converters in place of rotary and mercury arc converters; decentralised power generation from renewable energy sources; limit harmonics (refer Section 4.8); good construction and maintenance practices, viz. jointing, jumpering, tapping, etc.; good operational practices. Shaving of system peak during seasonal peak load months will curtail losses at a higher rate as losses at a time are proportional to the square of the current flowing in the circuit/system at that instant (see Figs. 2.4 and 2.11). Consider, for example, a grid or main sub-station having two power transformers, say a 66 kV sub-station with 2 × 8 MVA, 66/11 kV transform-

ers. The costs incurred in total transformer losses can be easily calculated. Up to a certain load, it will be found economical to run only one transformer, and above that load, the increase in the cost of load (copper) losses can be more than the saving in the iron losses in comparison to the two-transformer operation. Suitable controllers can be used to arrange an operation with one or two transformers in the most economical manner in terms of total transformer losses. The transformers can then be automatically switched in and out to minimise losses. In urban areas, transformer parallel operation is recommended to reduce losses at peak hours. To avoid unbalanced circulating current, the transformers' vector group, %age impedance and voltage ratio should be the same.

EXAMPLES: The no-load losses of 25, 63 and 100 kVA, three phase distribution transformers as per REC specifications for cold rolled grain oriented (CRGD) core (REC specifications 2/1997) and amorphous core transformers (REC specification 70/1993) are given in Table 4.6.

Table 4.6

No-load losses of distribution transformers

kVA capacity	*No-load losses (REC specification) watts*	
	CRGO core	*Amorphous core*
25	100	25
63	180	45
100	260	60

In India, there are more than 3 million distribution transformers of up to 100 kVA of CRGO core type, taking the average size as 63 kVA.

If all the existing 3 million transformers are replaced by amorphous core 63 kVA transformers with 45 watts no load loss, the saving in energy marginal cost Rs 4/–kWh is:

$$3 \times 10^6 \times 8760 \times 4 \times \frac{180 - 45}{1000} = 14190 \text{ million rupees}$$

This is equivalent to adding an additional generating capacity of:

$$\frac{3 \times 10^6 \times 135}{1000 \times 1000} = 405 \text{ MW}$$

This is equivalent to saving in investment of Rs. 16200 million in generation capacity (assuming investment rate of Rs. 40 million/MW of additional generation capacity).

8. *Revamping the Existing System:* In the existing system, higher losses can be due to:
   - Improper construction and maintenance of lines/transformers
   - Uneconomic conductor size
   - Inadequate layout of the feeder
   - Overloading of distribution transformers
   - Low-voltage conditions in the network
   - Poor power factor due to inadequate reactive compensation
   - Use of poor quality of construction material
   - Uneven distribution of load on various feeders and sub-stations

The measures to be adopted to reduce losses and improving the voltage regulation are:

- Power factor correction be done by providing capacitors near to load centre
- Re-distribution of load among various feeders
- Re-distribution of feeders among various sub-stations
- Re-routing of feeders or adding of link lines
- Proper tap-setting of interconnecting transformers
- Providing proper sectionalizing of lines
- Shifting of transformers to load centres
- Overloaded transformers should be augmented or new transformers be added
- Conductor of higher size be used for heavily loaded feeders
- Use of good quality material and proper jointing (refer to sub-section 11.7.1) and good maintenance be carried out
- High level of vigilance be regularly carried out to check pilferage of energy and other unauthorized use of electricity
- System extension for release of new power connection be done after taking into account of voltage regulation at tail end, balancing of loads and not overloading of transformers.

Upgrading the operating voltage by conversion of the existing system of operation at the next higher voltage level. For example, it may be feasible to convert the existing LT lines and system operating at voltage below 11 kV into 11 kV lines and 11 kV lines into 33 kV using the same supports and conductors by changing the cross arms and insulators only.

4.13.3 Optimum Losses

There is an optimal level of network losses, where the cost of further reduction would exceed the cost of supplying the losses. In a developing system, the loss levels are usually well above the optimal. Loss reduction requires capital investment to improve the distribution system. While framing a scheme to enable loss reduction, capitalise the losses saving in comparison to the cost of the scheme. The typical practice would be to capitalise loss savings over a period of 10–15 years using long-term discount rates. Also, account for the load growth over this period. For a growing load, the loss saving will increase over time. There are other benefits in loss reduction such as environmental benefits from reduction in fuel consumption at the generating plant, marginal improvement in system voltage and marginal saving in system capacity which are not be considered in the analysis of optimum losses. State Regulatory Commissions must enforce the power utilities to minimise the cost of supply by reducing losses to the optimal value. The optimum system losses in the Indian power system as suggested in the World Bank Report No. 6 are as given in Table 4.7 (also see Fig. 4.18).

Table 4.7

Optimum technical losses

System, component	*Maximum to be tolerated*	
	Within	*Cumulative*
Step-up station	0.50	0.50
EHV transmission and station	1.00	1.50
HV transmission and station	2.50	4.00
Sub-transmission	4.00	8.00
Distribution station	0.50	8.50
Distribution primary	5.00	13.50
Distribution transformer and secondary	2.00	15.50

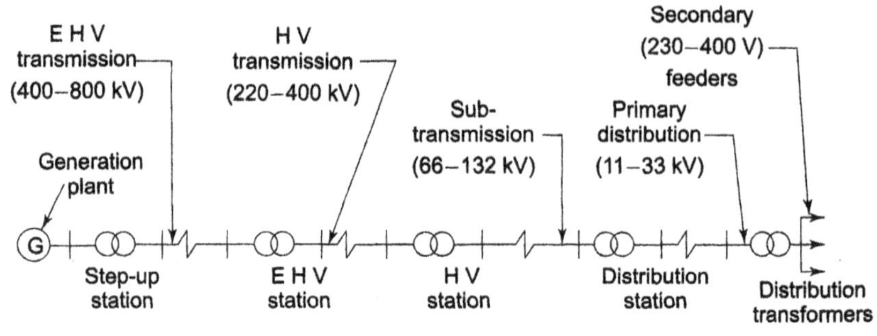

Fig. 4.18 *Segregation of system losses*

4.13.4 Cost Evaluation

In a system there are two types of losses: fixed, i.e. no load losses and variable, i.e., load losses. The cost of these losses differs in case of each organization because each utility has its own structure of fixed and variable costs. The annual cost of losses is assessed on the basis of the marginal cost of supply. The marginal costs should reflect not only the cost of production in terms of generation capacity and fuel consumption but also cost of capacity in transmission and the upstream distribution network [4]. In the absence of true marginal costs, bulk supply tariff or energy purchase rate (as per the contract) or incremental cost may be used to evaluate the cost of losses. The general principles of evaluating these losses are described below.

1. *No Load Losses:* Since these are essentially constant, extra equipment must be provided to supply the additional demand at peak hours. Efficient base-load units (firm hydro capacity, thermal or nuclear) can supply these losses most of the time. Hence, the per unit demand/energy cost for these is low.

2. *Load Losses:* These losses also constitute two parts: demand and energy. Load diversity complicates the evaluation of the demand-based cost of peak load losses. The annual peak of a system varies from year to year and the average loading varies from area to area and loading practices of individual utilities. The distribution system peak loading may not coincide with the total system (generation) peak. In these losses, the demand-cost depends upon peak responsibility factor (k) which is

the ratio of distribution system's load at the time of total system peak to its peak load. It generally varies between 0.2 and 0.8 for distribution systems and 0.8 and 0.95 for transmission systems. Since k is a ratio of loads, the losses are a function of k^2 which is used in determining the demand-cost. If a distribution system has P kW loss at its peak load, Pk^2 will be the loss at the time of total system (generation) peak, the cost of which is demand-based. Therefore Pk^2 kW losses demand must be provided by the generating equipment.

For energy based losses, the loss factor is important. It is the ratio of the energy loss in the system during a given time period to the energy loss that would result if the system peak loss had persisted throughout that period (see Section 2.9). One empirical approximation is [7].

Loss factor = (Load factor)$^{1.732}$

Another rule of thumb is:

Loss factor = c (load factor) + $(1 - c)$ (Load factor)^2

where $c = 0.3$ for transmission systems and 0.15 for distribution systems.

These two relations are considered accurate if the minimum demand during the period does not fall below 0.2 pu of the peak demand. In case the system load factor (L_f) is below 0.8, the following relation is more accurate [8].

$$L_s = L_f^2 + 0.273\,(L_f - K)^2$$

where, K = minimum demand in pu of peak demand; L_s = Loss factor

Total losses per annum = Loss factor × maximum losses based on continuous peak current

(i) In terms of the British experience, this factor fits the relationship:

$$L_s = 0.2\,L_f + 0.8\,L_f^2$$

(ii) The American experience gives:

$$L_s = 0.3\,L_f + 0.7\,L_f^2 \text{ for urban areas;}$$

$$L_s = 0.16\,L_f + 0.84\,L_f^2 \text{ for rural areas.}$$

(iii) The Australians use:

$$L_s = 0.2\,L_f + 0.8\,L_f^2$$

Such empirical relations can be established.

Once the loss factor is found, we can determine the capital energy cost of load losses. This cost equals 8760 times the product of the loss factor, distribution system peak load losses and energy cost. The number 8760 represents the number of hours per year. The peak load losses are

$I^2 \times R$, where I is the peak current and R is the resistance of the system circuits consisting of lines, transformers, etc. So the total losses capitalization can be formulated as follows:

Total losses (of all sections) cost per annum

= (8760 × No load losses × energy cost per kWh)
+ No load losses × demand cost per kW)
+ I^2Rk^2 × demand cost per kW + 8760 × loss factor
× I^2 × R × energy cost per kWh

The algorithm based on the above formulation can be prepared and with the help of suitable program and the system losses are computed and capitalized easily. Detailed system load-flow studies can be carried out right up to the consumer meter with a proper software to process the total system losses. For such studies, accurate field data of the network map, lines, equipment and loads including consumer services, is necessary.

In the above case, I is based on the actual network measurement. There may be the case when no measured data of peak load or peak current are available and the measured values are restricted to the electric energy consumption of low voltage consumers. In such circumstances, the available figures of energy consumption could be transformed roughly to peak load figures with the help of the following Velander's formula [10]

$$P_L = k_1 W + k_2 W^{1/2}$$

where,

P_L = peak load in kW
W = energy in kWh/year

Constants k_1 and k_2 depend on the nature of the load and should be taken as parameters while making calculations. Typical values of these constants for domestic and commercial consumers are as below [10]:

Category	k_1	k_2
Domestic	290	2500
Commercial (shops)	250	1900

Distribution system modelling of loads, lines, transformers, capacitors, etc. has been found to be an accurate tool for the calculation of losses [13].

EXAMPLE: In a power utility, the energy consumption during the year is 5000 GWh with maximum and minimum demands of 1000 and 100 MW respectively. Find loss factor and typical empirical relation.

Solution: Yearly load factor $(L_f) = \dfrac{5000 \times 10^6}{8760 \times 1000 \times 10^3}$

$$= 0.57$$

$$K = \frac{100}{1000} = 0.10$$

Therefore $L_s = 0.57^2 + 0.273\ (0.57 - 0.1)^2$

$$= 0.3852$$

$$L_s = cL_f + (1 - c)\ L_{f2}$$

$$0.3850 = c \times 0.57 + (1 - c)\ 0.57^2$$

$$c = 0.25$$

The empirical relation will be:

$$L_s = 0.25\ L_f + 0.75\ L_{f2}$$

EXAMPLE: A 2.5 MVAR, 11 kV capacitor bank has been commissioned at a 33/11 kV, 2 × 5 MVA transformer sub-station having the following data:

	Without capacitors	*With capacitors*
Peak load	8000 kW	8000 kW
Voltage	10.4 kV	11.4 kV
Power factor	0.846	0.99
Peak current on 11 kV	525 A	409 A
Peak current on 33 kV	175 A	136.3 A

33 kV line length is 35 km and conductor resistance is 0.584 Ω/km, load factor = 0.65.

Find saving in losses and sparing of system capacity after installation of the capacitors.

Solution:

(i) *Reduction in 33 kV line losses*

(a) Annual reduction in energy losses

$$= 3\,(I_1^2 - I_2^2)\ R.L.L_s \times \frac{8760}{1000}\ \text{kWh}$$

where,

I_1 = Peak current before capacitor commissioning = 175 A

I_2 = Peak current after capacitor commissioning = 136.3 A
R = Resistance of the line conductor = 0.584 Ω/km
L = Length of the line = 35 km
L_s = Loss factor = $0.3 \times L_f + 0.7 \times L_{f2}$
$= 0.3 \times 0.65 + 0.7 \times 0.65^2 = 0.49$

∴ Reduction in energy losses $= 3\,(175^2 - 136.3^2) \times 0.584 \times 35 \times 0.49 \times 8.46$

= 3,170,972 kWh per annum
= 3.17 GWh per annum

or Rs. 6.34 millions per annum @ Rs. 2/kWh.

(b) Demand losses saving

$$= 3\,(I_1^2 - I_2^2)\,\frac{RL}{1000}$$

$$= 3\,(175^2 - 136.3^2) \times 0.584 \times \frac{35}{1000}$$

= 738.741 kW

or say 739 kW

(ii) Sparing of system capacity

(a) For the same line current of 525 A, the new power available at 11 kV, due to improved voltage and power factor during peak hours is

$$\sqrt{3} \times 11.4 \times 525 \times 0.99 = 10{,}262 \text{ kW}$$

Therefore additional system power capacity available during peak

= (10,262 – 8,000) = 2,262 kW

(b) Additional energy distribution capacity available at load factor 0.65

$= 2{,}262 \times 0.65 \times 8760$
= 12,879,828 kWh per annum

or say 12.88 GWh per annum

Allowing 20% line losses, the net system capacity for energy distribution available will be

= (12.88 – 2.58)

= 10.3 GWh/annum

4.13.5 Measurement of Energy Losses

The losses can be measured by measuring total energy input to the power network, with good accuracy meters at the generating stations. Generating station power usage (auxiliary consumption) is also metered and deducted to give the net input of the transmission and distribution networks. Transmission system losses are determined by measuring the outputs from the transmission system, requiring that all meters in the generating and transmission stations and in the major industrial consumers directly fed from transmission, be read simultaneously on a fixed day in the beginning of the year. Aggregating and reconciling the inputs and outputs gives a good estimate of the transmission system losses. The total input to the distribution system is determined from these results, while total output is derived from the consumer billing data, and has to be made up for non-technical losses, mainly theft and metering-accounting errors. Flat rate consumption of agriculture can be estimated on sample basis. Theft can also be estimated, which could be between 5% and 15% of generation as per sample studies made in Indian power systems. In U.S.A., it is typically about 2%. Based on the studies, the consumer metering error is estimated as arising mainly from non-simultaneous reading of meters and reading estimation. Typically, utilities in some Western countries take this metering error of the order of +0.4 to 1% of generation. (see Fig. 4.19)

Detailed allocation of the energy losses across the different levels and circuits within the distribution system is a more difficult task, as extensive accurate measurement is difficult. The only feasible option is to calculate losses from circuit characteristics and network loading, requiring a 'bottom up' approach from the finest level of network detail, as opposed to the 'top down' approach for bulk (transmission) loss evaluation. The load and loss calculations use conventional formulae and parameters and may fully accommodate reactive power flow. The loss calculations can use in-house software running on the network data bases, as indicated in the Section 4.12.4. The energy losses in distribution system consists of no-load and load components and can be computed based on the formulae:

$$\text{Energy losses} = 8760\ (\text{No-load losses} + \text{loss factor} \times 1^2 \times R)$$

After calculations, there is need for correction upward by some intangible factors, including unbalanced phase loading and averaging inherent in load-data collection. Due to the relationship between the load and losses, losses due to load excursions on the high side are not balanced by the excursions on the low side. However, the approach of working from the 'top down' to determine bulk losses, and from the 'bottom' up to derive the detailed distribution losses, gives a very good picture of the sources of losses across the distribution system. However 'top-down' approach can be used in distribution system if accurate metering is made as shown in Fig. 4.20. Reconciling the two approaches gives a reasonably good level of accuracy.

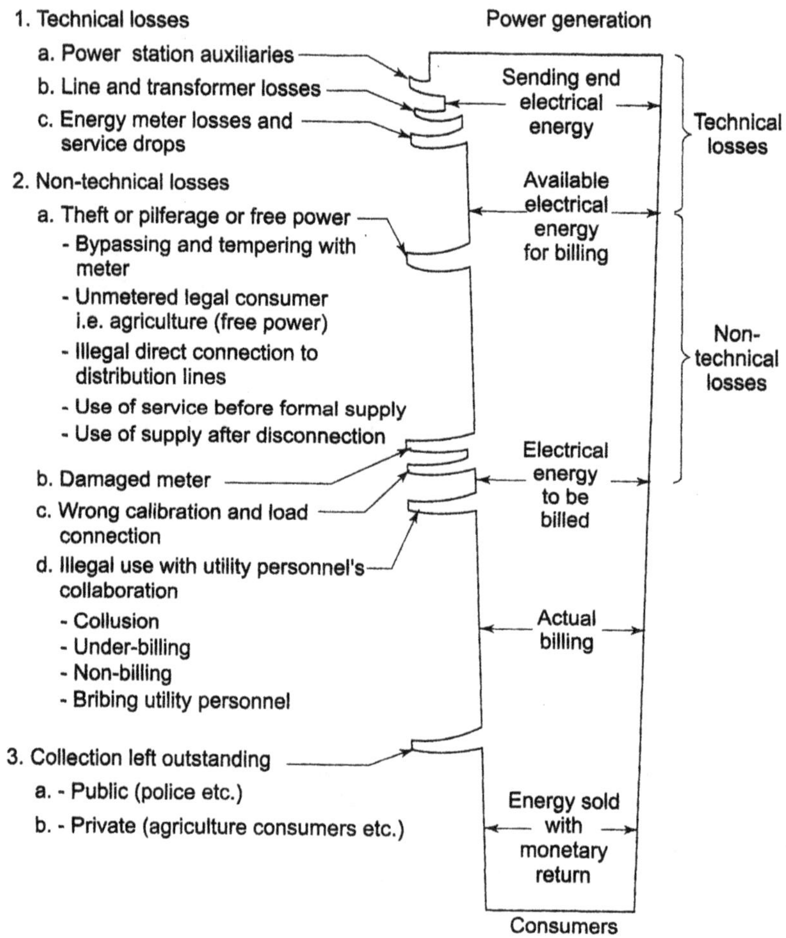

Fig. 4.19 *Sources of losses*

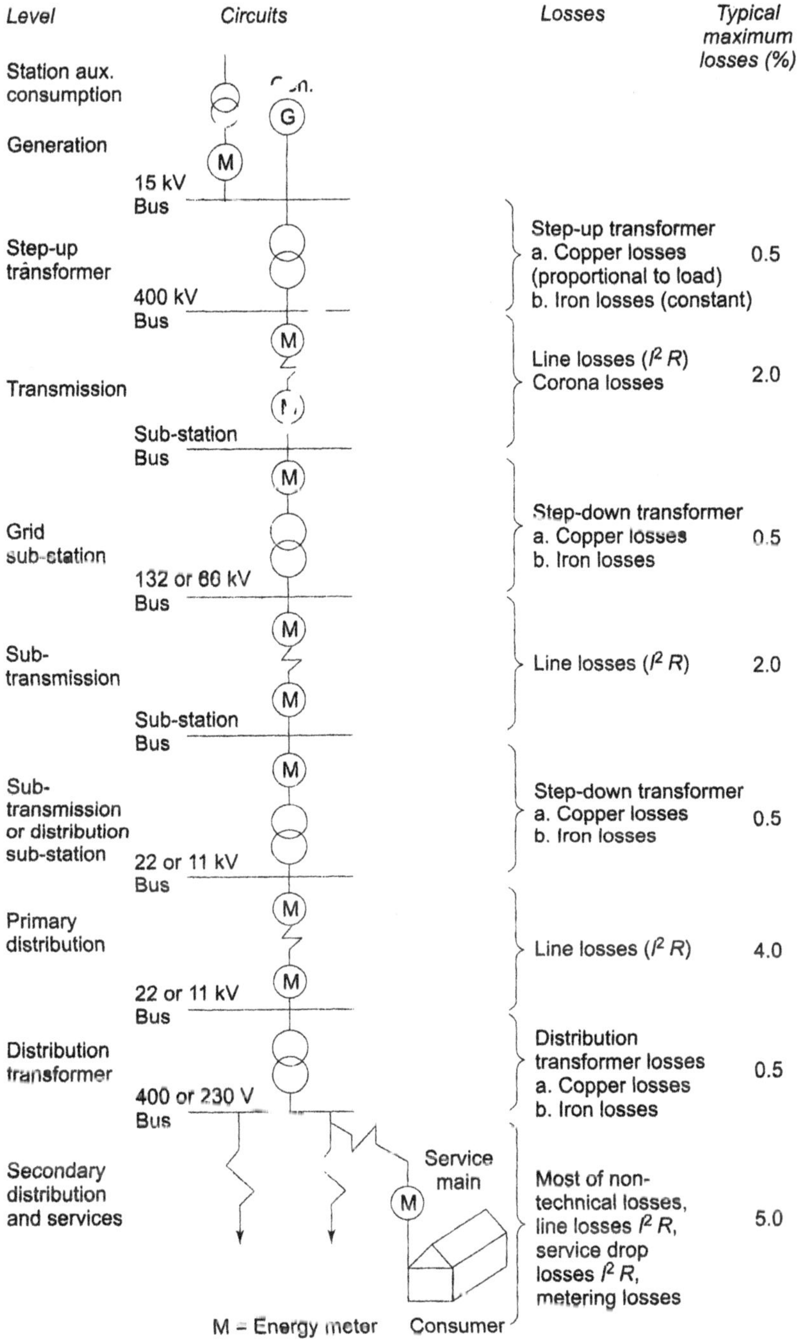

Fig. 4.20 *Breakdown of losses*

4.14 Energy Management

The main purpose of power Energy Management System (EMS) is to generate, transmit and distribute electric energy efficiently. EMS main function is to supervise, control and manage power networks in an integrated manner. The general scheme [14] is shown in Fig. 4.21. The Supervisory Control And Data Acquisition (SCADA) system involves generation and transmission system. The distribution, automation system/distribution SCADA supervise the distribution system for the purpose of:

- Improved overall system efficiency in use of both capital and energy
- Increased reliability of service to essential loads.

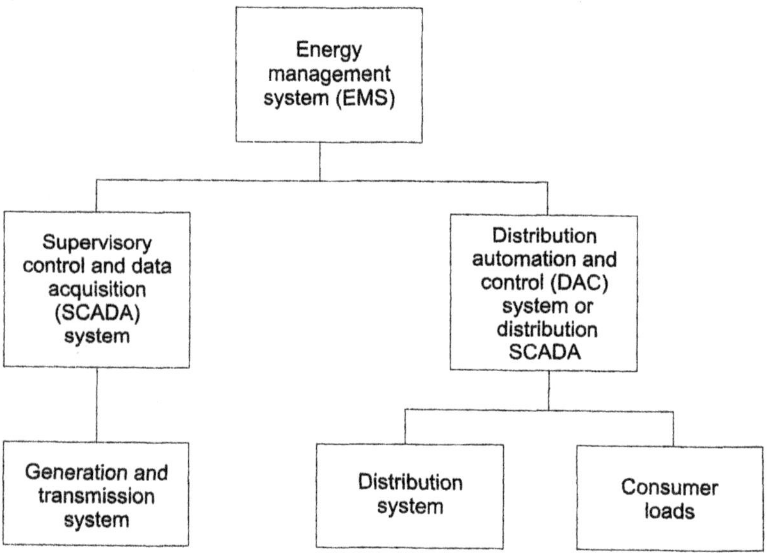

Fig. 4.21 *Energy management system scheme*

According to the Central Electricity Authority sub-committee, 56 number SCADA system load despatch centres are planned in the country. The works for construction of SCADA systems at National (at New Delhi for National Grid), Regional (Northern, Western, Southern, Eastern and Northeastern for regional grids), state and area levels are in progress. The complete National Grid is likely to be in complete working by 2007 AD.

DAC system has been installed in Mumbai, Chennai, Hyderabad and is needed urgently at Kolkata and Delhi. For example, the present (2002) peak demand of Delhi is about 2900 MW and is likely to rise to 4000 MW by 2007 AD. Effective power distribution without DAC or distribution SCADA system will be difficult in such case. DAC operating continuously on 24 h. basis could achieve 5–10% reduction in distribution losses. The main function of DAC or Distribution SCADA are:

4.14.1 Load Management

(i) *Load switching* It involves direct control of loads at individual consumer premises from a remote central station for purpose of reduction of overall system peak in case of interruptible load service contract, or to reduce the load on a particular sub-station or feeder due to load restrictions, etc.
(ii) *Peak load tariff* It is implementation of peak load tariff programme for industrial or bulk supply or grid supply consumers, by remote switching of meter registers automatically for the purpose of time-to-day metering.
(iii) *Load shedding* It permits rapid dropping of large blocks of load according a selected priority or staggering of rural/urban feeders supplies as per pre-programme.

4.14.2 Operational Management

(i) *Voltage regulation* It allows the remote control of selected voltage regulators, network capacitors switching to effect coordinated system-wise voltage control.
(ii) *Transformer load management* It enables the monitoring and continuous reporting of transformer loading data to prevent overloads and damage by timely improvement and replacements respectively.
(iii) *Feeder load management* It is monitoring of load of lines.
(iv) *Capacitor control* It permits selective and remote controlled switching of power capacitors for reactive power planning and loss minimization.

(v) *Fault detection location and isolation* Relays located in the system detect abnormal conditions. This information, in turn, can be used to automatically locate faults, isolate the faulted segment and initiate proper sectionalization, thus affording fast removal of faults.

(vi) *Load studies* It involves the automatic on-line collection of load data for analysis at the sub-station and further transmission to despatch centre for day-to-day planning and operation of power system.

(vii) *State monitoring* It involves real-time data gathering and status reporting to sub-station from which minute-by-minute status of the power system can be determined. Static state estimation is an intrinsic element of an modern EMS [6].

(viii) *Automatic consumer meter reading* It allows the remote reading of consumer meters for total consumption, peak demand, time-of-day metering.

The computer programs relevant to DAC or SCADA are available in various forms, with the various companies such as Electronics Corporation of India Ltd. (ECIL), Alstom, BHEL, CMC, ABB, etc.

4.15 Model Distribution System

An excellent distribution system can deliver electricity at competitive and affordable prices, provide excellent service and adopt appropriate new technologies to serve the consumers more efficiently. Modernisation and expansion of the existing distribution system can be *world class* or *a centre of excellence.* Improvements in the existing system should be models for development and expansion in other areas. Initially, the feeders, transformers and other network elements displaying a poor performance must be identified and selected. Accurate data should be collected and a detailed project report (DPR) prepared for the poor performing network in order to provide the following facilities:

- *Balancing Energy:* Provide improved (electronic) metering at sub-station bus at incoming/out-going feeder points, distribution transformers feeding agriculture pumpsets, selected distribution transformers feeding loads other than pumpsets. Import export type electronic meters measure: energy import; energy export; reactive energy lag; reactive energy lead; apparent energy; instantaneous load MW, MVAr, MVA; power factor; maximum demand; system frequency; load profile; etc. Provide 100 per cent improved (electronic) metering for all types of consumers. Prepare an energy balance-sheet feeder-wise and ensure energy audit at all levels.

- *Revamping:* Carry out renovation of power transformers, switchgear and instruments etc., at the sub-stations, lines-re-configuration, augmentation of line conductors, renovation of joints, jumpers etc., renovation of distribution transformers and LT lines, de-augmentation of distribution transformers, shifting of load from overloaded distribution transformers to lightly loaded distribution transformers.
- *Improving and Consolidating Revenue Collection:* Establish feeder-wise targets of revenue realization/assessment per kWh.
- *Capacitors:* Install proper capacitors (switched and non-switched) at sub-stations, primary lines and at the distribution transformer.
- *Agriculture Loadshedding:* Provide automated loadshedding for agriculture load.
- *Line Ratio:* Reduce HT/LT (11/0.415 kV) line ratio to nearly one. Install more or de-augment the existing distribution transformers. Introduce HV distribution.
- *Collecting, Collating and Calibrating Data:* Collect local accurate data of network elements including the meter at consumer ends and load profiles. Calibrate the data through a sample analysis. Prepare FM/AM/GIS for the distribution system management.
- *Segregating Technical and Commercial Losses:* (a) Compute technical losses with suitable software and co-relate the results with measured total energy loss values and billed energy. The difference between the total losses and technical losses is considered as commercial losses. (b) Explore power pilferage by providing electronic metering for 100 per cent of the consumers and at network input/output points including the distribution transformer in theft-prone areas and compare with the consumer billed energy. Estimate the technical and commercial losses feederwise. (c) Reduce losses through network re-configuration, line re-conductoring (augmentation), installation of capacitors and new distribution transformers, High Voltage Distribution System (HVDS), ABC system etc. Establish base level loss at different voltage levels.

- *Load Forecast and DSM:* Carry out feeder-wise spatial load forecasting for each category of consumer and global load-forecast for the next 5–10 years. Prepare a Demand Side Management Project Report and implement it in selected areas. Consolidate the DSM with load forecast projections.
- *Distribution Management System:* Provide an SCADA system, distribution automation, and computerized billing and collection.
- *Benchmarking:* Carry out benchmarking exercises as a regular feature.
- *Consumer Service:* Ensure consumer complaint call management and provide an Interactive Voice Response (IVR) telephone answering system; also set up a website providing information and a variety of services to consumers (e.g. billing).
- *Cost-Benefit Evaluation:* Benefits of load growth, loss reduction (technical and commercial), improved accurate energy metering (electronic metering), better continuity of supply and better quality of power will result in an increase in revenue and cost cutting. The cost-benefit report on the project should be prepared on completion.
- *Ultimate Success:* The entire scheme is a continuous improvement process and will require good governance, a professional approach, accountability. A world-class power utility is a dynamic organization that is always striving for improvement. Each distribution circle must have a ***planning and design cell*** under the control of the executive engineer.

PROBLEMS

1. A delta-connected three-phase load consisting of a resistance of 30 ohms and inductive reactance of 15 ohms in each phase is connected to a 415 V three-phase supply. Calculate: (i) the equivalent star-connected load; (ii) the line current; and (iii) the total real power consumption in the load.
2. What is the importance of a closed ring main system for a metropolitan city like Delhi? Define the precautions to be taken in designing this system.
3. A 132 kV sub-station has two power transformers each of 25 MVA, 132/11 kV and 10 per cent impedance. A 132 busbar has a fault level of 1250 MVA. One transformer is feeding 20 MVA, 11 kV induction motor having a starting current of five times the full load at zero power factor lagging. The other transformer supplies a domestic load

of 22 MVA at 0.8 power factor lagging. Calculate the dip in voltage on a domestic load 11 kV busbar when the induction motor is started.

4. Explain the design philosophy for the selection of various types of busbar arrangements in sub-stations. Which arrangement will you select for a remote rural area sub-station supplying a 10 MVA load through two 33/11 kV transformers? Give complete justification for your answer.
5. What is the usual X/R ratio of a distribution line and how does this ratio have a bearing on voltage regulation and losses?

BIBLIOGRAPHY

1. IEC 61000-4-15-CEM–part 4: Testing and measurement techniques–Section 15: Flickermeter–Functional and Design Specifications, November, 1997.
2. Pabla, A.S., "Power supply to arc furnaces", *Indian Journal of Power and River Valley Development*, Calcutta, June, 1975.
3. Yannucci, D.A., "Faults Influence Transformer Life", *Electrical World*, McGraw-Hill, New York, July 15, 1978, pp. 50–51.
4. Tobin Noel and Paul Shief "Managing to Reduce Power Transmission System Losses" *Power Technology International Spring*, 1996, pp. 83–86.
5. Coskerie, L.W., "The Prediction and Control of ferroresonant overvoltages in electricity distribution systems", *Electrical Engineering Transactions*, The Institution of Engineers, Australia, 1975.
6. Dy Liacco, T.E., "The Role and Implementation of State Estimation in an Energy Management System, *Electrical Power and Energy Systems*, Vol. 12, April 1990, pp. 75–79.
7. Wagner, C.L. and D.L. Nickel, "Choose correct transformer loss values", *Electrical World*, May 15, 1978, pp. 64–66.
8. Dewberry, Raymond A., "Loss Factor Evaluation", *Transmission and Distribution*, March 1980, p. 70.
9. IEC 61000-3-2: 1995 Limits for Harmonic Current Emission.
10. Lakervi, E. and E.J. Holmes, *Electricity Distribution Network Design*, Peter Peregrinus Ltd. London, 1995, pp. 213–263.
11. "*Computer Aided Distribution Management Utility Practices*", Publication No. 227, CBI&P, New Delhi, July 1992.
12. *Guide for Limiting Voltage Harmonics*, Publication No. 251, Central Board of Irrigation and Power, New Delhi, July 1996, pp. 14–19.
13. Sun, D.I.H. *et.al.*, "Calculation of Energy Losses in a Distribution System", *IEEE Trans.*, PAS-99, July-Aug. 1980, pp. 1347–1356.
14. Gönen Turan, "*Electric Power Distribution System Engineering*", McGraw-Hill Book Company, New York, 1986, pp. 529–533.

15. Pabla, A.S., "Energy Audit in Rural and Urban Distribution System Losses", *Proceedings of National Workshop on Distribution Systems*, New Delhi, 20-21, March 1986, pp 185–195.
16. Padbidri, M.S. and D.G. Pantulu, "Studies on Harmonics Pollution in Power Systems", *Proceedings of Sixth National Conference on Power Systems for the Year 2000 and Beyond*, Mumbai, Tata McGraw-Hill, June 4–7, 1990, p. 35.
17. Arrillaga, J., D.A. Bradley and P.S. Bodger, *Power Systems Harmonics*, John Wiley and Sons, Chichester, 1985, pp. 199–214.
18. Pabla, A.S. "Dirty Loads", *Proceedings of 2nd Afro-Asian Conference of Power Generation, Transmission and Distribution*, Vol. II, Kuala Lampur, September 5–8, 1994.
19. Xu, W., and others "Developing Utility Harmonic Regulations based on IEEE-519-B.C. Hydro's Approach, *IEEE Transac. on Power Delivery*, Vol. 10, No. 3, July 1995, pp. 1423–1431.

Chapter 5

Distribution Automation

The demand on electric power supply in India has changed drastically, both qualitatively and quantitatively. With increasing development, the dependence on electric power supply has increased considerably. While demand has increased, the need for a steady power supply with minimum power interruptions and fast fault restoration has also increased. To meet these demands, automation of the power distribution system needs to be widely adopted. All switches and circuit-breakers involved in controlled networks are equipped with facilities for remote operation (motor drives or actuators). The control interface equipment (RTU) must withstand extreme climatic conditions. Also, control equipment at each location must have a dependable power source. To cope with the complexity of the distribution network, the latest computer, communication and distribution technologies need to be employed. At the same time, a specialized software package developed in-house or out-house by the utilities and other institutions must be usefully selected.

5.1 Distribution Automation (DA)

It is an integrated system concept for the digital automation of distribution sub-station, feeder and user functions. It includes control, monitoring and protection of the distribution system, load management and remote metering of consumer loads. The distribution automation contains:

- Computer Hardware
- Computer Software
- Remote Terminal Units (RTUs)
- Communication Systems
- Consumer Metering Devices.

The benefits of DA are:

- Improved quality of supply
- Improved continuity of supply
- Voltage level stability
- Reduced system losses
- Reduced investment

The distribution automation system provides automatic reclosing of relays, automatic feeder switching and provides remote monitoring and controlling of distribution equipment (transformers, capacitors, breakers, sectionalizers, communication nodes etc.) from sub-station up to and including the consumer interface. It affords the utility in minimizing outage time and ultimately, better consumer service and lowering of the total delivered cost of electricity. It allows operation of the system with less capacity margin. The technical aspects of distribution automation are complex and need a thorough examination for their planning.

DA means something different to each utility, and functionally, there is no industry standard. The various functions can be:

- Electrical network analysis
- Work management
- Trouble call analysis
- Consumer load monitoring
- Intelligent remote metering e.g. automatic meter reading etc.
- Automated capacitor control
- Sub-station automation
- Intelligent electric devices
- Advanced remote terminal units
- Computerized power distribution relays
- Power quality monitoring
- Automated Mapping (AM)/Facilities Management (FM)/Geographical Information System (GIS).
- Energy Management

As work forces are being cut, and consumer expectations and demands are rising, automation becomes essential to system maintenance and improvement of consumer service. Some standards for various distribution automation functions such as data formats, communication protocols etc., are now beginning to emerge. The functions integration is becoming important as shown in Fig. 5.1. The Distribution automation system is also referred to as the Distribution Management System.

5.2 Project Planning

Rigorous project planning and management practices should be applied to complete the work in time and within the budget. The automation schemes should be constructed on a turn-key basis from a capable vendor with maintenance and training contracts for a year or two initially. The project plan should contain:

- A detailed schedule of all project activities and their estimated durations.
- A statement on the methods to be used to complete all the project activities.
- A quality statement which identifies all quality control and quality assurance steps to be applied.
- The structure of database is most critical to the integrated function of Distribution Management System (see Fig. 5.1). The choice of

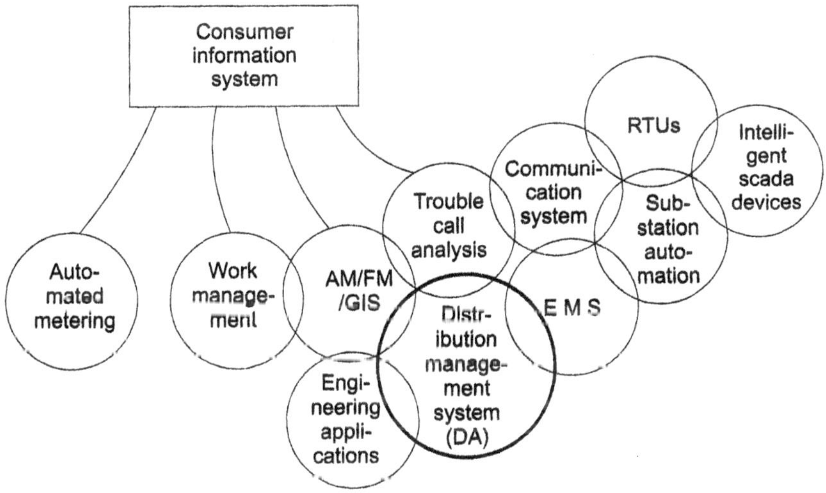

Fig. 5.1 *Typical distribution automation integration*

Data Base Management System is important for any SCADA/ EMS etc. to assure consistency of information over the telecontrol system [18].

* A statement on the organisation requirements and impacts within the utility organisation, to effectively manage the data capture/ conversion process, and to maintain the data in an upto date.

5.2.1 Cost Justification

The cost justification on distribution automation can be made on life cycle evaluation. Most utilities are realizing that automation technologies are both an expense and a strategic investment. Using this rational, they can justify. The spending on distribution automation is expected to increase dramatically in all continents.

Advances in engineering and mapping software, as well as the ability to integrate databases (along with increased computing power and computing speed), have enabled utilities to justify the expenses involved in installing, maintaining and upgrading computer-based automation systems.

In countries with mature power delivery systems, utilities need to maximize the value of existing automation systems while upgrading to more powerful information systems. This need will lead utilities to embrace open architecture, which will allow automation systems to be updated without incurring undue expenses. In developing countries where the existing automation infrastructure is minimal, utilities can build in automation technologies at the same time they build or expand their transmission and distribution systems. Vendors can be selected on the basis of the following points. Indicate the overall level of importance against each point on a scale of 1 to 5, where 1 = least important. The vendor with the highest total can be selected:

- Best life cycle cost
- Best overall performance
- Most value for money (cost-benefit ratio)
- Least risky offer
- Best technical approach
- Best maintenance environment
- Best solution
- Best open architecture system
- Best implementation schedule performance

5.3 Definitions

The following definitions will help in understanding Distribution Automation:

- **Automation Switching Controls:** Some of these are as follows:
  - (a) Outdoor Lighting Controls: Local electro-magnetic relay or electronic and/or thermal/time delay relay with photoelectric controls.
  - (b) Line or Capacitor Switching Controls: Local control by time, temperature, current, voltage and VAr. Remote control by VHF/UHF radio.
  - (c) Line Post Sensors: SCADA and local feeder monitoring for load switching and fault information.
  - (d) Faulted Circuit Indications: Fault location through use of on site LED, fluorescent flag, and remote SCADA indication for single-phase or three-phase overhead, underground or padmounted transformers.
  - (e) Radio Switches: Peak power demand reduction through VHF/UHF radio switching of consumer or feeder load.
- **Ethernet:** A popular network protocol and cabling scheme with a transfer rate of 10 megabits per second, originally developed by Xerox in 1976. Ethernet uses a bus topology capable of connecting up to 1024 PCs and workstations within each main branch. Network nodes are connected by either using thick or thin coaxial cable, or by twisted-pair wiring. Ethernet user Carriers Sense Multiple Access/Collision Detection (CSMA/CD) to prevent network failures or collisions when two devices try to access the network at the same time. The Ethernet standard has evolved into the slightly more complex IEEE 802.3 standard; the two standards are not exactly equivalent.
  - (i) IEEE 802.3 10 Base 2. An implementation of the Ethernet standard on thin Ethernet cable with a data transfer rate of 10 megabits per second, and a maximum cable-segment length of 185 metres.
  - (ii) IEEE 802.3 10 Base T. Establishes a standard for Ethernet over unshielded twisted-pair wiring; the same wiring and RJ 45 connectors used with modern telephone systems. The standard is based on a star topology, with each node connected to a central wiring centre, with a cable-length limitation of 100 metres.

(iii) IEEE 802.3 10 Broad 36. Defines a long-distance Ethernet with a 10 megabit per second data rate, and a maximum cable segment length of 3600 metres.

- **Information Technology (IT):** It includes administrative computing and all 'end-user' computing of a business and technical nature. The term extends to smart or intelligent, programmable electronic devices used in power operations, from the generation of electricity (with computer-based distributed plant control systems) to power distribution automation (including applications of such technologies as automated mapping systems, SCADA, EMS etc.
- **LAN:** A group of computers and associated peripherals connected by a communications channel capable of sharing files and the resources between several users.
- **Management Information Systems:** Most often defined in the power utility industry, as the department responsible for administrative computing systems and operations.
- **Man Machine Interface (MMI):** It is interface between man and technology for control of the technical process. The computer system at Master Control Centre or Central Control Room Integrates with RTU over the communication link with its transmission protocol, acquires the remote sub-station or distribution transformer/Feeder data and transfers the same to the computer system for MMI. Figure 5.2 shows the flow-diagram for Man Machine Interface for the power system. The software is configured in a manner that makes the single-line diagram of the distribution system. The entire system can be monitored and controlled from the screen.
- **Modem:** The word is a contraction of Modulator/DE Modulator. Modulation is the conversion of digital bit streams into analog telemetry suitable for transmission. Demodulation is the reverse of that process. The modem is a device that allows a computer to transmit information over a suitable communication link, such as a telephone line. It translates the digital signals that the computer uses to analog signals suitable for transmission over the communication link. A suitable communication programme is needed to operate the modem. When transmitting, the modem modulates the digital data on to a carrier signal on the communication link. When receiving, the modem performs the reverse

process and demodulates the data from the carrier signal. Modems usually operate at speeds ranging from 1200 to 19200 baud. The modem may be a separate device or may sometimes be integrated with the RTU.

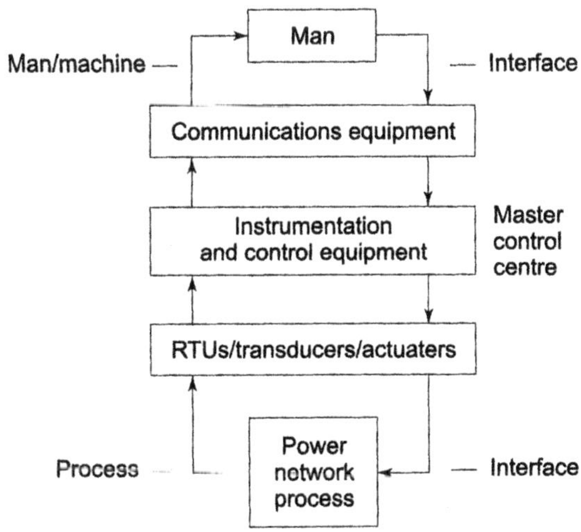

Fig. 5.2 *Flow-diagram for man-machine-power system interface*

- **Packet:** This includes any block of data sent over a network. Each pack contains information about the sender, receiver and error control information, in addition to the actual message. Packet switching is a data transmission method and simultaneously sends data packets from many sources over the same communication channel. Packet-switching networks are considered fast and efficient. Standards for packet switching on networks are documented in the CCITT recommendation X.25.
- **Protocol:** In computer lingo, a protocol is that which allows two computers to understand each other while transferring information between themselves. In networking and communications, it is the specification that defines the procedures to follow when transmitting data. Software products from different vendors can communicate on the same network if they use the same protocol. This is software handshaking.
- **Router:** In networking, an intelligent connecting device that can send packets to the correct LAN segments to take them to their destination.

- **Remote Terminal Units (RTUs):** Modern RTUs are microprocessor based devices and are designed to acquire data and transfer the same to the Master Station through a communication link radio, power line carrier, wire, fibre optic etc. RTUs collect data from transducers, transmitters, contact inputs from equipment/ instruments, meter readings etc., perform analogue/digital conversions, check data-scaling and corrections (typically at I/O card level), perform pre-processing tasks and send/receive messages from/to master station(s) via interfaces. The RTU is usually designed to monitor parameters such as: Bus-line volts, current, active power, reactive power, status of circuit breakers, switches and isolators, fault detection, temperature, level, pressure, flow etc. It can be mounted on the equipment, line etc. It can be a line pole/ tower mounted type, where a water-proof enclosure protects internal circuits such as the switch control circuit and the communication circuit from outside moisture. For example, The RTU for a transformer has the following functions:
  - (i) Data exchange between the communication line to the master station and low voltage line RTU for consumers.
  - (ii) Automatic data polling to RTU for consumers. RTU for the consumer has the following functions:
    - (i) Remote meter reading.
    - (ii) Load control.
    - (iii) Display of information for the consumer.

  A typical installation of the RTU for consumer automation is shown in Fig. 5.3. Various companies in India have developed the RTU for DAS such as Global, Shyam, ABB, GEC, CMC, EMCO, Siemens etc.
- **Workstation:** The role of the Workstation is to serve as an intelligent window to the database. It is a high performance computer. It is used to:
  - Provide a simple user interface for indexing the data.
  - Form the database access commands.
  - Communicate with the server database.
  - Accept the data are returned from the database.
  - Display the data graphically.

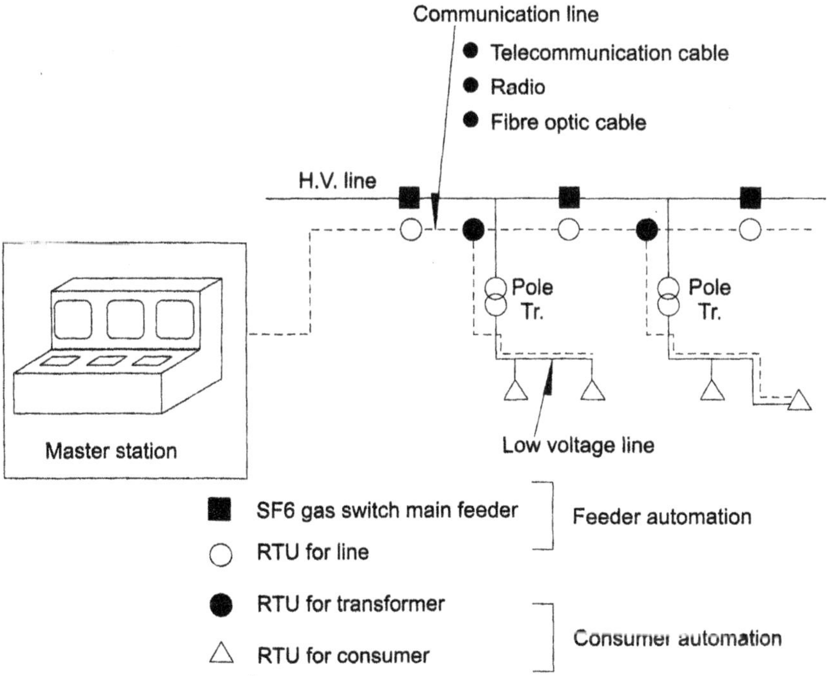

Fig. 5.3 *Typical RTUs installation for consumer automation*

Electric power utilities today, face a problem with processing large and growing volumes of data associated with the transmission and distribution system, and data associated with the substation and generation stations. In both these cases, drawings form the communicating media, since transmission and distribution data tends to be based on geography. The facilities are located on streets and on right-of-ways throughout the service territory. Substations, on the other hand, are described within limited bounds in a three-dimensional system. Engineering work-stations, useful in both these contexts, can be applied differently in each case.

The application of engineering work-stations to the transmission and distribution system is an example of the technology of Automated Mapping (AM), now generally termed Geographic Information System (GIS). The application of engineering work-stations is an example of Computer-Aided Design (CAD). Work-stations applied to the maintenance and operation of generating system belong to systems often referred to as Plant-facilities Management Systems or PFMS. While the tools GIS, PFMS and CAD appear to be similar (powerful multi-user CPUs, graphic displays, plotters, digitizers), the strategy for implementing these systems is different. The database structure and network could work very well for one, but may be totally inappropriate for another. It is important to look at the model of the data before implementation, to decide if the right tool is being used for the right application.

- **WAN:** A network that connects users across large distances and often crosses the geographical boundaries of cities and states.

5.4 Communication

There are many communication methods available. Evaluation of different communication systems for data communication between Distribution Control Centre (DCC) and any point on the distribution network is required at the planning stage. The fundamental requirements for communication infrastructure are:

(i) Determination of system average message rate;
(ii) If it can handle the requisite amount of data and multitasking;
(iii) Data throughput and system response times should meet various application requirements;
(iv) It should allow for network growth and added applications.

The communication methods may be used individually or combined.

5.4.1 Public Switched Telephone Network (PSTN)

Dial-up and dedicated leased telephone lines are often used for Distribution Automation. Telecommunication services through Packet Switching Network (PSN), cellular radio or rural radio have the advantages of providing cheaper and reliable services.

5.4.2 Power-Line Carrier (PLC)

It has an advantage of using the power network owned by the utility. It can reach any point on the network and extend automatically to newly added network elements. It has very limited bandwidth, in most cases less than 100 bits/second. This technology is useful for consumer load management and automatic meter reading of the consumer. PLC technologies are:

- Ripple control very low-frequency signals injected at high-voltage levels.
- Signalling by changing the power waveform on distribution feeders, so that the time of zero crossing in shifted.
- Distribution line-carrier-signals in the range of 23–98 kHz injected into distribution feeders as per IEC 61334-3-22.

(a) *Ripple Control*

Audio frequency ripple control systems are now working at more than 1000 utilities in thirty-one countries of the world. The term "ripple control" is employed to denote a system of remote control applied to electricity supply networks, whereby audio frequency signals are superimposed on the network, without the use of any pilot wires or modulation, but enabling control from a central point of an unlimited number of remotely connected relays.

These installations can send out a considerable number of command signals, each signal being characterized by the time interval which elapses between two impulses. The first impulse is known as the "starting impulse" and the second impulse as the "order impulse". Each receiving relay can respond to only those orders for which it has been tuned and set.

Two methods are currently used to inject audio frequency signals into a network. These are the parallel emission and series emission systems (Fig. 5.4). Depending on the signal frequency chosen and the method of injection adopted, a signal transmitter equipment can serve an extensive area. In the case of overhead networks, a radius of action of 150-200 km measured along the lines (typically 11, 33 or 66 kV) can be attained. Usually the parallel emission method is selected when operating at signal frequencies of 300 to 1500 Hz while at the lower signal frequencies of 150 to 200 Hz, the series emission method is preferred. It is a particular feature of centralized remote control systems that when a choice is once made, it is a matter of considerable expense to alter it. In this respect the series emission method offers greater flexibility.

The microprocessor based ripple control systems afford inexpensive flexibility, accuracy and are cheaper. Tata Power have developed successfully such a system called "remotrol" for remote control of power loads in large HT distribution. This scheme consists of installing a transmitter at the main sub-station and a receiver at each of the distribution transformers fed from this main sub-station. Each of the distribution transformers is provided with a box having one MCCB for LV control of the transformer and one MCCB for each LT feeder and a *remotrol* receiver. A coded message from the transmitter would be sent to desired transformer. This message would in turn be picked up by the corresponding receiver which will switch on or off the MCCB controlling the LT feeder. The Maharashtra State Electricity Board [19] has applied this scheme. It can be applied to LT networks such as urban street lighting control and switching off irrigation pump loads in rural areas. It is useful for improving the system load factor, remote energy metering control for off-peak tariff application.

(b) Cyclocontrol

This is the name given to new equipment and is at present controlling remote loads on distribution network (11, 22 and 33 kV) in a number of countries including United Kingdom, Sweden, Australia and North America. The remote control is achieved by connecting a cyclocontrol transmitter (thyristor wave-point control) to low voltage terminal of standard distribution transformer (also called the injection transformer, say, of size 500 kVA, 11 kV/415 V) to utilize the distribution network as a signalling medium. The transmitter causes a selected number of individual cycles of the main supply waveform to be altered in shape, in the region of zero voltage to form an intelligible transmission code. Signals propagate throughout the high voltage network to which the distribution transformer is connected and may be received on the secondary of all the distribution transformers. Practical tests have shown that signals can be propagated over considerable distances without significant attenuation. Rural HT spurs extending to 22 km at 11 kV and 96 km at 22 and 33 kV have been tested and shown to produce negligible signal attenuation. The denser loading in urban areas has not proved to be a problem. The peak value of the transmitter current depends upon the firing angle for a given system impedance. A firing angle of 25° is generally used, resulting in a transmitter peak of less than

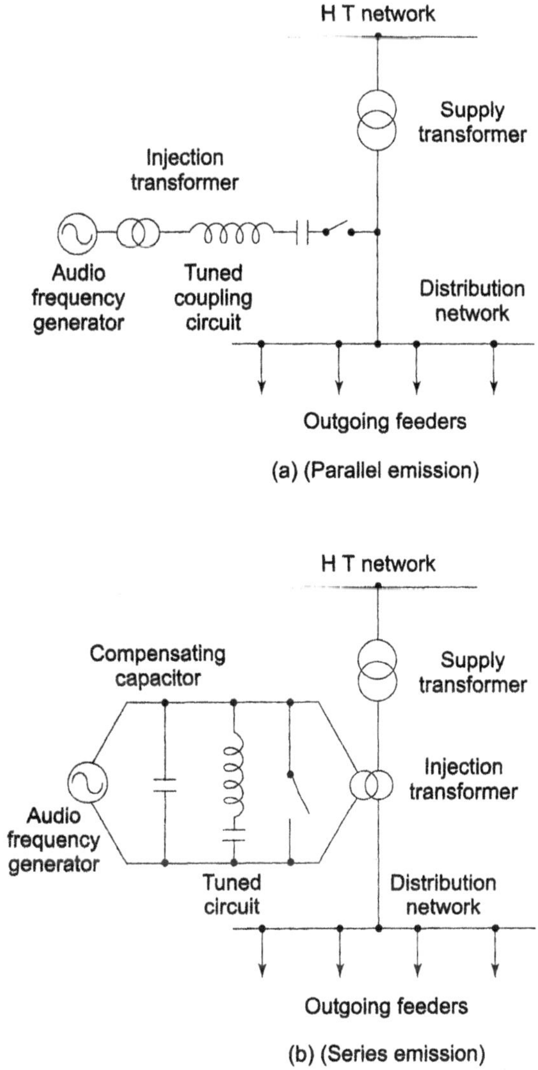

Fig. 5.4 *Ripple control systems*

10 per cent of the peak prospective current. At this angle the disturbance to other loads which are supplied from the same distribution transformer as the transmitter has been claimed to be less than 1 per cent in rms terms. Receivers, which may be connected to any low voltage circuit fed

from the same high voltage network as transmitter injection transformer (standard distribution transformer), detect and decode the signals impressed on the supply. One or two contactors within the receiver enclosure operate independently according to the instruction received for the purpose of controlling a variety of loads.

The comparative simplicity and low cost of this equipment and its installation in comparison with its "ripple" counterpart renders it a better solution to a variety of remote control problems in distribution systems such as street lighting, off-peak load, power-cut control, etc. Tata Power at Mumbai, Karnataka Electricity Board at Bangalore are using this system for street light control.

In cyclocontrol, outgoing signals from the sub-station are generated by drawing an accurately controlled current pulse from the transformer output (see Fig. 5.5).

This current pulse passing through the transformer's reactance distorts the voltage wave within a narrow window around the zero-crossing point. One bit of information is reliably defined by comparing one zero-crossing modulation with the next unmodulated waveform. Thus two cycles are needed to transmit one bit, giving an outgoing baud rate of 30 bits/second (bps).

The communication signal can be transmitted through any winding configuration in any transformer, as is part of the power sine wave. To date, the system has been used to distances of 50 to 80 km on distribution voltages up to 36 kV.

Signals are detected by transponders connected to the circuit anywhere on the feeders supplied by the transformer. Each transponder contains a microprocessor to provide remote intelligence, plus an inbound transmitter to generate inbound signals in response to commands.

One bit on an inbound message is defined by four pulses within a frame of four power cycles. This is done to ensure high reliability, even with the lower power of inbound signals. This gives a basic baud rate of only 15 bits/second. However, because of the large number of possible combinations of positive and negative pulses within four cycles, six completely independent channels per time slot can be piggybacked on a signal of 50 Hz power sine wave. This gives an effective rate of 90 bits/sec. per phase. Since zero-crossing points on the three-phase are separat-

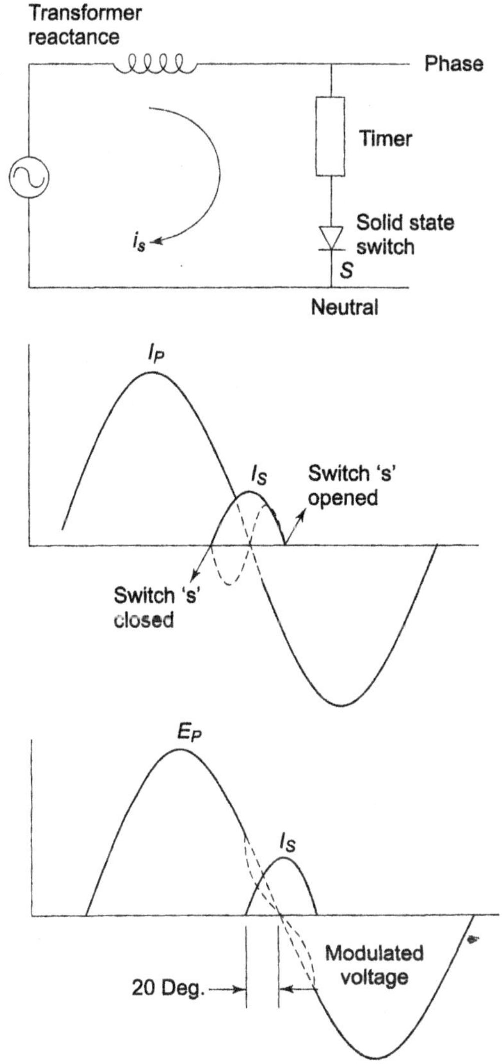

Fig. 5.5 *Cyclocontrol-voltage modulation by signal injection*

ed by 120 degrees, independent signals can be sent in on each phase, giving an effective rate of 270 bits/second.

(c) Carrier Frequency

A PLC channel consists of two PLC terminals at the two stations and connected together by the High Tension (HT) power line and coupling equipment like wave traps, coupling capacitors, coupling devices and High-Frequency (HF) lines as shown in Fig. 5.6. Information is accepted by one PLC terminal at the station A and passed on to its counterpart at the remote station B in the form of HF signals (10–450 kHz). A lower range of carrier frequency is used for distribution lines. For example, a frequency and 3 kHz to 148 kHz has been allocated [12] by CENELEC and IEC for data communication over a low voltage distribution network (PLC) for total metering. The remote PLC terminal delivers the same in original form to local users and passes it on to the adjoining PLC terminal after amplification, for transmission to the next station covers the next line section. Thus, two PLC terminals at different stations communicate with each other. PLC terminals use amplitude modulation, with Single Side Band (SSB) transmission to save the bandwidth. To adjust the voice signal (300 to 3400 Hz), the 4 kHz channel bandwidth is consumed. If F1/F2 are the HF carrier frequencies used by one PLC terminals as transmitting/receiving carrier signals, then F2/F1 are used by the opposite-end PLC terminal for transmitting/receiving the same.

By using Voice Frequency Terminals (VFTs), telegraphy signals are also transmitted over the power lines. Teleprotections signals are available from and delivered to protection couplers and protection relays connected to the PLC terminals at the two stations, meant to protect the power line between these stations against faults. Telemetry signals from electrical transducers and telecontrol signals for opening or closing of HT circuit breakers, are also sent through PLC.

5.4.3 Radio Communication

Radio communication systems, like PLC systems, are usually owned by the utility, but they have the important advantage of their operation being independent of the condition of the power system. Furthermore, many data channels, including high-speed channels, as well as voice communication can be provided. Radio technologies require licensing–generally UHF, and VHF radio is the most widely used medium. In VHF, two frequency slots for use in the power sector in association with other users are: 66–68 MHz and 136–174 MHz. The UHF band that

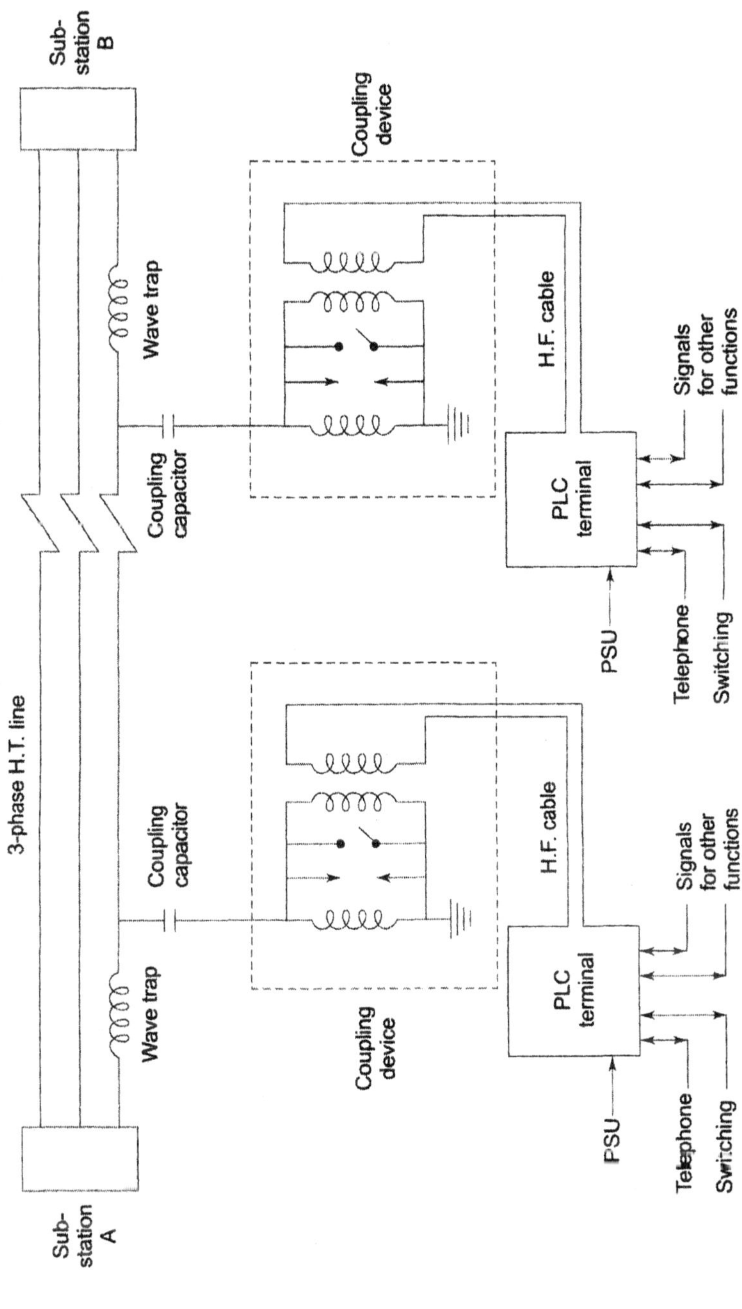

Fig. 5.6 *A schematic diagram for PLC channel*

have been allocated for communication for SCADA and Distribution Automation in power sector are: 314–367 MHz, 585–622 MHz and 915-935 MHz. The disadvantages of radio are the limited channels available from licensing authorities, the relatively low data rate possible (typically 1200 bits/sec.). The radio technologies generally used are:

(a) UHF Point-to-point Radio

UHF radio in the 915–935 MHz range is widely used by utilities for two way point-to-point communications. Systems in the 314–367 and 585–622 MHz range are also available. These systems usually support up to 15 voice and data channels operating at 4800 or 9600 bps. They require line of sight operation, in particular for the 915–935 MHz frequency bands. With flat terrain and antennas mounted at a reasonable height, reliable operation over a range of 30 to 40 km in practical. Dish antennas are used for point-to-point radios.

(b) UHF Multi Address System Radio

Multi address systems (MAS) were developed as a substitute for dedicated telephone lines, and are widely used for SCADA/EMS/distribution automation application. They consist of master radio station which communicates with several remote radios with IRTU (RTU interfaced with distribution equipment). Each system uses a pair of frequencies (314 MHz-935 MHz)–one for master-to-remote communications and the other for remote to master, so that duplex communication is possible. Coverage can be increased by providing remote more sub-master stations. An omnidirectional rod antenna is used at the master radio. This rod antenna can be mounted on a hill-top or on building roofs to increase coverage. At the remote radios, directional antennas provide better performance, but less expensive, less conspicuous and vulnerable "stick" antennas are also used. Typical MAS system has been shown in Fig. 5.7. This technology is useful for load management.

(c) VHF Radio

One-way VHF radio (typically, the 154 MHz band widely used in the USA) is generally used to send load control commands. Inexpensive radio receivers/switches are available for giving load control switching commands.

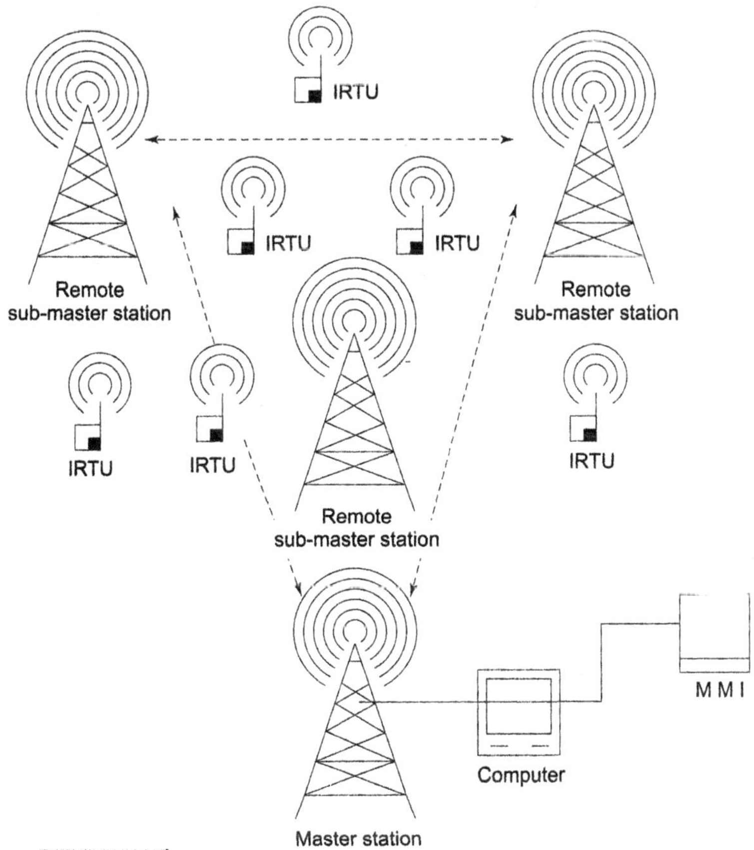

Fig. 5.7 *Typical UHF multi-address radio system (MAS)*

(d) Packet Switching Network (PSN)

A packet switching network uses low power UHF spread spectrum transmission (typical 935 MHz band) and does not require licensing of frequencies. It is a unique solution to frequency management using 'spread spectrum technology'. Due to its very low power density spectrum, it minimises interference to existing services. The packet switching send short burst of digital information over low power radios [14, 20]. These are data transmission networks with computerized nodes that perform specific communication tasks such as routing. Many power utilities in the world are using packet switching networks for many purposes, ranging from real-time SCADA, EMS to line and capacitor switching, AMR, administrative data; and from inter-utility to intrautility communications. The typical PSN system is shown in Fig. 5.8. for WAN and LAN through PLC. PSN technology is convenient for use in inter-utility communication, line switching in energy management systems and in data transmission between control centres. A number of utilities are using CCITT's X.25 protocol in these applications. X.25 is the ideal method to harmonize the different hosts.

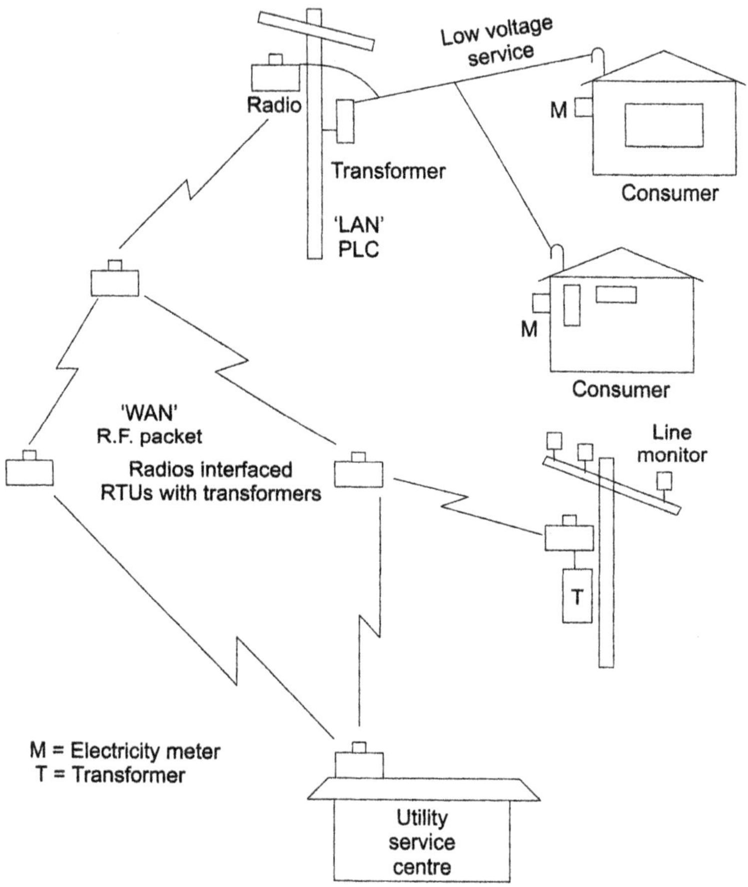

Fig. 5.8 *Typical packet radio UHF radio for WAN and power line carrier (PLC)*

(e) Cellular Radio

This system is widely used for AMR, DSM and DA. Its benefits include wide-area broadcast and monitoring capability, and independence from communication utilities. This technology offers very high-speed (9600 baud) real-time, bidirectional data gathering systems control, and should be ideal for field operations. The typical cellular radio architecture is shown in Fig. 5.9. The system has two levels of communication. A digital

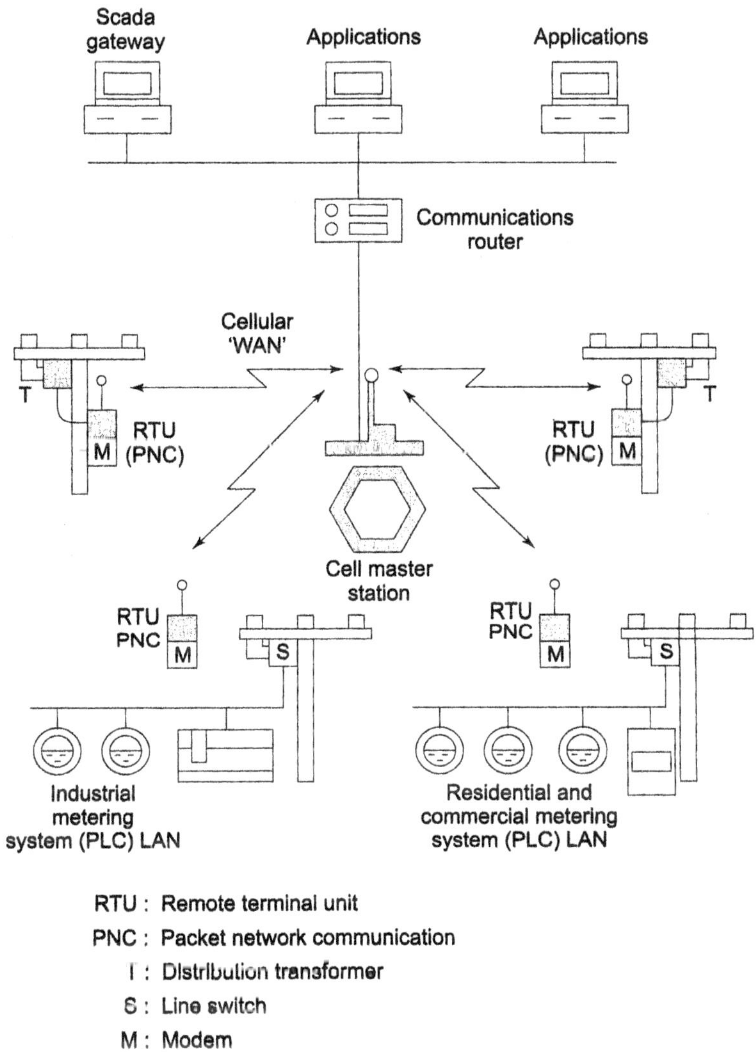

Fig. 5.9 *Typical architecture of the cellular radio HF (915 MHz) WAN*

cellular for wide area network (WAN) at 915 MHz range and a low power spread spectrum RF as well as PLC channel for Local Area Network (LAN).

5.4.4 Fibre Optics

Fibre optics, with its explicit downward cost trend in terms of product as well as installation and maintenance costs, has become a widely accepted choice, as it offers both technical and commercial advantages over conventional systems that use metallic cables and radio links. The communication basically consists of a transmitter, and a receiver for information signals coming from the user's device, which is connected through copper wires to the switching centre or exchange, where it is changed into a digital signal like 1s or 0s for easy handling. The signal is then transferred to the transmitter. In the transmitter, the information signal which is electrical, drives an optical source: Laser or a Light Emitting Diode (LED), which in turn, optically modulates the information signal, which get coupled into the optical fibre. The receiver located at the other side of the link detects the original signal and demodulates or converts back the optical signal to the original information signal (electrical). The signal is then connected through copper wires to the switching device or exchange for selection and connection to the proper user or user device. This type of communication is important for trunk routes power transmission lines. In such a situation, the problem of electrical interference is not encountered. This has the advantage of high data rate (9600 bits and much more) and immunity from noise, broad bandwidth and low attenuation.

Optical fibres are available in cable form. Such cables can be laid in cable trenches. There is a trend to associate optical fibres with power cables or conductors themselves. Optical fibres are embedded within the cables/conductors, either earth or phase conductor. Optical fibre communication is expected to play an important role in future protection signalling, either as a dedicated protective relaying link, or as channels within a multiplexed system which carry all communication traffic such as voice, telecontrol, telemetering and protection signalling. Typical fibre optic communication is shown in Fig. 5.10.

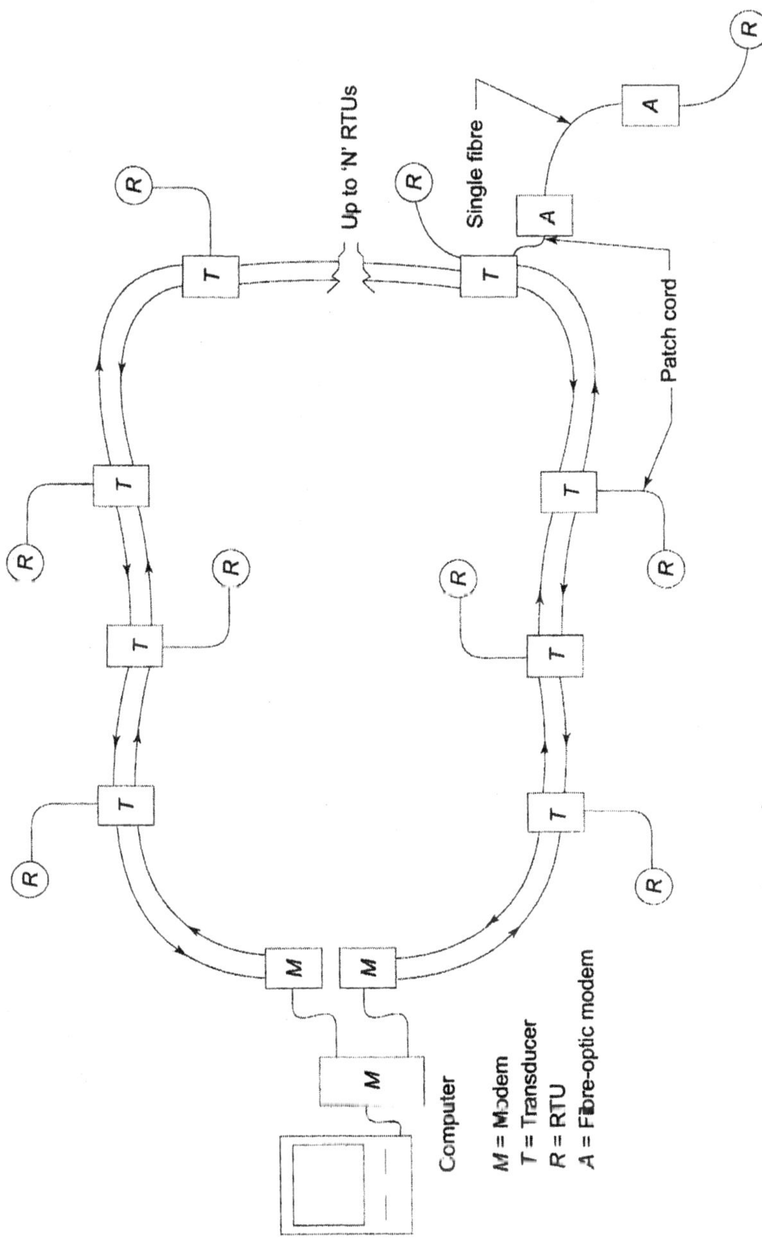

Fig. 5.10 *Typical fibre-optic communication system for underground distribution system for metropolitan area*

5.4.5 Satellite Communication

This technology is useful for load management controls. Satellite-based VSAT (Very Small Aperture Terminal) network communication is used for power systems. VSATs are only allowed to access Indian satellites. It has a 99.5 per cent reliability factor. A communication signal is sent from a VSAT to the satellite at a particular frequency (see Fig. 5.11). The signal is amplified, shifted to a different frequency and sent back to the VSAT at the destination. A VSAT network is distance insensitive. One has to pay only for the cost of VSATs and service charges which are very reasonable. The whole network can be monitored and controlled centrally. If there is any problem in the network, it can be diagnosed from the hub. Primarily, there are three types of technologies for VSAT networks. The 'Star' technology is approved by the Government of India. In this configuration, at the centre of the network is a big earth station known as the hub with a dish diameter of 6–11 metres. The control of the network is at the hub and it maintains a communication link between VSATs scattered all over the country or region. The configuration is therefore called the Star.

The other technology is the 'Mesh' technology. The configuration permits a group of VSATs to communicate directly with other VSATs going through the central hub. In this type of configuration, any one of the VSATs functions as a hub and controls the whole network. This configuration enables better voice and video communication than the Star configuration. This technology has also been introduced in India. In the third type of configuration, there is point to point connection between the VSATs. In such a network, data, voice and video signals are carried directly between the locations.

In satellite communication, an extended C-band (3.9–6.2 GHz) is used for communicating the data. HCL Commet, VSNL, Wipro BT, Hughes, Shyam, Telstra, Bharti BT, HFCL, AT&T, MAX, RPG satellite and communications have emerged as the large VSAT service operators in the country. VSATs are readily available in the market.

5.5 Sensors

An increasing number of electrical systems are being designed to provide monitoring and control from a central station within the installation. For this reason, products that include information and

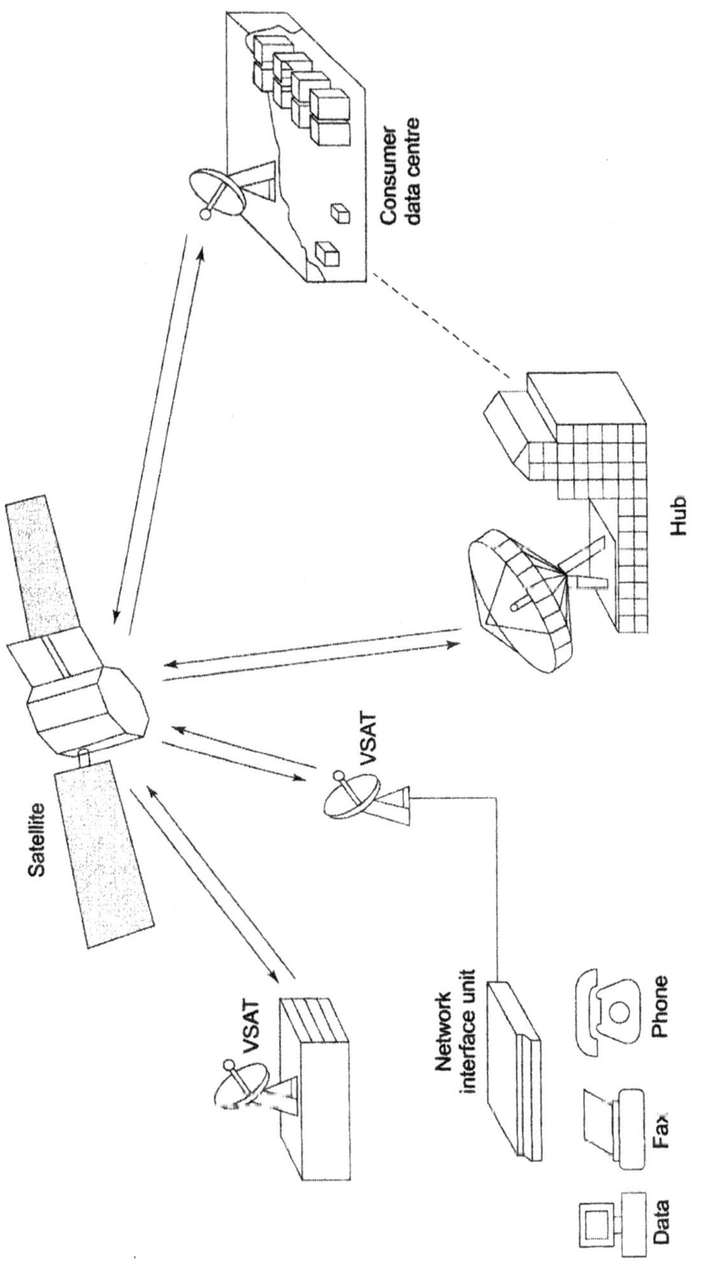

Fig. 5.11 *Schematic diagram of VSAT network*

communication features for use at sub-stations, consumer metering devices, switchgear, transformers, reclosers, digital fault recorders, sectionalizers, communication nodes etc. must be available. Real time data is acquired through transducers, transmitters and Remote Terminal Units installed at sub-stations, lines etc. [16].

Line sensors can be provided at line poles or towers which acquire data, such as phase currents and phase voltages and can also detect short circuit and grounding faults. By use of this data, the automation system can isolate the faulted section and monitor the real time data. Besides relays, there are transmitters/transducers, which gather data from basic sensors for telemetry. For example basic sensors may be current transformer, voltage transformer relay, level gauge, pressure gauge, flow metre, bimetallic temp. element etc.

Transmitter

This provides output (transmittable) signals after converting and amplifying low level signals of basic sensor elements. Modern transmitters are smart and microprocessor based, performing many other functions such as alarm, signal etc.

Transducer

This is a measuring element that senses the external action. It gathers parameters and supplies through remote telecommunication capabilities. Conversion components are often classified as *sensors* or as *transducers.* The difference is often blurred, but in essence, a sensor converts from one form of energy to another with no regard to efficiency and is generally used for measurement purposes. A transducer is used where the efficiency of transfer is more important, as in control systems. Transducers measure non-electrical quantities and convert them into electrical quantities. The commonly used transducers are photo-cell, hall-effect fluxmeter, microphone, thermocouple, thermistor, Resistance Temperature Detector (RTD), potentiometer, piezo-electric, piezo-resistive, capacitance, electronic bridge-circuit, strain-gauge and programmable power transducer (for real-time feeder telemetry data) etc. [5].

Receiver

In two way communication, the Receiver is used as terminal equipment along with the transmitter.

5.6 Supervisory Control and Data Acquisition (SCADA)

Distribution SCADA (see Sec. 4.13) supervises the distribution system.

The application areas of SCADA can essentially be categorized into the following groups:

- Small SCADA systems with a selected number of functions, e.g. for distribution networks and for electrical networks of industrial complexes.
- Medium-size SCADA and EMS with the full spectrum of functions for distribution and subtransmission networks, and selected functions for generation.
- Large-scale SCADA and EMS with an extensive and sophisticated range of functions for transmission networks and generation.

The various components of SCADA are shown in Fig. 5.12.

When interfacing and RTU to existing equipment, the following aspects may be considered.

- Availability of potential free contacts and motor drives or actuators.
- Wiring and marshalling.
- Interposing relays may eventually be required.
- Measurement transducers.
- Buffered power supply.

When building a new sub-station, the integration of the RTU functions into a microprocessor based station control system is desirable, since multiple cabling between the signal sources and the different systems for protection, local control and remote control can be avoided.

5.6.1 Hardware and Software

A typical SCADA system comprises the hardware and the software.

Hardware

The hardware may consist of:

- User-friendly man-machine interface
- Work-station
- Servers having a particular function
- Communication sub-system
- Peripherals
- RTUs.

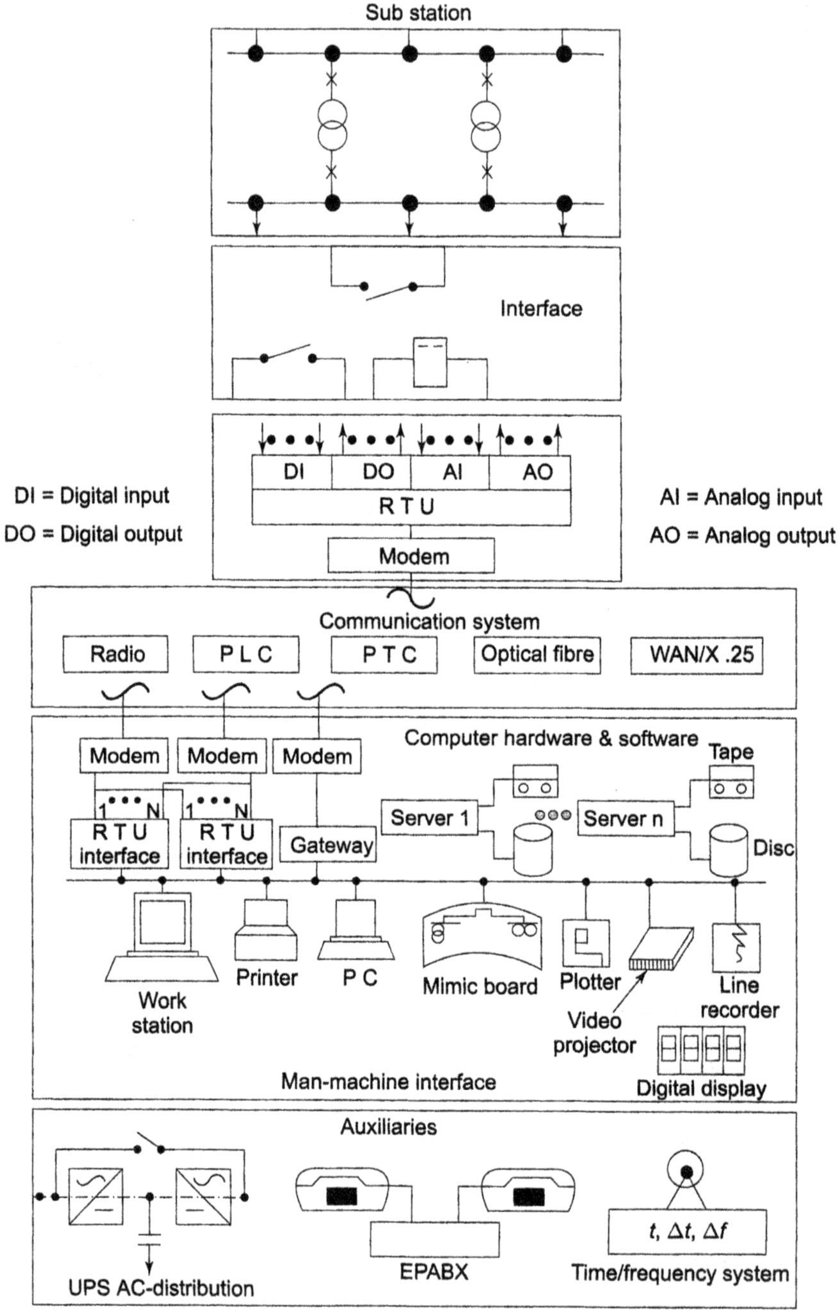

Fig. 5.12 *Typical SCADA system components*

All the above components communicate with each other via a Local Area Network (LAN) with internationally standardised protocols.

A flexible redundancy is provided, assigning hot stand-by servers to any server fulfilling time critical functions.

Software (5, 15)

A thorough understanding of available packages, operating systems, database access standards, user interface standards and systems integration techniques is essential to provide solutions to meet specific requirements. Also vital are high calibre staff and established policies, procedures and techniques for software engineering, project management and quality assurance.

Generally, experienced inhouse groups or external services are capable of maintaining and enhancing application programmes for distribution automation such as for SCADA, AM/FM/GIS, CIS etc. Procurement of new programmes and major upgrades are often assisted by specialized consultants. There are national and international suppliers' markets for a broad range of softwares. An international standard, SPICE, enables the purchaser to assess the relative capabilities of software suppliers and the risk involved in selecting them. Artificial Intelligence (AI) methodology is different from usual programming and normally, special skills are needed for applying AI tools or shells and setting-up and maintaining the knowledge base. This situation will be improved once the application of AI techniques becomes common and the skills of the maintenace teams are developed.

5.6.2 System Architecture

The modern trend for any automation system is open system architecture. The open system architecture can be defined as, an architecture in which equipment of different manufacturers may be functionally integrated in a manner that makes the particular characteristics of individual devices transparent to the system.

Among the number of hardware and software standards, the ones having the most significant effect on system openness have computer standards, which are gradually converging towards the framework set by the Open System Interconnection (OSI) model of International Standard Organisation. The application of well accepted standards enables the system's incremental growth and incorporation of products and devices of different vendors. This application is a prerequisite to a long-lived system and market driven rather than vendor-driven system evolution.

The other is distributed architecture. It is inherently flexible from the point of view of system extensions and upgrading. However, the system openness is not guaranteed by the distributed architecture itself, but rather by the use of industry and published standards throughout the entire design and procurement process. Hierarchical SCADA/EMS systems where communication between despatch centre (master) and regional centres (sub-masters) is obtained via Wide Area Network (WAN), while local communication between functional processing modes is obtained via Local Area Network (LAN). Figure 5.13 shows a distributed configuration, a regional centre is responsible for a subset of the entire RTU population and performs SCADA functions associated with that group only. However, during a regional centre failure, the despatch centre can assume the direct polling of RTUs associated with the failed regional centre. It is important to note that modern RTU architecture is distributed as well, with I/O processing accomplished through distributed nodes or processors located throughout the substation or power system or power plant and connected by LAN.

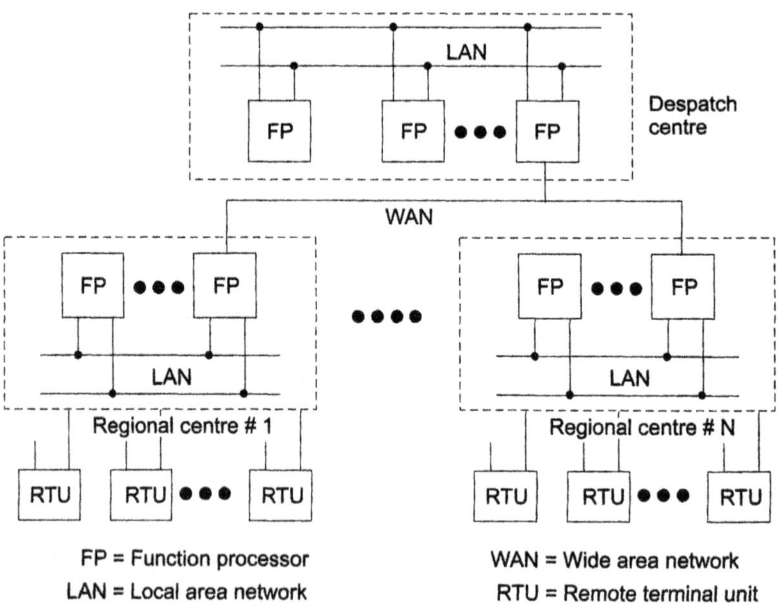

Fig. 5.13 *Distributed configuration of hierarchical (master/sub-master) SCADA/EMS system*

5.7 Consumer Information Service (CIS)

In India, consumer service is mostly done in the manner of 'fire fighting', i.e., it is complaint based. Consumers want fast, accurate and cost-effective service. CIS package services the following functions for better service to the consumer:

- Consumer information
- Account management
- Service orders
- Field service
- New business
- Meter reading
- Service rates
- Billing
- Accounts receivable
- Credit, deposits and budgets
- Collections
- Workflow management
- Website management

CIS application programmes are normally utility specific and should be developed inhouse or the standard package may require more of customization.

Account management maintains information describing service accounts and consumer accounts. CIS also maintains relationships between service accounts, consumer accounts and the consumers. Field service manages the scheduling, assignments and execution of field service orders. The field work includes new works, utility service and metre orders, complaint attendance orders, distribution service orders and investigation orders. The field service orders support the scheduling and execution of consumer requests for service, identification of available work dates for order scheduling, assignment of service personnel, dispatching of workteams and tracking the progress of each service order. Workflow management routes, analyzes, and distributes work in the system to appropriate functions, individuals and process. Work is broadly defined, encompassing orders, correspondence, information to be reviewed, revenue collection of defaulting amount cases, billing exceptions and various other activities. An extensive on-line help system can be available within CIS for each feature. With special software, the telephone technology is installed to facilitate the helpdesk operations.

5.8 Geographical Information System (GIS)

GIS stores distribution system records for the entire network, giving details of the age and position of sub-stations, lines and cables etc. Users can always know where their assets are and how they relate to each other.

Topological maps can be used for grid control i.e. load analysis, contingencies analysis; laying of cables, attending consumer complaints etc. The task of "digitising" power utility maps, and then to "vectorise" the digitised information so that roads, wires and other plant are recognised. Image processing and knowledge-based system techniques are used for automating this process.

Integrating geographical information systems into business support systems provides further benefits for an organisation. For example, emergency repairs would be carried out more efficiently if control room operators could send maps showing the location of equipment to maintenance staff.

The output of such analysis is always in bitmapping form. The GIS software package developed by the National Informatics Centre (NIC), is called GISNIC. The Geological Survey of India has a GIS mapping software which is capable of very high-level bit-mapping. GIS softwares have been developed by Infotech Enterprises, ISRO, RM Software India. GIS technology finds excellent use in facility mapping of power supply network.

5.9 Automatic Meter Reading (AMR)

AMR is the remote collection of consumption data from the consumer's power utility meter over a telephone line, radio system or power line carrier. AMR helps to:

- Improve billing accuracy
- Eliminate the need for estimating meter reading
- Shorten the time from consumption to billing and payment
- Improve load management
- Check tamper and leakage before it becomes a problem
- Increase the return on investment on a meter
- Provide remote connect and disconnect of the power connection
- Provide flexible tariff.

AMR comprises three elements:

- Automatic reading of meters
- A communication link between the meters and the central billing system
- A centralized information database on consumption.

Each AMR meter is provided with a transmitter/receiver module. It can read meters via a public switched telephone (PSTN), switched cellular, GPRS (Gross Packet Radio Service), GSM (Global Service for Mobile Communications) cellular connections etc. All these services are widely available in the country. About 10 per cent of AMR devices installed worldwide communicate by telephone and nearly 5 per cent by PLC. Radio is used in the remaining 85 per cent [17]. Radio is the least expensive technology for AMR. Radio AMR meter modules in the US generally operate at 900 MHz. In the UK, 184 MHz has been allocated for meter reading. In continental Europe, 433 MHz is the norm.

In the US, the radio band is shared with many users and spread spectrum modulation technique is encouraged. This system has been adopted in India. Certain environmental factors may require special considerations depending on where the meter is located. There may not be a clear path for the radio receiver in handheld computers, vehicles or pole-mounted units. Repeaters can be used to supplement such installations and are often needed in apartments, high-rise buildings or densely-populated city areas. Singapore Power uses a telephone line or GSM mobile telephone interface with RTU that reads the meter (records and stores pulses generated by the electricity meters) each day [11].

AMR software can have added value functions such as supervision of meters, operational conditions, detection of fraud attempts, consumer load profiles as per the standard: IEC 6206-31—data exchange for meter reading, tariff and load control.

5.10 Automation Systems

The required function of distribution automation system varies with each application. The typical distribution automation systems composed of master station and the distribution equipment for feeder and consumer automation system is as shown in Fig. 5.14. The detail of equipment is shown in Table 5.1.

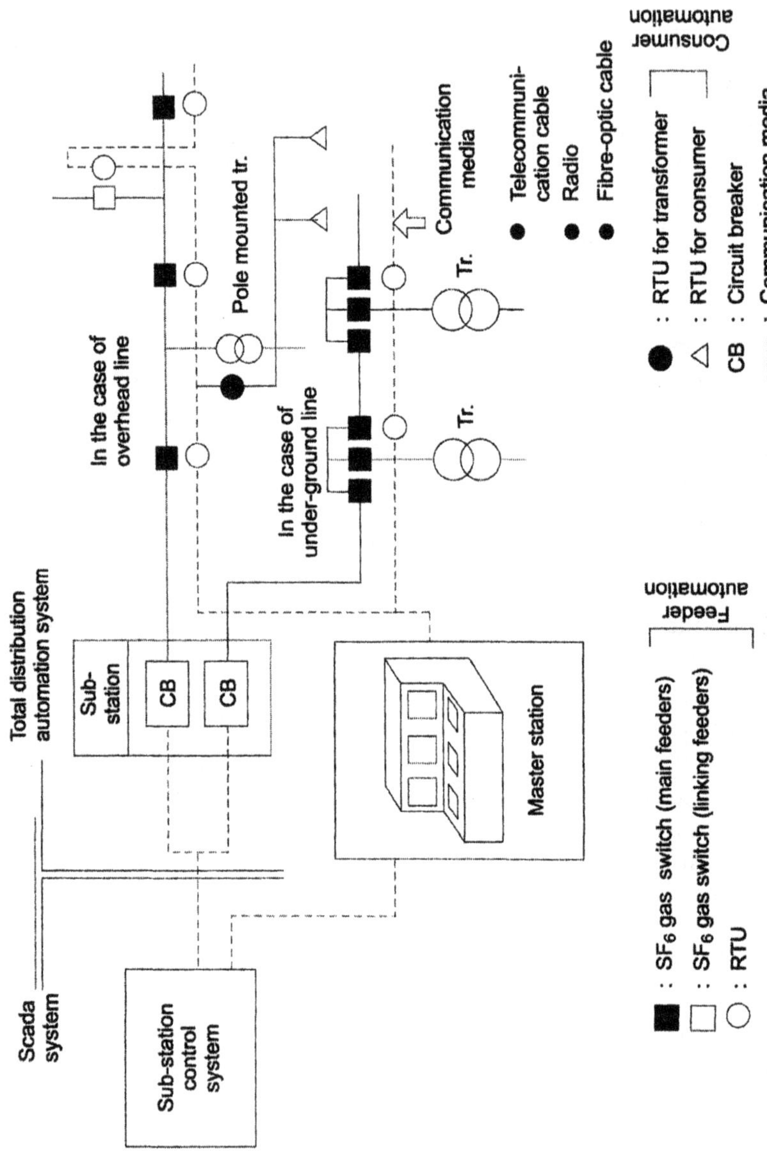

Fig. 5.14 *Typical distribution automation system*

Table 5.1

Typical equipment of distribution automation system

Master station	*Distribution equipment*	
	Feeder automation	*Consumer automation*
Work-station	Remote terminal unit (RTU)	RTU for transformer
Communication equipment	Remote-controlled SF_6 gas switch	RTU for consumer
Console or MMI	Line sensor	Communication media
	Communication media	

Typically, the following three types of communication can be linked with the master station and RTU for feeder automation or RTU for transformer and for consumer automation:

- Telecommunication
- Radio communication
- Fibre optic communication

The features of the above mentioned communications are given in Table 5.2.

The network for the Master Station is shown in Fig. 5.15.

Table 5.2

Features of communication methods

Feature	*Telecommunication (twisted pair cable or co-axial cable)*	*Radio Communication*	*Fibre optic Communication*
Suitable application	Semi-urban area	Rural area	Urban area
Data rate	600–1200 bps	300–1200 bps	1M bps
Interruption time by fault	Short approx. 1 min	Rather long, approx. 10 min	Very short approx. 1. sec.
Cost	Medium	Low	Relatively high

5.10.1 Remote Control Load Management for Agriculture Load (13)

In India, about 14 million electric run agriculture pumpsets are in service at present (2002). Pumpsets are controllable loads since 6–8 hours of operation is sufficient to meet crop irrigation requirements. It is desirable to defer pumping during peak load hours, and to spread pumping loads by dividing the pumpsets into groups controlled at different times. Load management can then be used to defer building of new generating and transmission capacity.

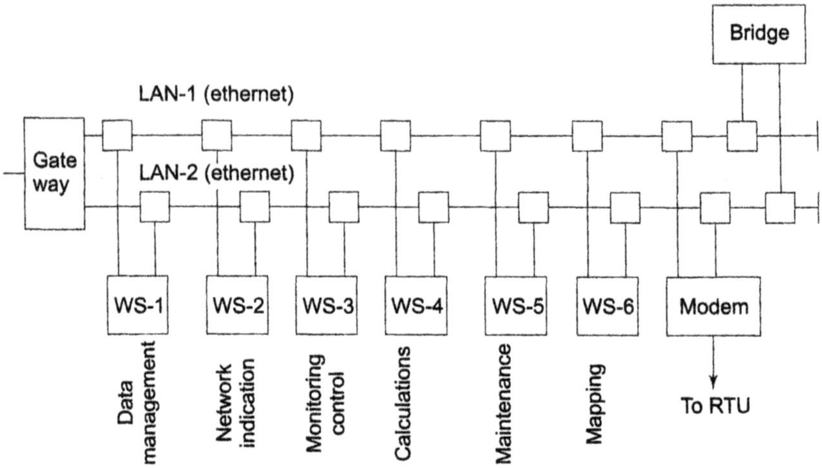

Fig. 5.15 *The typical network for master station: WS = work-station*

There are two options for remote controlling of switching distribution transformers exclusively feeding agriculture load:

(i) One-way load control at distribution transformers

One way VHF radio-operated pole top mounted radio switch interfaced with MCCBs can be installed on the low-voltage 230/400 V (LV) side of the distribution transformers. MCCBs are capable of breaking load and closing into on the maximum fault current of the transformer. This alternative requires a minimum number of low-cost commercially available, one-way radio switches.

(ii) Two-way load-control at distribution transformers

An MCCB is installed on the LV side of transformer. However, it is operated by a pole-top Remote Terminal Unit (RTU) with Multi Address System (MAS) radio communications (VHF or UHF) to and from the

DAS master station. Besides controlling the MCCB switching, the RTU will monitor the status of the switch and currents, voltages, power factor and capacitor control at the transformer.

- This alternative requires a minimal number of two-way MAS remote radios.
- Tampering with load-control equipment can be detected.
- Monitoring of pumping loads can assist in detecting power theft.
- General monitoring of the transformers is possible; switch status, currents, voltage and power factor.

This load management will ensure round the clock supply to all the consumers other than agriculture. Remote distribution transformers can be switched on or off by sending signals from master station to the pole-mounted advanced Remote Terminal Units (RTU) interface with MCCB.

The master station polls all RTUs, one by one, sequentially. Polled RTUs send their status of MCCBs and other values back to the master station. Every RTU has its own unique address code (field programmable on which it responds with the system capability say up to 999s RTU at the speed of communication of 30 bps). The typical scheme is shown in Fig. 5.16. The RTU is integrated with the distribution transformer. The feeder's transformers (see Fig. 4.9) are divided into two groups for staggering the load by switching on/off.

The Rural Electrification Corporation has approved a scheme as per REC Specification: 71/1993 for remotely controlling the agricultural load. It provides following features:

(i) Fail-safe transformers (see Section 6.2.3)

(ii) The on/off status of 11 kV switch, low oil level and the top oil temperature are brought out through transformer RTU to the master station. This scheme has been introduced in Punjab, Harayana and Rajasthan on a pilot basis. In Punjab, two 11 kV feeders having 144 distribution transformers (11/.433 kV) and feeding agricultural load in 21 villages, are controlled within a radius of 30 km from the master station at 33 kV sub-station Attowal, Distt. Hoshiarpur. The radio frequencies used are 849.05 MHz and 933.05 MHz respectively for transmitting and receiving signals. In Punjab, the scheme has been installed on new fail-safe silicon-steel core distribution transformers. In Haryana, the scheme was installed on a new 11 kV feeder having 105 new fail-safe amorphous core distribution transformers and also on the existing 105 transformers duly modified to a fail-safe position at another 11 kV feeder. The advantages are:

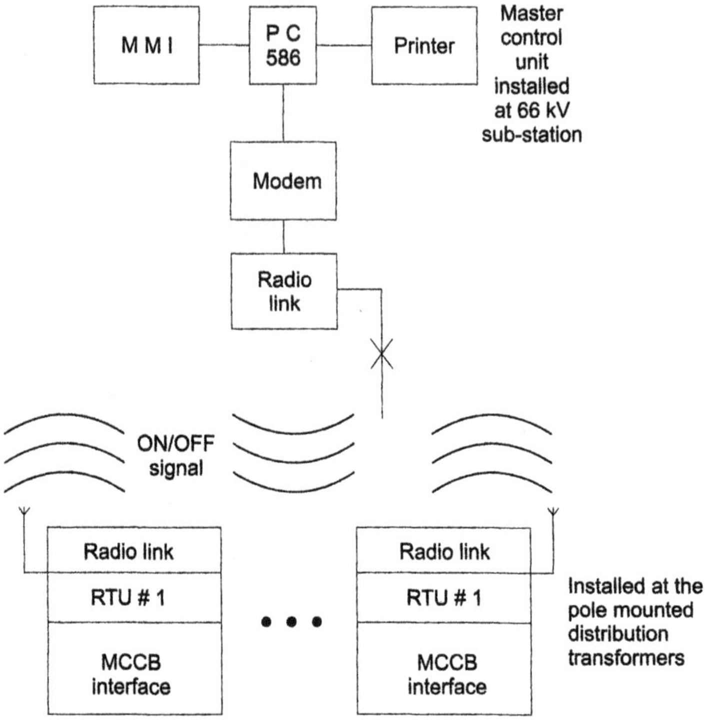

Fig. 5.16 *Typical automation for agriculture load control (VHF 136-174 MHz)*

- The agricultural supply can be staggered in two or three groups.
- Feeder overloading is curtailed (Demand-side management) by automatic load staggering.
- Transformer overloading does not occur due to split-phase supply manipulation.
- Due to online transformer overload protection, transformer damage is reduced.
- There will be substantial saving in system losses.
- This scheme is the best alternative for providing 24-hour supply to villages for lighting, industry etc. and for providing staggered supply for agriculture. In this case, the supply fed by separate transformers is maintained continuously.
- Improved voltage at the tail end.
- Peak shaving saves generation capacity.

The scheme in the three states has been constructed on a turnkey basis by Computer Maintenance Corporation (CMC) with a two-year maintenance contract.

EXAMPLE **Remote Control of Distribution Transformers in Punjab**

No. of distribution transformers exclusively for agriculture	: 90,000
Agriculture tubewells	: 0.73 million
Restricted yearly peak demand for agriculture (Two groups staggered)	: 2400 MW
Annual agriculture consumption	: 8,000 GWh
Approximate consumption per group	: 4,000 GWh
By switching off the transformers feeding agricultural load, a minimum of about 2% load losses and energy wastage are saved i.e. annual energy saving	: 80 GWh
About 5% saving in reactive power	: 80 × 5 = 400 million kVArh.
Annual energy saving @ Rs. 2 kWh	: 160 million rupees
Average cost per transformer for the management facility	: Rs. 40,000
Total cost	: 90,000 × 40,000 Rs. 360 million
Ignoring saving in reactive power, the pay back period is	: 2 years 3 months

PROBLEMS

1. Design the air conditioners that switch off automatically at the consumer premises in the town for shaving peak load.
2. What is telemetry and which type of data is telemetered to the Distribution Control Centre for proper functioning?
3. What is the difference between a 'sensor' and a 'transducer'? How do the thermo-couple, photocell and microphone convert signals into electrical quantities?
4. What is packet switching and what are its advantages?
5. Would you use a sensor or a transducer for the following:
   - Metering
   - Protection
   - Switching

BIBLIOGRAPHY

1. Proceedings of 'International Seminar on Modern Trends in Power Distribution Systems: Central Board of Irrigation and Power, New Delhi, January, 1993.
2. Shore Alan, 'How Intelligent Is Your System', *Electricity International*, vol. 3, no. 10, Dec./Jan. 1992, pp. 24–26.
3. Pollom Brian 'AMR—The Key to Enhanced Operation' Transmission and Distribution, September 1999, pp. 26–32.
4. Computer Aided Distribution Management Utility Practices, *Publication No. 227*, Central Board of Irrigation and Power, Malcha Marg, Chanakyapuri, New Delhi, July, 1992.
5. *Directory of Computer Software for Power Systems*, Central Board of Irrigation and Power, New Delhi, April, 1988.
6. Stevenson Alex, 'Which Communication Technology is Best for Automation?' *Transmission and Distribution*, June 1993, pp. 44–53.
7. Peri Reuven and M.V. Krishna Rao, 'Load Control for India', *Transmission and Distribution International*, Fourth Quarter 1994, pp. 26–30.
8. *Information Technology: A Review of Electric Utility Objectives, Plans and Goals*, Newton-Evans Research Company, January, 1996.
9. Bush Rick, 'T & D Spending Patterns', *Transmission Distribution*, January, 1996, p. 61.
10. Sehgal, B.K., *Computer-Aided Electric Power Control*, Power and Process Technologies, Incorporated Publication Division, New Delhi, 1989.
11. *Asian Electricity*, March 1999 pp. 22–24.
12. Newbury, John, 'Communications Field Trials for Total Utility Metering', *IEEE Transac. on Power Delivery*, vol. II, no. 2., April, 1996, p. 690.
13. Pabla, A.S., 'Future Technologies For Rural Electrification in India', *Proceedings of International R & D Conference*, CBI & P, New Delhi, 9–12 pp. 339–351.
14. Gelbeln Lawrence, 'Distribution Automation Increases Reliability', Transmission and Distribution World, October 1996, pp. 38–41.
15. *1997 U.S. Electric Utility Industry Software Directory*, PennWell Publication, U.S.A.
16. Barney George C. '*Intelligent Instrumentation*', Prentice-Hall of India, New Delhi, 1988, pp. 10, 118, 125.
17. Geiger Richard G, "Reading Meters by Radio" *Metering International*, Issue 2, 1998, p. 16.
18. Asian Power, September 2000, pp. 65–66.
19. Khare P.N., "Load Management by Judicious Load-shedding", *Electrical India*, 28 February, 1999, pp. 12–15.

20. Marihart, Donald J., "Communication Technology Guidelines for EMS/SCADA Systems", IEEE Transactions on Power, vol. 16, no. 2, April 2001, pp. 181–188.
21. Newbury John ad William Miller, "Multiprotocol Routing for Automatic Remote Meter Reading Using Power Line Carrier Systems", IEEE Transactions on Power Delivery, vol. 16, no. 1, January 2001, pp. 1–5.

Chapter 6

Optimization

6.1 Introduction

For any particular situation, there is a site, area and capacity best suited for an equipment that meets all criteria and real world conditions. The impact of this approach on the total system cost is minimum. This is '*optimization*' and is a challenge for distribution engineers. The solution has to reflect the right balance between the practical issues to addressed and the mathematical approaches that might be followed [1, 12].

Particular areas of interest in optimization are: (i) cost modelling of average networks, (ii) cost modelling of idealized networks, (iii) studies of different network configurations and (iv) optimization of a cable or feeder layout to suit a given pattern of the sub-station (v) siting and sizing of transformers.

An important factor to realize in network optimization studies is that the minimum cost attainable will depend on the configuration of the network to be used. This, in turn, is dictated by the minimum acceptable reliability for the network and the standard of the supply to be adopted etc. Clearly, if one system is to be more reliable than another, then its optimal cost is likely to be higher than that of a less reliable system. Similarly, if a higher maximum voltage drop on the network is to be tolerated, a cheaper overall system can be designed. Other factors which determine the minimum cost obtainable will be the number of voltage transformations in the system (if this is fixed by other considerations) and on circuit limitations which may exist in congested urban areas. These considerations have resulted in different design philosophies being adopted for high voltage transmission versus sub-transmission and distribution and urban networks versus rural networks. For example, in the latter case, a cheaper, inherently less reliable system will often be adopted because the cheapest system obtainable would still be too expensive if designed along the same lines as a network for a well-developed urban area.

6.2 Costing of Schemes

When we speak of network optimization we really mean the minimization of the overall network cost. Therefore, the method of costing, alternative network schemes is of great importance in optimization. The costing of network schemes is a complex subject and a number of recognized methods are available for this purpose. The method adopted should incorporate recognized financial procedures. Interest rate, discount, etc. will be affected by external conditions which will often vary the result over different periods. Usually the cost of establishing various items which make up an electric distribution system can be broken into two parts: capital (or installation) cost and annual (or maintenance) cost. The annual cost will cover such charges as interest, maintenance cost and costs of losses. As the losses in a power plant do not change with age and the maintenance charges are usually a constant percentage of the capital costs, the operating costs of a given network component will be related to the size of the component rather than to its age.

The capital cost may be either treated as a lump sum cost at the time of commissioning or be converted into an annual charge in perpetuity. This would cover the charges provided for the depreciation of the plant as well as interest for financing the loan which covered the purchase of the plant. The annual charges which represent the interest, depreciation and maintenance cost, will be constant each year during the life of an asset, provided that its 'book' life is not exceeded. For most items of plant, this will normally be of about 15–40 years.

A typical accounting procedure which is used when comparing schemes involving different expenditures over periods of time, is the Net Present Value technique (NPV) (see Section 3.5). In this method sums of money which are to be spent at sometime in the future, are converted into equivalent lump sums at the commencement of the study by calculating their present value.

6.2.1 Least Cost Analysis

For any investment to be least cost, it is the lifetime costs that are referred to. These include capital cost and operating and maintenance (O&M) costs. All these costs should be reconciled at the appropriate discount rate. That is, life-cycle costs are based on 'Net Present Value' (NPV) of all expenses: Capital, maintenance and operation over the life (see Section 3.5). Capital costs will be the cost of equipment purchased. O&M costs are generally taken as a certain percentage of capital cost.

It should be possible to draw up a schedule of these costs for several feasible alternatives over time. Alternatives could include different conductor sizes, the choice between installing one large transformer initially or on a phased basis, geographically extending a sub-station to give longer or shorter LT lines etc. All alternatives however, should be in line with a clearly stated objective and consistent design standards.

Once the lifetime costs of meeting a given objective have been determined, they can be compared on a present value basis. The present value is determined by discounting the lifetime costs (see Fig. 6.1).

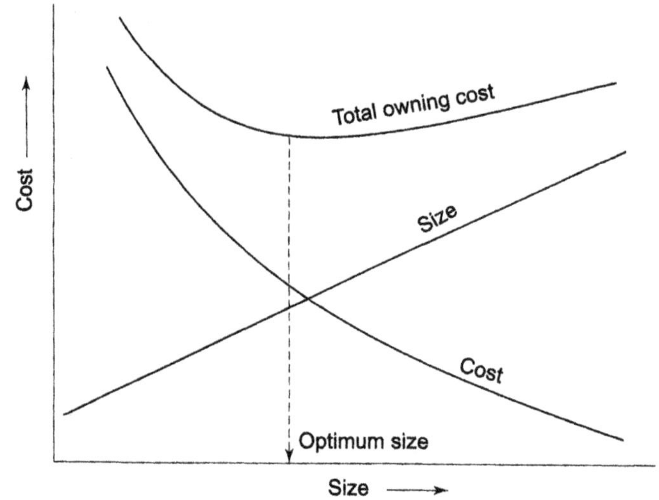

Fig. 6.1 *Optimum size of equipment (cables, conductor, transformers etc.)*

A least cost solution is a necessary condition for considering investments, but not a sufficient one. It is important to understand, that no consideration needs to be given at this stage to any resources saved.

For example, consider the case of reconductoring a line. The options to look at are keeping the existing line, or putting in a larger diameter conductor. The cost of the old line is simply the value of its power and energy losses. The new line will have capital cost associated with lower power and energy losses. These are the two time streams that are compared. It is irrelevant at this stage to think about what will happen to the saved power energy, and whether it will be used for supplying demand or saving capacity.

Equipment such as transformers, circuit breakers, computers etc. should be procured on the basis of lowest life cycle, which includes initial purchase cost and present value of running (O&M) costs over the span of life. For example, in case of distribution transformer, there are two kinds of costs—purchase price and cost of energy lost in losses during the active life of the transformer. The Total Owning Cost (TOC) of a transformer is the sum of the purchase price and net present value of energy losses. If the purchase is made on the basis of vendor rating (see IS : 12040), failure rate costs including repairs should be taken into the TOC account.

6.2.2 Technology Development Trends Considerations

Consideration should also be given to introducing new technologies into the system, e.g. insulated overhead conductors, aluminium alloy conductors, new types of circuit breakers, use of microprocessors devices in protection, local automation, network telecontrol and so on. Various electrical components should be considered bearing in mind their capabilities and also their energy losses in terms of cost, at the appropriate investment stage. One example is the introduction of low-loss distribution transformers where it may not be possible to justify replacement solely on saving in losses, although saving in losses may be a factor in determining any replacement of damaged or old transformers. Similarly, the use of the XLPE cable, with allowable higher conductor temperature, permits the use of a smaller cross-section area conductor than with the other cables. However, this reduced cross-section area results in an increased resistance and therefore higher I^2R losses. It is therefore necessary to take account of the capitalised losses as well as the initial investment cost when comparing the economy of XLPE cables against other cables.

6.2.3 State-of-art Technology

The optimum choice for adoption of the latest technology in a given condition, should be made on the basis of lowest life cycle costs or cost benefit analysis. The available technologies are:

(i) Transformers

A transformer is the work-horse of the power system. Foil or sheet-wound transformers have better capability to withstand short circuits, better thermal characteristics and better voltage distribution. They are at present produced up to 36 kV, 8000 kVA. The low voltage winding of this transformer is made with aluminium or copper foils/sheets of large width, as compared to rectangular or circular conductors used in conventional transformers. The improved performance results from the uniform stress distribution (mechanical, electrical and thermal) made possible by use of foil conductors.

Dry type transformer of resin-impregnated and/or resinencapsulated design, are in vogue up to 36 kV; 10 MVA in applications where fire hazard is a possibility and there is need for non-inflammable equipments. In addition, dry type transformers are mechanically more sturdy and occupy less floor space. (see IS: 2026 and REC Standard: 30/1983.)

SF_6 gas insulated transformers which are being made up to 66 kV, 30 MVA have major advantages as they have increased reliability, compact size, extended service life and they neither cause nor support fire.

The gas is inert with high dielectric strength and thermal stability. It is also inactive and harmless and not flammable. Any gas leakage is harmless. SF_6 is a non-toxic, inactive, so-called inert gas, which due to its physical properties is very well suited for use in transformers. Its density is five times that of air, so that filling of transformer tanks is simply done by slowly letting the gas flow in from below and thus displace air upwards. PD discharge, dielectric strength, and the cooling properties of SF_6 gas make these transformers superior in working performance. These transformers are in use in many highly urbanised cities such as Singapore, Hong Kong, London, New York, etc. SF_6 transformers are maintenance-free. The gas leakage rate is maximum 1 per cent per year. Thus an occasional check of the pressure guage is sufficient to monitor the operational condition of transformer.

The need to have complete reliability and continuity of service, mainly in distribution networks, has given rise to the development of completely fail-safe transformers which have an in-built circuit breaker, fuse and lightning arrester assembled along with the transformer.

Fail-safe Transformers (FST)

Fail-safe Transformers (FST) have the benefit of the lowest life cycle cost and more reliable service, useful for distribution automation. The Rural Electrification Corporation has standardised (REC standard 56/1993) this transformer with ratings: 11/0.433 kV, 16, 25, 63, 100 kVA. The protective devices in the unit are provided at the manufacturing stage, such as high voltage surge arrester, automatic reclosing low voltage circuit-breaker, HV fuse link, low oil level trip and lockout if the oil level drops to the minimum permissible limit. The indicating light shows 'Low Oil' which is automatically reset after the oil is topped.

The circuit breaker is controlled by an intelligent electronic control unit. It has inverse load-time characteristics and trips when the temperature of the top oil reaches 85°C. There is an indicating light to show that the maximum oil temperature has been reached and the light remains 'ON' till manually re-setting is done. The breaker closes automatically when the top oil temperature drops to 80°C. The breaker re-closes once after three minutes and there is a lock-out within five seconds if the circuit breaker re-trips. Another indicating light shows 'line fault' which gets automatically reset when the circuit breaker is successfully closed manually. Otherwise it will again indicate a 'line fault'. The total installed cost of a *Fail-safe Transformer* is about 10–15 per cent lower than the conventional transformer installation.

The distribution transformer provides the transformation from 11 or 22 kV to lower voltages down to the level of the final distribution network, three-phase, 415/240 or 433/250 V and single-phase 11/0.230 kV. These transformers tend to operate at poor load power factors. This means that the magnitude of the load loss is not too important. Iron losses are present all the time and it is therefore desirable to minimise its impact. Step-lapped joint construction for core arrangement allows the lower corner loss [4]. The Rural Electrification Corporation has standardized amorphous core distribution transformers as per REC Standard: 70/1993. No-load losses of an amorphous transformer is about $1/4^{th}$ of the CRGO core transformers. Madhya Pradesh, Haryana and Punjab have installed a few thousands of such transformers. The performance report of these transformers indicates a definite decrease in losses and reduction in the failure rate. Amorphous core transformers are being introduced all over the world to curtail no-load losses in the transformer.

(ii) Switchgears

Development has been in areas of arc quenching insulation media, contact materials and operating speed of mechanisms. SF_6 circuit breakers have today become the watchdog of power system protection and are used up to the highest voltage class. They have faster and efficient interruption characteristics, are maintenance-free and are more reliable. Advancements are being attempted to decrease the number of electrical breaks (from two to one) and still faster interruption by arc-management. For medium voltage applications, vacuum circuit breakers have been developed up to 36 kV voltage class. Vacuum contactors are popularly used for LT, 6.6 kV and 11 kV motor duty applications. Vacuum breakers and contactors have got very fast operating speeds, so fast that they sometimes generate undesirable, very high frequency surges in the system.

Gas Insulated Sub-stations (GIS) are more reliable, occupy less space—which is a critical criterion in metropolitan cities, do not require frequent maintenance and at 132 kV and above, are reported to be cost competitive with conventional sub-stations.

(iii) Protection

It determines the safety and stability level of the power system. Considerable developments have taken place in improving the speed of operation of relays, quick identification and isolation for maintenance. Digital relays operate fast and clean, allow flexibility in fixing trip settings, sequence of tripping and coordination with a large number of allied control relays. In addition, they occupy less space and afford computerized operations.

(iv) Zinc oxide lightning arresters

They have improved performance characteristics and are more consistent in operation. They have brought down the BIL of the system by at least one step, leading to a better safety margin.

(v) Condition monitoring

The need and desire to monitor the health of an equipment while in service, has forced the development of on-line monitoring systems. It is possible to monitore a variety of parameters such as analysis of dissolved gas in transformer oil, infra-red thermography, PD discharge measurement etc. on a continual basis, which affords cost effective maintenance of the equipment.

6.3 Typical Network Configurations

As was mentioned in the introduction, different types of network configurations are adopted for different networks—transmission, sub-transmission and distribution—reflecting the amount of power and security requirement of the networks. For example, a major transmission circuit carrying large blocks of power on which many consumers depend will be regarded more important than the low voltage distribution network in the street since the circuit failure of the former will inconvenience more consumers. As a result, a mesh network configuration is usually adopted for these networks. These networks offer a greater standby capacity than would normally be used for a low voltage distribution feeder to a domestic residence. In addition to this reliability aspect, the large numbers of consumers, and consequently a large number of supply points in the low voltage networks would mean that to provide a complete duplicate supply to each ultimate consumer would involve a prohibitive expense. On lengthy rural feeders, reliability can often be improved by introducing sectionalizing points in the form of auto-reclosers or cut-outs with air switches, which will disconnect one section of the line and prevent the whole feeder being disconnected from the supply.

Both mesh and radial configurations are used in sub-transmission networks. The actual configuration used depends on the relative importance of the circuit which, in turn, will normally be dictated by the load and/or number of consumers serviced. Some typical network configuration [6] are shown in Fig. 6.2.

Typical switching arrangements [6] at sub-stations are shown in Fig. 6.3 The simplest arrangement as shown in (a) to (c) is used for the distribution networks and the more complicated arrangements of (d), (e), (f) and (g) for sub-transmission networks. For comparative advantages and disadvantages of these switching schemes, any standard book [9], may be referred to.

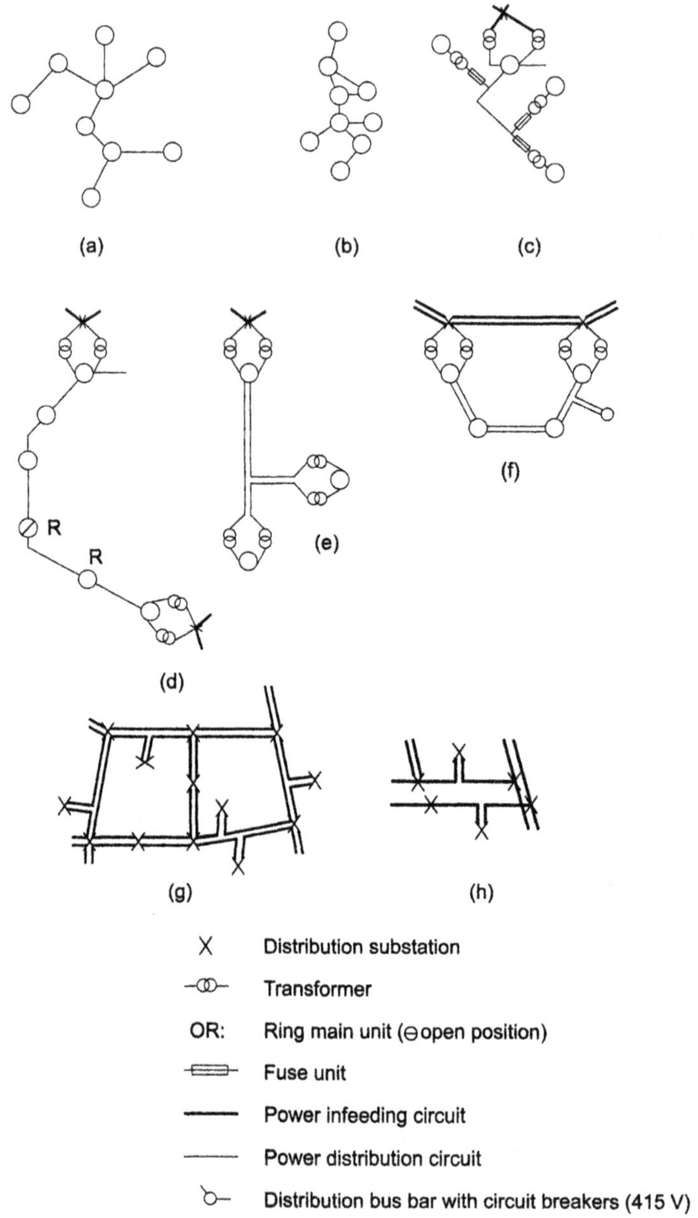

Fig. 6.2 *Typical networks:*
(a) Radial; (b) Meshed; (c) Radial supply; (d) Radial supply with switching back-up; (e) Transformer tee-off feeder; (f) Interconnection at lower voltages; (g) Meshed transmission; (h) Overlay (super voltage integrating network)

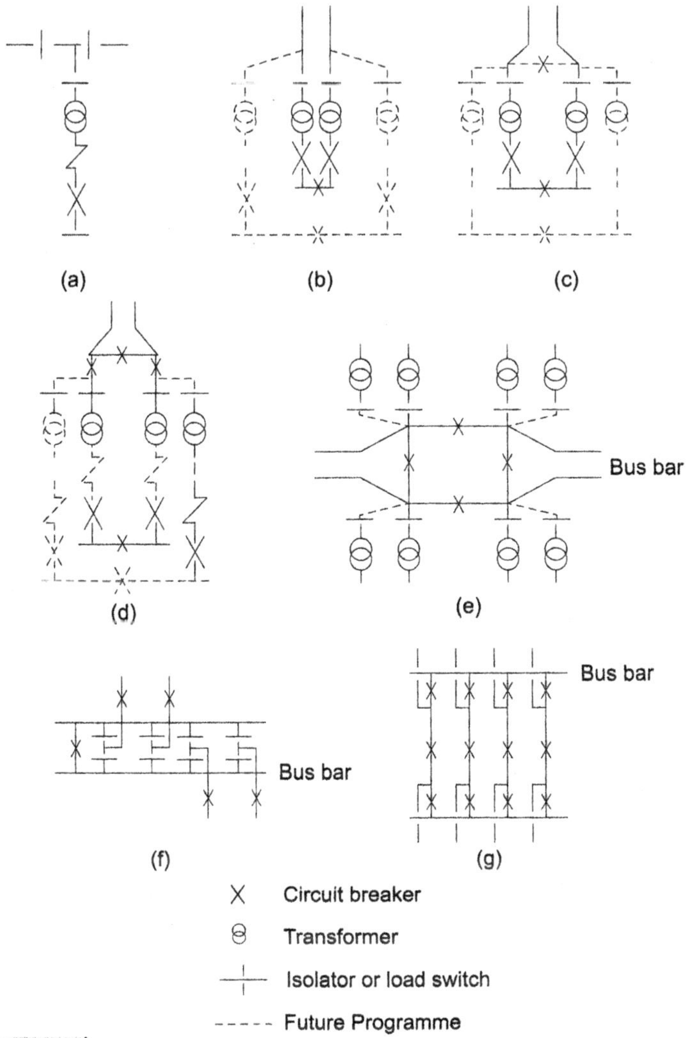

Fig. 6.3 *Switching arrangements:*
(a) Ring main; (b) Transformer feeder; (c) Single switch; (d) Three switch; (e) Four switch; (f) Double bus bar; (g) 1.5 switch

6.4 Planning Terms

Planning for the optimum power system can be classified into three main types: short, mid and long term. The procedures involved for each are somewhat different.

In a power distribution system 'long term' planning usually involves the 'horizon years' of twelve or more years ahead of the present: this span is greater for major transmission/generation studies. Unless very little growth is expected, the existing assets will normally have minor effect on decision making. Quite often the decisions are made with the aid of network standardization studies, involving cost modelling.

Short term or 'tactical' planning involves 1–2 years period ahead and generally involves only practical implementation of the recommendations of earlier long term studies.

In mid term (3–5 years) studies, the most economic methods of expanding the network are considered generally within the framework of the plant parameters set by the long term studies which would have usually considered the system development in a more generalized manner.

Thus, in network studies in the mid term, points of bulk supply, system voltage and choice of sub-station sites are considered and at each stage of network augmentation, only a limited number of alternatives are available. The engineer is faced with a series of 'decision trees' at each stage of augmentation. Impractical alternatives are quickly eliminated, leaving a small number of different augmentation strategies to analyse.

6.5 Network Cost Modelling

In the long term planning of power systems, it is often convenient to standardize network voltages and ratings. Given a set of standard network configurations (Figs. 6.2 and 6.3), costs of networks can be modelled using regression analysis and from the models, plant ratings and voltage chosen to minimize the overall cost distribution for the configuration adopted.

There are two basic approaches that have been adopted in the mathematical cost modelling of network. These are: (i) the cost modelling of an idealized network and (ii) the cost modelling of average networks. Both these approaches require basic cost models for the network plant, e.g. cables and transformers and a mathematical method of relating circuit lengths to the area and load supplied by the network.

6.5.1 Lines and Transformers

In distribution networks, the cost of two major items of the plant, i.e. sub-stations and lines (overhead and underground) can be closely idealized by straight lines. The cost of a sub-station consists of three variables—fixed cost (Cf), cost which depends on the rating and cost which depends on the voltage (see Fig. 6.4). This straight line relationship for sub-stations closely follows a similar straight line relationship for transformers. The fixed cost represents the cost of establishing a sub-station, e.g. the cost of land, civil works, building etc. and the variable cost (Cv) represents the cost of extra amounts of copper or aluminium required for increased rating and the additional amount of insulation required for higher voltages. A part of the fixed cost also includes the cost of basic components of the transformers, e.g. tank and tap changer.

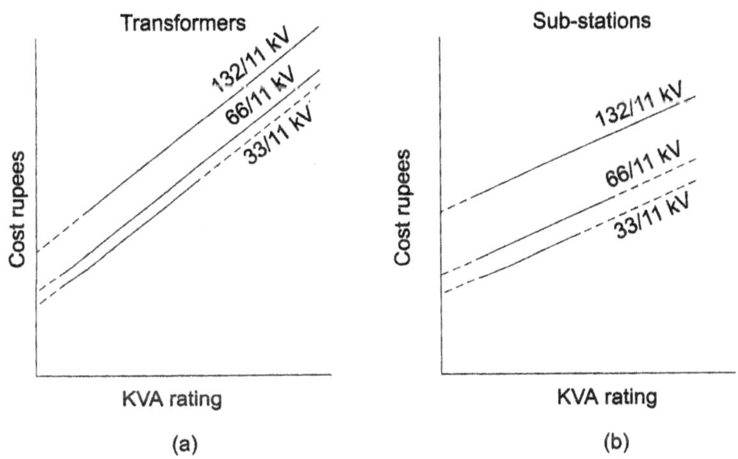

Fig. 6.4 *Typical cost modelling of network:*
(a) Transformers; (b) Sub-stations
The cost equation: $C_{ss} = C_{fs} + C_{vs} \cdot R$
C_{ss} = *Total sub-station cost*
C_{fs}, C_{vs} = *Fixed and variable components*
R = *Rating*

In a similar fashion, the cost of cables also varies in a straight line relationship with the cross-sectional area and voltage of the cable. This applies to both underground and overhead feeders. The cost '*K*' curves can be expressed in the following simple mathematical form:

$$K = a + bA$$

The price of cables, insofar as they are made of copper or aluminium, is reflected in the value of constant *b*; constant *a* represents fixed costs. As with substations and transformers, the variable component represents the increasing conductor material required for increased current rating and the increasing amount of insulation material required for higher operating voltages. *A* signifies rating the fixed cost represents the basic components in the cable (e.g. sheath and armouring in the case of an underground cable and poles and cross arms for an overhead line) as well as the labour cost for the installation of a cable which, for a given environment, is very nearly constant for different sizes of cables. The costing curves (see Fig. 6.5) can be found by regression analysis.

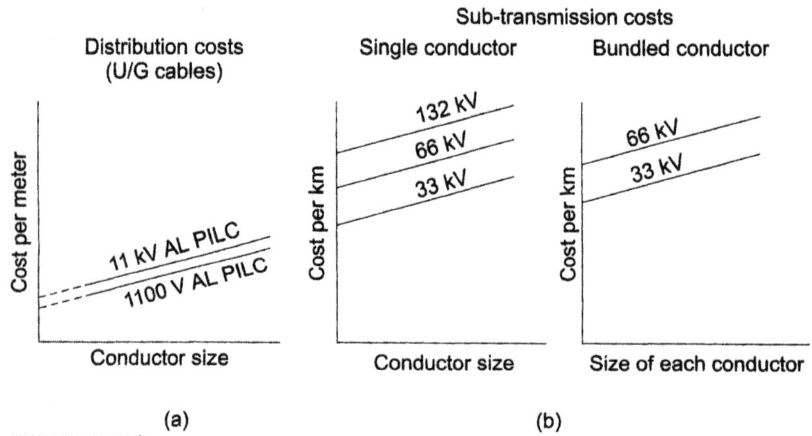

Fig. 6.5 *Typical cost modelling of lines:*
(a) Underground cables
Cost equation: $C_{dis} = C_{fdis} + C_{vdis}a$
(b) Sub-transmission:
Cost equation: $C_{sub} = C_{fsub} + C_{vsub}a$
$a = X$ *– area of conductor*

6.5.2 Relation of Circuit Length to Supply Area and Load

The end product of both cost modelling methods is to relate the cost per unit load to supply the complete area by the network. The approaches used to relate line lengths (and sizes) to supply area and load are as follows:

(a) The Cost Modelling of an Idealised Network

In this approach the network is idealised by an exact geometric shape. Normally the geometric figure is a hexagon. This is because the best of a supply area for any type of sub-station is a circle, but as circles cannot be laid out geometrically in continuous patterns without overlapping each other, some sort of a regular polygon must be used. The hexagon is the highest order polygon that can be arranged into continuous rows without overlapping and is therefore the closest to the optimum circular shape. The numbers and hence lengths of cables and positions of sub-stations are then automatically related to supply areas by the geometry of the supply area, allowing a complete cost model to be made.

There are certain weaknesses in this method that the set geometric shape usually in turn implies a fixed network configuration and a fixed ratio of ratings from one sub-station transformation level to another is not a realistic restriction in practice.

(b) Average Network and Load Density

At a given load density, the per kVA cost of a sub-station supplying the primary network decreases with an increase in the sub-station size and a corresponding decrease in their number. However, at a given voltage, the total length of the primary distribution circuits and, consequently, their cost increases with the size of the sub-station and the area served by it. Hence the need for optimization of distribution voltage and sub-station size when planning a new distribution system, the basic data available being the tract area to be served, its shape, i.e. the ratio of length to breadth and the anticipated load density in MVA/sq km for the various years. The latter can be estimated from the nature of load, e.g. residential, commercial or industrial, urban, rural, etc. and the extent to which the area will be developed.

The initial optimization study is carried out, starting from the primary distribution system consisting of trunk primary circuits, laterals (tee-offs) and distribution transformers with the main sub-station at the corner of the area. A similar study can be made for sub-transmission system. The lengths of the primary and lateral circuits in distribution are calculated from the following simple formulae, assuming square areas:

$$\text{Length of primary circuits (km)} = \left(N + \frac{3}{2}\sqrt{N} + 0.5\right)\sqrt{A}$$

$$\text{Length of lateral circuit (km)} = (\sqrt{A} - N\sqrt{X})\sqrt{\frac{F}{X}}$$

where,

A = tract area in km^2

X = unit distribution transformer area in km^2

N = number of primary circuits

The specific cost of an overhead circuit or underground length (calculated as above) may be considered for economic comparison between various alternatives with the help of the following empirical formula:

$$V_P = \frac{1}{P}\,[(1 + Y_L\, S_n\, F_P)K_L + \lambda\, L10^3\, k_1 \times 8760\, S_n\, F_P + CL\, 10^3]$$

where,

V_P = specific cost (Rs./MVA/km)

P = rated sending end power (MVA)

K_L = capital cost of line (Rs./km)

L = losses at rated power (MW)

L_F = load factor

λ = loss factor

$L_C = 0.2\, L_F + 0.8\, L_F^2$ according to British practice

k_1 = price of energy (Rs./kWh)

L_C = cost of installed capacity Rs./kW

Y_L = factor giving annual charge for line, i.e. the line annual cost

n = life of the equipment (year)

i = interest rate (%)

F_P = present worth factor $= \dfrac{1-(1+i)^{-n}}{i}$ [see Section 3.5]

S_n = annuity amounts, i.e. an annuity of 1 is payable each year for a period of n years, the annuity S_n after n years is

$$\left[\frac{(1+i)^n - 1}{i}\right]$$

Costs: From the above formulae, it may be seen that with an increase in load density and a corresponding increase in the number of primary circuits N, the length of laterals decreases where the length of the primary circuits increases, though not proportionately. The cost of primary circuits (6.6 ≃ 33 kV) as a percentage of the total system cost, therefore, decreases with an increase in load density. This decrease in circuit investment content makes room for the increase in distribution transformer investment content. By examining the various cost models and comparing different alternatives calculated from the above formula, we can select the distribution voltage, conductor size and line lengths on the basis of minimum costs. These studies can be made using computer techniques.

The optimization studies carried out at Mumbai for overhead sub-trans-mission and distribution system, found the following economic limits for various voltage levels:

Voltage	Limiting load and distance
11 kV	5 MVA, 3 km and below
22 kV	5 MVA, 3–10 km and for lower load levels
33 kV	5 MVA 10–30 km
	10 MVA up to 25 km
	10 MVA above 25 km
66 kV	20 MVA up to 30 km
110 kV	above 20 MVA

These limits were arrived at with $n = 40$ years
$Y_L = 0.12$, $F_p = 8.25$, $L_F = 0.6$, $\lambda = 0.4$, $i = 0.12$

(c) Economical Average Networks (8)

The optimum distribution system is the economical combination of primary line (HT), distribution transformer and secondary line (LT), which depends on area load density and diversity, system capital cost. These factors vary from State to State. The lower HT/LT line length ratio will have higher voltage drop and more line losses. There is trend towards increasing the HT/LT line ratio. Rural Electrification Corporation (REC) of India has also general directions to use smaller size of transformer, i.e. 25 kVA in rural area. This will increase HT line, decreased LT line in the system, thus increasing HT/LT line ratio. The following study was made to check up effect of this move, keeping in view firstly, that combined voltage drop in primary feeder, in distribution transformer, in secondary circuit and in service line should be within limits of 6 per cent, 4.5 per cent (impedance voltage), 6 per cent and 1 per cent respectively, with minimum system cost. Secondly, the economic level of load of distribution system is reached when the cost of the losses equals the cost of reducing the losses on annual charges basis, usually taken on 12 per cent of capital cost.

EXAMPLE: A case study was made in an area of 12 villages fed by one 11 kV feeder including spur lines of 20.5 km and LT lines of 41.5 km length, with 22 transformers with total capacity of 1577 kVA. Efforts were made to reduce the LT line and increase the 11 kV line and de-augmenting the various transformers, so as to increase HT/LT line ratio. The voltage regulation, losses and other relevant data were evaluated and were shown in the Table 6.1.

Table 6.1

Study of varying HT/LT ratio

HT/LT-line ratio:	1 : 2	1 : 1.8	1 : 1.7
Voltage drop			
11 kV line	4.93%	5.02%	5.1%
LT line	7%	4%	3%
Transformer	4%	4%	4%
Total drop	15.93%	13.02%	12.1%
Capital cost (Rs)	2,579,000	2,691,000	2,719,500
Losses			
11 kV line	1.022%	1.102%	1.178%
Distribution transformer	4.989%	5.364%	5.762%
LT line excluding service line	2.52%	2.32%	2.073%
Service line	0.877%	0.877%	0.877%
Total losses	9.40%	9.663%	9.890%
Annual cost of losses @ 80p/kWh (Rs)	133,200	136,930	140,150
Average transformer size	71.7 kVA	61 kVA	49 kVA
Total No. of transformer	22	26	31
Total transformer kVA	1577	1579	1530
Average LT km/Transformer	1.886	1.569	1.267
kVA-km (11 kV)	13,044	13,147	12,713

By quick look at above results, it is noted that capital cost for reduction of losses is well below economical limit. The capital cost per 1 per cent voltage regulation is less at 1 : 2 ratio, but regulation at LT is 7 per cent, which is more than 6 per cent allowed. Therefore, ratio 1 : 1.9 (interpolated) should be all right having lesser cost per 1 per cent voltage regulation as compared to ratios of 1 : 1.8 and 1 : 1.7. Also losses at

1 : 1.9 ratio will involve less cost as compared to these ratios. Losses are increasing with the increasing HT/LT line length ratio value. Predominantly, the losses are increasing due to increasing transformers. Therefore, it indicates that the solution to decreasing system losses is not that of de-augmentation of the transformers invariably, or using smaller size of transformers i.e. increasing the HT/LT line length ratio, but losses should be reduced by augmenting the size of line conductors and installation of low-loss distribution transformers with proper loss ratio. The HV distribution idea advocated by REC by using smaller size transformers (25 kVA) will of course be economical, if amorphous metal core transformers are used, which have one-fourth the core-loss of that of conventional transformers.

6.5.3 Developing Loads and Technology

In actual practice, the loading of a network varies with time so that the optimum network can only be planned for a certain load which, in turn, implies a design suited to a fixed point of time. It is fortunate indeed that the cost curves show a relatively flat base which means that a whole range of loadings can be accepted by the network yielding conditions which are close to the optimum. The network could, therefore, be designed so that in the initial phase of development, the sub-stations are loaded somewhat below the theoretical optimum load and towards the end of the period, are loaded somewhat above it, resulting in an average condition based on the optimum.

One important consequence of changing technology is the changing relative costs of cables and sub-stations. Over the last ten years a noticeable increase in cost has occurred in the manufacturing of transformers and switchgear, whereas especially the introduction of plastic insulation, the cost of cable manufacture has been kept relatively constant. This implies that the optimum rating of sub-stations built at the present time, will be higher than that of say fifteen yeas ago. The increasing load densities would also tend to reinforce this fact. This means that from time to timc, the network planner should be prepared to revise his concepts of standard rating in the light of changing load densities, changing plant costs and improving technology.

6.6 Voltage Levels

6.6.1 Selection of Voltage Levels

The selection of the voltage level is determined by the standards and safety regulations applying in various countries as well as by economic considerations. The loads in rural areas are generally scattered and the lines are very long. Thus the cost for rural systems is mainly determined by the cost of overhead lines. The cost of overhead lines increases slightly with the voltage, while the transmission capacity increases squarely. In addition to this, selection of a higher voltage and smaller conductor cross-sections brings about an appreciable reduction in costs. The most economical voltage for rural, medium or high voltage system will generally be in the region of the higher values. In an urban network, the conditions are different—the lengths of the cables or lines are much shorter as compared with the overhead lines in rural system, whereas the number of transformer stations is much higher.

The higher the rate of load growth in an area, the greater should be the difference in the stepping voltages, since it can be expected that the superimposed high voltage system will be economically loaded in the forseeable future, so that the capital outlay will pay for itself. Where the rate of development in the load is low, the ratio of 1 : 3 can be taken as a criterion for a suitable step from the medium to the high voltage. With normal growth in load, this ratio can rise to about 1 : 6. If the growth on load is above average and there are other reasons in its favour, the step voltage may be increased in the ratio of about 1 : 12.

A further decisive factor is the load density and its distribution in the area. The load density in an area is not constant, but may vary within wide limits. Thus, for instance, in the centre of a large city, the load density could be 20 MVA/km^2 in contrast to the value of only 0.01 MVA/km^2 of the rural outskirts. In the centre of the city, the concentration of power would make the highest possible voltage economical e.g. 132/11 kV. This might be uneconomical for the rural outskirts with a low load density in these areas. Generally, the following voltage levels are allocated. The steady growth of loads also makes it necessary that consideration be given to the load development, which may not only vary with respect to time but could also vary geographically. Matching of voltages to the standardized values is also important. The voltage application range is given in Table 6.2.

Table 6.2

Voltage application range

Voltage	*Range of application*
< 1 kV	Distribution systems for the feeding low voltage consumers such as houses, small workshops, commercial shops, tubewells, hotels etc.
1 to 36 kV	Distribution systems for feeding low voltage systems and large consumers such as shopping complexes, schools, hospitals, industrial plants, administration buildings etc.
36 to 150 kV	Distribution and sub-transmission system for feeding the medium voltage systems, for cities, large industrial works, railways main sub-stations.
> 150 kV	Transmission systems for large blocks of power and for inter-connected systems operations.

EXAMPLE: Small power consumers in Punjab are generally given supply in single-phase 230 volts for loads up to 10 kW, and for loads from 10 kW to 60 kW, three-phase 415 volts supply is given. Above 60 kW, generally, 11 kV supply is given. The basis for selection of supply voltage up to 100 kW is only connected load and above 100 kW, the basis is load demand. The supply pattern adopted in the Maharashtra State Electricity Board and the Andhra Pradesh State Electricity Board is under:

(i) For load demand between 100 kVA–1.5 MVA : Supply voltage is 11 kV.

(ii) For load demand between 1.5 MVA and 5 MVA. : Supply voltage is 33/66 kV.

(iii) For load demand 5 MVA above : Supply voltage is 132 kV.

In the Punjab State Electricity Board, the pattern adopted is as under:

(i) Connected load up to 60 kW : LT supply is given

(ii) For load demand between 60 kW and 5 MW : 11 kV supply is given

(iii) For load demand between 5 MW and 30 MW : 33/66 kV supply is given

(iv) For load demand between 30 MW and 50 MW : 132 kV supply is given

(v) Load demand above 50 MW : 220 kV supply is given

6.6.2 Number of Voltage Levels

The preferred standard distribution voltages in India are 240 V single-phase and 415 V, 11 kV three-phase. However, 33, 66 and 132 kV are considered sub-transmission voltages at present but are also being used as distribution voltage for very large consumers. In the near future these voltages will be commonly used for distribution purposes in India.

Many authorities have conducted network optimization studies for different number of voltages levels to ascertain the optimal voltage for a given level. It has been found that for transmission voltages, in the range of 300–400 kV stepping down to distribution voltages around 11 kV, three-tier voltage systems will be considerably cheaper than four-tier voltage systems.

For example, 400/132/11 kV is cheaper than 400/132/33/11 kV. The actual intermediate voltage chosen appears to have little effect on the overall cost. There is little overall difference in cost between a 400/132/11 and 400/66/11 kV system or between 220/66/11 and 220/33/11 kV system. The studies [11] indicate that the ratio of two voltage levels in the supply system in general, should be higher than a factor of 3 with regard to the transformation step from EHV level (220–500 kV) downwards. It should be higher than a factor of 5 with regard to the transformation step from HV level (50–150 kV) downwards. This means that it is desirable to get the lowest MV level in two or three steps from the highest voltage level used in the supply system—for example 400/110/22 kV or 765/400/132/11 kV. Reduction in the number of voltage levels by leaving out an intermediate voltage level between HV and MV (1 kV–36 kV) in general, will not only lead to a deeper penetration of HV systems, but also to an increased length of MV systems.

6.6.3 Adoption of Higher Voltages

Figure 6.6 shows that for increasing the voltage and conductor size of cables and transformers, the cost of these plant items increases. The capacity of a network can be increased by increasing the current rating at the same voltage or increasing the voltage at the same current. The variation of the cost of plant items with both current and voltage will often be such that it is more economical to go to a higher voltage rather than a larger current rating, especially in distribution networks, where there is a practical limit to conductor sizes above which the plant cost begins to rise very steeply.

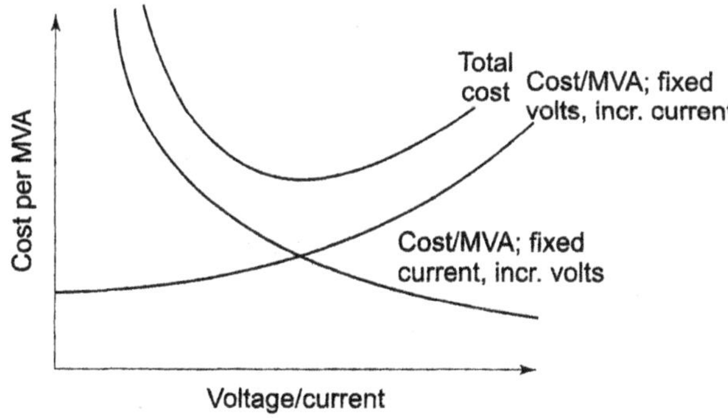

Fig. 6.6 *Cost versus current/voltage rating of conductors and transformers*

The insulation costs have of late tended to rise more slowly than the metal cost.

For a given load, therefore, the cost say, for a sub-transmission line, will vary as shown in Fig. 6.6 wherein at any point an optimum voltage and conductor size can be chosen for the load in question. Quite often the investment in the plant at the existing voltage level tends to restrict the opportunities available for voltage changes. However, when systems are being developed in new areas, a general rule can be applied: It is better to choose an operating voltage slightly above the optimum than below because in time, with increasing load, the system will become economically more viable.

In congested urban networks it is usual to adopt higher sub-transmission voltages by eliminating one transformation completely, e.g. conversion from 33/11 to 132/11 kV. The saving which results from this approach is because of the elimination of one voltage transformation. In a situation where large loads are being transported around the system, economic advantages of the higher voltage are obvious.

6.7 Synthesis of Optimum Line Networks

The procedure of network synthesis is sometimes described as the reverse of analysis. Instead of analysing a network for load-flow etc. to determine its suitability for the task in hand, the engineer is presented with certain specifications or requirements which a new network must meet while satisfying other criteria, including minimal cost and acceptable performance.

Basically, four approaches to the synthesis of network designs have been identified:

(a) Engineering judgement;
(b) Some optimization techniques to investigate certain aspects of the design, such as circuit ratings;
(c) Heuristic logic, simulating the designer's thought processes; and
(d) Mathematical programming techniques.

At present, (c) and (d) have been applied to some extent in long-range studies, especially mathematical programming.

In broad terms, the creative process undertaken in network synthesis by engineering judgement can be broken into the following steps:

1. A suitable horizon year (say, 10–12 years hence) is chosen, the projected load growth for the study period calculated and this is related to the capacity of the existing circuits, if any.
2. Depending on the extent to which the existing assets serve the area, a number of network layouts which could meet the anticipated loading requirements for the horizon years while satisfying other criteria, such as reliability, fault levels, etc., are devised. The designer will consciously try only those designs which involve minimal additional plant capacity while satisfying the general system standards. At this stage, according to the requirements of network augmentation, the designer may use the results of modelling (when, e.g. existing assets are minimal and the designer has a free hand to create a completely new network) to give a guide as to the sub-station circuit ratings and voltage levels. Alternatively or when the existing assets dictate the voltage levels available, he may decide whether or not a case exists for the elimination of a voltage transformation. On the other hand, he may be limited in the choice of voltage levels, points of supply, circuit routes and circuit ratings (as in the case of congested urban areas), and often, the range of alternatives may be restricted. Here, however, techniques such as mathematical programming may prove to be useful aid in the selection of line routes and sub-station locations. An improvement of reliability in the supply and elimination of plant deficiencies in older parts of the system may strongly dictate the form of network augmentations. In all cases, however, it is important to note that the phasing of the designs

considered are from the future to the present, for, once a certain network alteration or addition is made, the authorities are committeed to a course of action in the future by the very presence of this new plant. It is imperative that short term plans be part of an overall long-term plan, although this will gradually be modified as time goes on due to changing circumstances.

3. Various alternative designs are checked for load-flows, fault levels (and in the case of major transmission systems, machine stability) to verify their suitability; often minor changes or additions to the designs may be necessary.
4. Various alternatives are then compared on the basis of economics primarily; but often non-quantitative requirements, such as social acceptability, environmental impact or political expediency, may also affect the choice.

6.7.1 Optimum Size of Line Conductor and Transformers

(i) Sizing of Conductors

As per Kelvin's law, the most economical size of line conductor is that for which the annual charge on the investment is equal to energy losses. If these two values are not reasonably close, another size of conductor may be chosen. Either process may be repeated until the values are reasonably close. Conductor is often the biggest contributor to distribution system losses. Economic conductor sizing is therefore of major importance. If a conductor is loaded up to or near its thermal rating, the annual cost of the losses will exceed the annual charges of the line circuit. Therefore, line conductors are loaded below their thermal limit.

It has been seen that conductor over-sizing can yield substantial cost saving in spite of small penalties which may arise with incorrect projection of future load development. An increase in cross-sectioned area results in lower resistance and thus reduced losses. The extra capital cost associated with the larger conductor is paid for by the savings in losses. For example, in Punjab State Electricity Board, for rural area, initial 11 kV network with the trunk section from main sub-station is laid with ACSR 50 mm^2 and remaining spurs with ACSR 30 mm^2.

(ii) Optimum Size of Transformers

As per the Kelvin's law, the most economical size of transformer is that for which the annual cost of its copper/aluminium losses is equal to the annual carrying charges (usually 12 per cent of the capital cost) of the transformer installed plus the annual cost of the core-loss. Here if two values are not close, another size of the transformer may be chosen.

6.7.2 Load Centres

(i) Sub-stations

Sub-stations control power by means of transformers, regulating devices, circuit breakers, isolators, etc. and their location is based on the following criteria:

- For minimum losses and better voltage regulation, the sub-station should be close to the loads of its service area, so that the addition of loads times respective distances from the sub-station is a minimum.
- The location should provide proper access for incoming sub-transmission lines and outgoing primary feeders and also allows for future growth/expansion.

(ii) Distribution Transformers

For locating distribution transformer, the loads are treated as masses. The location point (X, Y) is established by moments equated round the reference point as origin of x- and y-axis. The LT lines length and loads position is plotted on graph/survey sheet to scale as per geographical orientation. Algebraic sum of all the moments around y-axis and x-axis are calculated and divided by total load (in BHP or kW) respectively, to get X, Y coordinates of the proposed transformer load centre and is marked as near as practicable at site.

i.e. $$\text{Total load. } X = L_1\, x_1 + L_2\, x_2 + \cdots$$

$$\text{Total Load. } Y = L_1\, y_1 + L_2\, y_2 + \cdots$$

$x_1, x_2, \cdots, y_1, y_2$ are distance of loads ($L_1, L_2, \cdots$) from reference point along x-axis and y-axis respectively.

EXAMPLE: Taking a simple case of three villages A, B, C, situated at equal distance of 500 m and their respective light load 8, 10, 16 kW having unity power factor, as shown in Fig. 6.7. The prospective load is almost negligible. Find the suitable location of their distribution transformer load centre.

Solution: For finding ordinates of load centre, taking moments:

$$X \times 34 = 16 \times 500 + 0 \times 8 + 10 \times 250$$

$$X = 308 \text{ m}$$

$$Y \times 34 = 10 \times 250 \tan 60$$

$$Y = 127 \text{ m}$$

Therefore load centre is at 308, 127 coordinates as shown in Fig. 6.7.

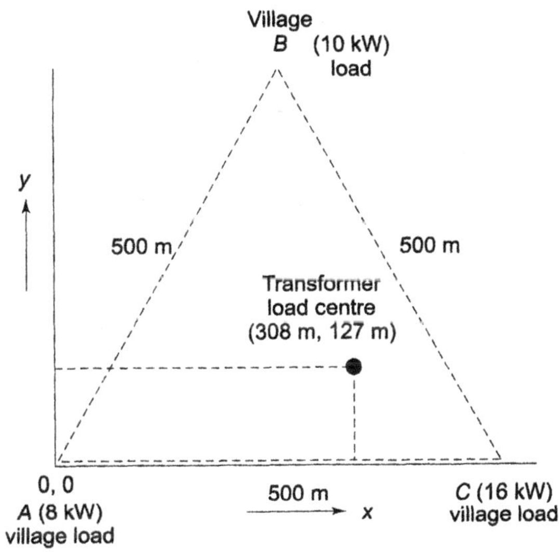

Fig. 6.7 *Typical distribution transformer load centre location*

6.7.3 Sub-station Size

The size of sub-station depends upon:

- Load density (e.g. MW/km^2)
- Load growth
- Utilization of transformer capacity
- Maximum fault levels
- Flexibility
- Siting

Normally, in city areas with higher load densities, the size of sub-station will be higher than in rural areas. This is mainly because of the fact that, the cost to distribute the power is lower when the load density is high. But the presence of intermediate voltage level influences the sub-station size considerably. Seen from the upper voltage level (for instance HV), the size will be bigger because of the fact that the density of sub-stations at that voltage level will be decreased. At the lower voltage level (for instance MV), the presence of the intermediate voltage level leads to a lower size, but a higher density of sub-stations.

In city areas, the values of the sub-transmission voltage levels can be selected, in order to make it profitable to redevelop with fewer intermediate voltage levels. In rural areas, because of much lower load densities this is a less likely option. This means that both the network structure and the sizes and densities of sub-stations significantly differ between urban and rural situations.

In most situations, the size and number or density of sub-stations is determined by a technical-economic analysis. The average distance between sub-stations (N):

$$N = \left[\frac{\text{Total areas}}{\text{Number of sub-stations}}\right]^{1/2}$$

The significant issues being the need for firm capacity within the sub-station itself, and the implications on the reliability standards to be met.

Due to the general scarcity of available sites and other local environmental aspects, it may be impossible to respect the optimal technical-economic solutions alone. Nevertheless, solutions should be looked for, so that the cost consequences are known and the optimum conditions may be approached as near as possible.

In general, an $(n - 1)$ design criterion dominates the optimization process of sub-transmission network planning. As a result of this, the majority of the sub-stations have one or two transformers. If load transfer is possible between feeding points, one transformer in a sub-station can often be found to be adequate.

From the economical point of view, the optimal successive transformer sizes between the same voltage levels, have a ratio of about 1.2 to 2.5. The resiting of transformers half-way through their life is sometimes part of the network philosophy. Reliability, quality criteria and loss reduction policies influence the time of changing transformers.

6.7.4 Optimum System Losses

The essence of optimization is the minimization of the net supply cost. The losses in the power system increase the cost of supplying energy to consumers. Figure 6.8 shows that the cost of losses increases and system investment cost falls, while economically, optimal loss level occurs at the minimum point 'O' of the curve.

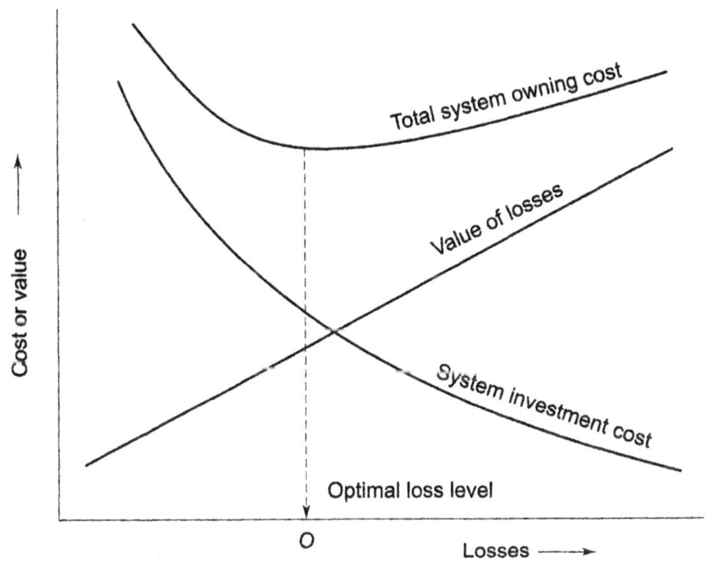

Fig. 6.8 *Optimum loss level*

6.7.5 Distribution System—Voltage

The distribution systems can be:

(i) Low Voltage Distribution System (LVDS)

It is based on the European practice, where three-phase distribution transformers of considerably large capacity are installed and a number of LV lines (secondary) of considerable length are extended to cater to the loads.

(ii) High Voltage Distribution System (HVDS)

It is based on the North American practice, where HV line (primary), is taken as near to the load as possible and small capacity single-phase distribution transformers are installed, minimizing or eliminating the LV lines. HVDS can be classified into four types:

(a) Phase-neutral system
(b) Phase-phase system and
(c) Phase-ground return system
(d) 3 phase system

The relative merits of the these types can be evaluated for a given situation.

Generally, the LVDS is best suited to meet the concentrated loads in small areas, whereas HVDS is best suited to meet the scattered loads with low load density.

A typical comparison of basic characteristics indicating relative merits of HVDS (11 kV phase neutral system) and LVDS (3-phase, 4 wire, 415 volts system) is given in Table 6.3. Taking the reference values for LVDS as 100 in the case of each of the parameters listed in the table, the corresponding values of HVDS are indicated:

Table 6.3

Typical comparison of HVDS and LVDS

	Item	*HVDS*
	Current	11.0
	Voltage drop	12.7
Losses:		
	Line	8.5
	Transformer load/kVA	115.0
	Transformer no load/kVA	138.0
Cost:		
	Line/kM	71.7
	Transformer/kVA	150.0

6.8 Applications of Linear Programming to Network Synthesis

Linear programming (LP) has found increasing uses in the synthesis of minimum cost networks. Computer program packages are available.

6.8.1 Minimum Length Criterion

This is often used for radial distribution systems where we have a single feeder emanating from a source sub-station joining a series of distribution transformers. The situation can be formulated mathematically as follows. For N points with routes x_{ij} between nodes i and j of length L_{ij} we have:

$$\sum_{\substack{i=1 \\ i \neq j}}^{N} x_{ij} \geq 1 \text{ for all } j\text{'s (reliability criterion)}$$

and

$$\sum_{\substack{i=1 \\ i \neq j}}^{N} \sum_{\substack{i=1 \\ j \neq i}}^{N} x_{ij} = N - 1$$

We wish to minimize

$$\sum_{\substack{i=1 \\ i \neq j}}^{N} \sum_{\substack{i=1 \\ j \neq i}}^{N} L_{ij}\, x_{ij}$$

with the x_{ij} integer.

This is the standard LP formulation.

6.8.2 Group Transfer Criterion (2, 6)

Again, we consider N points with routes x_{ij} between nodes i and j involving a circuit construction cost of C_{ij}.

We wish to minimize the expression

$$\sum_{\substack{i=1 \\ i \neq j}}^{N} \sum_{\substack{j=1 \\ j \neq i}}^{N} x_{ij}\, C_{ij}$$

Subject to constraints

$$\Sigma\, x_{ij} \geq v_g, \text{ where } i \in g, j \in N - g$$

for all possible groups v_g (excluding the slack bus), with the x_{ij} integer. The drawback of this method is that a system of N nodes will require $2^{N-1} - 1$ constraints. For this reason this particular method has not found extensive use in large networks. There are, however, ways of reducing the number of possible connections by, e.g., topological limits, i.e. sub-stations j and k may be limited to two circuits only, as well as allowing for the existing circuits, e.g., $x_{jk} \geq$ number of existing circuits.

6.8.3 Minimum Cost Approach using Transportation with Transhipment Model

If we consider the cost of shipping power from one sub-station to another as equal to the cost of installing the necessary lines/cabling and connections, we can set up analogy to the transportation model. This is a standard approach which is an extension of LP theory, and standard computer programmes are available for its solution. Here we let C_{ij} equal the cost of laying and connecting a cable from node i to node j, and N_{ij} represent the quantity shipped, i.e. the number of cables laid. The transportation with transhipment algorithm is used to minimize the overall cost.

6.8.4 Minimum Power Transfer and Distance Approach

In this approach the quantity to be minimized is the product of the overall power transferred and network length. If P_i is the power requirement of node i, P_{ij} the power transfer between nodes i and j, and L_{ij} the distance between nodes i and j, for a system of N nodes, the constraints are:

$$\sum_{\substack{i=1 \\ j \neq i}}^{N} P_{ij} = \mp P_i \text{ for all } i\text{'s}$$

We wish to minimize

$$\sum_{\substack{i=1 \\ i \neq j}}^{N} \sum_{\substack{j=1 \\ j \neq i}}^{N} L_{ij} P_{ij}$$

This again can be solved by using the transportation with the transhipment algorithm.

6.9 Optimum Phasing Sequence

The networks are usually designed to be optimal for a particular horizon year. Therefore, the network in its partly developed phase will not be optimal. However, an optimum development strategy can be formulated for a network. When the number of such strategies is quite limited, each can be economically appraised separately.

When a number of sub-stations, each (ultimately) comprising several transformers, are involved, the optimum strategy can be determined without the need for individual economic comparison of all possible alternatives. The general rules concerning the augmentation of sub-stations can be obtained from experience.

If, say, a sub-station consists of n transformers, each of rating R_1, an annual cost A_{R1}, and the cost of losses in $C^n_{R_1}(l_i)$ for a sub-station load l_i in a year i, then the cost of operating the n transformers each of rating R_1 for years j to k will be

$$F^j_i(R_1, n) = \sum_{i=j}^{k} [n\, A_{R_1} + C^n_{R_1}(l_i)]$$

during the period before the "book life" of the transformers.

For a given augmentation strategy covering the whole period of say, 1 to n years, the total cost would be the sum of the various cost functions (F) of each stage viz.:

$$F = f_1^{i_1}(R_1, n_1) + f_{i_1+1}^{i_2}(R_2, n_2) + \cdots + f_{i_m+1}^{i_n}(R_m, n_m)$$
$$\{l_k < n;\ m < n\}$$

Some practical results from the study of a sub-station are as follows:

(a) Three transformers (allowing first level emergency) are generally the most economical.

(b) It is usually cheaper to augment a fully developed sub-station by replacing transformers by larger ratings rather than by adding further transformers.

(c) It is usually most economical to replace a transformer by a unit which has a rating of about twice the original one.

6.10 Economic Loading of Distribution Transformers

Transformer loading is primarily limited by the aging or deterioration of insulation which is a function of time and temperature. Economic transformer loading can be determined by an evaluation of both aging and the characteristics of the transformer, i.e. it can be determined by analyzing the evaluation of loss of life along with transformer cost and the cost of operating the transformer. The changing relation of these costs makes it desirable for the utilities to determine the economic loading of transformers for today's conditions as compared with yesterday's conditions and to frame and review a particular policy, such as replacement of future damaged transformers, improvement of load factor, modification in the tariff structure, etc.

EXAMPLE: A 11 kV feeder to city area has 100 numbers of 11/0.415 kV transformers of sizes as given below:

Size	*Numbers*	*Total kVA capacity*
300 kVA	= 10	3000
200 kVA	= 24	4800
100 kVA	= 40	4000
63 kVA	= 20	1260
25 kVA	= 6	150
		13,210

The 11 kV line losses are 3 per cent. The yearly units supplied at 11 kV at outgoing point to consumers at secondary voltage are 28.7 GWh. Assuming power factor of 0.85, find the desirable transformers loss ratio.

Solution: Connected kVA of transformers = 13,210

Energy sold to consumers including core-loss and secondary line-loss

$$= 28.7 - 28.7 \times \frac{3}{100}$$

$$= 27.839 \text{ GWh}$$

$$\text{Average kW sales} = \frac{27.839 \times 10^6}{8760}$$

$$= 3176$$

Making allowance for power factor 0.85

$$\text{Average kVA sales} = \frac{3176}{0.85} = 3736$$

$$\text{Average transformer load} = \frac{3736}{13210} = 0.2828 \text{ p.u.}$$

i.e. 28.28%.

Average Losses:

The average losses will correspond to average load by the formula (see Section 4.12.4): Average loss = c (average load) + (1 – c) (average load)^2

$$= (0.15)\,(0.2828) + (1 - 0.15)\,(0.2828^2)$$

$$= 0.11 \text{ pu}$$

For maximum efficiency of transformers, no-load losses should be equal to losses on average operating load, i.e. No-load loss = 0.11

Full-load loss = 1 pu

$$\text{Loss ratio} = \frac{\text{Full-load loss}}{\text{No-load loss}} = \frac{1}{0.11} = 9$$

6.10.1 Estimating and Evaluating Transformer Loss of Life (4)

Transformer functional life is influenced by thermal, mechanical and dielectric stresses. It is most frequently terminated by one of the latter two stresses. The concept of insulation 'Thermal Life' is a useful one, in that it can set a standard level of physical condition below which insulation integrity may be prejudiced by mechanical or dielectric stresses, normally experienced in service.

IS: 6600–1972 guide for the loading of oil-immersed power transformers recognizes the normal life expectancy, if the transformer is run at its rated kVA with a yearly weighted average ambient temperature 32°C and hot spot temperature 98°C. This Guide is based on laboratory tests and 6°C insulation life rule. American specifications, Appendix C57.91 (1974), give a more realistic loading guide for distribution transformers with an average winding temperature rise of 55°C on the basis of life tests on functional units. In this the method for calculating the percentage loss of life for a small increment of time for an equivalent thermal daily load has been explained. To use the loading recommendations, the actual fluctuation load cycle must be converted to a thermal equivalent simple rectangular load cycle. A transformer supplying a fluctuating load generates a fluctuating loss, the effect of which is about the same as that of an intermediate load held constant for the same period of time. This is due to the heat storage characteristics of the materials in the transformer. The equivalent load for any portion of the daily load cycle may be given by the following equation in per unit value:

$$L_{eq} = \sqrt{\frac{L_1^2\, t_1 + L_2^2\, t_2 + L_3^2\, t_3 + \cdots + L_n^2\, t_n}{t_1 + t_2 + t_3 + \cdots + t_n}}$$

where L_{eq} is the equivalent load in pu and L_1, L_2, $\cdots$, L_n are the loads in pu and t_1, t_2, $\cdots$, t_n are the time intervals in hours for respective loads.

The procedure is as follows: Calculate the loss of life using the load cycle and average ambient temperature for a particular month. A summation of the loss of life for a particular month is calculated by multiplying the incremental daily loss by the number of days in the month assuming that the daily load cycle curve in the month is approximately same. Repeating the procedure for each separate month, the sum will estimate the loss of life for a complete year.

The percentage loss of life can be calculated by the following formula as per IEEE standard 345.

$$\text{Percentage loss of life} = \frac{100Z}{e^{[B/(\theta_g + 273) - A]}}$$

where Z is the interval of time in hours, B and A are constant obtained from the desired life expectancy curve given in Fig. 6.9. This curve is

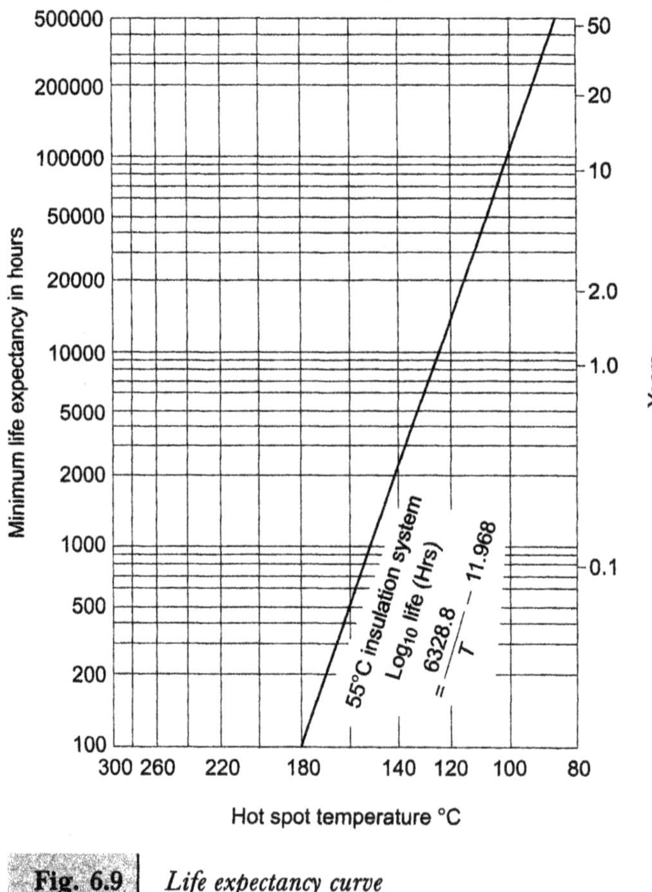

Fig. 6.9 *Life expectancy curve*

$$Log_{10}\ life\ (hours) = A + \frac{B}{T};\ T = \theta_g + 273$$

based on laboratory and model tests in which the decrease in mechanical and electrical strength has been measured and a series of functional life expectancy tests made under controlled conditions on production type distribution transformers under the auspices of the IEEE Transformer Committee [10]. The curve indicates a minimum life expectancy of above 15 years at 98°C or 0.0182 per cent loss of life/day for the transformer manufactured as per IS: 2026. θ_g is the hot spot temperature at any given time during either a temperature rise or cooling condition.

The ambient temperature is an important consideration in determining the loss of life when loading transformers. It is added to the temperature rise for any load in determining the hot spot temperature of the transformer, which in turn determines the loss of transformer life. Monthly ambient temperatures which are obtained by the averaging daily temperature over several years, are usually available from the meteorological department.

Knowing the no load and load loss standardized values, we can determine the hot spot temperature for the prevailing condition by the following formulae:

$$\text{Hottest spot temperature} = \theta_a + \theta_o + \theta_{gl}$$

where,

θ_a = average ambient temperature

θ_o = top-oil rise over ambient temperature

θ_{gl} = the hottest spot conductor rise over top-oil temperature

Ultimate top-oil rise for load $L = \theta_u$

$$= \theta_{fl}\left[\frac{(K^2R+1)}{(R+1)}\right]^n$$

where,

θ_{fl} = full load top-oil rise

K = ratio of load L to rated load

L = equivalent load pu under consideration

R = ratio of load loss at rated load to no load loss.

The hottest spot conductor rise over top-oil

$$\theta_{gl} = \theta_{gfl}\,K^{2n}$$

Here,

θ_{gfl} = the hottest spot conductors rise over top-oil at full load conditions

$\theta_{gfl} = 10°C$, temperature gradient allowed for distribution transformers as per IEEE Standard 462/1973.

$n = 0.8$ for ON transformers

Now we know the ambient temperature (θ_a), top-oil rise θ_o and the hottest spot conductor rise over top-oil (θ_{gl}) for the equivalent load under consideration. We can determine the hottest spot temperature.

The next step is to determine the present worth of revenue expenditure requirement to obtain the value of the transformer additional life lost for replenishment purposes. This can be calculated from the equation:

$$P_w \text{ of loss of life} = \sum_{0}^{m} \left[T \times F \times \frac{(1+I)^{m-n+1} - 1}{I(1+I)^{m-n+1}} \right] \cdot \frac{L_n}{(1+I)^n}$$

where,

I = interest factor

L_n = pu loss of life in year n

F = fixed charge rate

P_w = present worth or value

m = number of years of study

T = transformer cost

If the transformer characteristics are not accurately known, the maximum top-oil temperature derived from Fig. 6.10 can be used as an approximate guide for loading which is based on the difference between the hottest spot temperature and oil temperature as 20°C at rated load. However, oil temperature alone should not be used as a guide for loading transformers. The hottest spot winding rise over top-oil temperature at full load should be determined from factory tests. It should be recognized that due to thermal lag in oil rise, sometime is required for a transformer to reach a stable temperature for any change in load. Therefore, higher peak loads may be carried out for a short duration.

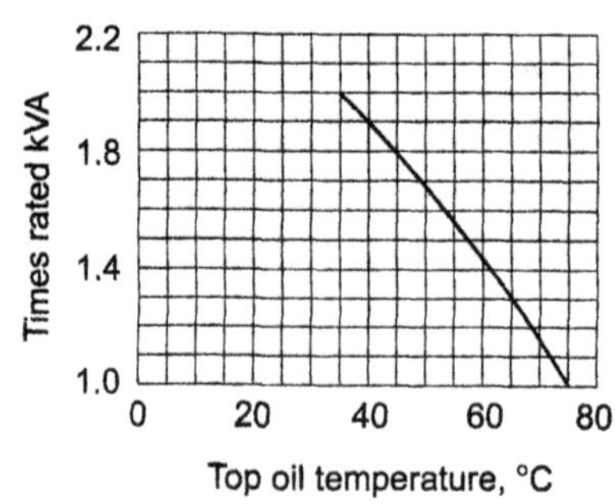

Fig. 6.10 *Approximate continuous loading for minimum life expectancy*

6.10.2 Cost of Transformer Operation

Utilities usually specify the values of no load loss and load loss for various sizes of transformers while procuring them. Indian Standards also specify the losses for various sizes of transformers as given in Table 6.4.

The efficiency of a transformer is maximum at a certain average loading (over most of its operating life) that is when average load losses are equal to no load losses. The optimum loading, expressed as a percentage of the rated load varies inversely with the loss ratio under-root i.e. the pu average load at which the efficiency is maximum

$$= \left(\frac{\text{No load loss pu}}{\text{Full load loss pu}} \right)^{1/2}$$

It is necessary to determine the present worth of year by year cost of the aging transformers and the cost to supply the losses and annual charges. The annual losses can be:

Losses (kWh) per annum

$$= \left[\text{No load loss} + \text{Full load loss} \times \left(\frac{\text{kVA}_{\text{peak load}}}{\text{kVA}_{\text{rated}}} \right)^2 \times \text{loss factor} \right] \times 8760$$

Table 6.4

Standard specification for maximum losses in distribution transformers

Size kVA	Voltage ratio	No load loss W	Full load loss W	Specification reference (max. values specified)
16	11/0.433 kV	80	475	REC specification 2/1971/97 and IS: 1180-1989
25	11/0.433 kV	100	685	IS: 1180/1989 and REC specification 2/1971/97
63	11/0.433 kV	180	1235	IS: 1180/1989 and REC specification 2/1971/97
100	11/0.433 kV	260	1760	IS: 1180/1989 and
		290	1850	REC specification 2/1971/97
400	11/0.433 kV	980	5500	CBI & P
1000	11/0.433 kV	1800	11000	specification for distribution
1600	33/0.433 kV	3000	16000	transformers
1600	33/11 kV	-do-	-do-	
1600	33/11 kV	3000	15000	REC specification 7/1974/93

Utilities have various methods for evaluating the characteristics of distribution transformers. Some utilities estimate the cost of feeding the energy necessary to supply the kWhs used by no load and load losses of the transformer plus a demand charge for the demand created on the system by these losses. Other utilities evaluate these characteristics plus the cost of supplying the transformer reactive requirements and loss of revenue due to regulation.

Also, a computer programme can be developed to calculate the year by year costs of any or all of these characteristics and the present worth of these values so that this could be added to the additional loss to obtain the present worth of the transformer.

Since the annual maintenance expense is common to all transformers, it may be ignored for any economic comparison.

EXAMPLE 1: Assuming 63 kVA distribution transformer with fully-developed agriculture tubewell load as:

5 HP motor = 7 nos., 7.5 HP motor = 2 nos., 3 HP motor = 7 nos.
and auxiliary light load = 1.280 kW
Load diversity factor = 1.10
Load factor = 0.35
Power factor = 0.85

Calculate the desired transformer loss ratio.

Solution:

Connected load = 71 × 0.746 + 1.280 = 54.246 kW

$$\text{Average operating load} = \frac{54.246 \times 0.35}{1.10} = 18.97 \text{ kW}$$

Average operating load in terms of percentage of transformer rating

$$= \frac{18.97 \times 100}{63 \times 0.85} = 35.42$$

Assuming full load losses = 1 pu

$$\text{Average operating losses} = \frac{(35.42)^2}{100^2} = 0.125 \text{ pu}$$

For most efficient transformer operation, the average operating load losses should be equal to "no load losses".

$$\text{Therefore, the desired transformer loss ratio} = \frac{\text{Full load loss (pu)}}{\text{No load loss (pu)}}$$

$$= \frac{1}{0.125} = 8.$$

EXAMPLE 2: An augmentation of 50% over-loaded 100 kVA, 11/0.433 kV distribution transformer in rural area is made with 200 kVA transformer. Assume loss factor $Ls = 0.3$. Find annual saving in losses.

Solution:

(a) Losses with 50% overload 100 kVA transformer:

$$\text{Load-loss} = \left[\frac{\text{Peak load}}{\text{Full load}}\right]^2 \times L_s \times \text{Standard load-loss} \times 8760$$
$$= (1.5)^2 \times 0.3 \times 1.85 \times 8760$$
$$= 10{,}940 \text{ kWh per annum.}$$

where, 1.85 kW is REC standard load-loss for 100 kVA transformer at full-load.

Iron-loss per annum = 0.29 × 8760 = 2,540 kWh per annum where 0.29 kW is REC standard iron-loss

Total losses = 13,480 kWh per annum ... (i)

(b) Losses with augmented 200 kVA transformer with load 150 kVA:

$$\text{Load-loss} = \left[\frac{150}{200}\right]^2 \times 0.3 \times 3.1 \times 8760 \text{ kWh/annum}$$
$$= 4582 \text{ kWh/annum}$$

where 3.1 kW is standard loss for 200 kVA transformer at full-load.

Iron-loss = 0.45 × 8760 = 3,942 kWh per annum

where 0.45 kW is standard no-load loss of 200 kVA transformer.

Total losses = 8,524 kWh per annum ... (ii)

Total annual savings (i)–(ii)

= 13,480 – 8,542

= 4,956 kWh per annum.

or Rs. 9912 per annum @ Rs. 2/kWh.

6.11 Worst-Case Loading of Distribution Transformers

6.11.1 Rural Areas

In rural areas, for agriculture load, the transformers with very few accessories are usually used. These transformers do not have load monitoring device except the maintained connected load register. The loading is done by worst case occurrence, which may be rare, such as maximum ambient temperature, full-load, the diversity and demand factors both unity. It is a costly proposition. The evaluation is shown in the following example:

EXAMPLE: Assuming transformer capacity as 100%, agriculture motive load power factor is 0.85. The total load on transformer will be:

(i) Motive load of tubewells in HP.

(ii) Losses of the tubewell motors, i.e. 100 per cent minus motor efficiency of average 83 per cent = 17 per cent of motor input.

(iii) Transformer losses (kVA) is generally 7 per cent of transformer capacity.

(iv) LT line losses (kVA) 3 to 5 per cent, i.e. 4 per cent of an average of consumer load.

(v) Light load of a tubewell @ two bulbs each of 40 Watt = 2 per cent of HP load, considering average tubewell 5 HP.

Thus connected motive load justified on transformer is as follows:

100 per cent as rating of transformer
minus 17 per cent due to losses on motors in kVA
minus 7 per cent due to transformer losses in kVA
minus 4 per cent due to L.T. line losses in kVA
minus 2 per cent due to auxiliary light load of tubewell in kVA = total of 70 per cent

Therefore, the transformer should be loaded for motive power up to 70 per cent capacity of the name plate rating. For example, 63 kVA and 25 kVA transformer be loaded respectively up to

$$\frac{63 \times 0.85 \times 0.7}{0.746} = 50 \text{ HP and } \frac{25 \times 0.85 \times 0.7}{0.746} = 20 \text{ HP}$$

motive load respectively. This practice has been adopted in Punjab.

Besides, there are other factors such as extra loading due to low voltage and/or low power factor at the consumer terminals. In some criminal pockets, such as international border areas/smuggling area, high rated motors may be used by unscrupulous consumers. In areas where ground water exploitation is high (see Sec. 14.10.3), the agriculture load may rise by up to 30 per cent due to lowering of the ground water level, especially during the summer paddy crop season. The farmers either scale down the level of pump system or install higher size bore-motor pump system without sanction to make up the reduced water output.

6.11.2 Urban Areas

In the urban areas, the load is composite and transformers can be loaded on the basis of optimum use or worst case.

(i) Optimum loading is based on the ageing of transformer, i.e. when overloaded, the transformer is assumed to age at accelerated rate and when no-load or underload the transformer ageing is slow. The overall life of transformer is supposed to be normal (see Sec. 6.10.1).

(ii) Worst case loading need transformers in the urban area equipped with oil and/or winding temperature gauge indication. Allowable temperature be taken as guide for loading of transformers. In case temperature gauges are not provided, the transformer needs to be regularly checked for ampere loading during the peak hours. The transformer is considered fully loaded when the peak amperes are almost equal to the transformer ampere rating.

6.12 Discussion

1. The subject matter of optimization covers the basic theory of design and operation of optimal networks. Often, in practice, the engineer is not able to undertake studies such as cost modelling covered in this chapter and is largely restricted by the existing assets, policy and local restraints. In addition; we have seen how differing standards in reliability, differing levels and load densities strongly dictate the type of system which will evolve.
2. In all cases, however, the engineer will be working with the aim that the end-product of his design should satisfy the requirements of technical performance, be safe and economical, and also be socially and environmentally acceptable.
3. The line of computer techniques is desirable for optimizing the design and studying a wide range of variables and alternatives. The programmes and cost models/data can be periodically up dated and plans reviewed to take into account the latest prices, load forecasts and other design policies. Computer Aided Designs (CAD) could reduce the distribution losses to the extent of 8–20% [7].
4. Each power utility should evolve certain planning model for laying new/extension systems and revamping of the existing system.

The work may be taken in three parts: (i) distribution transformer and its LT distribution system; (ii) primary feeders such as 11 kV or 22 kV and (iii) sub-transmission 33, 66, 132 kV and main sub-station.

The geographical mapping with location of loads of each part be done. The various organizations such as CPRI, Bangalore [3] have developed the computer programmes to obtain minimization of costs and losses by proper loading of the feeders, by re-routing of networks, by re-distribution of load, by strengthening of conductor, etc.

PROBLEMS

1. What is the goal of linear programming? Also indicate the steps desired in the application of linear programming.
2. For the purchase of 63 kVA distribution transformers, quotations were received from three vendors: *A*, *B* and *C* for Rs. 70,000, Rs. 48,000 and Rs. 47,000 respectively. Their respective no load losses and load losses were quoted as 65 W and 1130 W, 175 W and 1180 W, 180 W and 1235 W. Find out the most economical offer of the three.
3. A 100 kVA, 11/0.415 kV transformer has no load losses of 1 kW and load losses of 1.5 kW. Determine the kVA loading at which the transformer efficiency is the maximum. What will be this maximum efficiency? Assume the load power factor as 0.9 lagging.
4. Calculate the all-day efficiency of a transformer having a maximum efficiency of 98 per cent at 25 kVA at unity power factor and loaded as follows: 12 hours-2 kW at power factor 0.5 lag; 6 hours-12 kW at 0.8 power factor lag; 6 hours at no load.
5. Four industrial consumers, *a*, *b*, *c* and *d* apply for 11 kV supply with respective loads of 700 kW at power factor 0.8 lagging, 6 MW at 0.9 power factor lagging, 2 MW at 0.95 power factor lagging and 1250 kW at unity power factor, are situated at the corners of *abcd* square with 1 km side. Find out the optimum location of the 66/11 kV distribution sub-station to feed these loads.

BIBLIOGRAPHY

1. Belighter Charles S. and others, *Foundation of Optimization*, Prentice Hall of India, New Delhi, 1982.
2. Masud Enver, "An interactive procedure for sizing and timing of distribution sub-stations using optimisation techniques", *IEEE Trans.*, 93, Sept.-Oct., 1974, pp. 1281–1286.
3. 'Rationalization of Primary Distribution Networks', *Technical Report No. 137*, CPRI Bangalore, May 1986.

4. Heathcote J. Martin, '*The J&P Transformer Book*', 12th Edition, Newness, 1998, pp. 621–710 and 113–115.
5. '*Electric Power International*' ERA Technology, London, 1981, pp. A-19 to A-21.
6. Knight, U.G., *Power System Engineering and Mathematics*, Pergamon Press, London, 1972, pp. 126–127.
7. Willis, H. Lee., J.E.D. North Gategreece 'Comparison of several computerized Distribution planning methods' *IEEE Trans*, PAS, Jan. 1985, p. 233.
8. Pabla, A.S., "Average Power Distribution System" *Proceedings of Power India Technology in Transition Conference*, NPC, New Delhi, March 7–8, 1984.
9. *Standard Handbook for Electrical Engineers*, 12th edn. 1987, p. 17/3–7.
10. Petterson, L., "Estimation of the remaining service life of power transformers and their insulation," *Electra*, CIGRE, No. 133, Dec. 1990, pp. 65–71.
11. *Electra*, No. 162, October, 1995, p. 149.
12. Frauendorfer and others *Optimization in Planning and Operation of Electric Power Systems*, Verlag—Heidelberg-Germany, 1993.

Chapter 7

Reliability and Quality

7.1 Introduction

Two questions that often confront the power system design engineer are: (i) What degree of reliability does the system has? (ii) What do the varying degrees of reliability cost?

The answer to both these questions must necessarily involve the gathering and evaluation of considerable statistical data to achieve compatibility between an acceptable degree of reliability and minimum economic cost. The question that assumes importance is: What is an acceptable degree of reliability?

Reliability engineering with regard to distribution systems involves gathering outage data and evaluating system designs. System designs are evaluated and compared with alternative circuit configurations, sectionalizing provisions, protective schemes and automation. Therefore, the pressing need is for data on failure rates and repair times of components used in the distribution system so that reliability calculations can be more meaningful.

It is well known that distribution systems are affected by stochastic events such as faults on lines, sudden failures of power plants and random fluctuations in demand. Probabilistic methods are therefore essential for a sound assessment of the reliability of power systems. It is clear that competition in future will increase the level of uncertainty that affects operation and planning of the distribution system. To increase the reliability, it is necessary to understand the causes of outages and types of equipment failures. The most typical causes of outages are:

- Power utility's equipment failure
- Consumer's equipment failure
- Dig-in-for cables
- Trees
- Pollution
- Storm
- Flood
- Lightning
- Wear and tear
- Accident
- Power shortage
- System inadequacy
- Power theft
- Lack of consumer care.

The tropical environment (high temperature, dust, high humidity, heavy rainfall, high wind velocities, severe thunderstorms etc.) in most parts of the country accounts for the major outages. Another contributing factor is poor workmanship in most of the power utilities. For example, a well managed utility like Reliance Energy Limited has surpassed the expectations of its consumers by maintaining a reliability of 99.99 per cent in Mumbai. In India, about 25 per cent of the distribution system is 20 to 35 years old. These old networks are operating beyond their life expectancy, making the system less reliable. During the last few years, the unreliable grid power in the country has prompted large industrial consumers to invest in captive facilities. Most of these consumers have 60–80 per cent of the grid power as back up, even though the cost of captive power is 1.25 to 2 times that of the grid power. The captive power capacity in the country is about 22,000 MW ending March 2001, almost 25 per cent of the total power grid capacity.

The information technology sector is not taking any chance. Most of them have installed diesel generating sets and UPS machines to take care of sudden disruption in supply. Domestic and commercial consumers have also installed portable generating sets and/or UPS systems and *inverters*. In the year 2000-01 alone, nearly 1,50,000 portable generating sets and about 5,00,000 inverters were sold. Inverters used in residential houses have an efficiency of not more than 50 per cent, thus reducing efficiency at the consumer's end. This step is against the energy conservation policy. Also, UPS and inverters adds harmonics in the system. However, the advantage to the power utility is that these devices act as a demand-management system. Inverters charge batteries during the non-peak period and save the transient load at the peak period (during power cuts). The reliability of consumer supply has increased with the added cost [1].

7.1.1 Outage Data

Statistical information to determine the average failure rate of each component and its down time or repair/maintenance is based on data collected and averaged over many years. This requires an efficient fault recording system with complete details of time and location of each fault, information regarding which component in an item of equipment has failed, possible cause of failure, the age of the component, any previous failures of the component and the influence of external factors, such as weather, action of utility staff, general public, animals, birds, etc. Regular updating of the long-term average fault data information is desirable as this helps assess any improvement achieved due to maintenance procedure and the effect of removing equipment from a system prone to failure.

7.2 Definition of Reliability

From the engineering point of view a simple definition of reliability is that it is the probability that an item or a collection of items will perform satisfactorily, under specified conditions during a given period. The period assigned is thus an essential part of reliability specification. The period may be the complete lifetime of the item or any permissible period during maintenance.

7.3 Failure

When a device is put into operation there are initially a large number of premature failures which diminish with time as the device is debugged. During the working life of a mature design there are few failures. These will generally be of a random nature with a constant failure rate.

Finally, during the wear-out phase the failure rate increases until it is uneconomic to keep the device in operation. Three typical life cycles are illustrated in Fig. 7.1.

The failure rate is defined as:

$$\lambda = \frac{\text{Total number of failures}}{\text{Total of unit test or operating times}} = \frac{N}{\Sigma T_t}$$

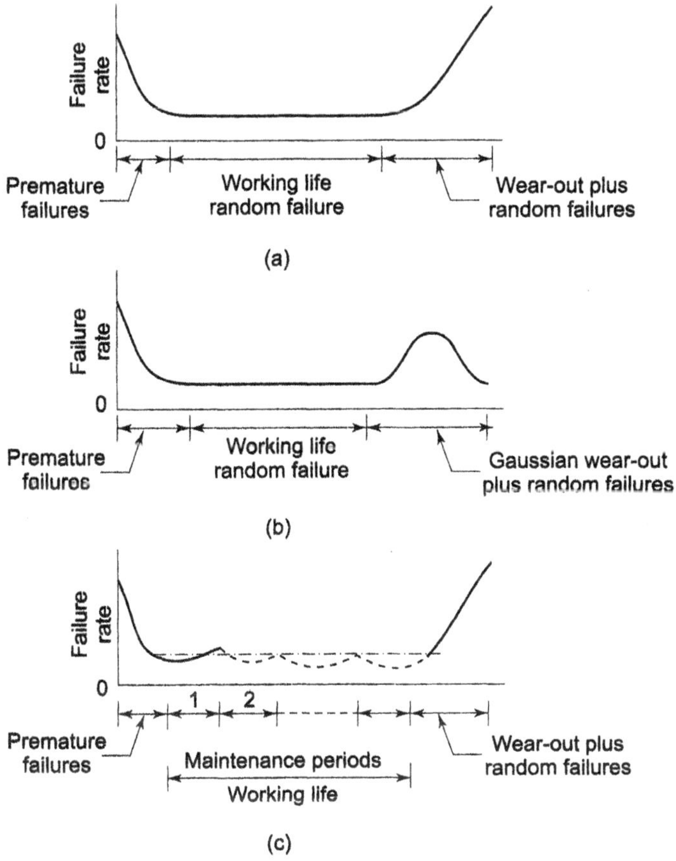

Fig. 7.1 *Life failure rate curves:*
(a) General nature curve showing constant failure rate in working life;
(b) General nature curve showing Gaussian distribution of wear-out plus random failure and constant failure rate during working life;
(c) Curve showing failure rate as constant and random in real life.

The reciprocal of λ i.e. $1/\lambda$ is defined as the mean time m between failures, i.e. $m = 1/\lambda$. In Fig. 7.1, the curves a and b show that during the working life, the failure rate is random and constant. This is not necessarily so. However, for a large collection of devices, random or constant failure is a fairly correct assumption. For general reliability assessment, curve c approximates closely to real cases.

If during the working life of a device failures are assumed to be random or constant, they can be described by an exponential or Poisson distribution. This greatly simplifies reliability assessment particularly where the failures are infrequent.

7.4 Probability Concepts

7.4.1 Distributions

Assuming very elementary concepts of probability as being generally understood, we consider the case of a device which has a probability of failure p after a given number of years of operation. The probability that the device will still be operating after this period is $q = 1 - p$. If there are n such devices, the probability that x of the devices have failed for mutually exclusive events is given by Bernoulli's theorem, i.e.

$$f(x) = n![x!\ (n - x)!]p^x\ q^{n-x}$$

This relationship is also called the binomial distribution since the probability of obtaining x out of n trials corresponds to the terms in the binomial expansion of $(p + q)^n$, i.e.

$$(p + q)^n = p^n + np^{n-1}\ q + \frac{n!}{[x!(n - x)!]}\ p^x\ q^{n-x} + \cdots + q^n$$

From the binomial distribution it is possible to find out how often the estimated probability is equal to the true probability. Then if estimated probability is x/n, it is correct if $x/n = p$. The probability that the estimated probability is the true probability is then given by

$$f(n\,p) = n!/[(n\,p)!\ (n - np)!] \times p^{np}\ q^{n-np}$$

When $n \to \propto$ and $f(np) \to 1$.

The binomial distribution is a discrete distribution and has points only where x is an integer. The standard deviation σ and mean μ of a binomial distribution are given by:

$$\sigma = \sqrt{npq} \text{ and } \mu = np$$

If p is very small and np is very large as compared to p, the binomial distribution can be shown to be:

$$f(x) = \frac{\left[e^{-np}(np)^x\right]}{x!} = \frac{e^{-\mu}\mu^x}{x!}$$

This equation is called the Poisson distribution for which Standard deviation

$$\sigma = \sqrt{np} = \sqrt{\mu}$$

Mean $\mu = np$.

If p is large and n is also large, e.g. when $p \geq 1/2$ and $np \geq 5$, the binomial distribution can be shown to be:

$$f(x) \simeq \frac{1}{\sqrt{2\pi}\cdot\sqrt{npq}} \times e^{-1/2[(x-np)(x-np)/npq]}$$

$$= \frac{1}{\sqrt{2\pi\cdot\delta}} \times e^{-1/2[(x-\mu/\delta)(x-\mu/\delta)]}$$

where,

$$\mu = np$$

$$\delta = \sqrt{npq}$$

This equation is called the Gaussian or normal distribution.

The Poisson distribution which covers rare occurrences, such as failures of devices, is extremely useful in reliability since a knowledge of the mean $\mu = np$ is sufficient to apply the distribution.

If a device fails with an average failure rate of λ with respect to time, the mean number of device failures in time, t is $\lambda\ t$. The probability for x failures in time t is then given by the Poisson distribution:

$$\frac{e^{-\lambda t}(\lambda t)^x}{x!}$$

Putting $x = 0$, the probability of no failures is $f(0) = e^{-\lambda t}$.

For any number of failures with a frequency λ/unit time:

$$f(n) = e^{-\lambda}\left[1 + \lambda + \frac{\lambda^2}{2!} + \frac{\lambda^3}{3!} + \cdots + \frac{\lambda^n}{n!}\right]$$

The probability that no failures occur for period $0 \rightarrow t$ is called the *survival* or *reliability function* $R(t)$, i.e.

$$R(t) = e^{-\lambda t} = e^{-t/m}$$

This standard reliability curve is plotted in Fig. 7.2. It is important to note that if we consider a period equal to the mean time between failures m, a device has only a 37 per cent probability of surviving this period.

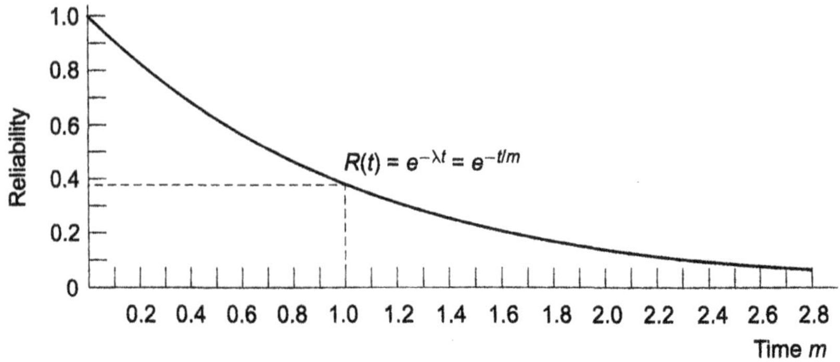

Fig. 7.2 *Reliability function*

7.4.2 Failure Distributions

In determining a particular failure distribution there are no absolute rules as regards assigning a particular mathematical relationship to the distribution. Also, as in all statistical problems, this must inherently be based on the concept of "best fit". The commonly used forms of failure distribution with a brief discussion on their particular applications are given below.

Exponential or Poisson Distribution

This distribution has been discussed in the previous section. It has a constant failure rate and is most readily applicable for distributions where the failure rate is small and the sample large.

Weibull Distribution

This type of distribution was originally developed empirically to fit increasing failure rates and as can be seen for $m > 1$ (m is a measure of skewness of the distribution curve), the failure rate increases, and for $m < 1$ the failure rate decreases. If $m = 1$, the Weibull distribution becomes an exponential distribution. Professor Weibull, after whom the distribution is named, used this form to describe the fatigue life of materials. Voltage aging of ac power supply is governed by this distribution [6]. The time for the breakdown of air gaps for a given applied voltage can be fitted to this distribution [8].

Gamma Distribution

This is skewed family of distributions. The failure rate increases for $m > 1$ and is bounded by λ for $m < 1$ with the failure rate decreasing. The gamma distribution can be considered dual to the exponential or Poisson distribution in that the Poisson distribution uses time as the variable while the gamma distribution uses rate as the variable. However, both the functions characterize the same process.

Normal Distribution

This type of distribution is of great use in reliability studies of distribution systems. Switching surge overvoltage strength of air insulation and switching surge overvoltage magnitudes in a power system generally obey this law. In other words, the distribution of breakdown probability as a function of applied overvoltages for air gaps obey the normal law of probability distribution. Distribution of mechanical strengths of poles, X-arms, insulators, conductors of overhead power lines follow this law [9].

Log Normal Distribution

In this distribution, the failure rate increases at first but then decreases and eventually approaches zero. This distribution has been found suitable for describing the length of time to repair a piece of equipment successfully. The occurrence of lightning overvoltages and overcurrent magnitudes in a system obeys this distribution.

7.4.3 Reliability, Availability and Maintainability Engineering Techniques (RAM) (7)

The power distribution system comprises equipment which is identified with a specific consumer or group of consumers, and the effect of consumer habits on distribution-equipment failures must be well defined. Also, the reliability characteristics of distribution system are easy to analyse, because most of the outage risk arises from single contingency events on account of being the radial feeder supply usually. Therefore, the "series system" model is generally applicable. In the case of large industrial plants, where a dual-feed arrangement is supplied, the reliability analysis becomes more complicated but it is still well within the capability of existing techniques and computer programmes are available to carry out the calculations.

The following RAM relations can be used to measure the field performance of equipment supplied by different suppliers:

1. Reliability can be determined by a formula that provides the Mean Time Between Failures (MTBF) of equipment (or its inverse–the failure rate of equipment). This formula provides a reliability 'figure of merit' and is given by:

$$\text{MTBF} = 1/\lambda = \frac{2T}{\chi^2\,(\text{DF})\,(\text{CONF})}$$

where λ is the failure rate, T is the operating time, χ^2 (chi square) is a value obtained from statistical tables, DF is the degree of freedom (numerically equal to twice the number of failures plus one), and CONF is the confidence level (usually about 90 per cent). The reliability of an electrical system is limited by the weakest element which fails first. Generally, the failure rate of a component depends on four main factors:

- Quality
- Temperature
- Stress (e.g. voltage)
- Environment (e.g. humidity, salt, dust in the atmosphere, vibrations etc.)

It is desirable to obtain MTBF and MTTR estimates for systems' components on the basis of field experience and experimental testing. These data will facilitate calculating the reliability of the system.

2. Maintainability is a measure of the Mean Time To Repair (MTTR) and is given by the formula:

$$\frac{\Sigma(\lambda R_t)}{\Sigma\lambda}$$

where R_t is the repair time.

3. Availability is a measure of the design availability for equipment comparison and trade-offs and is given by:

$$\frac{\text{MTBF}}{\text{MTBF} + \text{MTTR}}$$

Availability of distribution system represents an average fraction (per unit) of time during which power and energy could be delivered. In a well-developed system, the figure reaches a value of up to 0.995 so that average down time per year is as low as 15 hours.

It is desirable to specify MTBF and the availability figures in purchase orders for equipment along with the failure rates of major transmission and distribution equipment. The purchases should be based on the lowest life cycle cost, i.e. the lowest initial cost plus the present cost of future annual costs and outages of the system due to the failure of a particular component. Suitable life testing of the equipment is imperative to calculate and quantify the system or component reliability, availability and maintainability.

7.5 Limitations of Distribution Systems

The limitations that must be taken into account for the correct and reliable and qualitative working of a power distribution system are given below.

Thermal

Temperature determines the limiting load currents. It is significant that the limiting loads in this regard for typical system elements are set by thermal rather than mechanical considerations, and that such limits vary appreciably for different patterns of loadings and climatic conditions. Thus a range of ratings will be available to coordinate with loadings under various conditions.

Economic

Some cost is involved in the heat dissipated in the system components, i.e. in terms of losses. The economic level of loading is reached when the cost of the losses equals the cost of reducing the losses (on an annual charge basis).

Even in densely loaded urban situation, the normal component loadings are usually well below the economic limit. This justifies the exceeding of this particular limit for a relatively short time in an emergency.

Voltage Drop

The thermal and economic limits are directly related to the magnitude of the load, i.e. "quantity of supply". An indirect limit is also set by permissible voltage drops. Voltage limits as seen by consumer, are set, which then become one of the determining factors in system component design. It is usual to allocate a permissible voltage drop through each system component which becomes, in part, a limit of the magnitude of the load flow.

Unbalanced LT supply limits are defined to avoid undue overheating and losses in the system and malfunctioning of motors.

Fault Current Capability

The fault levels in urban areas have reached a very high magnitude with the continuing augmentation of transmission and distribution systems. Depending on system arrangements, the switchgear must be capable of either making, or making and breaking the available fault levels. With regard to feeders, duration of the fault flow is equally important, and is set by the switchgear capabilities and relay grading requirements. This quite often causes problems in a developing urban area, where more than one step of overload relay grading may be necessary within a particular feeder which incorporates associated small section spurs or tee-connections. Similarly, auto-reclosing which may be justified on reliability grounds can unduly stress line conductors from a fault capability view point.

Over-Voltages

Apart from voltage drops, components must be capable of withstanding abnormal voltage surges generated from within the system or by external sources.

Voltage Flicker and Dips

The magnitude and frequency of fluctuations which can result in annoying disturbances are well documented. Where equipment is operated under unacceptable levels, special supply arrangements are necessary and the operating staff must be aware that no switching can be tolerated which will parallel such loads with general network consumers. For example, separate LV distribution and sometimes transformers may be installed for the supply of welding loads. Supplies to arc furnace loads are usually from separate power transformers and feeders at a major supply point. Services rules are necessary to protect the consumers from interference caused by devices such as thyristor controls. Other obvious causes of voltage disturbances or dips (> 10 per cent) are system faults and heavy current testing.

A standard developed by computer manufacturers in the U.S.A. based on the IEEE Standard 446–1987, outlines minimum over and under side through capabilities for computer as a function of the duration of the disturbance as shown in Fig. 7.3. For example, a computer should be able to withstand a dip to zero volts for 0.8 cycles, or a dip to 87 per cent of rated voltage for 100 cycles.

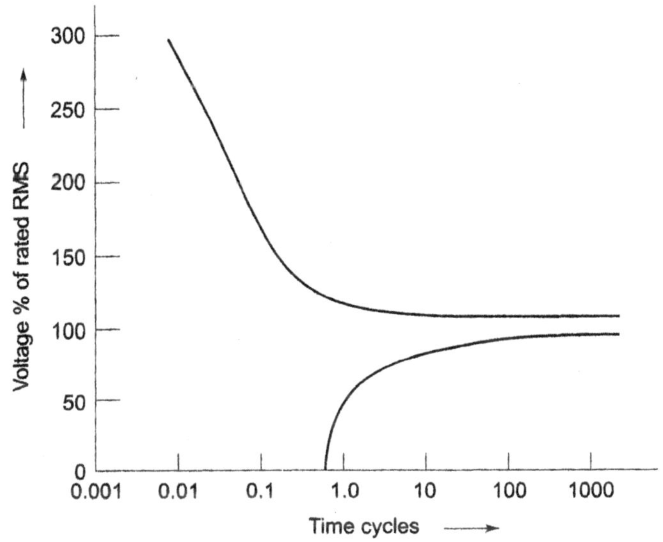

Fig. 7.3 *Voltage tolerance for computers*

Harmonic Interference

Electronic equipment manufacturers are quite vociferous in increasing their demands for reasonably pure sine wave supply to ensure optimum operation of computer installation, X-ray equipment, hi-fi audio and video equipment. They are very critical of the control tones or frequency injections imposed by the supply authorities and yet the same manufacturers make and market such devices as thyristors. It is essential that the equipment which generates harmonics higher in magnitude than the accepted levels are isolated from the general supplies. Such supplies may be traction and other rectifier loads and must be treated like arc furnace loads.

The generation of subharmonic frequencies by an induction furnace may cause considerable flicker to fluorescent lighting.

Frequency

(a) As per IE rule 55, a power frequency variation of 50 ± 1.5 Hz is permissible.
(b) The target for maintaining synchronous time within ± 3 s may be an important consideration, frequency should in which case be normally kept within ± 1 Hz. In the integrated system, the frequency limits could be achieved by installing sensitive automatic switching pump storage hydro-generation schemes.

Risk Analysis

Experience indicates that most distribution service interruptions are initiated by severe weather-related interruptions with a major contributor being inadequate maintenance. The coordination of preventive maintenance scheduling with reliability analysis can be very effective. System reliability can be improved by timely identification and response to failures. Contingency analysis helps determine the weakest spots of the distribution system. The special form of contingency analysis in which the probability of a given contingency is clearly and precisely expressed is known as *risk analysis.* Such analysis is desirable for important and essential supply feeders every year well before the start of summer/rainy season.

7.6 Power Quality

Reliability defines a number of interruptions in supply, their duration and frequency. The term *power quality* refers mainly to maintenance of the ac bus voltage waveform at rated frequency, undistorted, balanced and at rated voltage. The main quality problems are: Harmonics and momentary events (e.g., sags/swells). Voltage sags at one bus are caused by faults elsewhere in the system. They have received special attention now because spectacular failures have highlighted the problems that result from the performance of computer-controlled loads and adjustable speed drives during these voltage sags/swells. Sags/swells not only cause considerable productivity loss but they are also hard to control. Interestingly, sags/swells are events of short duration but high impact. The total sag duration is measured in cycles—not minutes or hours.

Harmonics, on the other hand, are continuous and omnipresent. Their effect should be integrated over time with additional magnetic circuit and I^2R losses. Other aspects of power quality are:

- Frequency variation
- Voltage imbalance

Frequency: The frequency of supply depends on the total balance between energy production and consumption and is generally regulated by a grid system operator. In India, there is large variation of grid frequency—47.5 to 53 Hz. (see Section 2.1.2). *Voltage imbalance:* (see Section 4.13.2).

The concern for power quality is due to continuing increase in the overall electronic loads. The industrial operations of modern plants are built around sophisticated electronic controls that are directly affected by voltage surges, momentary interruption in service and harmonic distortions. Such electronic controls are computers, adjustable speed drives and programmable logic devices. Some of the industries prone to power quality disturbances are:

Category	*Typical industry*
Continuous process	Paper, textiles, plastics
Precision machinery	Automobile parts manufacturing
Information technology	Data processing centres, banks, telecommunications, broadcasting
Safety and security-related	Hazardous process, Chemical processing, Hospital and health care facilities, Military installations, Power plant auxiliaries, Large transmission sub-stations.

7.6.1 Quality Problems

For a consumer, quality problems are caused by:

(a) The distribution system which creates disturbances travelling to the consumer's end.
   - The system transformers' magnetising current causes harmonics.
   - Heavily loaded transmission and distribution system and system faults produce voltage sags, swells and magnify imbalance in the system [4].

(b) Other consumers: Non-linear loads of other consumers such as welding sets, rectifiers, variable speed drives, arc/induction furnaces requiring variable high reactive power, large motor starting, inadequately designed wiring, heavy unbalanced loading, voltage swings and imbalance. These are transmitted to nearby consumers through the distribution system.

(c) Consumer's own equipment: Consumer's own non-linear loads create harmonics and voltage problems as listed under (b) above.

The symptoms of poor power quality are wide-ranging, of which the more common ones include:

- Hot neutral connections;
- Equipment malfunction of failure;
- Power factor correction capacitor failure;
- Frequent replacement of consumables, e.g. lighting; and
- Flickering lights.

EXAMPLE: As per the study carried out in India during the year 2001 by the World Bank [10], the voltages of power supplied to agriculture fluctuated frequently and over a wide range while the three-phase power supply was subjected to significant imbalances leading to large pump motor burn-outs. The minimum voltage level was recorded as low as 60 per cent of the rated requirement and farmers faced maximum phase imbalance of more than 30 per cent.

7.6.2 Power Quality Variations

The variations in power quality and possible power conditioning solutions, are listed in Table 7.1. These variations are shown in Fig. 7.4.

Table 7.1

Power quality variations [17]

Type of variations	*Method of characterizing*	*Defining characteristics*	*Causes*	*Power conditioning solution*
1. Impulsive transients voltages	Magnitude duration	Unidirectional, typical duration < 200 micro sec	Lightning, Load switching	Surge arresters Filters Isolation transformers
2. Oscillatory transients voltages	Waveforms	Decaying oscillation Components < 500 Hz Components 500 Hz-2000 Hz Components > 2000 Hz	Lightning Line/Cable switching Capacitor switching Transformer switching Load switching	Surge arresters Filters Isolation transformers
3. Voltage sags/surges	Waveforms RMS vs. Time	Duration 0.5-30 cycles Typical Mag. 0.1–0.9 pu Typical Mag. 1.05–1.73 pu	Remote faults	Constant voltage transformers
4. Undervoltages/ overvoltages	RMS vs. Time	Duration > 30 cycles	Motor starting Load changes Compensation changes	Voltage regulators, Constant voltage transformers
5. Interruptions	Duration	Complete loss of voltage Duration < 2 sec. Duration 2 sec. – 2 min. Duration > 2 min	Breaker operations (fault clearing) Maintenance	UPS systems Backup generators
6. Harmonic distortion	Waveforms Harm. spectrums THD	Continuous distortion (V or I) Components up to 50th harmonic	Nonlinear loads System response characteristics.	Filters Isolation transformers (zero sequence)

Contd.

Type of variations	Method of characterizing	Defining characteristics	Causes	Power conditioning solution
7. Voltage flicker	Magnitude, Frequency of modulation	Intermittent variations in 50 Hz voltage magnitude Frequency components < 25 Hz	Intermittent loads Arcing loads Motor starting	Static var systems Series capacitors
8. Noise	Magnitude Coupling method Frequency	Continuous high frequency Components of V or I Frequency components > 3000 Hz	Power electronics Switching Arcing Electromagnetic radiation	Wiring and grounding improvements, Chokes, filters, Shielding
9. Electrostatic discharge	Peak voltage Peak current		Discharge of static electric charge	Humidity control filters, shielding

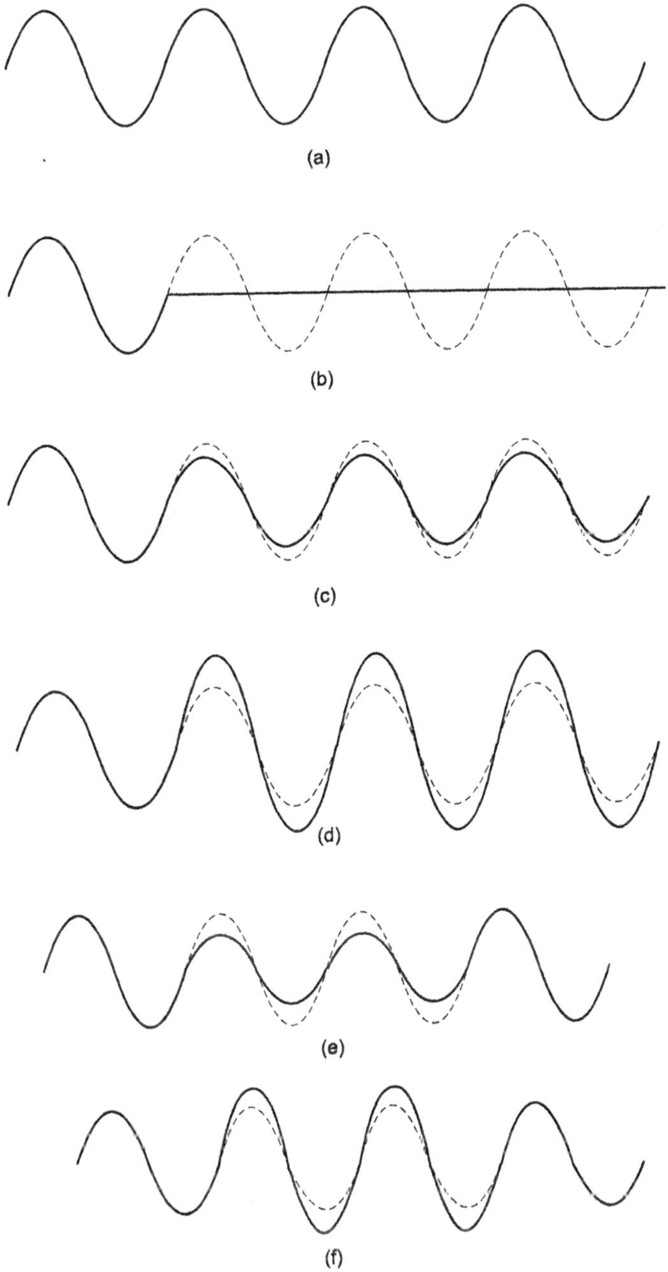

Fig. 7.4 *Power supply voltage disturbances (Contd)*
(a) Normal power supply (b) Power cut (c) Under voltage (d) Over-voltage (e) Sag (f) Swell

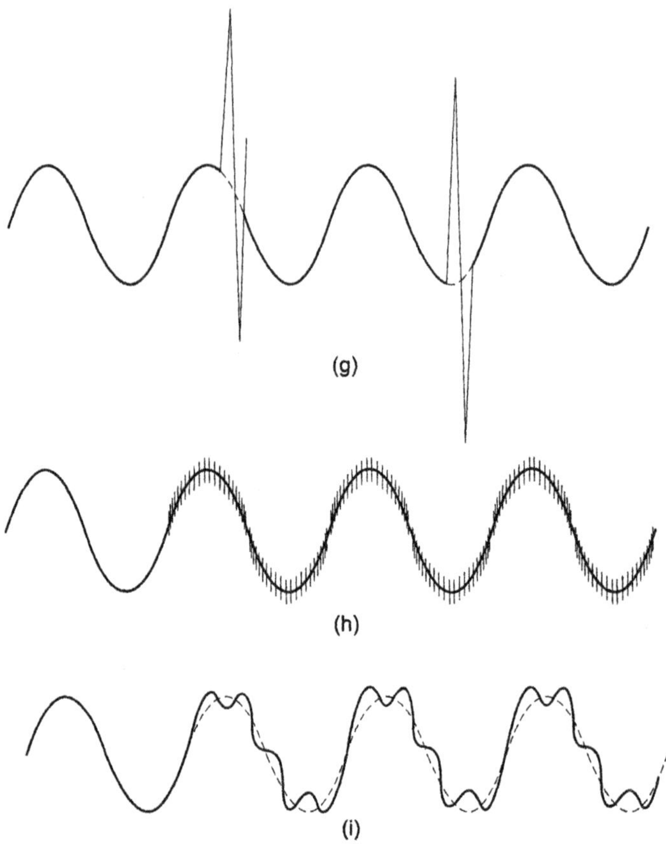

Fig. 7.4 *Power supply voltage disturbances*
(g) Transient over-voltages (h) Radio frequency interference
(i) Harmonics

7.7 Reliability Measurement

7.7.1 Method

The measurement of reliability in quantitative terms have been increasingly used in recent years to achieve consistency in system performance, assist in making a weighted choice between alternative planning schemes to improve the performance of a particular network, and, generally, to provide the background information which can become a basis for evaluating day-to-day planning [5].

The reliability indices generally applied are:

- **SAIDI:** *System Average Interruption Duration Index*, which is the average total duration of interruptions of supply that a consumer experiences annually. For example, this index is 1.5 minutes in Singapore (1998), 88 minutes in U.S.A., 52 minutes in France [2].
- **SAIFI:** *System Average Interruption Frequency Index*, which is the average number of interruptions of supply that a consumer experiences annually. It is average 1.5 in U.S.A. [2]
- **CAIDI:** *Consumer Average Interruption Duration Index*, which is the average duration of an interruption of supply for a consumer, who experiences the interruptions of supply annually.
- **ASAI:** *Average Service Availability Index* is

$$\frac{\text{Consumer hours service availability}}{\text{Consumer hours service demand}}$$

The usual measures of reliability are frequency of interruptions and their total duration. It may be in terms of expected number of days per year in which there is insufficient generation to meet power peaks of the days or energy of the days. The more common measure as a planning tool is the total duration of the outage time per year. As far as reliability as seen by the consumer is concerned, the total supply system is subdivided, in general, into the following three parts.

(a) Power stations, main transmission and sub-transmission systems down to the 11 or 22 kV busbar of the sub-station.

(b) The HV distribution system, i.e. from the 11 or 22 kV busbar at the sub-station to the 415 V at the distribution centre.

(c) The 415 V distribution system which also contributes to the total system.

Assessment of System Reliability

The following steps are required for assessing system reliability.

1. Draw the reliability model diagram of the system, i.e. in series or parallel or a combination of both. The system redundancy is an important consideration for higher reliability [12].

- *Series system:* Two components in series having reliability R_1 and R_2 will have a total reliability of $R_1 \times R_2$. For n components in the series, the total reliability is given by:

$$R_s = R_1 \times R_2 \times R_3 \times R_4 \cdots R_n$$

It is called the *product law of reliabilities.*

- *Parallel system:* If the above two components are in parallel, their total reliability will be

$$1 - (1 - R_1) \times (1 - R_2).$$

For n components in parallel, the total reliability is given by:

$$R_p = 1 - [(1 - R_1) \times (1 - R_2)(1 - R_3) \times (1 - R_4) \cdots (1 - R_n)$$

This is called the ***product law of un-reliabilities in parallel operation.***

- *Series-parallel combination:* This is equivalent of interconnected power distribution system. Figure 7.5a shows a series-parallel system, having a high level of redundancy. The equivalent reliability of the system (R_{sys}) with n parallel paths and each path containing m components, can be expressed as:

$$R_{sys} = 1 - (1 - R^m)^n$$

Figure 7.5b shows another series-parallel system (m series banks and each bank having n parallel components) which has a comparatively low level of redundancy. The equivalent reliability of the system will be:

$$R_{sys} = [1 - (1 - R)^n]^m$$

where R = reliability of a component, and

R^m = equivalent reliability of series path of m components (Fig. 7.5a).

$1 - (1 - R)^n$ = equivalent reliability of a bank having n components in parallel (Fig. 7.5 b)

The above formulae can used for reliability analysis and fault-tree analysis in the system.

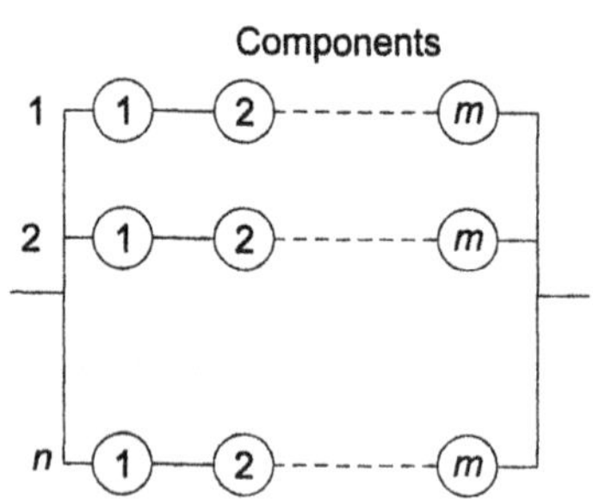

Fig. 7.5a *Series-parallel system*

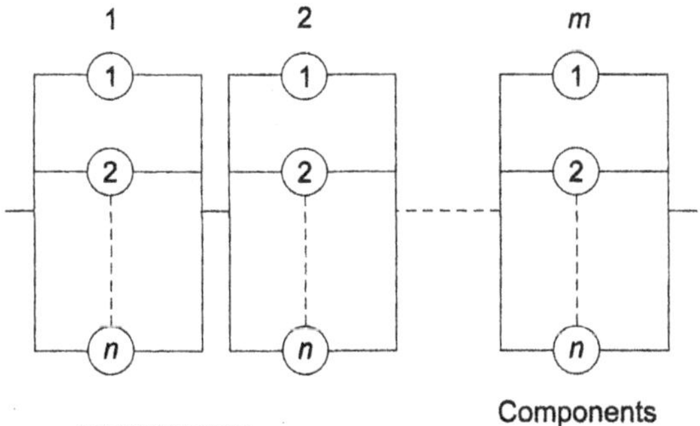

Fig. 7.5b *Series-parallel system*

2. Ascertain the data regarding the failure and repair rates of the individual components of the system, fault and maltripping interruptions of switchgear, outages due to scheduled maintenance, power restrictions and strike threats of field staff or trade union.

To convert the performance figures to outage times, the average of time taken to restore the supply for each type of failure is required. A variety of restoration methods are available and, naturally, the easiest one available in the circumstances should be selected.

3. The performance data in terms of outage time are synthesized into a reliability model to find out the system outage time as alternatives to the restoration of the working system.

7.7.2 Consumer Interruption Costs

There is increasing interest in determining the cost of unreliability. The determination of interruption costs involves an understanding of the nature and variety of consumer impact resulting from electric supply interruption. It is related both to consumer and interruption characteristics. (i) Consumer characteristics are: type of consumer; nature of consumer activities; size of operation and other demographic data; demand and energy requirement; energy dependency as a function of time-of-day, etc. (ii) Interruption characteristics are: duration; frequency and time of occurrence of interruptions; whether an interruption is complete or partial; if advance warning of the interruption is given by the utility; whether the area affected is local or widespread.

Finally, the impact of an outage is partially dependent on the attitude and preparedness of consumers which, in turn, is related to the existing reliability standard. These points can be checked by examining data collected from consumer survey. There are a wide range of methods [15] to estimate consumer interruption costs. Such costs can also depend on the consumer's willingness to pay for the supply interruption. As a simple example, in a developing country, interruption cost for commercial consumers is virtually nil as consumers usually carry out their operations despite the outage. Similarly, domestic consumers may complain of missing TV programmes, loss of comfort and so on. In a developed country, the considerations may be different including additional aspects as those of safety, personal injury, losses arising from civil disobedience, arson, etc. However, in case of industrial consumers, there is production loss, manpower loss, bearing the stand-by generation cost (capital as well as running). The interruptions costs per consumer of a category are determined in terms of Rs/kW or Rs/kWh. Electrical energy is not, however, a simple commodity. Unlike other forms of energy, it cannot be stored easily in large quantities. Continuity of supply thus has a value that can be much higher than the cost of the energy consumed. For India's businesses, power interruptions translate to annual costs of approximately Rs 5,000 crores in lost data, materials and productivity. Consumer's power interruption cost varies from one region to another. In most cases, this cost is lower if prior information/ notification is given. Other factors are time-of-use, consumer cultural level and economic standing. As per CIGRE Study Committee-39 (Bangkok), November 1989 report, the consumer end losses in terms of cost of undelivered energy is taken 20 times the cost of supply.

7.7.3 Optimum Reliability

Higher reliability can be achieved by the installation of better equipment or by providing more redundancy. Capital and the operating costs are associated with both solutions, and the capitalized value of these investments is the price the utility has to pay for an intended higher level of reliability. Actual values of these costs, as a function of system reliability, can be represented by an increasing curve, C in Fig. 7.6. The reliability measure on the abscissa may be any suitable index, such as LOLP for generating system studies, or the system availability, or kWh actually supplied divided by the kWh which would be supplied if there were no service interruptions for transmission and distribution system.

The cost of interruptions can be considered from the point of view of the electrical utility, or from the point of view of the entire consumers. In the first case, the cost consists mainly of the net loss of revenue from load not served; in the second, the losses suffered by all the consumers must be taken into account. For industrial consumers, these losses can be quite extensive; they may be due to lost production, spoiled products, idle labour, and so on. For residential consumers, the losses arising from serving interruptions are rather intangible; they are mostly associated with comfort and convenience. It is evident that capital cost of all interruptions is decreasing function D of system reliability. When C and D are known, total cost associated with various levels of system reliability is given by the sum of two curves, and optimal reliability cost at point 'O' evaluated.

7.7.4 Reliability and Quality Goals

Each power utility must set reliability goals based on local conditions, particularly local consumer requirements. A power utility must prepare operation codal instructions to undertake steps to restore supply. Mobile sets of a sub-station, a distribution transformer and a diesel generator should be constructed to restore supply in emergent conditions for certain consumers. A time schedule to restore power supply after a system fault may be given in the operation code. For example, in the United Kingdom, UK Planning Standard P.2/5 specifies the following times to restore supply after a fault in the system.

Consumer load	*1st interruption*	*2nd interruption*
Up to 1 MW	Repair time	Repair time
>1 MW $\leq$ 12 MW	3 hours	Repair time
12 MW $\leq$ 60 MW	15 minutes	Repair time
60 MW $\leq$ 3000 MW	Immediate	3 hours

Power cuts should be notified in advance in case of power shortages. The notice time in advance for any scheduled maintenance may be specified depending upon the importance of the consumer as an industrial process or public importance. For example, the notice period can be instantaneous, 1 hour, 24 hours, 48 hours, one week no interruptions for consumers with a connected load of 50 kW, 1 MW, 10 MW, 20 MW, 50 MW and above 50 MW respectively.

Another example of Energy Australia (ES2-Electricity Supply Standards—December 1998) is as follows:

If Energy Australia: (i) fails to provide to the consumer at least two working days notice of any planned interruption, or (ii) interrupts the consumer supply for longer than indicated in the notice, then Energy Australia will give the consumer 20 dollars compensation for the interruption of supply.

In India, the Orissa Electricity Regulation Commission has taken the lead in setting standards of service for distribution utilities. They are required to bring down voltage fluctuations within specified limits within 15 working days of the complaint in 60 per cent of the cases, and have to restore supply within 24 hours of the fault in 80 per cent of the cases. These standards are backed by provisions for the imposition of penalties for violations. There are procedures for grievance redressal. The regulatory commission in Andhra Pradesh has also taken initiatives to ensure quality. In Himachal Pradesh, the Electricity Regulatory Commission has prescribed standards for the restoration of power supply within 6 to 12 hours in the urban areas and within 24 hours in the rural areas after the fuse failure complaint is lodged by the consumer. Besides, for the interruption of electricity supply to carry out

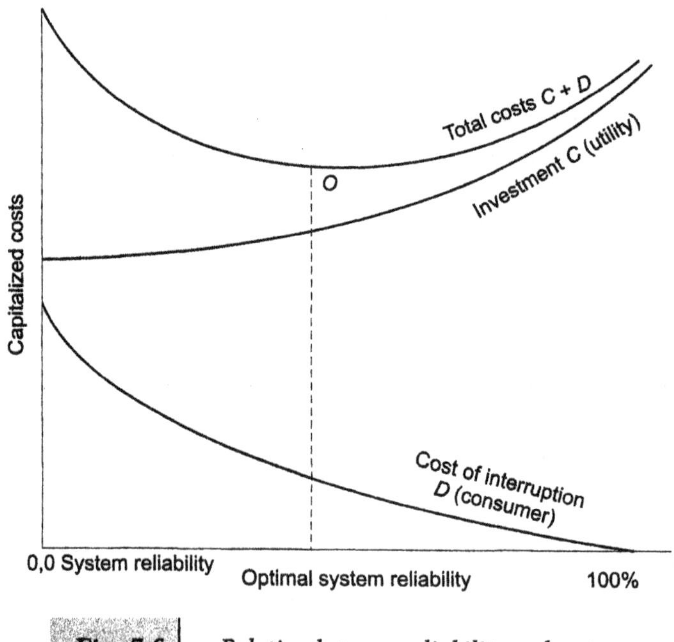

Fig. 7.6 *Relation between reliability and costs*

maintenance works, a 14-hour notice has to be given. Violation of these standards will entitle the consumer to a token compensation.

7.8 Power Supply Quality Survey (3)

A systematic power quality survey of the distribution system is of fundamental value before releasing the power connection for sensitive loads. It is common for consumers to use a power electronic system for power supply in the case of sensitive equipment. A power electronic system consists of a power source, filter, a power converter, a load and a control circuit. The block diagram is shown in Fig. 7.7.

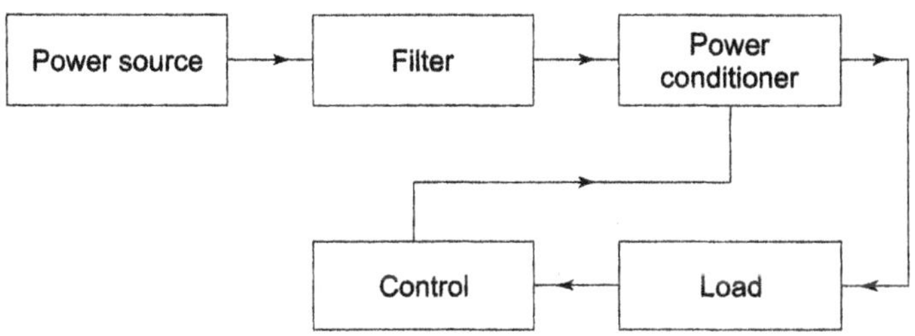

Fig. 7.7 *Clean power process*

The 'power source' could be three-phase or single-phase ac mains or it could be a portable supply such as a dc battery. A 'filter' may be necessary to prevent harmonics generated by the converter from being fed back to the mains or from being radiated into space (radio interference or noise). International standard: IEC 552-2 for electromagnetic compatibility (EMC) is applicable for manufacturers of power electronic equipment. The 'control' circuit monitors the condition at the load, compares this with present values and then adjusts the converter drive as necessary. A 'power conditioner' is an arrangement of semiconductor devices all operating in the switching mode.

7.8.1 Power Quality Monitoring

Power Quality (PQ) Monitoring keeps a check on the following PQ parameters (disturbances) of delivered power:

- Sags/surges
- Swells

- Dips
- Flicker
- Transient over-voltages
- Harmonics
- Frequency profiles
- Voltage and current unbalance

The measuring devices are installed on nodes in the network and/or the consumer end and provide the recording of every disturbance coming on the network as per IEC: 61000-4-7. A PQ contract can be made with consumers having sensitive load.

The internet provides an excellent opportunity to involve consumers in the monitoring process at a minimal additional cost to the utility. The data and analysis can be made available without the user having to install and learn specialized software. Security protocols implemented in the web server can ensure that only authorised users can access the data.

In addition, the server is so structured that users have read-only access to the data, helping to preserve its integrity. The basic system can consist of:

- Monitoring instruments at the nodes on the network and consumer end.
- Data collection computers that access and download data from instruments.
- A master server that processes the data and makes it available to authorised users.
- A power utility intranet or the internet that connects the user to the master station via file service or web pages.

Symptoms of various disturbances in distribution system and the possible remedial measures are given in Tables 7.2 and 7.3 respectively. 'Yes' in the tables, indicates a positive symptom and positive remedy respectively.

7.9 Reliability Aids

7.9.1 Mastering Information

Information is a resource that can be utilized to use all other resources in the best possible manner. The power systems are expanding and the

Table 7.2

Symptoms of power disturbances

Symptom	Disturbance				
	Short term voltage SWELL/SAG	Low nominal voltage	High nominal voltage	Transient	Harmonics
Computer data corruption	Yes	–	–	Yes	–
Electronic control malfunction	Yes	–	–	Yes	–
Flickering Lights	Yes	–	–	–	–
Low Lamp Life	–	–	Yes	–	–
Equipment overheating	–	–	Yes	–	Yes
Protection malfunction	Yes	–	–	Yes	Yes
Neutral currents	–	–	–	–	Yes
Motor failure	–	–	–	–	Yes
Power supply failure	–	–	Yes	Yes	–

Table 7.3

Power disturbances—remedial measures

Remedial measures	Disturbance				
	Short term voltage SWELL/SAG	Low nominal voltage	High nominal voltage	Transient	Harmonics
Utility					
Upgrade LV circuits	Yes	Yes	–	Yes	Yes
Install new transformer	Yes	Yes	–	–	–
Change to other LV circuit	Yes	Yes	Yes	Yes	Yes
Change on other HV circuit	Yes	Yes	–	–	–
Change transformer tap	–	Yes	Yes	–	–
Correct lightning arrester deployment	–	–	–	Yes	–
Install feeder automation scheme	Yes	–	–	Yes	–
Underground the supply	Yes	–	–	Yes	Yes
Stage switched cap. bank	Yes	–	Yes	Yes	–
Consumer					
Check grounding system	Yes	–	–	Yes	–
Check all connections	Yes	–	–	Yes	–
Check proximity of data to power cables	–	–	–	Yes	Yes
Upgrade LV circuits	Yes	Yes	–	Yes	Yes
Upgrade neutral cables/conductors	–	–	–	–	Yes
Install dedicated LV circuits	Yes	Yes	–	Yes	Yes

Contd.

Remedial measures	Disturbance				
	Short term voltage SWELL/SAG	Low nominal voltage	High nominal voltage	Transient	Harmonics
Replace worn contactors/switches	Yes	–	–	Yes	–
Change motor starting characteristics	Yes	–	–	Yes	–
Install PF correction capacitors	Yes	Yes	–	–	–
Install harmonic filter	–	–	–	–	Yes
Install electronic filters (wave tracking)	–	–	–	Yes	–
Install isolation transformer	–	–	–	Yes	–
Install voltage regulator	Yes	Yes	Yes	–	–
Install power conditioner	Yes	Yes	Yes	Yes	–
Install UPS	Yes	Yes	Yes	Yes	Yes

activities becoming complex and diversified. New tools for handling data are word processors/personal computers, optical scanners, microfilm storage, computer networks and Internet are aimed to lighten the information overload. This is a cycle: more diversity and change equals more information equals more technologies to handle information and may lead to still more diversity and change. That is the dynamics driving the information revolution. Furthermore, as we organize the information, we deepen our scientific understanding of the system and use information for transforming the processes of production and services. The successful deployment of information technology will make the enterprise dynamic, integrated, effective and responsible. Managing information (see Fig. 7.8) successfully mean processing information to add value in at least one of three ways by increasing:

- Productivity,
- Competency,
- Innovation.

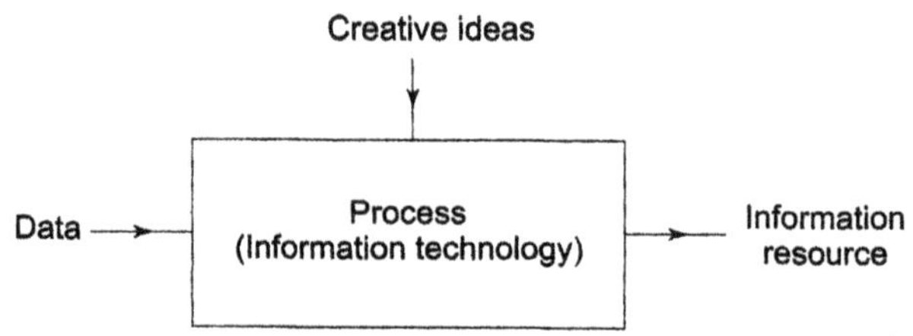

Fig. 7.8 *Information process*

The following points must be checked for the useful deployment of information technology:

- Does the information system unify the whole organisation, connecting various activities?
- Is the information system centrally managed and compatible throughout?
- Has the information system been designed by people who are going to use it?
- Is the information network widely accessible?
- Does it capture the information you need?
- Does it provide the support you need to make rapid, fact-based, accurate decisions?
- Does your information system allow you to know your consumers well: Track record regarding defaults, theft of energy, energy conservation, load profile, metering history etc.?

To avoid information overload, it is necessary to keep redefining the context in which knowledge is required so as to avoid storing everything and then not having any use for it. The significance of the information is only realised when it is used.

7.9.2 System Configuration

System configuration plays a very important part in reliability. For service continuity taking into account both long outages and momentary interruptions, the better system reliabilities are rated in the following order of merit (see Fig. 1.3):

1. Grid network
2. Spot network
3. Secondary selective
4. Primary selective
5. Primary loop
6. Radial

7.9.3 Standardization

It is imperative to evolve planning, design, operation and maintenance, material, constructional and instructional standards, codes in respect of the following with regular reviews and updating:

(a) Specification for the equipment of mass utilization such as lightning arresters, distribution transformers (three-phase and SWER system), cables, etc.

(b) Constructional standards for installing various equipment/connections, such as tapping arrangement for 11 kV lines, sagging and fuse mounting arrangement, etc.

(c) Manuals for the guidance of the field staff, such as for the repair of distribution transformers, design of 33/11 kV rural sub-stations, voltages drop calculations, erection of lines, cable laying, transportation and installation of energy meters in the field, etc.

(d) Training institutes need to be well established to bring the necessary awareness, skills and adroitness among the personnel working in the distribution system. The training may be laboratory or field through audio-visual aids, simulators and lectures and practical demonstrations. For example, it is essential that the operating staff be aware of the voltage ranges under various conditions of work so that assistance in either maintenance or emergency in the form of diesel generation, capacitors, and/or extra paralleling points can be arranged if the voltages are observed below minimum. It is desirable to prepare training manuals for linemen and operators.

(e) Software packages for different computer studies should be standardized. PERT/cost network analysis programmes for construction and maintenance works be evolved and standardized. Knowledge of Prima Vera and MS Project softwares is important.

(f) Accurate system diagrams are essential for reliable operation. Geographical plans of HV feeder routes enable pin-point a damage in a line. Estimated load peak records for winter and summer should be available in a chart form for lines and sub-station. Rating information should also be displayed, e.g. for sub-station transformers (sub-transmission system) and main attended type 11 kV sub-station.

7.9.4 Electrical Safety Rules (see Appendix V)

Minimum safety standards are mandatory and are framed in such a manner that the supply authorities can adopt them as working procedures. Supply authorities, in addition, frame their own rules which cover such aspects as requiring that all equipment must be regarded as alive until proved dead, protective clothing to be worn, demarcation of dangerous areas, safe working distances, use of safety equipment and insulated tools, and associated testing requirements. Methods of isolation, earthing, labelling, proving dead, use of tape barriers and the use of an access permit system are given. Special precautions to be observed in the use of equipment, such as tong tester, ammeters and working on the secondary windings of current and voltage transformers are also included in the rules.

7.9.5 Recent Developments

Recent technological developments are concerned with either cost saving techniques or improving performance. The benefits of any improved performance must, of course, be justified. The problem concerning transfer of technology or appropriate technology must be analysed according to the prevailing technological culture in a country.

PROBLEMS

1. A radial rural feeder has a reliability of 0.75. As per the schedule, this feeder is to supply power to agricultural consumers for 8 hours continuously everyday. The reliability of the back-up feeding sub-station, sub-transmission and transmission system reliability is 0.99. The generating plant has a Mean Time Between Failures (MTBF) of 362 days and a Mean Time to Repair (MTTR) of 3 days. Find the percentage of availability of the supply system for agricultural consumers on this feeder.
2. Draw the general life-curve showing failure rate and failures during the warrantly period, working life and worn-out period for a supply of distribution transformers. How are the failures statistically distributed during the working life and worn-out period?
3. (a) A well maintained 11 kV feeder has a 0.98 reliability. How can we increase the reliability to 0.9996 and at what cost?
   (b) A worst 11 kV feeder has a 0.38 reliability. How can we increase the reliability to 0.8 and at what cost?
4. Single phase parallels connected with automatic switching uninterruptible power supply (UPS) are increasingly being used by domestic consumers due to frequent power cuts in the power utility grid supply. The UPS has a reliability of 0.99 while the power utility power supply system has a reliability of 0.88. What is the total reliability for the consumer?
5. In a radial circuit, the controlling HV circuit breaker, overhead line, distribution transformer and service line, each have a reliability of 0.95, 0.92, 0.85 and 0.8 respectively. Find out the system reliability.

BIBLIOGRAPHY

1. Keerthipala W.W.L., and others, "A Single-Phase Parallelly Connected Uninterruptible Power Supply/Demand-side Management Systems", *IEEE Transactions on Energy Conversion*, vol. 15, no. 1, March 2000, pp. 97–101.
2. 'A Survey of Distribution Reliability Practices in US", *IEEE Transactions on Power Delivery*, vol. 14, no. 1, January 1999, pp. 250–255.
3. *IEEE Standard 1159-1995-IEEE*: 'Recommended Practice for Monitoring Electric Power Quality'.
4. Pabla A.S., "Power Quality Problems in India," *Technical Report, 15th International Conference on Electricity Distribution*, CIRED, Session 2, June, 1999, Nice, pp. 45–49.

5. *IEEE Standard 1366-1999-IEEE*: Trial Use Guide for Electric Power Distribution Reliability Indices.
6. Pattim, G. and L. Simoni, "Voltage life of dielectrics", *Electrotecnia.* vol. 58, Jan. 1971, pp. 25–32.
7. Parascos, Edward, and Johan Arceri, "Eliminate early failures", *Electrical World*, McGraw-Hill, New York, July 1, 1977 pp. 62, 63.
8. Jones, B., and R.T. Waters, "Air insulation at large spacings", *Proc. IEE*, vol. 125, no. IIR, London, Nov, 1978.
9. Orawshi, G.J. Bradbury, M.J., Vanner 'Overhead lines—some reflections on design', *IEE Proceedings*, vol. 133, No. 7, Nov. 1986, p. 142.
10. *World Bank Report No. 22171-N: 'India Power Supply to Agriculture'*, South Asia Regional Office, Energy Sector Unit, June 15, 2001.
11. Featheringill, William E., "Power Transformer Loading", *IEEE Transac*, vol. 1A-19, Jan., Feb. 1983, p. 21.
12. Gonen Turan, *Electric Power Distribution System Engineering*, McGraw-Hill Book Company, New York, 1986, pp. 17, 596–616.
13. Miura, N., H. Sate, H. Narita and M. Takaki, "Automatic Meter-reading System by Power Line Carrier Communication", *IEE Proceedings*, vol. 137, no. 1, January 1990.
14. Lakervi E. and E.J. Holmes, *Electricity Distribution Network Design*, Peter Peregrinus Ltd, London, 1995.
15. Wacker, Garry and Roy Billinton "Customer Cost of Electric Service Interruptions," *Proceedings of IEEE*, vol. 77, no. 6, June 1989, pp. 919–930.
16. *M4202 PD Detector*, Robotron Messelektronik, Dresden, Germany.
17. Power Quality Assessment Procedure, *EPRI, Report*, U.S.A., December 1991.

Chapter 8

Consumer Services

8.1 Supply Industry

The electricity supply industry is usually a very heavily capitalized sector of activity. There are, however, three important differences between the electricity supply industry and the usual business enterprise. First, the electricity supply authority has an obligation to serve all those who apply for service; second it does not have direct competition, and third, in place of direct competition, is subjected to government regulations.

8.2 Natural Monopoly

The business of electricity supply is what is called a natural monopoly i.e., the business is inherently monopolistic. In view of the high investment required, which gives a high ratio of investment to revenue, and the fact that the supply of electric service is undertaken in the general interest and welfare of the public, direct competition is wasteful, uneconomical and undesirable. Further, the available space in the streets limits the number of supply lines which can be accommodated. Natural monopoly thus results from practical considerations rather than statutory laws. The business so classified is given certain advantages under regulations, such as protection from direct competition and the right to use streets and highways.

It should be pointed out that along with these advantages, there are certain obligations, such as: (i) limitation on earnings and (ii) obligation to serve all who apply for service and restriction against withdrawal of service without regulatory approval.

However recently wheeling services for selling power to the selected consumers have been allowed in competitive environment by some power utilities and IPPs in U.K., U.S.A., Australia, India, Singapore and Argentina under the government regulations (refer Sections 8.3, 9.8.2, 9.8.3 and 9.8.4).

8.3 Regulations

The supply industry is regulated by the following statutory public controls.

The Indian Telegraphic Act, 1885

This act covers the privileges and powers of the government to place the telegraph lines and posts. Penalties and certain other supplementary provisions regarding electric power lines, are also covered by this Act.

The Indian Electricity Act, 1910

This is an act relating to amend the law relating to the supply and use of electrical energy. It regulates:

(a) *Licences:* grant of licences; revocation or amendment of licences; purchase of undertakings; annual accounts of licensees.
(b) *Works:* provision as to the opening and breaking up of streets, railways and tramways, notice of new works; laying of supply lines; notice to telegraph authority; overhead lines; compensation for damage.
(c) *Supply:* point of supply; powers of licensee to enter premises; restrictions on licensees; obligation on licensees to supply energy; powers of the state governments to give direction to a licensee; power to control the distribution and consumption of energy; discontinuance of supply to consumers; meters.
(d) *Transmission and Use of Energy by Non-licensees:* sanctions required by non-licensees in certain cases; control of transmission and use of energy.
(e) *General Protective Clauses:* protection of railways, aerodromes,

canals, docks and piers; protection of telegraphic and electric signal lines; notice of accidents and inquiries; prohibition of connection with earth and power for government to interfere in certain cases of default.

(f) *Administration and Rules:* advisory boards, appointment of electrical inspectors.

(g) *Criminal Offences and Procedure:* theft of energy; penalty for maliciously wasting energy or injuring works; penalty for unauthorised supply of energy by non-licensees; penalty for illegal or defective supply or for non-compliance with order; penalty for interference with meters or licensee's works and for improper use of energy; offences by companies; institution of prosecution.

(h) *Supplementary Provisions:* Exercise in certain cases of power of telegraph authority; arbitration; recovery of sums; delegation of certain functions of the state government to the inspection staff; protection for acts done in good faith; amendment of Land Acquisition Act, 1884; repeals and savings.

The Electricity (Supply) Act, 1948

This Act rationalizes the production and supply of electricity and generally provides for taking measures conducive to its development. It enacts:

(a) The Central Electricity Authority: Constitution; powers to require accounts, statistics and returns; direction of Central Government to the Authority; power of Central Government to make rules; powers of Authority to make regulations.

(b) State Electricity Boards, Transmission Companies, Generating Companies; State Electricity Consultative Councils and Local Advisory Committees; Constitution and Composition of State Electricity Boards; inter-state agreement to extend Board's jurisdiction to another state; formation, objects, jurisdiction, etc. of transmission companies, generating companies.

(c) Power and duties of state electricity boards, transmission companies, and generating companies.

(d) The works and trading procedure of the board, transmission company, and the generating company, coordination with the Regional Electricity Board.

(e) The board's finance, accounts and audit.

(f) Miscellaneous items such as effects of other laws; water-power concessions to be granted only to the board or a generating company; coordination between the board's and multipurpose schemes; powers of entry; annual reports, statistics and returns; arbitration; penalties; cognizance of offences; direction by the state government; provision relating to income-tax; members, officers and other employees of the board to be public servant; protection to persons acting under this act; saving of application of act.

The Electricity Regulatory Commission Act, 1998

The Act provides for the establishment of the Central Electricity Regulatory Commission and State Regularity Commission, Central and State Transmission Utilities, framing grid code and grid standards, rationalization of electricity tariff, transparent policies regarding subsidies, promotion of efficient and environmentally benign policies and for matters connected with it.

The Electricity Act, 2003

It's content supercedes and consolidates the provisions of the Electricity Regulatory Commission Act, 1998, the Indian Electricity Supply Act, 1948 and the Electricity Act, 1910. The Act provides for

- National Electricity Policy and Plan
- Promoting competition, trading and developments of electricity markets; open access and parallel distribution;
- De-licensing, setting up and operating new generating stations, captive generation, and dedicated transmission lines;
- Licensing for transmission, distribution and supply companies;
- Duties of generating, transmission, distribution and supply companies;
- Directions to generating, transmission, distribution and supply companies;
- Compulsory consumer metering;
- Reorganisation of state electricity boards;
- Rationalizing subsidies;
- Constitution and functions of the Central Electricity Authority;
- Constitution and functions of central and state regulatory commissions, appointment of advisory committees;
- Works; Consumer protection: Standard of performance;
- Appellate tribunal; investigation and enforcement; special courts, dispute resolution;

- Tariff regulations;
- Offences and penalties for theft of electricity, electric lines and materials, wasting electricity, extinguishing public lamps;
- Protection of railways, highways, aerodromes, canals, docks, wharfs and piers; telegraphic, telephone and electric signalling lines;
- Notice of accidents and inquiries;
- Appointment of electrical inspectors; others and miscellaneoeus provisions.

The Energy Conservation Act, 2001

This Act concerns any form of energy derived from fossil fuels, nuclear substances or materials and hydro-electricity, and includes electrical energy or electricity generated from renewable sources of energy or bio-mass connected to the grid. It delineates:

- The establishment and incorporation of the Bureau of Energy Efficiency;
- Transfer of assets, liabilities, etc., of the Energy Management Centre;
- Power and function of the Bureau;
- Power of the Central Government to facilitate and enforce the efficient use of energy and its conservation;
- Power of the State Government to facilitate and enforce the efficient use of energy and its conservation;
- Finance, accounts and audit of the Bureau; and
- Penalties and adjudication.

Miscellaneous issues: Power of the Central Government—to make rules, to issue directions to the Bureau or the State Government, to supersede the Bureau, if need be; default by companies; exemption from tax on income; protection of action taken in good faith; delegation; power to obtain information; power to exempt; Chairperson, Members, officers and employees deemed to be public servants; power of the State Government to make rules; power of the Bureau to make regulations; rules and regulations to be placed before Parliament and the State Legislature; application of other laws not barred; provision of Act not to apply in certain cases; power to remove difficulty.

The Indian Electricity Rules, 1956

As per Section 37 of Indian Electricity Act, 1910; the Central Electricity Board has framed the Indian Electricity Rules, 1956.

It contains 143 rules along with detailed annexures and covers:

(a) Authorization to perform duties.

(b) *Inspection of Electric Installations:* creation of inspection agency; entry and inspection; inspection fees; appeal against an order; submission of records by supplier or owner.

(c) *Licensing:* application, contents and form of draft licence; advertisement of application and contents thereof; approval of draft licence and a notification for grant of licence; commencement of licence; amendments of licence; preparation and submission of accounts and model conditions of supply.

(d) General safety precautions regarding construction, installation, protection, operation and maintenance of electric supply lines and apparatus; service lines and apparatus on consumer's premises; identification of earthed conductors; accessibility of bare conductors; provisions applicable to protective equipment; instructions for restoration of persons suffering from electric shocks; intimation of accidents; precautions to be adopted by consumers, owners, electrical contractors, electrical workmen and suppliers; periodical inspection, testing of consumer's installations.

(e) *General Conditions Relating to Supply and Use of Energy:* testing of consumer's installation; precaution against leakage; declared voltage and frequency of supply; placing and sealing of energy and demand meters; point of supply; precautions against failure of supply.

(f) *Electric Supply Lines, Systems and Apparatus for Low, Medium, High and Extra High Voltages:* testing of insulation resistance; connection with earth; voltage tests systems; approval by inspector; general conditions as to transformation, use and control of energy; pole-type sub-stations; discharge of capacitors; supply to neonsigns; supply of X-ray and high frequency installations.

(g) *Overhead Lines:* materials and strength; joints; clearances and supports; erection of or alteration of buildings; structures; conditions to apply where telecommunication lines and power-lines can be carried on the same supports; lines crossing; service lines; protection against lightning; unused overhead lines.

(h) *Electric Traction:* additional rules for electric traction; voltage of supply; difference of potential on return; current density in rails; size and strength of trolley wires; records.

(i) Additional precautions for mines and oil-fields.

(j) *Miscellaneous Provisions:* rules relaxation by the government; relaxation by the inspector; supply and use of energy by non-licensees and others; penalty for breaking seal and other penalties for breach of rules; repeal.

8.4 Other Legal Provisions

The electrical power engineer would be expected to have a working knowledge of the following legal provisions besides those indicated in Section 8.3.

The Industrial Dispute Act, 1947

This Act makes provisions for the investigation and settlement of industrial disputes, between employees and employers, employees and workmen or workmen and workmen with the right of representation by a registered trade union or federation of trade unions or association of employees, etc. Besides this, it also aims at securing amity and good relations between the employers and workmen and lays down measures for provisions of illegal strikes and lock-outs, relief to workmen in the matter of layoff, retrenchment and closure of an undertaking, etc.

The Factory Act, 1948

This act abolishes the distinction between seasonal and non-seasonal factories. The Act lays down definite minimum requirements regarding inspecting staff, health, safety, welfare, working hours of adults, employment of young persons, annual leave with wages, special provisions and penalties and procedure, etc. State governments are empowered to make rules requiring the association of the workers in the management regarding arrangements for the welfare of workers. Prior approval of the state government has been made necessary for every new installation of a factory or for the extension of an existing factory. It is also obligatory for the state government to see that all the factories are registered and obtain a licence for working which should be periodically renewed.

The Trade Union Act, 1926

This Act covers for the registration of trade unions after fulfilling the requisite conditions. It deals quite in details about the registration of trade unions; rights and liabilities of the registered trade unions, their recognition, powers to make regulations, various penalties and procedures, collective bargaining and unfair labour practices, etc.

The Workmen Compensation Act, 1923

This Act was designed for the payment of compensation by certain class of employers to their workmen for injury by accident. Any claim for compensation has to be ascertained in accordance with the provisions of the Act and rules made thereunder by the Provincial Commissioner for Workmen's Compensation.

Indian Contract Act, 1872

It covers mainly the communication, acceptance, and revocation of proposals, contracts, voidable contracts and void agreements, contingent contracts and performance of contracts. In addition to this, certain relations resembling those created by contracts, consequences of breach of contracts in sale of goods, indemnity and guarantee, bailment and agency etc. have also been covered.

Regulations for Power Line Crossings of Railway Tracks, 1987

The Consumer Protection Act, 1986

Town and Country Planning Acts

These Acts are of interest before erecting a sub-station or an overhead line. It is necessary to seek approval of planning authorities whenever these Acts are applicable. Pollution Control Act application is important when releasing power supply corrections to certain types of industrial consumers.

Pollution Control Acts

The Environment (Protection) Act, 1986; Water (Prevention and Control and Pollution) Act, 1974; Air (Prevention and Control of Pollution) Act, 1981 and the Forest (Conservation) Act, 1980 are important from the point of view of environment clearance at various stages for the construction of power plant and lines, and for the release of power connection to the ultimate consumer.

8.5 Distribution Code

These are the rules of a particular supply authority for the supply system, provision for metering, earthing and other installation matters, etc. Each supply authority produces a manual for these rules such as sales manual, etc. in accordance with the Sections 49 and 79 of the Electricity (Supply) Act, 1948. A summary of the points generally covered by such a manual is given below.

(a) Administrative information for grant of connections, billing, connected load, contract demand, disconnection, recovery of dues from defaulting consumers and dealing with theft of energy cases, etc.
(b) The system of supply and requirements for balancing single phase loads on a three-phase system.
(c) The limitations placed on water heaters, air-conditioners, welders, electric motors, steel rolling mills, arc furnaces, etc.
(d) Information regarding service lines, point of attachment of supply and fuses, change-over-switch, etc.
(e) Provision necessary for all metering including the location, accessibility and protection.
(f) Requirements for consumer's installations such as compliance with the wiring rules, inspection and testing of an installation.
(g) System of earthing prescribed and any special earthing requirements.
(h) Special requirements, such as limits to low power factor, limits on capacitor installations, harmonics limits for individual consumer at the point of common coupling and rules regarding high voltage installations. Use of capacitors at the consumer's end should be in such a manner that leading power factor does not occur, particularly at no-load or low-load conditions. The leading power factor causes switching oscillations which are harmful to sensitive electronic equipment.
(i) Information of demand, load and diversity factors of various types of consumers and plants.
(j) Tariff application for different categories of consumers.
(k) Settlement of consumer disputes etc.

8.6 Consumer Care

8.6.1 Consumer Satisfaction

An electricity consumer aspires for power supply at a reasonably low price and of acceptable quality. The consumer is satisfied if the following other needs are met:

- Continuity of power supply
- Good quality of power
- Correct billing
- Better collection method to reduce hassles and time spent in queues
- Power on demand
- Faster redressal of complaints
- Providing metered supply for agriculture and domestic consumers with guaranteed service
- A system of penalty to power utility in case of deficient service

A *Satisfaction Index* is actual satisfaction divided by expectation. It cannot exceed one. The ORG-MARG Survey (2002) focusing on availability, accessibility of affordable electricity was carried across 15 states in India [13]. The *satisfaction index* was found around 0.5 and particularly agriculture consumer had lowest score at 0.38.

8.6.2 Consumer Values

The values are benefits received by the consumer in terms of the electricity provided. These are services beyond the meter. Adding values delights the consumer. The consumer becomes enthusiastic and excited about the service. Some of these values are:

- Staff courtesy
- Timely response
- Good behaviour
- Positive, helpful attitude
- Keeping appointments
- Providing metering data to the consumer: Load profile, cost-saving tips in electricity consumption
- Anticipation of consumer needs
- Any tariff hike proposal must succeed a public debate through consumer consultative/advisory committees, and the proposal should be hosted on the website for public study and feedback.

8.6.3 Guaranteed Standard of Performance

Every state government or state electricity regulatory commission needs to issue a notification so that an agreed standard of service is provided to the consumer. Any deviation from the standard will attract penalties and some compensation to the consumer. The time limit for the following services needs to be specified.

- Time response to consumer for electricity complaint
- Maximum time for restoring power supply after the fault for different categories of consumers
- Time schedule for providing meter and power supply for new connection
- Advance notice of supply interruption
- Responding to meter/billing problems
- Investigating voltage problems
- Making and keeping appointments

8.7 Standards

National as well as international standards for electrical equipment are framed, the aim of which is to attempt obtaining a uniform set of equipment on a national and international basis for the benefit of consumers and public. For example, in order to promote international unification, International Electrotechemical Commission (IEC) is doing active work on 'world plug and outlet systems' and could be of tremendous technical import if a unified system is evolved. Bureau of Indian Standards has prepared a handbook which lists the various standards issued. IEC: 60038 specifies an international standard for power supply of 50 Hz, 230 V+/– 10 per cent to be adopted in most parts of the world by the year 2003. ISO: 9001-2000 should be adopted by power utilities to develop a quality assurance system for distribution of electricity and consumer services. ISO: 14001 for Environmental Management System needs to be adopted to improve the utilization of surrounding resources namely air, water, land, natural resources, fuel, flora, fauna and humans. This will require improving all-round efficiency and reducing wastage.

8.8 Consumer Load Requirements

A consumer may ask for service in any amount, time and place, and the electricity supply authority is expected to give that amount of load. The consumer expects to receive continuous service with properly regulated voltage. Consumer loads vary in size, time of use of service (diversity), period of use (load factor), service voltage, power factor, instantaneous peaks and location.

The consumer's requirement, that load be supplied at any time, makes it necessary for the supply authority to provide facilities for his maximum requirements; energy storage may be needed. The consumer cannot be made to wait. He must be supplied instantly with the full amount of service he demands. Since the service requirements of many consumers coincide with respect to time, this results in "peaks" and "valleys" in the load curve. There are periods when the equipment is fully loaded, while at other times, the same facilities may remain almost idle.

The length of distribution lines is determined by the consumer's location. Population density also influences this. The amount of spare capacity to be maintained by the supply authority is determined by the consumer requirement for uninterrupted service.

8.9 Consumer Factors

Maximum Demand

The instantaneous load variation of a system is familiar to all power engineers. The capacity of an electric equipment is fundamentally dependent upon its ultimate operating temperature; thus a transformer with considerably less than 1000 kVA capacity could take care of a peak of this magnitude which is imposed for a period of a few seconds or even a few minutes. The power station capacities, however, depend also upon the prime-mover rating and, therefore, have limited overload capacity.

In practice, electricity is bought and sold on the basis of demands which are in fact the average demands registered over a fixed time period usually 15, 30 or in some cases 60 minutes. Figure 8.1 illustrates that the magnitude of maximum demand varies with the time period over which it is measured.

The instantaneous peak is always the highest; as the time interval increases, the demand value decreases. A 30-minute period is frequently recommended for the following reasons:

(a) There is no undue penalty for inadvertent peaks of short durations.
(b) From the consideration of the various thermal time constants of electrical equipment such as electric motors.
(c) Most available metering equipment provides for registration of a 30-minute demand as per IS: 8530.

The maximum demand or peak load of an installation or system is usually expressed as the largest value of level of the 30-minute demand during a given period, such as a month or year. The ratio of maximum demand to connected load (rated capacity) of a consumer is called the *demand factor. Similarly utilization factor* is the ratio of maximum demand of a system or part of a system to the rated capacity of the system or part of the system.

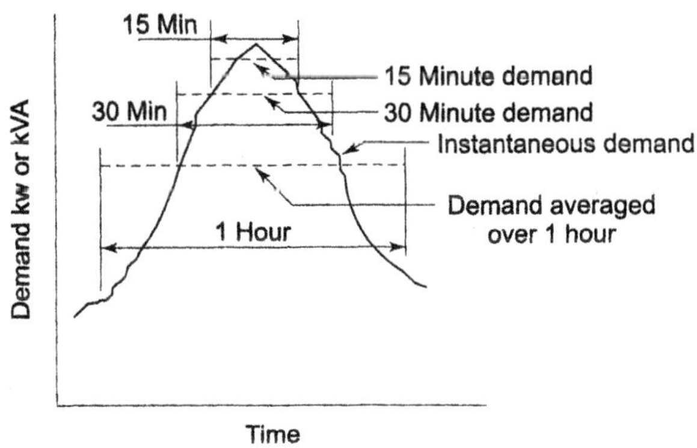

Fig. 8.1 *Variation of maximum demand versus time*

Load Factor

This is ratio of the average power to the maximum demand. It is important to note that a statement of load factor is meaningless unless the interval of time of the maximum demand and the period of time over which the average is taken are specified; for example, 30 minutes, monthly, yearly load factor, etc. thus

$$\text{Load factor} = \frac{\text{Units consumed in a given period}}{\text{Maximum demand} \times \text{hours in the period}}$$

Diversity Factor

This is the ratio of the sum of the maximum power demands of the consumers or the sub-divisions of any system to the maximum demand of the whole of the system under consideration.

Thus

$$\text{Diversity factor} = \frac{\Sigma d}{D}$$

where $\Sigma\ d$ is the sum of the maximum demands of the sub-systems or consumers and D is the maximum demand of the whole of the system or the consumers under consideration.

Stated in this form the diversity factor is a number greater than unity. For convenience, the *coincidence factor* which is the inverse of the diversity factor may be used.

As example, in general, the diversity factor usually falls within the following limits in Punjab:

At distribution transformers	1.00–1.55
At distribution primary sub-stations	1.08–1.60
At grid sub-stations	1.05–1.25

The annual diversity factors of various regions of the country have been estimated as per the 16th Electric Power Survey of India Report by CEA [11]. Annual diversity factor is the ratio of the aggregate of annual system peak loads of the constituent states/union territories of the region to the annual simultaneous peak load of the region. The following are the annual load factors:

Northern region	:	1.08
Western region	:	1.04
Southern region	:	1.05
Eastern region	:	1.05
North-eastern region	:	1.08

The diversity factor between different states in the northern regions [8] is 1.15. The country's five regions comprise the following states:

Northern region: Delhi, Haryana, Himachal Pradesh, Jammu and Kahsmir, Punjab, Rajasthan, Uttar Pradesh, Uttaranchal.

Western region: Goa, Gujarat, Madhya Pradesh, Chhattisgarh, Maharashtra, Dadra & Nagar Haveli, Daman and Diu.

Southern region: Andhra Pradesh, Karnataka, Kerala, Tamil Nadu, Pondicherry.

Eastern region: Bihar, Jharkhand, Orissa, Sikkim, West Bengal.

North-eastern region: Arunachal Pradesh, Assam, Manipur, Meghalaya, Mizoram, Nagaland, Tripura.

Islands: Andaman and Nicobar and Lakshadweep are separate and do not fall under any of the above categories.

Power Factor

The ratio of power in kW to kVA is called the power factor. The power factor at which an equipment operates is an economically important feature. Low power factor means more current and greater I^2R losses in the power system and equipment installation. The power factor generally lags due to the following main causes:

- Nearly 50 per cent load in the system is of induction motors and most of these are running under-load
- Other loads: Domestic, commercial, industrial (arc welders, arc furnaces, induction furnaces etc.) are lagging (see Appendix IX).

As the power factor drops, the system becomes less efficient. A drop from 1.0 to 0.9 lagging power factor results in 15 per cent more current being required at the same load. A power factor of 0.7 lagging requires approximately 43 per cent more current.

Power systems and consumer power installations are designed and constructed assuming lagging power factors. Installations with leading power factors should be designed taking into account additional safety margins. Difficult switching conditions occur as equipment with leading power factor are characterised by large inrush current and switching oscillations. The worst case is when a part of the installation with capacitive character is energised and the other part is switched on. This leads to larger oscillations, reducing the life time of installation including capacitors. When a leading power factor is allowed in a part of the electrical circuit, this part is seen by the supply system as capacitive load with all technical consequences. A good engineering practice is not to allow leading power factor in electrical power installations.

EXAMPLE: A power plant supplies the following loads with maximum demand as below:

Industries	100 MW
Domestic load	15 MW
Commercial load	12 MW
Agriculture	20 MW

The maximum demand on the power station is 110 MW. The total units generated in the year is 350 GWh.

Calculate, (i) yearly load factor (ii) diversity factor.

Solution:

$$\text{(i) Annual Load Factor} = \frac{\text{Total Units generated in the Year}}{\text{Maximum demand} \times 8760}$$

$$= \frac{350 \times 10^6}{110 \times 10^3 \times 8760}$$

$$= 0.36$$

(ii) Diversity Factor:

$$\frac{\text{Sum of maximum demand of individual consumers}}{\text{Station maximum demand}}$$

$$= \frac{100 + 15 + 12 + 20}{110}$$

$$= 1.34$$

8.10 Least Cost of Supply

For any electricity supply undertaking, all costs, including interest on loans, depreciation, provision for loan repayments, general reserves and running expenses, must be covered by the total income from consumers almost entirely by way of electricity tariffs. This is essential if financial stability is to be maintained. But a power generating or distributing authority, being a monopoly, has a duty not only to cover costs, but also to do its best for its consumers, and it must be able to defend itself against criticism. Integrated resources planning; appropriate system voltages; optimizing sizes of the line conductor and distribution transformer; optimum design of sub-station site, size and service area; enhancing efficiency and reducing operating costs will lower electricity prices. Lower electricity costs will have a profound effect across the entire economic spectrum. Lower electric bills for the military, post office and other major government entities will save taxpayers millions of rupees. Indian businesses, both large and small, will see one of their highest operating costs significantly reduced, making them more competitive in international markets. Think about its impact on schools, hospitals, nursing homes and shopping markets. The mission of a well-managed power utility should be to '*supply electricity at low costs*'.

8.11 Revenue and Return

Revenue is the lifeblood of a concern. Insufficient revenue usually results in insolvency or at least instability unless steps are taken to generate sufficient revenue. Sufficient revenue results in a profitable, stable and progressive organization providing satisfying employment and good working conditions to those concerned.

The generation of appropriate revenues to balance the continually rising costs, when selling in markets more competitive than ever before, is a constant battle in every commercial and industrial organization. It is most certainly so in the electricity supply industry. However, as per Indian Electricity (Supply) Act, 1948, (Section 59) the electricity boards are under obligation that as far as practicable, they will not run at a loss commercially and will earn a net surplus of 3 per cent on capital employed.

The sale of electricity is similar to the sale of other manufactured goods that is the whole cost of production and distribution must be recovered in the price charged to the consumer if the undertaking is to remain financially stable. It differs, however, from sale of ordinary goods in the important respect that is electricity must be generated as required and cannot be satisfactorily stored to meet future demands. This characteristic of electricity supply introduces the main problem to be surmounted while framing tariffs and is the principal reason why rates for certain classes of consumers must be higher than those for others.

The Rajadhyaksha Committee (1980) suggested that every electricity board should give a return of 15 per cent annually on its average capital base. This 15 per cent return includes a 6 per cent profit also. Few boards have achieved 15 per cent return even after allowing for over-stated 'profits because of the practice of using the historical cost of assets for depreciation purposes instead of evaluating the assets based on the present worth of future operating margins.

As per the Electricity Supply Act, power utilities in the private sector are allowed the tariff to recover all its costs, including interest, depreciation, tax, statutory reserve and reasonable rates of return.

8.12 Load Management (2)

The importance of load management has increased world-wide. In the developing countries, it meets gap between power demand and availability. The power utility activities designs to influence the timing and amount of electricity that the consumer may use.

Load management's benefits are system-wide. Alteration of the electric energy use patterns will effect the demands on system generating equipment and the load on distribution system. The load management is mainly concerned with those actions which deals with the patterns of energy consumption. These include differential metering, external control of energy supply such as consumer supply interruption, modification of consumer process characteristic to change energy consumption patterns, use of alternate energy sources by consumer during the critical supply periods, energy conservation which may improve the process efficiency or may eliminate altogether some elements of energy consumption by consumer or within the distribution system.

8.12.1 Successful Programme

(i) Power utilities will benefit from the load management only, to the extent that the action taken will move energy consumption from the system peak demand period. The programme will be effective only when applied to controllable loads, for which the intrinsic energy consumption characteristic of the average load may be modified to move energy consumption to off-peak and without adversely affecting the end-use consumers service.
(ii) It must be able to reduce demand during critical system load periods.
(iii) It must result in a reduction in new generation requirements, purchased power and/or fuel costs.
(iv) It must have an acceptable cost/benefit ratio.
(v) Its operation must be compatible with system design and operation.
(vi) It must operate at an acceptable reliability level.
(vii) It must have an acceptable level of consumer convenience.
(viii) It must provide a benefit to the consumer in the form of reduced tariff or other incentives.

8.12.2 Strategies

There are many methods of load management which should be considered in an integrated manner.

(a) Differential Tariff

The production cost tariffs are based on the argument that if all pricing were on the basis of producing the next unit, then the consumer would automatically choose the particular unit to be produced which will give him the best value. This can be in the form of three-step time-of-day tariff:

(i) A high charge rate reflect the large share of the demand related costs during peak hours.
(ii) Off-peak rate, for supply during night time throughout the year and/or on holidays with a low charge, that would exclude demand related cost.
(iii) Intermediate rate for supply during the remaining period, at a rate between the (i) and (ii).

It encourages the spread of electrical load and reduce the impact of demand peaks, which means fewer power cut problems. In European countries, by differential metering, the ratio of average system maximum demand to average system minimum demand has been reduced from 2.5 to 1.3. In India, this ratio varies between 5 and 3 in different utilities. This requires the use of differential or multi-rate meters.

Differential meters with demand metering are almost three times costlier as compared to conventional two-part meters. The meters are fitted with micro-computer pre-programmed for 5 to 10 years. Time-of-year element can be added by varying energy charges for different months. For example, there may be extra surcharge for summer peak months. Accounting for day and night times with options of holidays, seasonal variations, festivals of the year can also be done. With this, industries could be included to be staggered by providing peak rate tariff for weekly staggered off-days. These meters are desirable for large industrial consumers, bulk supply and grid supply consumers who utilize sufficient block of power. For example, in Punjab, these consumers (numbering about 3100) consume about 30 per cent of total power consumption (2000) and a good relief in peak demand can be expected by this action.

(b) Load Staggering

Load staggering improves system load factor and voltage regulation by various means:

- Manually switching off-on feeders or consumer mains by field staff.
- Time switch arrangement: Time switch can be used to give control signal to circuit breakers at specified time for switching off-on the feeders or service mains.
- Setting of two time zones differing by one hour in the eastern and

western states in a bid to a stagger the clock-related demand in these zones will positively reduce peak demand in the country.

- Shift shop timings during evening peak hours by 1/2 or 1 hour. Doing this in the critical months saves peak significantly. For example, in Punjab, advancing the closing hours of shops from 8 p.m. to 7 p.m. saved the summer peak by about 80 MW during the year 2001.
- Stagger weekly holidays of the industry and shops on different days of the week in different sectors of the distribution area. This arrangement in Punjab during peak months in summer (2001) saved peak by about 150 MW.
- Simple automation techniques are effective tool for load rostering. To perform various switch on-off functions, the central equipment (control room) can send commands to feeder circuit breaker terminal through radio or telephone or optical fibre line or distribution power line carrier, cyclo-control or ripple control.
- Segregation of Rural Supply:

(i) *Segregation of Rural Supply by Two-Phase:* In some States like Punjab, Haryana feederwise three-phase system in rural area is converted into the two-phase system for load management in rural areas. The circuit arrangement is shown in Fig. 8.2. Village single-phase loads,

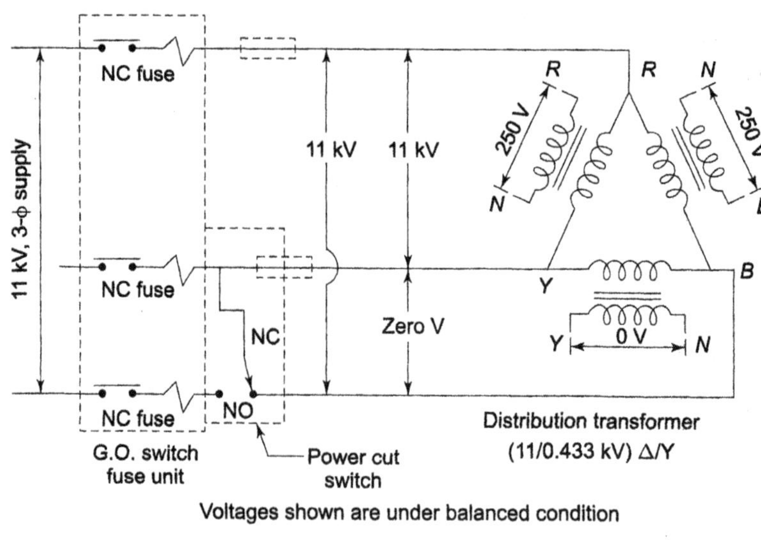

Fig. 8.2 *Power cut switching arrangement: 3 ϕ converted to 2 ϕ system*

lighting are permanently connected on the two phases only. There are some serious disadvantages of this practice [2]:

- Sustained overload on two phases of the transformers and may lead to accelerated damage of the transformers.
- Transformer losses increase by 50 per cent during this period.
- Due to unbalancing of two phase supply, the neutral of 11 kV winding is shifted. Due to this, HT line to neutral voltage on one limb of transformer and one LT line-to-line voltage increases by 16 per cent. It may saturate one limb and increase no-load current on the phase. The load current also increases, resulting in heating a part of the core and may damage the winding.

The neutral shift of HT winding generates negative sequence component, which has been observed to slowdown the three phase, three wire HT kWh meters (of 1 per cent accuracy class) on 11 kV feeders by 8 per cent.

With the agricultural load staggering by 11 kV, two-phase supply is not reliable as most of the consumers during peak period, particularly during paddy crop season, run their motors on two-phase, 250 volt supply (see Fig. 8.3) by connecting a suitable capacitor on LT side. The capacity of the capacitor can be determined by the following formula:

$$\text{Capacitor uF} = \frac{\text{Motor phase current (I)}}{3 \times 2 \times 50 \times 250 \times 10^{-6}}$$

$$I = \frac{\text{H.P.} \times 0.786}{\text{P.F.} \times \sqrt{3} \times 400 \times \text{efficiency}}$$

This arrangement increases the overload on the distribution systems including transformers.

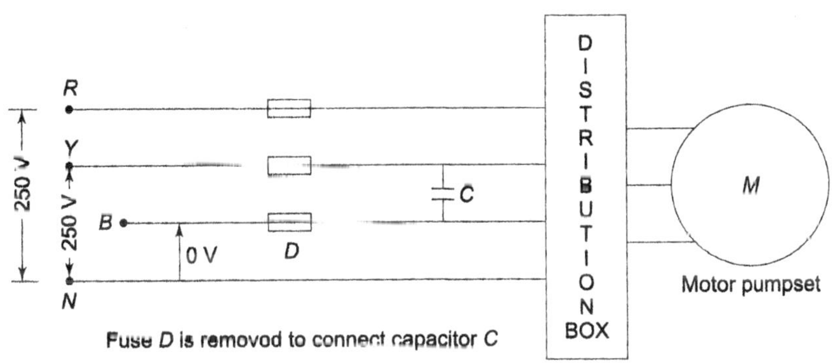

Fig. 8.3 *Running arrangement of agriculture pumpsets under two-phase supply*

(ii) *Automatic Transformer Disconnection Switch Scheme for Agriculture Transformers:* The Rural Electrification Corporation has approved the Automatic Transformer Disconnection Switch Scheme as per REC Specification 36/1984. In the scheme, the solenoid operated LV, single-pole switch conforms to Indian Standard: 9920. The switch is installed at the LV side of the distribution transformer feeding agriculture load only. The typical connections are shown in Fig. 8.4. A pole mounted 11 kV power cut switch is installed on an 11 kV outgoing feeder at the distribution sub-station (e.g. 66/11 kV) end for manual load shedding operation at the sub-station end. The feeder has dedicated distribution transformers for agriculture load and village load. Under load shedding/ power cut conditions, the power cut switch contacts are normally closed (NC), phase 3 of the primary and secondary sides is dead, and then the controlling solenoid gets de-energised, automatically opening the switch and making all the phases of the transformer dead. Other transformers (where this switch is not installed), i.e. the transformers supplying village

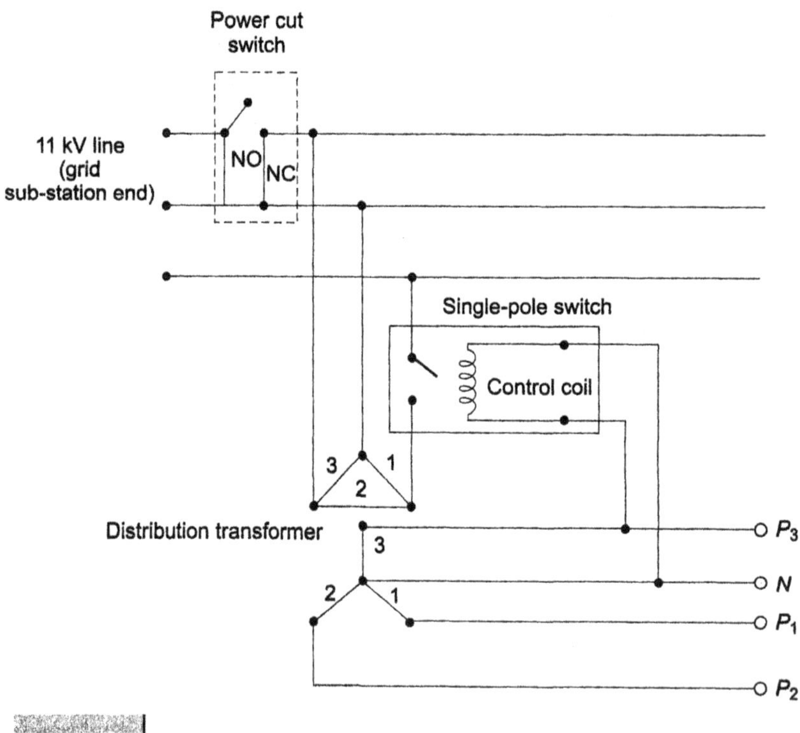

Fig. 8.4 *Alutomatic transformer disconnection switch-power cut arrangement*

domestic loads, remain energised and continuous supply is maintained. Also no-load losses for agriculture transformers are saved and the possibility of split phase supply (see Fig. 8.3) is eliminated. With the restoration of normal supply from the grid sub-station (when the power cut switch NO contacts are closed and NC contacts are opened), the controlling solenoid of this LT switch gets energised from the LV side. It closes the contacts of the control switch, resulting in the provision of normal power supply to the agriculture consumers.

(iii) *Two-phase Supply with Additional Phase Conductor:* Another 11 kV line configuration has been used in Punjab for giving 24-hour supply to villages as shown in Fig. 8.5. All the distribution transformers on 11 kV rural feeders in Punjab are segregated for agriculture and village loads. The arrangement of power cut as shown in Fig. 8.2 has been improvised by providing an extra phase wire on the same poles to reach the village transformer on the feeder. The only condition is that the village panchayat has to deposit a nominal amount as a deposit for obtaining 24-hour power supply. In the power cut arrangement, only one-phase supply goes to agriculture transformers, while the two-phase supply goes to village transformers, as in the case of the arrangement seen in Fig. 8.2. But in this new arrangement, agriculture supply is only one-phase and cannot be used to run agriculture pump-sets with the help of capacitors as shown in Fig. 8.3. The villages can be given 24-hour supply regardless of whether there is a power cut or not for agriculture supply.

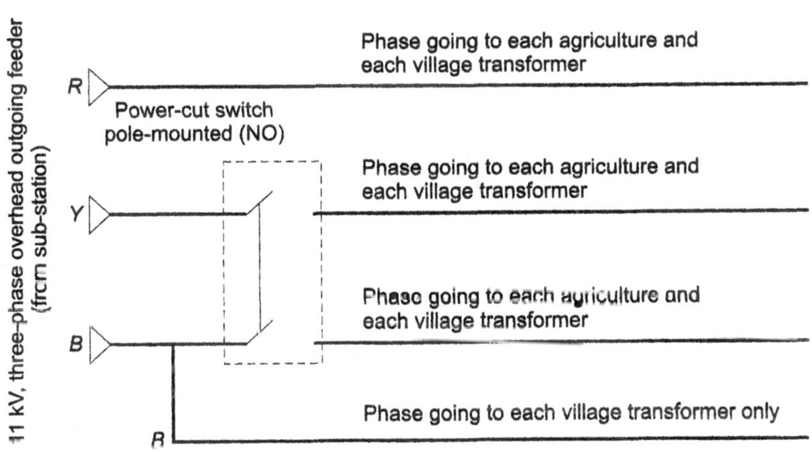

Fig. 8.5 *Agriculture load power-cut arrangement without interruption of supply to villages*

(iv) *Automatic Switching of Agriculture Transformers:* On a selected 11 kV feeder for automation, agriculture and village loads are fed from separate distribution transformers. Load shedding by automatic-switching of individual distribution transformer feeding agriculture load is done in two or three groups of transformers on the selected feeder. The Automatic Load Shedding Scheme is described in Section 5.10.1. The scheme facilitates a capacity saving of 11 kV feeder, as at one time, only one group of transformers feeding agriculture load is working out of two or three groups on the feeder. However, if any farmhouse wants supply, it should be given at 11 kV single-phase or three-phase on a self-finance basis. This scheme facilitates 24-hour supply to villages.

(v) *Automatic Switching of LT Feeder from Agriculture Transformers:* In this scheme also, 11 kV feeders are identified, for setting up dedicated distribution transformers for agriculture load and village load. The LT feeders from each agriculture load distribution transformer are evenly distributed in two groups as far as possible and are connected through an automatic change-over switch (MCCB type). Through automatic programming or remote control through 'Remotrol' (see Section 5.4) of this change-over scheme, each group of LT feeders from each agriculture load transformer on the feeder can be given 24-hour supply on alternate days. However, if any farmhouse wants continuous power supply, it should be given at 11 kV single-phase or 11 kV three-phase on a self-finance basis. This scheme facilitates 24-hour supply to villages.

Of the above five methods, methods (i) and (iii) are highly inefficient, increase losses to more than 50 and create gross metering errors. Any one of the remaining three methods may be adopted for giving 24-hour supply to villages as per the requirement and according to economic considerations.

(c) Interruptible Load Supplies

Some selective HT consumers should be governed by the service contract for interruptible power supply, such as stone crushers, battery charging plant, sport industries, hosiery, big sawmills, flourmills, foundries, etc., which can be given supply with the condition of interruptible supply in case of power restrictions. The supply control of such consumers may be from the sub-station by cyclo-control or ripple-control, which can be used for switching off-on individual consumers (see Sec. 5.4.1). As an example, ripple-control system used in Australia is shown in Fig. 8.6. Audio frequency of 750 Hz is used and the basic requirements are a minimum of two-motor generators at each station, the rating being such that with both machines available, injection of all groups can be carried out simultaneously, and with one being out of service, sequential injection of the groups could still be carried out. Coupling equipment is associated with each power transformer and to avoid incorrect switching, the operator is required to regard a transformer and its coupling equipment as a composite unit, and to ensure that both are in or out of service together. Wave traps or ripple filters are installed at station supply points to ensure that upstream impedance as seen from the coupling cell is constant. The ripple filters which are series resonant circuit at 750 Hz prevent spillover from one area to another.

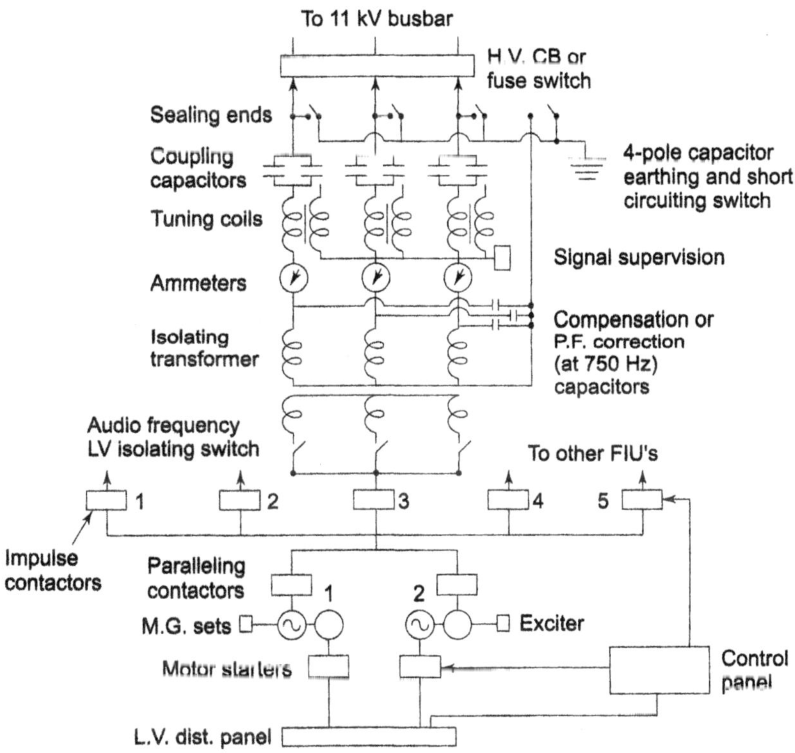

Fig. 8.6 *Typical ripple control system used in Australia*

(d) Maintenance of Essential Services

Protection is given to certain types of essential consumers. These may include hospitals, defence establishments, airports, city inner area, major continuous process plants, such as oil refineries, textiles, etc. These essential services must be maintained during power shortage period or during system instability or disturbances.

During a severe system disturbance, the maintenance of system frequency and hence stability is more important than providing all consumers with continuous supply. Automatic under-frequency load shedding relays should be provided at strategic locations at various grid sub-station to safeguard against total interruption or blackout of the system due to desynchronization, because of disturbances. Such provisions on the power system as given in Table 8.1 need wide consideration in India, as it is most important for the fast expanding power system.

Table 8.1

System emergency load-management measures

Cause of system emergency	*Load control measures*
System reactive-power deficit	Strategic load shedding
Severe transmission-line overload	Fast strategic load shedding
System reactive-power surplus	Fast automatic switching off capacitor banks by voltage rise control or fast switch-in energy storage batteries charging at the points of high voltage rise
System generation deficit	Shed load near lost generators
System generation surplus	Fast switch-off nearby storage water generating units or peaking hydro units

In this scheme, the first shedding stage at 49.6 Hz instantaneously, should be some major industrial consumers by contractual agreement. Other stages may be the switching off of selected outgoing feeders (132, 66, 33, or 11 kV, etc.) at 48.8, 48.5, 47.75, and 47.25 with a check relay at 49.0 Hz with a time delay of approximately one second. The priority of consumers depends upon various considerations, such as the regional importance of some consumers. The general priority of consumers/ feeders to be switched off may be the following:

(i) Major industrial consumers by contractual agreement
(ii) Agriculture consumers
(iii) Industrial and commercial (staggered weekly off-days)
(iv) Rural feeders
(v) Major industries
(vi) City area
(vii) Major continuous process industries
(viii) Hospitals, Aerodromes, Defence establishments

(e) Integrated System Operation

It has been a recurrent feature in the country that deficit states have to impose power restrictions on essential loads, whereas the neighbouring surplus states/zones are able to meet even the non-essential load. Thus, inter-transfer between surplus and deficit states/zones is necessary. It has been seen that the western and eastern systems are predominantly thermal and cannot make effective use of their thermal capacity during peak periods, while northern and southern regions with substantial hydro-capacities suffer from the recurring problems of lean monsoons. These problems have indicated the need for strong inter-regional ties, which will integrate the regional grids to lay the National Power Grid. Since the coming of pit-head super thermal power stations, the National Power Grid establishment has become more important.

(f) Use of Captive Generation and Co-generation

One of the measures for load management is to induce the consumers to develop and use captive generation.

CEA has estimated the 25000 MW captive generation capacity by 2010 AD. Large users of steam and hot water and other factories such as sugar mill, textiles, paper, distilleries, chemicals, cement, aluminium, fertilisers, iron and steel, petroleum, etc., have found it economically desirable to co-generate electricity and to use waste heat, etc., to meet their steam and hot water requirements. Such potential in India is estimated to be 10,000 MW.

Large captive, co-generation/other alternative power plants should have useful interconnection to grid power, as good tool for the total utility system peak demand reduction. Sale price of energy fed into the grid can be utility's avoided cost or plant's incremental generation cost or in between these two as economically settled. The plants should have facilities of (i) automatic synchronizing (ii) relaying for inter-locking of circuit breakers operation of power utilities and the plant, (iii) relaying for the plant protections of overcurrent, phase current balance, reverse power, under- and over-frequency, under- and over-voltage, (iv) a microcomputer energy management (EMS) package to keep account of export/import of energy. Such parallel operation has been widely accepted as an alternative to system expansion of the power utility and to achieve lower system operating cost.

(g) Distribution System Measures

(i) *T & D Losses:* Feederwise/transformerwise losses energy accounting is necessary to pin point higher losses and theft.

Losses can be computed on the basis of actual measurement of kWh, sent and received on each section of the power system through installation of meters. This requires very accurate solid-state metering and simultaneous meter reading. The efficacy of the data would largely depend on the quality of meters employed in the system for the purpose of energy audit. Typical electronic metering arrangement is shown in Fig. 8.7.

(ii) *Eliminate or Reduce Energy Subsidies:* Subsidized low energy tariff encourage intensive energy consumption. It has been observed that countries that do not subsidize energy, use it more efficiently than those that do, therefore, are less susceptible to energy price shocks. In India, power to agriculture sector is highly subsidized, due to which energy management in power boards has become difficult.

If, however, subsidy is to be given to tubewells consumers for agriculture development, it should be only for demand charges tariff. In such cases, the government should give free grants to the power boards for capital laying of the distribution network for agriculture sector or the consumers should be given power connection on self-finance basis.

(h) Conservation

Energy conservation reduces demand and energy consumption by efficient use.

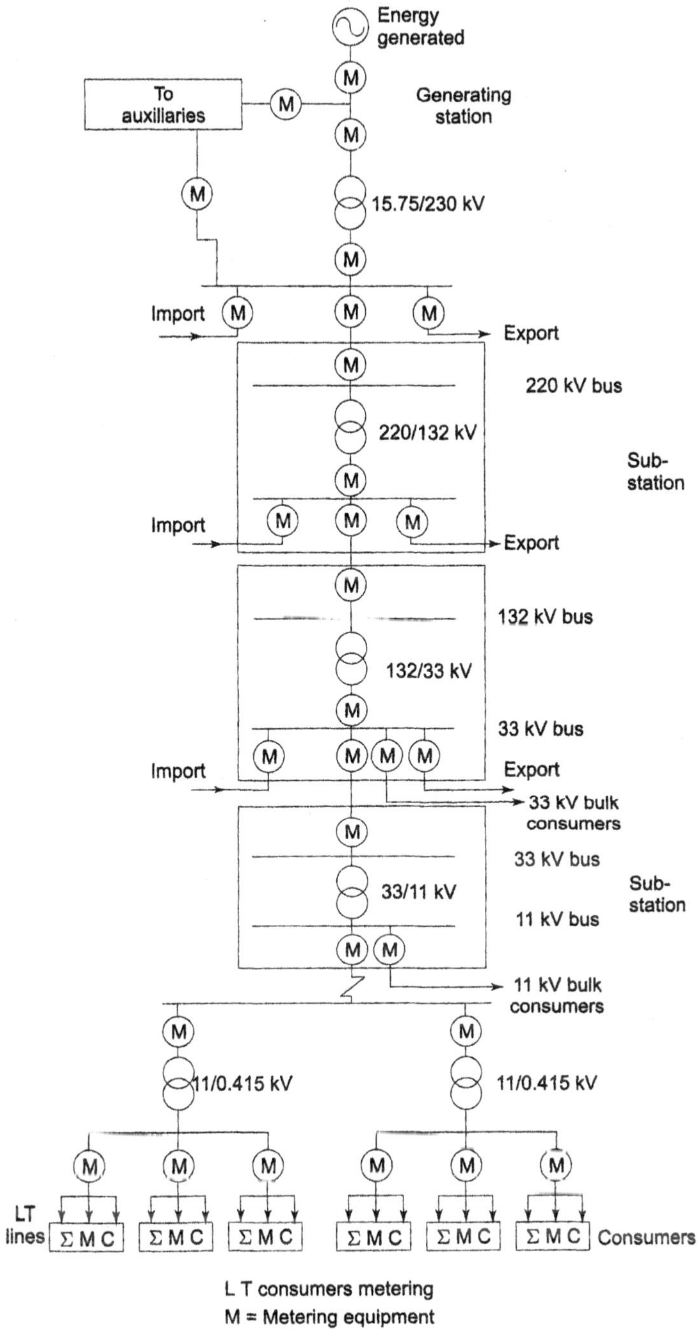

Fig. 8.7 *Metering arrangement of energy losses*

(i) Consumer Awareness Programme Consumers should be educated about the simple rules of energy savings in respect of efficient use of lighting, motors, motor-pump system. Use of motors and pump sets with ISI marks be made mandatory. In India, 75 to 80 per cent of energy generated is consumed by motors. Use of modern efficient (above 85 per cent) motors will save minimum 5 per cent energy. Large number of these motors are run at loads well below their optimum. Motors are usually designed for optimum performance almost at their rated full-load. If a motor operates at reduced load, then its efficiency starts to fall, first slowly and then sharply at lower loads. Motors operating at loads substantially lower than rated, result in increase in energy consumption. Electricity efficient use should be part of primary school curriculum.

(ii) Efficient Energy Use in Motors

(a) **Efficiency** Motors are a fairly efficient device at rated load. In general, three-phase motors are more efficient than single-phase motors, and larger motors are more efficient than small ones. There is only minor improvement in efficiency above 20 hp, and the knee of the efficiency versus size curve occurs at about 10–15 hp. The peak efficiency of a motor occures at full load with about 100 per cent (of name plate) balanced voltage at its terminals. However, as load is reduced from name plate rating, the optimum efficiency occurs at a lower voltage.

Motor voltage unbalance will increase motor losses due to a negative sequence voltage that causes a rotating magnetic field in the opposite direction of motor rotation. A 2 per cent voltage unbalance will increase losses by 8 per cent, a 3 per cent unbalance will increase losses by 25 per cent, and a 5 per cent unbalance will increase losses by 50 per cent.

The power factor of most three-phase motors is between 80 per cent and 90 per cent at full load, and decreases as load is reduced. The installation of capacitors for power factor correction (to 95 per cent or so) at the motor terminals will accomplish two tasks—improved power factor will decrease current requirements, thereby reducing I^2R losses in the supply line; and, more importantly, the use of capacitors at the motor will improve voltage regulation, by increasing the voltage level when the motor is used. If large banks of capacitors exist (or are planned), they should be switched as a function of plant load.

(b) **Over-size Motors** Over-size motors have higher starting current, lower running power factor and higher capital cost. A large number of motors have been observed to be running at loads well below their full load capacity. If a motor operates at a reduced load, then its efficiency starts to fall, first slowly and then sharply at lower loads, as illustrated in the following example:

EXAMPLE

	Case-I	Case-II
Motor load (kW)	15	15
Motor rated (kW)	19	30
per cent loading (approx.)	79	50
Typical motor efficiency at actual loading	89	77
Units consumed/Year		
(assuming run of 400 h/year)	67,600 kWh	78,000 kWh
Cost @ Rs. 2/- kWh	Rs. 135,200	Rs. 156,000
Cost saving for energy	Rs. 20,800	–

Saving in kVA demand will be extra.

Also, the motors at low load conditions have a low power factor as given in a typical case below.

Industrial Induction motor power factor:

- No load 0.18
- 25 per cent full load 0.56
- 50 per cent full load 0.74
- 75 per cent full load 0.81
- 100 per cent full load 0.85
- 125 per cent full load 0.86

The over-size motors should not be used as these lead to wastage of energy, unless efficient use technology is developed.

(c) **Soft-Starters for Induction Motors** These regulate the voltage at the motor terminals, so the magnetizing forces just meet the load demand. This boosts the efficiency of the motors operating below their rated outputs. Energy savings are significant for motors operating at less than 50 per cent load for about 50 per cent of the time.

(d) **Efficient Motor Designs** Efficient motors are efficient designs. These motors consume 5–8 per cent less electricity than standard motors when driving the same load. Their construction is similar to standard motors, but more material is used to reduce copper and iron losses. These motors achieve a higher efficiency by using higher grade steel, special low friction bearings, added copper windings, closer tolerances and small air gaps. These motors have the added benefit of a longer life because they run cooler than low efficiency models. There is a price premium on these motors.

(e) **Delta to Star Connection** The winding of any under loaded three-phase motor can be reconnected in star, rather than delta connection. This reduces the voltage across each winding to give 58 per cent of its rated value, so only motors constantly running at less than 58 per cent of full load will benefit. This is a cheaper option than buying a smaller motor and according to studies, the difference in performance between star-connected standard and star-connected efficient motors is minimal.

(f) **Variable Speed Drives** Variable speed drives adjust the speed of the motor, replacing constant speed motors. Electronic controlled variable speed drives and electro-mechanical variable speed systems are available. The variable speed motors are energy-efficient at reduced load with reduced speeds to meet the different load requirements.

Variable-speed drives are well established over the complete power range in all areas of industry. These include basic industries, material handling plants, transport systems and utility companies for mechanical equipment, e.g. machine tools, extruders, pumps, fans, compressors, as well as for all forms of transport such as ships, railways, elevators and conveyors, drives for paper machines and hot and cold rolling mills. Spinning machines and roller table drives are typical applications for multi-motor drives.

The penetration of variable-speed drives into these sectors has been further accelerated due to the development of new component and drive concepts, which allow new functions and performance characteristics to be realized. These include, for instance, user-friendly man-machine interface, comprehensive diagnostics, parametre assignment, low reactive power and harmonics etc. and thus can fulfil the increasing market demands on drive technology.

Economic aspects, which support the use of variable speed drives, essentially include:

- Energy saving for part-load operation, especially for fans and pumps.
- Less capital tied up with flexible production.
- Low investment costs.

The user can also benefit from a number of technical advantages, such as:

- Optimum solution of process-related tasks
- Reproducible product quality
- High operation reliability
- Low-maintenance and low wear operation

- Fast and accurate diagnostics
- Fast communication with higher-level automation systems using powerful interfaces and buses
- Simple, user-friendly operator control and display
- Environmentally safe operation.

(g) **Motor Rewinding** It has been observed, that every year, about one million motors get burnt and are rewound in India. Motor rewinding leads to loss in motor efficiency from 2 per cent to 10 per cent, even when rewinding the motor, a wire of adequate size is used. On an average, a new motor burnt-out on rewinding loses about 5 per cent efficiency due to heat damage to steel laminations. Also, heating before and after rewinding in the workshop should be controlled to avoid further damage to steel laminations.

(h) **Agriculture Motor Pumpsets** According to studies carried jointly by Punjab Agricultural University and Punjab State Electricity Board (1984), also separately by Rural Cooperative Institute of Management, Ahmedabad (1982) on performance of tubewell pump system, the replacement of inefficient pump was found to increase pump efficiency by 20 per cent, 10 per cent energy was saved by using low resistance foot-valve, 10 per cent energy was saved by use of rigid PVC pipes in suction line, 10 per cent saving was made in use of rigid PVC pipes in delivery line, instead of steel pipe. Total energy saving resulted was 50 per cent. In the country, there are about 14.5 millions of electric tubewells consuming about 27 per cent of the total energy consumption in the country (2000). This consumption could be reduced to almost 15 per cent if strictly the pump system for each tubewell is modernized [12]. The power utility can undertake Energy Saving Performance Contracting (ESPC) with an expert company such as Kirloskar, Jyoti etc. to carry out efficiency improvement works. M/s Kirloskar has done such jobs in Noida with the Noida Power Company Limited.

(iii) *Demand Reduction* Power utilities should have tariffs with in-built mechanism to discourage wasteful energy consumption. The tariff should be adjusted once for a consumer for a certain load/demand which has been assessed/sanctioned and is being utilized for a certain period. Any reduction in power consumption, achieved due to energy conservation, should be charged at progressive lower power rates. Alternately, metering (time-to-day) and tariffs should be framed to encourage the large supply consumers to arrange the operation of loads, so as to hold down maximum demand and improve load factor.

The levelling of demands will decrease the maximum current flow, though the demand current flow may continue for a longer period. As losses (I^2R) vary as the square of the current, the lower current will result in reduction in the total energy requirements of the consumer and reduction of system losses. Demand Controller units, now indigenously available, based on the microcomputer system could be used to supervise the operation of the consumer's equipment. The first step usually is to obtain the consumers demand profile, and if the profile shows only a few sharp peaks, it is a relatively easy matter to identify the equipment causing those demand peaks and take remedial measures. In most cases, such as cement and steel plants, the demands can be reduced, leading to elimination of some unnecessary operations and improved methods of operation. This generally calls for classifying the consumer's major loads into four categories, those which (i) can be re-scheduled (ii) can be deferred (iii) can be curtailed or eliminated and (iv) are essential base load.

(iv) *Energy Storage* Energy storage is a load management application that results in a load shift from peak to off-peak period. It can be in the industrial processes such as waste heat recovery, co-generation, charging of energy storage batteries, etc. Other feasible applications could be heating of water for bathing/washing during off-peak period in cold weather, etc.

(v) *Shunt Capacitors Installation* It has been seen that the shunt capacitors of industrial and agriculture tubewell consumers are not working properly. For example, mortality of capacitors installed on Tubewells in Punjab has been found to be 75 per cent in some areas. On an average, 50 per cent capacitors do not work on tubewells. The shunt capacitors should be invariably of ISI mark to ensure quality working, to reduce the energy and demand losses and maintain stable voltage. Faulty shunt capacitor should be considered as pilferage of energy and connection should be disconnected for specified period, besides charging of penalty.

(vi) *Standardization* Bureau of Indian Standards should revise various standard of electrical equipment to give invariably the actual consumption of energy on the name plate of the equipment by manufacturers. The Bureau of Indian Standards has issued almost 14,000 standards and almost 200 of these are related to energy efficiency. The various relevant Indian Standards concerning conservation are:

(a) *Improvement of power factor* Use of shunt capacitors reduces the line currents and thus reducing losses in the system. IS: 13925 (Part-I)-1998 'Shunt-capacitors for power system' specifies the maximum dielectric loss of the given capacitor to eliminate use of inferior type of capacitors. IS: 7752 (Part-I)-1975 'Guide for improvement of power factor in consumers low and medium voltage installations', provides guidance on proper selection and use of shunt capacitors to effect system efficiency.

(b) *Maximum limits on energy consumption* For fans, IS:374-1979 'Electric ceiling type fans and regulators' and IS: 555-1979 'Electric table type fans and regulators' specify the limits of minimum air delivery per watt consumption of electrical energy to ensure proper utilization of power. For lamps, IS: 418-1978 'Tungsten filament general service electrical lamps' and IS: 2418 (Parts 1 to 4)-1977 'Tubular fluorescent lamps for general lighting service' specify the maximum consumption in watts for each rating marked on them. The lamps have also to meet the lumens output for each rating marked on them. This helps to control the efficiency of conversion of electrical energy into light energy.

(c) *Minimum efficiency* Electrical appliances namely, IS:365-1983 'Electric hot plates' and IS: 2994-1992 'Electric stoves' specify the minimum value of thermal efficiency. IS: 325-1996 'Three-phase squirrel-cage induction motors for centrifugal pumps for agriculture application' and IS: 996-1979' Single-phase ac and universal electric motor'—specify the minimum value of product of efficiency and power factor, the maximum rated current and the minimum torque at full load condition. These parametres help in controlling the quality of motors which should be properly designed to achieve the desired values. Performance characteristics for three-phase induction motor of 2-, 4- and 6-pole have been covered in IS: 8789-1996 and IS: 12615–1989. Inclusion of minimum efficiency requirement in the standards for horizontal centrifugal pumps, IS: 6595–1993 and 9094–1979, would lead to improvement in the performance of the pumpsets.

(d) *Maximum power losses* IS: 2082–1993 for 'storage type electric water heaters' specifies the limits of maximum standing losses per day. The storage water heaters are switched on under normal conditions of use and water is withdrawn only at the time of need, may be in the morning during the winter season, the temperature of water being controlled most of the time by thermostatic control. During the idling period of storage, water heaters are not expected to have more losses than those specified in the standard. The limits of maximum losses on no-load and full-load conditions of distribution transformers up to 100 kVA have been laid down in IS: 1180 (Parts 1 & 2) and for power transformers in IS: 2026. Standards on cables and conductors lay down upper limit of resistance of the conductor, thus keeping the line losses under limits.

(e) *Optimum illumination levels* IS: 3646 (Parts 1 to 3)-1992/1966/ 1988 is the code of practice for interior illumination which provides the choice of light sources and the basis of most economic designs, keeping in view the optimum illumination levels necessary for different occupancies. The efficiency of incandescent lamps and fluorescent tubelights is 10 per cent and 40 per cent respectively. Overall lighting efficiency in India is 15 per cent. The Bureau of Indian Standards must bring a standard on 'Code of Practice on Energy Lighting'.

EXAMPLE: A 40 W fluorescent tubelight having conventional ballast consumes 15 W additionally. The electronic ballast consumes only 5 W. The electronic ballast converts power supply to high frequency and the current in the tube circuit is reduced for the same illumination. The total wattage for a 40 W tube will reduce from 55 W to 32 W. Thus, there is a saving of 23 W per tube, which translates to a 40 per cent saving.

8.13 Energy Audit

An energy audit ensures that every unit of energy and power gives the maximum in terms of production, just as a financial audit ensures broadly that every rupees goes the farthest. Norms should be worked out for each industry and the performance of a unit should be judged against these norms (see Appendix VII). The Energy Conservation Act, 2001, envisages setting up a bureau of energy efficiency that will introduce stringent energy conservation norms with a view to wipe out substandard equipment or appliances from the market.

A major reason for the high consumption of energy in the industry in India is the inadequate modernization of plants and continued use of obsolete technology. Inter-country and inter-firm comparisons of energy consumed per unit of specific industrial products in India and other developed countries clearly indicate that there is wide scope for conservation of energy in the industrial sector. For example, energy consumption in industries like paper, cement, petrochemicals and aluminium is 15 to 30 per cent higher than what is normal. In the Indian steel industry, specific energy consumption per tonne of steel is nearly twice the current figure in Japan. In lighting, a large amount of energy could be saved if efficient lighting lamps are used.

There is an immediate need to review energy consumption patterns in power-intensive industries like metal and alloys, textiles, pulp and paper, cement, chemicals, fertilizers etc., and to evolve industry-wise norms and database. For example, in Punjab, 10 per cent of large industries (about 3,050 consumers, each having a load of above 100 kW), consume about 75 per cent of the total large supply industrial consumption (2001). These consumers should be motivated to save energy and an energy audit should be made compulsory. Some states namely Punjab, Kerala, West Bengal, Tamil Nadu etc. have made energy audit mandatory once a year for high tension industrial consumers with maximum demand exceeding 500 kVA. The government should establish mobile units to undertake energy audits of medium and small-scale industries and other types of consumers.

Audit Studies

Electrical energy audit at the plant level must undertake the following activities:

(i) Electrical energy monthwise vis-a-vis the finished product (kWh/finished product)

(ii) Power bill study for each month regarding kWh, kVA, Power Factor (pf), production in numbers

(iii) Analysis of the load curve to curtail or shift some loads to off-peak periods

(iv) Monitoring energy consumption of various equipment separately to check efficiency, harmonics, starting currents, pf etc. of the equipment and taking remedial measures to achieve higher efficiency.

The process of modification, leading to power conservation, are specialized studies taking more time and needing more investments. Their payback periods are much longer. However, housekeeping measures needing lesser capital achieved payback periods of less than three years. The case studies of some high tension consumers in Tamil Nadu give the following common observations [4]:

1. In HT supplies, reducing monthly consumer demand on the power utility saves power system investments and energy losses for the power utility.
2. Attaining overall power factor nearer unity by installing appropriate shunt capacitors at optimal locations saves much energy and power.
3. The electronic energy meter provided by the Power Board for metreing HT supply also helps in conservation. These meters show up consumers with a lagging PF.
4. Staggering the working hours of non-process auxiliary and service loads keeps the demand low. Also, provision of peak demand controllers gives an alarm at the appropriate time either for manual switching off of non-essential loads or for their automatic switching off.
5. Reduction of power consumption and marginal reduction of power demand by improved and more energy-efficiency lighting systems is substantial. Filament lamps, especially those burning for longer hours, can be replaced by energy-efficient tubelights with electronic ballasts or SL lamps or LP sodium lamps etc.
6. Induction motors taking fluctuating leads or those considerably under-loaded would save power consumption if fitted with electronic motor controllers with soft starters. The motors are supplied automatically with lesser voltage at lower loads, improving both power factor and efficiency of operation. A marginal reduction of demand is also possible.
7. Medium-capacity DG sets consume nearly 1.5 times more oil to deliver the same units at half load. At quarter load, it is nearly twice as much. It is always preferable to load them at over two-thirds of their rated capacity. Hence, where varying doses of power demand cuts have to be managed, it is optimal to order two or three captive sets which are also capable of running parallelly, rather than one large set to feed entire loads.

Case study

In a fertilizer factory: The permitted peak demand of the factory was 531 kVA. The peak demand reached so far was 475 kVA. It is possible to bring it down to 425 kVA and make the balance available to others without the power board having to invest anything. Measures to be taken were detailed. There was a monthly saving in demand charges of about Rs 3000. The factory had installed two transformers of 500 and 250 kVA, 11 kV/400 volts. Both fed separately into two neighbouring buses. It was recommended that after interconnecting the two buses, the 250 kVA transformers could be dismantled along with their 11 kV side AB switch and HG fuse, and also the secondary cable. These dismantled equipment could be sold for thousands of rupees, which was more than the cost of interconnecting the two buses. This would be a one-time investment saving. The no-load loss of a 250 kVA transformer is about 1 per cent for such old units. At Rs 3 per unit of power saved, there would be a monthly saving of about Rs 4,000 worth of power.

The preliminary questionnaire for electrical auditing has the following pattern:

1. Name of company/firm
2. Full postal address
3. Name of contact person
4. (i) Telephone numbers
   (ii) Fax numbers
   (iii) e-mail/voice mail
5. Name of chief executive
6. Type of industry
7. End products
8. Date of factory commissioning
9. Approx. annual energy consumption: Equipment-wise (kW and kWh)
10. Purchased electricity:
    (a) Average power factor:
    (b) Price per unit (Rs)
11. Own Generation:
    (a) Installed capacity kVA:
    (b) Annual diesel consumption:
    (c) Total units generated/year:
12. Details of capacitor installed (kVAr):
13. Energy consumption/unit production:
    Load (kW)/unit capacity:
14. Energy consumption/unit production norms:
    Load (kW)/unit capacity norms:

8.14 Theft of Electricity

At present, about 20 per cent of power produced is stolen by unscrupulous people in the country. The theft varies between 5 and 37 per cent of the power delivered in different states. According to a study (2000), the theft as a percentage of total power delivered in some states is as below:

Punjab = 10%
Haryana = 15%
Delhi = 37%
Gujarat = 10%
Uttar Pradesh = 25%
Jammu and Kashmir = 30%
Orissa = 25%

With the increase in consumers, persons indulging in illicit abstraction of power and tampering with the supplies equipment have increased. Anti-social elements exploit the inability of electricity undertakings to remain vigilant at all times. Apart from the material benefit by way of reduction in the electricity bill, the consumer is indirectly benefitted by stealing energy. The income-tax, excise and commercial tax departments review the consumption of energy to assess the production in industry, such as rice mill, flour mill, etc. Less electricity consumption is directly taken as an index of less production. Thus there is a strong tendency on the part of certain consumers to indulge in the theft of power. However, during power cuts or restrictions, consumers indulge in thefts for obvious reasons.

Poor Paying Capacity

Another reason is that the increasing price of electricity is going beyond the paying capacity of low income group consumers. Nearly 25 per cent of the Indian population lives in slums (jhugi-jhompry). They either tap electricity directly from the mains unauthorized (popularly called '*kundi*' connections) or have unmetered supply. The remedial measure of installing prepaid meters or load limiters will help reduce power theft. Since a load limiter is pole-mountable, any attempt to bypass the device will be difficult as it is clearly visible to the electricity company.

Consumption Variation Check

There should be regular checks on the variation of current consumption of various consumers during the billing cycle. The consumption for a particular billing cycle should be compared with the consumption of the same period of the preceding year. Variations above the given limits should be listed for complete investigation to ascertain the reason for the variation.

Consumer Category	*Variation more than*
Large industrial, bulk supply	+/–10 per cent
Medium and small industries, street lighting	+/–20 per cent
Domestic, commercial	+/–25 per cent

Regular analysis of billing data to highlight the gap between supply and consumption from different grid sub-stations and from each outgoing feeder will indicate power theft or incorrect billing. In each case, an energy balance-sheet should be prepared.

Theft-prone Areas

Theft-prone areas should be identified by means of field checks or by a study of the feeder/distribution transformer/business area energy balance-sheet or a variation in individual consumer consumption. As indicated in Section 8.12.2(g), complete electronic metering upto LT (240/415 V) feeders need to be installed in the identified theft prone areas. The LT feeder/main and the electronic meter together should be protected by an MCCB. This will help to catch the culprit.

The legal position against theft of electricity is covered by sections 126, 135 and 138 of the Electricity Act, 2003. Theft of electricity is a cognizable offence.

Modes and Remedies

The methods and means mostly employed in theft of energy and remedies necessary to prevent its further escalation are as follows:

(a) Supply is tapped directly from the mains by hook rods.
(b) Supply is tapped from the suppliers cut-out which is before the meter. Most power utilities are now connecting the meter directly after aerial fuse at service pole and before suppliers cut-out (see Fig. 19.1).
(c) Meters should be installed at one entry point at ground level (even if multi-storey building(flats) in a proper safety enclosure appropriately sealed. In slum areas, meters should be installed in a group at the street entry point in a safety enclosure with sealing arrangement, PT and CT chamber must have excellent sealing. In case of PTs, the PTs fuses be sealed properly.

(d) In some cases, the current coil is by-passed with a loop. Some introduce a switch in the loop, so that they could allow the meter to run or stop at will.

(e) A fine small hole is made with the sharp edge or drill in the meter cover, and foreign matter like sticks are inserted through the hole to arrest the discs. The sticks are not permanently left in the meter and are inserted as and when required to regulate the meter reading. In some cases, the meter viewing glass is loosened due to handling. A plastic film, similar to a camera film, is inserted through the loose end of the glass to stop or to brake the rotation of the meter disk.

(f) Strong magnetic field near the disc, can slow or stop meter. Electronic meter readings are corrupted if a strong magnetic field is brought near it. The electronic meters must have proper magnetic and electrostatic shielding.

(g) Often unscrupulous consumers are helped by organized gangs with know-how and expert knowledge. There are persons who have sealing pliers with inscriptions similar to the inscription used by undertakings. These persons have some consumers as their regular customers. They frequently break open the seals in the meter, reset the meter reading backward and then reseal it.

(h) In the three-phase service, one of the elements is connected in the reverse, so that the metre would record only one-third of the actual consumption.

(i) In single-phase services, the phase and neutral to the meter are connected *vice versa*, so that the current coil is in a neutral circuit. Any load connected between the phase and earth goes unmetered.

(j) When supply is available in the same premises at two different tariffs, the consumer can easily misuse by connecting equipments intended for one to the other. The suppliers should evolve the tariff in such a way that supply at two tariffs to one premise should not be given.

(k) Meters are tilted as some meters become slow or dead stop when tilted vertically. Meter plumb fixing nuts-bolt or screws should be under sealed cover.

(l) Other measures are: keeping joint-free service lines; proper sealing of metering equipment; effective checking of connections/meters.

(m) Theft proof systems: There are places where large-scale theft of energy from power utilities is common. It is almost impossible to exercise proper vigilance and enforce administrative measures to curb theft of energy in such regions because they are crime and terrorist affected areas. Such places are international border areas strips in north-eastern states and the J & K. The possible remedy is to develop a theft proof system which is difficult to tap and which also involves the vigilance of the consumers themselves. For tubewell electrification within a 10 km belt of such international borders, only H.T. distribution should be undertaken as shown in Fig. 14.15. For each tubewell motor, a 5 kVA or 10 kVA transformer should be installed and a connection should be released on H.T., i.e. 11 kV or 22 kV. Power utility should install a 11/0. 48/0.24 kV transformer at the cost of the consumer. If the transformer is damaged, the consumer is liable to pay for the cost of replacement. The tariff may be based on L.T. metered supply or on the basis of transformer rating if flat rate tariff is applicable. In this case the motors on the tubewell can be single-phase 240 V or 480 V and the service cable should be laid by consumer himself. Such systems are in operation in western countries for rural electrification.

(n) Development of electric cooperative [6]: In rural areas, there is a large power theft from utility mains. The theft could be reduced if the entire responsibility of distributing the power to individual consumers is undertaken by village electric cooperatives. The cooperative could draw the metered supply from the utility local transformers at a tariff and then supply to their members in the village on payment of a certain tariff. The bills would then be collected by the cooperatives. Cooperatives can curb theft and improve distribution efficiency as has been observed in the working of about 50 of them working in Andhra Pradesh. It will be also easier to apply different period-of-day tariffs by cooperatives for consumers other than agriculture consumers.

(o) Totally electronic meters (microprocessor based) are tamper/fraud-proof and must be put in service. Meter software's excellent quality of monitoring of load consumption data and anti-fraud features are important (see Section 8.15). There are reported attempts by industrial consumers to damage the electronic meter by heating-up the meter enclosure or by injecting very high frequency current in the neutral circuit of the meter.

(p) In-built provision of '*Monthly Minimum Charges*' should be made in the tariff applicable to each category/type of consumer based on realistic consumption estimate.

(q) Pilferage of energy can be detected through analysis of individual consumer load curve, comparing it with similar consumers operating the same type of business during the same period of time.

(r) Large power consumers must be offered power at time of day (TOD) metering, coupled with an adjusted tariff system consisting of three rate levels–night, peak and day time.

8.15 Metering of Energy

About 50 per cent of the consumers in the country do not have energy metering such as agriculture, domestic and small commercial consumers. Most of these consumers are in Punjab, Haryana, Jammu and Kashmir, Uttar Pradesh, Madhya Pradesh, Chhattisgarh, Tamil Nadu, Gujarat etc. Energy accounting is necessary to improve efficiency. Energy meters must be provided for all consumes. However, for consumers among the lower strata of society (single-point connections given under REC's *kutir-jyoti* scheme etc), prepaid meters or load limiters may be provided. Modern electronic metering and communication technology facilitate meter data transfer between the consumer and power utility. It also enables consumers to control and understand his/her use of electricity. Advanced energy meters for large commercial and industrial consumers are designed for power quality monitoring (voltage sags, surges, interruptions) and load profile recording.

Electricity meter is a device which measures the amount of electricity a consumer uses. The primary types of meters are:

- Energy meters
- Demand meters

8.15.1 Ferraris Meters

Since 1883 a salient feature of the electrical energy meter has been a part rotating or oscillating, geared to a counting train. It is an induction meter in which the flux provided by currents in fixed coils react with the induced eddy currents in the disk and rotate it. The present form of the induction meter has reached a high degree of sophistication and reliability. The rotor can be suspended magnetically with pins and ring stones giving constraints to maintain the rotor spindle vertical while nylon gears and self-lubricating spindles have reduced friction in the counting train. Modern domestic meters are capable of operating for 20 years without losing their accuracy beyond legal limits. These meters have moving parts and are easier to tamper. Their accuracy is lower but they are cheaper compared to solid-state meters. These will remain in use for another 10–15 years. The main points of technological progress of the modern meters are:

Bearing Application

Nowadays, a magnetic bearing is being utilised in increasing numbers in addition to the classical jewelled bearing. The specific surface loadings of the bearing parts in the magnetic bearing are many times less than in the jewelled bearing. For this reason the guide bearing need not be lubricated. This ensures better stability of the bearing friction with respect to time and thereby better measuring properties in the low load range. As a result of the small specific surface pressures of the bearing parts, meters with magnetic bearings are less sensitive with regard to transport influences.

Brake Magnet

The strong brake magnets required in the present-day high load capacity meters are manufactured from high coercive Alnico magnet material. Thanks to their large specific volume of energy, they are excellently protected against demagnetization by other magnetic fields and their breaking force remains practically constant over long periods of time. Furthermore, their temperature dependency is relatively low. In order to stabilize the braking force it is sufficient if the flux of the magnetized magnet is weakened by a certain amount with the aid of an ac magnetic field, operating at mains frequency.

Electric Steel

If the electric steel applied in the metering element is not competently manufactured and treated, changes can take place in course of time, thereby leading to an unfavourable influence on the long-term characteristics of the apparatus. The aging characteristics of electric steel are decisively influenced by the manufacturing process. From experience, any aging is chiefly effective in the sphere of the initial permeability, i.e. with relatively low magnetic induction. The low load characteristics of the meter are hereby influenced. The only way of guaranteeing perfect long-term stability of the apparatus is to subject the utilized material to a critical test when it reaches the factory.

8.15.2 Solid-state Meters

New ways have been opened in the field of electrical measuring techniques with the advent of the kWh meter using solid-state metering elements. A solid-state meter is one without any moving parts. In this metering, two technologies are used: The 'shunt' or 'Hall Effect' technology [3]. The latter is more common. The output voltage of the Hall cell is digitally processed by the Application Specific Integrated Circuit (ASIC). This provides a calibrated pulse stream with appropriate status signals LED/LCD and/or pulse output (to port for remote metering). Metering data up to 100 events can be stored in the non-volatile memory of the meter for 10–15 years. The non-volatile memory is one which can retain stored information in the absence of power. The non-volatile memory mechanism is maintained through the magnetic induction jackpot system. However, a battery may be required to read the meter when there is no power at the meter.

These meters are generally calibrated once in their life time and have high accuracy in low-load conditions also, unlike ferraris meters. Normally, these meters have a 5-year warranty. They are now widely used for all types of consumers in the country. Use of electronic meters in the Punjab State Electricity Board has shown a nearly 10 per cent increase in consumption in electricity on account of higher accuracy (0.5 per cent) of these meters on full range of load. According to a study, the electronic meter even records the very low level of electricity consumption in light indicators, bells, inverters, stabilisers etc. This forms an average 100 W of consumption in a house. This consumption generally remains unrecorded with induction type meters having an accuracy of +/–3 per cent.

Also an analysis of some domestic consumers of villages and a town in Punjab where old electro-magnetic meters were replaced with electronic meters, had shown that the consumption pattern varied from – 20 to 183 per cent after the installation of electronic meters. A few consumers reduced consumption with some of them touching upto – 20 per cent. This is perhaps due to the fact that their earlier meter (induction type) may have been running faster. The overall rise in consumption was found to be 52 per cent. The major reasons for this rise are:

- The electronic meter has a higher accuracy (0.5 or 1.0 accuracy class) on a full range of load.
- The induction meter tends to slow down with time.
- Induction meters can be easily tampered with. Electronic meters are tamper-proof.
- Many consumers manipulate to install higher size induction meters which do not record low load consumption and replace these with the electronic meters wherein accurate consumption is recorded.

The application of electronics has made it possible the production of meters with a higher precision. These instruments thus serve both as a means of measurement for consumers and in the high power grid networks as well as for calibration and checking purposes. It advantages are:

(a) Very narrow error band over a large load range.
(b) Less dependency of the measuring properties on network and environmental influences.
(c) High reproductibility of the measured results due to the absence of moving components in the measuring circuit.

At feeders on the sub-station end, import/export type electronic meters are installed to measure: energy import/export; reactive energy lag/lead; apparent energy; instantaneous MW, MVAr, MVA, power factor, system frequency. For these meters, the CTs of accuracy class 0.2S and 0.5S as per IS: 2705 (part 2)-1992 should be used. These CTs have better accuracy even at lower loads, up to 1 per cent of the value of rated current.

However, the expenditure to achieve these improvements is also substantial and leads to correspondingly high costs for the electronic instruments. Integrated circuits/microprocessors are now at the stage of awaiting a profound reduction in cost that will come with mass production. The factor in favour of new high-voltage metering is the very low burden that the static meter presents to high-voltage instrument transformers. Recent developments of low-power output PTs and CTs will bring down the total cost of the meters. Moreover, the solid-state meters have much less internal loss as compared to conventional meters thus decreasing the system overall losses. These are also not vulnerable to pilferage of energy by consumers. Another factor that may surprisingly affect the future of this meter is its ability to measure total energy—both fundamental and harmonic unlike in the induction meter.

The application of solid state metering elements for household, commercial, industrial and other types of consumers is increasing. The added advantage is that the high technology meters can be used to give consumers price signals and encourage them to switch their consumption to off-peak period [7].

Modern electronic meters collect energy information in digital form. They make this data available for analysis, such as for demand side management, consumer service. Ironically, as the electronic meters become more complex and technologically sophisticated, they become simpler to use. Electronics have resulted in feature-packed meters, where a single device now performs many measurement needs: energy, demand, time-of-use, load profile, power factor, reactive power etc. [3] Utilities must consider compatibility issues when purchasing new meters. All the utility's electronic meters should share a common software platform, processing and reporting data the same way for simplified training, installation and system integration.

Since there are no moving parts, tampering of meters, as in the case of electromechanical meters, is not possible. The microprocessors give the opportunity to introduce multirate tariffs consumer service, data facilities, and many anti-fraud measures. They can be more easily linked to communications networks for the purpose of centralised automatic reading facilities and load survey or management. Very high precision class could be manufactured in electronic form. The meters are manufactured in accuracy class of 1, 0.5, 0.2 [3].

A menu driven software package, installed on compatible PC and running under windows, generally provides the means of programming the microprocessor-based meters. Access to the programming, software is controlled by secure arrangement of passwords. Programming and reading can be performed through the optical port directly from PC, or by using a Hand Held Unit (HHU). Alternatively, it can be achieved remotely through the RS 232 port or RS 485 port.

8.15.3 Overall Accuracy for kWh and Demand Meters in Service

As per Sec. 26 of Indian Electricity Act 1910, meters for energy and demand metering at consumer premises are to be kept up to correct level of accuracy by the power utilities. These electromagnetic kWh meters (Class-2) are generally manufactured as per IS: 722 [1] and are tested from 1/20th of full-load to full-load. Generally kWh meters tend to minus error on lighter loads and plus error on full loads. As per IE Rule 57, any meter at the consumer premises shall be deemed to be correct if its limits of errors do not exceed 3 per cent above or below of full absolute accuracy at all loads in excess of one-tenth load.

In most power utilities in India, when the kWh meter accuracy is disputed by the consumer, the meter is tested at Meter Testing Laboratory at the load points one-tenth, one-fourth, one-half, three-fourths and full load. In case the error of meter exceeds ± 3 per cent at any load, even though the average error may be less than ± 3 per cent. Then the overall average accuracy (OA) is calculated to overhaul the consumer accounts by giving equal weightage to each test point error (e), i.e.

$$OA = \left[1 \times \frac{1}{10}e + 1 \times \frac{1}{4}e + 1 \times \frac{1}{2}e + 1 \times \frac{3}{4}e + 1 \times 1e\right]$$

In the USA as per ANSI, American National Standard Code for Electricity Metering, IEEE, 1982 (P-182), the overall accuracy equation for a domestic consumer is given by:

$$\text{Overall accuracy} = \frac{4 \times \text{full-load error} + 1 \times \text{light-load error}}{5}$$

Light load is usually taken as 10 to 25 per cent.

In case of demand meters (Class-1), the equal weightage test points usually in practice are three-fourth and full contract or sanctioned demand. Errors should always be applied judiciously [10].

It has been seen that sufficient portion of energy is consumed at the light load, i.e. around 10–25 per cent of full-load by some type of consumers, such as domestic consumer. It has been seen that for domestic consumers the light load conditions remain for about 75 per cent of the time and consume about 45 per cent of the energy. So from the revenue point of view, implication of meter accuracy and accuracy stability at light load is as important as at full-load for the consumer. So it can be said that weightage of each testing point should be according to the contribution of energy at the relative load values. There is need to determine this power consumption pattern for residential, commercial and industrial loads to find average overall accuracy equation for the meters in different power utilities.

According to a survey of load data of some typical consumers in Punjab, the consumption contribution is given in Table 8.2.

Table 8.2

Energy contribution factor

Type of load	*Proportion of rated load/demand*				
	1/10	*1/4*	*1/2*	*3/4*	*1*
Industrial:					
(i) Small and medium loads	0.2	0.2	0.2	0.2	0.2
(ii) Large industrial loads	—	—	—	0.5	0.5
Domestic	0.1	0.35	0.1	0.1	0.35
Commercial	—	0.3	—	—	0.7
Agriculture	0.1	—	—	—	0.9

Overall average accuracy is summation of products of energy contribution factor and respective relative load test errors.

EXAMPLE: A commercial consumer when received high energy bill challenged the meter's accuracy. The induction type kWh meter when tested at the Meter Testing Laboratory was found to have following errors:

At $\frac{1}{10}$ load = − 1 per cent

$\frac{1}{4}$ load = + 0 per cent

$\frac{1}{2}$ load = 2 per cent

$\frac{3}{4}$ load = + 3 per cent

Full-load = + 6 per cent

Find the overall accuracy.

The meter needs correction as one of the test value has more than 3 per cent error. By the conventional equation we get

$$OA = \frac{-1+0+2+3+6}{5} = 2 \text{ per cent}$$

From Table 8.2, the more fair overall error will be

$$= 0.3 \times 0 + 0.7 \times 6 = 4.2 \text{ per cent}$$

That means, the consumer should be given benefit/refund of 4.2 per cent of the power bill consumption.

8.15.4 Demand Meters

The function of a demand meter is to measure the maximum demand of a load, taken by a large user of electricity. The reason or a maximum demand component in a large power user's billing is based on the utility's needs to recover the cost of providing generation, transmission and distribution facilities to meet the consumer's maximum load which might be considerably larger than the average monthly-load measured by the energy meter.

The maximum or peak load that a utility can deliver to a consumer, is limited by the thermal ratings of the utility's equipment. As generators, transformers, transmission lines, cables etc. do not overheat instantaneously, demand meters are designed to measure the maximum average load over a short time interval, or to simulate the thermal response of the utility's equipment.

Although the principle of maximum demand billing appears simple, its implementation presents a number of problems. There is no universally accepted definition of maximum demand or agreement as to which quantity should be measured. Integrating time can vary from 15 minutes to one hour from consumer to consumer, depending on past and current practices of each utility. For example the time is 30 minutes in India and 15 minutes in Australia.

8.15.5 Power Factor Metering (Induction Type)

Most of the electrical devices when connected to distribution system draw two components of power from the system. One is active component (kW) for converting energy into useful work in the form of heat, light or mechanical power. The second is reactive component (kVAr) and is responsible for producing the magnetic flux necessary for operation of equipment. The ratio of active power (kW) to the apparent power (kVA) is called *Power Factor.*

At sub stations, bus spot power factor is metered and it poses no problem in measurement with the power factor indicating meter. However, the consumer power factor measurement is desirable on an average basis for the billing period. The finding of average power factor from separately measured quantities of kWh and kVArh will not be correct, in case the power factor is varying with time, which most likely is the case of motive loads.

EXAMPLE: The case is shown in Fig. 8.8 regarding two loads:

(a) For constant power factor loads, both point-in-time and separate metreing values are equal, as shown below by calculation of referred loads in Fig. 8.8(a).

By point-in-time measurement of kWh and kVArh:

$$\text{kVAh} = \sqrt{2^2 + 3^2} \times 6 = 21.6 \text{ kVAh}$$

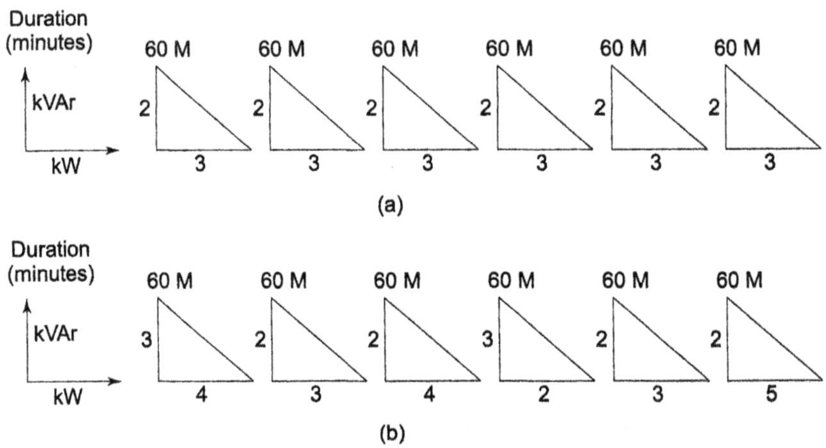

Fig. 8.8 *Effect of variation of power factor on kVAh metering*
(a) Constant power factor
(b) Variable power factor

By separate measurement of kWh and kVArh

$$= \sqrt{(3 \times 6)^2 + (2 \times 6)^2}$$
$$= 21.6 \text{ kVAh}$$

(b) With varying power factor loads, the calculations for load data in Fig. 6.8(b) are:

By point-in-time measurement:

$$\text{kVAh} = \sqrt{25} + \sqrt{13} + \sqrt{20} + \sqrt{13} + \sqrt{13} + \sqrt{29}$$
$$= 5 + 3.6 + 4.47 + 3.6 + 3.6 + 5.385$$
$$= 25.655 \text{ kVAh}$$

By separate measurement:

$$\text{kVAh} = \sqrt{14^2 + 21^2}$$
$$= \sqrt{196 + 441}$$

$$= \sqrt{637}$$
$$= 25.238 \text{ kVAh}$$

$$\text{Difference} = \frac{25.655 - 25.238 \times 100}{25.655} = 1.62 \text{ per cent}$$

Measurements are made therefore by conventional measuring of kWh with special kVAh metering. Special kVAh metering is dependent upon method of addition of active and reactive power.

The active and reactive power are measured individually by suitable polyphase measuring instruments (active and reactive energy meters). From this the geometric addition of apparent energy is established.

In modern electronic meters electrical impulses coming from transmitting contacts on each meter, can be further processed electronically to give apparent energy VA. This is called FAF metering. This comprises an active and reactive energy meter (inductive) and an FAF telecounting apparatus. It has push button-controlled LCD and storage capacity for 12 monthly kWh, kVAh, kVA values based on microprocessor design. It is reliable and measuring error corresponds to the errors of the retransmitting meters. Over-running of MD1 due to voltage outage in a phase is impossible.

8.15.6 Prepaid Electricity Meters

Electric energy is largely sold on credit. The sooner a bill is sent out and paid, the greater the profit for the utility. In this context, the fitting of prepayment meters for consumers with poor credit records can be a considerable saving to a utility.

Billing and revenue collection of the vast number of domestic consumers, especially at the poor strata of society, has always been a difficulty for the electricity utilities. A prepaid electricity metering system allows the consumers to buy credit from the Electricity Board or Company to use a specified quantum of energy in the form of an encoded token which, when inserted into an appropriate slot in the meter, allows the consumer to use the energy at his/her credit. There are two types of such meters: One is a magnetic card version where the consumer inserts the card in a slot and the other is a keypad version where the consumer enters the identification card number through the keypad on the meter. On receiving the credit information through the keypad or magnetic card, the meter credits the electric energy (kWh), corresponding to the amount of electricity purchased by the consumer. The meter then dispenses electricity till the credit lasts. The meter alerts the consumers sufficiently in advance when it is imminent that the balance energy to his credit is going to be exhausted, so that the consumer can get the token recharged or buy a new token depending on the system. In case the consumer fails to buy a further quantum of credit, the meter disconnects the supply as soon as the credit store of energy is used up. Such type of energy meters are in use in countries like U.K., New Zealand, U.S.A., South Africa etc. These meters can save the power utility a lot of operating costs in reading, billing and revenue collection of a considerable segment of the domestic category of consumers, and also improve their finances substantially. By means of these meters, utilities can avoid losses due to deficiencies in meter reading and billing.

Prepaid metering benefits both the utilities and consumers. The utility receives electricity payment about 45 days earlier than in the conventional monthly billing system. There is a direct link between the money spent and value obtained. The consumer should be given a 5 per cent discount for earlier payment as an incentive. Prepaid metering systems help to make the consumer conscious of the cost of electricity. In Western and African countries, the installation of such meters has resulted in energy conservation of 5 to 15 per cent depending upon the category of the consumer. As per life cycle costing studies carried out by ESCOM (South Africa's power utility), the prepaid metering system has proved to be far more cost-effective compared to the current conventional billing system. In Argentina, prepaid meters are being used successfully to supply power to co-operatives which are responsible for distribution of electricity to several communities. This meter is also suitable for giving temporary connections.

8.15.7 Site Checking of CT/PT, Three-phase Meters

The meter should be checked by shorting CT terminals one by one. In case of three-element meters, disc should slow down by one-third with every short circuiting. If the speed doubles with shorting of one element, it means that CT polarity of that element is reversed. In case of two elements meters if short circuited of one current element, it starts running in the reverse direction; which means that power factor is less than 0.5 lag. PT connections are checked by phase sequence indicator. Some of the cases of wrong connection intercepted in the field are given in Table 8.3.

Table 8.3

Defective connection CT/PT with meters

Defect	Energy measured as % of actual energy
Three element meter:	
Any one-phase CT not connected or shorted	66 %
Any one-phase CT polarity reversed	33 %
Interchange of phase CTs	0 %
One-phase PT fuse missing	66 %
PT phase leads interchanged	0 %
Two-element meter:	
Red-phase CT not connected or shorted	$\frac{\cos(30-\theta)\times 100}{\sqrt{3}\cos\theta}\%$
Blue-phase CT not connected or shorted	$\frac{\cos(30+\theta)\times 100}{\sqrt{3}\cos\theta}\%$
Red-phase CT polarity reversed	$\frac{\tan\theta\times 100}{\sqrt{3}}\%$
Blue-phase CT polarity reversed	$-\frac{\tan\theta\times 100}{\sqrt{3}}\%$
Interchange of phase CT leads	0 per cent
Red-phase PT fuse missing	$\frac{\cos(30-\theta)\times 100}{\sqrt{3}\cos\theta}\%$
Yellow-phase PT fuse missing	50 per cent
Blue-phase PT fuse missing	$\frac{\cos(30+\theta)\times 100}{\sqrt{3}\cos\theta}\%$
Any phase PT leads interchanged	0 %

Note: θ is power factor angle.

8.16 Periodical Testing of Meters

As per the Indian Electricity Rules, meters in service need to keep to the prescribed accuracy range. With time, the accuracy of the Ferraris meter degrades and the meter tends to slow down. For that, periodical testing and recalibration of all meters installed in the field for all categories of consumers needs to be carried out as per the meter standards. On recalibration, if the meter is found to be slow or fast, the consumer accounts are adjusted and the calibrated meter re-installed. The following are the practices of periodical testing in different countries.

Canada

The meters are tested in the 6th year.

Germany

Domestic meters are tested every 16th year.

India

Three-phase induction meters are tested every 2 years.
Single-phase induction meters are tested every 5 years.

Japan

Domestic meters of rating 30 A, 120 A, 200 A are tested every 10 years. Domestic meters of rating 20 A, 60 A are tested every 7 years. Other meters are tested after 5 years.

Romania

Single-phase induction meter are tested every 5 years.
Single-phase digital meter are tested every 8 years.
Three-phase inductive kWh power meter are tested every 5 years.
Three-phase inductive kWAr power meter are tested every 6 years.
Three-phase digital meter are tested every 7 years.

South Africa

For consumers having power connection of more than 100 kVA, the calibration cycle varies from 2 to 5 years depending on the size of the metering point.

United Kingdom

Domestic meters are periodically tested every 20 years.
Industrial meters are periodically tested every 15 years.
Commercial meters are tested every 10 years.

U.S.A.

Commercial meters are periodically tested in the 1st, 2nd, 4th, 5th, 8th to 16th year cycle. The size of the load determines the period:
Meters with a load of over 500 kW are tested once a year.
Meters with 300 to 500 kW load are tested every two years.
Meters with 100 to 300 kW load are tested every four years.
Single-phase meters are tested at 8 to 16 years interval depending on the size.
Demand meters under 100 kW are tested at 5 to 6 years intervals.

8.17 Consumer Load Monitoring

State estimation intelligent devices are available either incorporated in the electronic meters or separately. These devices record watt and VAr values typically three times per second. It looks for edge transitions, step-wise changes that are higher than a programmed threshold to determine exactly when a particular appliance turns on and off. Then it records, calculates and stores the date and time of each on/off event along with voltage, current and real and reactive power. The recorder also computes and stores kWh and kVArh for each recording interval. The particular appliance is identified by a stored software library of appliance signatures.

PROBLEMS

1. How does a consumer feel about the value added. Is giving timely energy saving tips to the consumer a value or a matter of satisfaction?
2. A 63 kVA, 11/0.415 kV distribution transformer has 9 kVAr LT capacitors (in star) on the secondary side just after the electronic metering (kWh & kVArh) point and supplies continuous metered power to the following 10 agriculture motor pumpset consumers. Their consumption for a particular month taken simultaneously with the main meter at the transformer was as follows:

Consumer	*kWh consumption*	*KVArh consumption*
Consumer A-5 H.P. motor	437	152
Consumer B-5 H.P. motor	358	105
Consumer C-5 H.P. motor	320	230
Consumer D-5 H.P. motor	350	112
Consumer E-5 H.P. motor	470	130
Consumer F-7.5 H.P. motor	610	35
Consumer G-7.5 H.P. motor	673	332

Consumer H-3 H.P. motor	215	45
Consumer I-3 H.P. motor	232	112
Consumer J-3 H.P. motor	216	33
	Total = 3881	160
Main meter (consumption reading)	4639	1080

Pump-set motors 3, 5 and 7.5 H.P. motors have a rated power factor 0.8 lagging and have LT capacitors of 1, 2 and 3 kVAr respectively. On an average, the LT line losses per H.P. have been computed as 4 kWh per month. (i) Balance energy on this transformer; (ii) Calculate losses; or (iii) Any theft; (iv) Detect any non-working of any consumer LT capacitor; and (v) Reactive power supplied by capacitors at the transformer to motors and beyond the transformer; and (vi) Is there any over-compensation at the transformer?

3. What is the relation of the power factor of induction motor running at rated or at partial load? How does the large industrial consumer 0.8 power factor leading affect the consumer's own equipment or that of other consumers connected on the same bus?
4. Prepare the monthly electricity bill for the industrial consumer having a consumption of 88,500 kWh with the maximum demand during the month being 145 kW. The consumer's connected load is 240 kW with sanctioned contract demand of 135 kW. Calculate the demand factor and load factor of the consumer installation.
5. What steps would you take to reduce the domestic consumer bill at no cost and/or low-cost measures?

BIBLIOGRAPHY

1. IS: 722, ac electricity meters induction type class 2, 1, 0.5.
   IEC: 521-1988, a.c. watthour meters induction type
   IS: 13010-1990 class 0.5, 1 and 2
2. Pabla, A.S., "Load Management in Punjab," *Proceedings, Vol. II of International Symposium on Adoption of New Techniques for Power Distribution Systems*, Calcutta, 17–19 December, 1987, pp. 369–374.
3. Orchard Nigel, "The Advance of Metering", *Utility Automation International*, October 2000, pp. 31–32.
4. Pabla A.S., *Electrical Power Systems Planning*, Macmillan India Ltd, New Delhi, 1998, pp. 314–327.
5. Jürgen Burchka, Klaus-Peter Panzlaff, "Methods and experiences in improving competitiveness through cost reduction resulting from the restructuring of internal processes, quality management and other methods in distribution networks", *Proceedings of the 15th International Conference on Electricity Distribution, CIRED*, Session 6, Nice-99, pp. 127–132.

6. Nanda, J. and D.P. Kothari, *Recent Trends in Electric Energy Systems*, Prentice-Hall of India, New Delhi, 1988, pp. 13–18.
7. *Electricity UK*, April, 1990.
8. *Annual Grid Report* 1997–98, Power Grid Corporation of India Limited, Northern Regional Load Despatch Centre, New Delhi, pp. 37.
9. *Static Energy Meters Specifications*, Power Finance Corporation Ltd., New Delhi, August, 1990, IEC: 1036–1990, B.S.: 7400–1991 and IS; 13779-1993.
10. Pabla, A.S., "Metering Computations", '*Electricity*', Punjab State Electricity Board Journal-1992.
11. 16th *Electric Power Survey of India*, Central Electricity Authority, Govt. of India, New Delhi, September, 2000, pp. 17, 24.
12. *Annual Report on the Working of State Electricity Boards & Electricity Departments*, Planning Commission, Govt. of India, June, 2001.
13. *The Economic Times*, 21 December, 2002.

Chapter 9

Tariffs and Billing

9.1 Objectives

The reason for having a differential tariff structure, as the reader may be aware, is that electrical energy is not, economically speaking, a uniform product. The cost of supplying 1 kW load at 2 a.m. is different from that at 4 p.m. on a cold winter day. For example, during off-peak night hours when the most efficient base load thermal plants are operating, the cost of energy may be, say only one rupee a kWh. However, during peak hours in the evening when even the most inefficient plant may have to be put into operation, the cost may be as high as three rupees a kWh. Further, the cost of supply at high voltage is different from that at low voltage. The high and low load factor also affects the cost of supply. Again, the costs of supply from different generations, such as hydro, nuclear and thermal differ, depending on regional resources and government policies. The generation cost of thermal plants (super thermal plants generation average cost is about 175 paise a kWh and of remote units it is in the range of Rs 2.20–3.40 a kWh at the 2000–01 price level) is higher than hydro-generation but lower than nuclear generation and much less than gas turbine or diesel power generation. The present cost (2000–01 price level) of installation of generation plants is as under:

(a) For thermal plants it is Rs 38,000–54,000/kW at new sites. The cost is slightly less if the project is at a developed site.

(b) The cost of hydro-electric project is subject to considerable variations depending on the type of project and load factor for which they are designed. Their cost range is of Rs 44,000-56,000/kW.

(c) The cost of a nuclear power plant is about Rs 60,000/kW.
(d) The cost of a gas turbine plant is approximately Rs 41,000/kW.

The average supply cost per kWh (at 2000–01 price level) at LT consumer terminal in paise in different states is given [10] in Table 9.1.

Table 9.1

Supply cost [10]

State	Cost of supply Paise per kWh
Andhra Pradesh	353.31
Assam	541.76
Bihar	350
Delhi	459.53
Gujarat	335.59
Haryana	359.98
Himachal Pradesh	218.75
Jammu and Kashmir	341.01
Kerala	275.96
Karnataka	320.65
Madhya Pradesh	272.38
Maharashtra	273.13
Maghalaya	267.66
Orissa	177.45
Punjab	247.17
Rajasthan	334.45
Tamil Nadu	275.64
West Bengal	342.71
Uttar Pradesh	293.59
Nagaland	377.89
Pondicherry	184.48
Sikkim	224.19
Tripura	285.96
Arunachal Pradesh	599.68
Goa	345.98
Manipur	379.72
Mizoram	466.05
All India Average	303.75

The main objects in framing a tariff structure are to ensure that:

(a) The total annual revenue covers the total cost of generation, distribution and transmission.

(b) As far as practicable, each class of consumer meets the total cost of supplying that particular class.
(c) The increased use of electricity by any group of consumers will not result in the additional energy being supplied at a significant loss.
(d) The use of electricity is encouraged in such a manner that the economy of the undertaking is improved.
(e) The tariff must be readily understood by the consumers.
(f) The tariff must be reasonably equitable as between different consumers.
(g) The tariff should also be such as to encourage consumers to improve their load factor or to transfer their demand from peak to off-peak periods or to conserve energy or to discourage pilferage of energy.
(h) Tariff modifications from time to time are made to establish good consumer relationship and to effect economies in tariff administration. This concerns largely the structure and metering arrangements, and also to enable the industry/agriculture to make the maximum contribution towards the general well-being of the community.

9.2 Costing

9.2.1 Costs

For the proper management of any electricity supply undertaking, it is important that reliable methods of costing are used to run it on a sound financial basis. It can be justifiably claimed that satisfactory tariffs cannot be framed unless they are properly related to the cost of supply.

At a time when increasing attention is being paid to the conservation and efficient utilization of energy resources, any approach towards the economic optimization of the electricity supply system must surely accept that it is imperative to have a realistic costing basis for tariffs. Generally, all costs are passed on to the consumer. Therefore, a level playing field for all power distribution developers is desirable. Income tax, depreciation, rate of return on equity etc. should be applicable as a uniform criteria. Depreciation rates should be rational and match the equipment/plant life. For example, correct depreciation rates allowed to

IPPs, NTPC, NHPC etc. have no relation to the life of assets and are intended to recover 90 per cent of the cost of assets within 12 years. The consumer pays higher tariff in this period. This has added greatly to cash reserves of these power generators. Similarly, a 16 per cent return on equity is allowed which directly adds to the price. There should be no cap on the rate of return. Efficiency should be the sole criteria. State electricity boards pay full fixed charges between 62.4 to 68.49 per cent generation plant PLF. Thereafter, they earn so-called incentives which adds to cost. Fixed charges should be considered on 75 per cent PLF.

9.2.2 Total Costs

Total costs are simply the aggregate of all costs incurred by the electricity companies or boards in generating, transmitting and distributing electrical energy throughout the year. These would include, for example, all operating expenses, including depreciation, duties/taxes and interest on loans plus an allowance for return on investment. The factors usually used in costing are discussed below.

Average Cost: The average cost to the utility per kWh output for the year would be obtained by dividing the total cost by the total number of kWh produced during the year.

For example, the average unit cost of energy during 1999–2000 in the country is given in Table 9.2.

Table 9.2

Unit cost of supply (%) during 1999–2000 [10]

Item	% of cost/kWh
Fuel	17.2
Power purchase	44.5
O & M	3.4
Estt./Admn.	13.6
Miscellaneous	1.8
Depreciation	6.6
Interest	12.9

The typical average cost of supply at different levels is shown in Fig. 9.1.

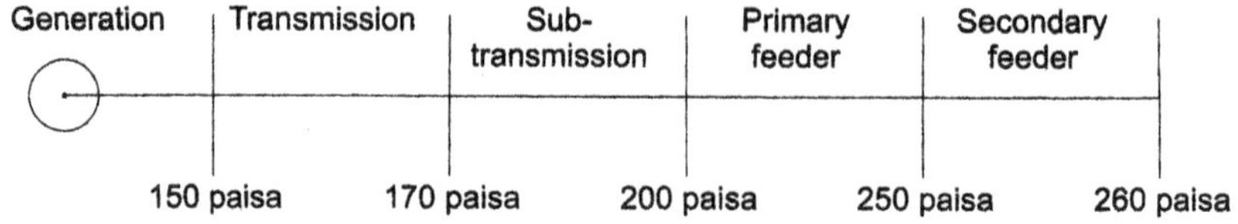

Fig. 9.1 *Supply costs at different voltage levels*

Fixed and Variable Costs: Fixed costs are costs that do not vary, and consist of consumer related costs and capacity costs. Consumer related costs are costs such as those of meter reading, billing, collection, providing service mains and submains, etc. Capacity related costs are made up of annual interest and depreciation on capital costs of the board's assets used in the generation and distribution of energy. The cost of administration, salaries, wages and other expenses that do not vary directly with the output and number of consumers are included in these costs. Variable costs are those that vary with output such as fuel cost which varies with increase in generation. Variable costs are also known as energy related costs or running costs. Variable costs include operation and maintenance costs. The formula for fixed and variable costs should be decided while framing the contract or project estimate. These costs decide the power utility's total annual revenue requirement as shown in Fig. 9.2.

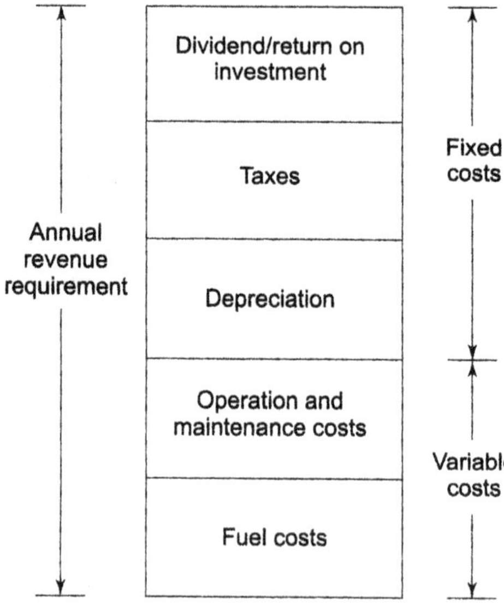

Fig. 9.2 *Power utility annual revenue requirement*

Other factors in costing are:

- *Market-based:* Interactions between the wholesale power supply market and demand (transactions) in the electricity market will determine the cost of power.
- *LRMC:* (see Section 9.4.10): It serves the economic objective of equating price with marginal cost.
- *Allocated cost basis:* Costs are allocated to classes of consumers in a fair, reasonable and stable manner (see Section 9.3).
- *Production-based cost:* These costs depend upon when and where the consumer requires power (see Section 9.3). These costs are not affected by the purpose for which the consumer uses power and by the class to which one belongs. Therefore, the cost depends upon the time at which the power is used, location and voltage. These costs are applied in the United Kingdom in the electricity spot market.
- *Social costing:* It is based on subsidisation of cost by the state government as a social obligation for certain types of consumers. For example, dwellings of those from the lower economic strata of society and agricultural consumers may have cross-subsidisation or subsidisation from the state government to reduce price/tariff.

9.2.3 Marginal/Incremental Costs

Marginal cost is the additional cost that is incurred in generating one more unit or conversely the cost that is saved if one unit less is generated (1 kW or 1 kWh). Marginal cost theory recognizes that no plant capacity cost is chargeable outside the potential peak period [2], [6]. Marginal cost is the gradient (cost per unit) of the cost function curve at the current operating point 'A' of the plant/system (See Fig. 9.3). This is called Short-Term Marginal Cost (STMC). The gradient of a curve at the operating point 'A' is the gradient of the *tangent* T-T′ drawn at A. [2, 4]. Marginal cost is used for the evaluation of system losses. STMC is the specific cost of squeezing more firm power from the existing system. Firm power is characterized by high availability to consumers. Another marginal cost *Long Term Marginal Cost (LTMC)* or *Long Run Marginal Cost (LRMC)* is the firm power cost of the next project to be developed. For practical use, LRMC is calculated as the weighted average of the firm power cost of the projects to be developed during the next five to twenty years. It can be proved that to achieve the minimum present value of all future costs, all new energy projects (generation or demand-side-management) should be implemented in the order of increasing firm power cost [13]. Incremental cost is the cost of serving one specific significant increment of power or energy (kW and kWh) requirement. It is defined by a gradient or slope of increment cost function which is beyond the current operating point as shown in Fig. 9.3. Here, the increment is $A'B'$ and the gradient of line AB is increment cost i.e. $(BB' - AA'/A'B')$. Marginal cost is used for evaluating system losses. Incremental cost is generally useful in determining the purchase/sale price of power and energy. Incremental costs measure the added costs of producing a significant increment of output.

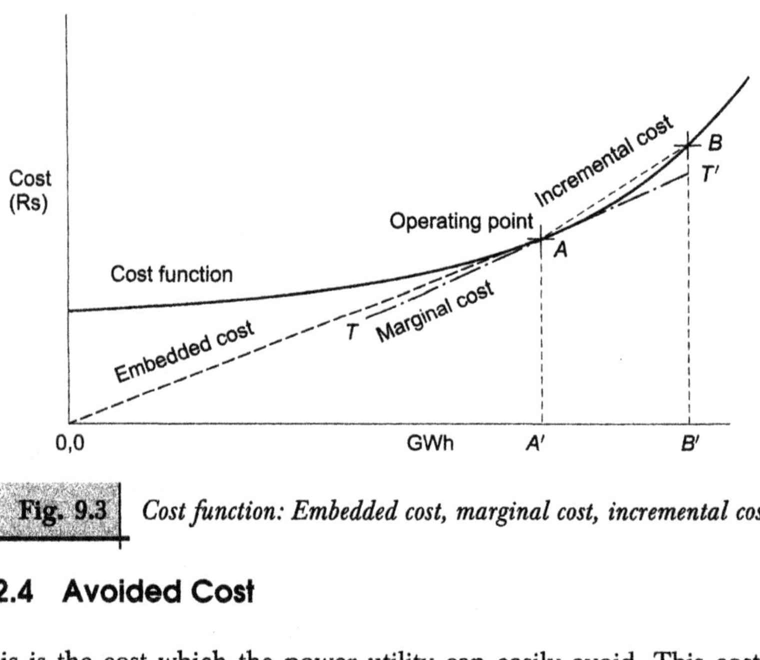

Fig. 9.3 *Cost function: Embedded cost, marginal cost, incremental cost*

9.2.4 Avoided Cost

This is the cost which the power utility can easily avoid. This cost is applied to power supply to the nearest power's grid point from newly-developed schemes: Co-generation, captive power generation or non-conventional resources of generation (photovoltaic, micro-hydro, wind power etc.). The sale price of energy fed into the grid from these co-generation or captive power or non-conventional resources is usually fixed as the power utility's avoided cost. The avoided cost is taken as a cost of generation which is predominantly installed by the utility. For example, it may be thermal or hydro. In U.S.A., captive/co-generation power plant companies sell electricity to utilities at a price equivalent to the 'avoided-cost' of the power from utility-owned plants [1].

9.2.5 Opportunity Cost

The cost of supplying power to a consumer may be decided on the basis of the opportunity cost. This is the cost that the consumer will be incurring per unit of energy in case of the best alternative available to them. The best alternative may be available in the form of wood, diesel, kerosene oil, etc.

9.2.6 Embedded Cost

It is the initial fixed cost plus variable operating cost per unit of electricity. It is cost that exists at the present level of operation of the plant. It is defined by the gradient (Rs/kWh) of the cost function curve up to the operating point 'A' as shown in Fig. 9.3. The cost function for operating a power project is usually non-linear (quadratic equation). The gradient of the curve up to the operating point is the gradient of line 0A i.e. $AA'/0A'$ = total cost/total units of electricity generated [2, 4].

9.2.7 Depreciation

Most of the fixed assets used by the power boards last for 15 to 40 years. In India, we generally follow a straight line method for calculating depreciation i.e. depreciation is equal for every year of the asset's life. The new thinking is that it is better to write down the cost of an asset when the operating margin or earning power is high, which will be the case when an asset is new, rather than to wait for later years when the operating margins may not be sufficient to bear the depreciation for the old asset. As an item of a plant gets older, its earning power or operating margin tends to decline for the simple reason that its operation and maintenance and repair costs will increase due to aging of the plant. Charging of depreciation as per Section 68 of the Electricity (Supply) Act, 1948, is notified from time to time.

Depreciation enables the correct cost for economic use of assets to be charged in a balance sheet before the profit is declared. In any one year, the cash flows should be represented by the margin between the operating income and operating expenses excluding depreciation and this represents the 'operating margin'. The value of the assets should be the present worth of the future operating margins taking into account the interest, etc. The provision for depreciation should represent the annual decline in the present discounted worth of future operating margins that can be earned by the assets currently in use.

9.3 Pricing Approach

The price is what you charge from consumer.

Allocated Costs Basis

There are three essential steps in this approach:

(a) The division of consumers into classes according to their pattern of use of electricity.

(b) The allocation of total annual costs between these classes in a fair, reasonable and stable manner.

(c) The drawing up of tariffs which are acceptable in form and which will bring in revenue from each class approximately equal to the total cost allocated to that class, so that one class is not made to subsidize another class. The cost elements, previously described, would theoretically call for rates containing components representing consumer, demand and energy charges. However, it is not practicable to apply technically correct rates to all consumers or to measure all load elements. A rate which is satisfactory for the large power user is too complicated for the residential consumer. A measurement of demands and power factors for small loads is often too costly to be justified in view of the amount of service involved. Therefore, at this point expediency enters, which modifies the tariff structure from the form in which it would most closely reflect the allocated cost.

Production Cost Basis

The approach based on production costs seeks to derive the income in a manner which would tend to produce the most economical system by improving the load factor. This is based on the principles that:

(a) The basic assumptions of the allocated cost approach are not valid, and may very easily result in an uneconomic system if this approach is followed.

(b) The best solution is a time-of-day tariff, i.e. fixed hours tariff depending also on the region and supply voltage.

It must be noted that the demand related cost of supply to a consumer essentially depends on whether one requires power at the time of system peak load (in which case plant must be installed to meet his demand) or off-peak (in which case the plant already installed can supply one's needs). Whether the supply at a given time is to a residential, commercial or industrial consumer is quite irrelevant, since it does not effect the production cost.

It is then claimed that the system peak demand should grow less rapidly and the valleys of the load curve would fill up, since the consumers would try to avoid the high charges of the peak period. The extent to which this will happen can be determined only through experience. If it seems likely that the valley of the load curve will fill up so rapidly that the time of the system peak will shift, then the peak rate would be reduced and the intermediate rate increased. The total income should, however, remain the same. On the other hand, if the load factor is not increasing fast enough, a greater differential between peaks and intermediate rates should be tried.

If a time-of-day tariff is implemented, it would improve the system load factor, and hence the overall system economy.

The nature of the tariff gives an indication to the consumers as to which are the expensive and cheap supply periods. In theory, the consumer's efforts to reduce his own costs will automatically produce a more economical overall system, thus reducing the costs. However, it has been shown that the price of an essential commodity has only a small influence on its demand. That is, the demand for an essential article is virtually constant regardless of its price. This would tend to indicate that since most people would consider electricity as an essential commodity, the consumption habits of the majority of consumers would remain constant and only a limited amount of load factor improvement could be expected from the time-of-day tariff. This fact could render this form of tariff unsuitable for general use such as domestic or commercial.

Market Price

The following are the parametres which influence pricing in the market regime:

- Spot market price
- Transmission and distribution charges.

9.4 Classifications

9.4.1 Consumers

No two consumers are absolutely identical in their cost characteristics. Therefore, for workable costing, some suitable averaging must take place which should be acceptable for a group of those consumers with broadly similar costing characteristics, such as domestic, commercial, industrial, etc.

9.4.2 Loads

To enable demand costs to be apportioned to various classes, it is desirable to identify their load profile. The load for different types of consumers is generally derived on sampling a number of selected consumer load curves from a class of population. The typical profile made on sample basis for commercial and industrial loads in a metropolitan area is shown in Fig. 9.4. These load curves besides being useful for tariff framing can also be used for calculating system losses for different types of consumers.

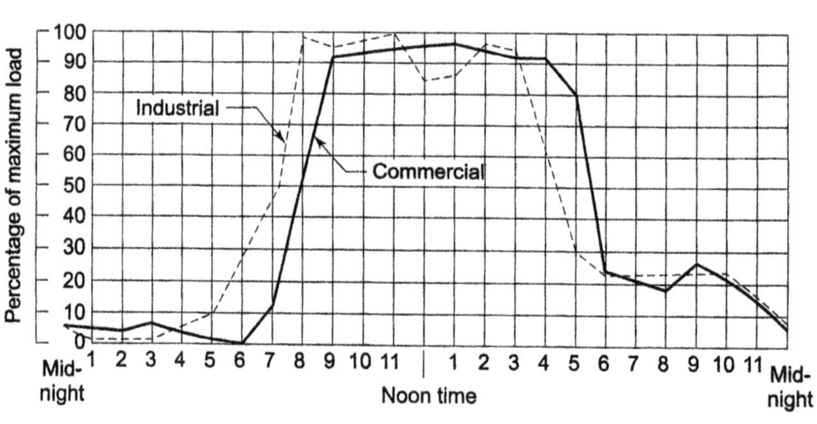

Fig. 9.4 *Typical comparative load curves of a metropolitan city*

9.4.3 Trinomial Apportionment of Costs

Energy Related Cost

This cost is largely represented by the cost of fuel consumed together with other basic production expenses such as on lubricating oil, chemicals for water treatment, cost of system losses and maintenance costs. In computing energy costs, the question of sharing of losses arises. Obviously losses differ for, say, a class of consumers taking supply at high voltage as compared to say, a rural class of consumers. Any modification in tariff can be justifiable made to account for this variation. As a simplification, losses could be shared equally between different classes.

Consumer Related Cost

These costs are those incurred by virtue of consumer's existence as such, quite independent of the electricity purchased. These are part of the base costs before the consumer presses his switch.

Annual costs of meters and their servicing, consumer services, meter reading and billing are clear examples of costs that have to be recovered whether the consumer uses electricity or not. A proportion of administrative cost is also consumer related.

Experience has indicated that some bias might have been earlier given to apportioning the demand component. Earlier costs associated with high voltage lines were considered as almost wholly demand related and similarly with sub-station costs. This is still so in urban areas, but the high voltage spur to an individual rural consumer, say with a small load, is seldom designed to meet that specific load but is determined on the basis of a minimum standard construction. The importance of this analysis is apparent when it is realized that the consumer-related costs can be directly obtained from the consumer in the form of a fixed charge, with the remaining rate kept down to a minimum.

It can be seen from Fig. 9.5(a) that although the capital cost (and thus the associated annual charge) is of an ascending order when related to distribution transformer size, the zero intercept indicates that there is a large proportion of capital cost (nearly Rs 40,000) that can be classified as consumer related. Where the distribution transformer installation serves a large number of consumers, this is not important but it can assume significant proportions in the case of individual rural consumers.

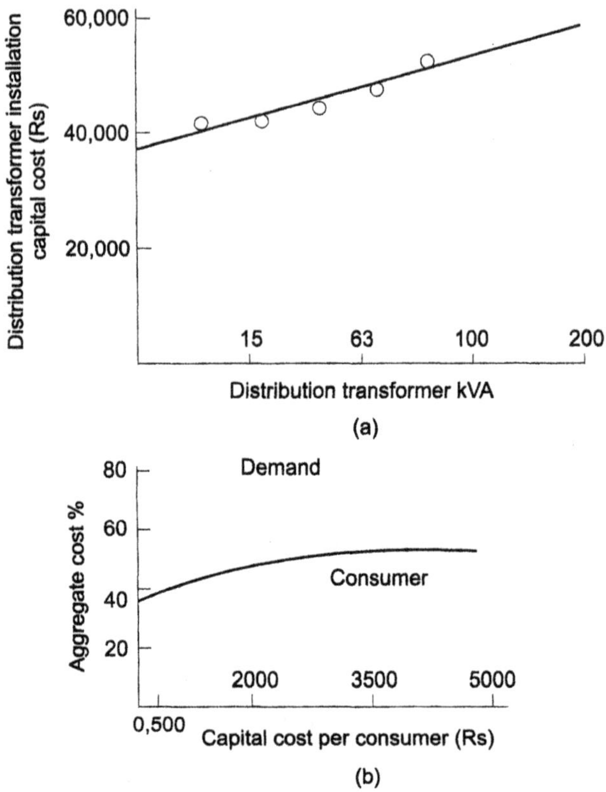

Fig. 9.5 *Consumer and demand related costs: (1996)*
(a) Consumer costs;
(b) Demand and consumer costs

Studies reveal [Fig. 9.5(b)] that though the cost of supplying a low consumption urban type consumer is substantially demand related, there is an increasing consumer related component of cost for rural consumers as the average cost per rural consumer increases.

Demand Related Costs

The demand related cost is the extent to which system capacity may be attributed to the consumer, for example, generation plant, transmission line and sub-station or transformer installed capacity costs. The demand component of the bulk supply charge is by far the largest single item of the demand related cost. For those undertakings with their own generating plant, a further apportionment is necessary. As an example, a typical apportionment is shown in Table 9.3.

Table 9.3

Typical allocating costing

Item	*Demand cost %*	*Energy cost %*	*Consumer cost %*
Hydro-Electrical Power Stations			
Water Storages	—	100	
Balance	100	—	
Steam Power Stations			
Fuel	10	90	
Generation stores, oil, water and sundries	50	50	
Management expenses and wages	75	25	
Interest, depreciation and insurance of coal handling plant	—	100	
Interest, depreciation and insurance excluding that of coal handling plant	100	—	
Diesel Power Stations			
Fuel	0	100	
Generation stores, oil, water and sundries	50	50	
Management expenses and wages	75	25	
Interest, depreciation and insurance	100		
Purchased Electricity (Inter-State)		*As Purchased*	
Transmission and HV Systems (excluding final LV sub-stations)			
Total charges	100		
Distribution Systems (excluding transmission and HV systems)			
Management expenses	50		50
Wages (linemen, etc.)	70		30
Wages, revenue staff and meter readers	—		100
Maintenance and repairs of mains and distributing stations	70		30
Maintenance and repairs of apparatus on consumers premises:			

Contd.

Item	*Demand* cost %	*Energy* cost %	*Consumer* cost %
(a) General			100
(b) 'Off-peak' meters and time or staggering switches			100 direct to 'off-peak'
Maintenance and repair of street lighting			100 direct to street lighting class
Interest, depreciation and insurance on:			
Street lighting	—	—	100 direct
Mains and sub-stations	70	—	30
Meters and services	—	—	100
'Off-peak' meters and time switches			100 direct to 'off-peak'
General			
Sales control or power cut expenses			100
Management and general expenses not otherwise allocated	Prorata to sum of all above items		

Another cost element which is 100 per cent demand related is sub-transmission cost and is an item where some misunderstanding may arise. Since the sub-transmission system carries kWh as well as kW, it may be assumed that it has an energy related component of cost. Nevertheless, if the demand were to be doubled with energy sales remaining the same, it would mean doubling the capacity of the sub-transmission lines and grid sub-stations. Usually, therefore, sub-transmission cost is considered as wholly demand related.

9.4.4 Allocation of Annual Costs Between Classes

The trinomial apportionment of costs identifies three elements of costs.

Energy Cost

These are allocated among the classes proportional to the kWh sales, with losses being either shared or more accurately apportioned.

Consumer Related Cost

This cost is usually allocated broadly on a consumer basis weighed to take into account the greater average consumer related cost for some classes compared to others.

Demand Related Cost

The joint demand cost is allocated by the *average excess demand* method (AED). Here the system demand cost is divided into two components, one proportional to the average demand of the system peak, and the other the balance or excess system demand as shown in Fig. 9.6.

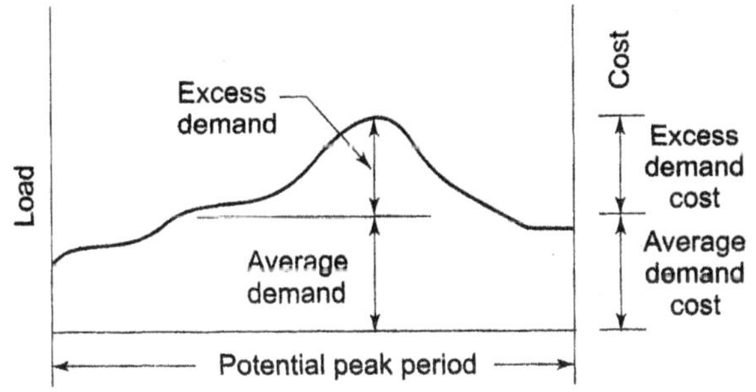

Fig. 9.6 *Apportionment of demand costs*

These two demand cost components are then shared between the classes in proportion to their individual 'average' and 'excess' loads. The method takes into account the demand and load factor of different classes of consumers. It, however, overlooks the time of occurrence of the peak load and, therefore, does not reflect the cost of supply.

Economists have advocated the use of the 'peak responsibility' method, which allocates these costs in proportion to class loads at the time of the system peak load. In this method the consumer's demand during the potential peak period can be related to his individual maximum demand by means of the peak responsibility factor. But this system also has two drawbacks:

(a) It is not equitable to allocate large demand cost to a class merely because the combined loads happen to produce a system peak at the same time as this class load becomes large.
(b) The allocations of loads are unstable, because a change in the time of the system peak would drastically alter all the class allocated costs.

Finally, we see that reconciliation of the cost incurred and revenue earned from each class is the main object of the analysis of cost allocation. There is need to identify the surplus or deficit in relation to the cost of supply to each class while reviewing tariffs.

9.5 Economically Efficient Tariff Structure

The National Grid Code stipulates least cost despatch by system operators on merit order scheduling. There is one method to reduce generation costs, i.e., by energy exchange between adjoining systems. When the power plants of the interconnected systems are operating in "merit order", the least efficient (i.e. the most costly) plant should be backed down first irrespective of its location in the integrated system. In the economic order of energy transaction, energy from the seller's system displaces more expensive energy of the buyer's system. Such transaction normally takes place during off peak hours. The rate is such that each utility equally shares the affected savings. For example, say system A has in operation a generating plant at the margin during off peak hours whole marginal energy cost (fuel cost/kWh) is 58 p/kWh. System B correspondingly has in operation a generating plant whose marginal energy cost is 130 p/kWh. Therefore, system A sells to B at 58 + (130 – 58)/2 = 94 p/kWh, i.e. each system makes a "profit" of 36 p/kWh. System B backs its generating plant producing 130 p/kWh and saves cost to the extent of 36 p/kWh.

More reliable supply will cost the consumer more on account of the provisions to be made such as standby capacity to ensure continuous availability of power, duplication of lines, switchgears, transformers, cables and to provide adequate capacity to ensure correct voltage. It may also be necessary to install capacitors at certain vantage points to improve low voltage conditions. Further, rigorous specifications for components call for higher investments which result in higher capacity costs. Thus the power cost to consumers will increase.

The location and density of load also affect the power cost. The longer the transmission and distribution lines, the higher the costs. Similarly, the lower the density of load, higher the costs. Hence it is cheaper to supply urban areas which have high load density than rural areas where load density is very low. The averaging process which the electricity boards follow to arrive at uniform tariffs for rural and urban areas means that urban areas subsidize the rural areas, resulting in income redistribution. The tariffs then do not reflect the true costs.

9.5.1 Long Run Marginal Cost (LRMC) Based Tariff

Electricity pricing in India has been based on the broad guidelines of the Electricity Supply Act, and the basis of pricing is average cost plus a minimum return of 3 per cent on invested capital for public power utilities and 16 per cent for private utilities and NTPC, NHPC etc.

The cost of power of the new private producers would necessarily be more than the pooled power provided by the State Electricity Boards (SEBs), the bulk of which would be from old, depreciated plants. Such average low-cost supply should mean that SEBs can supply power more competitively than the private distributors, who are more likely to have a larger proportion of their supply from new plants.

The growing peak demands in various regions in the country have been putting pressure on capacity expansion. These expansions are costly and the resources necessary for investments are not easy to come by. It is therefore desirable, that those users who desire additional capacity must really pay for it. Thus, if prices based on operating costs (variable charges) as well as capacity costs (fixed charges) are charged to the peak time users, they will continue to operate during peak periods, whereas others who can manage, will shift their loads away from peak to off-peak periods, when they do not have to pay the additional capacity charge. This would improve the daily utilisation of the power system capacity and at the same time would also conserve resources by releasing pressure on investments as the peak time demand growth is checked. A merit of LRMC based tariffs is that it recognises the differential nature of electricity supply costs across different periods of the operating day and hence avoids losses on operations during any part of the day.

The fixed charges consist of interest on capital, depreciation, operating and maintenance expenses (excluding fuel), taxes on income, return on equity and interest on working capital. The variable charge is the fuel cost recoverable for each unit of energy supplied. Given such a stream of expenses (to be recovered as tariff) over a number of years, the total present value of the fixed and variable charges can then be calculated. This can be done via the practice of discounting future receivables to arrive at the present values of fixed and variable charges (by discounting on an annual basis). To determine the LRMC based tariff, one has only to average out the present values. Figure 9.7 gives the process of calculation of LRMC tariff.

This method allows front-loaded tariffs with fixed costs to come down in real terms with time as debts are paid.

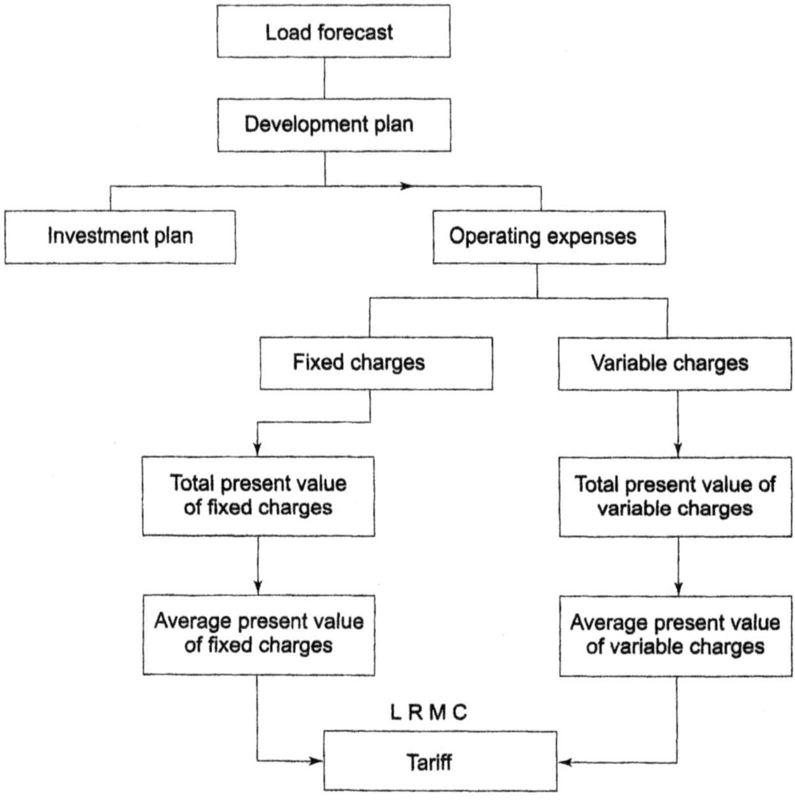

Fig. 9.7 *LRMC tariff calculation process*

9.6 Electricity Market

The Electricity Act, 2003, provides for the development of the electricity market to facilitate reduction of electricity prices. But there can be no free market in electricity when there are shortages. Electricity cannot be treated in the manner of exotic commodity like apples. If apples are in short supply, people will buy other items, leaving apples aside.

Electricity is a basic necessity of life. We have to use it even if there is a high price. Free market of electricity requires adequate availability of generating and system network capacity. There are certain problems in a free market.

- The market economy increases greenhouse gas emissions due to the tendency of increasing the generation capacity in a short time, that is, installing coal/gas/liquid-based thermal power stations. In Australia, greenhouse gas emissions have increased after the introduction of the market economy [9].
- The market economy has reduced energy conservation programmes as the emphasis is on increasing the revenue by selling more electricity.
- Market power abuse [12] is usually associated with a higher price above cost. It can also lower quality of power or service. Market power abuse is not limited to sellers only. Buyers too have market power, for example, large consumers have more ability to affect pricing than smaller one.

Strict regulatory measures can minimize the above effects. The ultimate free electricity market will be a 'power exchange,' almost similar to the stock market and can be managed by the Power Trading Corporation. The pricing may be structured in four different ways:

- Power at a scheduled price as spot market
- Power at a fixed price
- Time-of-day pricing (peak and off peak prices)
- Seasonal pricing: Summer, winter, monsoon, harvesting etc.

The internet can be a primary marketing tool (e-commerce). IPPs and power utilities can join to launch an internet portal for procuring supplies on-line and for other transactions. Power marketing will afford the following benefits:

- Through retail wheeling, reducing the need for expensive capital expenditure for new generating capacity, thereby satisfying demand through increased utilisation of existing assets.
- Online auction of power and energy will reduce per unit cost.

There is the chance of manipulation in a power exchange model, such as creating artificial power scarcity at the last-minute to hike prices based on opportunity cost. A mix of free market power purchase and long-term bilateral contracts between generators/suppliers and eligible consumers is desirable. According to an estimate, this mixture may be 35 per cent of bilateral contract purchase and 65 per cent free market purchase. In the competitive electricity market, the role of the Power Grid Corporation of India will be as a *Network Service Provider* for reactive power, normal system frequency, normal system voltage, system stability etc. [5].

9.7 Rational Tariffs

There are two basic tariff-making philosophies—cost-based, and market-based. The factors used in developing cost-based tariffs are identified as capacity related, energy-related and consumers-related. These factors vary for different classes of consumers (residential, agricultural, commercial, industrial etc.), and require an analysis of much data in order to properly allocate costs (as explained in Sec. 9.4.4). Cost-based tariffs are generally preferred because they are less likely to be criticised by consumers. However, political or social considerations sometimes override the inherent fairness of cost-based tariffs, especially in developing countries. When this happens, the tariffs are said to be market-based. However, to recover costs, cross-subsidization between various classes of consumers and/or some subsidization by the government is inevitable. For example: (i) At present, most of the cross-subsidisation of domestic consumers comes from high tariffs imposed on industrial consumers and (ii) Most of the state governments are subsidizing supply to agriculture.

(i) Cost-based Tariffs

Tariffs should have sufficient rates to raise adequate revenue to meet the financial requirements of the utility. However, attempts should be made to bring down the cost of energy by integrated development, optimization, reactive compensation, conservation, reduction in T&D loss and control on pilferage (see Sec. 9.5).

- The tariff should be based on supply cost for each category of consumer. However, urban consumers will subsidize the rural consumers to some extent. Profit oriented consumers i.e. commercial or large industry, should contribute to earn the some rate of return for utility.
- Peak consumers should pay both capacity and energy costs, whereas off-peak consumers (such as agriculture) should pay only the energy costs.

- The lower the service voltage, the greater the costs consumers impose on the system. Therefore, higher tariffs for low voltage consumers is desirable.
- Tariffs should be based on marginal costs of serving demands.
  (a) for different consumer categories
  (b) for different seasonal industries i.e. rice sellers, ice industries etc.

Marginal capacity costs are basically, the investment costs of generation, transmission and distribution facilities to supply the additional kW.

(ii) Market-based Tariff

Marketing is adding value to the electricity service, making it more attractive to consumers. This tariff helps the consumer to tailor his/her electric use to times when energy costs are low. Typically, off-peak tariff is lower than that of peak time. Time-of-day or time-of-use tariff levies are higher for peak hours, medium during the day outside peak hours and lower at night. There may be some incentives by the state government for a particular class of consumers as a measure of social justice or to encourage growth in a certain sector. Following are some examples of market-based tariffs. They may be more prevalent when sufficient justification can be provided:

- Certain industrial classes may be subsidized to attract new industry to an area.
- Residential rates may be subsidized by other classes for social/ political purposes.
- Agricultural tube-well services may be subsidized to encourage increased food production. The agricultural sector is a non-peak consumer and that too only for limited hours. the agriculture demand is seasonal and mainly confined to non-peak hours. Therefore, a lower tariff is justified. The agriculture consumers are mostly unmetered in India and their consumption is based on field study estimates.
- Inverted block rates have been used extensively to encourage energy conservation.

9.8 Tariff Applications

9.8.1 Types of Tariff

The more common tariff types are given below.

Flat-Rate Tariff

Tariff comprising a single kWh or HP charge only.

Two Part Tariff

Tariff comprising a kW or kVA charge and kWh charge.

Block Tariff

Tariff in which the price is based on a series of rates per kWh or kVAh applying to successive blocks of kWh/kVAh. Each block is of fixed size, though the size is not necessarily the same for each block. Successive blocks rate may be in increasing order for conservation of energy.

Variable-Block

A tariff similar in form to a block tariff but in which the number of kWh or kVAh of each consumer block depends on the size of some feature of its installation, e.g. connected horsepower. A variable block tariff is not recommended for general use but have application in special cases.

Restricted Hours

Tariff applicable to supply which can only be used during certain specific hours.

Off-Peak

Restricted hours tariff in which a lower price is charged for electricity used during prescribed off-peak hours.

Controlled Load

Tariff applicable to loads, supply to which may be interrupted at the discretion of the supplier, in order to regulate the peak loads. A method of such control is the frequency/injection system or time switching. This tariff is sometimes called *restricted hours tariff.*

Bulk Supply

Tariff chargeable from large consumers of general or mixed loads, such as military establishments, railways, public work departments, institutions, hospitals, etc. where further distribution of load is to be done by the consumer.

Fixed or Service Charges

A charge pertaining to a prescribed period but independent of the number of kWh supplied and the fluctuations in demand such as periodical meter testing charges, re-connection charges, re-sealing charges, meter rental charges etc. and other charges of taxes, O&M etc.

Capital Contribution Charges

A cash contribution by a consumer towards the capital cost of supply either by a single initial payment or by a number of payments over a period, such as service connection charges, transformation charges etc.

Guarantee of Minimum Revenue

A guarantee by the consumer that he will pay a minimum amount in each accounting period irrespective of consumption such as *monthly minimum charges* provision in tariff.

Power Factor

Power factor as near unity as practicable lowers the cost of energy supply. To ensure that average monthly power factor of a consumer load is not below a certain minimum, say 0.95, the tariff provides for extra charge if power factor is less.

Grid Tariff

This is applicable to primary power distributing licensees for resale to ultimate consumers.

Time-of-day Tariff

As stated in Sec. 8.12, differential tariff encourages the spread of electrical load, thus improving the system load-factor and energy conservation through the application of time-of-day or/and time-of-year rates. Such tariffs are suitable for large industrial consumers, bulk supply, grid consumers, who utilize sufficient block of power, so that it is possible for them to restrict their peak-load demand to an appreciable extent. In the country, there is a severe problem of power shortages of peak load capacity. Therefore, it is urgent and important to introduce time-to-day pricing where shifting consumption to off-peak periods is feasible. In such cases, charging a high tariff during periods of peak demand can result in a significant shift in demand and energy consumption. Even medium and small industrial consumers will shift consumption from the peak period if this tariff is applied.

Demand Conservation Tariff

Power utilities should have tariffs to encourage the existing consumers to hold down the load maximum demand and improved load-factors. Micro-computer based demand controller units are indigenously available, which supervise the operation of the consumer's equipment. In most cases, demand can be reduced by elimination of some unnecessary operation and adopting improved methods of operation. To discourage wasteful consumption of energy, the demand conservation tariff should be levied at progressive lower rates for a consumer, for whom certain demand assessed/sanctioned is being utilized for certain period. Then later on, demand as well as consumption is reduced due to energy conservation measures [see Sec. 8.12.2 (h)].

Wheeling Charges

Power transferred through the area from one utility to other utility is called '*wheeling power*' and use of this area transmission facilities for wheeling power is charged on monthly bases is called '*wheeling charges*'. These are based on percentage cost of the power transferred. For example, most of the state governments/regulatory commissions have approved wheeling charges as 2 per cent of the energy wheeled by the state, private sector, wind power or micro-hydro projects. Realistically, wheeling charge should be calculated on the basis of annual charges on the capacity of transmission and distribution used for wheeling of power and line losses. As electricity is not transported like a commodity, electricity flow follows the least resistance path. When one utility or control area transports power to another, the resulting power flows along paths (interconnected transmission and distribution lines) joining the two areas, regardless of ownership of lines. The amount of power flowing on each path of the transmission/distribution system depends upon the impedance of various paths. The impedance of the line depends upon the line length and design detail of the line. A low impedance path attracts a greater part of the total transfer than a path with high impedance. The load flow study under normal load conditions will be helpful to determine the path taken by the electricity delivered. In actual cases, the power will always flow from the nearest power plant instead of from the generator with whom the wheeling contract has been made. Due to this, system losses may be reduced and system efficiency improved. When power utilities enter into a wholesale power transaction with other utilities or consumers, they designate a pro-forma 'contract path' of transmission system through which power is expected to flow. The actual power will flow through parallel paths depending on loading conditions at the time when the transfer occurs.

Banking

It is the balancing of accounts of the co-generator for supply to grid and supply from grid in certain specified grace period. It is arrangement of supplying surplus power to the grid with the provision of withdrawing an equivalent amount at a later date within 12 months usually.

Buy-back

It is arrangement of allowing the consumer to offset the electricity bill by using his own grid-connected power generation (e.g. solar PV, wind power etc.). Power generated by the consumer is measured and credited against the consumer's electricity account. When the consumer's generation is more than his requirement, the power utility will buy-back the excess power at the rate settled.

Advance Consumption Deposit

It is the consumer deposit as security for supply of electricity, which is deposited before commencing the supply to the consumer. The amount of deposit is equivalent to consumption for the billing period, generally of 2 to 3 months, as the consumer consumes electricity on credit for the period.

Transmission Tariff

It is the tariff for the transmission of power. It is equal to annual fixed charges, which consist of O&M expenses, plus depreciation, plus interest on loan and working capital, plus return on investment, plus any other taxes annually payable. The annual fixed charges are based on fixed assets of transmission system.

O&M charges, depreciation charges, rate of return, fixed assets are notified from time to time, applicable for 5 years at a time, by Government of India, Ministry of Power for Grid Power Corporation of India, for each transmission line.

Electricity Pool

Electricity pool is not a physical location. It is a set of rules and procedures managed by the Load Dispatch Centre. The power pool operates the wholesale market under mandatory trading arrangements, bidding and settlement procedures. It is not possible to distinguish which generator produces the electricity consumed by a particular consumer. Hence, this is the concept of a central pool of generation to supply total consumer demand. Typically, pool rules require generating and supply companies to submit a day-ahead bid package and demand reservations for each half-hour period during the next seven days. The pool administration uses this to prepare a seven-day generating unit commitment plan, ranking bids in the order of merit to determine the system marginal price at which offers match demand. The pool price paid to generating companies generally comprise:

- The system marginal cost;
- Start-up and no-load adjustment and
- Ancillary services payment, black-start capability and spinning reserve.

The pool price, paid by the supplier, needs to be hedged or capped. It generally comprises:

- Price paid to generating companies:
- An adjustment for transmission losses;
- Pool administration charges; and
- Uplift, (e.g. for constrained-on units).

Figure 9.8 shows the transfer and supply of electricity via a pool. As mentioned before, electricity is a necessity. Therefore, the electricity free market must always have some minimum surplus in its generation, transmission and distribution capacity. The pool system is expected to bring down costs because of increased competition.

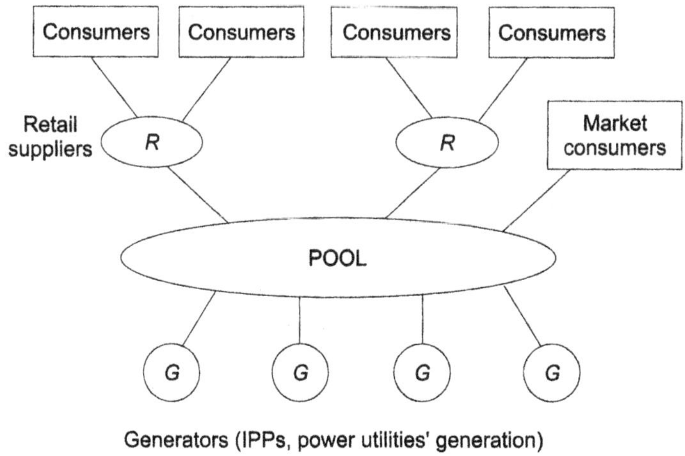

Fig. 9.8 *Pool electricity marketing*

Availability-based Tariff

This tariff scheme, applicable to central sector generators and power utilities, has been approved by the Central Regulatory Commission. It is based on the rational of correcting the grid frequency to bring it within the permissible band. There is a provision for fixed charge when power is available (peak and non-peak separately) and a separate energy charge when power is delivered. State power utilities need to pay a fixed price for the generation capacity allocated to them a day earlier irrespective of their consumption. However, variable cost will be paid only if energy is drawn. The tariff is linked to unscheduled interchange (UI) of power. UI charges will be paid or received under four conditions:

(i) When a generator generates more than the schedule, increasing the system frequency;
(ii) When a generator generates less than the schedule, decreasing the frequency;
(iii) When the beneficiary state overdraws power, decreasing the frequency;
(iv) When the beneficiary underdraws power, increasing frequency.

The UI charge accruse to the party adversely affected on account of grid indiscipline. In all the instances, there is sufficient incentive for both the generator and the beneficiary states to maintain a system frequency close to 50 Hz. UI charge also offers greater incentive to hydro power capacity in the system. On the supply side, there is a strict penalty for wrongly declaring availability. If the power utilities are saddled with surplus power, they will be forced to look for buyers. This will encourage power trading and better rates for the consumer.

Reactive Power Tariff

This tariff is on kVArh consumption. It should apply to all commercial consumers having a connected load of more than 20 kW on per kVArh basis. The charge should be such that it gives the consumer the incentive to install proper capacitors to reduce his/her reactive power consumption. Electronic meters are best suited for this tariff application. For example, BEST company in Mumbai has levied this tariff on large commercial consumers like hotels, show-rooms, shopping centres etc.

kVAh Tariff

With the availability of electronic metering, this tariff application has become urgent. Previously the cost of providing induction metering for kVAh tariff was prohibitive and therefore only kWh metering was adopted for general connections and small industry. Now electronic metering has brought the cost of kVAh or kWh or kVArh meters at par and made them affordable. A consumer was supplied reactive power free of cost. Reactive power is either generated at the generating plant or by power capacitors installed in the system at suitable points. The kVAh tariff takes into account the reactive power consumed and therefore the consumer is fully charged. This tariff needs to be provided for domestic, commercial and small/medium industrial consumers having a connected load of more than 15 kW.

9.8.2 Open Access

It offers producers and suppliers thorough access to the grid network and distribution system to enable the wheeling of power directly to consumers (see Fig. 9.9). Clear norms for open access and wheeling of power at the least cost needs to be established in the grid code. Bulk consumers should be allowed to access power directly from producers by paying suitable wheeling charges to the transmission and distribution utility. Open access on a non-discriminatory basis is critical for a competitive environment in the power sector as per the Electricity Act, 2003.

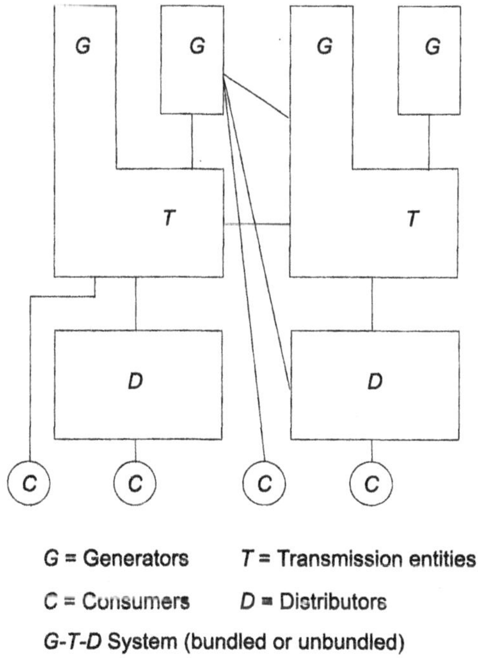

Fig. 9.9 *Open-access market structure*

9.8.3 Wholesale

This is power that is bought and sold in the electricity market among utilities, non-utility generators and other wholesale entities such as municipalities. The electric products of the wholesale market are:

- Power/energy capacity
- Reactive power
- Frequency regulation
- Spinning reserve
- Cold reserve

9.8.4 Retailing

It is the reverse of a power utility's natural monopoly to sell power. It is related to the sale of electricity in a free-market economy. The objective of retailing is to make a commercial margin by purchasing electricity in wholesale and selling it to the consumers. Another type of retailing can be Independent Power Producers (IPPs) selling electricity to large consumers directly on term contracts. Ultimately, electricity retail market development will facilitate the choice of retailers in the country by any category of consumer, possibly within the next one decade.

EXAMPLE: The relation used by Grid Power Corporation of India for determining the transmission tariff is as below:

Recoverable fixed charges every month from the beneficiary state:

$$= \frac{\text{Total annual fixed charges}}{12} \times \frac{\text{monthly average kWh sales to the beneficiary state}}{\text{Total monthly kWh sales to all beneficiary states}}$$

9.8.5 Suitable Tariffs

The following tariff forms suggested are a general guide for different consumers on the basis of economical costing. Each tariff should have added elements of '*minimum monthly charges*' and levy of '*fuel surcharge*' (based on fuel price increase liability from time to time).

Domestic or Commercial

From a costing viewpoint, domestic or commercial tariffs may be represented by a minimum charge and block energy rate. The cost of supply for small domestic consumers from the lower economic strata may be reduced by prepaid metering or by load limiters or through investment grant from state government.

Agriculture

Agriculture load requirement is only off-peak and power supply is required for maximum 8–10 hours continuously daily. Also this load is seasonal in character and the demand goes up only for 3-4 months in summer during the year. The annual load factor does not exceed 15 per cent. Therefore lower tariff is required. The agriculture produce is affected by the vagaries of nature and prices are also fixed by the Central Government. The measures for cutting cost for agriculture supply are:

- A good option is to allocate hydropower for agriculture. Hydropower is generated from land resources and will be used for the land. For example, if we allocate hydropower in Punjab for agriculture, the cost-based tariff for agriculture will be about 50 paisa/kWh;
- Metered supply for agriculture will help to remove wastage of electricity use and detect power theft;

- Improve the voltage profile and remove voltage fluctuations on the worst feeders;
- Reduce trippings and breakdowns on rural feeders;
- Balance energy and reduce losses on feeders and distribution transformers;
- State Government may give grants for capital works relating to agriculture supply, and
- Monitor the healthy working capacitors at the consumer premises.

These measures will reduce the cost of electricity/kWh and thus make reduced tariff, for agriculture feasible. A block tariff needs to be levied. The billing can be estimated every month with the final billing done six-monthly on the basis of actual reading of the meters.

Heavy and Large Supply

This class is generally composed of consumers with load factors in the higher range, say above 40 per cent. The magnitude of load for these consumers is generally large, and the number of consumers low compared to other classes. Two-part tariff with block energy rate along with power factor tariff suits.

Small Supply

This class is with a limited degree of homogeneity, and considering that the large number of consumers involved with relatively low consumption do not justify demand metering, a block tariff is suitable.

Off-Peak

A flat rate charge for some non-continuous process industries can be applied after coming to an agreement with the consumers.

Grid Supply

This tariff may be represented by the demand and a flat energy rate.

Bulk Supply

This is applicable to non-industrial (general or mixed load) large establishments, such as railways, government departments, hospitals, etc. The tariff based on flat energy rate and minimum charges is desirable.

Street Light

In urban areas, this tariff may be represented in demand per lamp point and flat energy rate. In rural areas may be based on flat rate per consumer meter in the village preferably for 40 W lamp point.

EXAMPLE: Given the following data of a power system, calculate the marginal cost tariff at 11 KV and at 400 V.

Thermal Power Plant	
Capacity construction cost	: Rs 30000/kW
Plant Load Factor	: 58 %
Auxiliary consumption	: 10 %
Coal consumption	: 0.65 kg/kWh
Coal cost at site	: Rs 1.50/kg
Fuel oil consumption	: 5 ml/kWh
Fuel oil cost	: Rs 4000/kl
O&M cost	: 2.25 %
Hydro Plant	
Capacity construction cost	: Rs 40000/kW
Auxiliary construction cost	: 1 %
O&M cost	: 1 %
Load factor	: 40 %
Hydro-Thermal generation mix	: 35:65
Transmission construction cost	: Rs 15000/kW
Subtransmission to primary distribution at voltage 11 kV cost	: Rs 10000/kW
Transformation 11/0.4 kV and LT network cost	: Rs 8000/kW
Line losses: Total	: 15 %
Transmission	: 4 %
Sub-transmission to primary feeders	: 6 %
11/0.4 kV transformation and LT network	: 5 %

Solution

(i) Thermal Plant:

Construction cost/kW of thermal plant	: Rs 30000
Depreciation @ 7 %	: Rs 2100
Interest @ 8 %	: Rs 2400
O&M cost @ 2.25 %	: Rs 675
Total	: Rs 5275

Number of kWh generated/kW capacity/annum at 58 % LF

$$= \frac{1 \times 8760 \times 58}{100} = 5081$$

Auxiliary consumption	= 508 kWh
Net kWh delivered at generation busbar per annum	= 4573
Fixed cost/kWh	$= \frac{5273}{4573}$
	Rs 1.15
Fuel cost per kW capacity/annum (Coal)	= 5081 × 1.50
	= Rs 7621.5
Fuel oil cost per kW/annum	$= \frac{5081 \times 5 \times 4000}{1000 \times 1000}$
	Rs 101.6
Total fuel cost/kWh	$= \frac{\text{Rs } 7723.1}{4573} = \text{Rs } 1.688$
Total generation cost/kWh	= 1.15 + 1.688
	= Rs 2.83

(ii) Hydro generation:

Cost/kW	= Rs 40000
Depreciation @ 3 %	= Rs 1200
Interest @ 4 %	= Rs 1600
O&M cost @ 1 %	= Rs 400
Total	= Rs 3200

Number of kWh generated annually at 40 per cent PLF = 1 × 8760 × 40

	= 3504
Auxiliary consumption	= 35 kWh
kWh delivered at generation busbar per annum	= 3469
Generation cost/kWh	$= \frac{3200}{3469} = \text{Rs } 0.92$

(iii) Combined Thermal-Hydro supply cost/kWh

	= 0.65 × 2.83 + 0.35 × 0.92
	= Rs 2.16
(iv) Transmission cost/kW	= Rs 15000
Interest @ 8 %	= Rs 1200
Depreciation @ 3 %	= Rs 450
O&M cost @ 1.5 %	= Rs 225
Total	= Rs 1875

kWh supplied per kW capacity
(thermal + hydro) per annum $= 0.35 \times 3469 + 0.65 \times 4563 = 4186$

kWh transmitted $= 4186 - \frac{4}{100} \times 4186 = 4019$

Transmission cost/kWh $= \frac{1875}{4019} = 0.47$ Rupee

(v) Cost of subtransmission down to

11 kV feeders/kW capacity	= Rs	10000
Interest @ 8 %	= Rs	800
Depreciation @ 3 %	= Rs	300
O&M @ 2.5 %	= Rs	250
Total	= Rs	1350

kWh delivered at 11 kV $= 4019 - 4019 \times \frac{6}{100} = 3778$

Cost/kWh of 11 kV $= \frac{1350}{3778} =$ Rs 0.36

(vi) Cost of transformation 11/0.4 kV and LT network/kW

capacity:	= Rs	8000
Interest @ 8 %	= Rs	640
Depreciation @ 3 %	= Rs	240
O&M @ 2.5 %	= Rs	200
	= Rs	1080

KWh supplied to consumers $= 3778 - 3778 \times \frac{5}{100}$

$= 3589$

Cost/kWh $= \frac{1080}{3589} = 0.3$ Rupee

(vii) Marginal cost

Keeping 3 % as minimum profit margin as per provision of Indian Electricity Supply Act, 1948.

Marginal cost/kWh at 11 kV $= 1.03\ (2.16 + 0.47 + 0.36)$
= Rs 3

Marginal cost/kWh at LT $= 1.03\ (2.16 + 0.47 + 0.36 + 0.3)$
= Rs 3.39

9.9 Billing

Correct Billing

In 2001, about 50 per cent of energy consumed was not billed in the country. The picture of total energy in the system is shown in Fig. 9.10. Item 1 shows the system losses and constitutes about $1/3^{rd}$ of the energy produced. There is a large scope to reduce these losses. Under item 2, energy consumed and not billed is due to under-billing, incorrect billing and no billing. Energy not billed is due to unauthorised connections (i.e. pilferage), inaccurate meters, unmetered supply and free electricity supply. Induction type meters tend to slow down with time, leading to under-billing for consumers. It is a known fact that energy consumed and billed (item 3) is nearly $1/3^{rd}$ the power produced in the country. Correct billing needs accurate metering and metering for all consumers.

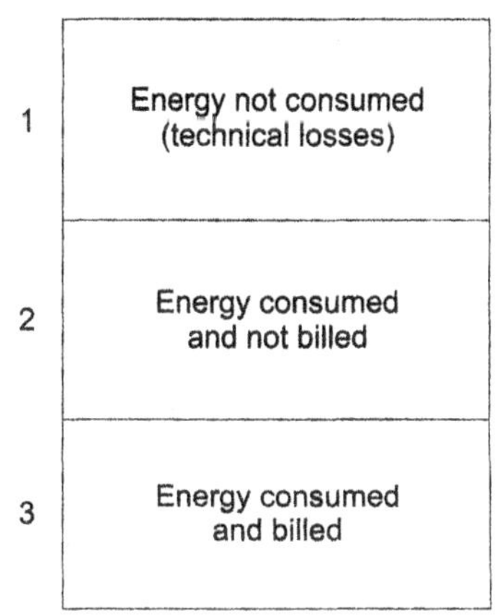

Fig. 9.10 *Energy accounting*

Bill Preparation

The bills are prepared on the basis of one of the following:

- Actual meter reading.
- Estimated reading in case of locked premises.
- No billing but pass-book for flat-rate rule payment in the case of most agricultural consumers in the country.
- Permitting no bill i.e. free electricity.
- Getting advance payment without billing by providing prepaid meters.

- No billing but fixed rate on a fixed number of lamps or by providing load limiters.
- Budget billing is consumer-defined, having flexible capabilities to set up payment plans for residential consumers. By averaging consumer energy cost over 12 months of the financial year, budget billing allows the consumer to take the high and low out of the electricity bill. The advantages are: Equal monthly installments; reduce bills to a manageable size; the bill shows a monthly charge and new balance but consumers pay a budgeted amount; once a year, the consumer account is reconciled for credit or under-payment.

Billing Cycle

The process of periodical electricity consumption, meter reading, billing and finally, collection completes a billing cycle. The billing period is counted in days taken for consumer energy use plus preparation of bill plus delivery of bill as per schedule plus due date period allowed for payment. There is a need to reduce this billing cycle period with modern technologies.

Revenue

Maximum attention should be paid to high revenue-generating accounts and be based on the A-B-C principle. 'A' are the most revenue-earning consumers and constitute nearly 20 per cent. 'B' are nearly 60 per cent of the consumers, bringing in low revenue. 'C' are the remaining 20 per cent consumers who are defaulters. Also, according to Pareto Law, about 20 per cent consumers in a power utility generate approximately 80 per cent of revenue and the remaining 80 per cent of generate only about 15 to 20 per cent of the revenue.

Tips

On every consumer bill, tips on electricity conservation and impact on environment of electricity delivered (e.g. emission/kWh) should be given.

9.9.1 Meter Reading and Bill Presentation

A meter-reading system must deliver accurate and timely bills to the consumer. The methods in use are:

- Manual
- Remote
- AMR
- Pre-paid
- Load limiters
- Bill online
- Electronic billing

Manual

In manual reading, meter readers are employed and involve cost and time. But the manual method is still the most cost-effective for most consumers. For modern electronic meters, intelligent handheld meter-reading devices can be used which are relatively cheap. The cost is soon recovered by saving on the elimination of data entry and expedited billing process. A cost-benefit evaluation revealed a pay-back period of one to two years in various utilities. Additionally, the devices provide useful meter-reader performance analysis features such as high or low consumption edits, notepad for recording information by the meter reader e.g. meter glass broken and messages for the meter reader e.g. beware of dog, key required and meter location details. A number of innovations can be made in the reading and billing system like:

(i) *Spot billing:* For industrial consumers, the meter reader visits the consumer's premises and issues the bill there, after taking the reading. This saves time in issuing the bills.

(ii) *Annual meter reading:* For certain consumers, meters can be read yearly but bills can be issued bi-monthly, quarterly or biannually on the basis of estimated consumption. Any discrepancy in reading is adjusted in the final billing when yearly reading is taken. This method can be used for agriculture and remote domestic consumers.

(iii) *Monthly, bi-monthly, quarterly or biannual reading:* This method is for most of the domestic, small commercial and industrial consumers. For example, some utilities in England take a biannual reading but billing is done every month on a average basis and adjusted biannually. Some utilities do a quarterly reading but bill monthly.

(iv) *Outsourcing of the Billing Function:*

- For villages, meter-reading and billing can be entrusted to panchayats at a commission of 2.5 per cent of the bill amount, or

- In villages, bill distribution and collection may be entrusted to panchayats on the basis of commissions, or
- The village chowkidar may be entrusted to deliver the electricity bill at a lumpsum payment with the approval of the state government. For example, in Haryana, the village chowkidar delivers electricity bills on a monthly honorarium payment of Rs 100 and Rs 50 as travelling allowance.
- *Self billing:* The Tamil Nadu Electricity Board allows small domestic and commercial consumers to read their own meters on a monthly basis. They prepare their bills from a ready reckoner pasted on the meter.
- Bill distribution and collection in slum areas (jhugi-jhompari etc.) should be entrusted to one responsible person on a commission, based on the collection made.

Remote

Remote/AMR (see Section 5.9): It is suitable for apartment buildings, shopping complexes, large commercial and industrial consumers. The benefits are: Reduced meter-reading costs, ability to read difficult access meters, facilitates flexible frequency of meter-reading, improved consumer service, quicker response to resolving consumer inquiries i.e. eliminates the need to physically visit the site for meter-reading required at the time of the inquiry, improved meter-reading safety and reduced read-to-bill time.

Prepaid Meters (see Section 8.15)/Load Limiters

For poor paymasters such as single-lamp point/kutir-jyoti scheme connections, revenue collection are not significant for supplies to rural consumers who use very little electricity (see Section 14.7.1). For such supplies, the cost of conventional meter-reading, billing, collection and administration is comparatively high. The cost of revenue collection is reduced substantially by using load limiters instead of meters since payments are fixed and there is no need for meter-reading or billing, and accounting is easier. Bills for load limiter consumers can be made payable in advance since amounts are fixed. This improves cash flow for the power utility. The fixed monthly payments for a Load Limiter supply make budgeting easier for the consumer and reduce the likelihood of defaulting on payments. An annual advance payment option can be offered to enable rural consumers to pay for their supply when income is generated at harvest time. A lower connection fee can be charged with load limiters to reflect its much lower capital cost and the reduced supply-side cost due to peak demand limitation.

Online Bill

In cities, electricity bills can be delivered to consumers via the Internet. Consumers can log on to the power utility's website with a specific account number and password. In Mumbai, BEST delivers electricity bills as well as a graphical representation of their annual consumption to consumers via the Internet.

Electronic Billing

Consumers will be able to receive and pay electricity bills at the *touch of a button*. In Internet-based billing and payment service, the consumer can receive and pay bills by logging on to the utility website. That information includes billing and payment history, day-by-day or month-by month electricity usage and useful weather statistics. Consumers can see usage statistics in both numeric and graph form and compare usage and costs on a month-by-month and year-by-year basis. Consumers can link quickly to frequently-asked questions on billing, payment and service. The power utility website management can be interactive, leading to enhanced revenue collection. For large consumers, presentation and payment of bills can be made daily. AMR technology is suitable for this system.

Meters (ABB, L&G etc.) are available in the country for billing remote metering and internet communication. The open architecture is shown in Fig. 9.11. This system is already in use in Singapore and U.S.A.

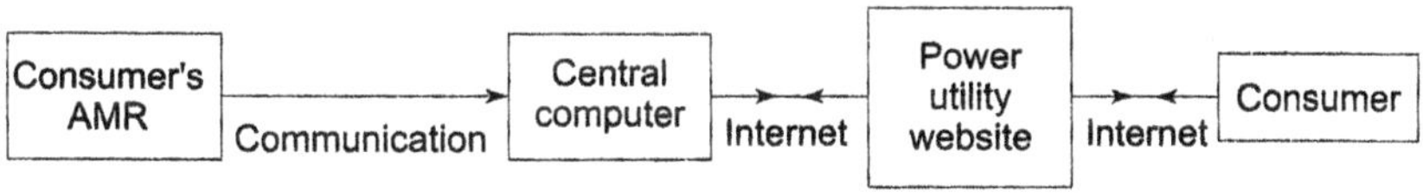

Fig. 9.11 *Electronic billing*

Bill Notification through e-mail.

9.9.2 Collection

The following methods may used in innovative ways for the convenience of the consumer:

- Collection boxes for bank drafts/cheques for electricity payments may be installed at strategic points.
- Creating one collection agency for water and electricity.
- Collection through banks/post offices.
- Collection on a fast track via the power utility website. This facility is introduced by North Delhi Power Limited recently.

EXAMPLE: In the Punjab State Electricity Board (2001), there were about 3,100 industrial connections, each having a connected load of more than 100 kW. Their total annual revenue was about Rs 2,000 crores, about 50 per cent of the total annual electricity revenue of Rs 4,300 crores. The Punjab State Electricity Board billing cycle for large industrial consumers is given below:

Billing period (30 days) $\longrightarrow$ meter reading (1 day) $\longrightarrow$ bill preparation (2 days) $\longrightarrow$ bill delivery (2 days) $\longrightarrow$ due date for payment (7 days) = 42 days.

Therefore, total days taken to make the payment after power is consumed = average $15 + 1 + 2 + 2 + 7 = 26$ days.

It takes 25 or 26 days on an average to get the payment after energy is consumed. If electronic billing and payment is made on a daily basis through the power utility's website, we save 25 days' interest (@ 12 per cent) $= 2000/12 \times 12/100 \times 25/30 =$ Rs 16 crores. The payback period for this facility will be nearly two years [8].

PROBLEMS

1. As per Pareto Law (the 80/20 Principle), in a power utility 80 per cent of the revenue comes from 20 per cent consumers. Give your comments.
2. How will you design the low cost agriculture tariff which is essential for promoting the agriculture in any state?
3. Define the following:
   - Average cost
   - Marginal cost: short-term marginal cost, long-term marginal cost
   - Incremental cost
   - Avoided cost
   - Opportunity cost
   - Embedded cost
4. The Bhilwara Group, a large industrial textile consumer from Ajmer district in Rajasthan state has installed an 86 MW hydropower captive power plant (PLF: 50 per cent) at Malana in Himachal Pradesh to supply power to their industrial complex at Ajmer in Rajasthan. They have signed an agreement with the Rajasthan State Transmission Company and the Ajmer Distribution Company to provide 14 per

cent electricity free of cost to the transmission company as wheeling charges. Make calculations to justify these charges. What will be the wheeling charges/kWh?

5. Estimate the generating cost per kWh delivered from the thermal power plant having the following particulars:
   Installed capacity = 1000 MW
   Capital (loan) cost = Rs 40 million/MW
   Plant load factor = 75 per cent
   Fuel and overhead cost = Rs 1.5/kWh
   Depreciation = 12 per cent
   Interest = 9 per cent

BIBLIOGRAPHY

1. Flavin Christopher and Nicholas Lenssen, *Power Surge–A Guide to the Coming Energy Revolution*, Earthscan Publications, London, 1995, p. 248.
2. Swift Loui, *Mathematics and Statistics for Business, Management and Finance*, Macmillan, London, 1997, pp. 237–257.
3. Maria Cugnetto, 'Metre Reading Systems for the Future', *Proceedings of the 5th International Transmission & Distribution Conference, Distribution 2000*, November, 1999, Brisbane, Australia, pp. 341–350.
4. Willis H Lee, *Power Distribution Planning—Reference Book*, Marcel Dekker, 1997, p. 190.
5. Gil Julian Barquin and others, 'Reactive Power Pricing : A Conceptual Framework for Remuneration and Charging Procedures", *IEEE Transactions on Power Systems*, vol. 15, no. 2, May, 2000, pp. 383–489.
6. Lescouerur, B. and J.B. Galland "Tariffs and Load Management—the French experience', *IEEE Transac. on Power Systems*, vol. PWRS-2, no. 2, May 1987, pp. 458–464.
7. *Economic Growth in the Future*, Edison Electric Institute, McGraw-Hill Book Company, New York, 1976, pp. 257–286.
8. *Asian Power*, July 2000, p. 18.
9. "Australasia Market Report", *Asian Power*, Dec. 2000/Jan. 2001, pp. 24–27.
10. *Annual Report on the Working of State Electricity Boards and Electricity Departments*, Planning Commission, Government of India, June, 2001, pp. 82–96.
11. Pabla, A.S., 'Modelling of Agriculture Electricity Consumption', *Proceedings of 59th R&D Session*, CBI&P, Calcutta, 1–4 February, 1994.
12. David A. Kumar, 'Market Power in Electricity Supply', *IEEE Transactions on Energy Conversion*, vol. 16, no. 4, December, 2001, pp. 358–359.
13. Pabla, A.S., *Electrical Power Systems Planning*, Macmillan India Limited, New Delhi, 1998, pp. 220–221.

Chapter 10

Grounding

10.1 Grounding System

One of the key factors in any electrical protection scheme is grounding. If any acceptable measure of safety is to be attained, correct grounding design and application must be followed. The principles and standards normally applied in grounding are found in the reports of several national organizations [1, 2, 3, 9].

A grounding system to be totally effective, must satisfy the following conditions:

(a) Provide a low impedance path to ground for personnel and equipment protection and effective circuit relaying.
(b) Withstand and dissipate repeated fault and surge currents.
(c) Provide corrosion allowance or corrosion resistance to various soil chemicals to ensure continuous performance during the life of the equipment being protected.
(d) Provide rugged mechanical properties for easy driving with minimum effort and rod damage.

In any discussion of grounding, the question always asked is "How low in resistance should a ground be?" In general, a low earthing resistance means high earth fault currents but low overvoltages during fault conditions. It is difficult to determine this in ohms. The lower the ground resistance, the safer the grounding. Further, for protection of personnel and equipment, it is worth the effort to aim for a ground resistance of less than one ohm. It is generally not practical to reach such a low resistance along a distribution system, transmission line or in distribution sub-stations. In some regions resistances of 5 ohms or less may be obtained without much trouble. In others, it may be difficult to bring resistances of driven grounds below 100 ohms.

Accepted standards stipulate that transmission sub-stations should be so designed as not to exceed a earth resistance of one ohm. In distribution transformer sub-stations the maximum recommended resistance is 5 ohms. In sub-stations (66 kV and above), the buried grid system of any sub-station will provide the desired resistance (IEEE Std 80–1996 deals exhaustively with sub-station grounding and fundamentals of personnel safety). The problem may arise in those systems where the voltages are 33 kV and below in which driven rods are normally used. The sub-station grounding grids depend on the dual action of the grid and connected grounding rods. From the viewpoint of sheer resistance, the rods here do not materially reduce the resistance of the system but have a separate and important function. The mat itself has an initial impedance which is several times its low frequency resistance to ground. This can reduce the effective dissipation of surges. The grounding mat serves to control within safe limits the step-touch potentials at higher fault currents. Ground rods, in contrast, which are simple resistance rods, can exhibit surge impedance that is about half of their simple low frequency resistance. This is so because of the breakdown of soil resistance under the high-voltage gradients at the surface of the rods. As a result of this characteristic, the rod should be located in the immediate vicinity of the station arresters. Ground rods, of course, will be required in high voltage lines (132 kV, 66 kV) where the maximum earth resistance of 15 ohms is acceptable and in distribution lines (33–0.4 kV) where maximum earth resistance of 25 ohms is preferred.

These parameters can usually be met with proper application of basic grounding theory. There will always exist circumstances which will make it difficult to obtain the desired ground resistance. When these situations develop, several methods of lowering the ground resistance can be employed. These include parallel rod systems, deep driven rod systems utilizing sectional rods and chemical treatment of the soil. Additional methods discussed in other published works are: buried plates, buried conductors (counter-poise), electrically connected building steel and electrically connected concrete reinforced steel. Due to the inherent property of absorbing and retaining moisture; a clay known as *bentonite* may be used for reducing the earthing resistance in areas where soil resistivity is high, of the order of 300 ohm-metres and above [4].

The artificial treatment of soils with fly ash (from coal based thermal power stations) offers promise in the reduction of earth resistance in high soil resistivity areas. The results conclusively prove that fly ash offers better grounding compared to conventional earthing of salt-charcoal-soft-coke (IS: 3043–1987).

EXAMPLE Bentonite powder and soil in ratio of 1:4 by volume was used in pits for earthing of 11 kV lines by electrode. Earth pit resistance, before and after mixing bentonite, was taken when water was poured into the pit and measurement was taken after two days, as given in Table 10.1 [13].

Table 10.1

11 kV line earthing with electrode earth pit resistance (ohms)

Location	Before using bentonite powder (earth resistance-Ω)	After using bentonite powder (earth resistance-Ω)
1	25	9
2	7.5	2.5
3	30	4.5
4	10	0.8
5	10	2.2

10.2 Earth and Safety

Now let us look at the other half of the ground connecting system, the earth itself. First, the area of contact between the earth and ground rod must be sufficient so that resistance of the current path into and through the earth will be within the allowable limits of a particular application. The resistance of this earth path must be relatively low and must remain reasonably constant throughout the year.

To understand why earth resistance must be low we need to apply Ohm's law i.e., $E = I \times R$ (where E is the voltage in volts, I is the current in amperes and R is the resistance in ohms). For example, assume a 415–volt supply (240 volts to ground) with a resistance of 4 ohms. Now assume that an exposed wire in this system touches a motor frame that is connected to the grounding system which has 20 ohm resistance to earth (refer Fig. 10.1). According to ohm's law, there will be a current of 10 amperes through the fault from the motor frame to the earth. If a person touches the motor frame and is solidly grounded to earth, he would be subjected to 200 volts (20 ohms times 10 amperes). This may be fatal depending upon the resistance of the person which varies with touch voltage. The relation of human (normal adult with dry skin) body resistance to voltage is nonlinear and for frequencies up to 100 Hz and for d.c. currents is as given in Table 10.2.

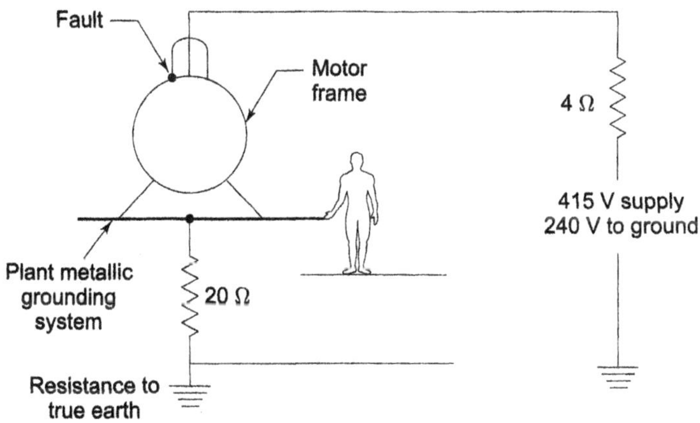

Fig. 10.1 *High earth resistance fault*

Table 10.2

Safe touch potentials

Touch voltage rms (V)	*Body resistance Ω*
25	2500
50	2000
250	1000
Asymptotic value	650

A let-go current of 10 mA for men and 8 mA for women have been accepted as standard. Currents of 100 mA and above are known to be fatal. The IEC publication 479 "Effects of current passing through a human body" includes a graph (Fig. 10.2) showing the zones of danger resulting from contact with a.c. supply 50/60 Hz on adults. If currents passing through the human body could be confined to the level and times specified in zones one, two and three, the danger of electrocution under normal circumstances would be practically eliminated. However, in some cases shock intensities could result in people being thrown off ladders which could result in physical injuries. IEC 364 gives the circuit disconnecting maximum times against touch voltages as reproduced in Table 10.3.

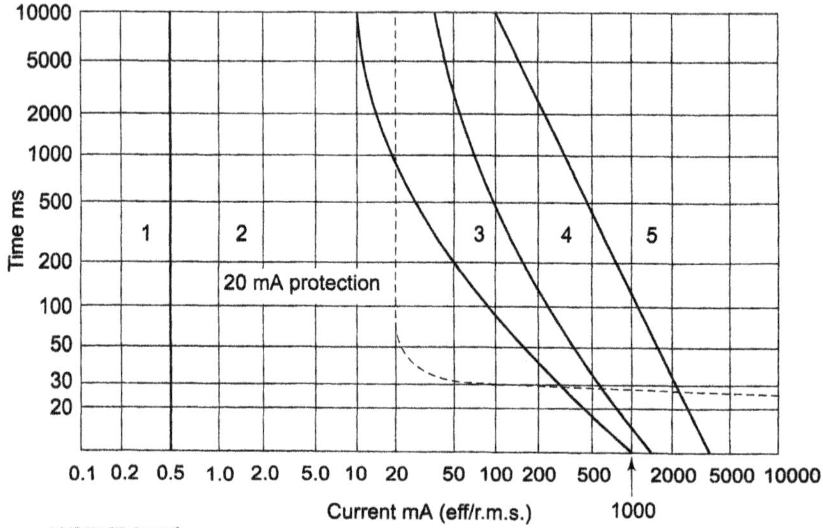

Fig. 10.2 *Effects of current through human body:*
Zones 1, 2 and 3: Danger of electrocution practically eliminated;
Zones 4 and 5: Danger of electrocution

Table 10.3

Maximum touch voltage duration

Prospective touch voltage (V)	Maximum disconnection time (s)
< 50	∞
50	5
75	1
90	0.5
110	0.2
150	0.1
220	0.05
280	0.03

The committee appointed by the Central Board of Irrigation and Power has recommended the use of the following equation to determine safe body currents for the purpose of sub-station grounding grids in India [14].

$$I = \frac{0.155}{[t]^{1/2}} \text{ amperes}$$

$I =$ Power frequency ac (r.m.s.) current that a human being of average body weight can withstand without ventricular fibrillation

$t =$ Duration of shock in seconds.

10.3 Nature of an Earth Electrode System

Resistance to current through an earth electrode system has the following three components:

(a) Resistance of the ground rod itself and connections to it.

(b) Contact resistance between the ground rod and earth adjacent to it.

(c) Resistance of the surrounding earth.

Ground rods, masses of metal, structures, pipe, strip or conductor, plate, cable armour etc. are commonly used for earth electrodes. These electrodes are usually adequate in size or cross-section so that their resistance is a negligible part of the total resistance.

The resistance between the electrode and earth is much less than what is normally suspected. If the electrode is free of paint or grease, and the earth is firmly packed, the U.S. National Bureau of Standards No. 108 has shown that the contact resistance is negligible.

A ground rod which has been driven into uniform resistivity earth conducts current in all directions. Let us consider the electrode as being surrounded by shells of earth all of equal thickness (Fig. 10.3). The earth shell closest to the ground rod has the smallest surface area and consequently offers the greatest resistance. The next earth shell is somewhat larger in area and offers less resistance and so on. Finally, a distance from the ground rod will be reached where inclusion of additional earth shells does not add significantly to the earth resistance surrounding the ground rod. This distance is known as the *effective resistance area*. It is mainly dependent on the depth of the ground rod. Of the three components involved in "resistance", the resistivity of the earth is most critical and most difficult to calculate and overcome.

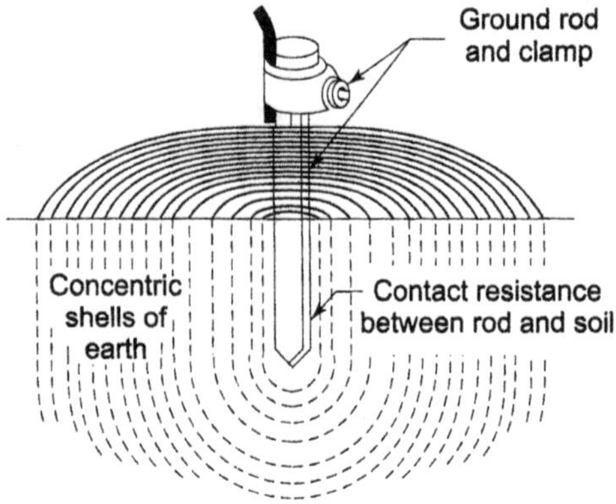

Fig. 10.3 *Resistance components of earth electrode*

Calculating Earth Resistance

The equations for earth resistance from various systems of electrodes are quite complicated and in some cases may be expressed only as approximations. All such expressions are derived from the general relation $R = \rho L/A$ and are based on the assumption of uniform earth resistivity throughout the entire soil volume although this is seldom the case. A commonly used resistance to earth formula for a single ground rod developed by Professor H.B. Dwight* of Massachusetts Institute of Technology is

$$R = \frac{\rho}{2\pi L}\left(\ln \frac{4L}{a} - 1\right)$$

where,

$\rho =$ average soil resistivity (ohm-cm)
$L =$ ground rod length (cm)
$a =$ ground rod radius (cm)
$R =$ resistance of ground rod to earth (ohm)

Effect of Rod Size on Resistance (6)

Whenever a ground rod is driven deeper into the earth, its resistance is substantially reduced. Generally, speaking, doubling the rod length reduces the resistance by an additional 40 per cent (Fig. 10.4a).

*H.B. Dwight, "Calculation of Resistance to Ground", *AIEE Transac.* vol. 55, December, 1936.

Increasing the diameter of the rod, however, does not materially reduce its resistance. Doubling the diameter for instance, reduces resistances by less than 10 per cent (Fig. 10.4b).

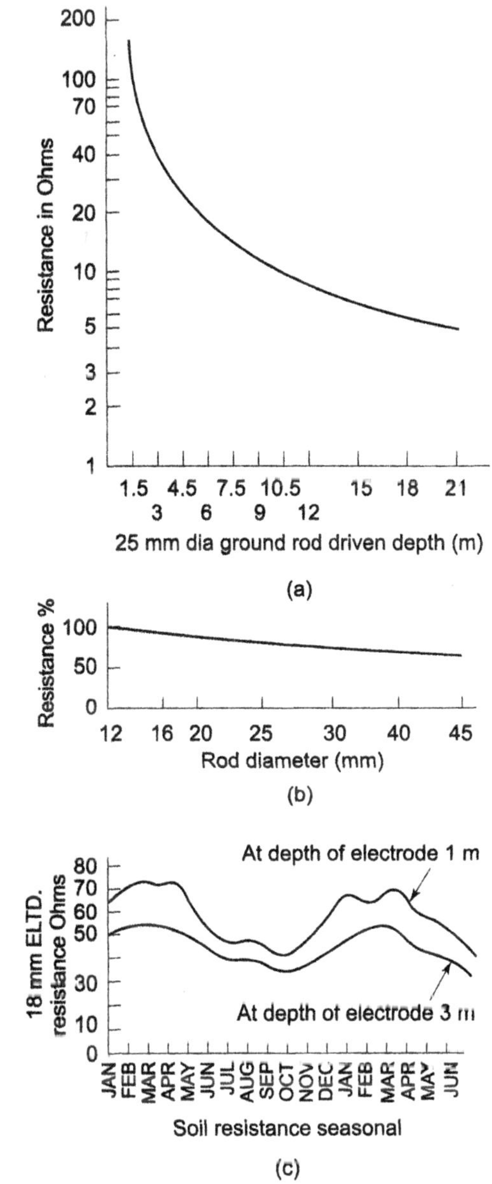

Fig. 10.4 *Variation in earth resistance:*
(a) with depth; (b) with diameter; (c) seasonal

Effect of Soil Resistivity on Earth Electrode Resistance (6)

Dwight's formula shows that the resistance to earth of grounding electrodes depends not only on the depth and surface area of the grounding electrodes but on soil resistivity as well. Soil resistivity is the key factor that determines the resistance of a grounding electrode and the depth to which it must be driven to obtain low ground resistance. The resistivity of the soil varies widely throughout the world and changes seasonally. Soil resistivity is determined largely by its content of electrolyte, consisting of moisture, minerals and dissolved salts. A dry soil has higher resistivity but a wet soil may also have a high resistivity if it contains no soluble salts.

Because soil resistivity is directly related to moisture content and temperature, it is reasonable to assume that the resistance of any grounding system will vary during the different seasons of the year. Such variations are shown in (Fig. 10.4c). Since both the temperature and moisture content become more stable at greater distances below the surface of the earth, it follows that the grounding system should be constructed with the ground rod driven down a considerable distance below the surface of the earth to be most effective at all times. Best results are obtained if the ground rod reaches the permanent moisture level.

10.4 Earth Conductor Sizes

Steel conductors and electrodes are used for distribution line and substation grounding. The minimum conductor area required can be calculated from the following empirical formula [3]:

$$\text{Area in mm}^2 = 12.15 \times 10^{-3}\, I\sqrt{t}\,,\ \text{for welded joints}$$

$$= 15.7 \times 10^{-3}\, I\sqrt{t}\,,\ \text{for bolted or compression joints}$$

where,

I = fault current in amperes

t = duration of fault current, usually taken as 3 s.

While selecting the conductor for thermal stability according to the above formula, mechanical strength and corrosion of conductor are given due consideration. For mechanical ruggedness, the minimum size of steel strip conductor should not be less than 10×6 mm^2 or G.I. wire

4 mm dia and extra corrosion allowance may be considered on the following lines:

(a) Very mildly corrosive soils, having resistivity above 100 ohm-m, no allowance for corrosion be made.

(b) Mildly corrosive soils, having resistivity 25–100 ohm-m, corrosion allowance of 15 per cent of thermal designed conductor area is made.

(c) Severally corrosive and treated soils, having resistivity less than 25 ohm-m, corrosion allowance of 30 per cent of the thermal designed conductor area may be allowed.

As compared to bolted/compression joints, there may be little preferential corrosion on welded joints of welding materials. Tack welding should be avoided and instead continuous welding is recommended.

In general, mean soil resistivity values for various types of soils: marshy, loam, wet sand, wet gravel, dry sand/dry gravel, rocky are 25, 100, 200, 500, 1000, 3000 Ωm respectively. The fault current in distribution transformers [3] is generally of the order as given in Table 10.4.

Table 10.4

Fault currents in 11/0.433 kV distribution transformers

Transformer kVA rating	*Order of fault current (A)*
16	200
25	400
63	1100
100	2000
200	3700
400	6500
500	7600
750	9300
1000	10,000

The fault current will be almost double if two transformers of the same size are run in parallel. The conductor sizes may be calculated with due corrosion allowance and nearest standard size of steel strip or MS rod or GI wire may be selected.

10.5 Design of Earthing Electrodes

10.5.1 Electrode Design

The earthing electrodes up to 33 kV distribution systems are usually of a minimum size of MS rod 20 mm dia or GI pipe 25–40 mm dia and 2.5–3 m long (with due consideration for mechanical strength and corrosion) buried in the ground at a depth of 0.5–0.75 m from the ground level. Deep or long driven rods have been found to be very useful in reducing the earth resistance. 8 SWG G.I. wire spiral shape 'coil earthing' electrode has been specified by REC (standard J_1/1972) for small transformers 25 kVA and below.

At places (of high soil resistivity) where the grounding resistance obtained with this arrangement exceeds the prescribed limit, multiple electrodes should be used. In case of two electrodes the interconnection should be made with a MS strip of the same size as the grounding conductor and the distance between them should not be less than twice the length of the electrode. If the use of a third electrode is also necessitated, it should be so placed that the three electrodes when interconnected form an equilateral triangle with sides not less than twice the length of the electrode. The resistances of two or three rods can be considered in parallel for practical purposes and the total earth resistance will thus become half or one-third of a single rod earth resistance respectively. Sometimes, it may be difficult to keep such large spacings and so it becomes necessary to determine what reduction in the total resistance can be obtained by connecting rods in parallel. Empirical relations for various rods in parallel are given in the form of relations with reference to the earth resistance of single rod as given below [5]:

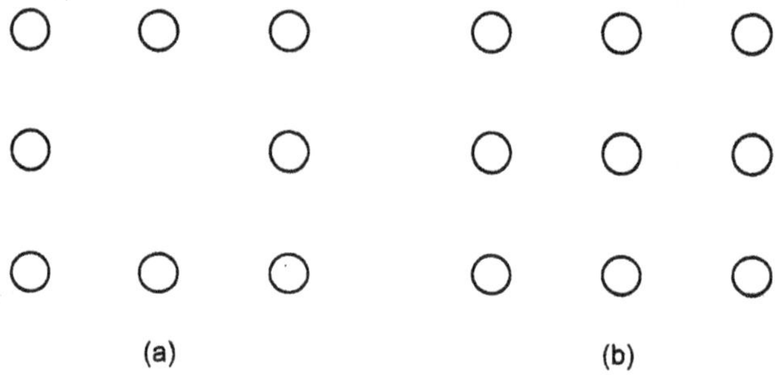

Fig. 10.5 *Earth electrodes:*
(a) Rods in parallel in hollow square; (b) Rods in parallel in solid square

(a) Two rods in parallel

$$\frac{\text{Resistance of two rods in parallel}}{\text{Resistance of one rod}} = \frac{1+x}{2}$$

where, $x = \dfrac{L}{d\left(\ln \dfrac{4L}{a} - 1\right)}$, d being the distance between the two parallel rods.

(b) Three rods in parallel arranged in the form of equilateral triangle of side d,

$$\frac{\text{Resistance of three rods in parallel}}{\text{Resistance of one rod}} = \frac{1+2x}{3}$$

(c) Rods arranged in a hollow or solid square as shown in (Fig. 10.5): If N is the total number of rods, the ratio is expressed as

$$\frac{\text{Resistance of } N \text{ rods in parallel}}{\text{Resistance of one rod}} = \frac{1+kx}{N}$$

where k is constant and dependent on the number of rods. The values of k are given in Table 10.5.

Table 10.5

Number of rods inside of a square	Total number of rods	Value of k
Hollow square		
2	4	2.7071
3	8	4.2583
4	12	5.3939
5	16	6.0072
6	20	6.4633
7	24	6.8363
8	28	7.1479
9	32	7.4195
10	36	7.6551
Solid square		
3	9	5.8917
4	16	8.5545
5	25	11.4371
6	36	14.0650
7	49	16.8933
8	64	19.5003
9	81	22.3069
10	100	24.9587

In general, number of rods or pipes may be connected in parallel and the resistance is then practically proportional to the reciprocal of number of electrodes used, so long as each is situated from the other not less than twice the length of the electrode.

Interconnections between the electrodes of a multiple electrode system should be made at a depth of 0.5–0.75 m from the ground level.

At locations where the multiple-driven electrodes system cannot be adopted because of space limitations or some other reasons, the treatment of the soil is desirable as discussed earlier. Two separate and distinct connections to earth as risers are desirable. Each connection should be capable of carrying the full fault current.

10.5.2 Grounding Nomogram (6)

To assist the engineer in determining the approximate ground rod depth required to obtain the desired resistance, the grounding nomogram given in Fig. 10.6 can be used. For example, to obtain a grounding resistance of 20 ohms in soil with resistivity of 100 ohm-m, an 18 mm dia MS rod must be driven 6 m deep. It should be noted that the values indicated on the nomogram are based on the assumption that the soil is homogeneous and, therefore, has uniform resistivity.

10.5.3 Electrode Earth Resistance

In practice, as per Indian Standard IS: 3043 Code of Practice for earthing, the following formulae are used for calculating electrode resistance of various types:

(i) Rod or Pipe Electrode

$$R = \frac{\rho}{2\pi L} \ln \frac{4L}{d}$$

where ρ = resistivity of soil in ohm-cm

L = length of rod or pipe in cm

d = diameter or rod or pipe in cm.

(ii) Strip Electrode

$$R = \frac{\rho}{2\pi L} \ln \frac{2L^2}{\omega t}$$

where

ω = depth of burial of the electrode in cm.

$t =$ width in case of strip, twice the diameter in case of round conductor in cm.

$\rho =$ resistivity of soil in ohm-cm.

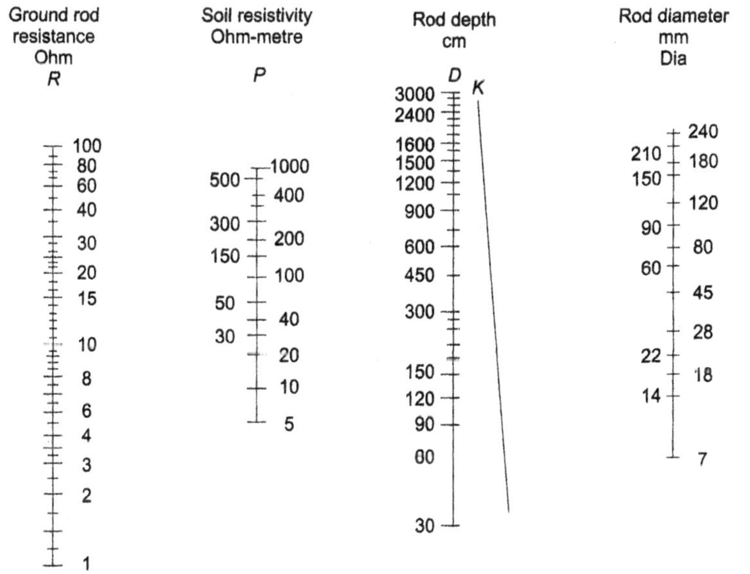

Fig. 10.6 *Ground nomogram:*

1. *Select required resistance on R scale;*
2. *Select apparent resistivity on P scale;*
3. *Lay straight edge on R and P scale and allow to intersect with K scale;*
4. *Mark K scale point;*
5. *Lay straight edge on K scale point and dia scale and allow to intersect with D scale;*
6. *Point on D scale will be rod depth required for resistance on R scale.*

(iii) Plate Electrode

$$R = \frac{\rho}{400} \times \left[\frac{\pi}{A}\right]^{1/2}$$

where,

$A =$ area of both sides of plate in m^2.

$\rho =$ resistivity of soil in ohm-metre.

(iv) Resistance of sub-station grounding grid by the method of Laurent:

$$R_{\text{grid}} = \frac{\rho}{4r} + \frac{\rho}{L}$$

Maximum grid potential rise = $I \times R$

where,

ρ = soil resistivity in ohm metres

r = radius of circular plate of same area as grid (m)

L = Total conductor length of grid including the length of cover perimeter, cross connections and connections to equipment and structures (m).

10.5.4 Temporary Grounding

Temporary grounding is required when doing maintenance and construction jobs on the lines and sub-stations. Properly applied temporary grounding creates an equipotential work zone guarding against possible threats:

- Voltage induced from adjacent energized lines
- Fault current on adjacent systems
- Lightning stroke
- Accidental re-energizing.

It provides a parallel path around the worker with low resistance by earthing sets. Almost all the current passes through the parallel path. A frequently-used value of human body resistance is 1000 ohms. This requires the parallel path to have a resistance measured in milli-ohms. For example, if it is desired to maintain a maximum 50 volts across a worker whose body resistance is 1000 ohms during a fault with 10000 amperes (assuming that clamp body and contact resistance is negligible comparatively), by Ohms law, it can be seen that, a parallel protective temporary ground resistance of 5 milli-ohms or less is required.

10.6 System Earthing

A power system is earthed at appropriate points by a suitable scheme to achieve certain advantages such as: reduced equipment costs, reduced operation and maintenance expenditure, greater safety, improved service reliability, better system and equipment overcurrent protection, improved lightning performance and reduced radio interference.

10.6.1 System Neutral Earthing

Most three-phase electrical system neutral points are either unearthed, or effectively (solidly) earthed, or earthed through the impedance of a resistor or reactor (non-effectively earthed).

Unearthed

A system is considered unearthed, if no intentional connection is made between any part of the system and earth. In reality, a system is always capacitively earthed through the inherent shunt capacitance that exists between the conductors and earth. In a balanced system with no loads connected between line and neutral, the neutral is held at earth potential because of the presence of this shunt capacitance. Line to neutral voltage equals line to earth voltage, and the neutral to earth voltage is zero.

Under balanced conditions, each phase current is displaced by 120 degrees and the vectorial resultant is zero. When a fault occurs, the faulted phase assumes the earth potential and the neutral potential is displaced from earth by an amount equal to the line to neutral voltage (V_{ph}).

The voltage between each of the healthy phases and earth rises to the line to line value (line to line voltage $= 3 \times V_{ph}$). The resulting increase in current across the shunt capacitance causes the current between each of the healthy phases and earth to increase by a factor of 3; since the currents are only displaced by 60 degrees, the result of the vector diagram is an earth fault current of $3I_c$.

An ungrounded system can continue to operate with an earth fault, provided that the earth current ($3I_c$) does not rise above a few amperes, and the phase-to-earth voltage of the healthy phases do not rise above the line to line voltage, and the fault can be located, when shut down can be arranged at convenience.

Operating experience with ungrounded systems has shown that substantial over voltages can develop. The presence of capacitance and inductance in the system leads to arcing and intermittent spluttering type faults, which cause the system voltage to escalate to dangerously high value (up to six times the normal value has been produced in laboratory tests). Over-stressing of the insulation and failures in system components can be caused by voltages of this magnitude. Transient over-voltages, especially in the medium voltage systems, increase the probability of failures and generally outweigh any advantages an ungrounded system might offer.

Effectively (Solidly) Earthed

Effectively earthed systems have a direct connection between the neutral and earth, with no intentional impedance. Protective multi-earthed utility distribution systems have a neutral wire which parallels the three-phase conductors for the length of the distribution line, and is earthed at intervals. This system has been widely adopted in India.

Effectively earthed systems are not subject to transient over-voltages which result from intermittent earth faults. The solid earth connection stablizes the neutral voltage and prevents elevation of the phase-to-ground voltage.

If the neutral is available for connection, effective earthing involves no additional equipment and is the least expensive method. However, the earth fault current is limited only by the arc and stray return path impedances which are small, resulting in high fault current. Extensive and time consuming repairs may be needed after an earth fault current.

Resistive Earthing

Resistive earthing systems have an essentially resistive impedance inserted between the neutral and ground. This method of earthing has the advantages of both the above unearthing systems, while eliminating most of their disadvantages. The potentially dangerous system over-voltages caused by arcing type earth faults are suppressed by dissipating the energy in the resistor, so significantly improving system stability and personnel safety.

Resistive earthing techniques have been proved to be economical, safe and more reliable—ultimately improving system performance. Indonesia, Malaysia and Singapore have moved towards resistive earthing systems.

The mitigation of an earth fault's damaging effects, and resulting hazards to personnel are even more pronounced when compared to solid earthing. The energy released and the resultant damage by a fault, are approximately proportional to the square of the fault current multiplied by the fault duration. If the fault current is reduced from 10,000A with solid earthing to 100A with resistance earthing, the magnitude of the fault energy is reduced by a factor of 10,000 (assuming equal fault durations).

The sizing of the resistor allows the system designer to modify the fault current magnitude to suit the system's protection scheme. The exact magnitude of resistance is often not critical, as long as it satisfies several criteria. Low resistance earthing provides an earth fault current typically between 25A and 2000A. The magnitude of the earth fault current must be at least as large as the current flowing through the system's shunt capacitance, for the resistance to adequately limit the transient over-voltages. The protection scheme is optimised by selecting a fault damage, while still allowing enough to operate relays selectively and reliably.

Because the fault is cleared quickly by relays, damage to equipment is minimized. The fast response time also improves personnel safety, prevents additional faults from occurring and limits overheating and mechanical stress on conductors.

High resistance earthing uses a resistance, which is sized to limit the earth fault current to slightly higher than the capacitive current and typically, no more than 10A ($R < X_c/3$). If possible, the earth fault current should be large enough to allow for system growth. The fault current in high resistance earthed systems is low enough to permit continued operation while the fault is located, and a scheduled shut down is arranged for repairs. Fault tracking in high resistance earthed systems can be accomplished without circuit interruption.

In India, the system is effectively earthed. The coefficient of earthing depends upon the system zero phase sequence and positive and negative phase sequence reactance and resistance (see Fig. 13.1). Studies reveal that in a distribution system, the coefficient of earthing is nearly less than 80 per cent. On systems 110 kV and 132 kV, this coefficient generally does not exceed 75 per cent.

Future Practice

The Indian power system is expanding rapidly. Grid network inter-connections are increasing and regional grids have developed. The national grid will soon become operative. This will result in increasingly high fault levels in the system. Therefore, it is desirable to introduce resistance earthing in the Indian distribution system at the earliest to contain earth fault level. For example, earthing practices in highly developed power systems are:

(i) France: Neutral impedance earthing.

(ii) Canada: HV neutral grounding is through lightning arresters and LV neutral grounding is through 1.5 ohm reactors.

(iii) Australia: Steel reinforcement of PCC poles is used as an earth electrode for multiple neutral earthing and earthing of other metallic parts of the line for 11 kV lines.

10.6.2 Lines and Sub-stations

The earthing of overheads lines and sub-stations is important due to following considerations:

(a) To have low zero-phase sequence impedance to keep the higher fault current and enable operation of relays for circuit breaking under fault conditions.
(b) For the safety of personnel and livestock.
(c) To reduce communication interference by keeping the earth potentials low.

The touch voltage is more critical than step voltage. In outdoor sub-stations 33, 66, 132 kV, a person is not likely to be exposed to dangerous touch voltage when he is more than 2 metres away from any metallic equipment or structure in switchyard. A surface layer (75 cm) of crushed stone around the metallic equipment, and structures up to a distance of about 2 metres will provide the necessary high resistivity layer below the feet to enable a person to withstand higher voltage.

Every pole of primary distribution, three-phase, three-wire system with or without earthwire should be earthed, having earth resistance not more than 20–25 ohms and laid as per the Indian Standard code of practice for earthing [2]. This practice ensures safety in case there is an earth fault. For better overvoltage protection, it is desirable to install (on lines without earth wires) an earth wire on the approaching span of the distribution sub-station. This reduces the intensity of the incoming impulse waves and distributes the ground fault current by providing a parallel path and thus decreases local ground potentials under fault conditions.

According to I.E. Rule–90:

(a) All metal supports of overhead lines and metallic fittings attached thereto, shall be permanently and efficiently earthed. For this purpose a continuous earth wire shall be provided and securely fastened to each pole and connected with earth ordinarily at three points in every km, the spacing between the points being as nearly equidistant as possible. Alternatively, each support and the metallic fitting attached thereto shall be effectively earthed.
(b) Each stay wire shall be similarly earthed unless an insulator has been provided in it at a height not less than 3.0 metres from the ground.

In lines, maximum earth fault impedance should be such that under fault conditions, the fault current must not be less than 2.5 times the relay setting or fuse rating of the distributor or main. Zero sequence impedance of the line and footing resistance of pole can be calculated and this fault current stipulation checked. In city areas, a bottom earthed guard must be provided for maintenance as well as for the safety of personnel and livestock. In rural areas, the lines with or without an earth wire should be constructed in a way such that a breakdown in the conductor must result in earth short circuit, i.e., it must touch the earthed structure pole or earth wire. Snapped conductors of 33 kV, 11 kV and LT lines on high resistance soil or dry soil with moderate resistivity have been found to give a feeble fault current, i.e. less than the full load or setting of relay or fuse rating due to high contact and soil resistance. The feeder's breaker in this case does not trip or the fuse will not blow out. Numerous accidents have been observed on such conductors snapping even on loamy soils in dry season. Such accidents are unavoidable unless special measures are taken, such as the breakdown creates a short circuit to earth with a special mechanical design of X-arm or line, etc. which is prevalent in the southern states of India (Fig. 10.7). Other measures include a reduction of the earth resistance of electrodes. Adoption of static digital relays, as explained in Sec. 10.6.6 is a good remedial measure.

There are two types of earthing systems in overhead line supports—one is 'counterpoise' earthing, and the other one is 'pipe type' earthing. In case of counterpoise earthing, horizontal earth rods of several arms of strip or round steel stock are buried radically around the tower or pole to a small depth, and connected to the earthing down leads. It is usually provided in hard rocky soils. Earthing conductors are made of galvanised steel wire stocks no less than 6 mm in diameter and at least 35 mm^2 in area. The earthing conductors are buried at a depth of 0.6 m.

Pipe type earthing is used in normal soils. Normally, supporting structures of overhead lines are earthed with the aid of deep driven vertical earthing devices. These are in the form of round or angle steel stock or pipes driven into the earth to a depth of 3–20 m. Generally, a 25 mm diameter hot dip galvanized M.S. pipe is used.

Underground System: When an extensive underground cable system is available, the armour and/or lead sheath form a most effective earth-electrode. In majority of the cases, the resistance to earth of such a system is less than 1 Ω. A freshly installed cable serving deteriorates according to moisture content and nature of soil. The cable armour/sheaths are used to provide a metallic path to fault current returning to the neutral.

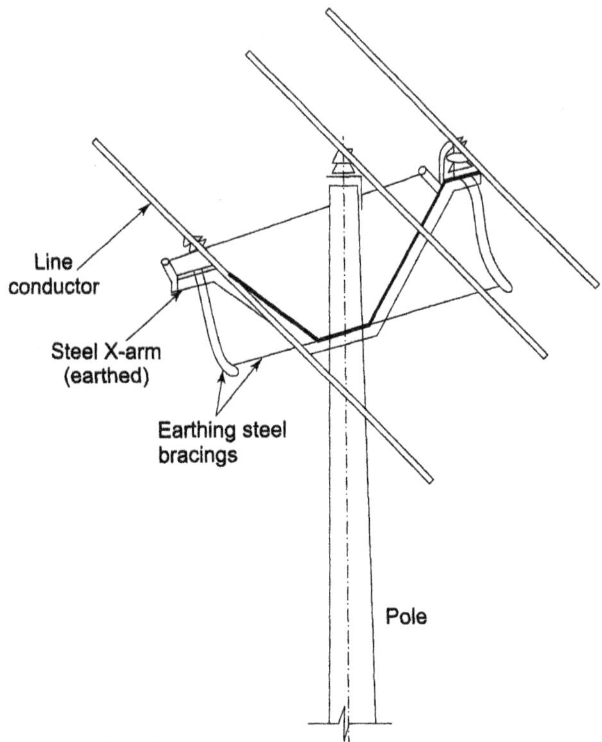

Fig. 10.7 *X-arm earthing arrangement*

As per I.E. Rule 67, the distribution transformer neutral point should be earthed by not less than two separate and distinct connections with earth, each having its own electrode. All non-current carrying metalwork of installation be connected to earth as per I.E. Rules 51 and 61. Separate earth connection to separate electrode should be made for lightning arrester installed at the distribution transformers. The lightning arrester inter-earthing connection to transformer tank/neutral be made as given in Sec. 13.11.

Reinforced Concrete Footing as Earthing Device

In the case of LT and primary distribution lines and sub-stations (415 V, 11 kV, 22 kV or 33 kV) there is ample scope for savings by avoiding conventional earthing electrode system and using instead steel reinforced concrete footing of line, sub-station supports such as steel tubular or braced channel poles with RCC muffs and PCC and RCC poles buried portion. This is still an area of research in India and proper standards have yet to be formulated. The concrete does not loose its conductivity when used in water-bearing soils even after setting. Suitable earthing terminal of extended reinforcing steel can be made at the top of the reinforced concrete footings or top of PCC or RCC poles.

Because of their conducting nature, steel reinforced or pre-stressed concrete poles present effective grounding (MEN) for distribution overhead lines in Australia to provide safety for the general public, utility employees, and to protect nearby telecommunication installations on a cost-effective basis [16]. A concrete pier of four bars has approximately one-half the earth resistance of simple 1.5 mm dia driven electrode [18].

For the calculation of the earthing resistance of the footings, reference should be made to various published data [11].

EXAMPLE According a study [11], the ground resistance of the reinforced concrete foundations in sub-stations yards is good grounding element. When all the foundations are connected in parallel through the laid ground conductor and overhead aerial wires, the resultant ground resistance of the foundation in the sub-station yard may be comparable with the ground resistance of the laid conductor. The study showed the individual ground resistance of the power transformer foundation; 115 kV lead end tower footing; 115 kV, 230 kV, PT, CT supports as 11.1 ohms, 11.6 ohms, 11.78 ohms, 20.5 ohms respectively.

10.6.3 GIS Sub-stations

The area occupied by a GIS sub-station is typically only 10–25 per cent of that of the equivalent air insulated conventional installation which makes the achieving of low ground resistance difficult. It is necessary to resort to additional methods to achieve the required value of earth electrode resistance [15, 17].

- *Increasing* the length of conductors laid within a single loop will reduce the resistance of the grid mesh, but not in direct proportion to the additional length laid.
- *Connecting* the reinforcing steel mesh of concrete floor slab and structural steel to the earthing grid will certainly serve to reduce the total earth electrode resistance.
- *Using* the deep driven ground rods or chemical treatment may be beneficial.
- *Bonding* of metallic enclosures of GIS assembly is important, as these enclosures carry induced currents of significant magnitude which must be confined to specific paths.

10.6.4 HV Neutral Grounding

One neutral grounding is normally provided in at each voltage level between generation and distribution. If a neutral point is not available (e.g. delta connection, bus bar points etc.), a zig-zag transformer (interconnected star) is used. It is a three-phase auto-transformer with no secondary winding as shown in Fig. 10.8. Each limb of the transformer has two identical windings wound differentially, such that under normal conditions, the total flux in each limb is negligibly small and therefore, the transformer draws very little magnetising current. The transformers are typically, of short time rating, usually 10 seconds to 1 minute. Therefore, the size of such a transformer is small as compared to the power transformer of the same rating.

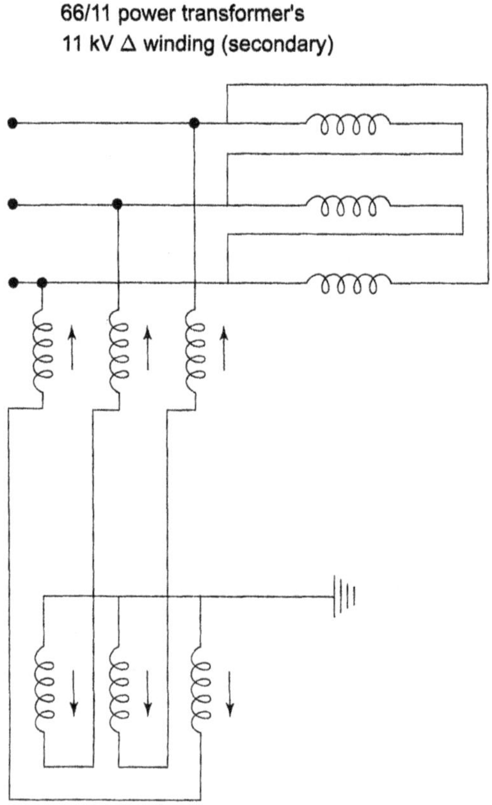

Fig. 10.8 *Zig-zag transformer for neutral grounding*

10.6.5 LT Lines and Consumer Permises

With the increasing use of electricity for heating, cooking and domestic appliances, the number of electrical accidents has increased as a result of inadequate earthing for the dispersal of large fault currents. Of the various methods of earthing in vogue, the protective Multiple Earthing of the line Neutral (MEN) [2] with equipotential bonding has been found to be the best as it affords maximum safety. It also avoids the use of expensive earth wire and thus reduces the loading and height of the supports, resulting in considerable economy. The neutral in LT line construction in vertical formation is placed below phase conductors and is earthed at various points along the line. The adoption of this method of earthing, however, necessitates the use of a minimum of three earth electrodes per kilometre of the line length and the installation of earth electrode system in consumer premises, insulating the stay wires, etc. The resistance to earth faults has to be kept low enough so as to enable the fuses to blow off quickly. This value of the resistance should not be more than 6 ohms. This is normally provided by three 20 ohms electrodes per kilometre of the line or more electrodes if electrodes of a higher ohmic value are used. The resistance of the ground electrode may be lowered in vulnerable positions by the use of the local multiple rods, preferably triangular spaced minimum at double the electrode length.

I.E. Rules 32, 61 and 90 do not specifically provide multiple earthing for LT system. An amendment in this respect is desirable due to increasing fault currents and accidents in systems. The advantages of Protective Multiple Earthing (PME) or Multiple Earthing of Neutral (MEN) are that it is possible for a supply industry to provide to all consumers a safe and efficient system of earthing at lower cost than by any other means.

Consumer Installation

As shown in Fig. 10.9, the supply system PEN conductor is earthed at several points. The power utility should provide the consumer, on his premises, a suitable earthed terminal in an accessible position at or near the point of the commencement of supply. A separate earth electrode system is necessary at the consumer's installation for medium, high and extra-high voltage levels as per IE Rule 33. However, no separate consumer earth electrode is required for low voltage i.e. a single-phase consumer (230 V supply). All exposed conductive parts of the consumer's installation are connected to the distribution system PEN conductor via his main neutral link with the main earth terminal. The consumer installation's TN-S type of earthed system is shown in Fig. 10.10.

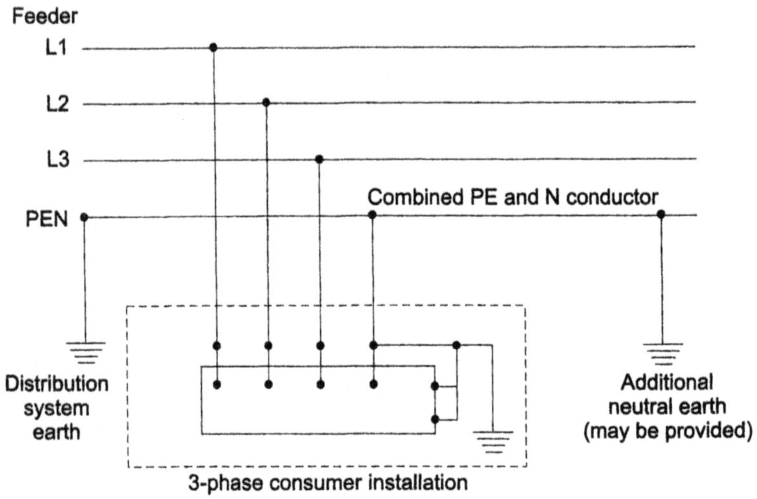

Fig. 10.9 *Multiple-earthed system (TN-C supply system)*

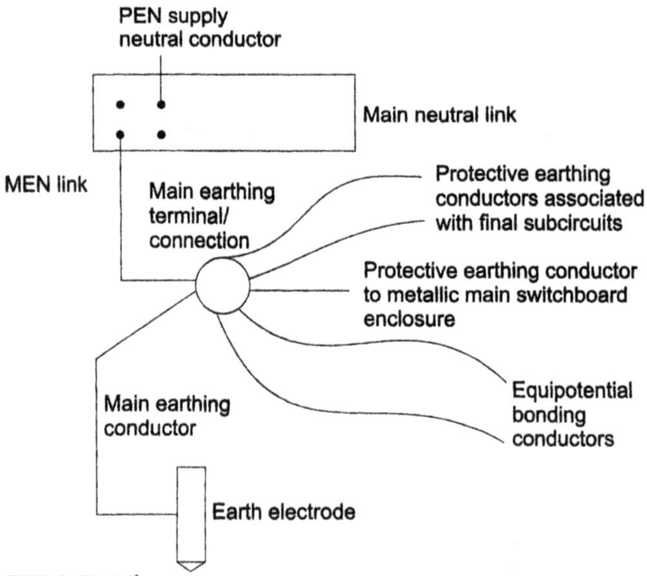

Fig. 10.10 *TN-S consumer installation with MEN link*

10.6.6 Earth Fault Protection of Feeders under Feeble Fault Current Conditions

The usual earth faults in the distribution system are:

(a) Phase conductor making contact with metallic cross arm earth through line support or separately.

(b) Phase conductor making contact with multi-earthed neutral wire in case of LT.

(c) Snapping of phase conductor on low, medium, high resistivity soils or in any of the surface contact e.g. dirt, grass, tree etc.

(d) Arcing earth faults.

(e) Snapping of multi-earth neutral.

The electromagnetic relays are generally set at the lowest setting of 10 per cent of the full load currents of the concerned lines. Their operation and operating time depends on the magnitude and duration of the earth fault currents. These relays fail to operate in the event of high resistance or feeble earth faults and conductor snapped conditions. Conductor snapping or dry hard surface soil or high resistivity soil including rocky soil is most risky. Such faults continue undetected by electromagnetic relays even after many hours of their occurrence and constitute a potential danger to human life and livestock. Such conductors falling on networks up to 33 kV systems have caused many accidents in various parts of India.

The resistance of the fault points does not necessarily cause a short circuit, especially in LT overhead lines. The fault current is considerably less usually under snapped conductor conditions. This type of fault resulted in an accident in Maharashtra [13]:

There was an 11 kV line across the river. One conductor accidently fell in the river. A few cattle entered the river to drink water. As soon as they did, they experienced shock and lost their lives. A farmer in the nearby fields noticed this and immediately informed the state electricity board (MSEB) office. The line was then switched off by the board's staff.

As a result of research studies in the USSR [7] the earth resistance measurements across a bare conductor fallen on the ground under different surface layers of earth are available and are given in Table 10.6.

Table 10.6

Earth resistance of bare conductor falling on ground (30 m long, X-section 16 mm^2)

S. No.	Description of ground	Weather condition	Mean earth resistance (ohms)
1.	Loamy soil with sparse grass	Very damp	101
2.	Black (fertile) soil covered with thick grass	Damp	167
3.	Moist earth in garden	Dry	583
4.	Tamped gravel road	Dry	690
5.	Ditch with water, black earth, loamy surface	Dry	28
6.	Asphalt yard (road)	Damp	653
7.	Snow at – 12°C	Dry	1000

Also, the earth resistance of a bare copper conductor 35 m long, laid on soft ground, has established that this resistance extended from 120 to 200 ohms. After the same conductor was pressed into the ground by automobile tyres, the earth resistance dropped to 8–14 ohms and only when placed into a ditch with water, did the resistance drop to about 3 ohms.

Earthing may not give protection against faults which are not essentially earth faults. For example, if a phase conductor of an overhead spur line breaks, and the part remote from the supply falls to the ground, it is unlikely that any protective gear relaying on earthing, other than the current balance protection at the sub-station, will operate, since the earth-fault current circuit either includes the impedance of load which would be high relative to the rest of the circuit, or it is open circuit where there is no load.

Causes for Non-operation of Electromagnetic Relays Under Feeble Fault Conditions

The principal causes of non-operation of the electromagnetic relays under feeble fault conditions are:

(a) High resistance earth faults other than conductor snapped condition can create earth fault currents of magnitude of the range of one to few amperes which is much lower than the earth faults relays even at their lowest current setting.

(b) Conductor snapping on high resistivity soil or even on hard surfaced dry soil of moderate resistivity results in poor contact with the earth surface. Consequently, the contact resistance of the conductor with the surface becomes so high that the fault current can be lower than the setting of the relay.

(c) Poor contact of dropped conductor with very hard soil can create arcing ground which set up very feeble earth fault current, as low as a few 10 milli-amperes.

Static/Digital Relays

Studies reveal that static/digital earth fault relays will successfully identify the above faults whose selection is based on the following properties of earth faults.

(a) Voltage signal is created due to earth faults current across the transactor or electromagnetic earth fault relays in the neutral lead of wye connected CT secondaries of the concerned line.

(b) Magnitude of the voltage increases as the magnitude of the earth fault increases till the concerned CT saturates.

(c) Voltage signals during strong earth faults are of fundamental frequency except during the first few cycles.

(d) Voltage signals due to high resistance earth faults are of fundamental frequency superimposed with strong harmonics. The harmonic content depends upon the resistance of fault, i.e. resistivity of soil/soil surface.

(e) The voltage signals created by arcing faults ae not continuous. Sometimes they reappear at random intervals and are lost from one to many seconds.

(f) The frequency band on harmonics is between 100 and 850 Hz and in some cases, in addition to this band, the harmonics will be in the kilocycle range. The harmonic content is in sufficient magnitude during high resistance faults.

(g) Voltage signals created by flow of earth current due to SWER line connected to three-phase feeder are mainly of fundamental frequency with or without lower harmonics.

(h) Spurious signals of strength 5 to 10 mV are found on the secondary of the transactor when there is no earth fault. These signals either disappear or become very insignificant.

10.7 Earth Testing

10.7.1 Methodology

Earth testing is done using the earth megger method based on the principle of fall of potential. Let us consider the case of an earth electrode consisting of a pipe driven into the ground and suppose that a potential is applied between pipe E and rod R driven in the ground at an infinitely large distance away as shown in Fig. 10.11. The current flow is measured by the ammeter A. If a second rod P is now driven into the ground at

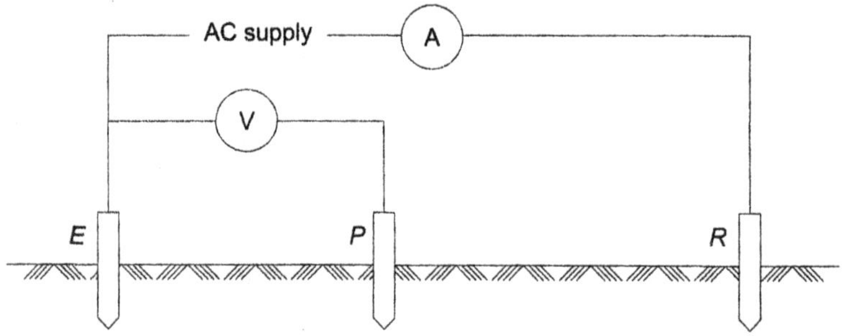

Fig. 10.11 *Fall of potential method*

various points surrounding the pipe E, the voltmeter will measure the potential difference between the pipe and various points in the surrounding soil. By Ohm's law, this potential difference will be directly proportional to the resistance of the earth up to the point measured and hence the relationship between the resistance and distance from the pipe E under test can be plotted as shown in Fig. 10.12. It will be found that the resistance increases as the rod P is placed further away from the pipe under test and that the rate of increase rapidly diminishes so that at a certain distance from the pipe, the rate of increase is so small as to be almost negligible. In fact, the resistance represented by this distance will be about 99 per cent of the total resistance to an infinite distance. Similarly, curves can be plotted in other directions radiating from the pipe under test and a series of points obtained enclosing a specific area. For all practical purposes this area may be considered to contain the whole of the resistance of the pipe to earth and is known as *resistance area.* The requirement as regards the spacing of these electrodes for measurement are:

(a) The current electrode R must be sufficiently far away from the electrode under test, E, so that the two resistance areas do not overlap.
(b) The potential electrode P must be placed outside the two resistance areas, i.e. within the horizontal portion of the resistance curve.
(c) The potential electrode P must be between electrodes R and E in straight lines joining them.

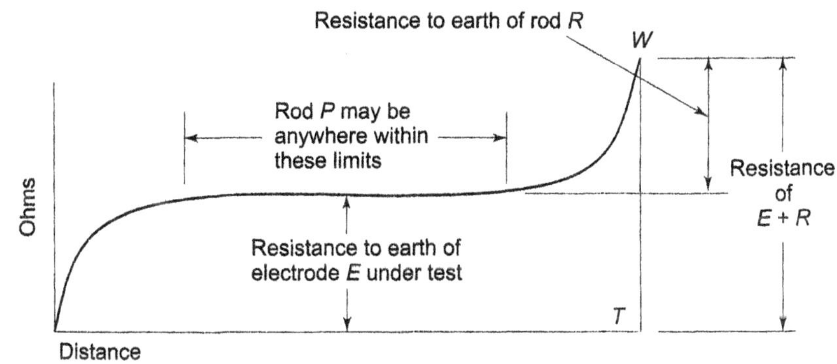

Fig. 10.12 *Effect of the resistance area of the distance rod "R" on the fall of potential curve*

10.7.2 Earth Tester

This instrument is used for measuring the resistance of earth electrodes and resistivity of soil. The preferred model is the 4-terminal type. It gives ac output. It is a combined cross-coil ohmmeter (ratio-meter) and hand driven generator of a special type so designed that alternating current is passed on to the soil while direct current is passed through the measuring instrument. This instrument usually has the voltage range of 10–110 V with a current range up to 5 mA, frequency of about 92–95 Hz, ohmic range from 0–2 up to 10,000 ohms. Indigenous designs are available in the test voltage range of 50–200 V approximately.

The ohmmeter consists of two coils mounted at a fixed angle to one another on a common axle and moving in the field of a Permanent Magnet (PM). A current proportional to the total current flowing in the testing circuit passes through the Current Coil (CC), while the Potential Coil (PC) carries a current proportional to the potential drop across the resistance under test. The cross-coil system makes the instrument a true ohmmeter giving readings in ohms which are independent of the applied voltage, speed of rotation and contact resistance of the ground rods.

The connections of the earth tester are shown in Fig. 10.13. Direct current from the generator passes through the current coil of the ohmeter to a rotating current reverser driven from the generator handle. Alternating current is thus delivered to the current terminals (the common terminals C_1 and C_2 in the diagram) of the instrument which are connected to the contact under test and to the "current" temporary earth connection. The potential coil of the ohmmeter obtains its supply from the potential terminals (common terminal P_1 and the terminal P_2 in the diagram), the latter being connected to the 'potential' temporary earth connection.

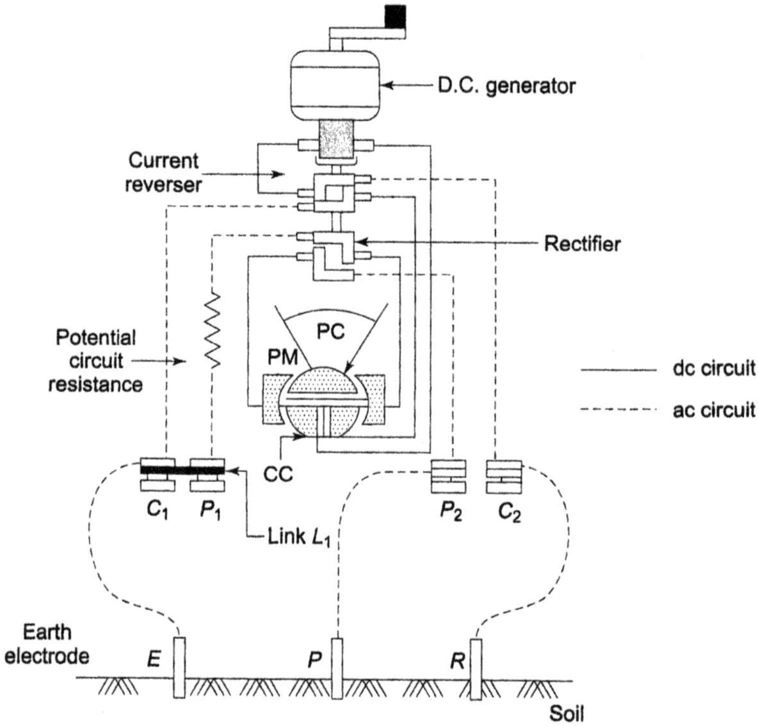

Fig. 10.13 *Working diagram of earth tester*

Since this supply is taken from the "soil" section of the current circuit and is therefore, alternating, it must be made unidirectional before passing through the potential coil. A commutator mounted on the same shaft as the main current reverser and synchronized with it, is therefore, interposed as a rectifier between the potential terminals and potential coil. The latest instruments are provided with electronic synchronized commutator system.

In this manner both the current and potential coils of the ohmmeter are supplied with direct current. But the 'soil' section of the testing circuit is supplied with alternating current.

Method of Use

When testing with an earth tester, the spacing of the temporary electrodes is of utmost importance as the two resistance areas must not overlap.

As a rough guide the following figures may be of assistance. When testing an earth electrode consisting of a single driven pipe or a single earth plate, the current electrode should be about 30 m away from the electrode under test, the potential electrode being about 15 m away. If the earth electrode is grounding grid, the distances should be further increased.

The temporary current electrode is driven into the ground at a distance considered sufficient from the earth electrode under test. Three readings are taken with the potential electrode driven in at three points in turn, one midway between the earth electrode under test and current electrode, second 3 m nearer the earth electrode under test, and a third 3 m nearer the current electrode. If the three readings so obtained agree with one another within the accuracy required for measurement, the mean of these three values may be taken as the resistance of the earth connection. If, however, the agreement is not sufficiently close, the current electrode should be moved and driven into the earth at a greater distance from the electrode under test. The three readings should then be taken again. The whole process should be repeated until the three readings obtained with one setting of the current electrode agree within the limits of accuracy required for the measurement.

The potential electrode must be placed in a direct line between the current electrode and electrode under test.

Soil resistivity measurements [12] are made by the following four-probe method of Wenner.*Link L_1 is removed. The outer two electrodes are used as current electrodes connected to terminals C_1 and C_2 and the

*F. Wenner, "A method of measuring earth resistivity", Report no. 258, *Bulletin of Bureau of Standards*, vol. 12, no. 3, Oct. 11, 1915.

inner two as potential electrodes connected to terminals P_1 and P_2. All the electrodes are spaced at equal distance in a straight line. The meter gives a direct reading in resistance and the soil resistivity is calculated by the relation

$$\rho = 2\,\pi\,a\,R$$

where,

ρ = resistivity in ohm-m
a = electrodes spacing in m
R = resistance in ohms

The spacing between electrodes indicates the soil depth up to which soil resistivity is measured [1].

PROBLEMS

1. A lineman working at a 33 kV sub-station yard was deputed on a sunny day to rectify a light point at the top of the switchyard steel structure. The switchyard is laid with 7.5 cm thick gravel layer as a safety measure. While fixing and positioning the ladder in the switchyard manually, it went out of control, causing the lineman to fall on a 33 kV overhead busbar. This caused an earth fault with the bus and people on the scene heard an explosive sound. The power supply tripped instantly. The lineman fell on the ground after getting a severe electric shock. He sustained some burns on both his hands and both feet causing bleeding. But his life was saved. Explain what are the factors that saved his life. Assume his body resistance to be 1000 ohms with a contact resistance of 100 ohms (shoes) and 33 kV bus fault level 750 MVA. Make any reasonable assumptions if required.
2. (a) What is the formula for determining the neutral point earth resistance?
   (b) What is the significance of a distance of 8 m between electrodes when measuring the earth resistance with the earth tester?
   (c) How will you reduce electrode resistance of 30 ohms to approximate 10 ohms at a spot?
   (d) What is the contribution of earth in the earth electrode resistance?
3. For a 100 kVA, 11/0.415 kV distribution transformer what is the (i) maximum earth resistance permissible, and (ii) minimum size of the earth conductor required?
4. What are the three components of earth resistance and which of them is the most critical?
5. Answer the following:
   (a) What is the general value of earth resistance of a pole in 11 kV line?

(b) What is the general value of earth resistance of 11 kV line?
(c) As per IE Rule 33, is a separate consumer earth electrode required for a low voltage single phase 230 V supply consumer?

BIBLIOGRAPHY

1. (a) "IEEE guide for safety in sub-station grounding", *IEEE Std. 80*, New York, 1996.
   (b) Pabla, A.S. and V. Singh, "Design of sub-station earthing for voltage systems of 66 kV and above", *Irrigation and Power Journal*, Central Board of Irrigation and Power, New Delhi, April, 1977.
2. IS: 3043-1987, Code of practice for earthing, p. 11.
3. "Steel grounding systems where grounding mat is not needed", *Technical Report No. 5*, Central Board of Irrigation and Power, New Delhi, July, 1976.
4. Bentonite earthing for high soil resistivity areas, *Technical Report*, CPRI, Bangalore, Nov., 1977.
5. Tagg, G.F. *Earth Resistances*, George Newness, 1964, pp. 99–111, 123–124.
6. "A modern approach to grounding systems", *Data Folder 7303.* ITT Blackburn, Missouri, 1977.
7. "Fundamental of electrical safety", Mir Publishers, Moscow, 1975, p. 230.
8. VAST-Vattenfall (Swedish Power Association-Swedish State Power Board), "Earthing of stations and switchyards (Design recommendations and measurement methods)", *Earthing Committee Report*, Zurich, August, 1973.
9. Manohar, V.N., R.P. Nagar, "Design of steel earthing grids in India", *IEEE Transac.*, PAS-98, Nov.-Dec. 1979, pp. 2126–2134.
10. Gohokar, V.N. and V.V. Gohokar "High Impedance Faults—Identification on Power Distribution, *Proceedings of Symposium* on *Electricity Distribution in the Developing Countries*, CBI&P, New Delhi, 20–21 January, 2000, pp. 254–259.
11. Thapar, Baldev, Omar Ferrer and Donald A. Blank, "Ground Resistance of Concrete Foundations in Sub-station Yards", *IEEE Transactions on Power Delivery*, vol. 5, no. 1, January, 1980, pp. 130–136.
12. Sakis Meliopoulos, A.P., *Power System Grounding and Transients*, Marcel Dekker Inc., New York, 1988, p. 312.
13. Jeyaraman, B., "Earth Resistance Improvements," *Journal of Institution of Engineers (India)*, April 1993, pp. 39–40.
14. Arunachalan, R. and others, "Investigation of Flyash for Effective Earthing", *Proceedings of 58th Annual R&D Session*, CBI&P, Bangalore, April 19–20, 1993, p. 133.

15. "Earthing System Parameters for HV, EHV and UHV sub-stations", *Technical Report 49*, CBI&P, New Delhi, September, 1985.
16. Postelethwaite, J. Derek and W (Bill) Greenland, "An Effective Solution to Concrete Distribution Pole Grounding, *Transmission and Distribution*, December 19, 1971, pp. 44–50.
17. "Earthing of GIS—an Application Guide, *Working Group 23.10 Report*, Electra, December 1993, pp. 31–51.
18. Beaty H. Wayne "*Handbook of Electric Power Calculations*" McGraw-Hill, New Delhi, 2001, pp. 14.5.

Chapter 11

Overhead Lines

11.1 Choice of the System

The distribution system can be either overhead or underground. It is usually overhead, though for higher load densities in cities or metropolitan areas, it is underground. The choice between overhead and underground depends upon a number of widely differing factors such as the importance of service continuity, improvement in appearance of the area, comparative annual maintenance cost, capital cost, and useful life of the system.

The exact point at which the underground system becomes more economical varies with individual cases. Sometimes a combination of overhead and underground systems is desirable. In a mixed system, the primary distribution may be underground while the transformer and secondary distribution overhead or vice versa. There are certain merits and demerits of each system. The voltage regulation of the underground cable system is more efficient as compared with the overhead. Therefore, in city areas it is becoming increasingly necessary to employ high voltage cables even up to 132 kV. The cable system voltage in chosen generally very close to the economic optimum with respect to the envisaged power transfer. The ranges of approximate power transfers generally found economically acceptable are given in Table 11.1. The main differences in overhead bare wire and underground cable lines are:

(i) Number of joints

The standard length of bare wire conductor wound on drums varies from 2 to 15 km depending upon the size of the conductor. Similarly, the manufactured length of power cables varies from 0.5 to 2 km. Therefore, cables having shorter lengths will result in comparatively higher number of joints in the underground system compared to the overhead bare wire for the same length of line. The joint is the weakest link in the system.

(ii) Inductance per km

The inductance in overhead bare conductor line is about four times than in cable line for the same conductor size. Higher inductance means more voltage drop. Typical three-phase line inductances are:

Voltage rating and conductor size	Overhead (ACSR) (ohm/km)	Cable (Al) (ohm/km)
11 kV, 50 mm^2	0.4	0.1
415 V, 50 mm^2	0.28	0.07

(iii) Charging current

The charging current in a underground cable is many times more than the overhead bare conductor line for the same voltage rating. This is due to the cable having comparatively more capacitance.

$$\text{Charging current per phase } (I_C) = 2\pi fCV$$
$$\text{Charging reactive power} = 3\,VI_C$$

where,

C = cable capacitance in farad (F) per phase/km
V = phase voltage in volts

Typical values of charging reactive power for 66 kV, 132 kV overhead and underground lines are as below:

Line voltage rating kV	Overhead charging MVAr/km	Underground charging MVAr/km
66	0.0122	0.941
132	0.0525	2.429

Table 11.1

Economic transmission capacity

Cable system voltage kV	*Approximate range of economic transmission: MVA per circuit*
11	3–10
22	7–12
33	12–30
66	30–80
132	80–300

11.2 Optimum Design Considerations

The distribution lines for a large supply complex must be based on optimum design. The most economical design of a system relies on minimizing not the total installed cost but the sum of installed cost, estimated present worth of losses, present worth of damage to lines, and revenue loss due to outages in its life time. This method involves the line loading criterion/practice, accurate statistical data on weather conditions and a probabilistic evaluation of potential damage. Line service life is also needed to be established for estimation.

Life tests can be carried out to analyze the service behaviour of different insulation materials, effect of extrusion techniques and types of various construction designs. The accelerated life tests at increased voltage on actual cable test lengths can be performed. A mathematical relation for the life of cables is based on the maximum electrical stress and real life. The voltage life of cables obeys Weibull distribution. Mathematically the relation is

$$E_{max} = Kt_d^{(-1/n)}$$

where E_{max} is radial stress kV/mm, t_d is the breakdown time, and K is a constant depending on the dielectric material and the type of field in it and n depends on the dielectric material only. Similar mathematical relation for accelerated temperature life testing is

$$t = ab^{-bT}$$

where,

t = time to disqualify the dielectric material when maintained at temperature T

T = temperature (in degrees Kelvin)

a = constant of proportionality

$b =$ constant for a given dielectric (normally b is approximately 0.085 near normal rated temperature).

The typical voltage and current aging curves obey the above laws, as given in Figs. 11.1 and 11.2 for cables. Similarly, chemical degradation (thermal as well as voltage stress aging) can be represented by the Eyring Model [3].

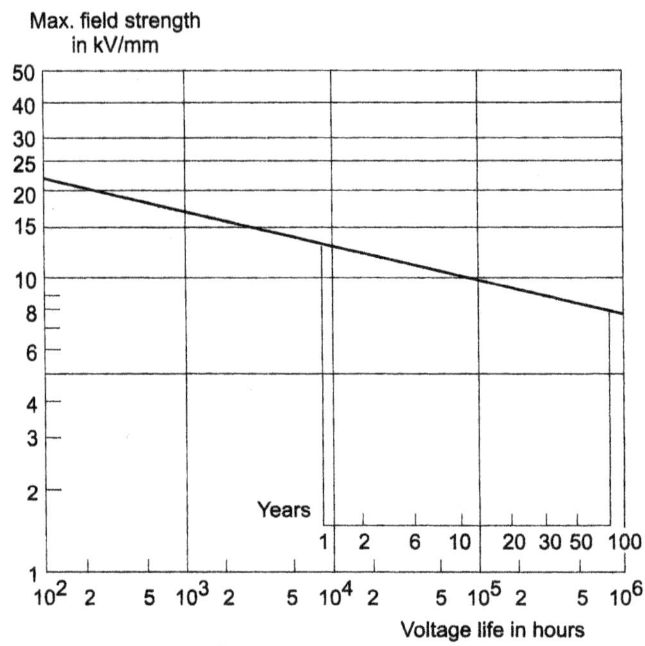

Fig. 11.1 *Voltage life (XLPE cables)*

11.3 Overhead Lines

An overhead power line is intended for transmission of electric power, by a bare or covered overhead conductor supported by insulators, generally mounted on cross arms near the tops of poles.

(a) The supports used for overhead line construction vary in design and the purpose they have to perform. These include ordinary (intermediate), angle, anchor, tee-off, and terminal (dead-end) supports. Ordinary supports, spaced suitably, serve to carry wires at a definite height above ground and must be capable of enduring the load due to the conductors and insulators, including the ice and wind loads on the conductors. Ordinary supports account for

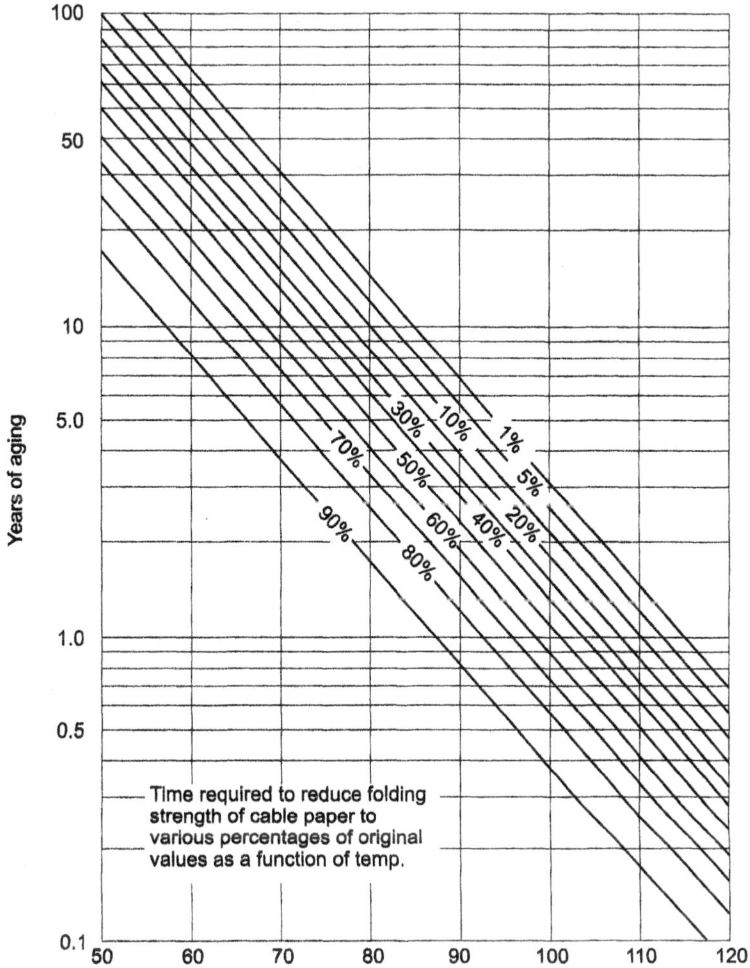

Fig. 11.2 *Thermal aging-paper cables*

nearly 80 per cent of the entire number of supports used for an overhead line. Angle supports have to carry the tension due to all the conductors and also, the tension due to the pull among the bisector of the internal angle formed between the wires of the adjacent spans. They are set on curves and at corners where the line run changes direction.

Anchor supports take up the load resulting from the difference in wire tensions along the overhead line. They are set in straight portions of the line at its main supporting points and also at each side of a crossing over various structures etc. The anchor span is the distance between two anchor supports carrying wires rigidly fastened to insulators. Anchor supports can function as intermediate, angle, branch, or terminal supports. Tee-off line supports are used for lines branching off the main line to supply power to communities remote from the line route. Terminal supports erected at the start and the end of overhead lines take up the loads due to the undirectional tension of wires along the line.

(b) Several factors determine the span lengths and the number and types of supports required, for the erection of an overhead line. These are the lay of ground, the route adopted, the type of conductors used and the number of conductors to be suspended, weather conditions in the area in question, the population density of the territory where the line is to be laid out, and the requirement placed on the continuity of service and operation safety of the line.

(c) The height of supports depends on the number of conductors they are to carry, the arrangement of conductors on a crossarm(s), the suspension height required, and the depth to which the supports are to be set in the ground as per IE Rules. The suspension height is the sum of the vertical clearance from the lower conductor to ground and the sag of the conductor. The sag of the conductor is the distance along the vertical from the lowermost point of the conductor in the span to the straight line, connecting the points of fastening of the conductor on two neighbouring supports.

(d) The vertical clearance of an overhead line conductor above ground is the distance along the vertical from the point of the highest sag to the ground. The vertical clearance to obstructions is the shortest distance along the vertical clearance from the conductor to an obstruction.

In the following sub-sections the rating of the overhead conductors from thermal considerations is discussed. While applying these ratings, due consideration should be given to voltage drop limitations and to the statutory requirements concerning clearances, stresses, etc.

11.3.1 Permissible Operating Temperature

The permissible operating temperatures of an overhead conductor depend on the maintenance of adequate clearances and limitations of loss of strength through annealing. Generally, the maximum current which any aerial conductor is designed to carry must not cause it to be heated such that it may result in annealing of the metal of the conductor or a reduction in the clearances specified. Usually for normal day-to-day loading, a maximum operating temperature of 75°C is permitted which is allowed to reach up to 100°C for emergency loading. (see Appendix II)

11.3.2 Loss of Strength due to Annealing

Annealing of metal conductor takes place to some extent at all temperatures and is a function of both temperature and time. It is accompanied by a reduction in the ultimate tensile strength on a cumulative basis. Appreciable annealing and loss of tensile strength will not take place if the conductor operates within prescribed temperature limits. In special circumstances, ratings on an estimated life basis can be determined. For this the engineer needs to know the relationship between the operating temperature, time of operating temperature, and the corresponding loss of strength of a conductor. A typical relationship of the thermal life of aluminium conductors is shown in Fig. 11.3.

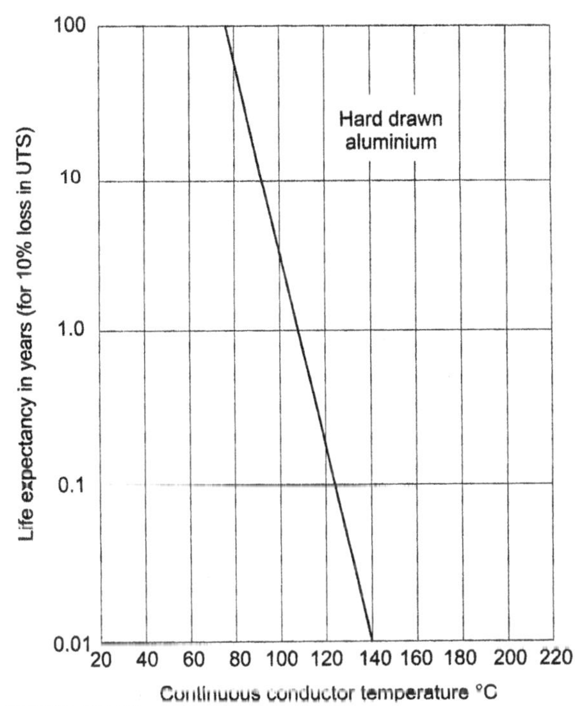

Fig. 11.3 *Thermal life of aluminium conductors*

11.3.3 Basis for Determinating Continuous Rating

The rating for an overhead conductor is calculated from the condition of heat balance which gives a relation between the heat received from various sources and heat emitted by the conductor. The formulation of the rating can be made by neglecting the temperature gradients within the conductor.

Heat Sources

Electrical Losses

The heat input due to electrical losses H_{cond} is obtained from the expression:

$$H_{\text{cond}} = I^2 R \text{ watt/cm}$$

where I is the current in the conductor in amperes and R is the ac resistance in ohm/cm length. The resistance corresponding to the permissible conductor temperature is used.

Heat Absorbed from the Sun

Heat absorbed from the sun H_{sun} is given by the expression:

$$H_{\text{sun}} = \alpha E d_c \text{ watt/cm}$$

where,

$\alpha =$ solar absorption coefficient
$E =$ intensity of solar radiation in watt/cm^2
$d_c =$ diameter of the conductor in cm.

The solar radiation intensity depends on the time of the day, season of the year, direction of the rays and atmospheric conditions. The solar absorption coefficient has a low value for bright surfaces. It varies between 0.2 and 1.0, depending on the material. The higher value is for the old conductors.

Heat Dissipation

Heat dissipates from the conductors due to radiation and convection:

(a) *Radiation:* Heat dissipated by radiation is based on the Stefan-Boltzmann law. It states that the rate of heat dissipation due to radiation H_r is proportional to the fourth power of the absolute temperature of the conductor. It can be calculated from the following expression:

$$H_r = E_c\,\sigma[(T_{amb} + \theta_c)4 - T_{amb}^4]\ \pi d_c \text{ watt/cm}$$

where,

E_c = emissivity coefficient of the conductor surface

T_{amb} = ambient temperature in degrees absolute

θ_c = rise in conductor temperature in °C

σ = Stefan's constant (5.7×10^{-2} watt/cm^2)

The emissivity coefficient has a low value for new surfaces and high for old surfaces, falling within the range of about 0.2–1.0. An empirical formula [20] relates E_c and conductor aging in years Y by the relation:

$$E_c = 0.23 + 0.7Y/(1.22 + Y)$$

(b) *Convection:* Many empirical relations are available for calculating the heat dissipation from a conductor H_c on account of convection. These relations are based on the conductor diameter, wind velocity and temperature rise above the ambient temperature. These relations differ from each other mainly because of variation in test conditions. The relation most commonly used is given by:

$$H_c = 13.8 \times 10^{-4}\ \theta_c\,(Vd_c)^{0.448} \text{ watt/cm}$$

where,

θ_c = conductor temperature above the ambient temperature °C

V = effective wind velocity cm/s

d_c = conductor diameter in cm.

Equation for Continuous Rating

This relation is obtained by combining the heat losses and absorption gains from the sources given above. For steady temperature conditions, the following relations give the continuous rating for heat conditions:

Electrical losses + Heat from sun = Heat dissipation by convection + Heat dissipated by radiation, i.e.

$$H_{cond} + H_{sun} = H_c + H_r$$

or

$$I^2R + \alpha Ed_c = 13.8\ \theta_c\,(Vd_c)^{0.448} \times 10^{-4} + E_c\,\sigma\ [(T_{amb} + \theta_c)^4 - T_{amb}^4]\pi\ d_c$$

Current in conductor is:

$$I = \sqrt{\frac{13.8\ \theta_c(Vd_c)^{0.448} \times 10^{-4} + E_c\ \sigma[(T_{amb} + \theta_c)^4 - T_{amb}^4]\ \pi d_c - \alpha Ed_c}{R}}$$

11.4 Design of Overhead Power Lines

11.4.1 Design Factors

An overhead power line is normally designed keeping in view the electrical and mechanical requirements.

The electrical design involves selection of voltage, selection of insulators and conductors, voltage control and selection of protective control gear.

The mechanical design involves stress and sag calculations and the design of supports and cross arms. The supports must be strong enough to withstand ice, wind loads acting on the supports, conductors, insulators, cross arm, etc. Wind pressure on poles and conductors are calculated as per the basic wind speeds given in IS: 802 (part-1, section-1)-1995:

$$P = 0.6\,V_d^2$$

where,

P = design wind pressure in N/m^2

V_d = design wind speed in m/s

Wind load is calculated as per IS: 5613. Basic maximum wind pressures including winds of short durations as in squalls, for three zones in the country is given in IS: 5613 (part-1, section-1)-1985.

It is desirable to have multi-circuit line design configuration for the 21st century electricity market. Open access to distribution may require additional lines for which right-of-way may be difficult otherwise [10].

As for the routing of an overhead line for single- and H-pole lines, easy accessibility for patrolling and maintenance work determines the layout, i.e., they are laid near the roads on one side as far as possible.

11.4.2 Selection of Voltage

The line voltage is determined by the load and distance over which it is to be transmitted. The most economical voltage is given by the following empirical formula for three-phase system.

$$V = 5.5 \times \sqrt{\frac{L}{1.6} + \frac{P}{100}}$$

where,

V = line voltage in kV

$L =$ distance in km

$P =$ load in kW.

The standard voltage nearest to the value of V is adopted (see Sec. 6.6).

11.4.3 Selection of Conductor Size

Both aluminium alloy conductors and ACSR are commonly used in distribution. The size of conductor is selected in view of the following factors:

(a) Current carrying capacity.

(b) Allowable voltage drop or line regulation.

(c) Breakdown strength of the conductor.

The choice of conductor [8] from a mechanical viewpoint depends upon:

(i) *External loading:* Wind speed, ice loading and ambient temperature, and

(ii) *Internal characteristics:* stranding, modulus of elasticity, thermal expansion and creep. For example, considering creep and economics, AAC is used only for LT lines. The line characteristics: (i) voltage regulation is influenced by line parameters and system frequency, and (ii) current carrying capacity is assessed from the heat balance equation given in Section 11.3.

(d) Line voltage.

Finally, current carrying capacity is decided by maximum conductor temperature rise or operating temperature. Operating temperature is limited by mechanical aspects such as allowable span, mid-span sag, joints, creep in conductor and long-term mechanical effects. Generally, 85°C (AAAC), 75°C, 70°C, 65°C or 60°C (ACSR) maximum operating temperature is used. Lower temperature is used for longer span lines, particularly lines in rural areas where jumpers may give trouble at a higher loading [9]. The allowable voltage drop is considered a critical factor in determining the conductor size for 11 kV and LT distribution line with thermal (ampere) loading not exceeding 80 per cent of the normal thermal rating (A) based on the maximum operating temperature. Other factors considered are line losses and minimizing the total cost.

Current carrying capacities of conductors are given in (Appendix II).

11.4.4 Calculation of Strength of Poles

In determining the strength of a pole, the pole is considered as a cantilever beam anchored at one end, i.e., the pole is designed for bending strength. Let W be the load in kg applied, 0.30 m from the top of pole. Then

$$M = fZ$$

where,

M = sum of the bending moments of the loads on the pole at ground level

Z = modulus of section of pole [value available from structural reference book]

f = fibre stress maximum.

The maximum fibre stress being known, the working load can be obtained by assuming suitable factors of safety as given in Table 11.2. The wind load on the pole face and that on the conductor (on 2/3 projected area) in addition to the pull of the conductor, if any, constitute the working load. For wind load calculation of a structure, the full projected area is taken for poles—PCC, RCC, wooden—and 1.5 time the projected area for braced channel poles to account for the two faces. For wind pressure refer to Appendix II.

Table 11.2

Factors of safety as per IE rule 76

Pole structure	Factor of safety
For RSJs (rolled steel joists) and rail poles	1.5
Steel tubular poles, braced channel poles	1.5
Woold poles	3.0
RCC poles, spun poles and PCC poles	2
Hand moulded RCC poles	2.5
For steel X-arms and struts	2.5 on ultimate strength
For conductors	2.0 on ultimate strength
For stay, guard and bearer wires, bolts in tension and compression	2.5 on ultimate strength

11.4.5 Selection of Line Supports

The principal types of line supports are: wood poles, concrete poles of RCC and PCC, steel tubular poles, rails and RS joists and lattice type poles.

Wood Poles

Chemically treated wood poles are used for distribution lines. The advantage of using wood poles is that they are low in cost, but they are susceptible to decay. The specifications for wood poles are covered by IS: 876 and IS: 5978. According to this standard the timber which is suitable for poles has been classified in three groups depending up on the strength. Jointed wood poles (IS-6056 for jointed wood poles for overhead lines—sal, deodar, chir, kail, etc.) with wire bound lap joint are considerably cheap and found to be very suitable for LT and HT lines in rural areas.

Group A: Very strong timber with modulus of rupture in bending above 850 kg/cm^2 such as sal.

Group B: Strong timber with modulus of rupture 630-850 kg/cm^2 by teak, sisso, etc.

Group C: Moderately strong timber with modulus of rupture 450–630 kg/cm^2 such as chir, deodar, etc.

Wood poles have been further grouped in seven classes according to the ultimate breaking load as follows.

Class I:	Ultimate breaking load not less than	1350 kg
Class II:	Ultimate breaking load between	1100–1350 kg
Class III:	-do-	850–1100 kg
Class IV:	-do-	700–850 kg
Class V:	-do-	550–700 kg
Class VI:	-do-	400–550 kg
Class VII:	-do-	300–400 kg

Note: The above loads are assumed to be applied at 60 cm from the top of the poles.

Usually aluminium cap is provided on the top of these poles to protect the end grains.

RCC and PCC Poles

These are more expensive than wood poles but cheaper than steel tubular poles. They are used extensively for power distribution in urban areas and in remote, and inaccessible places as RCC can be manufactured at the site itself. They are also preferred where decorative considerations predominate. Concrete of ratio 1:1.5:3 with 15 mm size well-graded shingle is used for their manufacture.

RCC poles have an extremely long life and need little maintenance but they are bulky in size and comparatively heavy. They have shattering tendency when hit by a vehicle.

PCC poles take care of these shortcomings to some extent. However, the handling, transportation and erection of these poles is more difficult because of their heavy weight.

Steel Tubular Poles

These are of two types—stepped and swaged.

Stepped poles are manufactured from a single tube, the diameter being reduced in parallel steps by passing the tube through a series of dies.

The specifications for tubular poles are covered by IS: 2713–1967. Due to their light weight, high strength to weight ratio and long life, they possess distinct advantages over other types of poles. The use of a pole cap at the top, concrete muff in the ground and regular coating of paint will prolong their life.

Rails and RS Joists

Old and second hand rails and RS joists are frequently used as supports for overhead lines. The working load and other technical data for some sizes which are commonly used are given in Table 11.3.

Table 11.3

Working load of rails and RS joists

Section	Length (m)	Planting depth (m)	Distance of load from top (m)	Working load (kg)
Rail 24.78 kg/m	9.00	1.50	0.30	175
Rail 29.74 kg/m	9.00	1.50	0.30	261
Rail 37.17 kg/m	9.00	1.50	0.30	345
Rail 44.61 kg/m	11.00	1.80	0.45	377
ISMB 125 mm × 75 mm	9.00	1.50	0.30	164
ISMB 150 mm × 80 mm	9.00	1.50	0.30	222

Lattice Type Poles

These are fabricated from narrow-base steel structures. They are light in weight and economical and can be assembled at site if bolted construction is used. Normally both welded and bolted types are used.

The portion embedded in the ground should be protected by concrete muff and the remaining portion by regular paint unless galvanized steel is used.

Cross Arms

Cross arms of hard wood (sisso, sal), or creosoted soft wood (chir) or fibre glass are used mostly. Steel cross arms are stronger and last much longer. MS angle iron and channel iron sections are generally used for this purpose. Smaller sections are used for communication circuits. The shape and length of the cross arms depend upon the desired configuration of conductors.

Insulators (refer also to sub-section 11.7.2)

Insulators made of glazed porcelain, tough glass, polymer, are used for supporting the conductors. The principal types of insulators are described below:

Pin Insulators: These are manufactured for voltages up to 33 kV and are cheaper than the other types. IS: 1445 and 731 cover detailed specifications for these. Typical dimensions for them as per these specifications are given in Table 11.4.

Table 11.4

Typical dimensions of pin insulators

Voltage (kV)	*Maximum diameter (mm)*	*Height (mm)*
11	152	137
22	210	195
33	278	242

Shackle Insulators: IS: 1445–1977 covers shackle insulators for voltages below 1000 volts. The two standard sizes listed in this specification are 90 mm dia × 75 mm height and 115 mm dia × 100 mm height.

A shackle insulator is supported by either two straps and two MS bolts, or one U-clamp or D-strap two MS bolts as per IS: 7935.

Shackle insulators are used for distribution lines dead ending and supporting conductors laid in vertical formation.

Disc Insulators: Disc insulators are made of glazed porcelain or tough glass. They are used as insulators on high voltage lines for suspension and dead ending. They can be assembled to form string assemblies to suit a particular voltage level. Single stack polymer insulators are used in uprating the existing lines or for erection of new compact lines.

The specifications for disc insulators are covered by IS: 731–1971 and their dimensions have been standardized in IS: 3188–1980 as given in Table 11.5.

Table 11.5

Dimensions of disc insulators (Ball and Socket coupling)

Insulator	*Nominal diameter (mm)*	*Nominal spacing (mm)*
Type 1	255	145
Type 2	280	170

The smaller disc insulator is used for voltages up to 11 kV. Disc insulators are "ball and socket" or "tongue and clevis" type. A suspension clamp is used to support the conductor, if suspension configuration of the line is chosen.

In the tongue and clevis type insulator, a round pin with a cotter pin is used to hold the tongue of one unit in the clevis of the other. In the case of the ball and socket type, the insulators are assembled by sliding the ball of one insulator into the socket of the next insulator from the side. A cotter pin or 'W' clip is slipped in from the back of the socket so that ball cannot slide out. Disc insulators are manufactured with mechanical strength of 4500, 7000 and 9000 kg for use up to 132 kV lines.

Guystrain Insulators: Egg type procelain strain insulators are used for insulating stay wires, guard wires, etc. wherever it is not proposed to earth them. As per IS: 5300, two strength sizes are used: 44 kN and 88 kN respectively for LT and HT lines.

Polymer Insulators: Nowadays track resistant epoxy resin impregnated unidirectional glass fibre core insulators equipped with two metal end fittings have come in use in advanced countries for distribution lines. The core is covered by a polymeric weathershed for environmental protection and to provide required creepage path length [1, 11]. These have specific advantages over the porcelain or tough glass insulators, such as outstanding insulation characteristics, light weight, easy to install, higher vandalism resistance and better appearance. These are very suitable for contaminated environment. Further, they require practically no maintenance service. These insulators are very useful for uprating the existing lines. For example, existing 11 kV lines could be uprated to 33 kV on the same supports with a changeover to these insulators. Also, there is a possibility of uprating the present single pole lines 33 kV to 132 kV with the help of these insulators. Such upratings are in common practice in western countries.

Post Insulators: These insulators are used for supporting bus bars in switchgear, contact assemblies in HV air break switches and supporting points at substations. IS: 5350 covers dimensions of three types of porcelain post insulators, namely, indoor, outdoor cylindrical and outdoor pedestal for systems with nominal voltage greater than 1000 V.

11.4.6 Spacing of Conductors

The spacing of conductors is determined by the voltage and span. As a rule of thumb the spacing between the conductors should be 1% of the maximum span for HT lines. A spacing of 20–30 cm is sufficient for LT lines. The lighter conductors should have more spacing than the heavier ones. The spacings of aluminium conductors is given by the formula:

$$S = \sqrt{d + \frac{V}{150}}$$

where S is in m, V is voltage in kV and d is sag in m.

IS: 5613 Part I/Sec. I–1985 suggests the following formula.

$$D = 500 + 18V + \frac{L^2}{50}$$

where,

D = spacing in mm
V = line voltage in kV
L = span length in m.

The recommended spacing of conductors in general are given in Table 11.6.

Table 11.6

Recommended spacing of ACSR conductors (Refer Fig. 11.4)

Voltage (kV)	Size of conductor (mm)	A (cm)		B (cm)		C (cm)		D (cm)	
		60 m span	90 m span	60 m span	90 m span	60 m span	90 m span	60 m span	90 m span
11	6/1/2.59	75	—	75	—				
	6/1/3.35	75	105	75	105				
	6/1/4.50	60	90	60	90				
	30/7/2.79	60	75	60	90				
33	6/1/2.59	135	—	135	—	105	—	45	—
	6/1/3.35	120	150	120	150	90	120	45	45
	6/1/4.50	105	135	105	135	75	105	45	45
	30/7/2.79	105	120	105	135	75	105	45	45

Lines running on suspension insulators assume a position away from the vertical under wind load. This angle is assumed as 60° with the vertical under conditions of maximum wind load as shown in Fig. 11.4. Generally, the minimum clearance is kept at 30 cm for voltage up to 22 kV plus 0.85 cm for every additional kV of voltage.

Minimum ground clearances as per IE Rule 77 are given in Table 11.7.

Table 11.7

Minimum ground clearances along and across streets

Class of voltage	Across street (m)	Along street (m)
Low and medium voltage	5.8	5.5
High voltage	6.1	5.8

Minimum ground clearance for O/H lines including service lines erected elsewhere than the above locations shall be as follows:

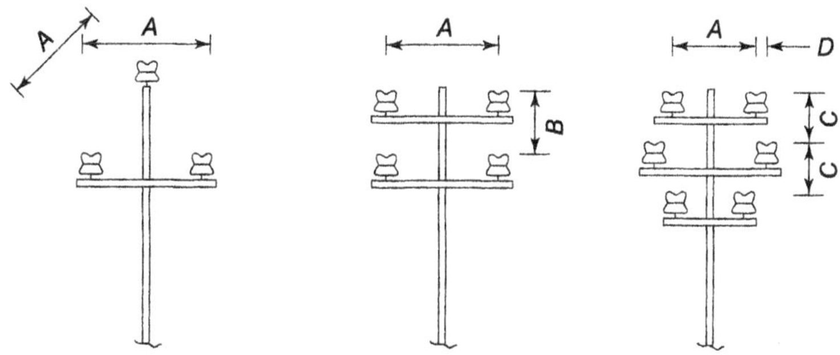

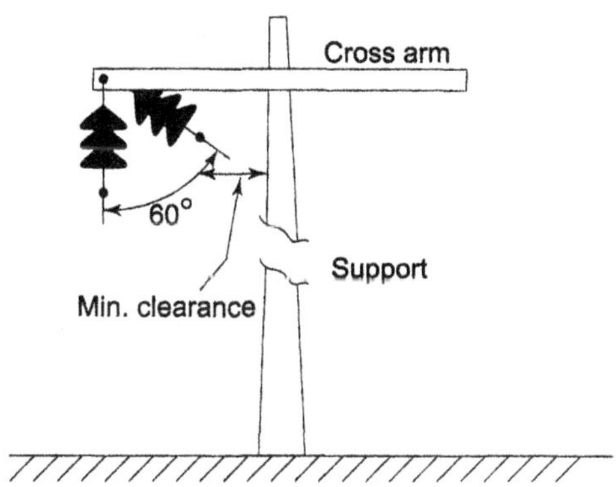

Fig. 11.4 *Minimum line clearances*

(a) Low medium and high voltage lines: 4.6 m up to 11 kV if bare.
(b) Low medium and high voltage lines: 4 m up to 11 kV, if insulated.
(c) For high voltage lines above 11,000 volts: 5.2 m.

For extra high voltage lines, the ground clearance shall be 5.2 m plus 0.30 m for every 33,000 volts or part thereof by which the voltage of the lines exceed 33,000 volts.

Minimum clearance from buildings:

(a) As per IE Rules 79 and 80, for LT and medium voltage, the vertical clearance should be 2.5 m and horizontal 1.2 m from the nearest point.
(b) Vertical clearance for high voltage lines up to 33 kV will be minimum 3.7 m. For extra high voltage lines the vertical clearance will be 0.3 m more for every additional 33 kV or part thereof.
(c) Minimum horizontal clearance

For lines up to 11 kV	1.2 m
For high voltage lines above 11 kV and up to 33 kV	2 m
For extra high voltage lines	2 m plus 0.3 m for every 33 kV or part thereof.

Minimum clearance between power and communication lines:

Low and medium voltage lines	1.38 m
11 kV lines	2.13 m

Minimum clearance between lines:

Low or medium voltage lines	1.30 m
High voltage lines up to 66 kV	2.44 m
Extra high voltage lines 132 kV	3.05 m

11.4.7 Other General Considerations

Route The proposed route should be shortest distances practicable, avoiding as far as possible the difficult terrain, urban development, high-amenity areas, difficult crossings, natural hazards. Working drawing be prepared for route plan, indicating position and type of supports, span lengths.

Configuration Horizontal configuration is not favourable in terrain having strong winds, as a short-circuit may occur between conductors. The chances of a short-circuit can be reduced by increasing the space between the conductors, or by adopting a vertical or intermediate configuration. In cold regions, vertical configuration is not desirable, as the ice getting coated around conductors may sometimes disappear in one conductor, but not in all. This causes an unequal sag in the different conductors and may result in a short-circuit between them. The chances of such a short-circuit can be reduced by increasing the space between the conductors, or by adopting the horizontal or intermediate configuration. The type configurations are shown in Fig. 11.5. It is preferable to have a triangular construction (Fig. 11.5b) for 11 kV and 33 kV as it does not create imbalance in the impedance among phases. Sub-transmission lines of 132, 66 kV level have vertical configuration for double-circuit lines and horizontal for single-circuit lines. In such cases, transposition of phase conductors is applied to reduce voltage imbalance and consequently reduce continuous negative phase sequence currents.

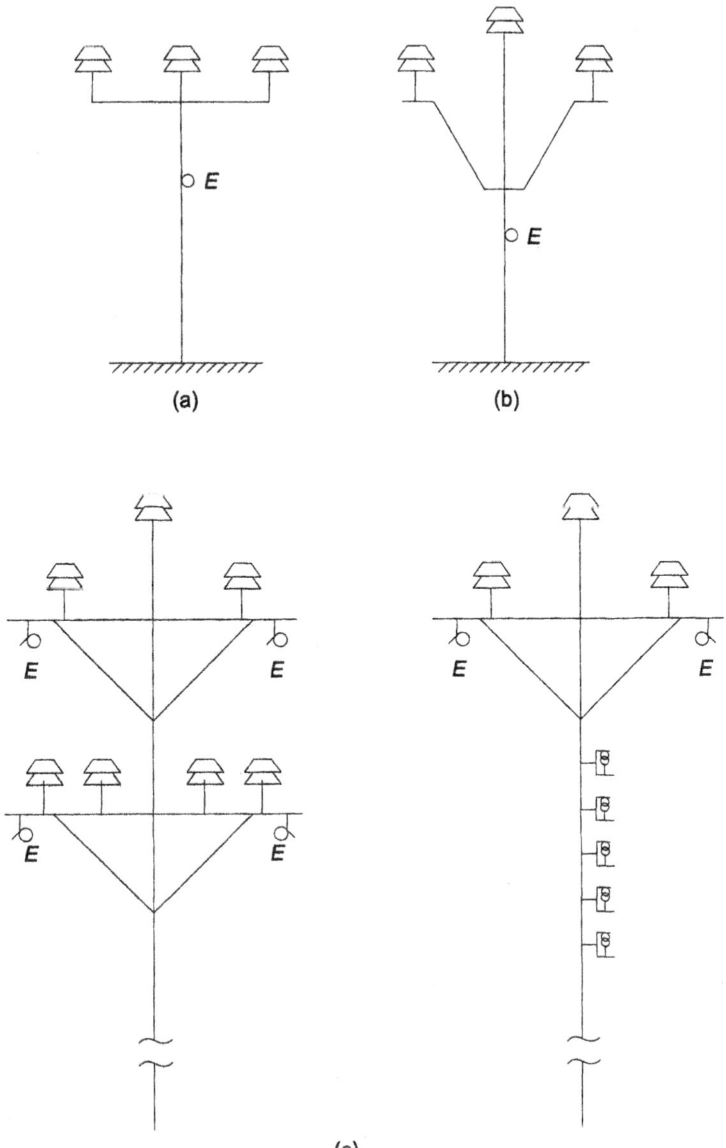

Fig. 11.5 *Typical configuration of lines*
(a) Shorter poles; (b) Longer poles with fewer bird fault
(c) Composite lines

Poles The forces acting on poles are the self-weight of the pole, weight of conductors/ice, the wind pressure action on the pole, cross-arms and conductors. Intermediate poles are designed for these acting forces and dead-end poles are designed extra for broken wire conditions. Burial depth of poles varies from one-sixth to one-tenth of total pole height, depending upon the nature of the soil and is usually between 1.5 to 1.8 m as per relevant IS–785, 876, 1678 and 2713. High depth foundation may be required at riverbeds, etc. Every 5th pole may be provided with wind stays or reinforced concrete muff (spun) at foundation for lines in high wind areas. However in low compact or sandy soils, each pole is desirable to be provided a concrete muff or reinforced foundation. A hole should be drilled in the ground with the use of earth-augers. However, if earth-augers are not available, a dog pit of the size 1.2 × 0.6 m should be made in the direction of the line. Poles be positioned in holes such that bigger section modulus of pole is always transverse to the length of the line, except the dead poles. The portion of wood poles below ground should be painted with bitumen to avoid deterioration. Steel-tubular poles, rolled steel Joists and Rails be provided at the bottom with cement concrete muffs, projecting 20 cm above ground.

Insulators On 11, 22, 33 kV lines, usually Pin insulators are used on straight runs and up to a maximum of 10° deviation. These disc insulators are intended for use at pole positions having more than 30° angle or for dead ending. For lines having a bend of 10° to 30°, either double-cross arms with double pin insulators or disc insulators should be used. For low and medium voltage lines, shackle insulators are used.

Sag and Tension: Normally all the spans cannot be kept equal because of the profile of the land and proper clearance consideration. The adjustments of tension would be necessary in adjacent span, since any alteration in temperature and loading would result in unequal tension in the various spans. This is obviously impracticable, as a constant tension must be applied at the tensioning position and tension must be uniform throughout the section. With suspension insulators, the tension inequalities would be compensated by string deflections, but for post or pin insulators, these inequalities would have to be taken by the binders which are not desirable. Therefore, a constant tension is calculated, which will be uniform throughout the section. For calculating this uniform tension an equivalent span (ruling span L_R) for the whole length of the line is chosen and is calculated:

$$L_R = \sqrt{\frac{L_1^2 + L_2^2 + L_3^2 + \cdots}{L_1 + L_2 + L_3 + \cdots}}$$

L_1, L_2, L_3, etc. are different spans in a section. Sag S can be calculated by

$$S = \frac{\text{Actual spans}}{\text{Ruling spans}} \times S_R$$

where S = sag for actual spans, and

S_R = sag for ruling span

The normal span is reduced by 20 per cent on road crossings. The sagging could be carried out by winch to save time and labour.

Conductor Jointing A joint should conform to I.E. Rule-75. For conductors up to 50 mm^2, crimped joints are made with simple hand crimping tools and for higher sizes compression type or hydraulic type crimping tools are used. A simple nozzling joint (electrically not efficient) may be cautiously made where crimping tools and accessories are not readily available. A joint should not be located within 3 m from the point of support. It should be at least 7.5 m away from the dead-end poles. Joints in all the conductors in a span should not coincide. One joint is allowable in one span, but in crossing span, there should be no joint.

Jumpering Conductor to be used for jumpering should be of the same size as that of the line conductor. There should be no joint in a jumper.

Guarding Guarding should be provided at power line crossings of: telecommunication lines, road/katcha paths, buildings, canals/rivers, along the road and public places. A guard wire should be galvanized steel wire of minimum 4 mm diameter and be fitted on the structure on either side of the crossing and be effectively earthed by providing earth electrodes. Guard wires should fully cover the phase wires and accordingly should not be less than 30 cm away from the outer phase wires. For power, telecommunication line crossings, the distance between two consecutive cross-lacings should not be more than 0.6 m over the entire portion of the crossing, with additional three bracings on either side of the crossing. For other crossings distance between two consecutive cross-lacings over the entire span length should not exceed 3 m and distances of cross-lacing from the supports should not exceed 0.75 m.

Sectionalizing/Tee-offs A primary spur line with length exceeding 2 km should be controlled by providing GO switch or sectionalizer. On trunk lines, sectionalizing point should be created at a distance not exceeding 1.6 km.

Trees Clearances No tree should be allowed to exist under the line and within 1.5 and 2.5 m respectively, from line and distribution transformer structure. It may be noted that leakage to tree touching all the LT line phases/per spot have been observed from 0.20 to 0.4 A, the maximum being during rainy season, observed for eucalyptus trees. Also different trees have varying touching resistances. Such as eucalyptus has touching resistance of about 25 per cent of that of dalbergia siss.

11.5 Overhead Line Construction

11.5.1 Supports

The different types of supports for overhead lines, as explained earlier, are prestressed concrete, spun, reinforced concrete, steel—tubular or braced channel and wooden. Their size and height depend upon the span length, conductor size and tension, phase and ground clearances. The single pole is generally used, though it transverse strength can be increased 3-4 times by using two poles arranged in H-formation. Presently for 66 kV and 132 kV lines, H-formation or tower structures are used but due to better awareness of over-voltages working and protective devices, it has been found adequate to use single pole structures as proposed in Section 14.10. The X-arm may be of steel or wood.

11.5.2 Conductors

Conductor represents 30–50 per cent of the installed cost of the line. All aluminium conductors (AAC), AAAC and ACSR conductors are generally used. Technical specifications of conductors are covered in IS: 398. These conductors are of standard construction and the ultimate tensile strength of the whole conductor is based on the total strand strength. For all aluminium conductors, the sum of the minimum average tensile strength for individual wires before stranding multiplied by a stranding factor is given below:

3 strand conductor	97%
7 strand conductor	96%
19 strand conductor	94%
37 strand conductor	90%

For ACSR, the ultimate tensile strength is 98 per cent of the sum of the minimum tensile strength of the individual aluminium wires plus 85 per cent of the sum of the minimum tensile strength of the individual steel wires before stranding. All aluminium alloy conductors (AAAC) use has now wide acceptance due to superior overall performance and longer life. AAC is used for LT lines.

11.5.3 Erection of Lines

When handling the drums in the field it is important to ensure that they are always rolled in the direction of the arrow marked on them. If they are rolled in the opposite direction there is risk of damaging the conductors. The drums are best mounted on drum stands or jacks for drawing out, the former being preferable since they are more rigid. It should be remembered that the drums of all aluminium conductors are considerably lighter than the corresponding drums of AAAC, ACSR. Care should, therefore, be taken to ensure that the drum is not pulled off its support when running out. This is unlikely to happen if a steady pull is maintained and the use of jacks avoided. It is best to pull off from the top of the drum, as this gives more play and avoids the risk of kinking if the drum over-runs.

AAC, AAAC and ACSR conductors are easier to pull out by hand or power operated hoists. If the conductor is being pulled out under tension with a braking device on the drum, it is advisable to interpose a swivel to allow for the twisting of the conductor during running out. The swivel is not required if the conductor is run out slack with the drums free.

Care should be taken to avoid grazing the outside wires by dragging over projected rock and the like, which should be covered by timber or rollers. (see IS: 5613).

Pulling Up

After the conductor runs out over the full length of the section being erected, it is fixed at one end to the permanent anchor clamps and a temporary clamp called come-along clamp is attached to the other end to pull it up to the required tension. These come-along clamps can be of automatic type of simpler bolted type.

In the automatic type, the jaws should be lined with soft aluminium or with a folded piece of emery cloth. In the bolted type, the conductor is usually held between aluminium alloy or hard wooden blocks and bored with grooves of the correct conductor diameter. In no case should aluminium be gripped directly by steel surfaces, and the gripping power should not be increased by serrating the surfaces.

An important point in erection is to 'take out the stretch' of the conductor, so that under working conditions there is no gradual increase in the sag due to the setting down of individual wires. There are two ways in which this can be accomplished—by either prestressing or over-tensioning.

Prestressing

In this method the conductor is pulled up to a tension considerably above the desired tension but not exceeding 50 per cent of the breaking load, for a short period, say 20 minutes. This is called *killing tension.* At the end of this period, the conductor is slackened off until it reaches the correct sag or tension for the prevailing temperature. The conductor can then be removed from the anchor clamps and transferred to the insulators or suspension clamps in the normal manner.

Over-Tensioning

This consists in pulling up the conductor to a tension which is a little above the theoretical value at the prevailing temperature and fixing it at that tension with a corresponding reduction in sag. After a certain time, it will settle down to the correct sag and tension due to creepage. A tension of 5–8 per cent above the correct figure has been found suitable for all classes of aluminium and steel cored aluminium conductors except in the case of large sizes containing more than 37 strands, when the over-tension should be increased to 12–15 per cent. IE Rule 76-C allows initial and final tensions of 35 and 25 per cent of UTS respectively at 32°C for unloaded lines and without external loading. Simultaneously, the working tension should not exceed 50 per cent of UTS with maximum wind loading or other external loading at average temperature of the coldest month of the year. A dynamometer may be used to check when the line has been pulled up to the required tension.

Checking Sag and Tension

It is usual to determine the correct tension for the prevailing temperature by pulling up the conductor until the correct sag is reached in one span (ruling span) of the section where it is kept under observation. Standard sag-temperature-tension charts can be referred to when necessary, e.g., REC standards are available.

Sight Method: The sag is best read by mounting boards on the line supports at each end of that span. These are mounted in such a way that the top edge corresponds to the lowest point of the sag of the conductor and a sight is taken between the top of the boards. For long span construction of over 250 m, it is sometimes better to use a sighting instrument at one end in place of the board. Except on the largest overhead lines, it is usually satisfactory to correctly sag one conductor and to adjust the others at the same sag without separately checking them.

The sag/tension of line can be calculated by the expression:

$$\text{Sag in m} = \frac{(\text{conductor wt. in kg/m}) \times (\text{span in m})^2}{8 \times (\text{tension in kg})}$$

After the conductor has reached the correct tension and sag, the permanent clamp is fitted in place of the come-along clamps and the conductor fastened to all intermediate supports by means of either suspension clamps or by binding it to the insulators in the case of pin-type insulators.

Pulse Method: When a conductor is struck sharply with a wooden stick at a point near one of its supports, a wave is initiated which travels along the conductor to the support at the other end, from where it is reflected back. This to and fro action continues till the wave is damped out. The time required for such a wave to travel to the next support and back is a function of the sag of the conductor in the span and is given by the formula:

$$D = 306.7 \times \left(\frac{t}{N}\right)^2$$

where,

$D =$ sag in mm

$t =$ time in seconds for N return waves

$N =$ number of return waves.

This method is reasonably accurate when the conductor is stationary and not touching any object and there are no joints in the span. The number of return waves is selected and the test repeated until three equal readings are obtained.

Tension can also be calculated using the following formula:

$$\text{Conductor tension in kg} = \frac{(\text{conductor wt. in kg/m}) \times (\text{span in m})}{26.41 \times (\text{period in s})^2}$$

11.6 Vibrations

Aeolian vibration damage has been reported on primary distribution overhead lines. Vulnerable spans are of open cross-country lines at water crossing such as canals, rivers, etc., and in valleys. Longer spans are also liable to suffer aeolian vibration damage. The damage occurs due to fatigue break in a progressive fracture following the natural planes of cleavage between crystals. The characteristic appearance of such a fracture, therefore, is sharp (i.e., without necking down) and exposes the natural crystalline structure of the material. Such fractures of a conductor strand occur at the support point (binding or clamp).

Spiral vibration damper is the optimum device for distribution lines up to 22 kV. It has got a flat response over a wide range of frequencies. It is a one piece PVC helical rod that can easily be applied on the conductor. It fits loosely on the conductor except at one end where the overall diameter is reduced to grip the conductor. The damper is very cheap and has a long life.

The damper is installed on line spans where the vibration level is more than 150 microstrain at 25 per cent of the ultimate tensile strength. One damper is installed at each span end. Typical length of dampers for different conductor diameters are given in Table 11.8.

Table 11.8

Typical length of spiral dampers

ACSR conductor diameter range	*Overall damper length*
6 – 8.5 mm	125 mm
8.5 – 2 mm	130 cm
12 – 4.5 mm	135 cm
14.5 – 9.5 mm	155 cm

On HV lines, 33 kV and above Stockbridge dampers are considered suitable. It is important to install the dampers on vulnerable spans of distribution lines to reduce the outage rate.

11.7 Line Accessories

Bolted/ Twisted This is an associated equipment required for fastening the conductor to supports and taking off the power or supply points such as joints material, clamps and compounds. These are discussed in this section and some are shown in Fig. 11.6.

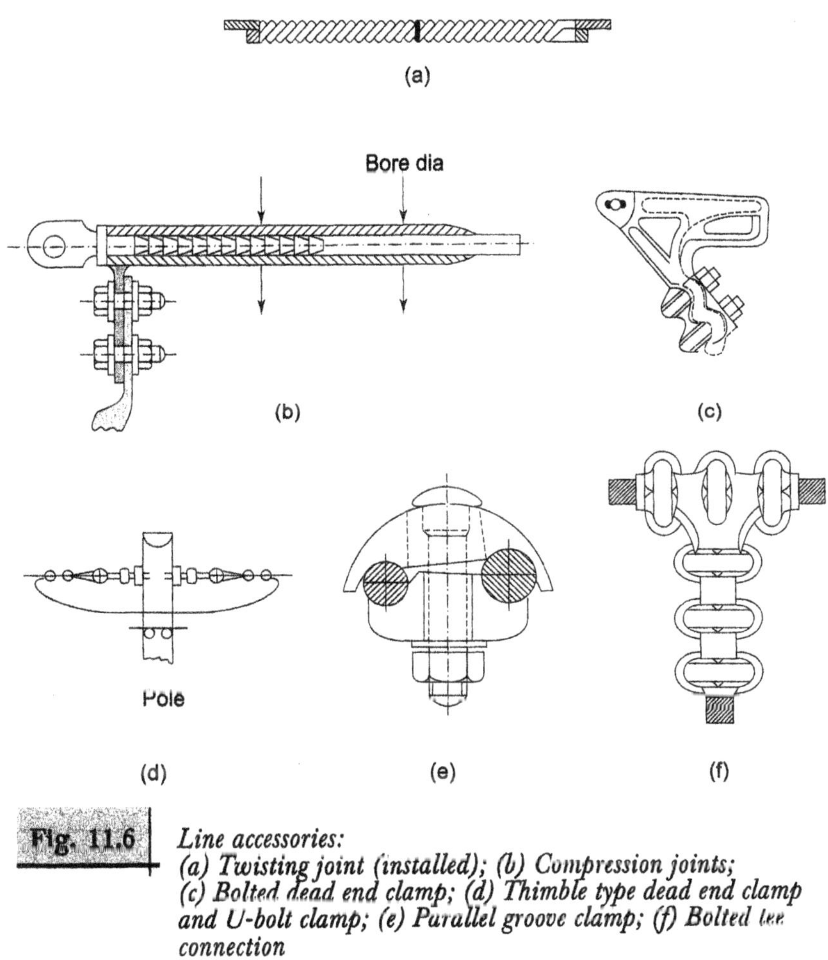

Fig. 11.6 *Line accessories:*
(a) Twisting joint (installed); (b) Compression joints; (c) Bolted dead end clamp; (d) Thimble type dead end clamp and U-bolt clamp; (e) Parallel groove clamp; (f) Bolted tee connection

Preformed Line Fittings These are helically formed fittings base on helical principle—the ability of the products to follow the reduction of the conductor/cable diameter, which occurs under tensile loading. This is achieved with very appropriately designed inside diameters of the rods for each item and conductor size. It ensures no ultimate relaxation or slippage due to compression set or cold flow, after the grips have been applied in the field. Also, the surface stresses are relatively low and are distributed along the length of the grip, thereby removing any points of severe stress concentration. This prevents the grip biting into the conductor and causing damage. Moreover, the elastic characteristics of the loop of the grip aids in absorbing shocks and compensating for small length changes due to elongation and contraction. See REC specifications: 25/1983 and conductor standards E-31 to E-33.

Grips, dead ends, grip guys, ties, service grips, splices for lines up to 33 kV are used. These are made of aluminium alloy. These are cost, labour and time saving. It eliminates chance for error or judgement. No tools are required. These fittings are fast and simple to apply and assure uniformity of application every time. These have excellent fatigue characteristics, resulting in longer life for the conductor. They offer possibilities of multiple guying or service connections and are capable of reapplication, so as to withstand severest conditions of vibrations. These eliminate the possibility of loose contacts/connections. Their use as dead-ends, maintains continuous run through the loop of the helical dead-end without any conductor cutting/joint and thus reduces the line losses. The typical application of these accessories is shown in Fig. 11.7.

11.7.1 Joints

Unijoints

These are one piece aluminium compression joints. When applied using the proper procedure and tools, they develop a minimum mechanical strength of 95 per cent of the conductor's rated ultimate strength and satisfactory electrical characteristics. These joints are usually supplied prefilled with the proper abrasive gripping compound. Crimping tools are used for compression. Such joints can be used in dead-end or mid-span.

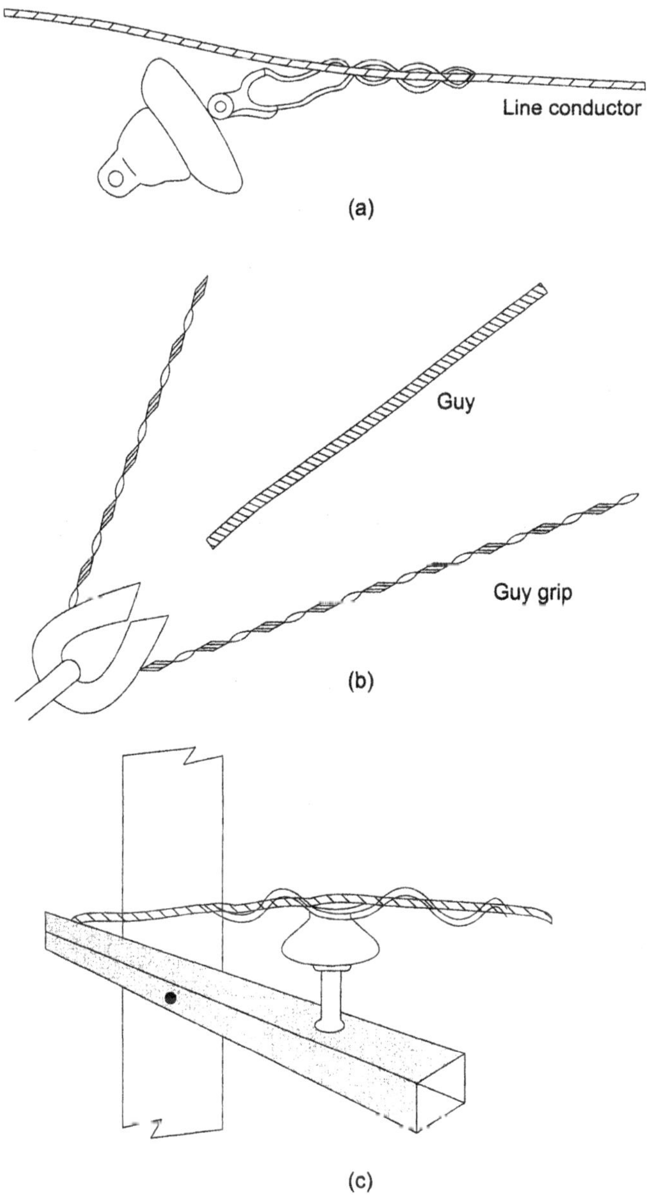

Fig. 11.7 *Helical preformed line fittings (Aluminium alloy)*
(a) HT conductor dead end
(b) Guy grip dead-end for use with guy insulator or thimble of stay-rod
(c) Distribution tie for pin insulator

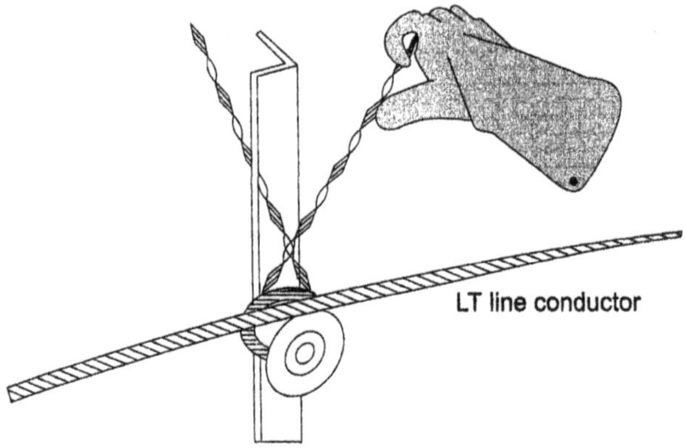

(d)

(e)

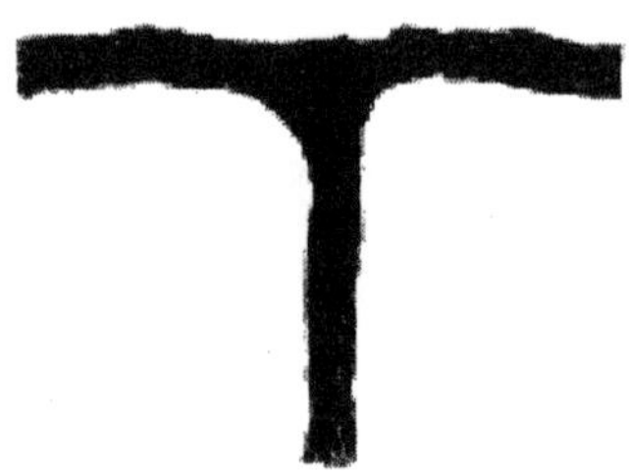

(f)

Fig. 11.7 *Helical performed line fittings*

(d) Spool ties for shackle insulator for tying the LT conductor to the side groove
(e) Lashing rod for bearer wire and PVC cable for service connections
(f) Tee-connector for service connection from LT conductor to the tap conductor or cable

Twisting Joints

These make the simplest mid-span joints. In this, the joints are made from oval-shaped aluminium tubes. Each joint covers a specific range of conductor. The joint will require one tube for AAC, two tubes for ACSR, AAAC and three tubes of high strength AAAC, ACSR. When applied correctly, the twisting joints develop 90 per cent of UTS of the conductor and satisfactory electrical efficiencies. Before a joint is made, it must be ensured that the insides of the tubes and ends of the conductor to be inserted in these are free from dirt and grease, and are bright and cleaned with a suitable abrasive. The electrical efficiency of the joints can, however, be improved without impairing their mechanical efficiency by incorporating a suitable metallic abrasive compound in the joint. Two twisting joint wrenches are required and for a conductor with right-handed outer lay, the wrenches should be rotated in a clockwise direction when facing the mouth of the tube.

Two-part Compression Joints

This joint consists of two parts: a steel tube and an aluminium tube for compression over the steel core and aluminium respectively. They are neat in appearance and particularly suitable for ACSR with multi-wire steel cores. Compression joints allow the development of substantially higher mechanical and electrical properties than these can be obtained on small conductors with twisting joints. A portable compressor and hexagonal oval shaped dies are used for applying these joints.

Dead-end Joints

Various types of joints for dead ends are discussed below.

(a) *Uniterms or Uniclamps:* It is one piece wire loop type and provides a suitable method for dead-ending distribution lines on spool or shackle insulators. It is a simple compression dead-end which employs the unijoint principle and crimping tools can be used for compression.

(b) *Dead-end Clamps:* These are snubbing type clamps, suitable for dead-ending on disc type insulators. They are made from aluminium alloy with galvanized steel U-bolts and are completely satisfactory in mechanical aspects. Power utilities in some states (e.g. Jammu and Kashmir) use dead-end snubbing type clamps made of malleable cast iron on 11 kV and 415 V lines. These clamps have 15–20 times more losses (hysteresis and eddy current) compared to losses (eddy current) in the aluminium alloy clamps. This practice of using malleable cast iron clamps must be stopped and those already installed should be replaced with aluminium alloy clamps.

(c) *Thimble Type:* These are used with smaller conductor sizes on disc insulators. One-and-half turns of the conductor are made around the aluminium alloy dead-end thimble (profiled to accept 1.5 conductor turns). The conductor can be secured with one tube of the twisting joint. Alternatively, two U-bolted clips may be used to secure the conductor.

11.7.2 Insulator Ties

For tying aluminium conductors on insulators, two types of ties are generally used. The *top groove* (pin insulators) tie is frequently used at tangent supports. In the case of shackle insulators in D-clamp, the *side groove* tie is used for angle as well as tangent supports. At angle poles with pin insulators, side groove is used.

11.7.3 Taps and Jumpers

Service taps are made by various accessories, which are not subjected to mechanical tension. Tapping should be taken off only at a point of line support.

Compression Line Taps

These are compression type line taps and give efficient service. Their success is due to the convenience with which they may be applied.

Parallel Groove Clamps

These are generally made of aluminium alloy or galvanized steel. The latter are used for steel earth wire or conductor. These are available in a range of sizes suitable for specified conductor diameters. Where connection to a copper conductor is required, i.e., a bimetallic joint, these have one groove tinned. Bimetallic connections should always be installed in such a manner that the copper conductor is situated below the aluminium conductor. This is to ensure that no copper contaminated water comes in contact with aluminium. Proper preparation of the contact surfaces prior to assembly will result in efficient connections. An abrasive compound may be used for this purpose.

Universal Parallel Groove Clamps

These accommodate a wide range of conductor sizes. For bimetallic connections both conductor grooves are tinned. The preparation and assembly is similar to parallel groove clamps.

Bolted Type T-Connectors

These are made of aluminium alloy and galvanized steel bolts and are widely used for jumper connection of equipment at sub-stations and heavy size conductor tee-offs for lines. Its mechanical strength is better as compared to parallel groove clamps.

11.7.4 Guys and Struts

Guys of stranded galvanized steel wire are used on all terminals, angles and other such poles where the conductors have a tendency to pull the pole away from its true vertical position. Transverse guys, perpendicular to the direction of the overhead line, are installed when the poles are located in sandy and water-logged soils of low cohesive strength. The guys are fastened to the pole near the load points centre with the help of pole clamps. The other end of the guy wire is secured to a stay rod embedded in the ground. The stay rod should be located as far away as

Table 11.9

Ultimate strength of stay wires

S.No.	*Item*	*Size*	*Ultimate tensile strength*
1.	Stay wire (galvanized steel)	7/2.5 mm	1610 kg
		7/3.15 mm	4130 kg
		7/4.0 mm	6450 kg
2.	Stay rods (M.S.)	16 mm^2 × 2 m with plate 23 × 23 × 0.6 cm^3	5500 kg
		19 mm^2 × 2 m with plate 30 × 30 × 0.6 cm^3 (larger plate of size 45 × 45 × 0.6 cm^3 should be used in soils of low cohesive strength)	8000 kg
3.	Turn buckle (M.S.)	9 mm	1750 kg
		12.5 mm	3250 kg
		16 mm	5500 kg
		20 mm	8000 kg

possible. Turn buckles are used for tightening the guys where the stay rods are not provided with a tightening arrangement. Storm guys are provided at intervals (say every 4th pole) on long lines in high winds areas. The common sizes of stay wires, stay rods and turn buckles are given in Table 11.9.

Struts may be used where it is not possible to use stay wires because of limitation of space. Tubular poles, pipes or rails may also be used for this purpose. Bow guy with horizontal strut (REC standard G-4) need less space.

11.7.5 Guard Wires

As per IE Rule 88, guard wires of galvanized steel having breaking strength not less than 635 kg should be used. Guard wires are required to be used at all points where a line crosses a street, road or railway line, other power lines, telecommunication lines and have to be earthed at all points where their continuity is broken.

11.7.6 Jointing Compounds

For all bolted connections in aluminium conductors approved compound should be used to provide sound low resistance connections and to protect joints from deterioration. First clean both the connector and conductor contact area with a stainless steel wire brush—to a bright shiny finish to remove the aluminium oxide film. Then immediately coat all contact surface with joining compound. Properties required to a good jointing compound for all connections are:

(a) Low electrical resistance in thin films.
(b) Non-corrosive to aluminium, copper, steel, zinc, tin and silver.
(c) Consistency not varying much with temperature from about 10°C to 100°C.
(d) Prevent oxide formation on aluminium.
(e) Exclude moisture with good protection from corrosion under all atmospheric conditions.
(f) Non-toxic and non-irritating.
(g) Ready availability at low cost.

11.7.7 LT Line Spacers

Mid-span clashing of conductors, causing phase-to-phase faults on LT lines is a common phenomenon occurring due to excessive winds, uneven or excessive sagging, use of longer span than required, etc. To overcome this problem, spacers are used (see REC standard 29/1987). There are two types (see Fig. 11.8) of spacers: one is spiral type made of PVC or polypropylene and is applied between two adjacent conductors by twisting the spiral ends around the two conductors manually. It automatically grips the conductor without any change in position. These can be used for horizontal or vertical configuration of lines. Normally, one set of spacers in the middle of span is adequate. The other type is composite spacer and is made of moulded polypropyline. It is in single mould except the clamping piece. A single spacer can hold a number of conductors simultaneously, as per requirement.

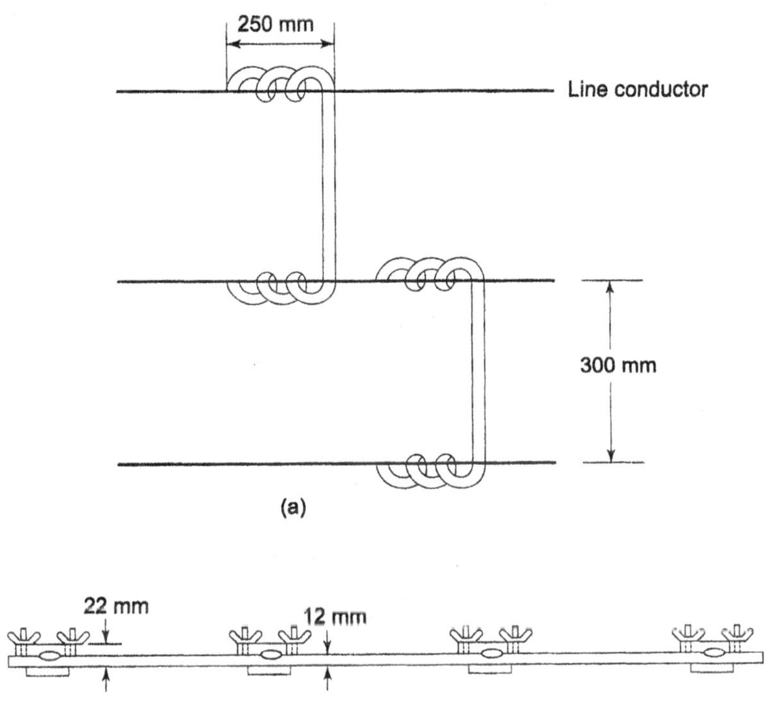

Fig. 11.8 *Spacers for LT lines:*
(a) Separate PVC/polypropylene spacers (6 mm diameter);
(b) Composite spacers

11.8 Installation of Distribution Transformer

11.8.1 Outdoor Installation

Transformers may be mounted outdoors in one of the following ways:

(a) Direct Mounting

Directly clamped to a pole by suitable clamps. The method is adequate for small transformers up to 25 kVA only.

(b) H-pole Mounting

Transformers are mounted on cross-arms fixed between two poles and are rigidly fastened to the poles. The method is suitable for transformers up to 200 kVA capacity [see Fig. 11.9].

(c) Platform Mounting

A platform is constructed on a structure made of four poles for placing the transformer of sizes: 250, 300, 400, 750 kVA. This method is

Fig. 11.9 *Transformer pole mounting arrangement: 2-Pole in metropolitan area*

recommended for places where it is hazardous to place transformers on the ground.

(d) Floor Mounting

The method is suitable for all sizes of transformers. The floor level must be higher than the surroundings to prevent flooding. The foundation should preferably be of concrete. If a number of transformers are located together closely, fireproof barrier walls should be provided to limit the damage arising from a mishap to any transformer. The enclosures of the floor-mounted transformer should be designed to permit free circulation of air on all sides. If possible, the outdoor transformer should be protected against direct sun rays. This will increase the life of paint and also prolong the life of the transformer.

The rollers of a transformer, after it has been placed in its final position, should be firmly locked to prevent any movement during a storm.

11.8.2 Indoor Installation

The building for housing a transformer should be large enough to permit free access from all sides and high enough to enable the dismantling of a transformer. The following minimum clearances from wall sides are considered satisfactory.

	Min. clearance (m) from wall sides
Wall on one side only	0.5
Wall on two sides	0.75
Wall on three sides	1.0
Wall on four sides (as in an enclosed room)	1.25

Adequate passage and doorways should be provided so that the largest size transformer can be conveniently removed for repairs, etc.

Adequate provision for ventilation is vital in the case of transformers located indoors. Free circulation of air on all sides of the transformer and within the building must be ensured. The air inlets located as close to the floor as possible and the outlets as high as the building will enable the hot air to escape. As a rule of thumb, a ventilation area of at least two square metres for the outlet and one square metre for the intel per 1000 kVA of transformer capacity should be provided. Where this is not possible, exhaust fans should be provided for forced circulation. Air inlets and outlets should be protected against entry of rain water, birds, etc. [see Fig. 19.6(g)].

Oil-filled transformers may not be placed in excessively damp locations. Dry resin encapsulated type transformers are better suited for such locations.

Arrangement of Cable Boxes

Transformers meant for indoor use may have cable boxes mounted either directly on them or provided with bushing when the cables are terminated in cable terminal boxes mounted separately and nearby. The latter arrangement permits the easy removal of a transformer without disturbing the cables.

11.9 Compact Lines

11.9.1 Upgradation of Existing Lines

(i) Several alternative approaches are available to help increase the capability of existing lines. The most obvious method of increasing the capability of an existing line is to increase the operating voltage. This method can sometimes be accomplished by modifying existing structures on a line, which is usually more economical and practical than trying to obtain right-of-way for a new transmission line.

(ii) It is feasible to upgrade any existing 11 kV line to 33 kV line or 33 kV lines to 66 or 132 kV lines as compact line.

(a) As per clause 7.2.2 of IS: 5613 (Part I/Section I)—1985, such 'V' or 'U' type cross arms made of mild steel are essential in situations where birds are found in large numbers, such as refuse dumping or animal burial grounds. To reduce the vertical distance between conductor and to accommodate the increase of sag for the upgraded 33 kV line, it will be advantageous to change the delta formation of the conductors to a horizontal (or near horizontal) formation. This can be achieved by fitting a 33 kV 'V' type cross arm to the existing 11 kV pole. The conductors and support normally employed for the existing 11 kV lines, can be used for 33 kV lines also. As a first step, 11 kV pin insulators are to be replaced by 33 kV pin insulators. The upgraded line will be adequate to be operated at 33 kV. For implementation, IE rules relaxation may be required with regard to ground clearance.

(b) 11, 22, 33, 66 or 132 kV lines on P.C.C. Poles (existing or new) in compact construction are becoming technical and commercially mature technology with overall economy and saving ROW. For example, polymer insulators can be used for upgrading the existing lines with porcelain insulators, without changing line structures and conductors with 10 per cent expenditure only.

11.9.2 New Lines

The dimensions of a transmission line can be compacted by restraining the movement of the conductor at the point of attachment to the insulators by certain arrangement and by installing ZnO arresters:

(i) Insulator Arrangement:

- *V String:* Suspension strings in 'V' configuration in all types of structures.
- *Horizontal Line Post:* The use of line post insulators horizontally.
- *Strut-Suspension Combination:* It also helps to reduce the overturning moment requirement of support due to transverse loading as effect of wind, because the transverse load due to wind is transmitted through the strut to the structure at a lower point. The strut remains in tension or compression, depending upon the direction of wind.
- *Horizontal V:* The insulator assembly is able to pivot about an axis, which is at an angle with the vertical. This arrangement creates a restoring force for the insulator, after it has been disturbed longitudinally due to any reason.
- *Oscillation Cross-Arm:* It is an improvement on the horizontal V-assembly, in the sense that it employs a single insulator, which is free to oscillate about an axis.
- *Insulator Cross-Arm:* The insulated cross-arms permit a more compact support design and can carry higher mechanical loads.

(ii) ZnO Arrester

Use of zinc oxide arresters have lower residual voltage (discharge voltage level) than silicon-carbide (SiC) designs. These can eliminate the need of shield wires and offer satisfactory phase-to-ground protection against lightning. The insulator manufacturing companies in the country are going in for the production of insulator-cum-arrester cross-arm assemblies for compact line supports.

11.10 Aerial Bunched Cables (ABC) System

The ABC system is comprising of insulated cable conductor, tension and anchor clamps, suspension clamps, tap connections for low and high voltage overhead line networks. The cables are usually cross linked polyethylene insulated, complying with IEC 502 having thickness in excess of the standard values, giving extra mechanical strength and incorporating carbon black colour to provide protection against ultraviolet radiation. Two to five assembled cores in standard conductor sizes up to 240 mm^2 along with the accessories and in rated line voltage levels from 240 V to 33 kV are in use.

The system is being increasingly used in Andhra Pradesh, Kerala, Maharashtra. The overall cost of the system is much less as compared to the bare conductor system. It requires less clearance, shorter poles, longer spans and less maintenance. Alternatively, insulated line can be strung at or under the eaves of buildings. This system has better safety and fault-free nature. The interruption of supply and transformer damage is reduced and the maintenance costs are considerably less. Chances of accidents is much reduced. The problem of tree trimming is eliminated, which is cumbersome job in hilly regions, in forest and in storm-prone areas. Voltage regulation is improved in this system as compared to bare overhead system, as the reactance is less. For example for low voltage system, the typical reactance is: 0.1 Ω/km, for ABC (four core) line compared with the 0.3 Ω/km, for bare four-wire line. The overall construction cost of the line is reduced for ABC lines as compared to bare conductor lines.

This system is most suitable for remaining rural electrification in the country. About 13 per cent villages (January, 2001) are yet to be electrified, which are mostly situated in difficult terrain, in the hills, forests and coastal areas, where the ABC system could speed up their electrification. In urban areas of old cities, where lines have to be taken through narrow lanes, this system could be easier to install and maintain.

The ABC system is used in the South Pacific, the South-East Asian countries, Australia, Papua New Guinea, Sri Lanka, China and New Zealand. A cheaper alternative to the ABC System is covered conductors, which are of two types: Covered Conductors Thick (CCT) and Covered Conductor Standard (CC). For any specific application, it is necessary to take into account, not only technical parameters, but also life cycle costs. Life cycle costs refer to all the costs incurred during the life of a system, such as initial construction cost, exposure to storms and accidents and assumed repair costs, the cost of regular maintenance and social costs in terms of unserved energy and public inconvenience. While all these insulated systems have individual advantages, covered conductors are comparatively light and cheap and suited to long spans in rural areas. CCTs have a superior abrasion resistance for forest areas, hilly terrain etc. ABCs are an alternative to CCTs in significantly reducing electromagnetic field (EMF). See REC specifications: 64/1993 for 11 kV and 32/1984 for LT.

11.11 Service Line

A service line (230/400 V) is a supply line tapped from the distribution main and taken through the shortest possible route to supply power to a single consumer or to a group of consumers on the same premises. The tapping is made from the nearest overhead distribution line or from the nearest underground cable junction. The service line is usually an insulated conductor which is stretched and clipped along the G.I. bearer wire (see IS: 8061). Service tee-off may be done in the following ways:

- Tapping overhead/underground distribution line.
- Connecting to a low-voltage distribution board of a sub-station situated on the premises.
- Connecting to a distribution pillar in case of underground distribution
- Looping it through the termination point of another service.

Entry of service cable to the building should be through a lead in pipe/conduit. It is terminated into the adequately ventilated enclosure containing the consumer supply unit. In multistory buildings, readily accessible vertical ducts or chases of adequate capacity are provided for service cables.

PROBLEMS

1. A three-phase overhead line with a phase impedance of 2 + j7 ohms, supplies a load of 30 MW at 33 kV and at 0.85 lagging power factor. Calculate:
   (i) the sending end voltage,
   (ii) voltage regulation of the line,
   (iii) transmission efficiency,
   (iv) voltage regulation if load power factor is 0.96 leading instead of 0.85 lagging.
2. What is compact line and where is it required? It is proposed to install four 22 kV lines into a route for a single-circuit 11 kV line with bare conductors operating in the city. What steps will you take to do so.
3. A three-phase, 11 kV line consists of three parallel conductors in the same horizontal plane. The two outer conductors are each located at a distance of 1 m from the central conductor. If the conductor diameter is 6 mm, calculate the inductive reactance per phase of a 10 km length of the line.
4. What are the dimensions of the right-of-way for 415 V, 11 kV and 33 kV overhead lines as per Indian Electricity Rules?
5. An 132 kV overhead line 10 km long, can be represented simply by a series inductive reactance of 4 Ω per phase. The line carries 100 MW with sending end voltage of 135 kV and receiving end voltage of 128 kV. What is the transmission angle?

BIBLIOGRAPHY

1. Worldwide service experience with HV composite insulators, *CIGRE Study Committee No. 22 Report, ELECTRA*, No. 130, 1990, 69-77.
2. *Power Sub-transmission and Distribution Systems—prevailing practices in India*, Central Board of Irrigation and Power, Publication No. 198, New Delhi, December, 1987, pp. 75–79.
3. Pattini, G., and L. Simoni, "Voltage life of Dielectrics", *Electrotecnica*, vol. 58, Jan. 1971, pp. 25–32.
4. IS: 5613 (Part I and II), Design installation and maintenance of lines up to and including 220 kV.
5. Willis H. Lee, *Power Distribution Planning Reference Book*, Marcel Dekker, New York, 1997.
6. Pansini Anthony J., "*Electrical Distribution Engineering*", McGraw-Hill Book Company, New York, 1983.
7. Krivada A., and D. Birthwhistle, "Test for Overhead Insulated Mains", *Proceedings of the 5th International Transmission & Distribution Conference—Distribution 2000*, Brisbane, November, 1999, pp. 790–803.

8. Jones, G.R., and others, "*Electrical Engineer's Reference Book*, 15th edition, Newness, Oxford, 1993, pp. 22/1-22/18.
9. *Manual on Transmission Criteria*, Central Electricity Authority, May 1985, p. 15.
10. Christensen Peter C., "*Retail Wheeling—A Guide for end-users*", 3rd Edition, Penn Well, Tulsa, U.S.A., 1998, p. 157.
11. 'Distribution Polymer Insulation', HUBBELL TIPS and NEWS, vol. 7, no. 1, May, 2002, pp. 4–6.

Chapter 12

Underground Cables

Underground distribution cables are a vital part of any power distribution system. This means that the selection of a particular cable must be based on many considerations ranging from the cost of losses to environmental concerns. Underground distribution costs are between 2 to 10 times that of the overhead system. It is preferred due to elimination of outages caused by abnormal weather conditions such as snow, rain, storms, lightning, fires, trees, accidents etc. This system is environment-friendly. Improved cable technology has reduced the cost of the underground system compared to the overhead. The improvement in materials and manufacturing process allows higher voltage stress in the design of modern cables. This enables reduction in the overall diameter of the cable. For example [8], the stress and overall diameter trend of the latest 3rd generation 132 kV XLPE cables is shown in Fig. 12.1.

Underground system is important for metropolitan cities, city centres, air-ports and defence services. For example, distribution system in Mumbai and Hong Kong is predominantly underground.

12.1 Circuit Design: Key to Reliability (6)

Circuit configuration for underground systems has a great influence on reliability. Circuit configuration describes the way in which circuit elements are connected and is particularly concerned with the location and type of switching equipment. The emphasis in underground system design is to provide necessary circuit redundancy and switching facilities to isolate the faulted system components and to restore service.

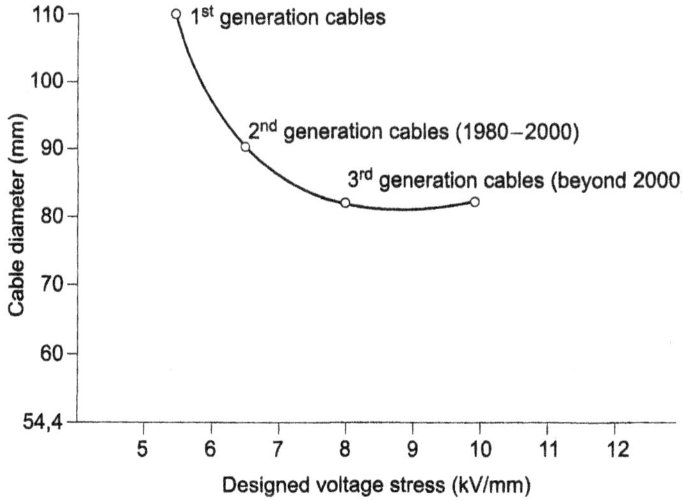

Fig. 12.1 *Typical reduction in the diameter of 132 kV XLPE cable with increasing insulation design voltage stress*

There are three basic circuit configurations used in the primary system: radial, open-loop and dual. There are many variations in circuit design, but generally they consist of one or more of the three basic elements described below.

1. Radial circuit configuration is the simplest to install and operate and is the least expensive. But because it has no redundancy and no sectionalizing, it has limited application in underground systems.

2. The open-loop configuration [Fig. 12.2(a)] for a circuit supplied from an overhead line is the predominant configuration used in underground distribution, specially in urban centres. The open-loop provides facilities to isolate any faulted components, without affecting services to the unfaulted portion of the system. The load break switching is provided by load switches which are also used to sectionalize the primary circuit at each transformer. The open-loop configuration is also used for total underground systems as shown in Fig. 12.2(e). The fused laterals are looped to the corresponding laterals from different equipment similar to Fig. 12.2(a) and the main feeders are looped to the corresponding main feeders from different sub-stations. Therefore, this plan provides the sectionalizing facilities necessary to isolate any fault. The looped-lateral circuits provide for service restoration following a lateral cable fault or a fault in a switch and fuse equipment. The loop main circuits provide for service restoration following a main cable fault or the outage of a substation. In the loop configuration, each half of the

loop must be capable of supplying the total load of both halves. Therefore, the maximum normal load on each half is limited to 50 per cent of the emergency capability. Expressed in terms of reserve capacity, an open-loop circuit must be designed with 100 per cent reserve. This requirement imposes an economic penalty on the open-loop configuration main circuits, but it continues to be preferred to lateral circuits.

This configuration has two disadvantages. It is somewhat difficult to operate and requires the highest reserve capacity of all the configurations. Service restoration following a fault is a complex procedure involving: (i) locating the faulted section, (ii) isolating the fault and (iii) restoring the unfaulted section.

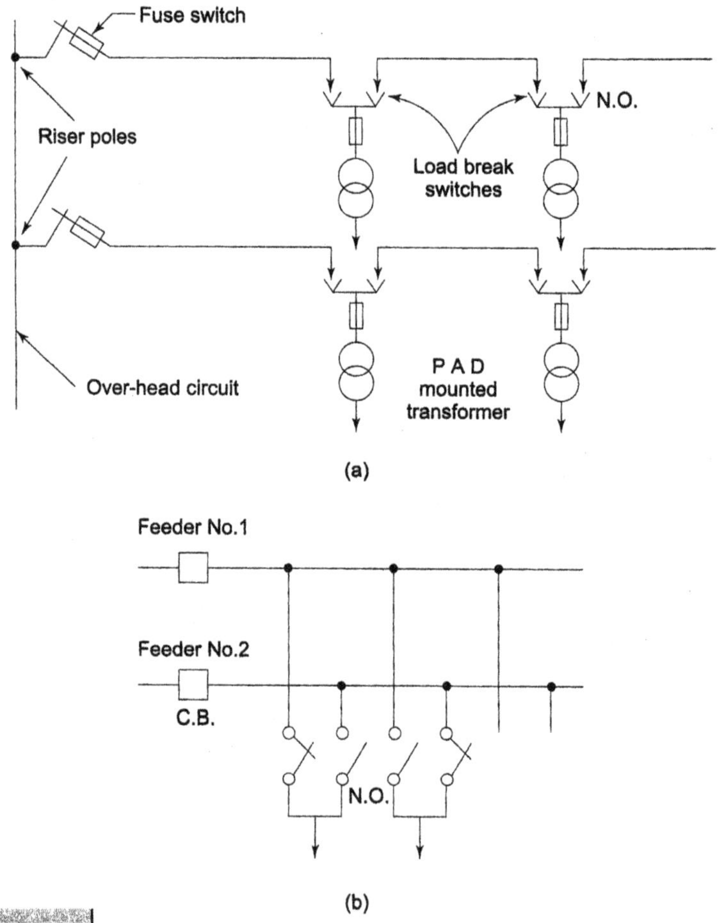

Fig. 12.2 *Underground circuits.*
(a) Open-loop, (b) Dual

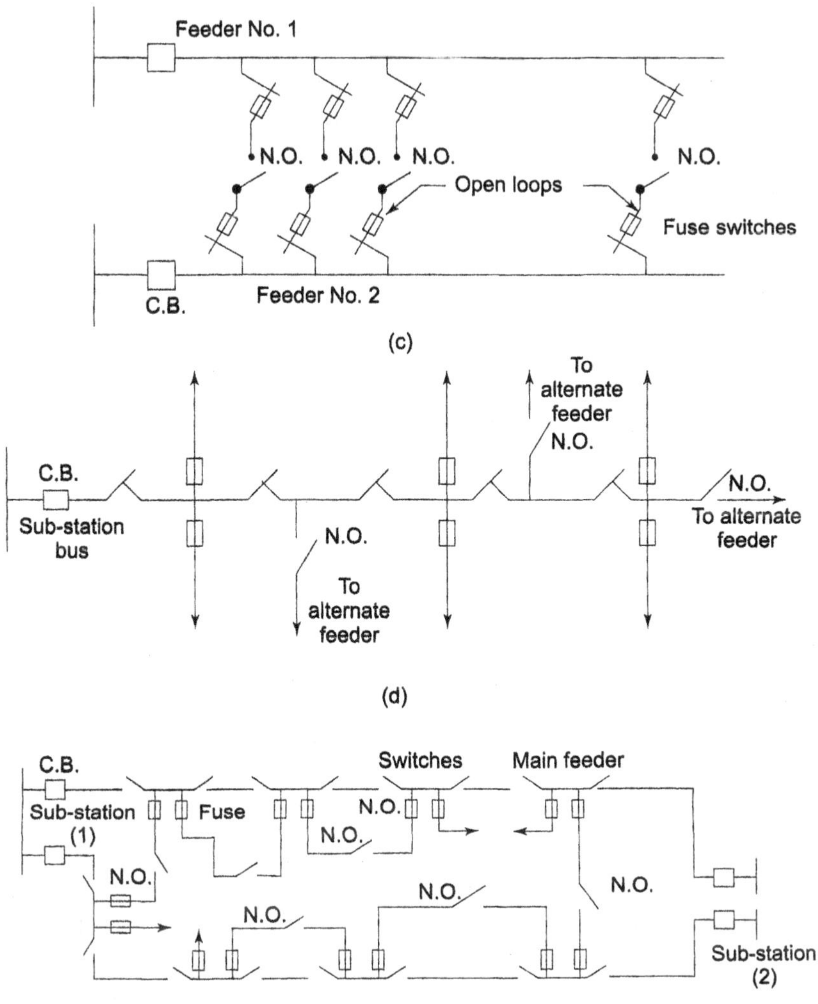

(e)

Fig. 12.2 *Underground circuits:*
(c) Ladder, (d) Open-tie; (e) Open-loop with laterals

3. Dual configuration [Fig. 12.2(b)] provides two primary circuits at each transformer. The main feeders are radial which eliminates the need for switching equipment. Although dual supply has been the traditional method of providing increased reliability for selected loads on both overhead and underground systems, it is usually used for serving a transformer in a given area. An advantage of the dual feeder system is that service restoration is simple and straightforward and can be initiated immediately without first determining where the fault is located.

The total time required to restore service on the dual feeder system increases linearly with the number of connected transformers, whereas with an open-loop system the total time tends to level off as the number of transformers is increased. The disadvantage of the dual feeder system is the need for duplicate primary circuits throughout the load area. The two circuits should be separated physically, perhaps placed on opposite sides of the street to avoid loss in both circuits. For this reason, the dual feeder arrangement is more expensive than the open-loop. An additional consideration with the dual feeder system is the possible damage to both the feeders due to a switch failure if the throwover switch is a common box. As regards reserve capacity this system is similar to the open loop configuration, although the dual feeder concept can be extended to three or four feeders. Each circuit then picks up only half to one-third of the additional load on the outage of one feeder and the normal feeder load can be 67 or 75 per cent of the circuit capacity.

This system is typically used for underground systems serving commercial loads, such as shopping centres and is not preferred for residential or general purpose underground circuits.

4. The ladder configuration [Fig. 12.2(c)] combines the looped lateral circuit with the radial main circuits with each end of each lateral connected to a different main feeder through a fuse switch unit. This arrangement is sometimes used for serving residential or mixed residential-light commercial load areas. For easy manual service restoration following the main circuit outage, the fuse and the switches of each looped lateral should be located at one place. The difficulty is that a planned outage of one main circuit will involve numerous switching operations for the transfer of each lateral to an alternative circuit.

5. The open-tie circuit [Fig. 12.2(d)] arrangement is one way to minimize the reserve requirements of the main circuits. The design incorporates a feeder circuit with three load centres or distribution transformer sub-stations and with open tie circuits arranged in such a manner that each load centre can be supplied from an alternate feeder. If this concept is applied uniformly throughout the underground system, each feeder would require a reserve capacity of only 33.3 per cent and, therefore, could carry a normal load of 75 per cent of the circuit capability—a significant increase over the simple open-loop configuration. The concept of providing a tie circuit for each load centre is applicable for any number of load centres. The relationship between the number of tie circuits, the required reserve capacity and feeder normal loading are summarized in Table 12.1.

There is always a practical limit to the number of load centres that could be installed on each feeder circuit and to the number of tie circuits that could be interleaved among the available feeders.

Table 12.1

Number of ties versus capacity and loading

No. of ties	Reserve capacity %	Normal loading %
1	100.0	50.0
2	50.0	66.7
3	33.3	75.0
4	25.0	80.0
5	20.0	83.3
6	16.7	85.7
7	14.3	87.5

Operation

The switching arrangement can be manual or automatic. The latter is fairly easy and cheap nowadays. Remote control schemes for operating distribution sectionalizing are receiving increased attention.

12.1.1 Insulated Cables

Aluminium rather than copper is largely used as a cable conductor for power cables as it is economical. Low-voltage cables up to 1.1 kV voltage rating are thermoplastic (PVC, polyethylene) or elastomeric (XLPE, EPR). These are manufactured as single, two, three or four core cables. Thermoplastic insulation material may be seriously damaged when subjected to, even for a relatively short period, temperatures higher than permissible for continuous operation. Therefore, current rating of thermoplastic cables is determined not only by the maximum conductor temperature admissible for continuous running (see Section 12.2) but also by the temperature likely to be attained under conditions of excess current. The cable rating is derated by the protection factor (see Appendix II). Standard cable rating is multiplied by the rating factor for excess current protection:

$$\text{Rating Factor} = \frac{\text{Thermal Rated (amperes)}}{\text{Design overload protection (amperes)}}$$

Cables with 3.3 to 33 kV rating are generally PVC or XLPE or paper-insulated lead covered (PILC). These are manufactured as multi-core cables. However, PVC cables are not manufactured beyond 11 kV grade. XLPE or PILC 66, 132 kV cables are usually manufactured as single core.

XLPE cables have about 20 per cent higher rating for the same conductor size compared to PILC and PVC insulated cables. The life of the XLPE cable is almost double that of PILC. XLPE cables are of unshielded construction upto 3.3 kV, above which it is shielded as per IS: 7098.

EXAMPLE A three-core aluminium conductor cable PVC insulated which is laid in the ground (buried) can carry a continuous load of 235 A. The rating is based on admissible conductor temperature. The excess current protection is designed to operate when current exceeds $1.3 \times 235 = 305$ A for four hours. If the cable is laid on surface in air, the excess current protection is designed to operate when the current exceeds $1.5 \times 235 = 352$ A. Find the cable rating actually allowed.

Solution:

Since the excess current is more than continuous rating, we should apply derating factors equivalent to the excess current factor.

Therefore, the cable rating allowed when cable laid is buried $= 235/1.3 = 180$ A. The cable rating allowed when cable is laid on the surface in air $= 235/1.5 = 150$ A.

Construction of Cables

Shape The use of sector and segmental or oval conductors in the place of concentric-stranded conductors affords a reduction in the cable diameter by 20 to 25 per cent, hence, effects savings in the materials used for insulation, sheathing, and protective covering. The conductors of single-core cables of all sizes and of multi-core cables up to 16 mm^2 are made round in section. The multi-core belted cables with insulation around, the cores, or screened 25 mm^2 and over in cross section, are shaped into sector or segmental forms. In belted cable, the multi-core cable is applied part of insulation to each conductor individually, and the remainder is applied over the laid up or assembled cores as a common belt or binder. In screened cable, the multi-core cable, the insulation of each conductor is separately enclosed in a conducting or semi-conducting layer, in order to ensure a radial electrostatic field surrounding the conductor. Belted cables in addition to radial voltage stress through layers of paper, have tangential stress along the surface of the paper, causing burns at the surfaces over the cores. Tangentially, electric strength of impregnated paper is one-tenth that of radial strength. Due to this, belted cables have a lower rating compared to screen type cables.

Insulation The common types of insulating material used to insulate cable conductors from one another and from outer metal sheaths are paper, plastics, and rubber.

(i) Paper impregnated insulated cables are extensively used due to high breakdown strength, but impregnated paper is hydroscopic which requires complete sealing by sheathing and cable boxes. The insulation comprises of electrical grade craft paper and is applied in the form of tapes laid helically with successive layers. The cables are manufactured as per IS: 692–1994 for voltage rating up to 33 kV.

(ii) Plastic insulation for power cables is produced from polythene and polyvinylchloride (PVC). Polythene is more correctly known as polyethylene (PE) .

   (a) Polyethylene features good mechanical properties in a wide temperature range, resists attack by acids, alkalies, and moisture, and displays high dielectric properties. Depending on the manufacturing techniques employed and the contents and types of additions made into the compound, polyethylene of both high and low density can be produced. The high density polyethylene excels the low density counterpart in mechanical strength and has a higher melting point. Introduction of organic peroxides into polyethylene with the subsequent curing, appreciably increases its melting point and renders it more stable to cracking. Cured polyethylene partially deforms at 150°C. Special compositions added to polyethylene impart it self-quenching properties. Polyethylene used for conducting shields of PE-insulated cables contains such additions as, polyisobutylene, acetylene soot, and streamline acid.

   (b) Polyvinylchloride is a solid product of polymerization that does not sustain combustion. Various additions are made to PVC to improve its characteristics: plasticizers render it more elastic and frost resistant; caoline (porcelain clay), talc, and calcium carbonate increase its dielectric strength. Colouring admixtures added to PVC impart it a definite colour. On exposure to heat, solar radiation and various media, PVC undergoes aging, i.e., looses elasticity and cold resistance since plasticizer volatilizes. PVC cables are manufactured as per IS: 1554–1988.

(iii) Rubber insulation is a compound consisting of natural or synthetic rubber, fillers, softeners (plasticizers), vulcanization accelerators, antioxidants (age resistors), pigments, and other ingredients. The rubber compound used for cable insulation is advantageous for use as its elasticity is combined with excellent moisture-resisting properties.

Outer Covering Cables are made complete with outer coverings applied over the cable insulation to protect it against exposure to light, moisture, and various chemical substances and also to guard it against mechanical injury. Lead and aluminium are the best materials for cable sheaths as regards the air-tightness they afford, moisture seal, moisture resistance, flexibility and thermal stability as used for paper insulated cables. General purpose plastic or rubber insulated cables not intended for exposure to mechanical strains do not require metal sheaths, and usually have plastic or rubber jackets employed as an outer protection against moisture. The thickness of protective covering depends on the type of protective material used, cable diameter and the service for which the cables are intended. A protective jacket can be a combination of bedding, armour and outer covering to provide protection against corrosion and mechanical injury.

The cables may be classified according to the following concepts:

(a) *Cores:* Single core, two core, three core and three-and-a-half core, four core, etc.

(b) *Shape:* Circular and sector type.

(c) *Arrangement:* Belted type, screened, oil filled, gas filled, armoured and unarmoured.

(d) *Dielectric:* Paper insulated (PILC), polyvinyl chloride (PVC), cross-linked polyethylene (XLPE) and oil or gas filled (nitrogen or SF_6), EPR (ethylene proplyene rubber), PE, etc.

(e) *Armouring:* Steel wire or tape or strip.

(f) *Sheath:* Aluminium, Lead, Rubber.

(g) *Voltage:* 650/1100 V, 1.1/1.1 kV, 3.8/6.6 kV, 6.35/11 kV, 11/11 kV, 12.7/22 kV, 19/33 kV, 66 kV, 132 kV.

12.1.2 Electrical Characteristics of Cables

Resistance

$$R_{ac} = R_{dc}(1 + Y_s + Y_p)$$

where,

R_{ac} = ac resistance
R_{dc} = dc resistance
Y_s = correction for skin effect
Y_p = correction for proximity effect

Capacitance

(a) For single core, three-core screened cables with a circular conductor, capacitance C is given by

$$C = \frac{0.024 \in}{\log \frac{d_{\text{in}}}{d_c}} \text{ uF/phase/km}$$

where,

d_{in} = diameter over insulation
d_c = diameter of conductor
$\in$ = permittivity of dielectric material (average 3.6 for impregnated paper insulation)

X_c = Capacitive Reactance = $\frac{1}{2\pi f C}$

(b) For belted cables, the capacitance can be best obtained by actual measurement.

Inductance

For three-core, circular conductor cables, the inductance L is given by the empirical relation

$$L = 0.460 \log \frac{\text{GMD}}{\text{GMR}} \text{ mH/phase/km}$$

where,

GMD = geometric mean distance is equal to axial spacing of conductors
GMR = geometric mean radius. Generally taken as $0.5\sqrt{\text{conductor area}}$

Armour increases inductance by about 10 per cent due to magnetic induction in 3-core armoured cables.

$$\text{Inductive reactance } (X_L) = 2\,\pi f L$$

The values of R_{ac}, X_L, X_c are usually given in the cable manufacturer's catalogue (see Appendix II)

12.1.3 Package Type Sub-station

It is compact, totally enclosed, safe, tamper-proof sub-station unit where HV switchgear, transformer and LT switchgear are mounted on a common base. The items are assembled at factory in a single unit, with underbase on rollers for easy handling. At site only the HV and LV cable connections are made for commissioning. Its maintenance cost is less. Package type sub-stations are generally used in rating from 315 kVA to 2 MVA for both indoor and outdoor installations in high load density urban underground distribution system. For outdoor installation, a housing or a kiosk is provided, made up of sheet steel. The doors are provided for different compartments to have independent access.

The transformers can be oil cooled or dry types. Dry type is preferred for indoor application. HV switchgear is generally vacuum or SF_6 type, with radial or ring main arrangement. The typical ring-main package sub-station single-line diagram is shown in Fig. 12.3 and radial one is Fig. 19.2.

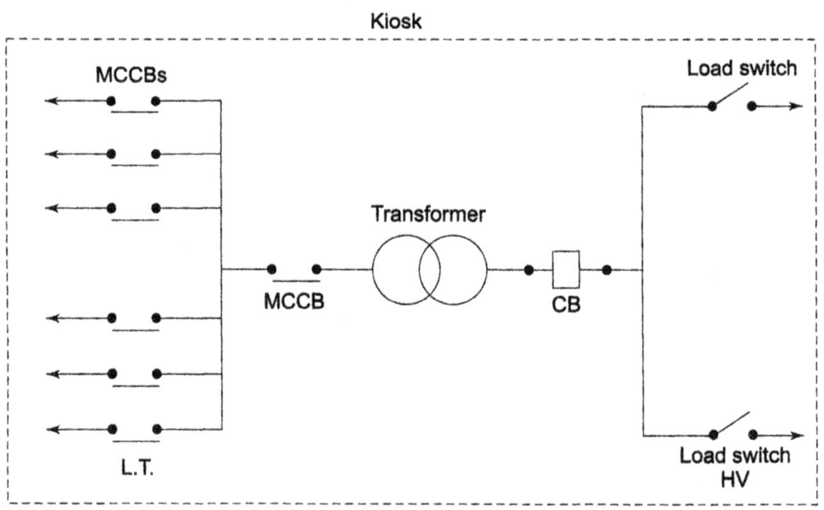

Fig. 12.3 *Typical single-line diagram of ring-main package type or pad-mounted sub-station*

12.1.4 Submarine Cables

Submarine cables are used in river or sea water and have enhanced current ratings compared with land cables and thermal resistivities as low as 0.3°C m/W are used. These values can only be justified under ideal conditions and if the cable is buried either in a controlled manner during the installation process or by silt during its operational life, it is prudent to use a figure of 0.5 or 0.7°C m/W. This still means that the majority of route will have superior thermal resistivity values than the land sections and this is particularly so, if the cable is installed inside closed steel tubes or risers on off-shore structures.

If the conditions are extreme, it may be necessary to consider methods of improving the thermal environment at the terminations or if this is not possible, by locally increasing the conductor cross section, thereby reducing its temperature to within the permitted continuous operating level.

The preferred conductor for submarine cables is high conductivity copper. The only possible alternative is aluminium, but this is susceptible to corrosion and hydrogen production when in contact with sea water. Individual strands are tinned for additional protection against corrosion. Water blocking is sometimes specified and is desirable with XLPE insulations which are susceptible to the water tree phenomenon.

The alternative insulation materiels are the traditional impregnated paper in conjunction with a lead alloy sheath and the extruded elastomeric materials XLPE and EPR.

Submarine cables must be designed to survive the installation process with a good margin of safety and this is generally achieved by the application of one or more layers of round zinc coated steel wires. These are applied over a bedding of polypropylene or jute roves and are between 4.0 and 7.0 mm in diameter. Particularly in sea-bed conditions with rocks and swift sea-bed currents with the danger of objects falling from platforms, additional protection either in the form of rock armours or by steel tapes applied under conventional armour is required.

Submarine cable terminations on land generally follow the conventional practice of either outdoor pole termination or termination into transformer switchgear. The land fall of the cable must be protected up the beach and the most usual method is a deep cut trench to below low water level, or a pipe or duct embedded in concrete.

12.2 Determination of Cable Rating

The method of computation is based on Ohm's law in thermal resistivity instead of electrical units. The formula gives the heat transfer across a layer between opposite faces, and a temperature difference which exists is expressed as:

$$\text{Heat flow in thermal watts} = \frac{\text{Temperature difference in } ^\circ\text{C}}{\text{Thermal resistance in thermal ohms}}$$

A thermal ohm is the difference in °C between opposite faces of a cm^3 produced by the flow of 1 watt of heat. It is expressed in units of °C cm/watt. Inherently, the thermal resistivity depends upon the material of the conductor, insulation, protective covering and ground. Taking all these factors into account, the ratings are determined by well established empirical formulae and assigned by the cable manufacturer along with the correction factor for use in different grouping tables. The general empirical relation for the determination of continuous rating is given below [5].

$$I = \left[\frac{\theta_c}{R\{G' + n\{G_b(1+\lambda) + (G_a + G_d + G_s + G_e)(1 + \lambda + \lambda_1)\}} \right]^{1/2}$$

where,

$I =$ current in conductors (A)
$\theta_c =$ Conductor temperature rise above ground ambient (°C)
$\lambda =$ sheath loss factor
$\lambda_1 =$ armour loss factor
$G' =$ thermal resistance between conductor and sheath (°C cm/watt)
$G_s =$ thermal resistance between armour and cable surface (°C cm/watt)
$G_a =$ overall thermal resistance between outer surface of cable and inner surface of duct (°C cm/watt)
$G_b =$ Thermal resistance between sheath and armour (°C/watt)
$G_d =$ Thermal resistance from inner and outer surfaces of duct (°C cm/watt)
$G_e =$ Thermal resistance of soil (°C cm/watt)
$R =$ ac resistance of conductor at conductor temperature (Ω/cm)
$n =$ No. of cores.

The thermal resistance between armour and sheath is considered negligible. The above expression is for the worst case of lowest current rating for PILC cable buried in the ground with some portion laid in a buried duct at a road crossing. The equivalent thermal circuit of the cable is shown in Fig. 12.4. There may be other situations as follows:

(i) When the whole cable is laid in air but shielded from the sun. In this case, terms $G_a + G_d + G_e$ are replaced by G_{air} which is the thermal resistance of cable in free air (°C cm/watt).

(ii) When the whole cable is laid in a duct in air but shielded from the sun. In this case, term G_e is replaced by G_{air} which is the thermal resistance of duct in free air (°C cm/watt).

Generally, cables laid in free air have a lower current carrying capacity than cables buried in the ground, because air offers comparatively more thermal resistance. Similarly, cables laid in water (submarine cables) have higher current carrying capacity than buried cables as water has better thermal conduction than soil.

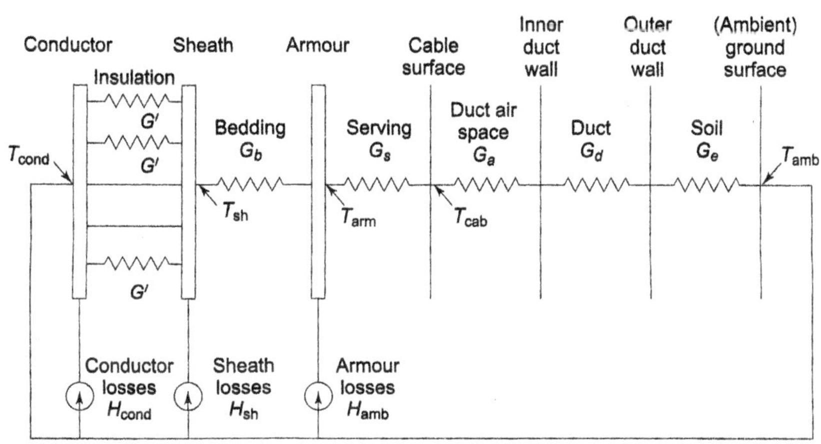

Fig. 12.4 *Equivalent thermal circuit of paper-insulated lead-sheathed cable laid in an underground duct*

12.2.1 Heating of Cables

Within the cable there are three sources of heat, namely, I^2R losses in the conductors, losses in sheathings and losses in armourings. In calculating I^2R losses it is usual to start with the resistances given in standard resistance tables. The dielectric loss is of importance only in super voltage cables, i.e. 132 kV and above. However, PVC cables of rating 6.6 kV and above have significant dielectric loss. In 11 kV PVC cables, the dielectric losses are 6–8 per cent of the conductor losses while XLPE cables having 0.1per cent. 3-core cables with metallic sheath and armour have sheath and armour losses of about 2 per cent of the cable conductor losses. However, it is significant for single core cables. Sheath and armouring losses, while not absent in three-core cables, are the dominating factors in single-core cables carrying alternating currents.

Grouping

When a buried cable *A* carries a load *I*, it will heat the surrounding soil, increasing the temperature above the general ambient. If a cable *B* is now laid in the soil adjacent to cable *A*, then cable *B* will assume the temperature of the soil it replaces. The temperature increase of *B* above the ambient is commonly referred to as Mutual Heating (MH). When a cable is being heated by more than one neighbouring cable, the principle of superposition is applicable, i.e., total heating is obtained by adding the heat generated from each of the neighbouring cable.

$$MH_{\text{total}} = \sum_{1}^{n} MH_n$$

12.2.2 Continuous Rating

The continuous rating of a cable is calculated by the formula in Section 12.2 under conditions of thermal equilibrium, i.e.,

Heat energy generated by cable = Heat energy dissipated by the cable into the environment.

The maximum continuous current ratings are recommended in IS: 3961–1967/1968 (see Appendix II). These ratings are based on the following assumptions and conditions of installation:

(a) Thermal resistivity of soil 150°C cm/W
(b) Thermal resistivity of PVC; XLPE; Paper 650; 350; 500°C cm/W
(c) Ground temperature 30°C
(d) Ambient air temperature 40°C

(e) Depth of laying the highest point of cables, laid direct or the top surface of buried ducts is taken to be as follows:

0.650/1.1 kV cables	750 mm
11 kV cables	900 mm
22 and 33 kV cables	1050 mm

(f) Maximum allowable continuous conductor temperature VIR, PVC insulated cables 70°C

XLPE cables	90°C
PILC cables:	
11 kV three-core screened, armoured	70°C
11 kV three-core belted armoured	65°C
6.6 kV, 3.3 kV three-core, 1.1 kV four-core armoured	80°C
22 kV PILC cables	
Single-core, three-core (screened) and three-core (solid type)	65°C
33 kV PILC cables	
Single-core, three-core (screened) and three-core (solid type)	65°C

Unarmoured cables laid direct in ducts are subjected to the additional regulations that the maximum permissible temperature rise above the ambient shall not exceed 35°C for 11 kV, 22 kV and 33 kV cables.

It is impractical to operate the cables continuously at the maximum temperature, on which the continuous rating is based. Cyclic rating and short circuit rating temperature rises are important for the actual operation.

EXAMPLE: Based on the above assumption, the comparative current ratings in amperes for three-core, 11 kV buried cables can be approximately as given below.

Size mm^2	XLPE	PILC	PVC
35	120	100	105
120	240	205	200
300	385	335	330

The current carrying capacity for screened lead-covered cables is higher by 5–10 per cent as compared to belted type lead-covered cables.

The current-carrying capacity of unarmoured and armoured cables is practically the same.

12.3 Stress Grading

Electric stress is greatest near the centre of the conductor and diminishes logarithmically towards the outside. The stress near the conductor can be reduced by grading the insulation since the stress is inversely proportional to the dielectric permittivity.

The stress G, in volts per mm at a distance x (mm) from the axis of the conductor, is given by the equation:

$$G_x = \frac{E}{x \log R/r}$$

where,

$E =$ voltage between the conductor and metallic covering (volt)
$R =$ outer radius of insulation (mm)
$r =$ inner radius of insulation (mm)

The maximum stress is ordinarily obtained when $x = r$ and the minimum stress is when $x = R$.

If the conductor is stranded but not compacted, the lines of force are not exactly radial, especially near the conductor, and a factor has to be applied to increase the stress calculated by the above formula. It is necessary to use a suitable factor if a comparison is to be made between stresses in a cable with a strandard conductor and those in a cable with a solid, smooth or shielded conductor. The stress is overcome by laying over cable conductor/insulation, the taped screen of metallised paper in case of paper-insulated cables and semi-conducting plastic screen in case of extruded cables. While procuring cables, the power utility must quote the average and maximum stress values in the cable insulation specifications. A safety factor of 2 to 5 is taken when designing for working stress.

Where the conductor is sector-shaped or oval, the calculation of the maximum stress becomes more tedious and it is usual to assume without fear of significant error, that it is equal to the stress at the surface of a cylindrical conductor of radius equal to the minimum radius at the apices of the "shaped" conductor.

12.4 Thermo-Mechanical Effects in Cable Systems

If a cable conductor is not free to expand and lateral movement of the cable is prevented by the soil, a force is developed in the conductor as the temperature is varied (see Fig. 12.5). For small temperature variations the effect is linear and reversible. With larger variations, however, inelastic deformation of the conductor occurs at a rate that depends on the material and temperature. A rigid joint at high temperatures will be subject to the same compressive force as the cable conductor. More usually, the conductors move into the joint which relieves the force to an extent that depends on the friction between the cable cores and sheath. Movements due to cyclic loading can disturb the joint insulation or any screen over individual cores.

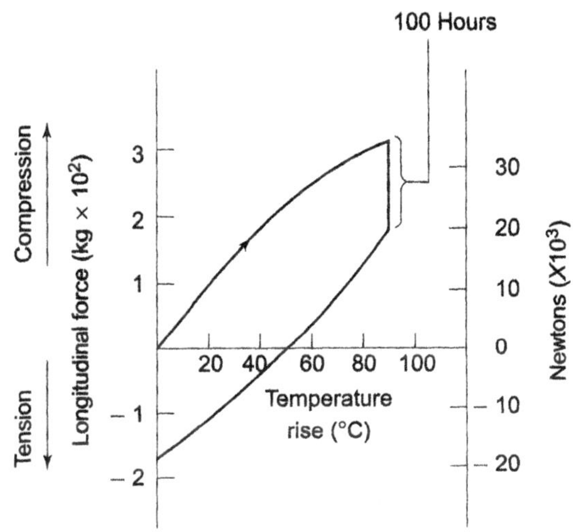

Fig. 12.5 *Variation of force in copper conductor of a cable tested under conditions such that movement was impossible*

The distortion of the cores in the joint that is the typical effect of the conductor expansion could be mistaken for bad jointing. Ground sinking and conductor shortening produce tension and pulled ferrules could, in practice, result from a combination of both these effects. The subject remains of interest particularly in the context of possible increased cable rating.

12.5 Cable Materials

12.5.1 Insulants

The main points regarding the properties of dielectrics used for cable insulation are:

(a) High insulation resistance.
(b) High dielectric strength.
(c) Good mechanical properties, e.g., tenacity and elasticity.
(d) No reaction with acids and alkalis at the operating temperature range.
(e) Nonhygroscopic, otherwise, watertight covering is required.
(f) Not too costly and easily processable during manufacture and in the field.
(g) Heat resistant for FRLS/FS cables (see IS: 5831).

The properties of insulating materials is given in Table 12.2.

12.5.2 Cable Oils and Compounds

Desirable Qualities

These are as given below:

(a) Low coefficient of expansion.
(b) Low viscosity at impregnation temperature.
(c) High viscosity at working temperature (solid cables only).
(d) Solidifying point below service temperatures.
(e) Some lubricating property.
(f) Low permittivity, temperature coefficient and high resistivity.
(g) High dielectric strength.
(h) Chemically stable and free from occluded gases.

12.5.3 Cable Sheathing

Also, the sheathing material has great bearing in determining the capacity of the cable, such as thermal conductivity, bending resistance, weather proofness, etc. The comparative properties of different sheathing materials are given in Fig. 12.6.

Table 12.2

Comparison of insulation materials

Item	*Oil impregnated paper*	*Natural rubber*	*Butyl rubber*	*EP rubber*	*Polyvinyl chloride (PVC)*	*Polyethylene PE*	*Cross-linked polyethylene*
Dielectric constant	3.5	3.0–4.0	3.5–5.0	3.0–5.0	5.0–7.0	2.3	2.3
Impulse level (Volts/mil)	1800	—	—	—	1100	—	2000
AC Breakdown voltage (kV/mm)	40–50	22–28	25–30	30–40	20–30	40–50	40–50
Tan δ at max. operating temperature (%)	1 or less	3	2	2	15	0.05	0.05
Permissible continuous conductor temperature (°C)	80	60	85	90	70	75	90
Permissible emergency overload temperature °C	100	—	—	—	120	—	130
Permissible short circuit temperature (°C)	160	—	220	250	160	130	250
Softening temperature (°C)	—	(–50)-(–60)	(–40)-(–60)	(–50)-(–60)	(–20)-(–30)	(–70)-(–80)	–60
Specific resistance (ohm-cm)	10^{15}	10^{15}	10^{16}	10^{15}	10^{14}	10^{17}	10^{17}
Thermal resistivity (°C cm/watt)	500	—	500	500	600	350	350
Flame retardance	—	—	Burns	—	Self extinguishing	Burns	Burns
Performance in moist environment	Poor	—	—	Good	Good	—	Excellent
Tensile strength (kg/mm²)	—	0.8–1.3	0.4–0.8	0.5–1.0	1.3–2.0	1.2–1.5	1.8–2.0
Elongation (%)	—	300–500	300–600	400–600	200–350	400–600	350–500
Specific weight	1.0 or less	1.2–1.5	1.2–1.5	1.1–1.3	1.2–1.5	0.92	0.92

Comparison of Sheathing Materials

Item	Lead	Aluminium	Vinyl	Chloroprene	Polyethylene
Specific Resistance (ohm cm)	—	—	$10^{12\sim15}$	$10^{7\sim12}$	10^{15}
Tensile Strength (kg/mm^2)	1.5	8 ~ 18	1.0 ~ 2.5	~ 2.0	~ 1.0
Elongation (%)	5.5	2.0 ~ 30	100 ~ 300	300 ~ 1000	~ 350
Bending Resistance	Δ	Δ	•	•	o
Heat Resistance	•	•	Δ	o	Δ
Cold Resistance	•	•	Δ	•	•
Weather Resistance	•	•	•	•	•
Ozone Resistance	—	—	•	o	•
Fire Resistance	•	•	o	o	Δ
Oil Resistance	•	•	•	Δ	•
Acid Resistance	×	×	•	Δ	•
Alkaline Resistance	×	×	o	o	•
Water Resistance	•	•	•	•	•
Specific Weight	11.34	2.7	1.4	1.5	0.92

Notes

- • Best
- o Deterioration takes place but is insignificant for actual applications
- Δ Somewhat greater deterioration but not sufficient to preclude use
- × Improper for practical use

Fig. 12.6

12.6 Cable Jointing

The overall reliability of a distribution system depends on the weakest link, namely, the joints. The joint is considered a weak point and, therefore, jointing accessories and techniques assume an important and critical role despite their comparative low value in the overall investment (see IS: 1255).

The main points of aluminium cables jointing are: Oxidation, galvanic corrosion, creep rate and coefficient of expansion. Oxidation can be broken down by mechanical and abrasion, applying the mixture of high temperature grease or compound. The galvanic corrosion may be eliminated by either tin-plating of the terminal or connector. Control of stresses, which exist at a point where the shield or screen is terminated, is necessary. Normally cable joints and terminations are made by soldering, crimping or compression (see IS: 8309). "Deep-notch crimping" and "hexagonal pressing" for aluminium conductors have gained importance due to their relative simplicity. Selection of proper sized/matched barrel of the furrule or terminal is important.

12.6.1 Types of Joints

Different types of joints shown in Fig. 12.7 for power cables can be simply classified as, branch terminations for indoor use, cable terminations for outdoor use and tee joints for LT cables.

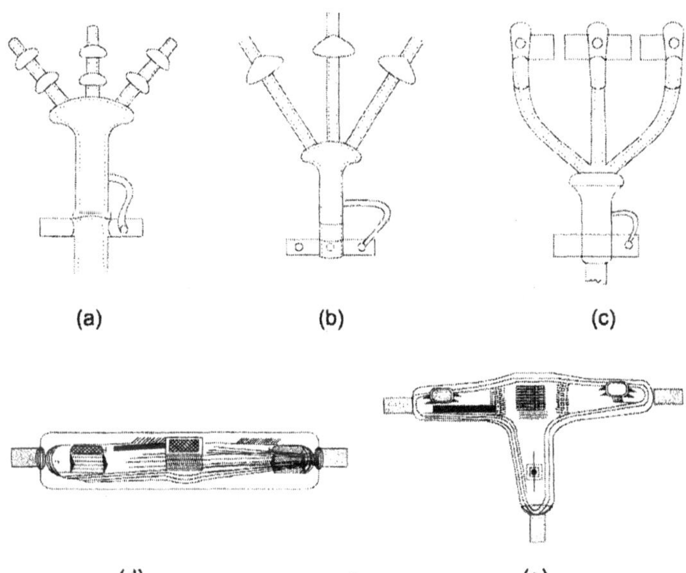

Fig. 12.7 *Typical cable joints:*
(a) Outdoor termination box (11 kV PILC cables); (b) Outdoor termination joint (LT plastic cables); (c) Indoor termination; (d) Straight through joint box; (e) Tee joint for LT cables

12.6.2 System of Accessories

Different types of cables accessories are:

(a) Conventional cast iron box hot pouring bituman filled system.
(b) Cold pouring cast resin system.
(c) Heat shrinkable accessories.

(d) Cold shrinkable sleeves.
(e) Taped techniques.
(f) Premoulded slip-on type accessories.

The contents of the kits, jointing and conductor jointing techniques used for each type of joint and system accessories depend on the voltage range of power cables. The range is generally classified as:

(a) For low tension cables rated up to 1.1 kV (refer IS: 8438).
(b) For cables of voltage rating 1.1 kV to 11 kV (refer IS: 7093).
(c) For cables of voltages above 11 kV.

A typical comparison of kit contents and jointing techniques is given in Table 12.3.

It may be seen, that in view of the sensitive nature of XLPE to partial discharges and its deterioration under stresses, the XLPE cable joint has to be so designed that the stress remains uniform and radial even after disrupting of semi-conducting layers and core screen. XLPE joints must provide for:

- Stress relief at screen termination area.
- Stress relief at ferrule region in straight through joints.
- Maintenance of proper insulation between core-to-core and core-to-earth under running temperature of 90°C.

12.7 Installation of Cables

Some of the important factors which should be borne in mind during installation of cables up to 33 kV rating are given as follows as per IS: 1255/1983. Cable-laying methods are (see Appendix III):

(i) *Direct burial* The cable is buried in the soil. Often, a narrow trench is dug and the cable is reeled in and covered. Alternately, the soil is bored with an automatic boring machine and also cable toed-in with that.

(ii) *In ducts* The cables are installed inside buried ducts, tunnels or surface-mounted ducts.

(iii) *Surface mounted* The cables are laid in open brick or cement concrete trenches or open channels or steel/aluminium trays/ ladders. In the trenches or chennels, the cables are laid on supports erected inside.

The grouping of multi-core cables can be horizontal or vertical together or spaced. Single-core cables can be in trefoil formation: Touching or spaced. The neighbouring cables have the effect of

Table 12.3

Comparison of typical cable jointing systems

Name of accessory	*Conventional hot bitumen-filled type (Only for PILC)*	*Cold puring cast resin type (For both PILC and plastic insulated)*	*Cold or Heat shrinkable (Both for PILC and plastic insulated) sleeves*	*Taped (For plastic insulated cable)*	*Pre-moulded (Only for single-core, round, armoured plastic insulated cable)*	*Conductor jointing system*
(1)	*(2)*	*(3)*	*(4)*	*(5)*	*(6)*	*(7)*
1. Straight Through Joints:						
(i) For cables up to 1.1 kV	Main kit contents: 1. C.I. box 2. Hot melt bituminous compound	Main kit contents: 1. Casting resin epoxy or polyurethane 2. Moulds 3. Earthing system	1. Cold or Heat shrinkable polymer sleeves for core insulation 2. Cold or Heat shrinkable sleeve for encapsulated cores 3. Earthing system	1. Insulation tape for core insulation 2. Insulation tape for encapsulation of insulated cores 3. Earth continuity	1. Slip-on-type premoulded EPR joint body 2. End caps	Compression or soldering technique; mechanical connectors are the latest technique
(ii) For cables above 1.1 kV up to 11 kV	Items 1 and 2 same; additional items: 3. Lead sleeve 4. Impregnated crepe paper insulation	Items 1 and 2 same; additional stress relief material given for screened cables of 11 kV and above	As above and with the kit of stress control tubing; metallic shielding braid	Items 1 to 3 same; additional items: 4. Stress control tape	Items 1 and 2 same; additional items 3. Stress relief EPR material	Soldering is used; compression jointing is also used

Contd.

(1)	(2)	(3)	(4)	(5)	(6)	(7)
(iii) For cables above 11 kV up to 132 kV	Items 1 to 4 same; additional items: 5. Semi-fluid oil resin compound 6. Copper stocking earth and screen	-do-	-do-	-do-	-do-	Only soldering recommended
2. Branch Tee or Service Joints:						
(i) For cables up to 1.1 kV	Same as for 1(i) with CI Box of suitable shape Additional item: T shaped ca. box	Same as with 1(i) with suitable casting mould and claw connectors	-do-	Same as for 1(i)	Same as for 1(i)	1. Married joint with soldering 2. Compression jointing 3. Mechanical connectors
(ii) For cables above 1.1 kV up to 11 kV	-do-	-do-	-do-			
3. Indore End Sealings:						
(i) For cables up to 1.1 kV	1. CI box 2. Hot puring bitumen	Same as with 1(i) suitable casting mould	Cold or Heat shrinkable polymer sleeves for core insulation	Same as 1(i)	Pre-moulded EPR slip on type end sleeve	Compressed and soldering: Compression jointing preferred for plastic cables
(ii) For cables up to 11 kV	Same as 3(i) additional item: 3. Porcelain bushing for indoor use	Same as with 1(i) suitable casting mould; additional material: For screened cable: stress relief material	Same as 3(i); additional material: Heat shrinkable SC tubing for stress relief for screened cables	Same as 2(i) additional material: Semiconducting Stress relief tape	Same as 3(i) additional material: Moulded stress cone	-do-

Contd.

(1)	(2)	(3)	(4)	(5)	(6)	(7)
(iii) For cables above 11 kV up to 132 kV	Same as 3(ii) additional stress relief cone; Bushing are used for a particular voltage	-do-	Same as 3(ii)	Same as 3(ii)	-do-	-do-
4. Outdoor End Sealings:						
(i) For cables up to 1.1 kV	Same as in 3(i) but with HT tape for core protection and bushings	Same as in 3(i) but with environmental protection for cores	Same as in 3(i) additional material: rain sheds	Same as in 3(i) additional tape for environmental protection	Same as 3(i) but with rain sheds plastic cable	Compression recommended for
(ii) For cables above 1.1 kV up to 11 kV	Same as in 3(i) additional material: Porcelain bushing with rain sheds	Same as in 3(ii) with: 1. Porcelain bushing with rain sheds 2. Stress relief material for screened cable	Same as above with additional sheds, additional stress relief tubing	Same as in 3(ii) additional items: 1. Special tape for environmental protection 2. Rain sheds	Same as 3(ii) but with EPR rain sheds	Both compression and soldering can be used for PILC
(iii) For cables above 11 kV up to 132 kV	Same as 3(iii) bushings have rain sheds	Same as with 4(ii) with resin cast bushings of appropriate rating	-do-	Same as above with additional sheds	-do-	-do-

additional heating and reduce rating. The manufacturer's catalogue must be referred to check up the proximity factor to calculate the actual rating of the cable installed a group (see Appendix II). At the time of installation, checks required are:

(a) Before laying the cables, their insulation should be examined with an insulation tester as a precautionary check against any probable damage.
(b) The drum should always be rolled in the direction marked on it by an arrow. In the absence of any such mark, the drum should be rolled in the direction same as that of the inside end of the cable and opposite to that of the outside end.
(c) The cable should be taken from the top of the drum with a supporting ramp if necessary, the drum being braked to avoid overruning.
(d) The cable drums, if required to be shifted at site should be moved by means of cable wheels. In case cable drums are rolled, it should be done in the direction as shown by the arrow mark inscribed on the flange of the cable drum.
(e) The installation radii should be as large as possible. Table 12.4 gives minimum installation radii as recommended in IS: 1255-1983 for lead or lead alloy sheathed cables.

Table 12.4

Minimum Permissible Bending Radii for Cables

Cable voltage rating kV	PILC		PVC and XLPE	
	Single-core	Multi-core	Single-core	Multi-core
Up to 1.1	20D	15D	15D	12D
Above 1.1 to 11	20D	15D	15D	15D
Above 11 to 33	25D	20D	20D	15D

D is the outer diameter of the cable

(f) In cold weather, cables should be warmed before handling. They should be installed when both cable and ambient temperatures are above 0°C (32°F).
(g) Moisture tests should be made on jointing material before jointing. In the case of PILC cables, the test should be done on the insulating papers taken out from the cable.

(h) Where the cables is to be jointed with the existing cables, the sequence of cores at the two ends to be jointed should be in the opposite direction, i.e., if at one end it is in the clockwise direction, at the other end it should be anti-clockwise. This is necessary to avoid the crossing of cores while jointing. This will also decide the direction in which the cable is to be pulled.

(i) A joint being the weakest point of the electric power distribution system, all necessary precautions should be taken to protect it. Jointing materials and accessories, e.g., conductor ferrules, solder, insulating and protective tapes, protective filling compounds, joint boxes, etc., of the right quality and sizes should be made skillfully for jointing work. In this connection it is always advisable that the method and working instructions of the supplier are strictly followed.

(j) About 3 m cable loops should be left spare near the sending and receiving ends of the cable. A layer of 10 cm thick sand should be laid at the bottom of the trench. After laying the cable over the sand layer, a 10 cm sand cover all round the cable must be given further. Top of the panned down sand, above the cable should be covered by placing bricks in flat position. After suitable intervals, cable route and voltage rating indicator should be provided. At road crossing, the cable should be laid through pipe or ducts having internal diameter not less than twice the outer diameter of the cable.

Armoured Cables

All bonding clamps at the joints, terminations and armour wires should be thoroughly cleaned. The clamps should be adequately tightened. This is necessary to ensure proper electrical contact, because the armour is the only return path for the earth fault current.

Unarmoured Cables

In the case of unarmoured cables, the external metallic earth bonding connector of adequate size should be used.

Earth

All joints, terminators, armour wires, metal sheaths and external metallic bondings should be connected to the earth at both ends. Precautions should be taken to eliminate the chemical and bimetallic corrosion of earth connectors or bonds.

Filling Compounds

(a) The design of the box and composition of the filling compound should ensure effective sealing against entry of moisture to conductor ferrules and armour connectors.
(b) If hot pouring protective compounds are used, the temperature of the compound while pouring should not exceed 150°C.

Tests

After installation (laying and jointing), the cables must be tested as per IS: 1255–1983. The insulation resistance test on cables should be made before as well as after jointing and recording the reading. DC high voltage tests should be carried out before the commissioning of cables. The duration of applied test voltage as shown in Table 12.5 should be 5 minutes. The voltage should be increased gradually to the full value maintained there continuously for 5 minutes. No breakdown of insulation should occur during the test.

Table 12.5

Commissioning test voltages

Voltage grade of cable (V)	Test voltage (dc) (V)	
	Between conductors dc for unscreen cables	Between conductors and sheath armour dc
	(A) Cables for Earthed Systems	
650/1100	3000	3000
1900/3300	9000	5000
3800/6600	18000	10500
6350/11000	30000	18000
12700/22000	—	37500
19000/33000	—	60000
	(B) Cables for Unearthed Systems	
3300/3300	9000	9000
6600/6600	18000	18000
11000/11000	30000	30000

12.7.1 Trenchless Excavation

The conventional method of digging trenches for providing underground cables services in thickly populated cities, such as metropolitan cities: Delhi, Mumbai, Chennai, Kolkata etc., is costly, risky, time-consuming and practically causes harm to the environment. Micro-tunnelling vis-a-vis trenchless technology is a solution to such problems. This technology is in use for digging bores up to 200 mm diameter up to 120 m. A directional boring machine drives the soil-displacement hammer through the soil pneumatically or hydraulically. After boring is carried out, this machine toes the cable/pipe duct along the underground bore. This method avoids public inconvenience in cable installation.

12.8 Principal Causes of Cable Failure

The principal temperature dependent causes of cable failure are thermal degradation, dielectric thermal instability, ionization due to void formation and fatigue failure of lead sheaths. These causes could occur simultaneously at different rates and the significance of any cause will depend on factors relating to the construction of the cable and loading conditions.

12.8.1 Temperature Degradation of Paper Insulation

Paper is manufactured from cellulose material which chemically consists of chains of glucose rings. When glucose is subjected to high temperatures, the mechanical strength of paper is reduced to a shortening of the chains. This is accompanied by a liberation of water. The presence of moisture accelerates the rates of chemical decomposition and reduction of mechanical strength. The temperature degradation of dielectric is designated by the general term aging. It has been shown experimentally that the life expectancy for paper insulation is a function of both temperature and time. The experiments also confirmed that the rate of deterioration of cable paper is measurable. It is insignificant at about 100°C but is appreciable at about 120°C. The maximum allowable temperature for an indefinite cable life would be of the order of the former value if other mechanisms of cable failure could be eliminated. In some circumstances it may be economically justifiable to load a cable for a planned life expectancy.

12.8.2 Dielectric Thermal Instability

Dielectric losses in a cable are dependent on the operating voltage and temperature. For a given voltage, the dielectric losses increase with a rise in temperature. The increase in dielectric losses causes the temperature of a buried cable to rise which further increases the dielectric losses. Under adverse conditions, the process continues causing a breakdown. This runaway condition is referred to as *thermal instability*.

12.8.3 Void Formation and Ionization

Void formation results because cables are made of different materials which have different coefficients of expansion. As a result of repeated expansions and contractions which occur during a daily load (heating-cooling) cycle, voids are created which become the sites for various discharges. In extruded cables, voids are created along the impurities. The discharges cause ionization and erosion and result in the gradual breakdown of the dielectric. The presence of voids with subsequent ionization is a chief cause for the deterioration of cable life.

12.9 Selection of Cables

In general, the factors to be considered for evaluating the suitability of a cable for a particular application are load, system voltage, cable insulation, short circuit rating, environmental condition, sheath and protective coverings, heat dissipation losses, etc., economic considerations and safety. The following points are important:

(i) *Maximum continuous current expected* While choosing the conductor size for a current rating, proper care should be taken of all the rating factors depending on actual conditions of installation.

(ii) *System voltage* The type of the system: Earthed or un-earthed? Is it solid grounding resistance/reactance grounding? The unearthed system will require full insulation from the core to the ground and the cable will be costly compared to the earthed system.

(iii) *Voltage drop* At full load, the voltage drop should be within permissible limits.

(iv) *Conditions of installation* Methods of installation, estimated thermal resistivity of soil, type of covering, type of armouring, the need, if any, for additional corrosion protection.

(v) *Expected short-circuit level of the system* On the basis of an expected short-circuit current and time of clearance, an appropriate conductor size for the cable may be selected.

(vi) Final selection is made on the basis of economic evaluation (see Section 3.5).

Distribution cable losses constitute conductor I^2R_{ac} losses. Dielectric losses ($2\pi fCV^2 \tan \delta$) are generally negligible but in case of PVC and EPR cables of 6.6, 11 kV rating, the losses are significant (see Section 12.2.1). The total losses are evaluated as:

Annual cost of conductor losses = Loss factor × Losses at maximum continuous peak current × 8760 × Cost/kWh

The dielectric losses may be added as some percentage of conductor losses for cables having significant losses. Two methods for economic evaluation are:

(a) Present Worth Method [2]:

This method recognises that the power utility must recover all costs of a cable facility over its lifetime. This is the initial installed cost plus the present value of the sum of the annual costs of losses. The present worth of lifetime losses is determined by multiplying the present value factor (P_v) to the total annual costs of losses.

Present value factor is derived by the formula:

$$P_v = \frac{(1+i)^n - 1}{i(1+i)^n}$$

where i is the discount or interest rate, n is the life of the cable in years. For example, if i is 10 per cent and n is 40, then $P_v = 9.779$.

The cable having the lowest Net Present Value (NPV) i.e. lowest installed cost plus present value of lifetime annual losses costs, will be selected.

(b) Capital Recovery Method

A series of annual charges is determined for the life of the cable by applying a certain capital discount rate (typically 12%). The annual charges are interest, depreciation, insurance charges, taxes, rate of return plus annual cost of losses, and annual operation and maintenance charges. The sum of these annual charges is called the *annual cost of capital.* The stream of annual costs over the lifetime is discounted by the capital recovery factor (typically calculated for 40 years of cable life) to arrive at the capital cost. The cable having the lowest capital cost is selected.

12.10 System Fault Location

12.10.1 Faults

If a fault occurs in a underground cable, it becomes essential that it is located as quickly and accurately as possible. Accuracy is necessary if excessive trenching work is to be avoided. The type of fault most likely to occur is single conductor fault-to-earth or to the protective metal sheath. In multi-core cables, the fault current is likely to give rise to excessive local heating at the fault, causing further breakdown of insulation and extending the fault to the remaining conductors. Open circuits may occur occasionally, which will usually be at the cable joints.

Various tests have been developed to locate either type of fault and most of them involve the application of a low dc test voltage. However, it should be noted that in some cases the fault may only be apparent when a high dc voltage is applied since a low voltage will give a meaningless result. In general, higher the test voltage applied, the more sensitive the test. The whole testing equipment can be accommodated in mobile van—*cable fault locating and testing van.*

12.10.2 Fault Identification

Prior to locating a fault, it is necessary to determine the nature of fault so as to make a better choice of the method to be used for fault location.

1. First, isolate the faulty cable and *test* each core of the cable for earth fault. One terminal of the insulation tester is earthed and every conductor of the cable is, in turn, touched with the other terminal. If the insulation resistance tester indicates zero resistance during any measurement, conductor-to-earth fault for that particular conductor is confirmed.
2. Then check the insulation resistance between the conductors. In case it is a short-circuit fault, the insulation resistance tester will indicate zero resistance.
3. After this, short and earth the three conductors of the cable at one end. Check the resistance between the conductors and earth and

between individual conductors (at the other end) to check open circuit fault.

4. In case there is any fault, the insulation test of individual cores with sheath or armour and between the cores is essential. The test should also be done by reversing the polarity of the insulation resistance tester. In case of any difference in readings, the presence of moisture in the cable insulation is confirmed. The moisture in the cable forms a voltage cell between the lead sheath and conductor because of the difference in the conductivity of these metals and the impregnating compound forms an organic acid when water enters it.

12.10.3 Fault Localization Methods

The following methods [6] are applied for fault location:

- Murray loop test
- Fall of potential test
- dc charge and discharge test
- Induction test
- Impulse wave echo test
- Time domain reflectometry test

The first four tests are conventional and are used for underground or control cables and the fifth and sixth tests can be used for both insulated cables and overhead lines. The details of these tests are given below.

Murray Loop Test

This method can be used for both low and high resistance faults in the following circumstances:

(a) When there is a fault in either one or two conductors, the third conductor in the cable remaining unaffected.

(b) When all the three phases are faulty, provided a core of adjacent cable is used for measurement.

(c) When three conductors are faulted, if the contact resistance of the conductors at the fault differs from each other by a factor more than 500.

(d) When the contact resistance does not exceed 5,000 ohms if working with a low voltage bridge and 1.5 megaohms if working with a high voltage bridge.

The Murray loop test is the most common and accurate method of locating faults and should be made use of whenever circumstances permit. It can precisely locate the fault if the fault current is more than 10 mA, e.g., if the battery voltage is 100 volts, then the fault resistance may be of the order of 10 kΩ. The sensitivity depends on the detector used in the test circuit. A high gain electronic dc amplifier can be used for a higher degree of sensitivity of the instrument. In its simplest form, the faulty conductor is looped to a sound conductor of the same cross-sectional area and a slide wire or resistance box with two sets of coils are connected across the open ends of the loop. A galvanometer is also joined across the open end of the loop and a battery or a dc hand generator supplies the current for the test. Balance is obtained by adjusting the slide or resistance. The fault position is given by the formula:

$$\text{Distance to fault} = \left(\frac{a}{a+b}\right) \text{loop length.}$$

where,

$a =$ length of the bridge arm joined to the faulty core

$b =$ length of the bridge arm joined to the sound core and loop length is equal to twice the route length.

In the case of high resistance faults, either the battery should be substituted by a high voltage dc supply obtained through a rectifier set or the galvanometer should be replaced by a sensitive electronic detector. When a sound return of the same cross section as the faulty conductor is not available, modified forms of the Murray loop test are adopted, such as (i) Murray loop test from each end; (ii) Murray loop test from the end using Fisher connections; and (iii) Murray loop test using overlap modification. The details of these tests are usually available in textbook.

Fall of Potential Test

In this test, the principle involved depends upon the measurement of the voltage drop on the cable conductor when a current is flowing through it. The only essential instruments consist of an accumulator, rheostat, ammeter and low range moving coil voltmeter. The measurement gives the voltage drop up to the fault and by comparing the voltage measurements made from each end, the position of the fault can be readily calculated. There are many different circuit arrangements, but the accuracy is not as high as that of the Murray loop test.

Charge and Discharge Tests

This method is used for broken cores with high resistance to earth. The usual method of measuring the charge capacity is to charge the cable under test to certain dc voltage for about 15 seconds and then discharge it through a moving coil galvanometer, the point to which the needle kicks being noted. For the purpose of locating breaks, it is usually sufficient to measure the relative values of capacity from each end of the broken core. To avoid false readings, it is necessary to earth all the broken cores at the far end except the one under test at the test point.

If C_1 and C_2 are the observed capacities at each end and l the length of the cable, then the distance d of the fault from the end giving the reading C_1 is:

$$d = \left(\frac{C_1}{C_1 + C_2}\right) l$$

Induction Method

The induction method can be used for the location of faults-to-earth in the case of a cable having no metallic sheath or armouring. It has the advantages of indicating the exact position of the fault without the necessity for any calculation or assumption regarding the cable resistance. The cable is supplied with intermittent pulses of current derived from a dc source and an interrupter as shown in Fig. 12.8. The cable route is then explored with a search coil connected to a telephone receiver, this coil taking the form of about 200 turns of fine gauge wire

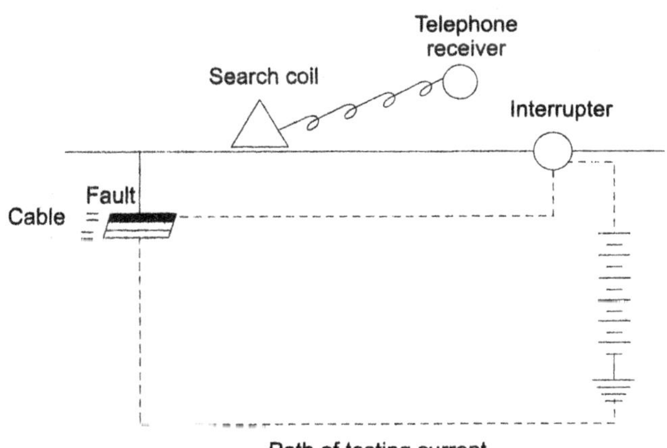

Fig. 12.8 *Induction method*

wound to form a triangle of about 1 m side. The coil is held near the ground with its plane parallel to the run of the cable as shown in the figure. Until the fault is reached, the cable will carry pulses of current and the intermittent nature of the magnetic field set-up will induce an intermittent emf in the coil producing a note which can be heard on the earphones. As soon as the fault is passed, the cable will carry no current and the note will cease. This technique may also be used to trace faults in buried cables, the precise route of which is not recorded. But as mentioned above, the presence of a metallic sheath will effectively screen the cable and nothing will be heard on the telephone.

Impulse Wave Echo Test

This method is based on the principle that a pulse propagating along a cable will be reflected when it meets with an impedance mismatch. This effect can be shown visually in a cathode ray tube. The pulse propagation velocity is inversely proportional to the square root of the dielectric constant of the cable. For a cable of uniform dielectric, the pulse reflected at the mismatch is displayed on a cathode ray tube at a time delay directly proportional to the distance of the mismatch from the test; irrespective of the conductor size. The fault position is given by

$$X = \frac{t_1}{t_2} \times \text{route length}$$

where,

X = distance from test end

t_1 = pulse time to fault

t_2 = pulse time to far end of cable.

This is the quickest and universally applicable method.

Nowadays portable digital fault locators are available embodying the impulse wave echo technique. It consists of a unit having a crystal controlled digital timing technique thus giving increased simplicity of operation and accuracy of more than 1–2 per cent of the range. Fault distances are displayed in metres on 5 digit LCD, eliminating the need for zero or scale setting. Pulse travelling wave reflection/transmission is shown on oscilloscope screen of the metre (Fig. 12.9). Such instruments can locate the fault at any distance between one metre and 25 km, with the ability to operate either from mains or nickle cadmium battery. This equipment can be used for LT/HT power cables, and control cables and overhead lines.

Time Domain Reflectometry (TDR) Test

The Impulse Wave Echo Test is a satisfactory method for single-cable runs. TDR is most accurate for the automatic cable fault location method for low-voltage distribution cable network [9] involving multiple tee-offs. Cable network impedance consists of resistance, inductance and capacitance. Reflections are caused by impedance changes. TDR can measure reflections caused by series and shunt impedances up to several hundred ohms.

Choice of method-general considerations

With the cable isolated from the energized system and free of safety grounds, lightning arresters, transformers, etc. measure the resistance on both the source and load ends with 1000 V megger and record the phase-to-neutral, phase-to-phase and phase-to-shield observations. If nearly zero readings are measured in any of these tests, use a volt/amp-ohm-metre method to determine the resistances using the lowest applicable scale. If the insulation test indicates a very high resistance to ground, high voltage fault (fault resistance greater than cable surge impedance of 10–100 Ω), burning equipment is used to burn the fault high resistance before using this test. This burn out produces stable low transition resistance by carbonizing the insulating material at fault. The burn out unit provides controlled current output at high voltages with the help of kenotrons, a high frequency generator and a transformer. The testing staff must rely on the accuracy of instruments, bridges and their ability to locate faults correctly. The information volunteered by others about the fault condition should always be checked.

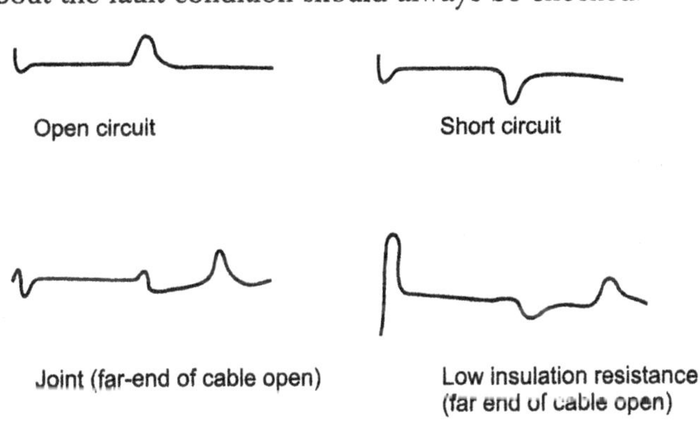

Fig. 12.9 *Oscillograms for cable faults*

12.10.4 Route Tracing

Self-contained instruments are available for tracing the routes and depth of hidden or buried wires and cables. The location of underground cables is based on the principle of the concentric electromagnetic field surrounding a current carrying conductor. To identify and locate a cable, a predetermined frequency current from a generator is transmitted along the cable. The resulting magnetic field is then explored by means of an inductive probe or detector rod with the integral search coil and receiver, which can be equipped to give both audio and visual signals. The transmitter (generator) can be omitted if design considerations enable the receiver to locate a cable carrying a power frequency supply.

PROBLEMS

1. A 1.1 kV, PVC cable has a thermal rating of 200 A at a permissible maximum conductor temperature of 70°C. If the over-current protection operates at 250 A, explain how this protection limits the cable rating and to what value.
2. A submarine cable is 350 km long with a conductor diameter of 0.5 cm having a 0.5 cm thick gutta-percha covering. Calculate the total capacitance of the cable taking the relative permittivity (ε_r) as 4.
3. Would you identify the distribution system as overhead line or underground which has.
   (i) greater X/R ratio
   (ii) greater value of capacitance,
   (iii) greater value of inductance,
   (iv) greater value of voltage regulation, and
   (v) higher effect of harmonics?
4. Why are HV cables laid deep into the ground? What precautions are taken while laying a cable through trench passing through the following locations:
   (i) road crossing,
   (ii) railway crossing,
   (iii) crossing of water mains, and
   (iv) crossing of bridge?
5. A 1.1 kV, two core cable, 50 Hz, has a capacitance of 380 nF/km, in a 2 km length of cable. How much capacitive reactive power is generated?

BIBLIOGRAPHY

1. "Loss evaluation for underground transmission and distribution cable systems", Task Group 7/39, Cost of losses, *IEEE Transac. on Power Delivery*, vol. 5, no. 4, November 1990, pp. 1652–1659.
2. Swift Louise, "*Mathematics and Statistics for Business and Finance*", Macmillan, London, 1997, p. 427.
3. IS: 1255–1983, Code of practice for installation and maintenance of power cables up to and including 33 kV.
4. IS: 1554 (Part I and II), Specifications for PVC insulated cables for working voltages up to and including 1.1 kV and from 3.3 kV up to and including 11 kV.
5. Cable Rating Concepts, *County Council Plant Rating Manual*, Part-3, Appendix-1, Sydney, December, 1971.
6. Easley, J.H., "Circuit design is key to reliability" *Electrical World*, May, 1978, pp. 52–53.
7. "Polymer solutions to contaminated environments", *Tips and News*, vol. 3, no. 2, April, 1997, pp. 4–6.
8. Bartlett A.D. and others, "Cost Reduction in Power Cable Systems Using Lean Cable Technology" *Proceedings of the 15th International Conference on Electricity Distribution*, CIRED, June 1999, Session I, CIRED, Nice, pp. 87–94.
9. Soraghan, J.J., and others, "Automatic Location for Underground Low Voltage Distribution Networks", *IEEE Transactions on Power Delivery*, vol. 16, no. 2, April, 2001, pp. 346–351.
10. Bozic Z., and E. Hobson, "Urban underground network expansion planning", *IEE Proceedings on Generation, Transmission, Distribution*, vol. 14, no. 2, March, 1997, pp. 118–124.

Chapter 13

System Overvoltages

13.1 Causes of Overvoltages

Internal Causes

Switching, faults and resonance are the main causes of internal overvoltages of a system. Theoretically, switching surge voltages may touch 4 p.u. but in actual system conditions, these voltages do not rise beyond 2.5 p.u. Studies reveal that statistically these overvoltages are normally distributed.

Under sustained phase-earth fault conditions, temporary power frequency overvoltages arise. Phase voltage of healthy phases at fault or arrester location rise from 80 to over 100 per cent of line-to-line voltage depending upon the earthing arrangement of the system. For effectively earthed systems, in single-phase to-ground-fault, the voltage rise of healthy phase (phase-to-neutral) is up to 80 per cent of the line-to-line voltage. In case of non-effectively earthing system the voltage rise is up to 90 per cent and for isolated neutral system it is up to 100 per cent or more of line-to-line voltage. These percentages are known as coefficient of earthing of a system and depend upon the system phase sequence resistances and reactances as shown in Fig. 13.1 [1].

Resonance overvoltages may come up to 3–4 p.u. and have been discussed in Chapter 4.

External Causes

These are due to atmospheric conditions of lightning discharges—interclouds, intra-clouds and cloud to ground/line. These may be in the

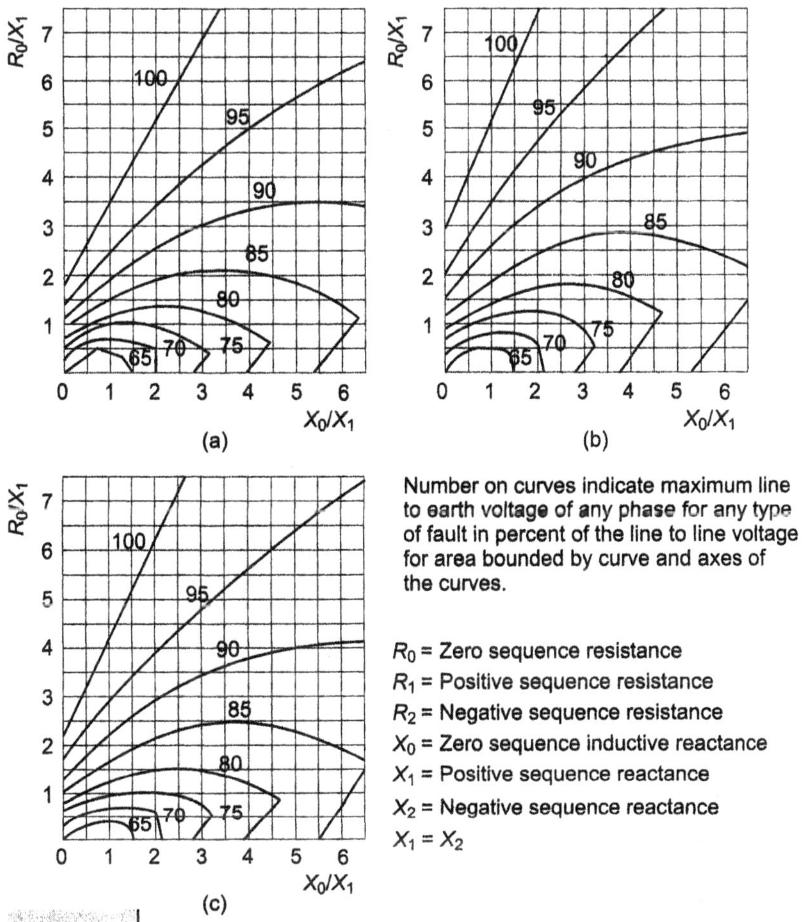

Fig. 13.1 *Maximum line-to-earth voltage at fault point for the grounded neutral system under any fault conditions:*
(a) Voltage conditions neglecting $R_1 = R_2 = 0$
(b) Voltage conditions for $R_1 = R_2 = 0.2\ X_1$
(c) Voltage conditions for $R_1 = R_2 = 0.1\ X_1$

form of direct lightning strokes, and indirect lightning, i.e., induced strokes.

The latter type of lightning mechanism is generally prevalent in the distribution systems. According to studies undertaken by F. Popolansky lightning currents/voltages is statistically log normal (distributed). The protection against these voltages is provided by lightning arresters. Sustained faults and switching overvoltages do not pose any problem in the distribution system as adequate power frequency short duration insulation strength as per IS is provided for the equipment at the design stage to look after these surges. Lightning overvoltages insulation protection is provided by lightning arresters and/or spark gaps. As per IS: 2165–1977 the power frequency and lightning impulse strength are stipulated as given in Table 13.1, which are the guaranteed minimum strength of the equipment/system under worst working conditions.

Table 13.1

Withstand test voltages

Normal voltage of the system kV	Highest system voltage kV	Power frequency strength kV rms	Impulse strength 1.2/50 μs as kV peak
0.415	0.450	2.5	5
11	12	28	75
22	24	50	125
33	36	70	170
66	72.5	140	325
132	145	275	650
		230	550
		185	450

Our knowledge of overvoltages and overvoltage protective devices has now increased and there is an awareness to reduce the insulation levels of the equipment for economical reasons without loss of reliability. Such reduced Basic Insulation Levels (BIL) are explained in Section 14.10. In a distribution system the predominant overvoltage faults are due to lightning, which are discussed in the following sections.

13.2 Lightning

13.2.1 Global Lightning System (8)

The phenomenon of lightning is now generally accepted to be a means of keeping in balance the 'global electric system'. The global electric system consists mainly of the lower ionosphere (that layer of atmosphere 50 to 75 km above the ground) and the earth surface forming a capacitor with the air between them acting as an imperfect dielectric. The lower ionosphere and the earth surface are, relatively speaking, highly conductive. It is estimated that the potential difference between these capacitor terminals is more or less steady at about 300 kV, the earth being negative and the ionosphere positive. The charge density at the earth's surface is around 1.1×10^{-9} C/m^2. The current transfer between the earth's surface and ionosphere is around 1500 A in fair weather. The current density at the earth's surface is estimated to be 3×10^{-12} A/m^2. These are average values and have been arrived at using measurements of air conductivity and potential gradients. The steady electric field at the earth's surface is about 3 V/cm. It is easy to infer from the potential difference and current values that the 'resistance' of the bulk of the atmosphere between the lower ionosphere and earth is 200 ohms. This steady leakage of this 'global capacitor' would lead to a fall in the potential difference between the plates. Lightning is nature's device of restoring the potential difference of the global capacitor. On the average, therefore, lightning must cause a charge retransfer to maintain the potential difference at about 300 kV. Measurements have shown that active lightning involves on the average a charge of 20 C over 10 seconds which means a current of 2 A. Therefore, to balance the leakage of 1500 A, there must be about 700 to 800 active thunderstorms every instant. Data indicate that about 2000 thunderstorms take place continually at any moment. Majority of lightning discharges occur between cloud to cloud. Worldwide isocernaunic level varies from 200 maximum at the Equator and decreasing to zero at the Poles.

In view of the above, it must be appreciated that a lightning discharge must transfer negative charges to the earth. This is mentioned here to emphasize the fallacy that a cloud existing between the plates of a capacitor must be charged as per the laws of induction. Thus, the lower end of the cloud will be negatively charged, which is also the polarity of the earth surface. The lightning current would, therefore, be negative in polarity. If, however, events were to take place which increased the potential difference between the earth and ionosphere, then thunderstorm activity may be expected to result in (i) the lower end of the cloud becoming positively charged and (ii) lightning current becoming positive. Positive lightning currents have also been measured but are relatively rare.

13.2.2 Mechanism and Damage from a Lightning Flashover

The principal mechanism of lightning flashover on HV, EHV and UHV lines are the shielding failure and the backflash events due to direct strokes. For the lower high voltage and distribution lines, the induced voltage accompanying strokes close to the line predominantly contribute to lightning overvoltages. Studies reveal that about 80 per cent of lightning overvoltages are caused by induced overvoltages on distribution lines. The induced overvoltages cause small arrester currents, and are of positive polarity.

Strike up to a kilometre away [5] can cause transient overvoltages or surges on main power, data communication, signal and telephone lines. They can cause data loss and corruption in computers and equipment damage.

On 11 kV lines lightning induced voltage in excess of 100 kV (BIL) have been recorded. The induced overvoltages can be calculated by the formula:

$$V = \frac{20\, hkI_0}{x}$$

where, V = voltage induced (kVP)
h = height of phase conductor above the ground (m)
k = 1.2
x = distance from phase conductor to striking point (m)
I_0 = stroke current (kA peak)

Studies indicate that lightning is responsible for about one-third of all faults on HV and distribution systems during storm days. Roughly 75-80 per cent of these lightning faults are of a transient nature and lines can be reenergized on reclosing the breaker.

Lightning damages to the distribution system are a serious problem to many utility-systems and account for the majority of consumer outages causing the highest expense in breakdown of distribution equipment. Inadequate lightning protection results in faults on distribution system that may cause through-fault failures of sub-station transformers. Pole-mounted distribution transformer failures caused by lightning have also been a longstanding problem on most of the systems. The economic justification for protection is tied to the cost of repairs which in turn is a function of the forced outage rate or the rate of lightning severity. The other considerations are consumer inconvenience and the loss in revenue due to outages.

The relation between thunderstorm days and lightning flash density is given in Table 13.2 as per British Standards BS: 6651–1992. Thunderstorms days per year at a place can be known from meteorological department.

Table 13.2

Relationship between thunderstorm days per year and lightning flashes per km^2 per year

Thunderstorm days per year	Flashes per km^2 per year: Mean	Limits
5	0.2	0.1 to 0.5
10	0.5	0.15 to 1
20	1.1	0.3 to 3
30	1.9	0.6 to 5
40	2.8	0.8 to 8
50	3.7	1.2 to 10
60	4.7	1.8 to 12
80	6.9	3 to 17
100	9.2	4 to 20

13.3 Protective Devices

The devices generally used for overvoltages protection are spark gaps and lightning arresters. The general voltage/time characteristics of these devices and the equipment insulation are given in Fig. 13.2 for power frequency or impulse voltage withstand voltage. A lightning arrester has characteristics similar to that of the equipment insulation. A spark gap has nearly hyperbolic characteristics.

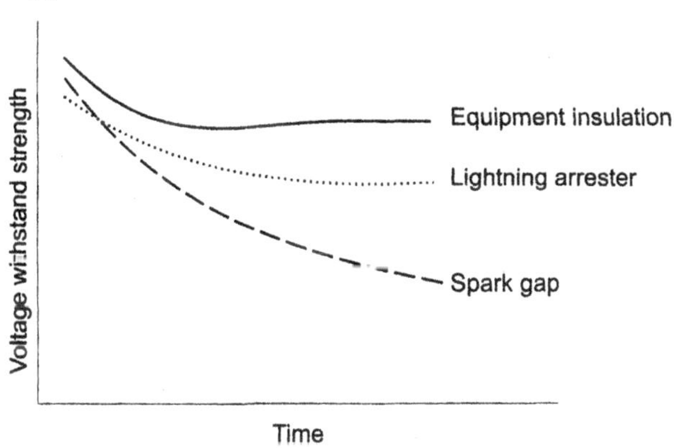

Fig. 13.2 *General power frequency or impulse voltage/time characteristics of the equipment insulation, lightning arrester and spark gap*

13.3.1 Spark Gap

The spark gap is a surge protective device which consists of an open air gap between an energized electrode and an earthed electrode. In case of a double-gap electrode system, one electrode may be un-energized and unearthed. A spark gap limits the overvoltage by sparkover. Hence the critical sparkover voltage is an index of the protection level. Every sparkover is accompanied by a power-follow current which could be interrupted by de-energization. This constitutes short circuit and currents cause electrodynamic stresses in apparatus. The sparkover voltage increases very steeply for higher rates of rise in the applied voltage. The non-uniformity of the electric field in the gap results in higher sparkover voltages for fast rising voltages of negative polarity than for the positive. The sudden collapse (chopping) of the voltage may cause additional voltage stresses in apparatus insulation.

This type of protection is comparatively more suitable for lines or terminal equipment having shielding earth wire in approach spans to attenuate the steep lightning waves or for regions having low lightning severity. The provision of rod gaps may be as back-up protection in case primary protection lightning arresters fail. Wherever possible, the gap horns (rod type generally) are desirable to be mounted horizontally or inclined at an angle of 45°. Gap horns in the vertical or in any other position unduly expose the insulators to arc effects. On power transformer bushing, the gaps provision is not preferred due to birdage fault problem.

Clearance and Gap Adjustment

The strength of an air gap in simple terms should be the dielectric strength of air/unit length multiplied by the gap length. This simple relation does not hold good as the varying factors are: gap electrode configuration, wave shape, polarity effect, gap length, i.e. smaller or medium or large gap. Compromise on these points has to be made to evolve certain formulation for gap setting. Some aspects in this respect are discussed below.

1. The minimum clearance depends on the configuration of the gap horns which are generally made of galvanized steel. A disposition of horns such that the arc tends to be blown away from the support is desirable. A value of about 0.75 d (d being the gap length)

between the arc and surface of the porcelain supporting the gap, is generally satisfactory for lower voltage. This decreases to 0.3 for higher voltages.

2. The rod/rod gap flashover voltage (generally quoted as the 50 per cent probability value with 100 per cent and 0 per cent values differing by about ± 15 per cent) is proportional to the gap distance to a power of about 0.9 and if the gap is vertical, it may be up to about 30 per cent greater for a negative surge than for a positive one.
3. A minimum ratio 1.25 of the withstand level U_w of the apparatus to be protected to the level U_i of the gap is assumed. In general, this ratio has been accepted for satisfactory protection provided the overhead shielding wires extend to some distance from the station to exclude the occurrence of very steep-fronted waves. The adjustment of the gap is to some extent empirical. A low value may be chosen first and increased if necessary within the limits consistent with satisfactory protection. The facility of adjustment of gap to local conditions is one of its characteristics. Table 13.3 gives some typical gap lengths for the electrode configuration shown in Fig. 13.3 as per IS: 3716–1978 and IEC Publication 71A.

Table 13.3

Typical distance between rods used for protective gaps

Highest of voltage equipment kV	*Impulse withstand level of apparatus to be protected kV peak*	*Configuration dimension cm*	*Distance d*	
			Normal cm	*Maximum cm*
12	75	(a) h = 30, 2r = 11, g = 6	4	6
	95	(e) R = 9, S = 3.2	2 × 3.2 (double gap)	—
24	125	(a) h = 30, 2r = 11, g = 6	9	13
		(d) r_1 = 18, r_2 = 12	10	11.5
		(b) h = 20, 2r = 15.5, k = 5	7	9
36	170	(b) h = 20, 2r = 15.5, k = 5	12	15
		(d) r_1 = 25, r_2 = 13	14	16.5
72.5	325	(a) h = 110, 2r = 35, g = 20	28	36
		(c) r_1 = 38, r_2 = 23	30	34

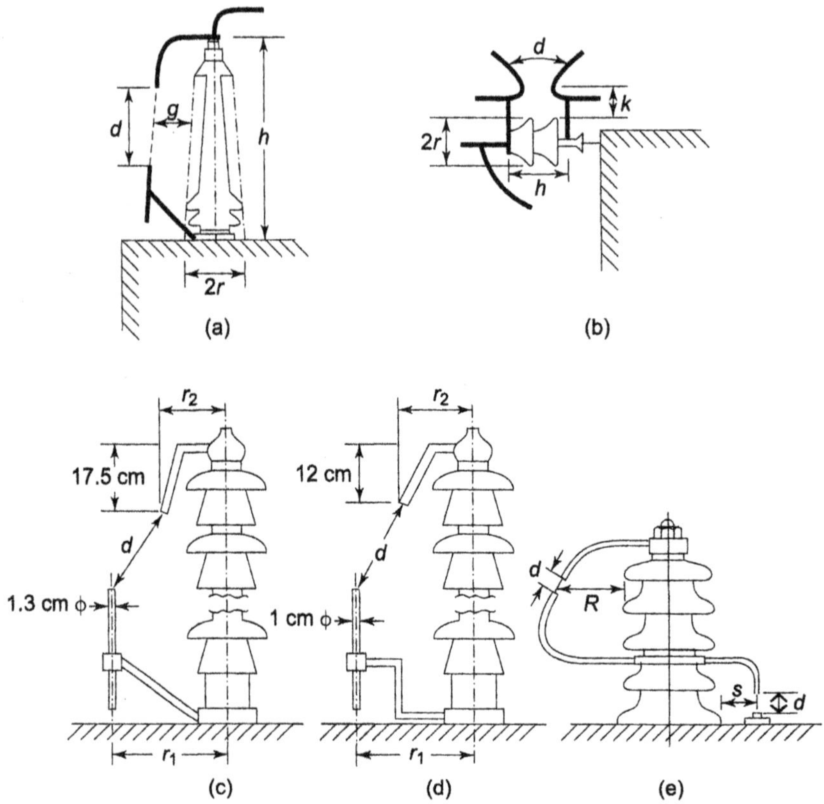

Fig. 13.3 *Configuration of protective arcing gaps at bushing of:*
(a) Indoor distribution transformer
(b) Isolating switch
(c) Power transformer
(d) Power transformer
(e) Distribution transformer

4. Equipment manufacturers generally specify the gap length adjustment for a particular sparkover gap which is based on the actual testing at factory works on the complete equipment or arcing gap in parallel with the bushing or insulator. Gap adjustment is based on establishing BIL of external insulation by a suitable margin of protection. The shape of the horn on the line side may be designed to control the distribution of the electric field with regard to the corona on high voltage equipment such as 66 and 132 kV ratings.

13.3.2 Lightning Arrester

Nonlinear Resistor type

The lightning arrester basically consists of a set of spark gaps in series with silicon carbide Nonlinear resistor (NLR) elements (see IS: 3070 (Part-I)-1985). Under normal system voltage conditions, the spark gaps are nonconducting and isolate the HT conductors from ground. However, whenever an overvoltage of a magnitude dangerous to the insulation of the apparatus protected occurs, the spark gaps breakdown and allow a current to flow to the earth through the NLR.

The volt-ampere characteristic of the NLR can be approximately described by the simple expression $V = KI^{\beta}$, where K and β are dependent on the composition and manufacturing process of the NLR. The value of β lies generally in the range of 0.3 to 0.45 for modern arresters. It may be easily verified that the current would increase approximately ten times when the voltage across the NLR doubles.

Therefore, with multiple spark gaps, arresters can withstand high Rates of Rise of Recovery Voltage (RRRV). The non-uniform voltage distribution between the gaps presents a problem. To overcome this, capacitors and nonlinear resistors are connected in parallel across each gap. In a higher voltage range, capacitor and nonlinear resistors are connected across the stock of gaps and NLRs (see Fig. 13.4). With steep

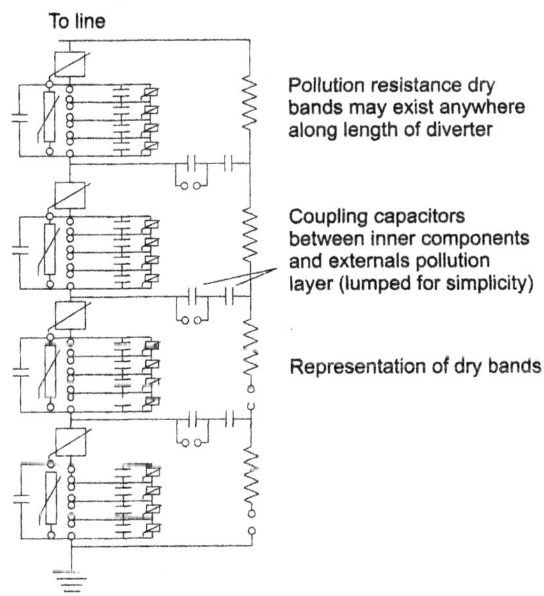

Fig. 13.4 *Equivalent circuit for polluted arrester*

wave surges the voltage is mainly controlled by capacitors and at power frequency by nonlinear resistors. It is obvious that when the overvoltages cause a breakdown of the series gaps, the current would be so high as to make the overvoltage subside very fast. The highest voltage that would appear across the lightning arrester would be either the sparkover voltage or the voltage developed across the NLR during the flow of surge currents. The lowest sparkover voltage of the arrester is called the *hundred per cent impulse sparkover voltage* of the arrester. The voltage developed across the NLR during the flow of surge current is called *residual voltage.* The lower the values for these two quantities, the better the degree of protection of the apparatus. However, these cannot be reduced indiscriminately as that would endanger the arrester itself for the following reasons.

First, the passage of the surge current through the arrester gaps leaves considerable ionization behind and the system voltage could drive a current through the diverter. This is known as *power follow current* or simply *follow current.* If the follow current persists, the NLRs get heated up and as they have a negative temperature coefficient of resistance, they may get permanently damaged in a few tens of cycles resulting in arrester failure and a short circuit on the system. The follow current must therefore be "quenched" at the earliest. The first current zero is a convenient time to quench the follow current. The quenching of the follow current essentially means that the series spark gaps become insulating as far as the normal power frequency voltages are concerned. The magnitude of the current that can be quenched is thus dependent on the structure of the gap. Also, the NLR must be so designed as to limit the follow current well within the quenching ability of the series gap.

Secondly, as the sparkover voltage of the gap is reduced below certain levels, the arrester would be sparking over for overvoltages that can safely be withstood by the apparatus insulation. Therefore, this reduction would be superfluous and at the same time increase the duty on the NLRs which may consequently deteriorate fast. It is, therefore, necessary to fix a minimum power frequency sparkover level. In the initial stages of arrester-development, it was believed that if the sparkover value of the gaps is 1.5 times the rated voltage, the gap would be able to reseal against the flow of the follow current at the first zero.

As already mentioned, the hundred per cent impulse sparkover level and residual voltage determines the degree of protection afforded to the apparatus. Therefore, for an arrester of given ratings, the hundred per cent impulse sparkover voltage and residual voltage will have to be below a certain maximum level. The general practice is to have a 50 Hz sparkover voltage of 1.5 to 2.5 times the rating. The maximum limits of hundred per cent impulse sparkover voltage and residual voltage at rated current (impulse current) lie in the range of 3.5 (HV) to 5. The increased degree of protection becomes essential at a higher voltage to economize on the enormous amount of expensive insulation. At higher voltage, quenching the very high magnitude of follow currents is achieved by elongating the arc path magnetically. The magnetic field can be derived from either permanent magnets or by the follow current itself flowing through a coil. The magnetic field would be perpendicular to the arc path.

Metal Oxides Arresters

Silicon carbide nonlinear resistors material is not ideal. It is not nonlinear enough and thus imposes design restrictions. Also, its characteristics call for a large number of spark gaps. A new class of nonlinear material was introduced in 1968 by the Japanese Matsushita Electric Industrial Company. The new materials are basically ceramics formed from zinc oxide together with additions of other oxides. These materials have since become generally available for protection of circuits operating at a few kV up to transmission voltages. Because of the high degree of non-linearity, the material allows considerable simplification in arrester design and yield improved performance and long life.

In principle, the material could be operated in a so-called gapless mode, i.e. without ancillary spark gaps, as a fully solid state device. The reduction of residual voltages at high currents is the main advantage of these arresters as shown in Table 13.4.

Table 13.4

Protection levels of NLR, ZnO arresters [IS: 3070]

System voltage (kV)	BIL (kV)	Arrester rating		Residual voltage (Protection level) (kVP)	
		(kV)	(kA)	NLR	Zinc oxide
11	75	9	5	46	30
33	170	30	10	108	100

The advantage of a fully solid state device is that conduction occurs continuously so that rapid changes in power frequency current with consequent undesired transient voltages are avoided. Again, in principle, the inherent difficulties of spark gap time lags and of chemical changes resulting from discharges within enclosed volumes, can be avoided. For certain specialized arresters applications, such as the protection of capacitor banks, gapless arresters appear to offer substantial advantages. A difficulty in such application with active gap arresters is that it is necessary but difficult to cause the arc to move rapidly from the point of initial sparkover. There are at least possibilities that gapless arresters may be less vulnerable to the effects of external pollution than the orthodox ones. These arresters are manufactured with polymer or porcelain housing as per IS: 3070-Part 3–1993.

13.3.3 Rating of Arrester

The arresters are rated generally in terms of the normal discharge current, power frequency voltage.

Field studies indicate that only about 5 per cent of arrester current are higher than 10 kA and more than 70 per cent are less than 2 kA. These currents are represented in laboratory tests by 8×20 μs waves. The magnitude of lightning overvoltages on lines is dependent on the height of the line: whether the line runs in the open space or is shielded by other tall grounded objects like trees, buildings, etc. The higher the level of the line structures and conductors and more open the lines the greater the likely magnitudes of the lightning overvoltages. Correspondingly, the lightning currents through the arresters are also dependent on the line voltage. The standardized ratings for lightning currents, given in Table 13.5, have been in practice for a long time. These have been arrived at laboratory estimates. These impulse currents are mostly responsible for the initiation of power follow currents. Therefore, the power frequency voltage rating (maximum system voltage $\times$ coefficient of the earthing of system) and the impulse current rating together form the major specifications for lightning arresters.

Table 13.5

Impulse current and voltage ratings

System	Impulse current 8 × 20 μs rating
(a) 230 V (400 V solidly grounded system) to 580 V representing domestic lines	1500 A and 2500 A
(b) Distribution networks (400 V to 33 kV)	5000 A
(c) Sub-station (station class)	10,000 A

Standard ratings and characteristics of arresters are given in IS: 3070.

Current Rating Considerations

The ground flash density roughly varies by a factor of ten between various regions in the world. It seems reasonable, therefore, that arresters of a higher current class should be used in high flash density areas rather than in the areas with low flash density, if no other means such as spark gaps, earthed cross-arms or ground wires are used to reduce the arrester current.

The arrester current has been observed to be higher on distribution lines with insulated cross-arms than on lines with earthed cross-arms by a factor of 2.4 to 3.0.

The higher the insulation level of the distribution line (within the practical range of BIL), the higher the arrester current. The insulation level has a great influence on the arrester current especially in the 20–50 per cent probability range.

13.4 Failure of Lightning Arresters

13.4.1 NLR Arrester

There are many possible causes of failure of NLR arresters but the predominant ones are: defective sealing, rough handling during transport, excessive stress caused by temporary overvoltages, lightning surges, faulty earthing, thermal ageing of NLR, and malfunction caused by pollution on the housing of the arrester.

High water content is an example of a defect which arises during manufacture. It arises either because of inadequate drying or because of defective seals. Rough handling before and during installation can result in damage to internal components and displacement of gap elements. The problems associated with pollution are much more complex. These arise principally because external contamination produces conductive regions which in effect act as electrodes on the outside of the housing. Such regions are connected by a capacitance to the internal components, and in the case of multi-unit surge arresters, are bodily connected to the intermediate flanges. The current carried by the external contamination causes drying-out and consequently dry bands are formed which become the seat of discharges.

A complex voltage distribution thus appears on the outside of the housing which changes with time and produces two classes of undesirable interaction. Radial electric fields can be set up between the inner assembly and external pollution layer. These fields can be quite large, even with a continuous conductive layer on the exterior. Very large fields can indeed exist once dry bands are found. The sparkover of dry bands generates high frequency currents which also are fed into the interior through the capacitance.

The effect of these radial fields is to produce discharges which can damage the internal components directly, and indirectly, by the formation of unwanted chemical products. These may lead to the failure of the device and corrosion of its components.

Another effect of exterior contamination is to modify the equivalent circuit and thus reduce the sparkover voltage at which the device operates. Figure 13.3 shows that the typical equivalent circuit of 132 or 220 kV class of a polluted surge diverter is complex, particularly because of the nonlinearity of the resistors and spark gaps.

Reduction in Sparkover Voltages

A defect in surge arresters is a reduction in the sparkover voltage to a value sufficient to permit continuous passage of large currents from the networks. These rapidly cause a build-up in the internal pressure which can lead to an explosion within a few cycles. Pressure relief diaphragm are fitted to accommodate this eventuality, but they cannot totally eliminate the hazards of damage. Research at Central Electrical Research Laboratories, U.K. has shown that arrester for 132 kV and high systems are prone to large reduction in sparkover voltage, even from quite modest value of external pollution.

Pollution inside the arrester such as silica dust and moisture in the gap reduces breakdown voltages. This increases with the number of operations the arrester has to make and may even lead to the passage of continuous current. Polymer insulator housed arrester survive better under contaminated conditions [5].

13.4.2 Comparative Reliability

A careful study [9] of analysing the causes of failures of 250 silicon carbide (NLR) arresters for distribution systems was carried out in United States. The causes of failures were found:

Moisture ingress	85.6%
Surges	5.9%
Contamination	4.5%
Misapplication	2.5%
Unknown	1.5%

There is little doubt that the frequency of failures for the reasons given varies from system to system because of differences in the frequency of lightning storms, application practices and a number of other factors including the quality of arrester designs in service; however, the results of most studies appear to fit a similar pattern. The failure causes can be grouped roughly into three categories:

(i) Moisture leakage and contamination.
(ii) Overvoltages including switching and resonance overvoltages.
(iii) Surges of excessive magnitude and duration.

According to other studies, moisture leakage and severe contamination have several times less effect on metal oxide distribution arresters than on silicon carbide arresters. Therefore, metal oxide distribution arresters can be considered more reliable than silicon carbide arresters from the important point of view of moisture leakage and pollution. Metal oxide arresters are more likely to fail as a result of system overvoltages because these conduct current in response to the overvoltages and for this reason, somewhat more care must be taken in application to match the magnitude and time duration of system overvoltage to the temporary overvoltage capability of the arresters. Metal oxide arresters absorb more energy than silicon carbide arresters on lightning discharges because their discharge voltage is a substantially constant function of current. On account of this, the probability of failure may be high in areas of high lightning intensity.

The advantages offered by metal oxide arresters in terms of superior reliability (when properly applied) should tend to favour their selection as the means of surge protection over gapped silicon carbide arresters.

13.4.3 Life Duration

The arrester's life is considered 25-30 operations of full rated discharges. It may be hundred to few thousands operations at lower discharges. Generally, it has been observed that arrester's life ranges from 15 to 30 years. The detections tests regarding malfunction of arresters are, therefore, recommended to be carried out on arresters after 15 years in service.

13.5 Detection of Arrester Failures in the Field

A portable testing kit has been developed by CPRI, Bangalore for detecting malfunctioning or failure of arresters up to 33 kV voltage and non-portable ones up to 220 kV system arresters.

Tests:

(i) Failure can be detected by using the portable kit to measure the impulse frequency flashover voltages of the arrester. Also a lower power frequency sparkover voltage will indicate that the arrester is defective. The kit consists of a 30–300 kV condenser and an air-cored pulse transformer. The impulse waveform is produced by discharging, the capacitor through a pulse transformer, the output of which is connected to the arrester. When the arrester discharges, a voltage develops across the high current shunt and this is applied to the indicating metre.

(ii) *Leakage current:* Leakage current is a good symptom of the condition of the arrester. High leakage current indicates the internal deterioration in the arrester and thus increases the temperature. A temperature rise of about 10–20°C can be detected by infra-red thermography or infra-red thermometer remote sensing. Also, the leakage current can be measured by a test where the power frequency voltage is gradually applied up to 90–95 per cent of the rms rating of the test unit. The measured leakage current under this voltage, when exceeding 2 mA for NLR arresters and exceeding 1 mA for ZnO arresters, indicates a faulty arrester. In the field, the arrester (33 kV and above) are now commonly installed with the surge monitor which measures the leakage current (grading as well as external) besides the registration of lightning strokes. The test button-on this monitor indicates the leakage current of the arrester and thus its condition.

(iii) *Insulation resistance:* An arrester with a reasonably large leakage current will show a lower insulation value when tested with a 2500-V megger between the line and earth terminals. Regular testing of arresters for insulation value then helps to sort out arresters with a high leakage current. The indicative is comparative degrading of insulation resistance in periodical tests. This occurs when non-linear resistors become defective.

13.6 Allowable Maximum Distance of Separation

13.6.1 Indirect Strokes

When the arresters are installed exactly on the terminals of equipment, lightning voltage impressed is residual voltage of the arresters. When arresters are physically separated from the equipment to be protected, additional voltage comes on the equipment. The increase in the distance between the arrester and protected equipment decreases the protection to the equipment. The increased voltage rise is mainly due to lead inductance. As a simple example, let us consider 10 kA, 8/20 μs current wave of the lightning stroke.

Lead or conductor inductance = 1.25 μH/m

$$\text{Voltage drop or rise/m} = L\frac{di}{dt}$$

$$L = 1.25\ \mu\text{H/m}$$

$$\frac{di}{dt} = \frac{10{,}000}{8 \times 10^{-6}} = 1.25 \times 10^{9}\ \text{A/s}$$

$$L \times \frac{di}{dt} = 1.25 \times 10^{-6} \times 1.25 \times 10^{9}\ \text{V/m}$$

$$= 1560\ \text{V/m}$$

Now considering 100 kA high current stroke of 4/10 μs wave, we have

$$\frac{di}{dt} = \frac{100{,}000}{4 \times 10^{-6}} = 25 \times 10^{9} \text{ A/s}$$

or

$$L \times \frac{di}{dt} = 1.25 \times 10^{-6} \times 25 \times 10^{9} \text{ V/m}$$
$$= 31.25 \text{ kV/m}$$

The surge impedance of the shunt path through the arrester to ground has an adverse effect as a result of the voltage drop in these leads. This shunt path includes, in series, the line lead to arrester and the ground lead from arrester to ground electrodes/mat in addition to the impedance of the arrester and of earth around grounding electrodes/mat.

As a rough guide, Table 13.6 gives the allowable maximum permissible distance in metres for station type arresters.

Table 13.6

Maximum permissible arrester's distance

Nominal system kV	*BIL kVP*	*Arrester rating*		
		80%	*90%*	*100%*
		Separation distance m		
11	75	12	—	—
33	200	20	17	15
	170	15	—	—
66	350	35	30	25
	325	30	—	—
132	550	35	—	—
	650	60	45	35
220	900	60	—	—
	1050	100	75	55

13.6.2 Direct Strokes

In the case of shielding failure, the function of the arrester is quite limited for direct strokes.

EXAMPLE: Consider a 10 kA lightning stroke of 4/10 μs wave, on an 11-kV line with 100-kV insulation level and 400-ohm surge impedance.

Voltage build-up (sparkover) at the point of strokes is given by

$$V = 1/2\ ZI$$

where

Z = surge impedance of line

I = stroke current (crest)

Hence surge voltage = 400 × 10 × 1/2 = 2000 kV peak.

This voltage of 2000 kV peak rising to the peak of value in 4 μs gives the rate of rise of 2000/4 = 500 kV/μs. The insulation level is reached in 100/500 = 0.2 μs. If the line is not to flash over, the time taken for travelling and consequently reflected wave from the lightning arrester should be limited to 0.2 μs and the distance of the arrester location from this point of stroke should be 0.2 × 300/2 = 30 m. Here 300 is the speed of the light wave in m/μs. (The actual speed of the travelling wave will be 5–10 per cent less because of line resistance attenuation.) Hence a lightning arrester of 10 kA rating even if located at a distance of 30 m from the equipment to be protected, would protect the same from a direct stroke of 10 kA. In such a case, where the lightning strikes, the phase conductor very near the lightning arrester, most of the stroke current will be discharged through the arrester. This is the theoretical value, but in actual practice the separation distance is kept much less due to the aforesaid adverse effects of leads, arrester impedance, etc.

13.7 Travelling Waves

The travelling wave is attenuated by corona loss, skin effects and distorted by reflections. The indirect strokes on equipment/power conductors are due to a stroke to the earth in the vicinity of line or a bound charge on the conductor due to a charged cloud above the conductor.

The amplitude of the travelling wave along a power conductor to the equipment connected to the overhead line depends upon the: (i) impulse withstand level of the insulators, (ii) type of supports (steel or wooden or RCC) and (iii) type of earthing adopted.

When a wave of crest value e travelling along a line of surge impedance Z_1 reaches a circuit of a different surge impedance Z_2, a part of the wave is reflected and a part transmitted.

The transmitted or incident wave at the junction is

$$e_i = \left(\frac{2Z_2}{Z_1 + Z_2}\right) e = ae$$

The reflected wave is

$$e_r = \left(\frac{Z_2 - Z_1}{Z_2 + Z_1}\right) e = be$$

where a is the coefficient of transmission and b the coefficient of reflection. From the above relations it follows that

(a) At the open end of the line, Z_2 is infinity, and the voltage will be $2e$.

(b) The high voltage winding of the transformer has a surge impedance of several times the surge impedance of the line and hence amplitude of the travelling surge on the transformer winding case for practical purposes is assumed to be $2e$. [6]

In addition to the reflected wave phenomenon, there may exist still higher peak voltage at the apparatus as a result of oscillations caused by the inductance of the line between the arrester and apparatus and the capacitance of the apparatus. Furthermore, the voltage drops in the leads from the line to arrester and from the arrester to ground which are effected by the rate rise of the surge current, add to the drop across the arrester. Any difference in the ground potential between the apparatus ground and arrester ground also adds to the voltage across the apparatus insulation. In view of this, the arrester lead as well as ground lead should as far as possible be kept short and ground of the arrester and apparatus be interconnected and low in resistance.

Another factor which affects the total voltage imposed on the equipment is the presence of multiple tee-offs or lines (incoming/outgoing) on the station bus. From the theory of the travelling wave, the voltages in such a case will be:

$$V_t = V_f\left(\frac{2Z}{n-1}\right) \Big/ \left(Z + \frac{Z}{n-1}\right) = \frac{2V_t}{n}$$

where,

V_t = voltage of the transmitted wave
V_f = voltage of the incidence forward wave
n = number of circuits of same impedance and more than 2
Z = surge impedance of single circuit

EXAMPLE: An overhead 11 kV feeder is connected to main 66/11 kV substation bus through an underground cable. Overhead line inductance is

0.96 mH/km and capacitance is 0.0075 μF/km. The cable inductance is 0.20 mH/km and capacitance is 0.25 μF/km. If a lightning surge of 100 kVP travels along the line towards the sub-station, estimate the voltage surge and surge current reaching the bus.

Solution: Surge transmitted (e_t) will be:

$$e_i = \frac{(2Z_2)e}{(Z_1 + Z_2)}$$

$$e = 100 \text{ kVP}$$

$$Z_1 = \sqrt{\frac{L}{C}} = \sqrt{\frac{0.96 \times 10^{-3}}{0.0075 \times 10^{-6}}}$$

$$= 358\ \Omega \text{ for overhead line}$$

$$Z_2 = \sqrt{\frac{L}{C}} = \sqrt{\frac{0.20 \times 10^{-3}}{0.25 \times 10^{-6}}}$$

$$= 28.3\ \Omega \text{ for cable}$$

$$e_t = \frac{2 \times 28.3}{358 + 28.3} \times 100$$

$$= 14.65 \text{ kVP}$$

$$\text{Cable current} = I = \frac{2e}{Z_1 + Z_2}$$

$$= \frac{2 \times 100000}{358 + 28.3}$$

$$= 517.7 \text{ A (crest)}$$

13.8 Protection Schemes

13.8.1 Design

The lightning protection design scheme for distribution circuits is basically arrived at by applying the proven line protective parameters based on:

(a) The acceptable outage rate for the purposed line
(b) The expected number of strokes to the line
(c) Knowledge of electrical characteristics of the line structures when they are subjected to the expected lightning surges.

The acceptable outage can be determined from the present outage rate. Few utilities have this information. The number of strokes to the line is determined by the area shielded by the line multiplied by the stroke density. For example, a line having a shielded area of 0.21 km^2 and stroke density of 10 strokes/km^2 would receive 2.1 strokes. The rule of thumb approach to determine the shielded area is to make the width of the shielded area four times the height of the line. This width is multiplied by the length of the line less any portion of the line that is shielded by other taller structures or trees. Further, accurate data on stroke densities to electric lines are not available. The few studies that have been done in this area are with limited data and only in widely scattered locations. The response of distribution lines to lightning surges need to be determined. A lightning surge is defined by three quantities—rate of rise of the wave front, crest value of the wave and time of half crest value for the wave (decay time).

13.8.2 Rotating Machines

Rotating machines require special attention because their insulation is about half of the power transformer of the same voltage rating. This is also for reasons of economy and performance. Typical insulation values (lightning impulse) of rotating machines are given in Table 13.7.

Table 13.7

Typical insulation values of rotating machines

Insulating class kV	*Rotating machine*	*Transformer*
11	35	75
15	50	95

In addition, for rotating machines with multi-turn coils, insulation between the turns is important. The turn-to-turn insulation in a transformer can be made to sufficiently withstand the turn-to-turn stresses that occur. Also, it is practical to introduce a favourable voltage distribution in the transformer windings. This is not practical in rotating machines.

The lower the rate of rise of voltage, the lower the turn-to-turn stress. Therefore, to limit the stress between turns is to limit the rate of rise of the surge voltage permitted to reach the machine terminals. This is achieved by shunting the arrester with a capacitor. The normal practice is to slope the voltage wave so that the time required to reach maximum voltage is 10 μs or more. Generally capacitors of the following values are used.

(a) 0.5 μF for 0.4–9.0 kV system.
(b) 0.25 μF for 11.5–13.8 kV system.

Normally 11 kV, three-phase system will have a minimum inductance of 50 μH. Hence the capacitance of 0.25 μF may be used. The time to peak rise is given as

$$\text{Time} = \pi\sqrt{LC} = \pi\sqrt{50 \times 0.25}$$
$$\simeq 10\ \mu s$$

13.8.3 Sub-stations

In some power utilities, arresters are installed only at or near the terminals of the transformers. These are provided to take care of the equipment covered in the separation distance allowed. In case, a circuit breaker is in an open condition and the arrester becomes isolated from some of the equipment in the sub-station, the arcing horns on the approaching line structures as well as at the sub-station equipment will protect the equipment from various surges. These arcing horns are set at 90–100 per cent of the value of the BIL of line, except on the transformer's bushings where it is set for 80–85 per cent of the value of the BIL of the transformer. If the line is without an earth wire, then on the approach spans of the line (1–1.5 km) it is essential to provide an earth wire to attenuate the incoming impulse wave.

In some organizations, there is a practice to install lightning arresters at the terminals of power transformers and also at the terminal towers of the incoming and outgoing lines. In all schemes, separate earth wires (with possibly the shortest line and ground leads) should be provided for each of the lightning arresters as per IE Rule 92 and IS: 3043.

The outdoor sub-stations are protected by properly positioned earth wires (called shielding wires) on the tops of the substation structures or column masts against direct lightning strokes. Protective angle (angle formed by a slanting line and the vertical) not exceeding 20° is accepted as complete safe protection and angle 20°–30° is considered average safe. If it is not feasible to run earth wires over the sub-station, it is necessary to erect masts or rods over the stations' structure columns so that the buses and equipment in the stations will fall within the protection cone of mast, which is considered to have base radius equal to twice the height of the mast or rod plus vertical column.

13.8.4 Lines

The protection of overhead power lines of voltage level 132 kV and above, against direct lightning strokes, is done by use of properly positioned earth wires on the tower/pole tops. The protection angle afforded by an earth wire is defined as, the angle between a vertical line through the earth wire and slanting line connecting the earth wire and outermost phase conductor. For good shielding, this angle should not exceed 30°. But a protective angle of 45° is satisfactory when the tower is on a hill side. In the areas where the isocrenoic level is less than 100, unshielded construction of lines up to voltage level of 132 kV is economical. The shorter structures are made possible by construction without overhead shielding wire. However approaching spans of 1 to 1.5 km from the approaching sub-station are shielded by earth wires, to reduce the magnitude of any travelling lightning wave entering the sub-station (as stated in Sec. 13.8.3).

13.9 Pole-Mounted Distribution Transformers

The predominant mechanism of lightning incidence of distribution network is through induction. The impulse level of 11 kV and LT equipment is as follows:

(a) 11 kV lines: 100 kV peak.

(b) Distribution transformer: 11/0.433 kV: 75 kVP, 95 kVP (11 kV winding)

75 kVP is reduced insulation level generally used. Therefore lightning arresters are necessary at the 11 kV terminals of the distribution transformer.

(c) LT (0.433 kV) equipment/lines: Generally of the order of 5 kVP.

13.9.1 LT Lines

The distribution transformer impulse level is generally reduced and to protect this insulation level, lightning arresters are desirable at the 11 kV terminals of the transformers for lightning protection. However, arresters on LT side/LT line may not be required because of the comparatively remote chances of lightning outages on the LT side on account of the following factors.

(a) The height of the LT line is less as compared to the HT line and, therefore, there is a lesser probability of building strokes.
(b) The LT lines pass through the populated areas where these are shielded by buildings, trees, etc.
(c) The continuous neutral earth wire or separate earth wire on LT lines attenuates the lightning magnitudes across the line insulators because of the capacitive coupling effect of the earth wire and also flattens the waveform, i.e. decreases the rate of the voltage rise.
(d) On the LT side, as there are comparatively a higher number of line tee-offs, the lightning stroke currents get distributed. Therefore, the strokes are diluted.
(e) The severe lightning strokes may be transferred on the HT side (roughly in the voltage ratio). The HT side arrester can help discharge the high voltage surge in that case, or these strokes can cause flashover across LT insulators without causing any damage. Recent studies have revealed that about 85 per cent of lightning strokes flashovers do not cause outage damage to the line and equipment.

However low voltage lines are more susceptible to lightning overvoltages due to their lower insulation level. LT arresters at the transformer LT side/LT line may, however, be desirable for wide open terrains, cross-country such as deserts in Rajasthan. A need for them may also arise in areas having high lightning activity, such as in some parts of Kerala, Assam where it is of the order of more than 100 thunderstorm days per year.

13.10 Cable Junction

Arresters are desirable at cable junction points on overhead primary distribution lines emanating from substations. At these junction points, in the absence of arresters, the lightning travelling waves can create abnormal overvoltages due to reflection because of changes in the surge impedance [8].

13.11 Lightning Arresters: Earthing

All provisions for lightning protection may fail if detailed consideration is not given to earthing. IE Rule 92 recommends separate earthing for LA on HV and LV system components. With this method of connecting lightning arresters, the entire voltage drop in the arrester discharge path is impressed across the parallel path from the primary line through the transformer to the grounded secondary neutral. In the event of the surge voltage drop in the arrester discharge path exceeding the impulse insulation strength of the possible parallel path through the transformer, a flashover from the bushings or a breakdown of winding insulation may occur. Cases have been noticed where the damage to the distribution transformer because of lightning have occurred in spite of the presence of lightning arresters [6].

The practice followed in Australia (as per stipulation of the Electricity Authority of New South Wales) with regard to the erection and connection on 11 kV/433–250 V transformer is that for lightning protection HV surge arresters should be fitted and located as far above the HV bushings as possible with necessary precautions as regards to the use of an appropriate size of fuse on the HV side. The earthing connection on the lightning arrester is then taken to the transformer tank earth connection and from there to the HV earthing electrode. The neutral of the LV side is separately earthed with the LV earthing electrode where separate HV and LV earth systems are used. With this arrangement, surge disturbances become shunted to the earth via the LV neutral earth connection as well as the lightning arrester earth connection. A coordinating neutral spark gap is also provided as close as possible to the neutral as shown in Fig. 13.5a. The gap discharging should be the surge potential of the tank which reaches a certain predetermined value. The LV neutral earth circuit should have as low a resistance as possible. Experience shows that these connections are quite safe and effective so far as consumers fed from the LV network are concerned.

In some systems [2] it has been seen that in order to provide maximum protection to the transformer against lightning, all earthing terminals (i.e., lightning arrester, neutral and transformer tank, etc.) would be inter-connected. Thus the grounding terminal of the lightning arrester is connected to the transformer tank and to the grounding conductor of the neutral of secondary of transformer as shown in Fig. 13.5b. A neutral coordinating gap may be placed for distribution transformers in urban or industrial areas where stringent conditions of supply are desirable. In rural areas where lines are scattered and longer, the gap may not be desirable. The surges transmitted to the LV side are attenuated and should apparently not adversely affect the equipment and consumers' premises. By such interconnection, the transformer tank

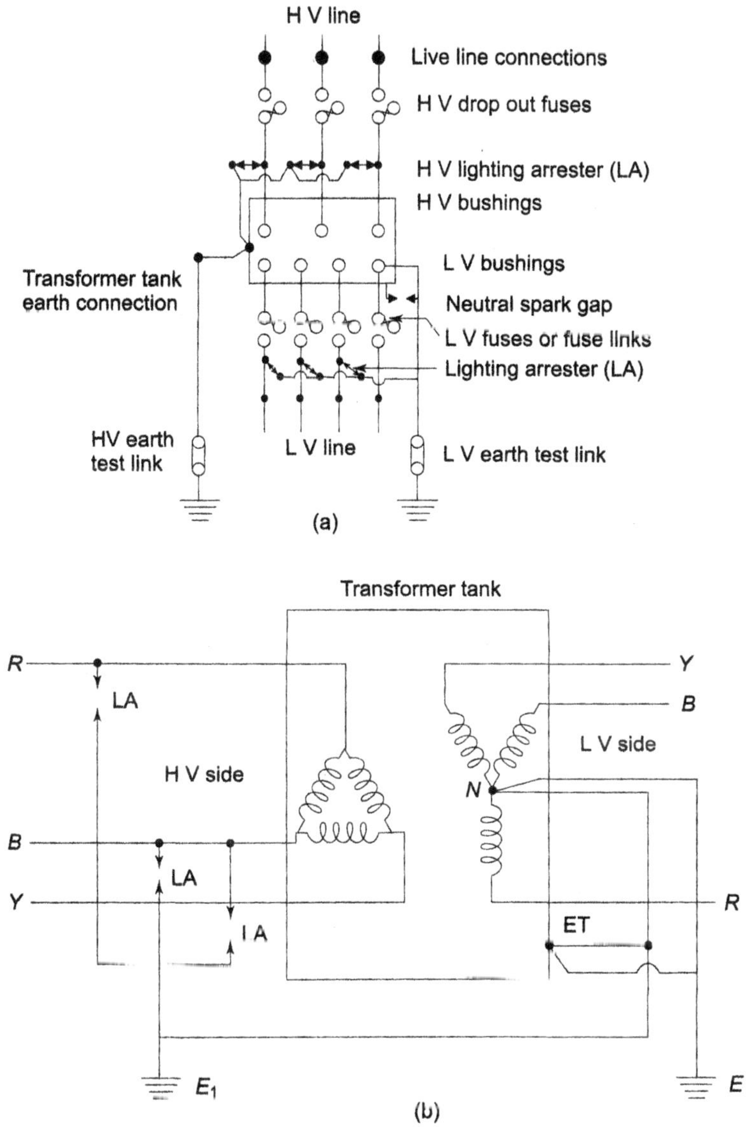

Fig. 13.5 *Grounding of lightning arresters at the transformer (a) Arresters on both sides; (b) Arresters on HV side*

and windings will be limited to the potential rise at the ground terminal of the lightning arresters. The latter method is fairly simple and can be applied in Indian conditions.

Shielding Earth Wires: For tower lines above 132 kV and outdoor sub-stations, properly positioned earth wires on tower tops and sub-station structure masts will protect these from direct lightning strokes. Protective angle (angle formed by a line through earth wire and outer phase conductor or equipment and the vertical) not exceeding 20° is accepted as complete safe protection and angle 20°–30° is considered average safe.

PROBLEMS

1. What is the basic difference of protection provided by spark gap and lightning arrester? How is the basic insulation level of equipment and lightning arrester residual voltage related? Apply this relation to zinc oxide lightning arrester protection of a 33 kV power transformer having a basic insulation level of 170 kV peak.
2. What is the predominant mechanism of occurring of lightning over-voltages on distribution lines? A 100 kA lightning stroke strikes one km away from a 11 kV line. Calculate the voltage induced in the line.
3. What are the internal causes of over-voltages in the distribution system? What are the minimum withstand values of these voltages for a 11/0.415 kV distribution transformer?
4. At a sub-station, the insulation level of the power transformer is kept at its lowest. Explain why and how it is kept at the lowest.
5. At a power transformer, a lightning arrester is installed with a line and ground lead conductor (inductance 1.5 micro H/m) of a length of 10 m. The lightning arrester has a residual voltage of 108 kV peak. Find the total lightning voltage impressed on the transformer when a lightning stroke of 10 kA, 8/20 μs occurs.

BIBLIOGRAPHY

1. (i) IS: 3716–1978.
   (ii) International Electrotechnical Commission Publication: 71–A. Application guide for insulation co-ordination equipment located in the exposed situations.
2. Rajatte Y., J. Fortin, B. Cyr, 'Lightning Overvoltages on LV Networks Fed by MV lines with Multigrounded Neutral, *15th International Conference on Electricity Distribution*, CIRED, Nice, June, 1999, pp. 103–109.
3. Lightning Protection, '*Asian Electricity*', *The Journal for Power Management*, September, 1995, pp. 68–73.

4. Beaty Wayne, H., "Researchers gather lightning data", *Electrical World*, McGraw-Hill, August 1, 1978, pp. 52–53.
5. 'Polymer Solutions to Contaminated Environments', *Tips and News*, vol. 3, no. 2, April, 1997, pp. 4–6.
6. Heathcote Martin J. The J & P Transformer Book, Newness, 12th Ed. London, 1998, pp. 523–539.
7. Golde, R.H. *Lightning Protection*, Edward Arnold, London, 1977.
8. Marshall, J.L., *Lightning Protection*, John Wiley and Sons, 1973.
9. Sakshaug, E.C., J.J. Burke and J.S. Kresge, 'Metal Oxide Arresters on Distribution Systems—Fundamental considerations', *IEEE Transactions on Power Delivery*, vol. 4, no. 4, October, 1989, pp. 2076–2089.

Chapter 14

Rural Supply

Rural electrification has two primary objectives. First, to improve the economic status of the rural population by increasing the productivity of human and animal labour and secondly, to promote rural welfare by providing an environment equal in comfort and convenience to that enjoyed in urban areas. Thus electrification is an important factor in preventing the drift of the rural population to urban areas, where higher wages and more comforts are available. The reliable supply to rural areas will help to reduce slum areas in cities. The World Bank Report India [9] indicates that in India, rural consumers appreciate lower prices, they also value and are prepared to pay for greater reliability of supply. As per the Census 2001, 72 per cent of the population of the country as a whole, lives in rural areas. The proportion of the rural population (in percentage) in the total population in various states as given below, indicates the quantum of rural electrification in the following states:

State	*Percentage of the rural population*
Tamil Nadu	56
Maharashtra	58
Gujarat	63
Karnataka	66
Punjab	66
Haryana	71
West Bengal	72
Andhra Pradesh	73
Madhya Pradesh	73

(Contd.)

(Contd.)

Kerala	74
Uttaranchal	74
Rajasthan	77
Jharkhand	78
Uttar Pradesh	79
Chhattisgarh	80
Orissa	85
Bihar	90

India is a predominantly agricultural country and has the highest number of electricity-run-motor-pumpsets as compared to other countries in the World.

The ultimate goal of rural electrification is:

- To electrify the remaining 80,000 villages in the country;
- To strive for better service in the villages already electrified;
  - (i) To provide continuous 24-hour electricity supply to villages and rural industry;
  - (ii) To provide minimum 8–10 hour continuous reliable electricity supply for daily agriculture though the agriculture load factor in any state will not exceed 15–20 per cent;
  - (iii) To provide electricity for all rural houses as nearly 70 per cent of the households in rural India are still without electricity; and
  - (iv) To improve voltage profile and remove voltage fluctuations on the worst feeders.
- To bring efficiency in the use of electricity which is very low at present and is not more than 15 per cent overall in rural India.
- To design low tariff for agriculture as high tariff in agriculture is not sustainable.

The State Regulatory Commissions must pay special attention towards proper rural electrification. They may insist on:

- Introducing metred supply for all consumers as this step will promote energy conservation, increase the reliability of supply and help in the detection of power theft;
- Keeping prescribed standards of electricity service for villages and for agriculture;
- Promoting dispersed generation from local renewable resources;

- Providing technical and managerial training;
- Administering subsidies judiciously; an innovative approach is required, the subsidy can be in the form of capital grants as one time payment thus reduced tariffs but they should cover only energy charges and not demand charges;
- Accrediting of service providers for specific equipment such as metres, energy audit, consumer complaint service providers, etc.;
- Ensuring that funds are allocated efficiently for new schemes and linked to well-defined performance criteria by power utility. Performance targets can be: demand (power and energy) growth, reducing distribution losses (technical and commercial), improving revenue collection, installing solar PV in houses, etc.

14.1 Rural System

About 35 per cent of the total power generated in India is consumed by over 70 per cent of the country's rural population. Concerted endeavour must be made to raise this consumption to 50 per cent of the total generation, which can serve as a good index of India's rural prosperity. Power engineers can play a greater role in rural India by devising economical and reliable power supply systems for integrated and balanced socio-economic developments of villages to make their economy self-generating. Small power cottage industries using local raw materials and manpower must be planned and agro-industries having the local market must be made progressive. At present, about 70 per cent of the population work on agriculture. It is imperative to reduce the pressure on agriculture by switching over to the industrial sectors in rural areas.

The present construction practice of supply system for rural electrification is given in Section 14.9. It is essential to provide low cost technology for rural electrification such as low loss distribution transformers without tape changer and breather; outdoor sub-stations up to 66 kV without a control room building; 11 kV gear: auto-reclosure (vacuum or SF_6 or air-break) with ac series trip with kiosk type for outdoor use, etc.

Tap-off sub-stations should be erected in the rural areas having sufficient load growth and have EHV lines of 33, 66, 132 kV passing by. The cost of tap-off sub-station is much less.

While rural electrification is an important input for rural development the fact remains that it is a financial burden on the electricity boards/companies and needs to be heavily subsidized. For example in the country, the average cost of power supply for agriculture at the LT consumer terminal during 2000–01 was 212 paise per unit, while the average tariff was 28.5 paise per unit [11]. The State Electricity Boards (SEBs) during 2000–01 incurred losses in supplying subsidized power to agriculture consumers. It is necessary to keep the cost of rural electrification low, so as to afford lower power tariff by establishing generating plants near load centres in rural areas. For this, micro/mini hydel power plants should be established on hill streams and canal falls, which have an estimated potential of 10000 MW in the country. Agro-waste based mini-thermal plants be set up. Solar and wind farms be developed. For example, Punjab has already commissioned some micro-hydel scheme on canal falls. Solar units for rural water pumping [3] are becoming economical. The small gas turbine generating plants can be installed, where small amounts of natural gas is available, such as in Tamil Nadu, Andhra Pradesh, Arunachal Pradesh, Gujarat, Assam, etc. The small schemes must have inbuilt economy.

According to available data, there are 5.87 lakhs of inhabitated villages in India of which 5.08 lakhs (86.53%) had been electrified by March, 2001. The rural electrification that is still to be accomplished is mostly in difficult terrains such as hills etc. Also fact remains that about 70 per cent of households in rural India still not electrified. To fulfil this task utmost economy in design and the use of correct construction material are required.

Vested interests have created a menace in rural areas by such methods as installation of spurious name plates on induction motors; fictitious LT capacitors; fuse replacements of the distribution transformers by the consumers themselves and theft of energy directly from distributors or service mains.

In general, the rural consumer is not well conversant with the use of electricity and electrical devices. It is advisable that the supply undertakings themselves provide viable consultancy/extension services to rural consumers in the selection/use of electric motors, capacitors, lighting and other electrical devices. They should be educated in conservation of energy measures, such as the use of efficient devices and of alternative energy resources such as solar energy, gobar gas plants, wind power, etc. and animal power as well. For example, an enormous amount of electrical energy is wasted in minor irrigation due to the capacity of the pumpsets not matching the irrigation needs and the low system efficiency of motor pumpsets. Pumps which do not attain at least a 70 per cent efficiency and 4.5 m manometric suction lift at the best efficiency point should not be installed. It is important to communicate convincingly the desired consumer tolerance for various system outages (forced or scheduled) [8].

In rural areas long scattered 11 kV lines and heavy peaks with predominant agriculture loads lead to the usual problem of bad voltage regulation, lower power factor, heavy distribution losses and lower load factor (0.2 $\cong$ 0.4). The general performance of the induction motors commonly used for agriculture pumping and industrial loads are normally designed for a power factor of 0.8 or 0.85 at full load. But this factor falls appreciably if the motors operate on partial loads which in practice is often the case. There are many factors because of which motors run at partial loads. A major one is the tendency to install a motor of a higher capacity than the one needed under given conditions. This may be due to lack of proper guidance in the selection of the motor pumping set or a tendency of the farmer/artisan to go for higher capacity motors because of the fear that lower capacity motors may get burnt. Mostly the power factor of the rural loads without compensation is about 0.7. As the losses are inversely proportional to the square of the power factor, an increase in the power factor, say, from 0.7 to 0.85 (being the statutory value as stipulated in the Indian Electricity rules), reduces the losses by as much as 22.5 per cent.

During the power shortage/cut period, there is a tendency to use autostarters by tubewell consumers, specially during paddy season. When feeder is switched on from sub-station, one time start-run of consumer motors imposes a high in-rush current (about four to six times the motor ratings) on transformer, simultaneously with the transformer own switching in-rush current (~ eight times the transformer rating). The net instant overloading on distribution transformer (11/0.433 kV) can be 10 to 14 times, which can do mechanical shifting of the windings, as mechanical shock is proportional to square of peak current. This may lead to damage of the transformers.

The problem of low voltage conditions due to long and over-loaded line is predominant. More than 50 per cent consumers in the rural areas have unmeterd supply in the domestic, commercial and agriculture sector. Most of them are charged a flat rate on a point in the installation or getting free electricity. Free electricity is supplied to domestic consumers in Himachal Pradesh, Punjab, Tamil Nadu, Madhya Pradesh, J&K and agricultural consumers in Punjab, Tamil Nadu. Most consumers misuse this facility and use extra appliances and heaters than what is permissible. There are nearly 39 lakh single-point (Kutir-Jyoti programme) consumers in the country enjoying free supply or paying a nominal price. After privatization of power distribution in Orissa, BSESC is installing meters on top priority. Similarly, in Noida, NPCL is installing meters for rural consumers after convincing them of better supply when it is done.

14.1.1 Rural Distribution System of Other Countries

Australia

- High voltage: 11 kV and 22 kV, 3-phase, solidly earthed; 12.7 kV and 19.1 kV single wire earth return.
- Distribution transformer: Delta/Star. Low voltage systems: 415/240 V, 3-phase, 4 wire solidly earthed neutral; 240/480 V, Single-phase 2/3 wire.

China

- High voltage systems: 10 kV, 3-phase, 4 wire impedance earthed.
- Distribution transformer: Star/Star.
- Low voltage system: 380/220 V, 3-phase, 4 wire earthed neutral.

FIJI

- High voltage system: 11 kV, 3-phase, solidary earthed.
- Distribution transformer: Delta/Star.
- Low voltage system: 415/240 V, 3-phase, 4 wire solidly earthed neutral.

Hong Kong

- High voltage system: 11 kV, 3-phase, solidly earthed.
- Distribution transformer: Delta/Star.
- Low voltage system: 380/220 V, 3-phase, 4 wire solidly earthed neutral.

Indonesia

- High voltage system: 20 kV, 3-phase, resistance earthed. (One Province has 3-phase, 4 wire multiearthed 20 kV system with star/ star distribution transformer configuration).
- Distribution transformer: Delta/Star
- Low voltage system: 400/230 V, 3-phase, 4 wire solidly earthed.

New Zealand

- High voltage system: 11 kV, 3-phase solidly earthed, 11 kV single wire earth return.
- Distribution transformer: Delta/Star.
- Low voltage systems; 415/240 V, 3-phase, 4 wire solidly earthed neutral.

Philippines

- High voltage system: 13.2 kV, 3-phase, solidly earthed.
- Distribution transformer: Delta/Star.
- Low voltage systems; 415/240 V, 3-phase, 4 wire solidly earthed neutral.

Thailand

- High voltage system: 22 kV, 33 kV, 3-phase solidly earthed.
- Distribution transformer: Delta/Star.
- Low voltage systems; 400/230 V, 3-phase, 4 wire neutral solidly earthed, single-phase 2 and 3 wire.

For Rural Electrification, distributed stand alone generation systems are progressively increasing in most of the countries.

14.2 Reliability

As mentioned in Sec. 7.5 on distribution system limitations, reliability and quality of supply refers not only to the number of complete outages the system may suffer but also refers to any deviations from the design limits of voltage, frequency, waveform, current carrying capacity, voltage flicker and system impedance up to the point of utilization. Any of these may affect industrial or commercial consumers, but the most relevant ones as regards rural consumers are voltage deviations and outages. However, some industrial consumers who are occasionally supplied from a rural supply system can find outages a serious hazard. For instance, an interruption without warning to a quarry crusher, has been known to block it to an extent where it could only be freed after great effort or sometimes by blasting within the crusher itself. The loss of one phase of a three-phase supply, or large voltage fluctuations in the system can have an equally adverse effect.

The relative reliability of different types of the rural supply systems can be disputed in the absence of comparative data on the interruption-frequency and durations on systems with similar exposure. For example, in the case of lightning magnitude and frequency, it might be argued that SWER (single wire earth return) system is likely to suffer from reduced reliability because of maintenance problems in earthing and the added vulnerability of an isolating sub-station. It must be admitted at the same time that the SWER line itself must be more reliable than an equivalent single phase two-wire line, as it has half the conductor length and half the number of insulators exposed to faults. Also, the phase-to-earth flashover voltage on SWER line pole is much greater. The augmentation of a rural system by a duplicate feeder or sub-transmission can improve the reliability of the supply in two ways. First, the system is generally broken up into a number of smaller systems, thus reducing the exposure of each consumer to the chance of a fault. Secondly, in the initial stages after augmentation, and when the load is less, an alternative supply is usually available by manually switching back to the other system. The latter facility becomes less useful as the load continues to grow after augmentation and the system eventually almost reverts back to a radial system. It is difficult to see that full alternative supply or duplication of supply will ever be widely used in rural areas because of the high capital cost involved. The best way to improve reliability is to investigate the causes of interruptions and take suitable action to overcome them with improved protection.

It is worthwhile to limit the size of the distribution transformer and the length and number of feeders emanating from it to a manageable size. For example, in the Punjab State Electricity Board, a practice has been adopted to use a maximum size of 63 kVA, 11/0.433 kV distribution transformer for rural agricultural loads with LT feeders not more than three and, at the same time, the length of a feeder not exceeding 0.8 km.

14.3 Faults and Protection

Lightning

It causes phase-to-earth or phase-to-phase insulation flashover, followed by switchgear or drop-out fuse or fuse operation and sometimes pole or crossarm fires. It can also cause failure in distribution transformers, and occasionally, direct strikes can destroy poles and break conductors. Lightning overvoltage protection is important and the installation of rod gaps or lightning arresters on HT side of pole-mounted transformers is desirable.

Animal and Bird Life

These are generally a cause of nuisance to transformer poles where phase-to-earth and phase-to-phase clearances are smaller than on normal line poles. Birdguards on transformer poles and insulations using polythene tubing or bushing caps, where clearances are less, can prevent most of these problems. Sometimes large birds, such as swans, eagles, etc. may fly into line conductors, or a large number of small birds may perch on the top conductor (where phase-to-phase clearance is vertical) causing conductors to flashover and be damaged or broken by the resultant arc. The remedy is to increase interphase clearance or use 'V' or 'U' type *x*-arms (as per IS: 5613) for lines near bodies of waters such as ponds, river, canals, refuse dumping etc. or to avoid poletop designs which provide vertical clearance on such places, and animal burial grounds nearby or to mount suitable PVC/XLPE sleeves on the effected portion of conductors or to use aerial bunched cable (see sub-section 11.10).

Mechanical Failures

These can occur due to defective conductors, and aeolian vibrations and loading conditions not provided for in the design, falling of tree or tree branches on lines during storm days. Cross-arms and poles can fail for similar reasons. Failure of guys can occur due to constant "back scratching" by cattle or because of ground corrosion.

Man-made Faults

Vandalism, such as stoning, shooting at insulators, throwing of chains/ wires on conductors etc., can cause troubles resulting in permanent faults. Broken insulators may not be traced for weeks and can cause pole or crossarm fires in the case of wooden structures.

Sectionalizing Devices

These devices at suitable locations functionally improve the services reliability by isolating faults to a portion of a circuit and by reducing the fault locating time. These are of the type of auto-reclosers, fuse cutouts, circuit breakers, load switches and air switches with fuses.

The calculation of system fault currents at salient points on the system is essential though tedious, before automatic or nonautomatic switchgear, such as reclosers, sectionalizers, drop-out fuses and horn-gap fuses with air switches can be applied with any degree of confidence (see Sec. 17.3).

Fault Indicators

These can be installed on rural tee-off points and can help to reduce fault location times by patrolling party and enabling the restoration of supply.

Power Transformers Operation

Generally, fault frequency on rural feeder is more due to long overhead lines and their vulnerability to natural hazards. At main or grid sub-station in rural areas having two or more transformers, the transformers should not be run in parallel, so as to keep the fault level low. It will facilitate to reduce the fault strain on breakers, cables, lines, etc. and will enhance their lives and improve the continuity of supply.

Network Protection

Rural network being lengthy and scattered needs proper protection, proper fault indication, proper sectionalizing for speedy removal of fault and continuity of supply. In most of the electricity boards, rural distribution transformers, primary protection is provided with isolating switch with HT fuse and secondary protection is with LT fuse cut-out. The transformers are usually mishandled by consumers who put oversize-fuses in case fuse blows. This practice has led to more transformers damage. As a remedial measure, isolating GO switch on each transformer be dispensed with. The isolator switch be provided at line tee-off points or on spur along with HT expulsion fuses for group of transformers with LT protection by MCBs or proper fuse cut-outs for each transformer.

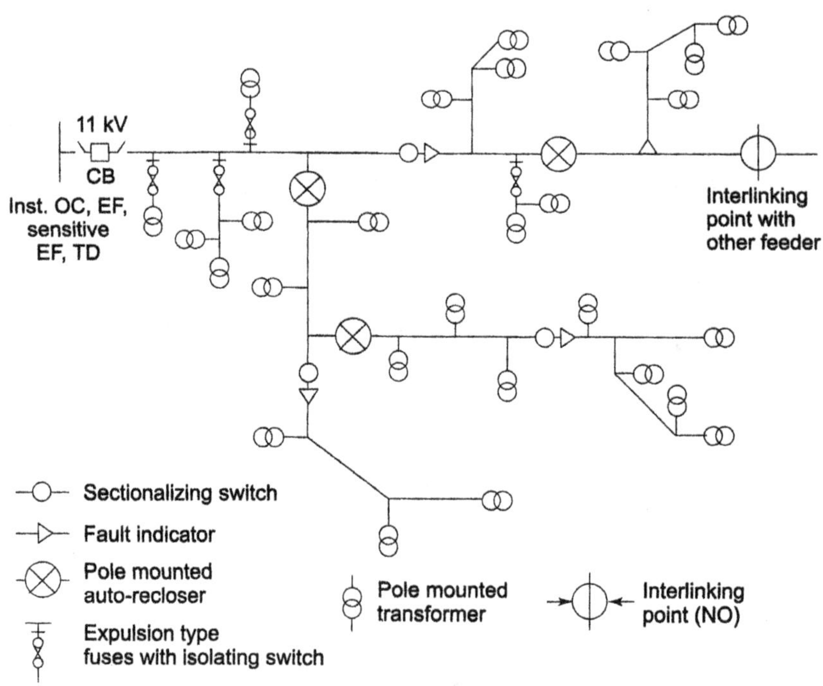

Fig. 14.1 *Typical primary rural feeder protection*

The important feature of protection should be:

- Close fault in case of lengthy feeder needs special attention.
- Snapping of conductors should be taken care, i.e. detection and elimination of low value earth faults.
- Proper sectionalizing points be provided to segregate faulty section and maintaining supply to remaining healthy part.
- Alternate supply link be provided in case main line or main sub-station supply becomes faulty.

Figure 14.1 shows a typical rural feeder protection. It is protected by a circuit breaker having both instantaneous and inverse time over current and earth fault relays to detect phase/phase and phase/earth faults. Instantaneous element is desirable for close heavy fault. These relays provide a coordinated protection scheme for the circuit as a whole, including spur lines from the main-line circuit, which are additionally protected by either pole-mounted HV fuses with isolating switch or autoreclosing breakers. Additional earth-fault protection is provided on the circuit breaker by a sensitive earth-fault relay capable of detecting low earth-fault currents and operating in conjunction with a definite time relays. To reduce the time taken to locate faulty section, particularly or lengthy circuits with fuses, use is made of pole mounted, *x*-arm fitted, fault current-passage indicators (see Fig. 14.3).

14.4 Improvement of Existing Distribution Systems

Any design for improving the system should at least cover the whole of sub-transmission and distribution in the given area and, therefore, comprise relevant levels of voltage, such as 132, 66, 33 and 11 kV besides LT network as a whole, sub-stations and distribution transformers including those initiated at consumer premises [see Sec. 4.12.2(8)].

14.4.1 Causes

Poor power factor, high distribution losses and poor voltage conditions in the rural system are generally because of the following factors:

(a) Long and overloaded 11 kV and sub-transmission lines.

(b) Low power factor of the agriculture tubewell consumer installations (prior to power factor corrections).

(c) Absence of shunt compensation in the sub-transmission and distribution system (required because of heavy inductive loads of agriculture).

(d) Location of distribution transformers centres away from the load centres.

14.4.2 Remedial Measures

The main consideration in voltage regulation which is kept within the usual stipulated limit as –9 per cent, +6 per cent for 11 kV and ±6 per cent for LT as prescribed by I.E. Rule 54. The corrective measures for this are:

(a) Rerouting the main link line of the feeder by the shortest route.

(b) If rerouting is not possible, then the feeder should be studied for bifurcation at suitable points with a view to transferring loads to lightly loaded feeders. Heavily loaded, long feeders should be rehabilitated as per the following World Bank guidelines to improve voltage profile, reduce losses and increase reliability of supply.

The trunk route conductor should be a minimum of 100 mm^2 ACSR and spurs should be a minimum of 50 mm^2 ACSR. 9 m poles of 400 kg strength for LT and HT are prescribed. A growth period of 5 years horizon is considered. Benefits of additional energy sale, saving in losses are evaluated in horizon year. Feeder improvement schemes are evaluated on life cycle cost-benefit basis as per REC norms. The typical evaluation proforma is as below:

(i) Additional demand due to load growth at the end of horizon (5th) year (MVA) on the proposed or existing EHV distribution substation: A

(ii) Expected sale of additional energy (kWh):

$$B = A \times 1000 \times LF \times pf \times 8760$$

(iii) Average weighted tariff of the SEBs or power utility: W

(iv) Value of benefits due to the sale of additional energy-B × W

(v) Expected net saving in line losses in HT (11 kV) lines in the horizon year (LU) after improvement

(vi) Expected net saving in line losses in LT (0.433/250 V) lines in the horizon year (LU) after improvement

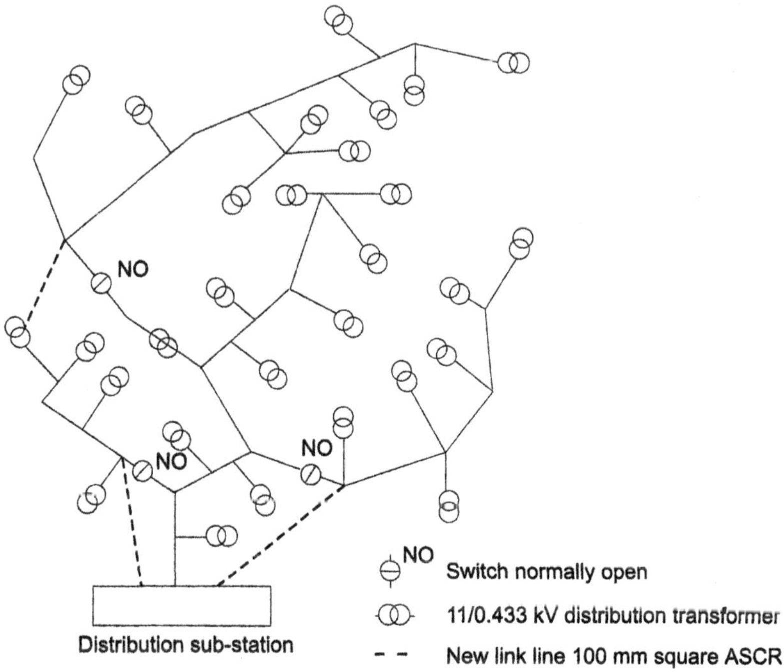

Fig. 14.2 *Rehabilitation of a typical 11 kV long and heavily loaded feeder into 3 feeders*

(vii) Expected saving in line losses due to introduction of capacitors
(viii) Total savings in LU = (e) + (f) + (g)
(ix) Benefits to the utility due to saving in losses = (h) × LRMC
(x) Total benefits to the power utility = (d) + (i)
(xi) Total scheme cost (including cost escalation, if any)
(xii) Percentage gross return = (j)/(k) × 100
(Here LF = load factor; LU = lakh kWh; LRMC = long range marginal cost).

The scheme giving gross return of more than 40 per cent is considered financially viable. In the above analysis, expenditure on additional energy generation and revenue increase due to better continuity of supply have not taken into account. Interest on loan during construction period is part of total cost. The typical improvement of a long 11 kV feeder, trifurcating into three feeders is shown in Fig. 14.2.

(c) Provision of shunt compensation on 11 kV feeders can be made. It is observed from the system studies that provision of shunt compensation directly on the line is only economical, when the load factor is fairly high (say, above 20 per cent) and the voltage regulation of the feeder is above 10 per cent for 11 kV. The problem of switching and over voltage can be avoided by compensating only 25 to 30 per cent of the kVAr on average load.

The pole mounted capacitor bank optimum electrical location distance (D) in km from sending end of the feeder should roughly be:

$$D = \left(1 - \frac{K}{2}\right) \times \text{feeder trunk length in km}$$

where

$$K = \frac{\text{Capacitor bank kVAr}}{\text{average load kW}}$$

Switched capacitors automatic control by a small capacitor switch of vacuum/SF_6 type with CT and PT pole mounted are in use. The capacitor switch automatically brings the capacitors into service, depending upon line loading and disconnects the capacitors during low load period. On very long and heavily loaded feeder, the automatic switched capacitors are very economical [10].

(d) Wherever the distribution transformer is overloaded and additional loads are anticipated, the existing transformer can be replaced by a higher voltage transformer, alternately, a new transformer centre may be provided to cater to the excess loads. Studies reveal that if the distribution transformer centre is shifted to the load centre, a considerable saving in LT losses can be achieved without much investment.

(e) When the power transformer in the HT sub-station is fully loaded, the augmentation of the transformer capacity can be carried out either by installing an additional transformer or by changing the existing transformer with a higher size unit. Whenever the transformers are augmented/added provision for OLTC transformer can be made.

(f) Provision of new 11 kV and LT feeders and distribution transformer can be made.

(g) Power factor at the load point can be improved considerably by installing LT capacitors at the consumer premises and at the distribution transformers.

(h) Sparse or isolated villages with one or two three-phase motor loads may be electrified with SWER system which poses no problems concerning voltage regulation.

14.5 Single Wire Earth Return System

For remote rural areas with sparse population, the Single Wire Earth Return (SWER) system has been found to be economical. Here, the earth is used as the return conductor and one wire as the phase conductor. This system requires auguring grounding rods deep into the soil at each consumer site in order to establish a sufficiently low impedance to the ground. This system is in use in India (Madhya Pradesh, Chhattisgarh), Australia, China and Canada.

In these areas, there may be one or two three-phase motors when not much addition to load for the next 5–10 years is expected. The main advantages of the SWER are:

(a) Economical to the extent of 30–40 per cent over conventional three-phase system under favourable circumstances. The galvanized steel line conductor system affords further economy.
(b) Better voltage regulation which is the main problem in long conventional feeders.
(c) Improved power factor as compared to the conventional system, being nearly 0.95, whereas it varies from 0.7 to 0.75 for the conventional system. The reason for this is that the inductive load such as that of irrigation pumpsets and flour mills, etc. in a SWER system is run in conjunction with the static phase converter which makes use of capacitors as one of the components.
(d) The outages on the SWER system are comparatively lesser than on the conventional system due to the less number of components per span.

The main limitations in adopting the system are:

(a) Maximum ground return current permitted is 7–8 A. This limitation is to avoid communication interference and safety hazards due to voltage gradients. A 1:1 isolating transformer is used to minimize the possibility of telephone interference.
(b) Maximum power limit of distribution of this system is limited up to 3–3.5 kVA/km of the line.
(c) Particular configuration and the maintenance of earthing electrodes is an important point while adopting this system.

(d) The SWER system without the isolating transformer might create extraordinary interference with telecommunication circuits. Detailed study must be done before its adoption (see REC Standard 14/1979).

14.6 Fault Locating

In accordance with the present practices in the rural distribution system, the main 11 kV feeders are protected by a circuit breaker with over-current and earth fault relays. Experience has shown that most of the faults are of a transient nature and the breaker can be usually reclosed successfully. In case of sustained faults, the circuit breaker does not close and the field staff is required to patrol the main feeder as well as the sectional lines to locate and isolate the faulty section before supply is restored. Though some of the electricity boards are using sectional fuses (drop out fuses, horn or HRC fuses) or auto-reclosers at the tapping points on the main feeder, experience has shown that in the majority of cases the fuses do not blow at all before the circuit breaker at the sub-station trips. Therefore, sectional fuses have practically no utility in the 11 kV distribution system unless the fuses are graded with reference to the time lag characteristics of the circuit breaker installed at the sub-station, which is rarely successful in practice.

With a view to eliminate the possible delay in locating and isolating the faulty section an 11 kV overhead fault locating equipment can be installed on the line at tee-off points. The usual fault locating equipment installed on HV or EHV lines are the pulse reflection locator and distance relays with distance measuring unit. The units for rural long lines must be cheaper and operationally viable.

A magnetic flux surrounds each conductor and induces a voltage proportional to the line current flowing in the detector head coil. The coil residually connected in series with resistor. Voltage developed across the resistor applied to the phase and earth fault measuring circuits. At the particular operating level, a signal is applied to the trigger circuit, giving physical indication.

The typical fault indicator manufactured by AB Chance Company, U.S.A., is shown in Fig. 14.3. The device has inverse time characteristics (see Fig. 14.4). This indicator can be reset manually from the ground level.

Fig. 14.3 *Fault Detector*
(Courtesy: Photograph reprinted with permission of A.B. Chance Company (Hubbel Power Systems, Inc.), U.S.A.

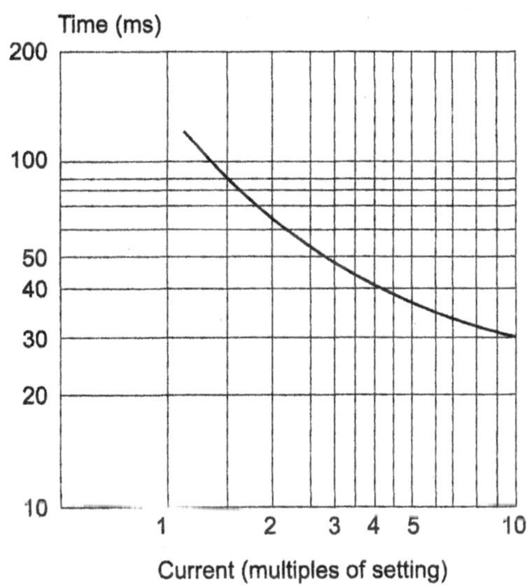

Fig. 14.4 *Fault indicator: Typical inverse time operating characteristics*

14.7 Auto-Reclosers

A typical rural distribution system comprises a radial 11 kV overhead line with several tee-offs covering a large area, much of which is inaccessible by road transport (see Fig. 14.1). Automatically-reclosing circuit breakers are available which increase the security of supply and reduce outages. These are called auto-reclosers and are small, usually vacuum or SF_6 circuit breakers. They can be programmed mechanically and/or microprocessor based to give up to four trips, either instantaneously or time delayed, or both, to suit the power system. The necessary energy for the opening and closing springs is obtained from the power line itself by an electromagnet shunt connected through auxiliary contacts across the incoming supply. When the auto-recloser it open (tripped), the electromagnet is energized and its armature partially charges the closing spring to a point where the armature opens the auxiliary contacts in the electromagnet circuit. The armature resets and the cycle is repeated until the spring is fully charged, at which point is closes the auto-recloser, and this action in turn opens the auxiliary contacts.

Phase-fault or overcurrent protection is given by three series trip coils which carry the lines currents, any one of which provides the energy to trip the auto-recloser (three-phase) when the line current exceeds the setting. Time delay in the tripping action and instantaneous tripping is obtained automatically with the help of electronic control unit or microprocessed based relays to program the minimum trip, number of trips to lockout, operating sequence and time delay. Both types of trip vary inversely with fault current. The instantaneous trip time decreases from about 0.35 to 0.06 s and the maximum delayed time trip from 40 to 0.7 s for currents of twice and twenty times the rated current respectively.

Earth fault protection is given by three residually connected (i.e. paralleled) CTs in the auto-recloser bushing insulators. IS: 7567–1993 specifies the requirements of automatic reclosing circuit breakers for ac distribution system of 11 and 33 kV.

The reclosers can be used for:

- As the primary feeder protection device at the rural sub-stations.
- On major tee-off lines to protect the main feeder from interruptions due to faults on the tee-off lines.
- For sectionalizing lengthy feeders in order to prevent outages on the entire feeder, when permanent faults occur near the end of the feeder.

14.7.1 Load Limiter

There are some countries in Asia, and Africa that use load limiters. They are very valuable as replacement for kWh metres particularly in relation to very small consumers, such as one lamp point connections or Kutirjoyti scheme of Rural Electrification Corporation of India for dwellings of poor sections of the society. In essence, the load limiter is simply an automatic circuit breaker. The only difference between load limiter and circuit breaker lies in the degree of accuracy in response to an excess current. The former, having a high degree of accuracy on over-loading. It enables electricity to be sold to consumers with low electricity consumption levels, on a secure and affordable basis. It provides significant capital and running cost advantages over kilowatt-hour meters and additional benefits including demand-side management. With this, the consumer is allowed to draw current up to a prescribed limit at all times, for which he pays a fixed monthly fee. If the current exceeds the prescribed limit, the load limiter automatically disconnects the supply. It automatically resets and can therefore be mounted in an inaccessible place, such as on the service pole. This reduces problems of electricity theft. Fixed monthly payments for a load limiter supply make budgeting easier for the consumer and reduce the likelihood of defaulting on payments. An annual advance payment option can be offered to enable rural consumers to pay for their supply when income is generated at harvest time. A lower connection fee can be charged for a Load Limiter to reflect its much lower capital cost and the reduced supply-side cost due to peak demand limitation. It is important for the consumer to see a clear financial benefit from subscribing to a load-limited supply. The approach taken in Zimbabwe is to charge a lower connection fee and set the tariff for the smallest load-limiter, in this case a one ampere supply, at the same level as the fixed monthly charge for metered supply. As the metered consumer is also charged for kilowatt-hours used, the load-limiter is clearly the cheaper option.

14.8 Determination of Rating of Induction Motor (Cage Type)

With the introduction of the flat rate tariff based on a horsepower of motor for agriculture supply by some electricity boards, many consumers have been tempted to install motors of higher ratings as compared to their sanctioned loads. Sometimes, industrial consumers also install higher rating motors as compared to their sanctioned loads. This is normally done when it is difficult to get load extension due to a restriction in the power supply imposed by the authorities. In such cases, either the name plates are deliberately removed or fictitious ones affixed in their place. To tackle this menace, it is desirable that the rating of the induction motor is determined to a reasonable extent. There is no fool-proof method to find out the ratings. The limited measures adopted for this are discussed below.

1. From the physical dimensions, class of insulations, etc. the ratings of the motors can be determined as per Indian standards. But the problem is that some manufacturers do not manufacture induction motors according to IS: 1231–1974 or other standards [1, 2]. If running of motor creates unbalance in the supply line voltages, the motor can be called substandard.

2. The other methods are based on two tests: no load and blocked rotor tests which have to be performed on the motor in the field. After the tests, the circle diagram is plotted (Fig. 14.5).

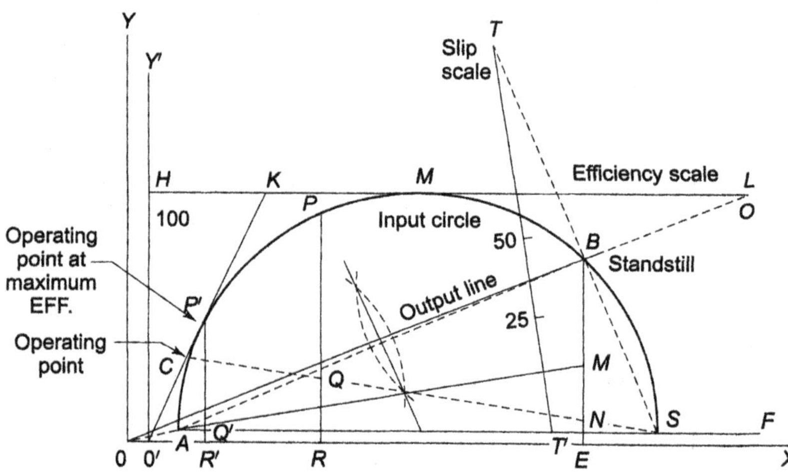

Fig. 14.5 *Circle diagram for induction motor (general purpose)*

This method is based on approximate equivalent network of the motor, neglecting the effect of the exciting current in causing a drop in the stator and the effect of stator resistance drop on the exciting current. The approximation errors are sufficiently small to permit a fairly accurate determination of characteristics of general single cage motor and not of double cage or deep slot type motors. Generally, agriculture motors or general purpose motors having single shallow cage type are used.

(a) The procedure of drawing circle diagram and making efficiency and slip scales on it are well explained in various textbooks [16]. The point of maximum efficiency is located. *K* is the point of maximum efficiency on the efficiency scale when the extended line from 0′ is drawn tangentially at *P*′. *P*′ *Q*′ represents the output and the *P*′*R*′ is the input to the motor. Output *P*′ *Q*′ at the point of maximum efficiency will give the rating of the induction motor because the continuous rated motors are designed for maximum efficiency at near or full load. The maximum efficiency rating is desirable to be given on the name plate which is normally in the range of 75–95 per cent, it being higher for higher unit ratings.

(b) The circle diagram also serves to determine the operating point of the motor from the slip scale which could be drawn on the circle diagram. As we know the slip can be determined from the full load rated speed of the motor, the value of which can be assumed reasonably of the order of 1440 rpm and 960 rpm motors respectively for synchronous speed, e.g. 1500 rpm (4-pole machine) and 1000 rpm (6-pole machine). To draw the slip scale, let the line SB be extended to any convenient point T such that TT′ line drawn perpendicular to the torque line AM (M intercepts the line NB such that BM/MN is the ratio of blocked rotor/stator I^2R losses) is made equal to 100, on per cent scale. For getting the operating point of the motor (thus H.P. of the motor), the line through the point S and the full load slip (percentage) point on the line TT′ is extended to the circle. The touching point C gives the operating point of the motor from which the H.P. rating of the motor can be deduced.

Drawing the circle diagram is time consuming and may be difficult to put into routine practice in the field except in some special cases for investigation. It is possible to develop self-contained instruments for these tests.

3. IS: 8789-1996 specifies following values of performance characteristics for 3-phase induction motors of various types and sizes:

- Output rating (kW)
- Maximum full load current (A)
- Minimum full load speed (r.p.m.)
- Breakaway torque in terms of full load torque (minimum percentage ratio)
- Minimum nominal efficiency (%)

The values checked during inspection and testing of motors in the field can be compared with standard values given in the above standard to find the size of the motor.

14.9 Constructional Practices

The construction methods and practices adopted must be continuously reviewed to reduce costs, to save times, and to have better performance. The important constructional practices recommended by the Rural Electrification Corporation Limited [5] are given below as typical cases.

14.9.1 Material Specifications

Line Conductors

(As per IS-398)

(a) All Aluminium Conductor (AAC) for LT lines:
 7/2.21 mm; 7/3.10 mm.

(b) ACSR conductors for 11 kV and LT lines:
 7/2.11 mm; 7/2.59 mm; 7/3.35 mm.

(c) AAAC conductors for 11 kV and LT lines:
 7/2.00 mm; 7/2.50 mm; 7/3.15 mm

Insulators and Their Fittings

(a) LT insulators and fittings: As per IS: 1445–1977
 (i) Schackle insulators with U or D clamp with nuts, bolts and washer fittings.
 (ii) Egg insulator for insulating guys points.

(b) 11 kV insulators and fittings; (refer IS: 731–1971 and 2486(1) - 1993). Porcelain pin insulator with galvanized pin, nuts, and washers, porcelain disc insulator (clevis and tongue type) with galvanized fitting of nut, bolt, cottor pin etc.

(c) Guy strain insulator as per IS: 5300.

Poles

(a) PCC, steel, spun, wooden: 200 kg
 7.5 m height for cross-country/along streets
 8 m for street crossings or along streets

(b) X-arm of channel iron (75 × 40 mm), or wooden.

(c) 9 m poles for H-pole structure for transformers.

14.9.2 LT Practice

System Configuration

Three phase, four wire for cross-country.
Three phase, five wire along streets, etc.
Single phase, two wire or three wire.

Conductor Support System

Vertical or horizontal or mixed configurations are used. The conductor clearance between phases and phase-to-neutral is kept as shown in Figs. 14.6 and 14.7. Guy points are insulated by interposing egg insulators.

(a) For 8 m height pole construction (vertical formation):
Embedded depth of pole = 1.4 m/1.5 m

Ground clearance of conductor	$\nless$ 4.6 m
Sag	$\ngtr$ 725 mm for 3 ϕ-5 wire;
	$\ngtr$ 925 mm for 3 ϕ-4 wire

(b) For 7.5 m height pole construction (Horizontal formation):

Embedded depth =	1.4/1.5 m
Ground clearance	$\nless$ 4.5 m
Sag	$\ngtr$ 1.10 m

Permissible spans are given in Table 14.1 and the regulations in Table 14.2.

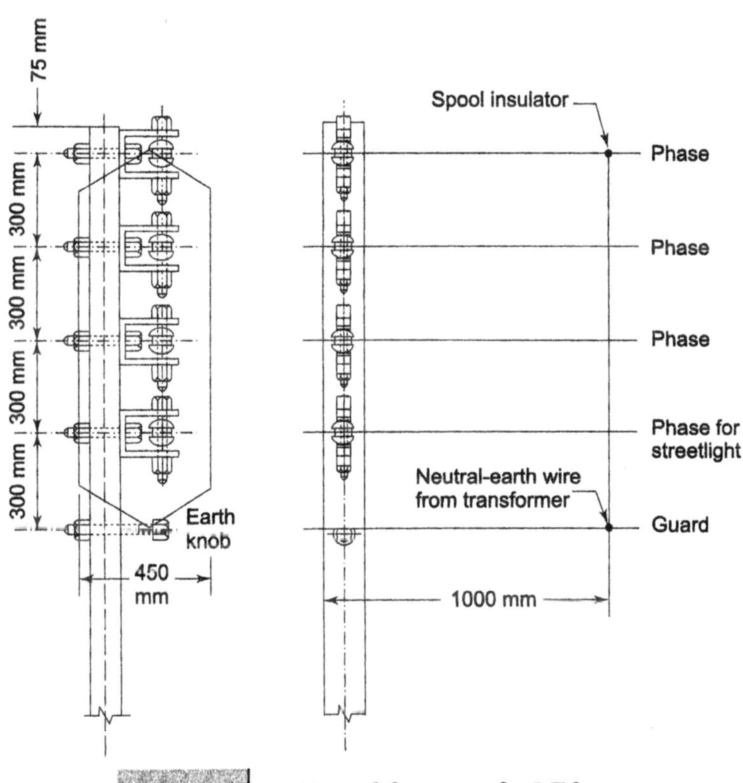

Fig. 14.6 *Vertical formation for LT lines*

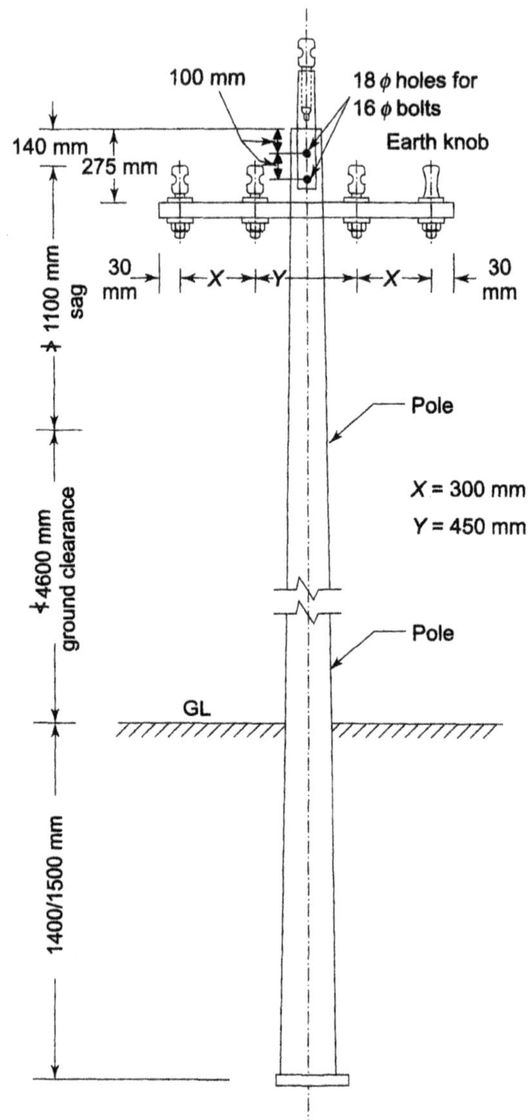

Fig. 14.7 *Horizontal formation for LT lines 415 V/240 V (3 ϕ, 5 wire)*

Protective Guard

It is used for lines at road crossings or in populated localities. Following are the points that should be adhered to for construction purpose:

(a) Cross-angle $\not<$ 60°

(b) Minimum clearance as per IE Rule 77 should be 5.8 m

(c) Guard wire is of GI wire of size 4 mm
(d) (i) Cage type guards for vertical formation (2 guards/span, see Fig. 14.6)
 (ii) Horizontal cross lacing type guard for horizontal formation construction (8 guards/span)
(e) Structures on either side of the road or street to be earthed.

Table 14.1

Permissible spans of LT lines in metres

Conductor	Working load of supports in kg	Wind zone 50 kg/m²	75 kg/m²	100 kg/m²
		Maximum permissible span in m		
(a) 415/240 V cross-country on 7.5 m support-conductor (Three-phase four-wire) horizontal formation				
AAC 7/2.21 mm	140	81	73	—
AAC 7/2.21 mm	200	—	73	66
AAC 7/3.10 mm	140	80	61	—
AAC 7/3.10 mm	200	—	75	66
ACSR 7/2.11 mm	140	107	83	—
-do-	200	—	107	90.5
ACSR 7/2.59 mm	140	107	71	—
-do-	200	—	107	77
ACSR 7/3.35 mm	140	93	57.5	—
-do-	200	—	88	62.5
(b) 8 m supports system — vertical formation				
AAC 7/2.21 mm	140	73	66	—
AAC 7/2.21 mm	200	—	66	60
AAC 7/3.10 mm	140	72	66	—
AAC 7/3.10 mm	200	—	67	63
ACSR 7/2.11 mm	140	100	91	—
ACSR 7/2.11 mm	200	—	100	97
ACSR 7/2.59 mm	140	100	77	—
ACSR 7/2.59 mm	200	—	100	82
ACSR 7/3.35 mm	140	99	62	—
ACSR 7/3.35 mm	200	—	93	67
(c) 415/240 volt lines along any street on 7.5 m support-conductor (Three-phase five-wire) horizontal formation				
AAC 7/2.21 mm	140	67	63	—
-do-	200	—	65	65
AAC 7/3.10 mm	140	67	50.5	—

(Contd.)

Conductor	Working load of supports in kg	Wind zone 50 kg/m^2	75 kg/m^2	100 kg/m^2
		Maximum permissible span in m		
AAC 7/3.10 mm	200	—	65	55
ACSR 7/2.11 mm	140	67	65	—
-do-	200	—	65	65
ACSR 7/2.59 mm	140	67	57.5	—
-do-	200	—	65	63
ACSR 7.35 mm	140	67	48.5	—
-do-	200	—	65	52.5
(d) 8 m supports of system — vertical formation:				
AAC 7/2.21 mm	140	57	53	—
-do-	200	—	53	49
AAC 7/3.10 mm	140	55	53	—
-do-	200	—	53	49
ACSR 7/2.11 mm	140	65	65	—
-do-	200	—	65	65
ACSR 7/2.59 mm	140	65	65	—
-do-	200	—	65	65
ACSR 7/3.35 mm	140	65	55	—
-do-	200	—	65	58

Notes: (1) Width of windward face of supports assumed as 120 mm.
(2) Applicable for states which have accepted 4.6 m as the minimum ground clearance along streets.

Table 14.2

Voltage regulation kW-km for 415 V, three-phase line for 1% voltage drop

Conductor	kW-km for 1% voltage drop (conductor temperature 60°C) 1.0 pf	0.9 pf	0.8 pf	0.7 pf
		Lagging		
7/2.21 mm AAC	1.37	1.21	1.14	1.08
7/3.10 mm AAC	2.69	2.18	1.98	1.81
7/2.11 mm ACSR	1.04	0.94	0.90	0.85
7/2.59 mm ACSR	1.49	1.31	1.22	1.15
7/3.35 mm ACSR	2.42	1.98	1.80	1.65

Note: For a conductor temperature of 50°C, the above figures would be about 3% above and for a temperature of 70°C, about 3% lower.

14.9.3 11 kV Lines

Configuration

Delta formation is usually used with 11 kV pin insulators with V type X-arm made of steel generally. The typical formation which is followed in practice is shown in Fig. 14.8.

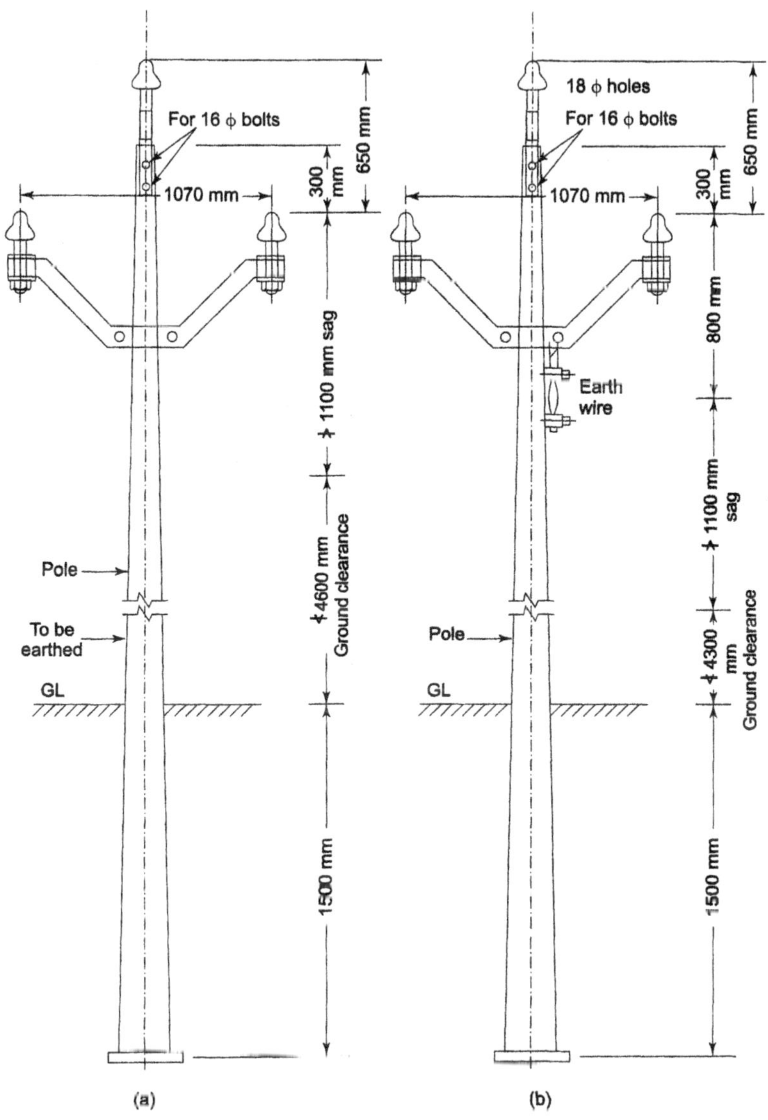

Fig. 14.8 *Formation for 11 kV lines (a) without earthwire (b) with earthwire*

Conductor Support System (see Table 14.3)

11 kV construction without earth wire:

Pole	7.5 m
Embedded depth	1.5 m
Ground clearance of lower conductor	≮ 4.6 m
Sag	≯ 1100 mm

Conductor Supports with Earth Wire:

Pole	8 m
Embedded depth	1.5 m
Ground clearance of conductor	≮ 4.3 m
Sag	≯ 1.10 m
Earth wire and lower phase conductor clearance	800 mm
Lower conductor and top end of pole clearance	300 mm
Clearance between lower and top phase conductor	650 mm

Guys are insulated by interposing disc insulators. Span and regulation are per Table 14.3 and 14.4.

Table 14.3

Maximum permissible span for 11 kV lines

Conductor	Working load of supports in kg	Wind zone 50 kg/m²	75 kg/m²	100 kg/m²
		Maximum permissible span in m		
(i) 11 kV lines, cross country on 7.5 m poles				
ACSR 7/2.11 mm	140	107	103	–
-do-	200	–	107	107
ACSR 7/2.59 mm	140	107	90	–
-do-	200	–	107	97.5
ACSR 7/3.35 mm	140	107	69	–
-do-	200	–	106	75
(ii) 11 kV lines, cross country on 8 m supports with earthwire 4 mm (55–95 kgf/m^2 quality)				
ACSR 7/2.11 mm	140	107	91	–
-do-	200	–	107	99.5
ACSR 7/2.59 mm	140	107	76.5	–
-do-	200	–	107	83
ACSR 7/3.35	140	99	60.5	–
-do-	200	–	93.5	66

Note: Windward face of supports assumed as 120 mm.

Table 14.4

Voltage regulation 11 kV lines (Fig. 14.8)

Conductor	kW-km for 1% voltage drop for a conductor temperature of 60°C			
	1.00 pf	0.9 pf	0.8 pf	0.7 pf
7/2.11 mm ACSR	706	674	640	605
7/2.59 mm ACSR	1150	970	900	840
7/3.35 mm ACSR	1920	1490	1330	1200

Note: For a conductor temperature of 50°C, the above figures would be about 3% higher and for a temperature of 70°C, about 3% lower.

Guarding

This must be provided for road crossing or lines in populated areas. Two earth wires on special X-arms with cross lacings are provided. Structures on either side of the road are earthed. Minimum groundwire clearance from ground as per IE Rule 77 $\nless$ 6.1 metres.

Special Crossing

Special arrangements are required for power lines crossing of railway tracks and telecommunication lines as REC construction standards J-3 and J-4 respectively [5].

Earthing Practice

GI pipe or MS rod is generally used for earthing. In case of high resistivity soil, chemical treatment (charcoal or coke and salt in alternate layers of 300 mm) is done. The pipe size is 40 mm dia, GI embedded vertically 200 mm below surface or ground. Riser of 8 SWG, GI wire is used.

14.9.4 Distribution Transformers

Distribution transformers, 11/0.433 kV, 16, 25, 63, 100 kVA are installed on H-pole structure of height 9 m with steel X-arms. Distribution transformer supporting X-arm is placed at a distance of

2.44 m approximately from the ground level. Each distribution transformer is provided with the following:

(a) 11 kV lightning arresters (5 kA) set.
(b) Either dropout fuse unit three-phase or horn gap fuse unit three-phase and air-break three-phase switch (Fig. 14.9).
(c) Distribution box.
(d) Earthing set.
(e) Danger board.
(f) Channel/angle X-arms with set of clamps, nuts, bolts, barbed wire, etc.
(g) LT cables as per IS: 1544 Part I.
(h) MCCB as per IS: 13118 on LT side of transformer

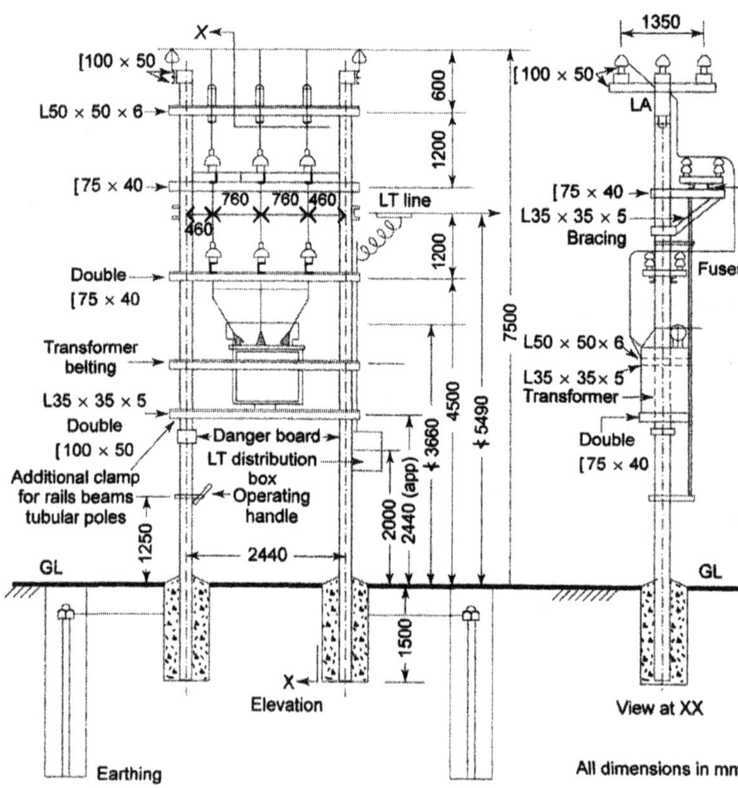

Fig. 14.9 *Distribution transformer installation (for transformer locations along line)*

14.9.5 Service Connections

The typical REC practice for a three-phase agriculture and industrial service connection is shown in Figs 14.10 and 14.11 respectively. The main features of the practice are:

(a) Maximum service span length ≯ 35 m.

(b) Twin-core (2 run) or four core PVC cables are used for different sizes of motors are follows:

   (i) Twin-core cable (two-runs)

2.5 mm^2	up to 3 HP motor
4 mm^2	3 HP-5 HP motor

   (ii) Four-core cable

6 mm^2	5 HP-10 HP motors
10 mm^2	10 HP-15 HP motors

(c) Bearer wire of 3.15/4 mm size of hard quality as per IS: 280–1962.

(d) Clamps are to be made from 40 × 6 mm^2 MS flat.

(e) Minimum ground clearance

Across road	5.8 m
Along road	5.5 or 4.6 m if relaxed
Elsewhere	4 m

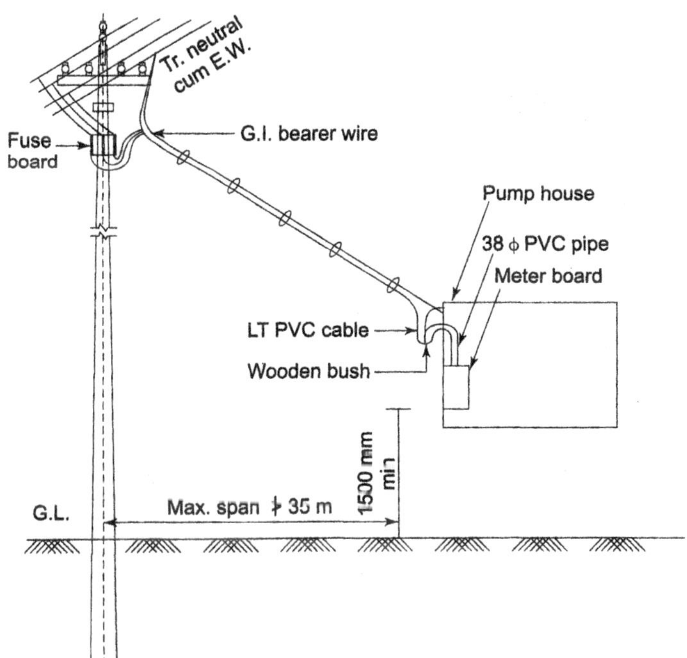

Fig. 14.10 *Typical agricultural pump service*

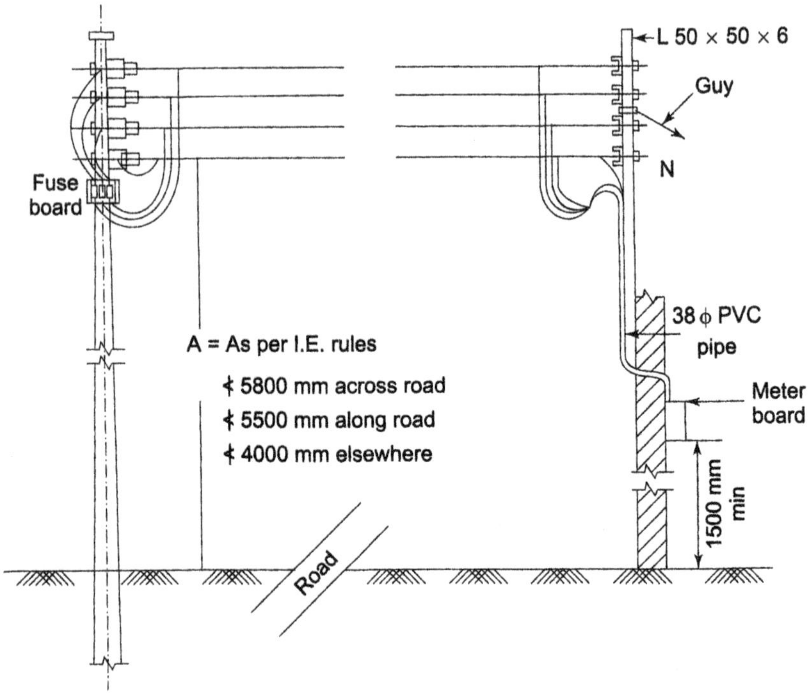

Fig. 14.11 *Typical industrial service*

14.9.6 33 kV Sub-stations

Two construction standards have been adopted by REC. One is with 11 kV outdoor gear and the other with 11 kV indoor gear. The general single-line diagram is shown in Fig. 14.12. A are used 33/11 kV transformers: 630, 1600, 3150, 5000 kVA are used.

The important features of construction practice are:

(a) 33 kV, vacuum or SF_6 circuit breaker (shown dotted in the diagram) is to be used when transformer, each with a capacity is 5 MVA or above. HRC fuses are provided for transformer less than 5 MVA.

(b) Circuit breakers on 11 kV side of the transformers will be provided if the transformer capacities are more than 630 kVA each. For better reliability, an electronic auto-reclosing (single reclosing) on 11 kV outgoing feeder circuit breakers will be provided.

(c) Conductors used for 11 kV and 33 kV jumpers and busbars should not be less than 80 mm^2 ACSR.

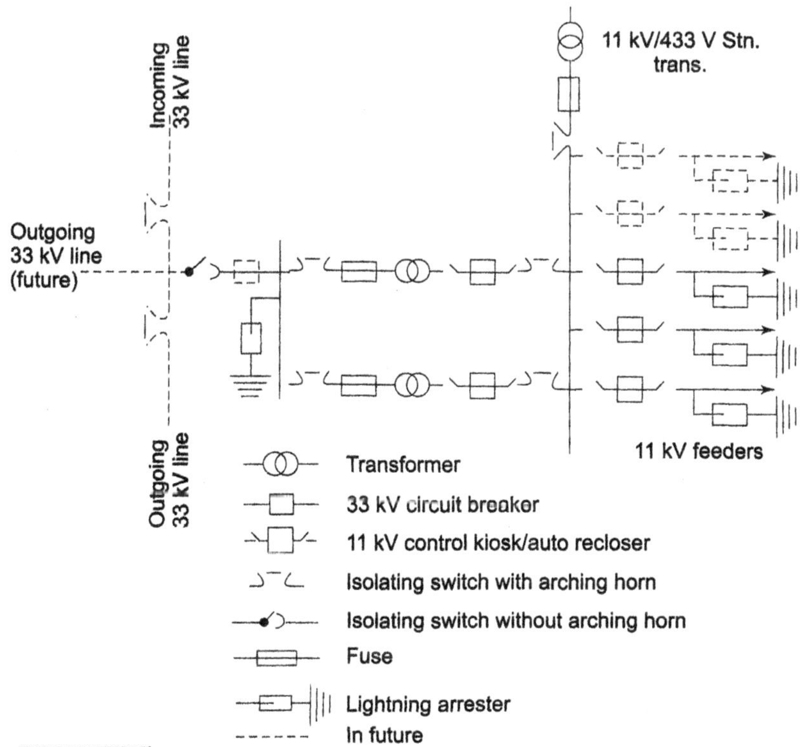

Fig. 14.12 *Key diagram of a 33 kV sub-station (outdoor 11 kV switchgear)*

(d) Expulsion type fuses (preferably employing tuffanol tubes) should preferably be used for proper protection of transformers.
(e) The supports will not be guyed but may be suitably concreted.
(f) 33 kV lightning arresters, 10 kA will be of station type.
(g) The station transformer should be located on the first H-pole of one of the outgoing feeders (preferably the shortest feeder). The capacity of the sub-station transformer would depend upon the load requirement in each case.
(h) Cables in the switchyard may be either buried or carried in concrete trenches depending on the prevailing standard practice in the power utility. 11 kV XLPE cables will be used.
(i) The joint at the bottom of the vertical support carrying the 11 kV cable box will facilitate the removal of the support if the transformer is to be shifted.
(j) Space for bus-coupler panel may be used for an outgoing panel at those sub-stations where no bus-coupler breaker is required or where it use may not be called for in the initial stages of load development.

Metering

It involves the following steps:

(a) Each of the incoming and outgoing 11 kV panels should be provided with a kWh metre and an ammetre for the proper assessment of losses. In addition, each incoming panel should have voltmeter. In case, there are no incoming panels a voltmeter should be provided on one of the outgoing panels.

(b) Only one PT and one power factor metre need be provided at each sub-station.

(c) Current transformers with separate secondary windings may be provided for (i) protection and instruments and (ii) metering or one dual purpose winding may be used.

A general arrangement for metering is shown in Fig. 14.13.

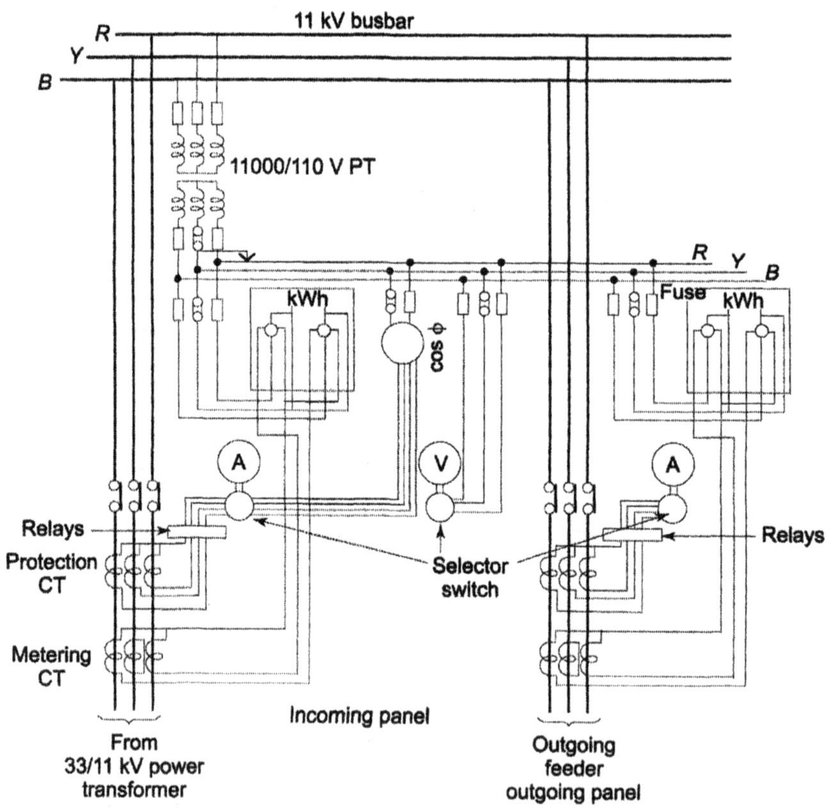

Fig. 14.13 *General metering arrangement*

14.10 Future Orientation of Rural System

14.10.1 Construction

The demand for power in rural areas is rapidly increasing as a result of techno-economic awareness and planned development. Time is not far when subtransmission or distribution lines of 132 or 66 kV line voltages will have to be laid in a reticulated arrangement. Economical designs as well as compact configuration are important for cheaper and viable rural electrification which could afford easier right of ways and less wastage of land. The practice of using single-pole for HV lines construction on tubular steel, rails and wooden poles is prevalent in advanced countries and has proved to be of adequate reliability. For example, the trident technology for 132 kV lines on wooden poles has been adopted in UK is shown in Fig. 14.14. A similar technique can be used for 66 and 132 kV lines in India thus saving capital costs by nearly 30–40 per cent.

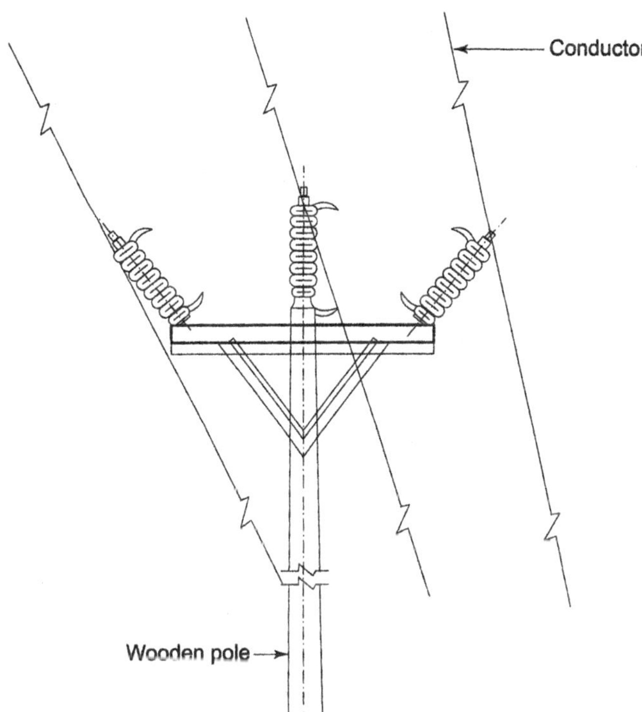

Fig. 14.14 *Trident construction for 132 kV line*

Research in the USA has shown that one can dimension 138 kV lines within the corridor limits normally assigned to low voltage distribution circuits. This work shows the feasibility of constructing 138 kV line with horizontal spacing of just one metre between phases with single pole supports.

Lines on wooden poles give lower profiles and better lightning performance. The existing 11 and 33 kV lines, with porcelain insulators and on single wooden poles can be uprated to 66 and 132 kV lines on the same poles with the help of either trident technology or polymer insulators as is already the practice in many Western countries. [14]

14.10.2 Adoption of Reduced Insulation Levels

The present insulation levels on 11, 33, 66 and 132 kV systems as given in Section 13.1 are much safer. Also, there is larger scope for reducing the BIL insulation level in general (both rural and urban) which could result in a tremendous saving in cost. Indian Standards should incorporate the downward revision. The present insulation levels were selected when early overvoltage protection devices were of inferior characteristics and their correct application was also not known.

The selection of the insulation level is generally based on the following factors:

(a) Neutral earthing conditions.
(b) Characteristics of the lightning arresters or spark gaps.
(c) Distance of arrester location from the equipment to be protected.
(d) Safety margin.

The distribution system in India is generally the effectively grounded neutral system. In this case the overvoltage, under sustained earth fault conditions, does not rise beyond 0.8 pu. Switching surge overvoltages also do not rise beyond 2.5 pu. Lightning arresters (ZnO) are now available with the improved maximum residual voltages occurring at the time of the lightning discharge and, therefore, reduced voltage stress is impressed on the equipment when lightning occurs.

The insulation requirements for line and sub-station insulation should be considered separately. The selection of the level for a line is based on several factors, such as its elevation above sea level, severity of lightning activity, soil condition, possible use of wood insulation and pollution level and is chosen to give an acceptable line outage rate.

Sub-station insulation mainly depends upon the protective levels afforded by the spark gaps or lightning arresters and shielding wires.

A power transformer is a very expensive item at the sub-station and the choice of lower insulation level is an economic incentive. Hence the transformer insulation has the lowest level and lightning and/or spark gaps are installed at the transformer to protect this level. Since each equipment, such as circuit breakers and isolators, cannot be provided with arresters individually, it is a normal practice to adopt a higher insulation level (above the transformer) to provide the equipment with good protection as is economically justified. The next level of about 10 per cent higher is given to the circuit breaker, isolator, station bus and line. The level of the line can be given an extra margin.

The protective levels of the chosen overvoltage limiting device must be lower than the equipment insulation levels by suitable margins. In the impulse region the margin takes into account factors such as higher magnitudes or faster rates of the rise of impulse currents through the arrester, possible reduction in impulse strength due to repeated application and higher crest voltages at the transformer due to the separation distance between the transformer and arrester. Margins of 20–30 per cent have been found satisfactory. The margin is taken as 15–20 per cent in the switching impulse region.

Based on the arresters residual voltages as given in Table 14.5, we can formulate the BIL levels taking into account suitable safety margins as shown in the following examples.

Table 14.5

Residual voltage of lightning arresters
Maximum residual voltage of arresters
(IS: 3070–1974 discharge current wave 8/20 μs)

Arrester rating kV rms	*10 kA heavy, light duty and 5 kA discharge peak kV*	*2.5 kA discharge kV peak*
9	32	32.5
18	65	65
30	108	108
60	216	—
120	400	—

EXAMPLES

11 kV System

Highest system voltage = 12 kV rms

(a) Highest system voltage phase to ground

$$= \frac{12}{\sqrt{3}} = 6.928 \text{ kVrms}$$

$$= \sqrt{2} \times 6.928 \text{ crest}$$
$$= 9.8 \text{ kV crest}$$

Maximum expected switching level (at 2.5 pu)

$$= 2.5 \times 9.8 = 24.5 \text{ kV crest}$$
$$= 17.32 \text{ kV rms}$$

Power frequency BIL = 17.32 × 1.20 = 20.78 say
(20% margin) ≃ 21 kV rms

(b) Impulse level with 30% margin

9 kV arrester maximum residual voltage = 32.5 kVP

$$\text{BIL} = 32.5 + 32.5 \times \frac{30}{100} = 41.6 \text{ kVP say } 42 \text{ kVP}$$

Transformer BIL desirable: 21 kV rms power frequency (pf)
42 kV crest impulse

Line circuit breaker, fuses should be approximately 10% higher:
24 kV rms pf
47 kV crest impulse

Similarly 22 kV System

Transformer BIL should be: 42 kV rms pf
84 kV crest impulse.

Line, circuit breaker, fuses, etc.: 48 kV rms pf
94 kV crest impulse

33 kV System

Highest system voltage = 36 kV rms

Highest system voltage

Phase to ground $= \frac{36}{\sqrt{3}} = 20.78$ kV

Expected switching surges

Max. (2.5 pu) = 20.78 × 2.5
= 52 kV rms

Power frequency BIL	$= 52 + 52 \times 0.20 = 62.4$ kV say 63 kV rms
Impulse level	
30 kV LA; max. residual voltage	= 108 kVP
BIL = $108 + 108 \times 0.3$	= 140.4 say 140 kV crest
Therefore, transformer BIL:	63 kV rms pf 140 kV crest impulse.
Line, breaker, fuses Isolators, etc. (10% higher):	70 kV rms pf 155 kV rms impulse.

Similarly 66 kV System

Transformer BIL:	125 kV rms 280 kV crest
Line, breaker, isolator etc.:	140 kV rms 310 kV crest

132 kV System

Highest system voltage	= 145 kV rms
Highest system voltage Phase to ground	$= \frac{145}{\sqrt{3}} = 84$ kV rms
Expected max. switching surge	$= 2.5 \times 84 = 209$ kV rms
BIL power frequency (15% margin)	$= 209 \times 1.15 = 240$ kV rms

Note: IS has specified 230 kV rms reduced level which should be adequate.

Impulse BIL

Max. residual voltage of 120 kV arrester	= 400 kVP
Therefore, BIL = $400 + 400 \times 0.25$	= 500 kVP
Line, breaker, isolator BILs:	270 kV rms pf 550 kVP impulse

Cost Saving

The saving of costs on transformers may be between 20 per cent and 50 per cent and savings may not be significant in line circuit breaker and isolators, etc. There may not be any tangible saving in system below 11 kV for any reduction.

14.10.3 Segregation of Rural Supplies

Segregation of rural supplies for agriculture, industry, village lighting etc. is necessary for (a) keeping continuity of supply to village lighting, (b) development of industry in rural centres, (c) to give proper supply according to agriculture crops. Power supply to agriculture in various states is subsidized. The estimates of subsidized supply are generally exaggerated to absorb part of the system losses and theft of power in agriculture consumption. The tendency is to show a good balance-sheet of power and this is taking place because of lack of accountability. Accurate estimates can be made if metres are provided on all agriculture connections. According to a survey carried out in various states including Punjab, farmers are willing to pay a higher price provided quality supply is assured.

The World Bank also carried out study [9] in India during the year 2001 and reported:

- Since a large part of the supply to agriculture is un-metred, power utilities under-report the system's actual distribution losses by ascribing a significant portion of non-technical losses and theft as agricultural supplies. The study found that actual transmission and distribution losses are significantly (27 per cent) above the official estimate.
- Farmers are willing to pay for improved quality.

(a) Agriculture Supply

In India, about 30 per cent electric power [11] is consumed in agriculture sector. The agriculture demand has large variation during the year, depending upon different crops sowing, growing and harvesting period. The maximum demand is for about a few months in the year and it is economical to cut off feeders for some hours every day during remaining lean period to save energy and system losses.

The power cut is carried out irrespective of intensity of agriculture, ground water availability of various areas. There is need to give justifiable demand based supply. Modern agriculture practices in the country have led to a tendency to over-exploit groundwater. The conditions of water level going down is serious in many states, like Punjab, Haryana, Karnataka, etc. For example, in Punjab, the dark, grey and white areas are identified respectively, where ground water exploitation is made over 75 per cent, between 65–75 per cent, less than

65 per cent of recoverable recharge. Further release of tubewells in dark, grey areas will make water level go down. The situation is likely to get out of control if immediate steps are not taken to increase the recharge of underground water by expansion of canal irrigation, artificial re-charging the ground water system by utilizing the surplus surface run-off by ponds, and small or micro-dams and/or by allowing river surplus water to stand for re-charging and running through chaes as drainage channels. In Punjab, 70 per cent land falls in such critical area. Two factors: One 'area factor' depends upon water strata depth and re-charging. This factor in white, grey, and dark areas approximately fit as 0.75, 3.5, 3.0 respectively. The other factor is crop water intensity (based on number of watering required normally for the predominant crops) factor is as given in Table 14.6.

Table 14.6

Crops water intensity factors

Crops	Factors
Wheat	5
Maize	4
Rice	18
Cotton	4
Sugarcane	12
Groundnut	3
Potatoes	10

Power cut rationing is desired on the basis of proportion of the product of above two factors. However, a suitable tariff surcharge could be levied for more hours of power supply and similarly a rebate be allowed for less hours of supply on flat rate tubewell consumers during the crops seasons. The other allied considerations are:

(i) The crops which are grown on ridges such as potatoes require intensive and narrow irrigation ditches, which is very difficult to irrigate during night/dark hours. Therefore, predominant potato growing areas be given agriculture staggered supply during day time.

(ii) In sub-mountain agriculture area, there is high risk of snake bites in summer, and, therefore, during summer, the power to such areas be staggered for day time.

(iii) For vigilance and security reasons as well as for international status, the areas close to international border should be given 24 h supply.

(b) Rural Industry

There is urgent need for balanced and integrated development of rural area in the country. The villages should be regrouped into six to seven villages for development of focal points or industrial growth centres. Power supply is main requirement to thrust the industry in the rural area. For each three or four industrial growth centres, a dedicated three-phase rural industrial feeder should be laid to cater the power demand for industries, such as dairying, agro-based industries, such as oilmills, ricesheller, fruits and vegetables processing industries, paper manufacture, industrial alcohol from sugarcane, manufacture of agro-waste briquette fuels from bagasse, groundnutshells, sawdust, wood shavings, jute sticks, wheat and rice husk and straw, dry tobacco-waste, etc., cottage industries and small welding repair shops, agriculture implements manufacturing, village water works, village artisan shops such as carpentry, black-smithy, etc. These feeders be given supply without any power cut. The industrial connections be released in growth centres without any restrictions. However, if any prospective consumer wants some industrial connection or essential service, such as on water supply, telephone, etc. from the rural industrial feeder but outside the growth centre, connection should be released on deposit cost basis. If any existing rural industrial consumer desire link with the industrial feeder, it should be connected on the cost deposit basis.

(c) Rural Lighting

Supply for lighting or other small industrial load for modern life in villages is required on regular basis without any power cut. The people in villages are now more dependent on power supply for day-to-day working, such as common use of computers, television, refrigerator, milk churning machines, fans, electric press, and other useful electrical appliances in the rural homes, lighting for students studies and for high security in villages, power supplies for public communication networks, such as telephones, wireless, etc. For this, above mentioned dedicated industrial-feeders should be tapped to provide village lighting with single-phase transformers, say 11 kV phase-to-phase tapping, with secondary mid-point (neutral) transformer 11 kV/480–240 V is desirable as shown in Fig. 14.15. This system investment will be small due to low cost of single-phase lines and the smaller rating distribution transformers. Such typical schemes for segregated agriculture, industry/lighting is shown in Fig. 14.16 [6]. REC specifications: 49/1987 and 50/1987 provides both single-phase distribution transformers resin encapsulated dry type and oil filled, $11/11\sqrt{3}/0.44–0.22$ kV (phase-to-neutral/single phase system) or 11/0.44–0.22 kV (phase-to-phase/single-phase system), in ratings of 6.3, 10, 16 kVA. The specifications provide in-built MCCB protection on the LT side.

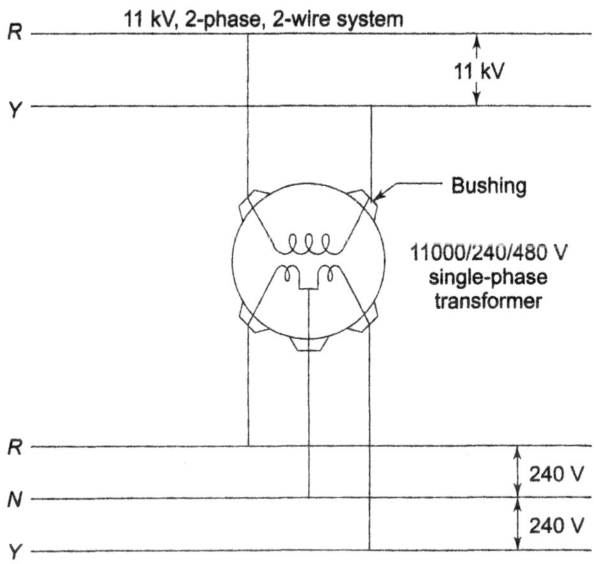

Fig. 14.15 *Phase-to-phase single-phase system*

If continuity of supply to villages for small single-phase loads/lighting is the only criteria, then the continuous earth-cum-neutral wire (ACSR 20 mm^2) multiple earthed can be run on LT insulators (for safety reasons) at each 11 kV feeder. And for load roastering on feeders, the three phases of each feeder can be inter-connected with one phase energised by a special switch installed at the sending end. To balance the load on the sub-station power transformer, different feeders can be connected to different phases. During power cut periods, all the villages will get normal 6.34/0.240 kV supply with single-phase (phase-neutral) transformers, and three-phase distribution transformers feeding for agricultural tubewells will be de-energised and thus save no-load losses. When there is no power cut, the phase inter-connection at the sub-station will be removed by the special switch to restore normal three-phase supply to tubewells and single-phase supply to villages.

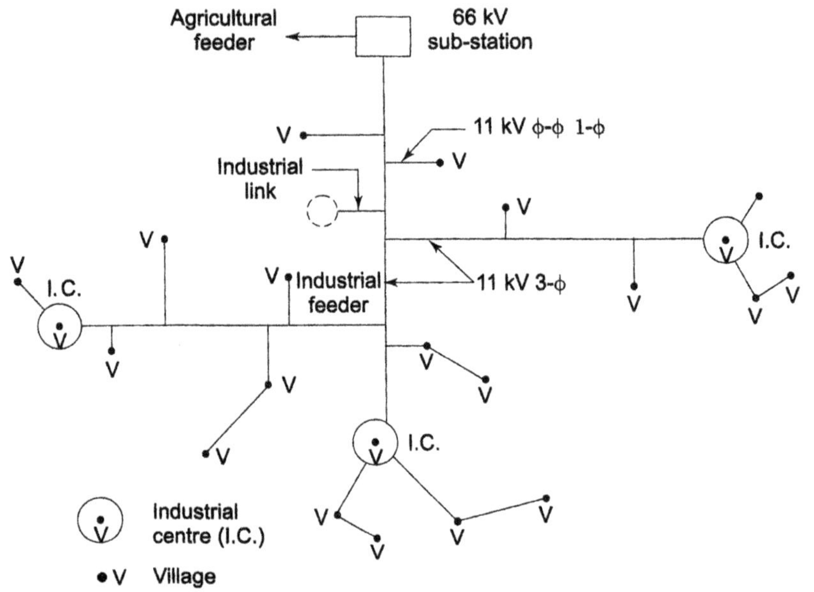

Fig. 14.16 *Typical segregated rural network (3 ϕ, 11 kV, phase-to-phase single-phase); I.C. provided initially with 100 kVA, 11/0.433 kV transformer. Each village provided with 11/0.240/0.480 kV, 10 kVA transformer (average size)*

EXAMPLE: During power shortage in Punjab, power cut is imposed either by shutting down the feeders in rural area or by giving two-phase 11 kV supply (see Fig. 9.2). The latter is generally used during summer, when the loading conditions are worst during paddy cropping and which leads to accelerated damage of transformers and additional losses.

On an average at every 10 km, the 220 or 132 or 66 or 33 kV grid or main sub-station exist in Punjab for taking out primary 11 kV, three-phase, outgoing feeders. Roughly on an average, every rural growth centre should have initially 100 kVA, 11/0.433 kV, three-phase transformers with one feeder of 3.3 km terminating at one growth centre. These are about 12,000 villages. Regrouping five to seven (on the average six) villages, about 2000-growth centres will be required. The requisite maping could be made on geographical survey sheets for actual location of growth centres selected with the consultation of Regional State Planning Authorities on the basis of geographical, socio-economic factors.

According to a study [6] to find out the best method to incur the least annual expenses in supplying the loads and with regard to requirement of power-cut and power regulation arrangement, it was found that it is necessary to segregate agriculture loads from other rural loads, so that agriculture cost could be reduced in seeking to lower tariff. Besides usual power-cut due to power shortage, supply to agriculture is desirable to be regulated on the basis of cropping pattern. Most of pumping sets run the maximum for four-months in summer for paddy season and there is lot of diversity for remaining eight months. The agriculture power cut is justified due to lowest agriculture tariff, and largest share to agriculture consumption is in Punjab as compared to other states in the country.

Segregation (see Fig. 14.16) of existing rural feeders as agriculture feeders could afford the benefit of switching off about 1100, 11 kV feeders and about 42,500 distribution transformers. Similarly, there is saving due to avoiding the two-phase supply (see Sec. 8.12.2) and consumer reduction in losses and decrease in damage of the transformers. The capital cost of laying network for industry and lighting incorporating 666, 11 kV dedicated industrial feeders along with phase-to-phase tapping for rural lighting to give continuous supply to the village single-phase supply system will cost about Rs 550 million and would be economically justified [6].

EXAMPLE: In Punjab, 24 hour continuous power supply on urban pattern to villages is progressively being provided. The scheme is financed by Punjab Government (85 per cent) and Punjab State Electricity Board (15 per cent). The village panchayat contribute Rs 10,000 to 20,000 depending upon the village population along with unskilled labour to execute the work.

A 11 kV separate feeder from the sub-station is taken to village or to group of villages considering load forecast for next 5 years. Except agriculture pumpsets, all other village loads up to 200 kVA are eligible for connection to this feeder. The loads up to 10 kW are released on single phase LT and loads 10 to 100 kW are given on three-phase LT supply. The loads beyond 100 kW are connected on 11 kV supply. The consumers existing beyond 500 m from the village Phirni (Ring Road) can avail this facility by the extension of the feeder at their own cost. In this case, loads more than 10 kW are supplied at 11 kV. The continuous supply system has been provided to about 3245 villages up to December, 2001.

14.10.4 Power Voltage Transformers

The voltage transformers are used for metering protection with very small power ratings in the range 100–1500 VA. These quite often are of the magnetic type, hermetically sealed units using a very small quantity of oil as insulation. The bigger units with power capacity of say 30–300 kVA with voltage levels of 33 kV/66 kV/110 kV/132 kV/220 kV on the primary and 440/220 V on the secondary, could supply power directly from the EHV line. The concept is used in the design of what are called Power Voltage Transformers (Power VTS), that are now gaining acceptance abroad for reaching electricity to places with earlier were considered inaccessible, technically or commercially. Typically, single-phase units with ratings of 10–25 kVA at voltage levels of 66 kV/110 kV/132 kV/220 kV are common. These can be connected together as a three-phase bank to reach higher ratings adequate for several locations.

The primary windings carry a very small current (1 A or less) and can be designed to withstand any electrical stress at the high voltage terminals, like standard VTs. As such, there is no need for high voltage fuses. For protection against surges, it is necessary to connect Lightning Arresters next to the primary terminals of power VTs. Disconnecting switches (Isolators) can be provided on the HV side, where necessary. The primary windings of the power VTs are highly resistive. As such, breaking and making the primary load current is not difficult for most of the standard isolators. These isolators can be manually operated. The secondary windings should be designed to be short-circuit proof. For protection against thermal overloads, suitable circuit breakers (MCCBs) can be provided. These may even have short-circuit release as usual. Where a bit of complexity is admissible, temperature monitoring (alarm + tripping) devices could be fitted into the tank. A typical arrangement is shown in Fig. 14.17.

It is the most suitable application for rural electrification along route of the high voltage transmission lines. Especially for hilly terrains with sparse population scattered over a large tract. A beginning could be made in India by identifying locations where reaching power supply is essential but difficult, and/or expensive by conventional methods. Many manufacturers abroad [4] are making Power VTs as a standard item. A few of these could be imported for use at various voltage levels up to 220 kV. Their performance could be monitored by the Central Electricity Authority or any other suitable organization.

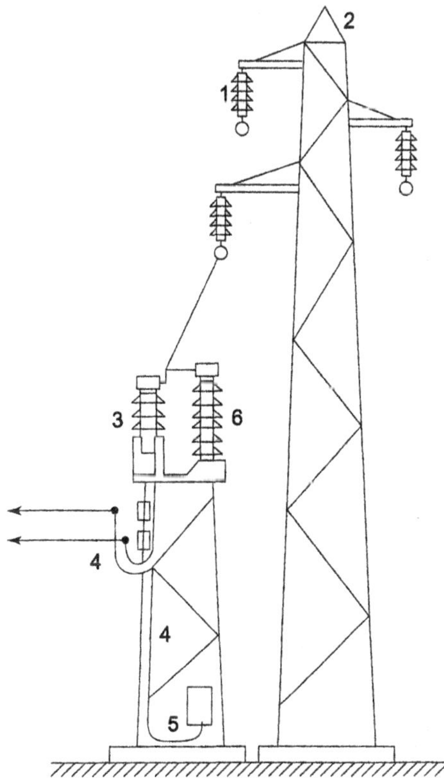

Fig. 14.17 *Typical mounting of power VT's; Legend: 1. EHV line; 2. Steel tower; 3. VT, 4. Low voltage line/cable; 5. Low voltage cabinet (MCCB); 6. EHV surge arrester*

14.10.5 Decentralised Generation and Distribution (DGD)

A policy of decentralised electrification, by using renewable energy sources can speed up the process of complete electrification. Classification of the remote villages and identification of locally available and suitable renewable energy sources i.e. solar, hydel, wind, biomass, etc., which could be harnessed either singly or in combination to meet the present and future demands. Solar Photovoltaic Power (SVP) street light system has been installed in more than 1000 villages. SVP systems have, however, been successful wherever NGOs or manufacturers had taken the responsibility of installation, operation of maintenance and collection of revenues. Active involvement of manufacturers in technical improvements suited to field conditions, arrangements for training, operation and maintenance, could promote larger demand and wider acceptance of SVP systems. Privatisation of generation and distribution of electricity in India, affords ample opportunities for SVP for pumpsets and household needs in remote areas away from grid systems. Thin film photovoltatic technology-based power generation has great potential up to 2500 MW in the Thar desert in Rajasthan (see section 3.4).

14.10.6 Task Ahead

The task ahead is the programme to achieve the mission of 'Electricity for All'. Strategies to be adopted will have to aim at:

- *Accelerating* rural household electrification.
- *Speeding up* electrification of tribal areas and provision of street lights in 'Harijan Bastis'.
- *Stepping up* rural industrialisation through adequate and assured electric supply.
- *Initiating* measures and special schemes for correcting problems of low voltages and frequent breakdowns in rural areas.
- *Providing additional resources* for expanding the special programme called 'Kutirjyoti', under which the cost of providing single point light connections to households of rural poor is given as grant, to serve a major social objective.
- *Continuing thrust* on pumpset energization programme. Three-phase HV (11 kV) distribution can be provided for tubewell electrification on a self-finance basis. The consumer shall pay for a 11 kV line and install his/her own 11/0.433/0.240 kV transformer along with other LT installations. Such a practice has been adopted in Punjab and Noida (NPCL). In Andhra Pradesh, a 11 kV single-phase distribution has been started on a pilot basis to accelerate the tubewell electrification programme. In Punjab, a 11 kV distribution transformer is provided to each consumer for the release of an agricultural connection on a self-finance basis.
- *Promoting* the use of innovative, cost effective and energy efficient technologies.

- *Promoting* small hydel generation, wind power generation, solar photovoltaic for augmentation of rural power supply and electrification of remote villages.
- *Promoting and evolving* suitable decentralised power distribution systems–strengthening the existing RE cooperatives and also promoting new viable cooperative and other distribution models at Panchayat level.

The present and future level of Rural Electrification is given in Table 14.7.

Table 14.7

Rural electrification in India 2020 AD [11]

Item	March 2001	March 2020
Energy consumption (GWh) per annum in rural sector	97800	547100
Tubewells (million)	12.8	19.5
Village electrified (lakhs)	5.08	5.87
Annual per capita rural consumption (kWh)	150	400
Non conventional resources generation (MW) (decentralised)	4500	3150 solar photovoltaic; 36000 wind power; 4500 small and micro hydel
No. of rural consumers (million)	35	100

PROBLEMS

1. A three-phase, 100 kVA, 11/0.433 kV transformer is working on two-phase, three-wire per supply to a village load of 72 kW, 0.9 lagging power factor as per power-cut supply arrangement (see Fig. 8.2.). The total village load has been put on the two LT line phases and neutral of the transformer. The transformer's winding resistance referred on secondary side (LT) and LT line conductor has a total resistance/phase of 2 ohms. Find the increase in transformer and LT line operating losses instead if a normal three-phase, four-wire supply is given and the village load is put on all the phases in a balanced manner.
2. A test was carried out on an agriculture tubewell three-phase induction motor. The measured values were: power input = 58 kW; stator losses = 1.1 kW; slip = 4%. Calculate the mechanical power developed and rotor copper losses.

3. The annual agriculture sector electricity consumption in India is about 2000 kWh/kW of connected load and this is increasing every year. This consumption is on the higher side as compared to that in other countries. Give at least eight reasons and one most important remedy for this situation.
4. Where will you use a load limiter? Give its advantages.
5. What are the different methods for providing 24-hour power supply to villages in agriculture-predominant states where there is a power shortage regime?

BIBLIOGRAPHY

1. (i) IS: 1231–1974 (third revision).
   (ii) International Electrotechnical Commission Publication: 72–1 (Part 1). Dimensions of three-phase foots-mounted induction motors.
2. IS: 2223–1983 (first revision), Dimensions of flange-mounted ac induction motors.
3. Mahmoud, M., 'Photovoltaic generators instead of diesel motors for water pumping from rural desert wells in Jordan', *IEEE Proceeding*, Vol. 137, No. 6, November 1990, pp. 391–394.
4. RITZ, *Merswandlerwerk GMBH*, Hamburg, West Germany.
5. "REC specifications and construction standards", Rural Electrification Corporation, New Delhi, issued in 1974/93.
6. Pabla, A.S., 'Effective Rural Distribution System in Punjab', *Proceedings of All India Seminar on Power Distribution Systems*, Institution for Innovation in Education and Technology, New Delhi, 4–5 February, 1987.
7. Cook H. Willis and others 'Distribution System Mapping for Rural Cooperatives, '*Transmission and Distribution*', February, 1992, pp. 22–25.
8. IS: 10804–1994, Recommended pumping system for agricultural applications.
9. *World Bank Report No 22171-N-India Power Supply to Agriculture*, South Asia Regional Office, June, 15, 2001.
10. *Guidelines for Formulation of System Improvement Programmes*, Rural Electrification Corporation, New Delhi, 1987.
11. *Annual Report on the Working of State Electricity Boards and Electricity Departments*, Planning Commission, Govt. of India, June, 2001.
12. Iliceto F. and others, 'Three phase and single phase electrification using insulated shield wire of HV lines energised on MV; 15th International Conference on Distribution, Nice 1999, pp. 229–236.
13. REC Specification–14/1979 (provisional), SWER distribution transformers.

14. *CIGRE Proceedings*, 'Design of Compact lines', subject 22.3, Paris 1990.
15. Meslier, F., P. Messager and F. Hemmer, "Rural medium-voltage network quality of supply and expansion planning", *9th International Conference on Electricity Distribution, CIRED*, 1987, AIM, Liege, paper a13.
16. Say M.G. 'Performance and Design of A.C. Machines,', 3rd edition, The English Language Book Society and Sir Issac Pitman and Sons, London, pp. 365–368.

Chapter 15

Power Capacitors

15.1 Reactive Power

Current passing through a wire creates an electromagnetic field (electric and magnetic fields). Physically, reactive power and energy is related to the magnetic field energy required in generators, transformers, power lines, motors, other loads etc. Reactive power energise the magnetic field. The operating power from distribution system is composed of both active and reactive elements. The typical daily load curves (reactive and active power) for a 66 kV feeder is given in Fig. 15.1.

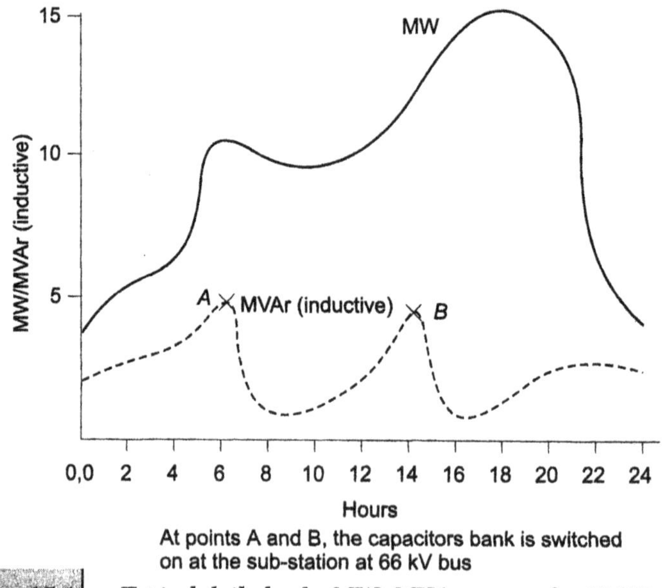

Fig. 15.1 *Typical daily load—MW, MVAr curves of a 66 kV outgoing radial Feeder*

The reactive power is consumed by overhead lines, transformers and loads. Its proper generation and control is important for maintaining the network voltage under normal and abnormal power system operation and to reduce system losses. The system voltage collapse due to lack of global control of reactive power flow [12] during crucial contingencies is emerging as a great problem. The system in northern grid has collapsed many times during the past few years due to lack of reactive power in the region. Cost per kVAr of reactive power generation varies between 2 per cent and 4 per cent of per kW installed generation costs, and will rise with increasing use of automatic and regulating controls. The system should be planned for carrying out suitable proportion of reactive energy.

15.1.1 Lines (11)

(a) The distribution lines consume reactive power (I^2X) depending upon the series reactance (X) and load current (I). The series reactance of line is proportional to the conductor self-inductance, which decreases as the spacing between conductors decreases. The minimum spacing is kept such as to prevent flash-over in foul weather conditions. According to a study (1985) of rural area in Punjab, the average reactive power consumed by primary HT (11 kV) and LT lines for each transformer at full-load is almost 1 per cent of transformer rating. The underground cable (11 kV and above) generate reactive power ($I_c^2X_c$) depending upon the capacitive reactance (X_c) and charging current (I_c) at rated voltage.

(b) Increasing the line voltage at transformer and motors above the rated voltage will increase the consumption of reactive energy, the extent of which depends upon their designs. Generally, increase of 10 per cent of the rated voltage will result in about 20 per cent reduction of power factor.

(c) In a distribution system supplying composite load, the general relation of reactive power with voltage is given in Fig. 2.2. The composite load may include some elements of lighting and domestic loads, motor load, inverter and rectifier load, as well as transformer and line losses. The curve is non-linear (parabolic) in nature. Reactive power is a function of series leakage and shunt magnetizing reactance of transformers, motors and other elements of lines and loads. While the series leakage components is load dependent and varies with the square of the load current, the magnetizing reactances vary with the voltage.

EXAMPLE: A case study was made at 132 kV grid sub-station, having 10/12.5 MVA, 132/11 kV transformer with 11 kV bus connected to 2.52 MVAr capacitors bank and about 5.5 MVA load (composite load of various rural, urban and industrial feeders through 11/0.433 kV transformers and motors rated 400 V). The 11 kV bus voltage was increased in steps with OLTC control and for line voltage 10.6, 10.9, 11.5 and 11.7 kV, at incoming panel, the lagging power factors 0.86, 0.86, 0.80 and 0.77 were found respectively.

This indicates that 6 per cent increase of voltage reduces power factor by 9 per cent. It is interesting to note that as capacitors reactive compensation increase with square of voltage, even then in this case power factor decreases with the increase of voltage. The observations without capacitors could not be taken due to non-operation of power factor metre in low range, i.e. below 0.65 power factor. However, with voltage increase, power factor reduction would be comparatively higher in that case.

15.1.2 Distribution Transformers

The reactive power consumption takes place in the series leakage and the shunt magnetizing reactances. The second component is voltage dependent, while the first component is proportional to the square of the transformer current. A completely unloaded transformer would be very inductive and has a very low power factor. The reactive power used by a transformer (up to 100 kVA capacity) at full load and at rated voltage is approximately 7–9 per cent of the rated power of the transformer. When unloaded, the amount of reactive power remains between 3 and 4 per cent of the rated power. The figures are applicable to conventional core construction in vogue. However, with wound-core construction reactive power, these figures will be lower by about 2 per cent. For higher size transformer (above 100 kVA capacity), the reactive losses are lower by 1–2 per cent. Numerically, the reactive kVA power consumed by a transformer is:

$$= \frac{\text{Percentage impedance} \times (\text{kVA load})^2}{100 \times \text{rated kVA}}$$

$$+ \frac{\text{Percentage no-load current} \times \text{rated kVA}}{100}$$

15.1.3 Loads

Loads have varying reactive power requirement. The low power factor can be a result of the equipment as in case of welders, or it can result from the operating conditions under which equipment is used as lightly loaded transformers and lightly loaded induction motors (see Sec. 2.1 and Appendix IX) are the worst offenders.

15.1.4 Monitoring

(i) Reactive power monitoring is desirable to maintain efficient system operation, keeping voltage and load stability. Some utilities exercise it by recording daily, the station bus spot power factor at minimum and maximum loads, or maximum and minimum power factor with respective loads. Efforts should be to keep the each station bus incoming power lagging power factors continuously not less than 0.95, 0.96, 0.97, 0.98 respectively at up to 22 kV, 33–110 kV, 132–220 kV, above 400 kV system.

(ii) Another method of reactive power check is made by noting daily hours of capacitor banks utilization. Total number of hours during a day or a month are successively compared to check the quality of reactive compensation. The emphasis should be to increase the hours of capacitors use.

(iii) Best course will be to install kVAr, kW energy consumption metering at 11 kV bus. Determine the ratio of kVArh to kWh on monthly or yearly basis. Efforts should be made to reduce this ratio. Circuit MVAr/MW metering is important at grid sub-stations.

15.1.5 Compensation

Mainly capacitors are used to develop reactive power near the point of consumption. For capacitor compensation at load, the user reaps the same advantage as the power utility for higher power factor on small scale. Also, if each load is compensated, the power factor remains relatively constant since in plants, loads are switched on and off and the dangers of over-compensation do not exist. If, however, power factor has been corrected only at the service entry, system power can make relatively wide swings as heavy loads are switched on and off. Suitable capacitor banks at grid or main sub-station are desirable to feed reactive requirement of lines, transformers and domestic consumers, etc., who have no capacitors at terminals.

At the point of capacitors over-compensation, the voltage rises and capacitive kVAr will export, i.e. it will flow to reverse side of the load. Under certain conditions, dangerously high transient voltage may prevail on the power lines at a point far removed from the actual load. Also, torque surges of over 20 times the motor rating can be transmitted to the motor shaft under rapid restart conditions. Over-compensation at consumer premises is not generally desirable. However, some over-compensation at the primary distribution bus (11 or 22 kV) at main or grid sub-station is tolerable to supply the reactive demand of local power transformers (say 66/11 kV class) of nearly 10 per cent of the transformer power kVA rating. This will allow bus power factor to cross up to 0.90 leading.

In European countries, fixed capacitor compensation on distribution transformer is provided (as shown in Fig 15.2). Minimum capacitor size is determined by the relation [2].

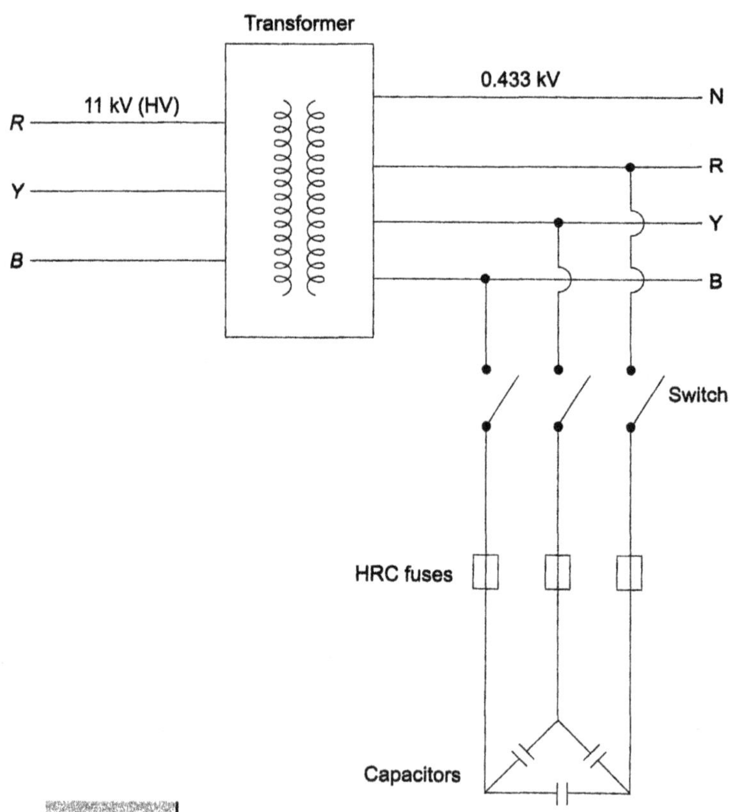

Fig. 15.2 *Typical capacitor connection to transformer*

$$\text{Minimum capacitors in kVAr} = \frac{\text{Transformer kVA} \times \%\,\text{Impedance}}{200}$$

The minimum capacitor size can be connected permanently on secondary terminals, irrespective of load. This is the value of capacitance required to compensate the no-load magnetizing kVA of the transformer. However, for minimum base load conditions, fixed LT capacitors of sizes 36, 18, 9, 3 kVAr, 440 V rating have been installed respectively on each 200, 100, 63, 25 kVA transformer in Punjab with an approximate total 1350 MVAr capacity on nearly 1,50,000 transformers. The annual failure rate is about 3–4 per cent. These capacitors comply with respective IS: 13340, 13341 and 13585 (part-1). The rated voltage of the capacitor unit is equal to the maximum service voltage of the network. Haryana has adopted the same practice. The Central Electricity Authority (CEA) has also approved this scheme. However, rigorous field studies of this practice need to be taken to find if any leading power factor condition occurs. Power supply at leading power factor creates current oscillations at the consumer's end. At a leading power factor supply, most of the electronic equipment will not operate properly. However, switched capacitor of higher sizes can be connected, depending upon the load, with automatic switching controls as shown in Fig. 15.3.

The capacitors are automatically controlled by a pre-programmed micro-processor based control unit, on the basis of VAr, current, voltage of distribution transformer or feeder. According to load, the switching operation of capacitor units are done in steps so as to closely follow the load curve. This gives a closer compensation and reactive load, and saves load losses in distribution transformers, 11 kV lines and upwards. According to studies, the investment on this control equipment and capacitors can be recovered in a period of less than a year, only from saving accrued in the line and transformer losses.

The following sizes of LT switched capacitors are being used in practice in India, by various State Electricity Boards by automatic switching control.

Distribution transformer kVA	LT switched capacitors kVAr
100	36
63	27
25	9

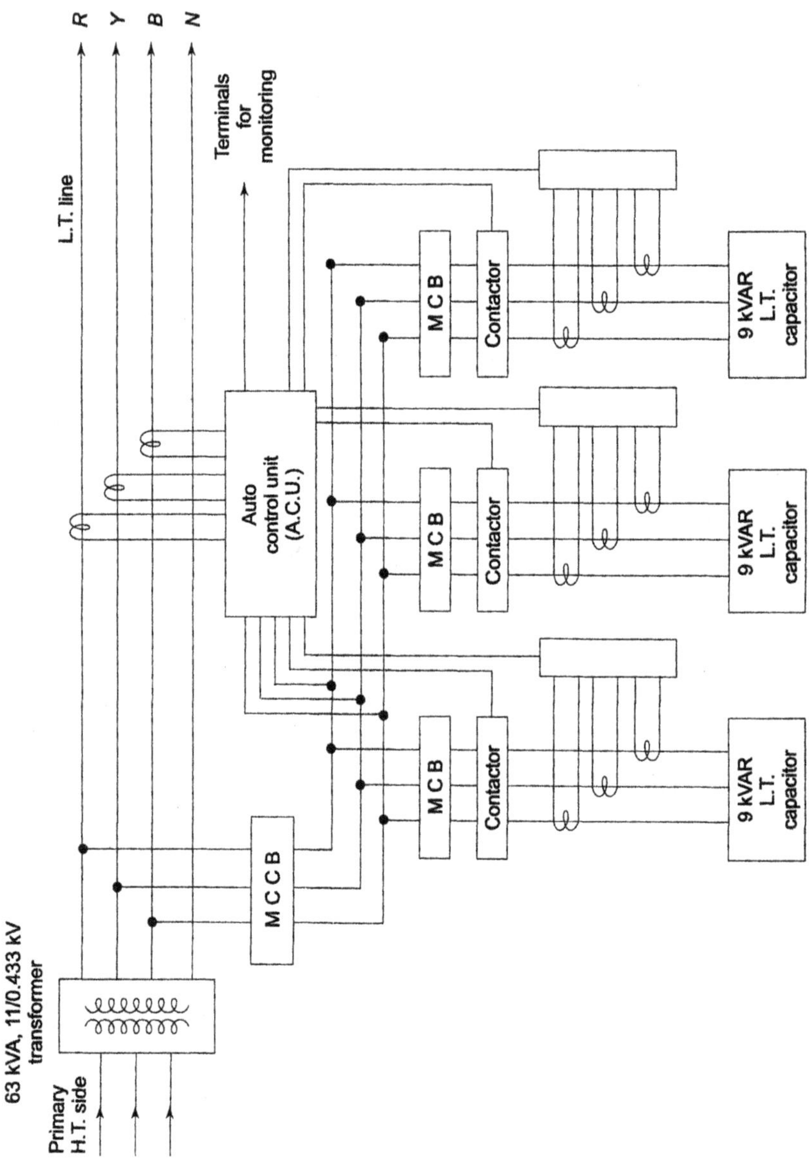

Fig. 15.3 *Typical scheme for automatic control of LT capacitors at distribution transformer*

REC has approved this practice as per their Specifications: 67/1993. Proper maintenance of the automation scheme needs properly trained field staff. The scheme along with an initial two years' maintenance and training contract should be given to a competitive vendor on a turn-key basis.

However, reactive compensation at 11 or 33 or 66 kV level is determined by computer-aided system studies. Adequate capacitors at 66, 33 and 11 kV level are provided at 220/66 and 132/33 kV sub-stations respectively on 66, 33 and 11 kV buses. Adequate steps (by installing isolating switches) in switched capacitors' banks may be provided for better utilisation of capacitors at low load conditions, specially for 11 kV capacitors. At distribution sub-station 11 kV bus, the power factor of the incoming load (incoming feeder) should be not less than 0.95 lagging and not beyond 0.95 leading. For any outgoing load (outgoing feeder), the power factor should not be less than 0.90 lagging. Fixed capacitors (usually bank sizes: 450 or 600 kVAr) can be provided on 11 kV feeders at suitable locations.

15.1.6 Maintaining of System Voltage

The management of MW requirement is far more easy than the management of MVAr requirement, in the sense that the MW capacity to be commissioned has to take care of the MW load requirement, the losses and auxiliaries. Once the balance has been struck, maintaining the declared frequency can be very well assumed. The solution is not that straight forward when the question of fulfilling MVAr requirement comes in the way. This is because, a simple balance of MVAr load requirement, and overall MVAr losses in the system with MVAr generating would alone not ensure maintaining of declared voltages, all over in the system. For maintaining declared voltages, within permissible tolerance at each busbar in the system, calls for balancing the MVAr inputs and MVAr outputs at each of the buses, keeping the voltage as an invariable function.

The drop in voltage due to MW flow alone, still requires to be made good by suitable means by adjusting taps on transformers. This, therefore, puts limits on tolerable voltage regulations on transmission lines, subtransmission lines and distribution feeders.

The operation of tap changers of transformers do not help the system when the system starves of MVAr. The MVAr loads mainly depend on the voltages. When the system is already short of MVAr and voltages fall, the operation of taps to increase the voltages, further aggravate the situation as it increases the MVAr requirement. The operation of taps, therefore, does not help in any way to the system, though it may give relief to the consumers to some extent.

It is argued that when there is a deficit in MW generation, the system demand is reduced and balanced by load-shedding and by imposing restrictions on drawal of power, and the voltages could be satisfactorily maintained with the available transmission system. However, experience has not always proved the above right. The generating units which are already on the bar, are being manoeuvred to generate as many MW to off set the shortage due to units, which are under outage, and this puts a further limit on MVAr generation from these units. This causes reduction in voltages on generating station busbars. The tap changers on the receiving station and sub-station transformers are further operated to even out these drops in voltages, causing MVAr loads to hang on the system. This causes a further drop in voltages, and the system starts deteriorating.

Figure 15.4a shows a single-line diagram of a simple distribution system. Fig. 15.4b indicates the typical voltage-load curve for a system. The voltage at the receiving end (V_r) is plotted against the total power delivered to these loads for a number of different power factors. As the power factor reduces i.e. 0.75 p.f. (say), the system is starved of reactive power. The load bus voltage shifts from point $A1$ to $A2$ on the downward portion of the operational curve. This causes voltage instability and leads to system voltage collapse. The solution is to add more capacitors at load bus and/or add more generating capacity at or near the load bus, to enable the stable operation of the system near unity p.f. for maximum power transfer capability.

15.2 Series and Shunt Capacitors

Modern *film capacitors* have packs consisting of sandwiches of aluminium foils and polypropylene films. A capacitor unit is made of groups of capacitor packs connected in series and parallel and sealed in a dry or oil-filled container. LT capacitors are usually dry. For protection of the unit, fuses in the series with each internal pack may be provided (see IS: 12672). Alternately, a single fuse may be provided externally for each

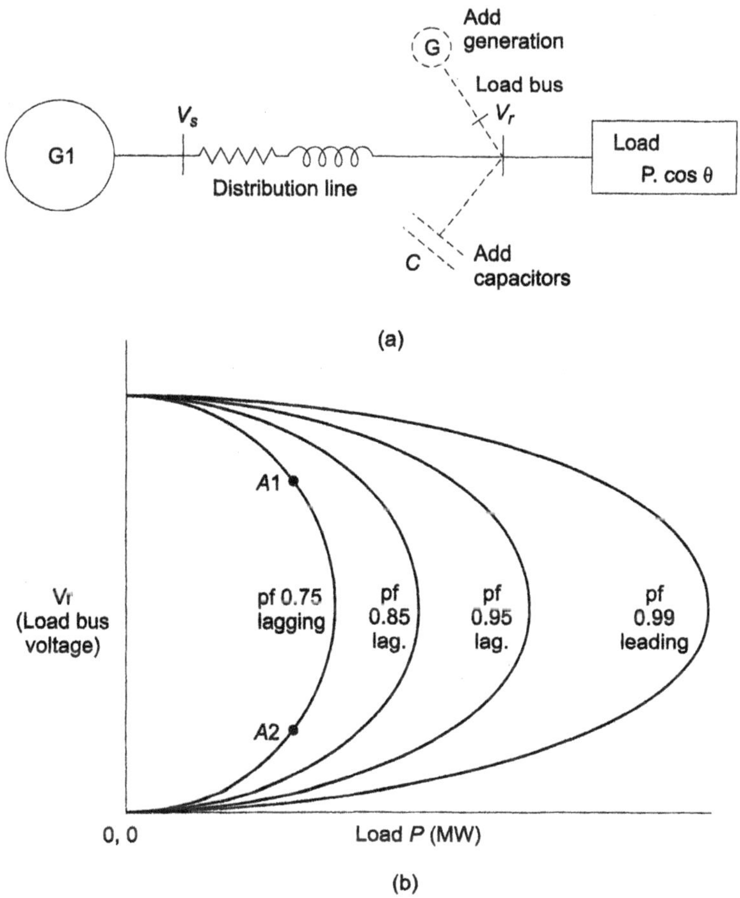

Fig. 15.4 *Effect of a capacitive reactive power starving and adding generation/capacitors at the load bus*

unit. Capacitors aid in minimizing operating expenses and allow the utilities to serve new loads and consumers with a minimum system investment. Utilities in the advanced countries have approximately 1 kVAr of power capacitors installed for every 2 kW of installed generation capacity in order to take advantage of the economic benefits involved. The extent of requirement of capacitor compensation to unity power factor with different power factors is shown in Fig. 15.5.

CEA has recommended 1.5 kVAr capacitors for every 2 kW generation capacity installed in the system.

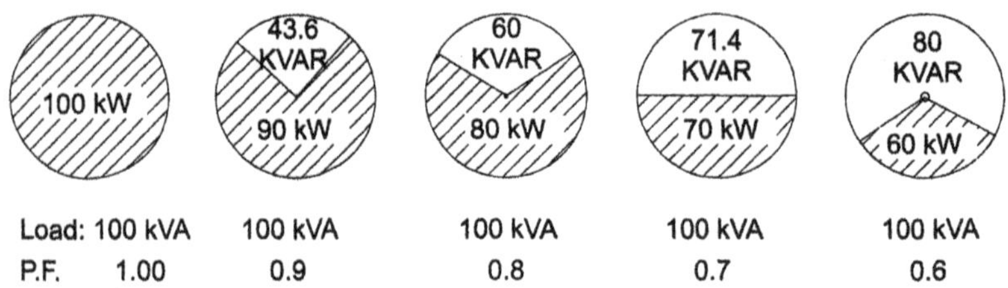

Fig. 15.5 *Active and reactive power change with power factor—keeping the load apparent power constant*

15.2.1 Comparison

Series and shunt capacitors in a power system generate reactive power to improve power factor and voltage, thereby enhancing the system capacity and reducing the losses. In series capacitors the reactive power is proportional to the square of the load current (I^2X_c), whereas in shunt capacitors it is proportional to the square of the voltage (V^2/X_c). There are certain unfavourable aspects of series capacitors. Generally speaking, the cost of installing a series capacitors is higher than that of a corresponding installation of a shunt capacitor. This is because the protective equipment for a series capacitor is often more complicated. In addition, it is usual that series capacitors have to be designed for a rather higher power than the one in force to cope with future increases in load (see IS: 9835).

Figure 15.6a shows an equivalent circuit of a distribution circuit along with its vector diagram. If we use *shunt capacitors*, the circuit and the corresponding vector diagram will be as shown in Fig. 15.6b. In this case, there is improvement in voltage and reduction in the line current (I′). Therefore, by installing shunt capacitors:

- The system's available capacity is increased.
- System losses are brought down by reducing the kVA flow and adding capacitive kVAr (see Fig. 15.7). By addition of CkVAr capacitors, demand on the system is reduced from kVA_1 to kVA_2.

Basically, capacitors should be placed as close as possible to the load requiring compensation. More specifically, capacitor banks should be located where the low voltage problem occurs. For this, load flow studies are required for maximum and minimum load conditions.

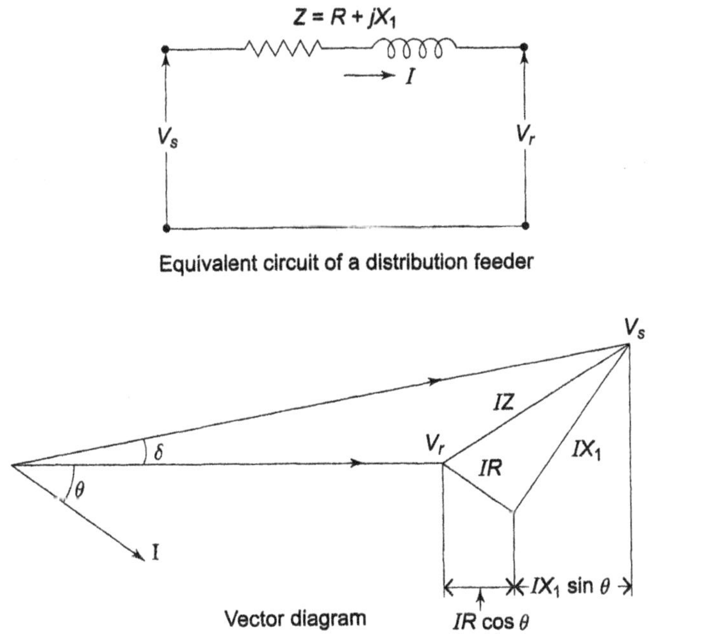

Fig. 15.6a *Equivalent circuit of a distribution feeder and vector diagram*

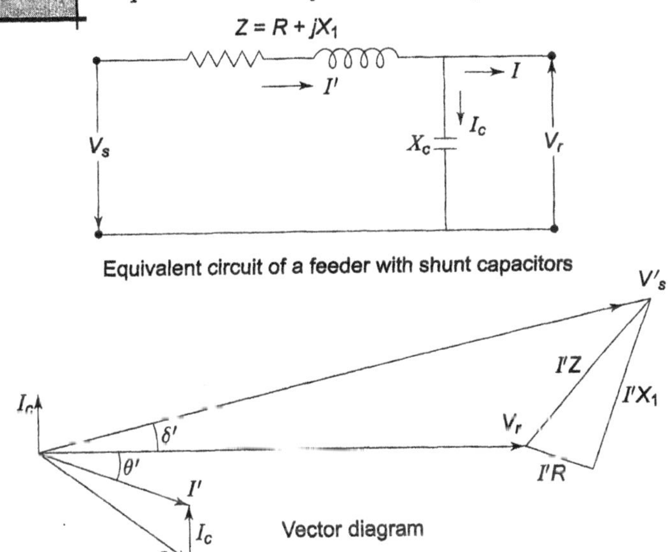

Fig. 15.6b *Use of shunt capacitors*

When series capacitors are installed in heavily-loaded long feeders (Fig. 15.6c), voltage drop in the feeder is reduced from $IR.\cos\theta + IX_l.\sin\theta$ to $IR.\cos\theta + I(X_l - X_c).\sin\theta$. Series capacitors are most effective on circuits with higher X/R ratio and for load variations involving a high reactive content. Series capacitor banks include capacitor, overvoltage protection equipment (metal oxide varisters or triggered bypass gaps), a bypass switch and a control system [6].

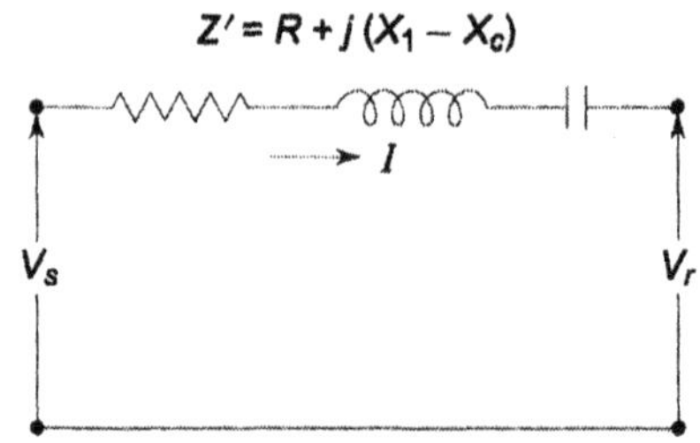

Equivalent circuit of series capacitors

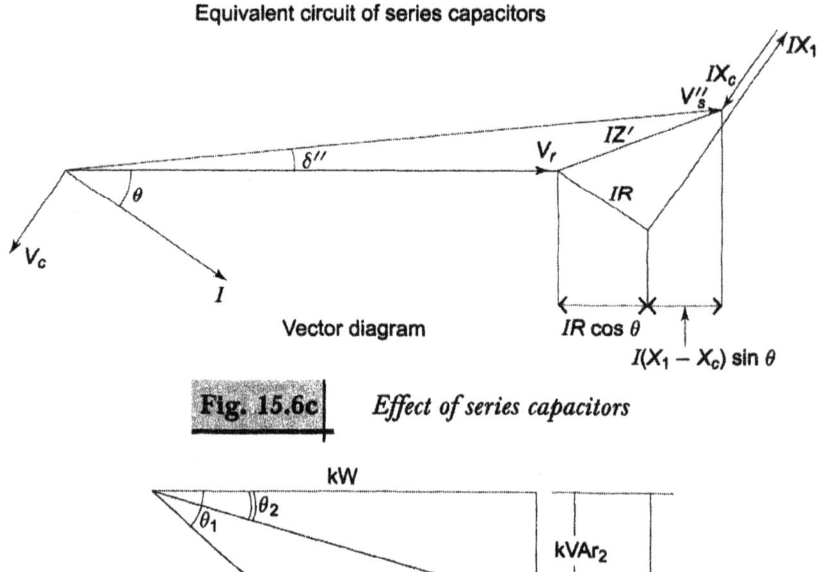

Vector diagram

Fig. 15.6c *Effect of series capacitors*

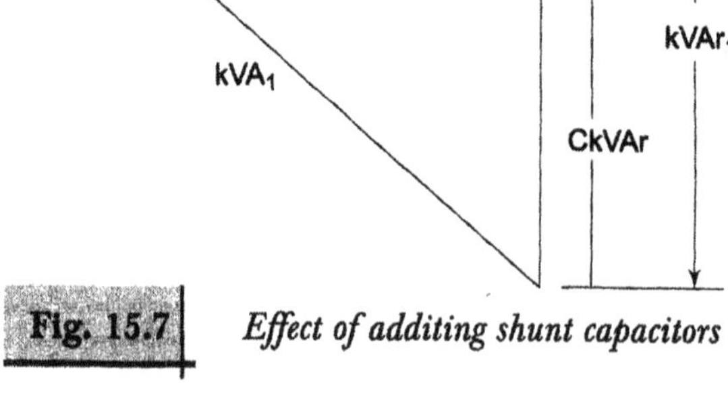

Fig. 15.7 *Effect of additing shunt capacitors*

The factors which influence the choice between the shunt and series capacitors are summarized in Table 15.1.

Table 15.1

Series and shunt capacitors

S.No.	Objective	Preference	
		Series capacitor	*Shunt capacitor*
1.	Improve power factor	Second	First
2.	Improve voltage level in an overhead line system with a normal and low power facor	First	Second
3.	Improve voltage level in an overhead line system with a high power factor	Not used	First
4.	Improve voltage in an underground line system with a normal and low power factor	First	Not used
5.	Improve voltage in an underground line system with a high power factor	Not used	Not used
6.	Reduce line losses	Second	First
7.	Reduce voltage fluctuations	First	Not used

Due to various limitations in the use of series capacitors, shunt capacitors are widely used in distribution systems.

For the same voltage improvement, the rating of a shunt capacitor will be higher than that of a series capacitor. Long, overloaded feeder hold better potential for voltage improvement with series capacitors. Up to 11 or 22 kV line, the line resistances are higher in magnitude than the inductive reactance and at 33 kV level, these are comparable. A part of the resistive drop is cancelled by increasing the degree of series compensation beyond 100 per cent up to 300 per cent. Series capacitors compensation may create certain disturbances: ferro-resonance in transformers, subsynchronous resonance during motor starting, shunting of motors during normal operation and difficulty in protection of the capacitors from system fault current.

Series capacitors are most effective on distribution circuits with higher X/R ratio and for load variations involving a higher reactive content.

EXAMPLE A 33 kV, 100 km feeder is heavily loaded and has low voltage problem at the receiving end during peak hours. Line impedance per phase is 10.5 + j35.3 ohms. How will you solve the low voltage problem?

Solution The most economical solution will be to install suitable series capacitors. We may provide 100 per cent compensation by installing series capacitors at the receiving end. The capacitors size (C) is determined as below:

For 100 per cent compensation:

$$jX_L = -jX_c$$

We know,

$$X_c = \frac{1}{2\pi fC}$$

Therefore,

$$\frac{1}{2\pi fC} = 2\pi fL = 35.3$$

$$C = 1/2\pi \times 50 \times 35.3$$

$$= 90\ \mu\text{F}$$

15.2.2 Shunt Capacitors in the System

Installation of shunt capacitors is important for reactive planning of a power system. It is apparent that it would be most economical if transmission lines are used to transfer only active power where the reactive power requirements of the loads are met within the distribution system at the consumer or at the most at the sub-station level. This would enable optimum utilization of transmission lines, improve their operational performance and reduce energy loss. This requires system studies and careful planning to meet the system reactive power requirement in the same way as active power planning is done and generator capacity additions are programmed [10].

The rated voltage of the capacitor unit should not be less than the maximum operating voltage of the network to which the capacitor is to be connected (see IS: 12360). The economic benefits that can be derived from shunt capacitor installation in the system are:

- Released generation capacity
- Released transmission and distribution system capacity
- Saving in system losses
- Reduced voltage drop and consequently improved voltage regulation

The optimum power factor at any point on the network is that at which the economic benefits of adding shunt capacitors equals the cost of capacitors.

Capacitors in the system are placed in banks and can be connected in any form—grounded star, ungrounded star, double star neutral floating, double star neutral grounded, delta, etc. The delta connected bank is used exclusively with only one series section per phase and is used up to 6.6 kV and in higher voltage, star connections are usually used. For star connection banks, in general, neutral of the capacitors is grounded only if the system or sub-station transformer is effectively grounded.

For the bank design, there are a number of constraints such as a minimum number of parallel units depending on the number of series sections and the need for more expensive current limiting fuses if either the available fault current or stored energy is excessive. For example, a 33 kV grounded star connected bank can be designed with one series section using 19 kV units and current limiting fuses. By using two series groups of 9.50 kV units per phase, the available fault current when a unit shorts, would be reduced so that expulsion fuses can be used. Also, 9.5 kV units are less costly than the 19.0 kV design. The loss of a unit increases the voltage across the group, thus increasing the possibility of failure of the remaining units. Any overvoltage may heat up a unit and cause a decrease in its life. The minimum number of units required to limit the voltage increase to 10% can be calculated from the following equation for the grounded star bank on its one unit loss.

$$N = 10\left(1 - \frac{1}{S}\right) - \frac{1}{S} + 1$$

where,

N = Minimum number of parallel units in one series section

S = Number of series sections in each phase.

It is better to have two levels of protection on large-banks. One relay is set to operate the alarm in the case of loss of one unit and the second relay to trip the bank circuit breaker when the allowable overvoltage levels (usually 10 per cent) are exceeded. A capacitor unit is protected by either internal '*wire*' fuses or an external fuse (expulsion type or HRC). Their relative advantages are given in Table 15.2.

Table 15.2

Merits of internal and external fuses

Item	*Internal fuse*	*External fuse*
Unbalance protection	Difficult and costly	Easy and economical
Occurrence of tank rupture	Sometimes	Rare
Fault detection, replacement time	High	Low
Regular measurement of units	Required	Not needed
Power losses	High	Low

Typical protection scheme at 132/11 kV, 66/11 kV, 33/11 kV substation may be comprising over-voltage relay; under-voltage relay; neutral voltage displacement relay; time-delay relay; over-current and earth-fault relay as shown in Fig. 15.8. When the reactive demand fluctuates frequently between wide limits, feature of automatic sequential switching of parallel capacitor banks be provided by overcurrent sequence switching relay.

High energy metal oxide arresters are required to protect capacitor banks, because:

- To prevent capacitor failures at a breaker restrike.
- To limit the risk of repeated breaker restrikes.
- To prolong the service life of the capacitors by limiting high over-voltages.
- To serve as an insurance against unforeseen resonance conditions, which otherwise would lead to capacitor failures.
- For overall limitation of transients related to capacitor bank switching, which can be transferred further in the system and causes disturbances in sensitive equipment in the system and consumer end [6].
- To serve as protection against lightning for capacitor banks connected to lines.

Arresters connected between phases give a superior protection, as compared with arresters between phase to ground.

15.2.3 Capacitors Operation

Capacitors are designed for 10 per cent over-voltage and 30 per cent over-current (see IS: 13925). Some capacitors may operate as fixed to

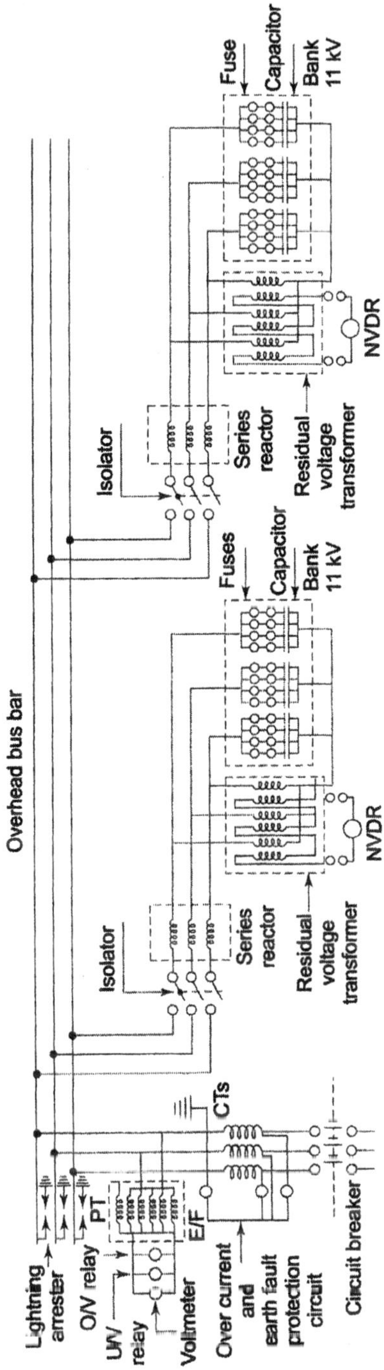

Fig. 15.8 *Typical protection for 11 kV capacitor banks at sub-station; NVDR: Neutral voltage displacement relay*

cater to the system's no-load and minimum load requirements. The other option is switched capacitor banks which are switched according to the requirement of the reactive power of the load in the daily load profile. The typical annual duration curve for use of capacitors is shown in Fig. 15.9.

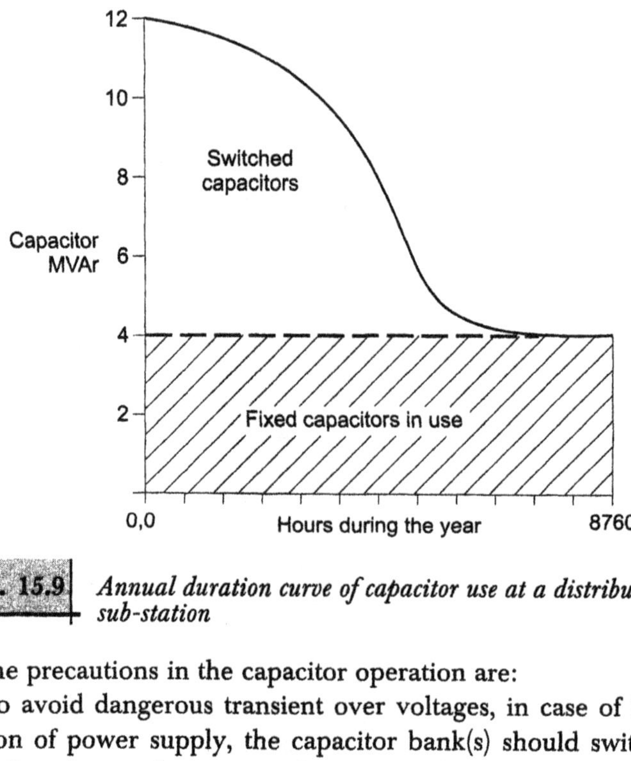

Fig. 15.9 *Annual duration curve of capacitor use at a distribution sub-station*

Some precautions in the capacitor operation are:

(i) To avoid dangerous transient over voltages, in case of interruption of power supply, the capacitor bank(s) should switched off before the supply is restored.

(ii) Power transformers at the sub-station and capacitor bank should not be charged simultaneously, while restoring supply after a trip out or shut down. Capacitor banks need to be put into service, one by one, as the situation warrants. However, reclosure of a breaker controlling capacitor bank may not be attempted before an elapse of 5 minutes from the moment of switch off, to ensure that residual voltage after switch off comes down to a safe value of 50 V or less.

(iii) If the voltage at the bus on which capacitor bank is connected reaches beyond 1.1 times or higher, the capacitors should be switched off.

15.3 System Harmonics

At present no standards are followed in India to limit the harmonic voltage in the supply system. However, with the increasing use of rectifiers, thyristor control shunt capacitors, etc. the harmonic generation can create problems with sensitive electronic equipment which can lead to errors in the protective gear, metering and control systems and also overload the capacitor banks besides causing resonance.

Reactors are sometimes needed to reduce the total current drawn by the capacitor banks if unduly large harmonics are present in the system voltage. As discussed in Sec. 4.8, the causes of harmonics in the power system may be due to:

(a) Over-excited transformers
(b) Mercury rectifiers and thyristors
(c) Arc furnaces
(d) Slot harmonics of rotary machine
(e) Corona loss
(f) Large motors running at low slip, fluorescent lighting, T.V. and radio equipment
(g) System switching and load chocking

The capacitors are designed to tolerate 30–35 per cent extra loading which may be caused by harmonics or overvoltage.

System study for harmonic contents is an important consideration before using reactors for capacitors installation. The third harmonic voltage in the supply system is generally eliminated due to the delta winding of the transformers. The fifth harmonic is the strongest and is followed by the seventh harmonic. The eleventh and thirteenth harmonics may also be present. The presence of capacitor banks can augment these harmonics and distort the voltage waveform. As the impedance of the capacitor bank reduces on higher harmonics, the current drawn increases appreciably even if a small amount of higher harmonic voltages are present. In the case of single bank installations, the system reactance also assists in reducing the higher harmonic currents. The location of the capacitor bank in the power supply network whether at the end of a long line or sub-station bus, or close to the transformer as well as the kVAr of capacitor bank can determine the extent of the harmonic generated. Figure 15.10 shows an equivalent circuit represented by a source impedance to the *n*th harmonic, Z_{sn} across which the capacitor bank is connected. The harmonic current and voltage due to the *n*th harmonic is given by:

$$I_{cn} = \frac{E_{sn}}{Z_{sn} + Z_{cn}}$$

$$V_{cn} = \frac{E_{sn} \cdot Z_{cn}}{Z_{sn} + Z_{cn}}$$

With the introduction of a series reactor in the capacitor circuit the condition of resonance can be prevented. The *n*th harmonic current and voltage will then be (neglecting resistances of the capacitor and reactors)

$$I_{cn} = \frac{E_{sn}}{Z_{sn} + j(X_{ln} - X_{cn})}$$

and

$$V_{cn} = \frac{E_{sn}\ j(X_{ln} - X_{cn})}{Z_{sn} + j(X_{ln} - X_{cn})}$$

where X_{ln} is the reactance of the series reactor to the *n*th harmonic and is normally greater than X_{cn}. The resultant impedance to the *n*th harmonic will be inductive. Thus a condition of resonance seldom occurs.

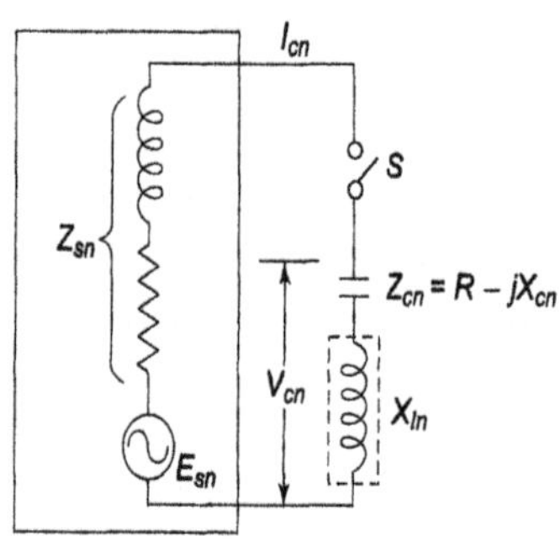

Fig. 15.10 *Equivalent circuit for harmonics:*
E_{sn} = nth-harmonic component of line voltage
Z_{sn} = nth-harmonic impedance of source side
Z_{cn} = nth-harmonic impedance of capacitor set
V_{cn} = Terminal voltage of capacitor
X_{sn} = Reactive component of Z_{sn}
X_{ln} = nth-harmonic reactance of series reactor

The series reactor can be so chosen as to eliminate a particular harmonic. The fifth harmonic being the most predominant for its elimination, we have the condition

$$Z_{c5} = j(5X_{l1} - X_{c1}/5) \geq 0$$

where,

X_{l1} = reactance of series reactor due to fundamental wave

X_{c1} = reactance of capacitor due to fundamental wave

Therefore,

$$X_{l1} \geq X_{c1}/25$$

Thus a reactor of 4 per cent value of capacitive reactance of the capacitor to the fundamental wave will eliminate the fifth harmonic. In practice, a 6 per cent and higher reactor is generally chosen as due to a fall in system frequency, the overall reactance may become capacitive. Due to source impedance, the actual % reactance needed may be much less. Series capacitors are manufactured as per IS: 5553(4)–1989.

15.4 HT Shunt Capacitors' Installation Requirements

The HT capacitors, 11 to 132 kV may be of the switched and non-switched type, depending on the minimum loading, maximum voltage conditions of feeders or substations. In case of nonswitched capacitors, the switchgear and damping reactors are not required. It has been found economical to install fixed capacitors (pole mounted) directly on long and heavily-loaded 11 kV feeders for compensation up to 30 per cent of kVAr of average feeder load. For switched capacitor banks, the switching and damping of inrush currents and the suppression of harmonics need special consideration. In the case of single capacitor banks, the damping reactor is not normally required from the consideration of inrush currents at the time of switching. The system reactance including that of the transformer at which the capacitor bank is installed is adequate enough to bring down the value of inrush currents within safe limits of the capacitor or switchgear. Even otherwise the duration of inrush currents is so small, of the order of a few cycles only, that their effect can be ignored. When a number of capacitor banks are used in parallel, it may become necessary to use series reactors for limiting the inrush currents. This is because the banks are not switched in at the same instant. The severest conditions of switching are when one bank is already switched in and the second bank is being switched with two poles of the second breakers closed and the third pole in transit for closing. The capacitor bank which is switched in a little later is also fed from the energy stored in the other bank and as such the inrush current may exceed the safe limits. Capacitors manufactured in India normally can safely withstand an inrush current of up to 50 times the rated current.

Assuming severest conditions of switching, the value of damping reactors in henry is given by:

$$L \geq \frac{Q_2}{Q_1}\left(\frac{U_n^2}{Q_1 + Q_2}\right) \times 1.27 \times 10^{-6}$$

where,

Q_1 = capacity of the bank to be switched in MVAr
Q_2 = capacity of the bank already switched in MVAr
U_n = rated voltage kV of the bank

Inrush Current

The maximum peak inrush current can be approximately given by the formula:

$$I_{max} = I_{c1}\left(1 + \frac{X_{c1}}{X_{l1}}\right)$$

where,

I_{c1} = capacitor's rated current (fundamental wave) rms
X_{c1} = capacitor reactance (fundamental wave)
X_{l1} = total inductive reactance of the system including capacitor bank (fundamental wave)

The inrush current comprises a steady component of forced oscillation at supply frequency and a free oscillation of frequency:

$$\text{Inrush current frequency } f_0 = \frac{1}{2\pi}\sqrt{\frac{1}{L_1 C_1} - \frac{R^2}{4L_1^2}}$$

Neglecting terms, $R^2/4L_1^2$ because R is very small as compared to L_1

$$f_0 \approx \frac{1}{2\pi}\sqrt{\frac{1}{L_1 C_1}}$$

L_1 is source inductance (H), C_1 is capacitance of single bank (F) (see Fig. 15.11a).

In case of parallel banks, which are already energized, the inrush current is predominantly governed by the momentary discharge from the energized banks and since the impedance between the energized capacitor bank and the capacitor bank to be energized may be small, it may result in high peak inrush current. The maximum peak current is given by the expression: [6, 9] (see Fig. 15.11b)

$$I_{max} = \sqrt{2} E_{n1} \sqrt{\frac{C}{L}}$$

where,
$$f_0 = \frac{1}{2\pi} \sqrt{\frac{1}{LC}}$$

$C =$ equivalent capacitance of the circuit in μF

$L =$ equivalent series inductance between the energized banks and bank to be energized in μH

$E_{n1} =$ line to neutral voltage

IEC publication 70 for capacitors recommends maximum inrush current not more than 100 times the rated rms current of capacitors.

Thus it may be desirable to install parallel capacitor banks with series reactors. The most important point to check is that such capacitors must have matched voltage rating with respect to reactors. Series reactors are normally installed to limit inrush currents and to prevent excessive harmonic voltages. Series reactors chosen with respect to harmonics are large enough that inrush currents cause no problems for capacitors and circuit breakers. Provision of the discharge devices is important to limit the residual voltage less than 50 V after switching off the capacitors [1].

Typically, distribution system capacitors' switching overvoltage ranges from 1.1 to 1.6 per unit. Transient frequencies due to switching usually fall in the 300–1000 Hz range [5]. Capacitor switching in the system causes transient magnification in a consumer installation having power factor correction capacitors. In addition, nuisance tripping of adjustable-speed-drives, consumer sensitive equipment damage or failure and computer network problems can occur, even if the consumer does not have capacitors. If analysis confirms such system conditions, the consumer can take the following remedial measures [5]:

(i) Applying MOV surge arresters

(ii) Converting power factor correction capacitors into harmonic filters (see Section 4.8).

EXAMPLE Two 11 kV capacitors banks, each of 2.5 MVAr, are installed in parallel at a 132/11 kV sub-station. Calculate the value of the series damping reactor with each capacitor bank.

Solution

$$Q_1 = Q_2 = 2.5 \text{ MVAr}$$
$$U_n = 11 \text{ kV}$$

By using the formula given in Section 15.4,

$$\text{Damping reactor inductance } (L) \geq \frac{121}{2.5 + 2.5} \times 1.27 \times 10^{-6}$$

$$\geq 31 \text{ micro henrys}$$

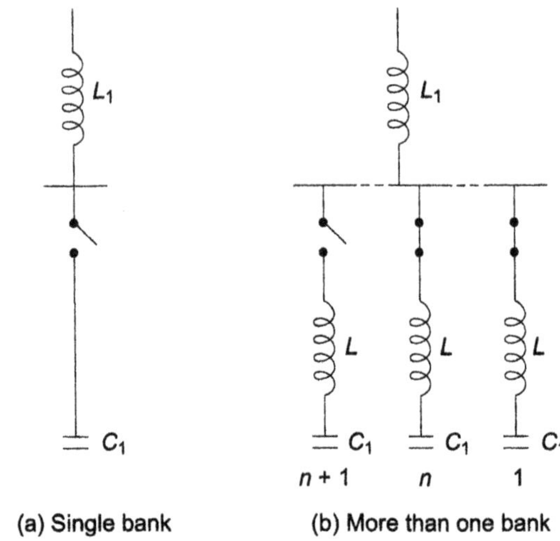

Fig. 15.11 *Switching of capacitors*

15.4.1 Circuit Breakers

The factors which control the design of a circuit breaker are:

(a) The speed of contact separation
(b) The speed of removal of ionized particles between the contactors
(c) The highest residual voltage that the contacts can withstand
(d) The highest induced voltages that the system can be permitted to withstand when used for controlling the transformers.

This last factor is important for the circuit breaker designed for general application and which may not be suitable for capacitor control. Capacitors switching very soon erode the contacts of a breaker. They leave fairly high discharge energies in a breaker while contact separating. These factors severely force down the breaker rating. For example, a general purpose Oil Circuit Breaker (OCB) which has a 250 MVA rupturing capacity can handle about 800 kVAr of capacitor load across 11 kV. A series reactor can help or even harm depending on the parameter of the circuit. When the circuit breaker has auxiliary contacts, these should be watched. They are the ones which are stressed most. Some circuit breaker manufacturers introduce a resistance between the main and auxiliary contacts for capacitor application. IS: 2834–1964 specifies the option of providing an auxiliary resistor for breakers meant for capacitor switching. Caution must be exercised in selecting circuit breakers which operate without causing excessive overvoltage due to restriking.

Restrike-free circuit breakers are recommended (see IS: 13118) for switching back to back capacitor banks. In that case high stresses owing to restriking currents are avoided and overvoltage to earth and across capacitors can be limited to 2 pu. It is important that breakers are designed for inrush current peak value and frequency and should withstand endurance test for 10,000 close-open operations (on the basis that the breaker is operated once a day for 30 years). Automatic SF_6 capacitor switches are found quite successful in controlling the pole-mounted capacitors on 11 kV lines.

15.5 Size of Capacitors for Power Factor Improvement

The size of capacitors to improve the power factor of the system at certain points can be computed with the help of the computer studies of the system. Manual calculations can also be made of comparatively small distribution system for the capacitor kVAr required to improve the power factor say from cos ϕ_e (existing) to cos ϕ_d (desired) with the following equation:

$$\text{kVAr} = \text{kW} (\tan \phi_e - \tan \phi_d)$$

or

$$\text{kVAr} = \text{kW} \times \text{MF}$$

where MF = multiplying factor

The nomogram shown in Fig. 15.12 solves this equation. With the help of this nomogram the MF for any improvement in the power factor can be read directly. Capacitor kVArs required for this improvement shall be the simple multiplication of MF and kW as shown in the following example.

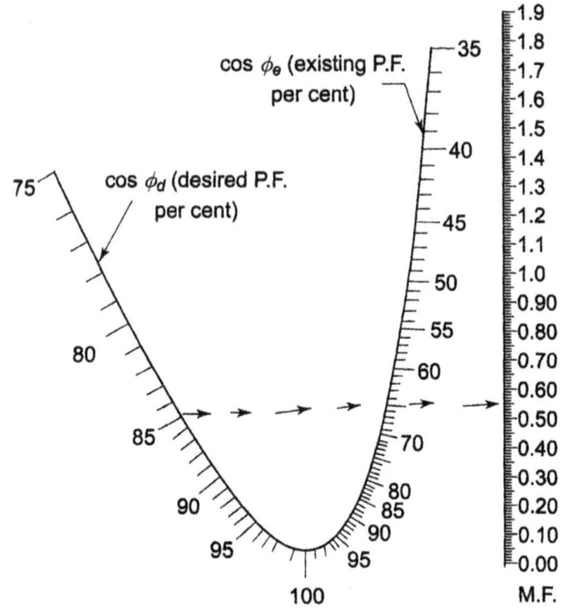

Fig. 15.12 *Nomogram for calculating multiplying factor required to determine capacitor kVAr; Multiplying factor (MF) = $\tan \phi_e - \tan \phi_d$.*

EXAMPLE: We are required to find out the capacitor rating to improve the power factor of 100 kW load from 65 per cent to 85 per cent. In Fig. 15.12, align 65 per cent (existing power factor) and 85 per cent (desired power factor) on the respective scales and extend to the multiplying factor scale to get MF as 0.55. Then the required kVAr rating of capacitor is $100 \times 0.55 = 55.0$.

15.6 LT Capacitors

15.6.1 Installation

LT capacitors are installed on the distribution system on individual lines, at distribution transformers or consumers motors to reduce system losses and improve the system voltage and capacity. In addition, they provide other advantages for the consumer, such as reduction in kVA demand, losses and stable voltage. The optimum benefit desired from the capacitors largely depends on the correct positioning of the capacitor in the system. However, the provision of LT capacitor on the individual service is generally preferred because of the following reasons:

(a) It can be controlled by the same control gear which controls the motor. IS: 7752 (Part-1)–1975 gives guidelines for improvement of power factor at consumer installation.

(b) If the capacitors are connected on the LT motor connection, the loading current flowing through the HT and LT lines will be reduced. Thus losses both on the HT and LT will also be reduced.

(c) The capacitors being directly connected to the motors come into operation when the motors are in service and are out automatically when the motors are not in use.

(d) The installation of capacitors at the sub-station and/or at the LT lines would require switching operations to bring in the capacitors in service during peak load periods and to cut out the capacitors during low loads, necessitating the provision of comparatively expensive switchgear.

(e) With the joining up of additional service connections, which is a normal feature, the location of the capacitors at the lines will have to be changed from time to time to give best results.

15.6.2 Economics of Capacitor Installation on Consumer Premises

The installation of capacitors at the load points substantially reduces the kVA demand. The tariff charges levied on the basis of energy consumed and the maximum kVA demand are accordingly reduced by a reduction in the kVA demand.

EXAMPLE

Assume load of consumer installation	100 kW
Present power factor	0.75
Required power factor	0.90
Maximum load in kVA at 0.75 power factor	133.33
Maximum load in kVA at 0.90 power factor	111.11

Reduction in maximum demand

133.33 − 111.11 = 22.22 kVA

Saving in monthly demand charges, say, at the rate of Rs 12.00 per kVA will be 22.22 × 12 = 266.64.

The kVAr capacity required for improving the power factor from 0.75 to 0.90 per kW load is 0.40 kVAr (refer Fig. 15.12) and therefore the capacity required for 100 kW load is 40 kVAr. The cost of a capacitor including fabrication charges for bank formation, say at the rate of Rs 130.00 per kVAr will be Rs 5,200.00.

Assuming the rate of depreciation to be 7 per cent (for an expected life of 15 years) and that of interest to be 9 per cent per annum, the total depreciation and interest charge will be:

$$5200 \times \frac{16}{100} \times \frac{1}{12} = \text{Rs } 69.33$$

Therefore, net saving per month is Rs 266.64 - Rs 69.33 = Rs 197.31. Thus the cost of complete capacitor installation is recovered in

$$\frac{5200}{197.31 \times 12} = 2.19 \text{ years}$$

i.e., the cost of the capacitor bank is recovered in about 26 months.

Also, from the system viewpoint there will be considerable reduction in transformer and line losses, resulting in maximum utilization of the installed capacity of the transformers, switchgears and lines.

15.6.3 Size of Capacitor for Motors

The size of capacitor permanently connected to a motor should be such that the capacitor current does not exceed no load current of the motor at normal voltage. Otherwise, dangerously high voltage will be generated when the motor comes to a halt because after disconnecting the motor from the power supply, the motor while still revolving acts as a generator by self-excitation. Generally the rating of directly connected capacitor is limited to 90 per cent of the magnetizing kVAr of the motor.

Recommended sizes for individual induction motors are given in Table 15.3 for general guidance. (see IS: 2993–1998).

15.6.4 Capacitor for Welding Plant and Arc Furnace

Welding equipment usually have a low power factor. The recommended capacitor rating for direct connection to the primary side of welding transformers, arc furnaces for power factor corrections are given in Table 15.4 for general guidance.

Table 15.3

Recommended capacitor rating for direct connection to induction motors to improve the power factor to 0.95 lagging or more at all loads

Motor HP 3 phase, 415 V	Capacitor rating in kVAr when the motor speed in rpm (approximately) is:		
	1500/1440	1000/960	750/720
3	1	1.5	2
5	2	2.5	3.5
7.5	3	3.5	4.5
10	4	4.5	5.5
12.5	4.5	5	6.5
15	5	6	7.5
20	6	7	9
25	7	9	10.5
30	8	10	12
35	9	11.5	13.5
40	10	13	15
45	11.5	14.5	16.5
50	12.5	16	18

Note: (a) For agriculture tubewells, ordinarily motors having speed of 1440 and 960 rpm are used with a working speed of pump of about 1400 rpm. The former is directly coupled and the latter is belt driven.

(b) Motors with a speed of 720 rpm approximately are normally used in industry.

(c) For motors up to 10 HP, inbuilt capacitors are desirable, keeping p.f. not less than 0.95 lagging at rated output.

15.6.5 LT Capacitor Connections

The connection of a delta-connected capacitor (three terminal) to an induction motor with slip rings and starting resistance or to a direct starting motor, involves no problem if the output of the capacitor does not exceed the consumed no-load power of the motor. If, on the other hand, the motor has a star-delta starter, certain problems may arise, viz. overvoltage due to self-excitation, high discharge currents and resonance.

Table 15.4

Recommended capacitor size [13]

(i) Welding plant Welding transformer continuous kVA rating	Capacitor kVAr
5	2
9	4
12	6
18	8
24	12
30	15
36	18
57	25
95	45
128	50
160	75
(ii) Arc Furnaces MVA furnace transformer rating	Capacitor MVAr (Usually in-built)
5	1.5–2
12.5	4–5
25–30	7.5–12
50–60	40–45

The overvoltage for a few cycles up to two or three times the rated voltage due to self-excitation (motor acts as a generator) could occur when switching from star to delta and the line is broken before the neutral. It may damage motor and capacitor.

If the switchover (star-delta) takes place such that the line and neutral point connections are broken simultaneously, no overvoltage arises but high discharge currents may be generated when the motor is again switched on. This is because the voltage across the capacitors remains the same during the time of disconnection but the line voltage on reconnection may be of a different phase, giving rise to harmful equalizing currents.

If during switching from star to delta, the neutral point connection is broken before the line, the series resonance between the capacitor and motor winding may occur. It amounts to a short circuit and the current momentarily rises to a very high value with corresponding high overvoltage across capacitor and motor winding.

The above difficulties are avoidable if a six-terminal capacitor is used and connected in single phase. The different connections for LT capacitors are shown in Fig. 15.13. With six terminal connections, the series resonance cannot occur since the capacitors and motor windings are always connected in parallel. The risk of self-excitation disappears. Equalizing currents are eliminated to a large extent, since the capacitor

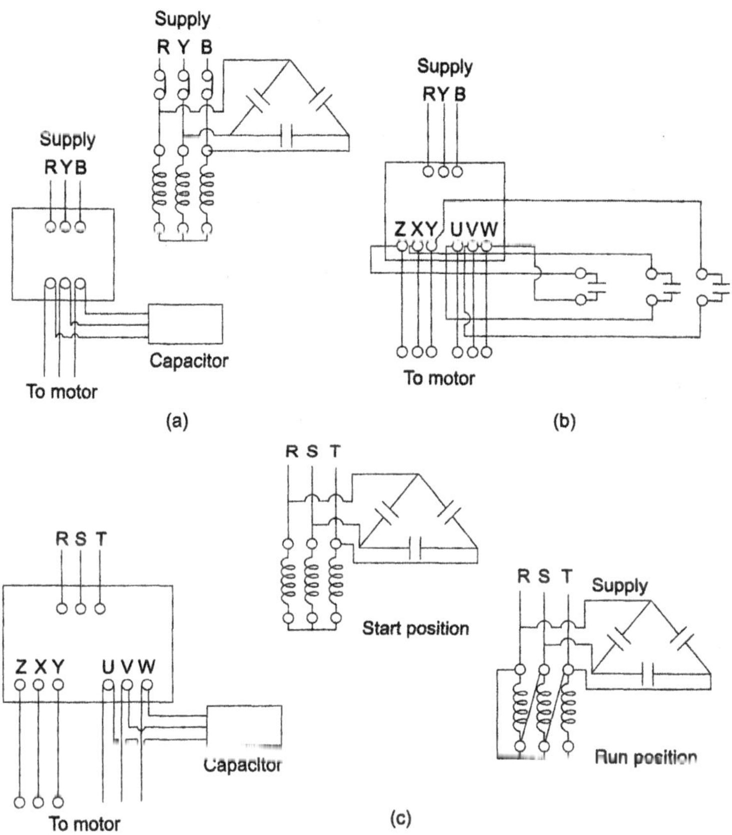

Fig. 15.13 *Wiring diagram for LT capacitors:*
(a) Direct start, auto-transformer start motors with three-terminal capacitor;
(b) Star delta start motors using six-terminal capacitors;
(c) Star delta star motors with three-terminal capacitors

discharges across the motor windings while it is disconnected from the line.

Therefore, the use of six-terminal capacitors is favourable for star-delta motors. It has been observed that by using three terminal capacitors there is a tendency for the line contacts of the starter to close before the star contacts open, thereby the capacitor gets connected in series with the stator phase winding. This gives rise to resonance and excessive voltage, thus leading to failure/damage of both motors and capacitors.

15.6.6 Spurious LT Capacitor Diagnosis

LT capacitors are being used by consumers with agricultural and industrial motors for improving the voltage, power factor, etc. A lot of spurious and substandard LT capacitors have come in the market, the diagnosis of which has become an important consideration. Following tests may be useful for verifying the genuineness of the 415 V capacitors. The portable clip on kVAr metre can also be used for field testing.

Measurement of Capacitor Current

The current drawn in each phase of an LT capacitor may be measured by means of a low range tong tester and these values are compared with the standard values for the capacities mentioned below in Table 15.5 at different operating voltages within a tolerance of – 5 to 10 per cent.

These values are based on the relation:

$$(\text{kVAr})_2 = (\text{kVAr})_1 \times \left(\frac{V_2}{V_1}\right)^2; \qquad (\text{kVAr})_1 = 3V_1 I_1$$

where, $(\text{kVAr})_2$ and V_2 are the rated values and $(\text{kVAr})_1$ is the kVAr at measured voltage V_1 and current I_1.

The difference in the amperage drawn from supply mains with and without capacitors at the normal operating load can be noted and the values for the following capacities of motors can be compared with the current values given against each.

3 HP	1 kVAr	0.65–0.92 A
5 HP	2 kVAr	1.0–1.5 A
7.5 HP	3 kVAr	1.45–1.75 A
10 HP	4 kVAr	2.5–3.4 A

Table 15.5

Standard value for capacities at different voltages

kVAr	390 V (A)	400 V (A)	415 V (A)	430 V (A)	440 V (A)
1	1.30	1.34	1.39	1.42	1.47
2	2.6	2.65	2.78	2.84	2.94
3	3.9	4.0	4.16	4.23	4.40
4	5.25	5.35	5.56	5.72	5.87
5	6.5	6.7	6.96	7.2	7.34
6	7.85	8.5	8.32	8.62	9.81
7.5	8.80	10.00	10.40	10.60	11.20
10	13.00	13.40	13.9	14.40	14.7
12.5	16.3	16.8	17.4	18.00	18.4
15.0	19.4	20.0	20.8	21.5	21.87
20.0	26.0	26.6	27.8	28.5	29.40
25	32.6	33.5	34.8	36.0	38.00

If the difference in amperage agrees with the above value for that particular rating of the motor, the capacitor may be taken as genuine.

15.7 Construction Features

The conducting system is usually made of high purity annealed aluminium foil or metal spray. The dielectric system can be made of:

(a) *Askarel-Impregnated Mixed Polypropylene–Paper Dielectric:* This dielectric in commercial designs frequently consists of two layers of 12.5 micron thick polypropylene film with a thin layer of low density (low loss) paper sandwiched between them and the whole impregnated with askarel. This construction has the advantage of: (i) Impregnation is made easy because the rough surfaces of the paper allow easy penetration of the impregnant into the paper polypropylene interface. (ii) The capacitor will have lower dielectric loss than the all-paper construction. Even with three layers, the total dielectric thickness is still quite small so that high discharge inception voltage stresses can be obtained. (iii) Because paper has a much higher permittivity than polypropylene, it is possible to work the combined dielectric at an economically high stress while applying stress to the paper layer than would be the case in an all-paper capacitor. Thus if the design stress in the polypropylene layer is about 40 V per micron, the stress in the paper layer will be about 18 V per micron. (iv) As an example of the resulting decrease in size, the dielectric systems of this kind allow the construction of 150 kVAr capacitors in the same dimensions as those for 100 kVAr conventional all-paper capacitors. The losses are within 0.15–1 W/kVAr.

(b) *Askarel-Impregnated Polypropylene (only) Dielectric:* In this construction, it is possible to produce high voltage capacitors with discharge inception voltages in excess of 100 V per micron, which makes it possible to have the working stresses in the dielectric of about 50 V per micron. This allows the dielectric volume to be reduced by about four times for the same all-paper capacitor kVAr rating, giving a reduction in the overall capacitor volume of about three times with concomitant cost reduction. Such capacitors when overheated are likely to behave badly on thermal cycling tests. The design practice working stress practice in India for these capacitors is around 30 V per micron. The loss for this type is less than 0.5 W/kVAr.

(c) *Mineral Oil-Impregnated Polypropylene Dielectric with Metallized Paper Electrodes:* In this system, the dielectric typically consists of a single layer of 6 micron thick polypropylene film, with electrodes of paper 9 micron thick and metallized with aluminium on both sides. Because of the use of metallized paper, it is necessary to use mineral oil impregnant. Impregnation is relatively easy, because the surfaces of the metallized paper are naturally uneven. These capacitors operate at stresses three times those for the paper dielectric types and are typically half the size of the latter for a given kVAr rating.

(d) LT capacitors of metallized plastic film dry type (loss < 0.5 W/kVAr) are now easily available from market.

15.8 Failures (7, 6)

The failures of capacitors due to insulation degradation are attributed to thermal, voltage and chemical processes resulting in changes of mechanical strength of insulation, degree of polymerization of cellulose insulation, evolution of gases, capacitance, tan δ value and partial discharges. Periodical measurements (as per IS: 13925) of capacitance, tan δ, partial discharge values and their progressive respective comparisons will give good indication of the working condition or degrading of capacitors.

Electromechanical forces and processes generated by a combination of electrical stress and temperature provide the means by which capacitor wearout takes place. The end results of all capacitor wearout is always dielectric breakdown. Several factors tend to speed up the normal wearout processes of a dielectric system. Among them are flaws in the material, improper processing or assembly and unusual operating conditions (refer Fig. 7.1).

Shortcomings in materials and processing procedures usually manifest themselves in very early failures. A poorly processed capacitor with dry spots will operate under conditions of partial discharges, causing material to degrade, with dielectric breakdown occurring with a period ranging from a few minutes to perhaps a few days. Early failures indicate that the materials are inadequate for the conditions of electrical stress and temperature to which the capacitors are subjected.

A second mode of failure in capacitors is the random failure mode. A random failure is not associated with early failures but is produced by a chance condition. For example, a failure occurring two years after commissioning that may have resulted from an unusual voltage excursion.

A third failure mode is dielectric system wearout. Ordinarily, the wearout mode becomes predominant after 20 years of operation. The normal wearout period is followed by an increasing failure rate.

A failed capacitor can be detected by observing the change in the capacitance of the unit. The change in the capacitance can be checked by calculating the actual capacitance from the measurement of applied voltage and current for the faulty unit and comparing it with the manufacturer's tested value for the unit.

16.8.1 Physical Fault Symptoms—General Precautions (7)

Fault symptoms	*Precautions*
• Leakage of oil from welded joints or terminals studs	Minor leakages should be stopped after thorough cleaning and applying araldite/m-seal.

(Contd.)

Fault symptoms	*Precautions*
• Capacitors bulging	Bulging takes place due to gas formation inside caused by internal sparking. The capacitor should be immediately disconnected, removed from service. For star-connected bank, phase balancing should be done by disconnecting equal number of capacitors from other phases.
• Capacitor overheating	It may be due to: (i) Poor ventilation. (ii) Excessive ambient temperature. Sun shade may be provided. (iii) Overvoltage, as voltage rises at the capacitor point. This voltage rise may be even greater, when harmonics are present or at time of light load. Therefore, nominal voltage rating of capacitor should be about 5 per cent above the nominal system rated voltage. (iv) High frequency oscillations caused by bad contacts in capacitor circuit.
• Cracking noise	This may be due to internal faults and hence the capacitor unit along with balancing units should be removed from service.

PROBLEMS

1. A three-phase, 50 Hz, generator supplies power to motor rated 60 kW load at 0.8 power factor lagging at 11 kV through a distribution line with a series impedance of 50 + j500 ohms per phase. If three capacitors each of 0.5 micro-farads, are connected in star across the load to improve the power factor, find the active power, reactive power and apparent power generated by the generator.
2. A three-core, 11 kV, three-phase, 50 Hz underground cable has a capacitance of 890 nF/km. How much reactive power is generated in a 1.5 km length of the cable?
3. Three capacitors, each of 7 μF, are connected in delta across a 415 V, three-phase line. If the system frequency is 50 Hz, calculate how much reactive power is generated by the capacitors. If the system

voltage falls by 10 per cent, what will be the corresponding fall in the reactive power generation?

4. (a) What is the formulae for reactive power generated by a synchronous generator and a power shunt capacitor?
   (b) What is the approximate cost of generating reactive power by a synchronous generator and a power shunt capacitor?
   (c) What happens at the node where a shunt capacitor is connected?
5. Three loads, each having an impedance of 20 + j15 ohms, are connected in star to a three-phase, three-wire, 400 V, 50 Hz supply. Determine the: (a) line current; (b) power supplied; and (c) power factor. (d) if three capacitors, each of the same capacitance, are connected in delta at the load terminals, find the capacitance of each capacitor to obtain as resultant power factor of 0.95 lagging.

BIBLIOGRAPHY

1. (i) IS: 13925 (Part 1)-1998, Part 2, 1994.
   (ii) International Electrotechnical Commission Publication: 70–1967. Shunt capacitors for power system for rated voltage above 1000 V.
2. Longland, T., T.W. Hunt and A. Brecknell, '*Power Capacitors Handbook*, Butterworths, London, 1984, pp. 93–98.
3. Srivastava Suresh, "Fuse Systems for Protection of Power Capacitors", *Ieema Journal*, vol. XVII, no. 8, August, 1999, pp. 17–22.
4. Ferracci Philipe, "Transient Over-voltages on Secondary Winding of MV/LV Transformer due to Capacitors Energisation-Correlation between computed values and experimental results", *Proceedings of the 15th International Conference on Electricity Distribution, CIRED*, Nice, France, June, 1999, pp. 91–96.
5. Grebe Thomas E, "Capacitor Switching and its Impact on Power Quality", *ELECTRA*, April, 2001, pp. 27–37.
6. Miske S.A., "Considerations for the Applications of Series Capacitors to Radial Power Distribution Circuits", *IEEE Transactions on Power Delivery*, vol. 16, no. 2, April, 2001, pp. 306–318.
7. Mehra, A.C., A.S. Pabla, and P.P. Garg, "Failure and life of shunt capacitors", *Electrical India*, Bombay, 30th, April, 1979.
8. Working Group 13.04 Report "Capacitors current switching—State of Art", *Electra no 155*, August, 1994, pp. 33–61.
9. IEEE Std. 341–1972, IEEE Standard: Requirement for capacitance current switching for AC high voltage circuit breakers—Rated on a symmetrical current basis.
10. Miller, T.J.E. "Reactive Power Control in Electric System", John Wiley and Sons, New York, 1982, pp. 353–361.

11. Pabla, A.S., "Distribution System Reactive Power", *IEMA Journal*, July, 1985, pp. 11–18.
12. Sachdeva, S.S., "Transformer Tap Setting and System Voltage Raise Impact on Optimum Planning of Static Capacitors", IEEE Transac., PAS, No. 5, May, 1984, p. 1024.
13. Tagare DM "*Electrical Power Capacitors—Design and Manufacture*", Tata McGraw-Hill, New Delh-2002, pp. 207, 301.

Chapter 16

Insulation Measurements

16.1 Insulation Supervision

The insulation of electrical apparatus and lines, etc. is designed and operated at a specific standard service voltage. It is also designed to operate continuously at voltages 10–15 per cent higher than the rated value to cover the possibility of such voltage appearing under certain conditions of operation. In addition, the electrical equipment must also be able to withstand short time overvoltages, such as switching and lightning surges. Clearly, the insulation of electrical equipment must have sufficient electrical strength to ensure the necessary factor of safety. While in service, the insulation may become soiled by deposits and absorb moisture. This somewhat reduces its electric strength.

In practice there are two main common forms of breakdowns in insulation—thermal and electrical. Thermal breakdown is mainly a result of heating brought about by the losses which arise when a dielectric is subjected to electric stresses. As the temperature of the dielectric rises, the losses increase. If they rise faster than the rate at which the dielectric is able to dissipate the heat, its temperature also rises. As a result, the dielectric may breakdown and conduct current.

Electrical breakdowns in solid dielectrics occur as a result of structural disintegration of the dielectric by the action of the electrical stresses applied to it. This form of breakdown begins when the voltage applied to the dielectric exceeds a limiting value. The nature of breakdown varies according to the temperature rise within the dielectric, frequency of applied voltage, cooling conditions, etc. The moisture content of the dielectric has a particularly important bearing on the value of the breakdown voltage. For example, the electrical strength of the hygroscopic materials drops very sharply with a rise in the moisture content.

In a liquid dielectric the breakdown is a result of the damage caused by electron and ion bombardment of the molecules, particularly of the dielectric. A liquid dielectric containing small bubbles of gas or particles of suspended matter will have a much lower breakdown voltage than one which is free of them. The presence of moisture, foreign matter, etc. will, therefore, increase the breakdown of the liquid dielectric.

During service, the insulation of electrical equipment will be subjected to the action of various factors which are harmful in nature and may sometimes endanger its normal operation. This is the reason why insulation is selected according to the conditions in which the equipment will operate. The service life of insulation will be reduced by aging which causes reduction in electrical and mechanical strength, changes in the structure of materials and cracking of surface. Deposits of dirt will also reduce its service life. In order to reveal defects in the insulation some regular measurements and tests are necessary. These are primarily the measurements of dc insulation resistance, leakage current of the insulation, dielectric losses, capacitance, voltage distribution, test for voltage discharges across parts of the insulation and test for electric strength. The preventive maintenance tests should be carried out on a schedule drawn up to coincide with the regular inspection and a repair schedule set up for the maintenance of electrical equipment.

16.2 Insulation Measurement—Non-Destructive Techniques (9)

16.2.1 Insulation Resistance

The insulation resistance is defined as the resistance (in megaohms) offered by the insulation to an impressed direct voltage. The resulting current is called *insulation current* and consists of two main components.

(a) The current which flows within the volume of the insulation and is composed of capacitance current, dielectric absorption current and irreversible conduction current.

(b) The current which flows in leakage paths over the surface of insulation, is termed as leakage current. It does not change with the duration of voltage application. This current in the primary factor on which the insulation quality may be judged.

Insulation resistance varies directly as the thickness and inversely as the area of the insulation being tested. A curve plotted between the insulation current and time (or insulation resistance and time) is known as *dielectric absorption curve.*

Unless made with a high degree of skill, insulation resistance and dielectric absorption measurements will show major fluctuations due to variations in the several factors discussed below. Each factor can cause large errors in the measurement of the insulation resistance which may not be chargeable to defects in the measuring instrument.

Effect of Previous Charge

One factor which can affect insulation resistance and dielectric absorption measurements is the presence of charge on the insulation. The charge may come from a previous measurement of resistance.

Effect of Temperature

Insulation resistance varies inversely with temperature for most insulating materials. To properly compare periodic measurements of insulation resistance, it is necessary to either take each measurement at the same temperature or convert each measurement to the same base temperature.

Effect of Moisture

Moisture will make a surprisingly large difference in the insulation resistance.

Effect of Age and Curing

Insulation with semisolid binder, such as asphalt and mica, undergoes a curing process with time. This curing process increases the dielectric absorption current taken by the insulation and thus the insulation test or the HV dc test measurement shows a decrease in the insulation resistance with increasing age. The more noticeable effect of age on the leakage current is mainly due to the development of cracks and contamination.

Polarization Index

The steepness of the dielectric absorption curve taken at a given temperature indicates the relative dryness of the insulation. This steepness may be expressed as the ***polarization index*** or ***absorption factor*** of the insulation resistance and at constant voltage is defined as follows:

$$\text{Polarization index} = \frac{R_{60}}{R_{15}} = \frac{I_{15}}{I_{60}}$$

where,

R_{60} = megohms insulation resistance reading after 60 s.
R_{15} = megohms insulation resistance reading after 15 s.
I_{15} = insulation current at 15 s.
I_{60} = insulation current at 60 s.

The stipulation of this polarization index varies with different equipments. For dry insulation, this index will exceed unity and reach values of 1.5–2. If the insulation contains moisture, this factor will be close to unity.

Insulating materials are used, in general, to insulate and support components of an electric network from each other and from the ground. For this purpose it is generally desirable to have the insulation resistance as high as possible, consistent with acceptable mechanical, chemical and heat resisting properties, though in many cases it need only be high enough to have no appreciable effect on the operation of the network. When adequate correlating data are available, resistance or resistivity may be used to indicate the suitability of a material in other aspects, such as dielectric breakdown, dissipation factor, moisture content, degree of cure, mechanical continuity, and deterioration from other causes. The insulation resistance test is of the greatest value when the test specimen has the same form as that required in actual use as it combines both volume and surface resistances.

The methods used for these measurements are covered by specification IS: 2259–1963. Equipments commonly used are:

(a) Tinsley insulation testing equipment.
(b) D.C. microvoltmeter.
(c) Megohm bridge.
(d) Insulation resistance tester–250 V, 500 V, 1 kV, 2.5 kV, 5 kV rating.

Insulation testers of various voltage rating are:

Rating (dc volts)	Used for nominal system voltage
250	240 V
500	415 V–1.1 kV
1000	6.6–11 kV
5000	22–220 kV

Insulation Resistance Values

There are no national or international standards for insulation resistance test values for power industry equipment. These values vary for different designs. Some manufacturers have their own standards. For example, Bharat Bijlee Limited, Bombay have standardized the following values as a criterion for passing the transformers at their works. These values are at room temperature, around 30°C and measured with 2.5 kV insulation resistance tester.

Transformer voltage class in kV	*Insulation value in MΩ (winding to earth)*
3.3	150
6.6 to 11	250–300
22	350
33	400–500
66	600–800
132	800–1000

However, for maintenance purposes, it is useful to compare the actual insulation resistances of equipment in service with the test certificate values at manufacturer's works to examine the insulation condition. The insulation temperature is very important while examining the values. For example, for transformers as a rule of thumb, the insulation resistance value reduces by half after every 10°C temperature rise.

As a general guide for transformer maintenance, IS: 1886–1967 specifies the insulation resistance values (Fig. 16.1) during the hot air drying out a transformer, i.e. it gives the insulation values for air-paper insulation system instead of oil-paper insulation in oil-immersed transformers.

For using air drying insulation values for the operation of oil-immersed transformers, the following conversion factor may be used:

(a) Insulation value in MΩ at 85°C in air

= Insulation value in MΩ at 40°C in oil

(b) Further, we can use 10°C as a rule of thumb for insulation change with temperature.

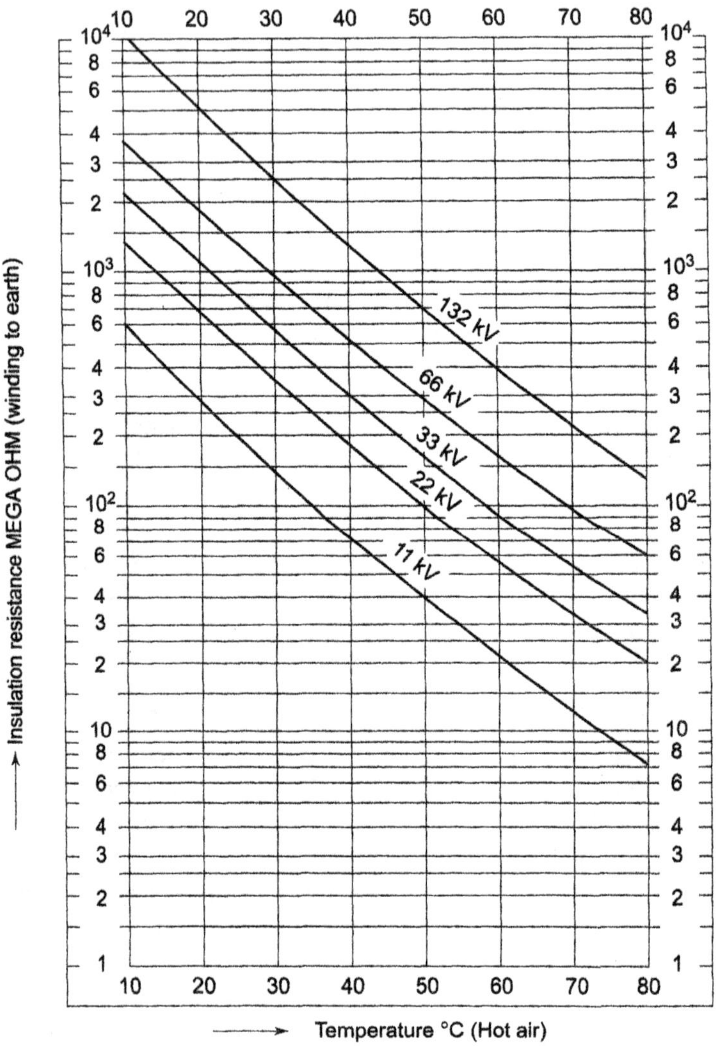

Fig. 16.1 *Minimum insulation resistance of transformers (without oil) in hot air drying process*

The curves given in Fig. 16.1 can be used to diagnose the causes of low *IR* values (i.e., less than 10 per cent of values specified in Fig. 16.1) of transformers in following steps: (i) Oil is checked thoroughly from top, middle and bottom samples, and if found in order, then the fault may be with the winding. Probably it may be having moisture. (ii) Take the *IR* value with temperature noted of complete power transformer, say it is *X*. (iii) Take its polarization index, its value near unity shows moisture. In that case, the transformer oil may be drained through drain valve up to about 0.3 m level from the bottom. From the remaining oil, take a little oil sample in a glass container and check if the mercury type bubbles of water are settling down in the container, which indicate the presence of water. (iv) Now drain the complete oil through bottom drain plug and measure *IR* value of transformer (without oil) with temperature noted, say it is *Y*. (v) The value of *X* and *Y* should be converted to same reference temperature (say 23°C) and compared. If *Y* is within 15 to 20 times of *X*, then oil is all right but the winding is defective. If *Y* is less than 15 times *X*, then both winding and oil is defective and need drying. If *Y* at the reference temperature is nearly equal to the value given in Fig. 16.1, then the winding is all right and oil needs replacement. But if it is lower by 30% or more, then there may be some doubt in the winding, in such a case the oil and winding both be dried up.

Motors/Generators

An empirical relation for a minimum one-minute value of insulation resistance (MΩ) for motors or generators at 75°C is

$$= \frac{E}{0.01\,\text{kW} + 1000}$$

where E = rated voltage of machine in volts
kW = rated power of machine in kW

Wiring in Buildings (3)

Wiring Rating (V)	Test Voltage (d.c. volts)	Minimum Insulation Resistance (MΩ)
LT wiring (250/440 V)	500	0.5
1.1 kV cable wiring	1000	1.0

EXAMPLE Figure 16.2 shows a test for a large 350 HP motor for an extended period of time. The test is based on the comparison of absorption characteristics of good insulation, versus the absorption characteristics of humid or otherwise contaminated insulation. A test voltage is applied (motorised insulation tester) for an extended period of time of 10 minutes. A reading every 15 seconds for the first minute and every minute up to 10 minutes is taken. The curves *A* and *B* are drawn as shown in Fig. 16.2.

A good insulation will show a continual increase in resistance as shown in curve *A*. Contaminated, moist or cracked insulation will produce a relatively flat curve similar to curve *B*. The slope is indicative of the polarization index (PI):

$$\text{PI} = \frac{\text{10 minutes reading}}{\text{1 minute reading}}$$

On large motors or generators, PI values as high as 10 are commonly accepted.

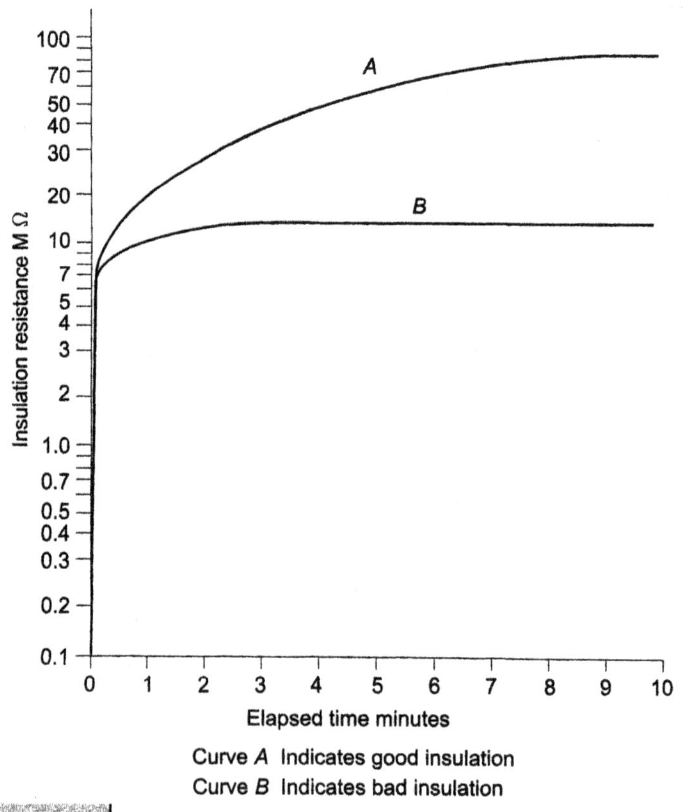

Fig. 16.2 *Typical absorption curve of 6.6 kV, 350 H.P. Motor*

16.2.2 Power Factor Measurement

Unless otherwise stated, the terms "power factor", "dissipation factor" and "tangent delta" are used interchangeably. Insulation pf is the cosine of the angle between the applied voltage and current and is obtained from the measurements of watts, volts and amperes usually with a specialized power factor test set. The dissipation factor is the contangent of the angle between the applied voltage and current and is usually obtained from a direct reading of the dial on a capacitance bridge. It should be remembered that pf or "dissipation factor" or tangent delta is a measure of insulation "dielectric power loss" and is not a direct measure of the dielectric strength. This loss in an insulation structure is associated with the oscillation of polar molecules trying to orient themselves with the alternating electric field. In an insulation system, the dielectric loss is given by $V^2 \omega C \tan \delta$ watts. Hence $\tan \delta$ is termed as the loss or dissipation factor. If some air spaces and cracks occur in the structure, the voltage distribution will alter and at a particular voltage the air will ionize depending on the dimensions and dielectric constant of the insulation. Thus ionization causes an extra loss which is manifested as a parabolic increase in the dissipation factor as the applied voltage is increased (Fig. 16.3). Hence, absolute value of $\tan \delta$ is of little importance than the rate of change of $\tan \delta$ or dissipation factor over a selected range of voltages. The conditions which cause abnormal power loss usually cause a reduction in the dielectric strength. The power factor values are independent of the insulation area or thickness and increase only with an increase of contamination by moisture and other foreign matter or by ionization and, therefore, are easier to interpret than insulation resistance values which additionally depend on the insulation area and thickness.

An increase in dielectric loss may accelerate insulation deterioration because of increased heating. But more commonly an increase in the dielectric loss is evidence of other factor leading to deterioration which also affect the dielectric strength. As in insulation resistance tests, the change in periodic test readings is more indicative of insulation deterioration than absolute magnitude readings. The insulation pf increases directly with temperature and may be as much as 10 times high at 80°C than at 20°C. Hence, temperature corrections to a base temperature must usually be made at 20°C for a meaningful comparison between values taken at different temperatures. American standards (C. 57.12.90) give the correction factor K for temperature effects as given in Table 16.1 for transformers.

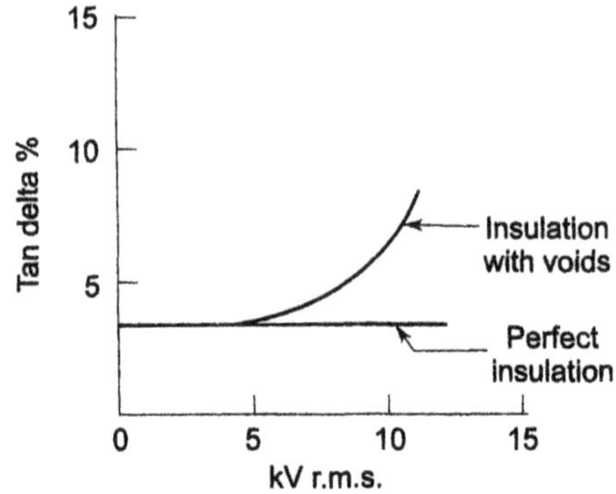

Fig. 16.3 *Variation of tan delta with applied voltage*

Table 16.1

Tan δ measurement correction factor K for transformers (tan δ_{20} = tan δT/K, where tan δ_{20} = dissipation factor after correction and tan δT = dissipation factor as measured)

Test temperature *T (°C)*	*Correction factor* *K*
10	0.80
15	0.90
20	1.00
25	1.12
30	1.25
35	1.40
40	1.55
45	1.75
50	1.95
55	2.18
60	2.42
65	2.70
70	3.00

As a nondestructive test, 10 kV rating Schering bridge equipment is considered for measuring the dissipation factor or power factors of capacitors, bushings, transformer oil, lightning arresters, cables, etc. A test voltage up to 1.5 times the nominal voltage rating of equipment when applied for 1 min. is considered nondestructive for new equipment. For equipment in service, 10 per cent nominal voltage or 10 kV, whichever is minimum, should be used as the test voltage.

16.2.3 Measurement of Leakage Time Constant of an Insulation

(i) One method is based on the charge absorption process that occurs in a nonhomogeneous dielectric. The permittivity of a dielectric is increased by the presence of free ions which are present in the insulation only due to contamination with moisture or other impurities. A testing set measures the capacitance of the dielectric by means of single-shot charge and timed discharge. The term capacitance versus time is applied to this method. The measurement of the amount by which the voltage rises within a fixed interval of time serves as a measure of the degree of penetration of moisture or other impurities into the insulation. The voltage rise is governed by the process of interlayer (slow) polarization in the insulation.

A typical diagram of this set is shown in Fig. 16.4. In measuring a rise in the capacitance of the insulation under test, the relay contact K_2 is made to close at the initial moment of discharge of the insulation to allow the charge to begin draining from the insulation. Specified several ms later, the relay will switch K_2 open and close K_1. This allows the slow-acting component of the insulation capacitance to begin charging capacitors C_N and C_M. Few ms later contact K_1 is thrown open to isolate the "memory" capacitor C_M. Following this (with the aid of a valve voltmeter) the voltage to which the capacitor has become charged is measured. It will be proportional to the rise in the capacitance of the insulation under test within the period.

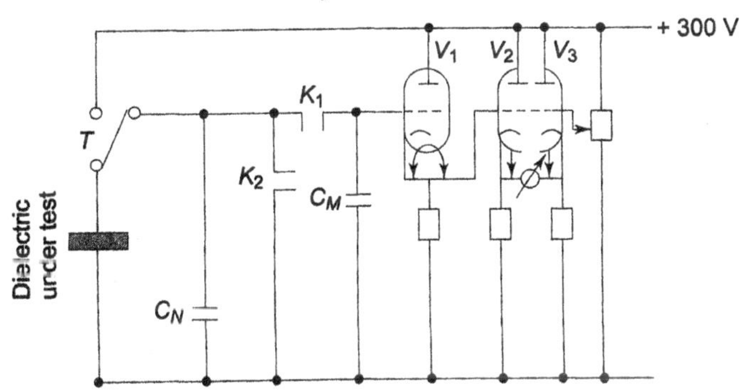

Fig. 16.4 *Leakage time constant measuring testing set:*
T = Changeover switch
V_1 = Buffer stage
V_2, V_3 = Valve voltmeter stages

(ii) Dispersion measurement: As moist insulation is found to have a higher dielectric dispersion as the test frequency is reduced, it is desirable to work at as low a frequency as possible to achieve high sensitivity. By working with square waves in place of sine waves, the dispersion may be determined directly. For most types of solid insulations used in oil, i.e. paper, pressboard, etc. the time constants when dry is less than 3 ms and when moisture is present it lies between 3 and 300 ms. This fact is made use of in selecting the charging and shorting pulse width in ERA dispersion metre. In case of very moist specimens, excessive leakage may invalidate the dispersion measurement. In such cases provision is made in ERA dispersion metre for the measurement of T, the leakage time constant, by the application of a square wave consisting of a charging voltage of duration 300 ms followed by a period of isolation, also of duration 300 ms. Assuming that the specimen consists of a simple shunt combination of capacitance and conductance, the leakage time constant is given by

$$T = \frac{t}{\ln(V_0/V_3)}$$

where V_3 is the voltage on the specimen at the end of the isolation period and V_0 the charging voltage. The metre is provided with a scale reading directly in leakage time constant. Here t is 300 ms.

There is evidence that the leakage time constant gives a more reliable indication of moisture in oil-impregnated-paper insulation than the measurement of the loss angle.

16.2.4 Partial Discharge Measurements

Conventional tests such as high voltage, impulse withstand and tan δ, reveal the gross defects in insulation. These tests give little assurance against slow deterioration and eventual failure caused by discharge due to voids and conducting impurities present in the insulation at the service voltage. The partial discharge test detects the incipient defects due to small voids in the insulation. The detection of discharges is based on energy exchanges which take place during the discharges. The exchanges are manifested as electrical impulse, em radiation (light), sound (noise), increased gas pressure and chemical transformations.

Discharge detection and measuring techniques are based on the observation of any one of the said phenomena occurring inside the solid insulation's cavities, cracks, gas bubbles etc. These include:

(a) Detection, determination of presence of discharges and voltage at which they appear.
(b) Location, determination of the site of discharges.
(c) Evaluation and magnitude of discharges and deterioration limit.

The object of a discharge test is to detect and measure the energy dissipated by the breakdown of a gasfilled cavity in solid insulation. For a lumped insulation, such breakdown results in the appearance across the terminals, a transient voltage having a waveform that may be taken to be a vertical step. The function of the discharge detector is to detect and measure the magnitude of this step voltage. The quantities normally measured with a discharge detector are: the inception voltage (U_i), extinction voltage (U_e), discharge magnitude (*pC*), discharge energy (1/2 *pC* U_i) and the quadrature rate of discharge. A CRO display on which the pulses can individually be counted and measured is normally used. To find the total discharge it has been suggested that the slope of the CRO trace can be used to estimate the total volume of the voids with the increase of capacitance of the equipment under test as the voids become short circuited by discharge [8].

A basic partial discharge detection system (Fig. 16.5) includes all the equipment required to measure the principal partial discharge parameters, i.e. inception voltage, extinction voltage, signal level in *pC* and system sensitivity. The system consists of three major components: partial discharge detector, power separation filter and calibration signal coupler. Also included are all the necessary interconnection cables, the power cord, and a comprehensive system instruction manual. It can be used with either ac or dc voltages.

The methods for measuring the partial discharge are illustrated in IS: 6209—"Methods for partial discharges measurements". Versatile portable non-contact PD detectors are now available for testing the equipment in service [10].

16.3 Insulation Testing: Destructive Tests

There are insulation high voltage proof tests for which various national and international standards are available for various items of equipment and insulation. These tests are considered destructive as their frequent application is liable to damage the equipment. Knowing the correct technique of applying the voltage with respect to time is very important

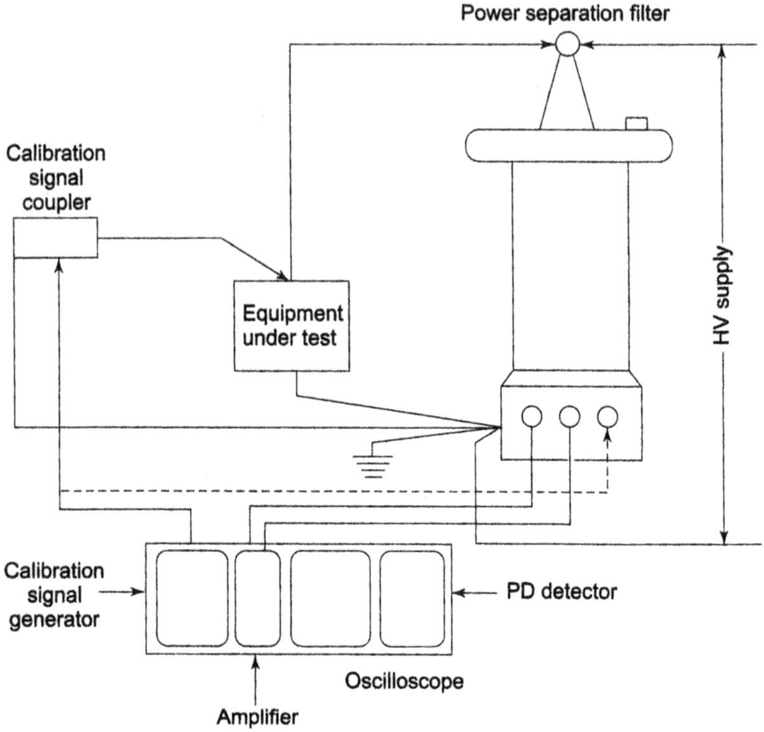

Fig. 16.5 *Partial discharge detection system*

in these tests. Withstand voltage are applied as per requirement on the insulation level. The different tests are classified as:

Power Frequency One Minute Voltage Withstand Tests

These tests are made to check the insulation strength for switching surges, overvoltage under sustained earth fault conditions and certain system resonant conditions.

Induced Voltage Tests

Tests made on transformers at double the frequency to check the inter-turn insulation strength.

D.C. High Voltage Withstand Tests

These tests are made on underground cables after installation. These tests are also performed in factory on capacitors and cables.

Impulse Voltage (Standard Wave 1.2/50 μS) Withstand Tests

Such tests are performed to test the insulation strength of outdoor equipment for lightning.

16.4 Transformer Oil Testing

16.4.1 Composition

The main contents of the insulating oil are naptha, paraffin and aromatic. The proportion of these contents vary for different sources of oil and decide its aging properties.

In the field, the factors most prominent in causing oil deterioration are the presence of air (causes oxidation of oil) and the effect of temperature. The complete exclusion of air may be achieved by the use of sealed transformers having a suitable oil expansion arrangement. However, the major exclusion of air is obtained by the use of an oil conservator with a breather system.

Some of the products of oil oxidation are catalytic for further oxidation. It is, therefore, important to avoid the mixing of oils having varying degree of deterioration. It is especially important in case of the transformer, to thoroughly clean the core and windings which have been immersed in severely deteriorated oil. The dielectric constant of paper and transformer oil is similar and is approximately 2.5 but very different from air or gas bubbles. It follows that if there are air/gas bubbles trapped in the oil or solid paper insulation, there is a much higher electric field in the air/gas bubble than the surrounding insulation. The bubble will breakdown at much lower applied voltage than oil/paper, producing additional gas and eroding oil/solid insulation in the transformer in contact with the bubble's perimeter. This can eventually lead to a breakdown of the complete insulation.

The oil for filling in the transformer should have the characteristics, as given in Table 16.2 (as specified in IS: 335–1993).

Table 16.2

Guidelines for the characteristics of oil to be obtained before energizing new transformers using new insulating oils according to IS: 335–1993

Characteristics	*Equipment voltage*	*Permissible limit satisfactory for use*
Electric strength (breakdown voltage kV)	Below 72.5 kV	40 kV (rms)
	72.5 kV and less than 145 kV	50 kV (rms)
	145 kV and above	60 kV (rms)
Specific resistance (resistivity ohm-cm at 90 degree C)	All voltages	1×10^{12} ohm-cm (min.)
Dielectric dissipation factor (tan δ) at 90 degree C (max.)	All voltages	0.05
Water content, ppm max.	Below 72.5 kV	25 ppm
	72.5 kV and less than 145 kV and	20 ppm
	145 kV and above	15 ppm
Interfacial tension at 27 degree C (min.) N/m	All voltages	0.030

16.4.2 Manufacture Standards

In India, the transformer oil is mainly supplied as per IS: 335–1993. The mineral oil supplied is generally treated further by equipment manufacturers so that it conforms to their standards of manufacture before it is used. The additional stipulations of individual are given below.

Bharat Heavy Electricals Ltd. (BHEL), Bhopal

tan $\delta \not> 0.0005$ at 27°C

Resistivity $\not< 700 \times 10^{12}$ ohm cm at 27°C

Water content should not be more than 50 ppm (as received condition). No inhibitor should be present in oil.

Before use, the water content of oil is brought down to about 15 ppm by filtration.

New Government Electric Factory (NGEF), Bangalore

Oxidation stability	(a) Sludge value should not be more than 1.2 per cent (percentage by weight). (b) Acidity after oxidation should not be more than 2.5 per cent (or 2.5 mg of KOH per gram of oil).
Dielectric strength	Should withstand 50 kV (rms) for one min when tested as per IS: 335.
Saponification value expressed as mg of KOH per gram of oil	It should not be more than 1.0.
Acidity	The total acidity should not exceed 0.03 mg of KOH per gram of oil.
Viscosity at 27°C	Should not be more than 27 centistokes.
Pour point	Maximum = – 10°C
Flash point	Minimum = 145°C
Copper strip corrosion (at 100°C for 3-hr)	There should not be any tarnish on copper.
Volume resistivity when tested with 500 V dc for one min at 90°C	Should not be less than 7.5×10^{12} ohm cm.
S.K. value (chemical test to find extent of refining)	Should not be more than 4 per cent.
Furfurol value (chemical test to find out aromatic content)	8 per cent maximum
Dissipation factor or tan δ value	For new oil at room temperature 0.001 or less; at 90°C it should not be more than 0.005.
Interfacial tension	This shall not be less than .04 newtons/m at 27°C.

16.4.3 Tests

In the new transformer oil from refineries as per IS: 335–1993, the moisture content should not exceed 50 ppm. There is general agreement that the water content of the oil in transformers should be limited as below.

(a) Transformers 132 kV and below: 25–30 ppm.
(b) Transformers 220 kV: 15–20 ppm.

The presence of moisture results in lowering the dielectric strength of transformer oil. Small dust particles of fibres contained in it readily absorb the water particles suspended as emulsion or dissolved in the oil and soon bridge the electrode gaps. Other solid insulation in transformers also get injuriously affected in the presence of moisture. It has been reported that the breakdown strength is very sensitive to moisture content in the range 0–150 ppm of the water content in the oil.

Measurement of Moisture Content

Karl Fischer titration (refer IS: 2362) is long acknowledged as a versatile method for performing routine moisture determination of oils. The reagent is a mixture of pyridine, SO_2 solution and iodine-methanol solution which reacts quantitatively with water giving a sharp electrometric end point. High accuracy (~ 5 ppm) and sensitivity can be obtained with this titration.

Apparatus Used

Karl Fischer titration attachment, titration potentiometer and transformer oil.

Procedure

(a) Fill the burettes with the Karl Fischer reagent and methanol solutions.
(b) Place a dry breaker with a stirring paddle on the instrument.
(c) Turn the potentiometer on and put the polarizing current switch in the on position.
(d) Turn the stirrer motor on and adjust the speed so that no air bubbles are formed or excessive splashing of the solution occurs.
(e) Turn the range switch of the potentiometer and scan with the cursor till the magic eye opens fully. Adjust the sensitivity so that a sharp opening of the magic eye is obtained.
(f) Wait for 10 min and start titrating with KF reagent. As the drops of the KF reagent fall, the magic eye closes and opens.
(g) Proceed with titration. At the end point the magic eye will remain closed for 20 s or more.
(h) Repeat the titration 2–3 times.

The rate at which the titrant is added is important.

Interfacial Tension

The interfacial tension between an electrical insulating oil and water is a measure of the molecular attractive force between their unlike molecules at the interface. It is expressed in newtons/metre (N/m). The main significance of the interfacial tension test of insulating oils is the fact that it provides a sensitive means of detecting small concentrations of soluble polar contaminants and products of oxidation.

Under certain conditions, when the interfacial tension drops below a certain value, it may be an indication that sludge precipitation in the transformer has started or is imminent. Interfacial tension is determined in accordance with the method of test for interfacial tension of oil against water by the ring method (IS: 6104–1971).

Dielectric Strength Test

The dielectric strength of an insulating material is the average voltage gradient at which electric failure or breakdown occurs under prescribed conditions. The dielectric strength of an insulting oil is important as a measure of its ability to withstand electric stress without failure. It can also serve to indicate the presence of contaminating agents, such as water, dirt or conducting particles in the oil, one or more of which may be present simultaneously when low dielectric strength values are found by test. However, a high dielectric strength does not necessarily indicate the absence of all contaminants.

Procedure: The dielectric strength can be determined in accordance with the method test for dielectric breakdown voltage according to IS: 6792–1992.

Aging Test

Transformer oils undergo aging during service causing an increase in acidity, saponification number, sludge content and viscosity. These changes in test values of the oil depend on the oxygen absorbed by it. Oxidation is the root cause of the deterioration of transformer oils.

Transformer oils are mixtures of several hydrocarbons. Very low aromatic content produces excessive acids and very high aromatic content produces excessive sludges. Optimum aromatic content is required and so "optimum refining" has been recognized as necessary for producing good oils. In addition, special chemicals (inhibitors) are added in small quantities. These chemicals decrease the rate of oxygen absorption.

With aging tests, an attempt is made to assess the suitability of the oil for several years of service in the field. As the aging process is accelerated during the tests, these tests are termed *accelerated aging tests.* They determine whether optimum refining has been carried out.

Procedure: Among the several tests for estimating the oxidation properties of oils, the oxygen absorption test is a simple, useful and suitable test for obtaining a comparative assessment between several samples. The amount of oxygen absorbed by the sample is measured as a change in the oxygen pressure when the sample is put in a container filled with oxygen and sealed.

Acidity Test

The following acidity limits have been adopted as a guide as per IS: 1866–1983, "Code of practice for the maintenance of insulation oils".

(a) When the acidity is less than 0.2 mg/KOH/g, the oil is in good condition from the standpoint of this test.
(b) When the acidity is between 0.2–0.5 mg/KOH/g, no action need be taken if the oil is satisfactory in all other respects.
(c) When the acidity is between 0.5–1.0 mg/KOH/g, the oil should be kept under observation and filtered, if necessary.
(d) When the acidity exceeds 1 mg/KOH/g, the oil should be reclaimed or discharged.

The portable kit for quickly determining the acid content of transformer oil has been developed by Central Power Research Institute. This kit consists of the following items:

(a) A rugged box with suitable compartments or partitions to accommodate various reagents and other components.
(b) A polythene bottle containing natural rectified spirit or preferably ethyl alcohol AR grade.
(c) A polythene bottle containing Q-0085N sodium carbonate solution.
(d) A drop bottle or an indicator bottle containing universal indicator.
(e) Clean dry test tubes.
(f) Three pipettes: one two ml capacity graduated to 0.1 ml division and two one ml capacity.
(g) Colour chart calibrated with neutralization number values.
(h) Suitable instruction booklet.

Specific Resistivity Test (7)

The measurement of resistivity is carried out as per IS: 6103–1971. The standard specific resistivity values at different temperatures of healthy oil are shown in Fig. 16.6. Useful information is obtained when the resistivity test at both ambient temperature and at 90°C is carried out. A satisfactory result at 90°C coupled with an unsatisfactory value at the lower temperature is an indication of the presence of water or cold precipitable material, without undue chemical deterioration or general contamination. Unsatisfactory results at both temperatures indicate a greater extent of contamination and that the oil is, therefore, less likely to be restored to a satisfactory level by drying and low temperature filtration.

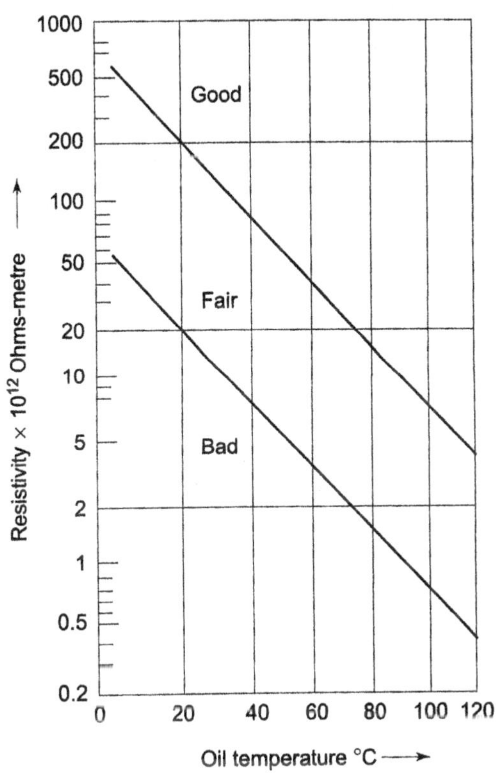

Fig. 16.6 *Condition of insulating oil as determined by dc resistance*

PROBLEMS

1. (a) What is the thumb rule for insulation resistance change with temperature rise (°C) for a transformer? (b) If the insulation resistance of both windings together to earth of 132/11 kV transformer is 200 MΩ at the room temperature of 25°C, what will be the insulation resistance at 75°C top oil temperature?
2. The polarization index (R_{60}/R_{15}) of a faulty 300 m paper insulated lead covered cable of an 11 kV outgoing feeder from the sub-station is 1.35 when measured with a 2.5 kV insulation tester. What does it signify?
3. (a) In the dehydration process of a power transformer, IR values are taken at fixed intervals. Initially the values reduce and finally the values increase. What are the reasons for this? (b) What are the minimum values of IR required for a 700 H.P. 6.6 kV motor and industrial wiring?
4. State the effect of the following values of the insulation:
   (a) A new transformer oil received from the supplier has tan $\delta = 0.08$ at 90°C.
   (b) Partial discharges in a 137.24 kVAr, 6.99 kV rated capacitor in service were measured to be 2 *pC* at the inception voltage of 7.2 kV.
   (c) The transformer oil of the transformer in the field was tested for specific resistivity and the measured values were 5×10^{12} at 20°C and 0.8×10^{12} at 90°C.
5. What is the significance of the power factor (tan δ) measurement test for a capacitor? What will be the tan δ of an HT power capacitor having losses of 0.2 W/kVAr?

BIBLIOGRAPHY

1. (i) IS: 335–1993.
   (ii) International Electrotechnical Commission Publication: 74.
   (iii) International Electrotechnical Commission Publication: 296. New insulating oil for transformers and switchgear.
2. Mashikian M.S., and others "Diagnostic Testing of Medium Voltage Cables by Partial Discharge Location—Recent Field Experience", *Proceeding of the 15th International Conference on Electricity Distribution, CIRED*, Nice, June, 1999, pp. 132–144.
3. Steward W.E. and T.A. Stubbs, "*Modern Wiring Practice—Design and Installation*", Newness, Oxford, 1995, pp. 322–323.
4. *The J & P Transformer Book*, 12th Edition, Newness, Oxford, 1998, p. 78.
5. Jones G.R. and others, *Electrical Engineer's Reference Book*, 5th Edition, Newness, Oxford, 1998, pp. 13/29.

6. New International Electrotechnical Commission Procedures for Insulation System Evaluation, Publication No. 610, 1979.
7. IS: 1866–1983 'Testing of transformer oil in service'.
8. Reynolds, P.H., "Partial discharge measuring techniques", *International HV Symposium*, Zurich, September, 1975.
9. Kuffel, E. and M. Abdullah, '*High Voltage Engineering*', Pergamon Press, Oxford, 1970, pp. 188–326.
10. *M 4202 Partial Discharge Detector*, Robortion Messelectronik, Berlin, 1992.

Chapter 17

System Protection

The protection requirement in an electric circuit depends on the anticipated hazards, relative degree of protection required against each and the capacity of the undertaking to bear the cost of the protective devices. Protection is provided to minimize the following hazards in the system:

(a) Sustained excess current, i.e., overloading

(b) Excess energy during transient overload, i.e., short circuit faults etc.

(c) Electric shock

(d) Fire

(e) Loss of discrimination

(f) Interruption of supplies

At the same time, protection is required to be effective with the least disturbance to the normal operation of the system, both in the branch protected as well as in the remaining branches. In other words, there should be no avoidable loss of discrimination at any level of fault current, the nuisance created by the interruption of supplies should be kept to a minimum and the restoration of supply on the clearance of a fault should be as rapid as possible. On immediate inspection, it is clear that some of these requirements are to a large extent contradictory and a compromise must be reached in a practical situation.

17.1 Time Current Characteristics

It is important to know the time current characteristics (TCC) of system elements, such as lines, cables, capacitors, transformers, loads (motors, furnaces and so on) etc., and their combinations so that time current characteristics of protection provided, are matched to avoid unnecessary operation of protection on switching surges. When a fault occurs, all equipment starting from the generation sources up to fault location are subjected to damaging effects caused by the flow of fault current, whose value is many times greater than rated current. Heavy stresses: magnetic, thermal and mechanical which are associated with the flow of fault current can result in serious damage. Whereas the peak value of fault current decides the magnetic and mechanical stresses, the thermal stresses are proportional to the area under the curve of the first major loop of fault current (see Fig. 17.1). The limitation of peak magnitude of fault current and the area under the curve of the first major loop of fault current indicate the degree of protection.

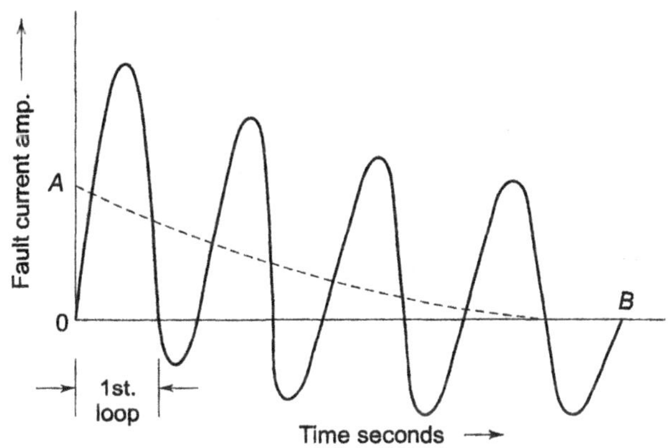

Fig. 17.1 *Fault current loop*

17.2 Fuses

17.2.1 Introduction

A fuse acts both a protective and disconnecting device. The fuse protection is based on the concept of thermal heating given by I^2t (Amp^2 s) where I is rms current during fuse operation and is deduced from I^2dt over pre-arcing, arcing or total operation time t as may be appropriate. Typical A^2s of HRC fuses manufactured as per IS: 13703–1993 are given in Table 17.1.

The time/current characteristic of a fuse is approximately $I^2t =$ constant for large currents, so that its operating time can be very small (fraction of a cycle) and thus the time grading is not a simple matter. Current-time characteristics of different fuse links and elements differ and need careful consideration for application in a system. Fuses can be of low or high rupturing capacity. Rewirable type are considered having low breaking capacity. All fuses having a breaking capacity of over 16 kA can be referred to as *high rupturing capacity* (HRC) fuses. HRC fuses provide better discriminative protection to a degree of sensitivity which is well within the requirements of general distribution. Their characteristics can be ascertained and declared accurately so that they can be applied in situations where the requirements are known. These remain consistent and stable in service without calibration or maintenance provided they are properly designed and manufactured. The design of the system in which fuses are used and other factors external to the fuse itself, affect the degree of discrimination which can be achieved by possible matching of current-time characteristics of the fuse link as well as of the device to be protected. An assessment of these factors is necessary for the most effective use of fuses, though the assessment need not be critical. Failure to achieve discrimination owing to the incorrect choice of fuses may result in nuisance.

Table 17.1

Performance of LT HRC fuses*

Rating amperes and type	I^2t (Amp^2 s)			Watts loss on full load
	Pre-arcing	Total 415 V	Total 550 V	
32 L	1050	2500	3300	3.7
40 L	1900	4400	5850	4.7
50 L	2600	6000	8000	6.0
63 L	5300	12000	16500	6.4
80 L	13500	31500	42000	7.5
100 L	27000	62000	82500	7.8
125 M	41500	96000	130000	11
160 M	74500	170000	230000	13
200 M	145000	340000	450000	15

**Source:* Catalogue ST/S, S & S Power Switchgear Limited, Chennai.

17.2.2 High Voltage Fuses

These are of two types as described below.

Noncurrent-Limiting Fuses

(a) Rewirable types fuses: horn gap type. These are of a lower rupturing capacity. The horns are generally made of steel. For effective quenching of the arc, the angle of dispersion of the arcing horns is designed for 60° and fuse assembly is installed in a horizontal position. Tinned copper wire is used generally as the fuse element (see Appendix VIII). The working of these fuses is effected due to pollution, wind and other atmospheric conditions and hence they do not afford complete protection, though they are cheap.

The element may be enclosed in a glass tube when it is liable to mechanical damage in an open air condition. The gap spacings are based on minimum clearances in air for live point to earth as specified in various standards. In practice the horn gap length is generally of about 200, 250 and 375 mm for 11, 22 and 33 kV systems respectively.

(b) Expulsion type: The arc is extinguished by the expulsion of gases produced by arcing initiating by the fusion of the element. The short circuit rupturing capacity of such fuses is higher because of the gases evolved inside the fibre tube which expel the arc products and quench the arc quickly. Although the arc may be extinguished when the alternating waveform of current passes through its zero value of amplitude within half cycle i.e. 10 ms, the arc may re-ignite as the recovery voltage rises across the ends of fuse links. Eventually, final extinction occurs and the fault is removed from the system after a few cycles. This process may create voltage dip in the system. REC Specification 53/1987 prescribes 11 kV, 100 A, 8 kA breaking capacity drop-out expulsion type fuse cutouts for 11 kV distribution transformers.

Current-Limiting Fuses

These fuses which are of high rupturing capacity are silica powderfilled, contained in a porcelain or resin cast envelope with an entire assembly of elements on the former and is impressed in the filler which may be air or oil tight depending upon the application desired.

A current-limiting fuse is constructed with silver ribbon element surrounded by fine granular silica sand housed in a fibre-glass tube. For low fault current, the fuse time-current characteristics are similar to that of the expulsion fuse. When high fault current flows, the silver ribbon element quickly heats and vaporises. The high temperature of the resulting arc melts the sand, forming a glass-like structure which restricts the arc and produces an increase in resistance. This high resistance changes the circuit power factor to near unity so that the time at which the current crosses zero is the same as the system voltage. This situation allows the fuse to clear the fault before pre-fault current zero and in less than half a cycle. There is no voltage dip in the system. Current-limiting action and fast operation reduces the I^2t (fault energy) let through into failed equipment, thereby reducing resultant damage. By suitable design and use of various metallic elements and filler material, the different time/current characteristics can be achieved to aid discrimination with other fuses and protective devices. The characteristics are given by the manufacturer in the catalogue. Relatively, cartridge type-HRC (high rupturing capacity) fuses are costilier compared to expulsion type fuses. These are used for protecting distribution transformers (generally 100 kVA and above) and LT feeders.

Table 17.2

HRC fuse link ratings*

Rated kV	Maximum current rating A	Assigned breaking capacity MVA	kA
3.6	100	150	—
	200	—	50
	225	—	50
7.2	100	250	—
	80	550	—
	160	—	40
	160	550	—
11/12	50	825	
	75	400	
	100	825	
	100	400	

**Source:* Catalogue of S & S Power Switchgear Ltd. Chennai

Rating

The current limiting fuses are available with superior low over-current breaking performance and very high breaking capacities. Typical rating of fuse links manufactured as per IS: 9385 are given in Table 17.2.

The expulsion type and horn gap type fuses are commonly used in distribution systems for protection of lines and transformers. The latter is extensively applied on 11 and 22 kV lines and transformers and the former mostly on power capacitors; 33, 66 and 132 kV transformers. Compared to the expulsion type unit, the horn gap fuse has certain disadvantages, viz. rupture of the arc takes place in free air which can cause damage to equipment close by, the fuse may blow out due to fast wind and pollution may change the characteristics of the fuse. Expulsion type fuses on the other hand, afford superior protection. For example, the 66 kV expulsion fuses developed and used by the Karnataka State Electricity Board withstood Table 17.3 test values.

Table 17.3

Test values

	Expulsion type	*Expulsion drop out type*
Rupturing capacity	360 MVA	503 MVA
Fault clearing time	3.5 cycle	1.25 cycle
Lowest pf of fault	< 0.1	< 0.1

For any type of fuse, the rated current of the fuse is usually higher than the normal service current. Important factors affecting the proper current rating of the fuse are overload currents possible in the circuit including sustained harmonics, transient overcurrents associated with transformer inrush currents, motor starting and capacitors and the requirement of discrimination with other protective devices. Fuses are referred to as general purpose or backup fuses depending on their application. The former are able to break any value of current up to the rated breaking capacity and can be applied as the sole overcurrent protection, while the latter are intended to be used in connection with other overcurrent devices and are so designed as to break currents above a rated minimum breaking current. Fuses carrying a current exceeding the rated current for a time longer than that associated with the minimum breaking current may be subjected to deterioration which could influence its characteristics, particularly the breaking capacity. Figure 17.2 shows the characteristics of expulsion fuse and HRC fuse in protection of transformer.

Since a fuse is a voltage sensitive device, the rated voltage of the fuse should be carefully selected, bearing in mind that the basis of the rating is generally related to the use on a three-phase earthed system. On such a system the voltage rating of the fuse link should be equal to the highest line-to-line voltage. On a single phase system, the voltage rating should be at least 115 per cent of the highest single-phase circuit voltage.

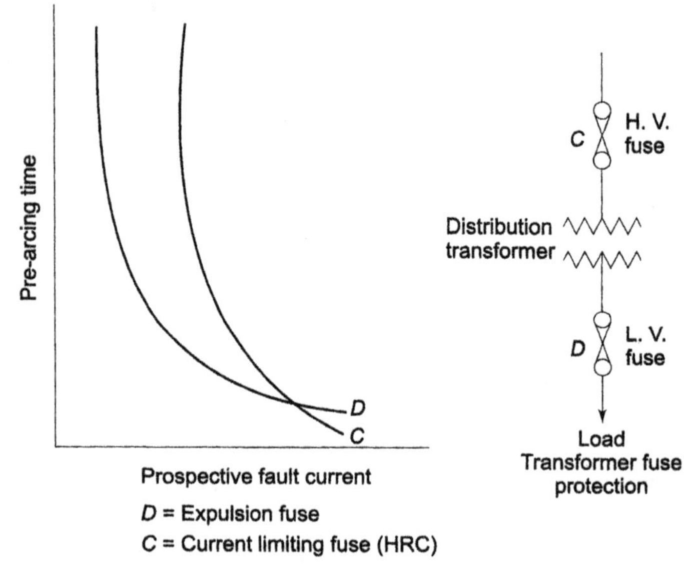

Fig. 17.2 *Typical time/fault current characteristics of a current-limiting fuse and an expulsion fuse*

17.2.3 Low Voltage Fuses

These fuses are for voltage ratings up to 650 V as per IS: 13703. Low voltage fuses are of the following two types.

Semi-Enclosed Type

It is very cheap rewirable fuse protection for circuits where a fairly low breaking-capacity is required. Copper tinned wire is generally used (for ampere rating see Appendix VIII). The performance of these rewirable fuses is not guaranteed. Hence, whenever proper discrimination is required, these fuses are not recommended. Switch fuse units using these fuses give only coarse protection. The minimum fusing current depends upon the material, length, shape and area of cross-section of the fuse element and type of enclosure employed. An approximate value of minimum fusing current (I) for a round fuse wire is:

$$I = kd^{3/2}$$

where k is constant depending on the material of fuse wire. k is 1870 and 2530 for aluminium and copper respectively. d is diameter of wire in cm. These fuses are used up to a maximum fault level of 4 kA. Generally, the following sizes are used for transformers (11/0.415 kV) and service lines (see Appendix VIII).

Transformer size (kVA)	LT fuse wire current rating (A) (tinned copper)
16	32
25	63
63	100
LT service single-phase load	
Up to 2 kW	16
Above 2 kW to 4 kW	32
LT three-phase service/sub-main/feeder load	
Up to 5.5 kW load	16
$> 5.5 \leq 11$ kW	32
$> 11 \leq 22$ kW	63
$> 22 \leq 37$ kW	100

HRC Cartridge Fuses Links

Fuse links of this type are usually in completely enclosed cartridges, containing one or more fuse elements and filled with an arc-extinguishing medium, usually silica sand. The design of the fuse element is capable of considerable variations. IS: 1370–1993 covers these fuses. The application of these fuses requires checking of voltage, current rating and class of fuses, i.e. quick action or delayed action, and adequacy of their breaking capacity. Apart from carrying the rated current continuously, fuses are generally designed to take a certain amount of overload for a pre-determined period of time before they operate. Installations which are designed to take very marginal overloads need protection even against small sustained overloads. In such cases an overload of only 25 per cent should be sufficient to blow the fuse. Fuses which are classified having fusing factors (the ratio of the minimum current required for fuse operation to the rated current) between 1.25–1.5 and 1.5–1.75, respectively, can be used in circuits where the initial currents are higher or where the fuses are integrated in the system such that they do not blow on smaller overcurrents. The data regarding this can, however, be procured from the manufacturer's catalogue. This is important to avoid unwarranted blowing of fuses.

These fuses are characterized by the property of 'cut-off' which means the ability to interrupt the flow of the fault current even before it reaches its maximum value. This property helps in achieving very high rupturing capacities and restricts the flow of the fault current only up to a quarter of cycle. Indian Standard: 13703–1993 prescribes breaking capacity having extremely low let through I^2t and cut-off currents.

17.2.4 Selection of Fuses

The important considerations are:

(i) It should be able to withstand momentary over-current due to starting a motor, and transient current surges due to switching on transformers, capacitors and fluorescent lighting, etc.

When the starting-current of the motor is known, it is assumed that the starting-current surge will persist for about 20 seconds, and a suitable cartridge fuse which will withstand the starting current for this time is selected. When the starting current is not known, it is assumed to be about five times the full-load current.

(ii) For capacitors, cartridge fuse links having rated current about 50 per cent greater than the rated currents of the capacitors, should be selected. Generally, fuses having rated current of about 25 to 50 per cent greater than the normal full load current for transformers and fluorescent lighting circuit are selected.

(iii) Its operation must be ensured when sustained overload or short circuit occurs.

(iv) It should provide proper discrimination with the other protective devices.

(v) Its selection should depend upon the load circuit; steady load circuits and fluctuating load circuits.

17.3 Switching Devices

Isolators

Isolators are switches that are meant for use at places where no load current or only a small charging current is to be interrupted. They are usually installed in a circuit, so as to enable a line or a piece of equipment to be isolated from the energized portion of the circuit as positive safety measures. As per IS: 9920 and REC Specification: 43/1987, air-break switch isolators are rated for the following breaking capacities:

Line-charging breaking capacity	= 2.5 A rms
Cable-charging breaking capacity	= 10 A rms
Transformer no-load breaking capacity	= 6.3 A rms
Load-breaking capacity	= 10 A rms

Load Switches

Switches are generally used to interrupt load currents, such as in primary feeders or in ring-main circuits or for interrupting the exciting current of large transformers or a group of smaller transformers. These are not designed to interrupt fault currents, but are suitable for closing on to a short-circuit. When subjected to severe switching demands, such as more than one operation per day, the circuit breaker should be used for economy. These are less costly and lighter in weight as compared to circuit breakers and reclosers.

Circuit Breakers

A circuit breaker is a switching device built ruggedly to enable it to interrupt/make not only the relatively large load current, but also the much larger fault current which may occur on a circuit. It contains fixed contacts and moving contacts. When the circuit breaker is closed, these contacts are held together. When the breaker opens, the moving contacts separate from the fixed ones and the arc at these contacts is extinguished. This arcing is due to dissipation of magnetic/electric field energy stored in the system (depending upon X/R ratio) beyond circuit breaker point.

A ring-main unit consists of circuit breaker or load switch with built-in isolators on either side. These are designed as per IS: 13118 (HV) and 13947 (LV) for asymmetrical faults, which are more severe than the symmetrical faults due to dc off-set of the fault current, depending upon *X/R* ratio of the circuit. Table 17.4 gives the asymmetrical factor, which are ratios of the asymmetrical to symmetrical rms fault currents at 0.5 cycle after fault initiation for different circuit *X/R* ratios.

Table 17.4

Asymmetrical factor as function of X/R ratios

X/R	Asymmetrical factor
2	1.06
4	1.20
8	1.39
10	1.44
12	1.48
14	1.51
25	1.60

As rule of thumb, for distribution system *X/R* ratio does not exceed 5 and for which asymmetrical factor is taken as 1.25. Different types of circuit breakers are discussed in the next section.

Auto-Reclosers

Reclosers are essentially circuit breakers of relatively lower normal current and short-circuit current carrying capability, and generally conform to IS 2516. These are light weight, less costly than circuit breakers and could be easily installed on poles. Maintenance-free vacuum or SF_6 type reclosers are now well established. Over-current and earth fault protection is based on electronic relay circuit built-in unit, giving a wide range of settings with a micro-processor based controlled switches located on the relay facia. The important feature of the modern design is that, the reclosers do not require any separate ac or dc control supply. These are totally self-powered. The auto-recloser is discussed in detail in Sec. 14.7.

Automatic Sectionalizer

A sectionalizer is a protective device which has no time-current characteristics and does not interrupt a fault. Sectionalizers are used in conjunction with auto-reclosers to disconnect parts of the system on occurrence of permanent fault. The sectionalizer disconnects the faulted line during the unenergized reclose interval of the recloser located nearer to the source of supply. In this way, parts of network with permanent faults can be disconnected without human intervention and without prolonged loss of supply to other parts of network.

A sectionalizer is a switch (vacuum or SF_6 type) which can carry short time fault current but cannot break them. In the event of a fault on the line, the sectionalizer remains closed, when the recloser or circuit breaker interrupts the fault. If the fault is permanent, the recloser or breaker will trip and reclose a number of times, thereby generating a series of pulses of fault current, which pass through the sectionalizer on the faulted spur.

The sectionalizer is fitted with overload and earth fault relays. Their function is to sense the pulses, count them and after a predetermined number (and while the line is dead), trip the sectionalizer. In addition, the sectionalizer is fitted with an external handle by which it can be manually closed or opened. It can thus be used as load break switch for manual line sectionalizing.

Figure 17.3 shows the recloser and sectionalizer coordination on a fault. The remote sectionalizer S_2 is set for two operations and the near sectionalizer S_1 is set for three operations. The recloser is set for four operations, which are instantaneous only. They need not be delayed, which means no problems are experienced because of overlap with the source breaker characteristic. The figure shows how a fault between sectionalizers 1 and 2 causes the recloser to trip three times and after the third impulse causes sectionalizer S_1 to open; sectionalizer S_2 and the other remain closed. The recloser closes and restores the supply to the rest of the lines. When the fault is repaired, the sectionalizer S_1 is reclosed by hand.

The two main advantages of using a sectionalizer are:

- The recloser does not have to be set for delayed trips and therefore the problems of coordination with breakers further back along the system are avoided.

- The sectionalizer can be fitted with electronic overcurrent and earth fault relay so that coordination is better achieved with earth fault as well as overcurrent faults [2].

REC has standardised 200 A sectionalizer for 11 kV lines as per REC specification 44/1993.

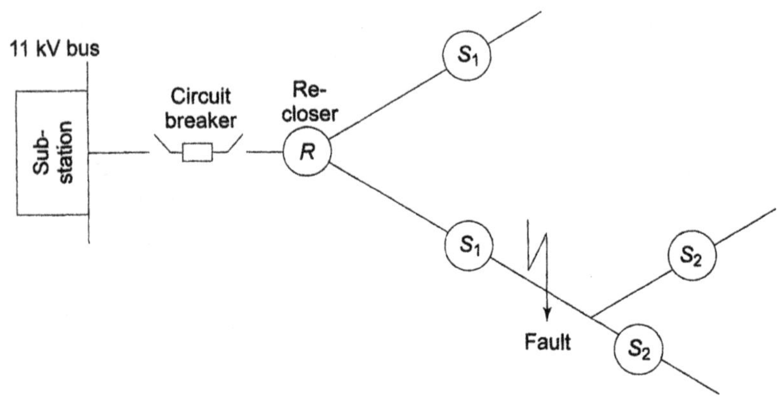

Fig. 17.3 *Recloser and sectionalizer coordination for feeder protection*

17.4 Circuit Breakers

A circuit breaker is a device which is not designed for frequent operation, but is capable of making and breaking all currents including fault currents up to its relative high rated breaking capacity. A wide variety of closing and tripping arrangements are available using relays with variable time delay and a number of operating mechanisms based upon solenoids, charged springs or pneumatic arrangements. The types of breakers used in a distribution system air: air break type, oil break type, vacuum type, and electronegative gas type (SF_6 gas breaker). One great advantage of circuit breakers is their speedy operation, comparatively speaking, on a small overloads and the considerable control of operating time under these conditions. Whereas a fuse characteristic is fixed in the design of the element, the operation of the breaker can be varied at will to suit variations in demand.

Table 17.5 shows that the various types of circuit breakers, their arc quenching medium and their voltage range.

Table 17.5

Type of circuit breakers

Type of circuit breaker	Arc quenching medium	Voltage range and breaking capacity
Miniature	Air at atmospheric pressure	400 V–600 V; for small current rating; up to 5 MVA
Air-break	Air at atmospheric pressure	0.4 kV–11 kV; 5–750 MVA
Bulk oil	Transformer oil	0.4 kV–33 kV up to 300 MVA
Minimum oil	Transformer oil	3.3 kV–220 kV; 150–2500 MVA
Vacuum	Vacuum	3.3 kV–33 kV; 250–2000 MVA
SF_6	SF_6 at 5 kg/cm^2 pressure	3.3 kV–765 kV, 1000–5000 MVA
Air blast	Compressed air at high pressure (20–30 kg/cm^2)	132 kV–1100 kV; 2500–6000 MVA

17.4.1 Oil Circuit Breakers

The mode of action of all circuit breakers consists in the breaking of the fault current by the separation of a set of contacts. An arc is immediately established on separation of the contacts. The means by which this arc is extinguished enables distinction of different types of circuit breakers. When the contacts are closed and carrying current, the problem of temperature rise and millivolt drop, etc. is common to all forms of circuit breakers and is a function of the contact material and contact force. Other features of the design also contribute but are of secondary importance.

For a voltage up to 11 kV, the plain break type of oil (bulk oil or minimum oil) circuit breaker is used to provide economical units of breaking capacities up to 250 MVA.

In bulk oil circuit breakers, oil performs two functions—it acts as an arc extinguishing medium, and it also serves as insulation between the live terminals and earth. The tank of bulk oil circuit breaker is earthed. Its main drawback is, that it requires a huge amount of oil at higher voltages. Due to this very reason, it is not used at higher voltages.

A minimum content oil circuit breaker does not employ a steel tank. Its container is made of porcelain or some other insulating material. This type of circuit breaker consists of two sections; namely, an upper chamber and a lower chamber. The upper chamber contains an arc control device, fixed and a moving contact. The lower chamber acts as an insulating support and it contains the operating mechanism. These chambers are filled with oil, but they are physically separated from each other. The arc control device is placed in a resin bounded glass fibre cylinder (or backelised paper encloser). This cylinder is also filled with oil. The fibre glass cylinder is then placed in a porcelain cylinder. The annular space between the fibre glass cylinder and the porcelain insulator, is also filled with oil. The number of interrupter units contained in a tank depends upon the fault current to be interrupted, and the system voltage. Up to 11 kV voltage, the minimum oil circuit breakers generally employ a single interrupter per phase. The typical figures for higher voltage are two per phase at 132 kV.

17.4.2 Air-break Circuit Breakers

Air-break circuit breakers are quite suitable for high current interruption at low voltages, as compared to oil circuit breakers which are not suitable for heavy current interruption at low voltages due to carbonisation of oil. In an air circuit breaker, air at atmospheric pressure is used as an arc extinguishing medium. It employs two pair of contacts: main contacts and arcing contacts. The main contacts carry current when the breaker is in a closed position. They have low contact resistance. When contacts are opened, the main contact separates first, the arcing contacts still remain closed. Therefore, the current is shifted from the main contact to the arcing contacts. The arcing contacts separate later on and the arc is drawn between them. The arc interruption is carried out by high arc resistance principle. The arc resistance is increased by lengthening, splitting and cooling the arc. The arc length is rapidly increased by employing arc runners and arc chutes. The arc moves upward by both, electromagnetic and thermal effects. It moves along the arc runner and then it is forced into a chute. It is split by arc splitters. A blow-out coil is employed to provide magnetic field to speed up arc movement, and to direct the arc into splitters. The blow-out coil is not connected in the circuit permanently. It comes in the circuit by the arc automatically during the breaking process. The arc interruption is assisted by current zero because arc high resistance is obtained near current zero, as the dielectric strength of contact gap increases to such an extent that it can withstand and voltage stress across it.

17.4.3 Moulded Case Circuit Breakers (MCCBs)

The moulded case circuit breaker is a simple, compact, economical air break circuit breaker which can be conveniently mass produced because of its moulded construction (see IS: 13947–2). Its cost lies somewhere between the cost of a switch fuse combination and conventional circuit breaker of an equally reputable make. However, for more complete comparison, it is necessary to consider the wholesome cost in terms of the size, weight, maintenance, assembly effort, speed of breaking, securing costs and other factors, such as down times and accuracy of protection. The moulded housing is made of heat resistant phenol-asbestos or plastics which are easy to install. Only low voltage types are available in ratings up to 660 volts ac and current rating up to 2500 A and breaking capacities up to 100 kA rms symmetrical at 0.20 power factor. Apart from their modest breaking capacity having the same advantages and disadvantages as large air break circuit breakers, modern MCCBs afford reliable protection and are also available with a trip-free mechanism. They are usually fitted with a built-in tripping system which can either be of a thermal-magnetic or fully magnetic type with inverse time-current characteristics respectively, for overload protection and instantaneous protection against short circuits. The typical inverse time-current characteristics for both the tripping releases are given in Fig. 17.4. These characteristics are specified for certain ambient temperatures 20, 40°C and a suitable correction factor is specified for other ambient temperature conditions. The tripping mechanism characteristics can be designed for specific requirements of the distribution circuit. Current transformers are not used with MCCBs. The thermal setting is usually adjustable between 50 and 130 per cent of the rated current for overload protection as defined by IS: 3842 and magnetic setting between 300 and 1200 per cent of the rated current for short circuit protection. In case of fully magnetic tripping mechanism, the inverse time-current characteristic overload setting is usually between the range of 75 and 200 per cent and instantaneous setting between 300 and 1200 per cent. Other protection provisions, such as shunt or undervoltage release and one or two auxiliary switches are available as options. Microprocessor based electronic tripping system option is now available as an alternative to thermal-magnetic tripping. The electronic tripping system is more accurate and immune to harmonics.

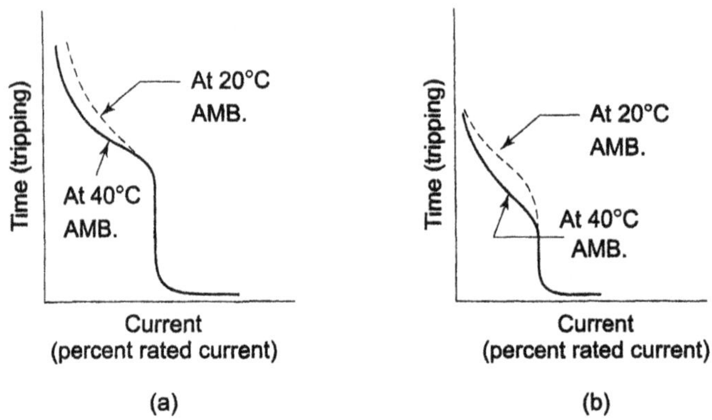

Fig. 17.4 *Typical inverse time-current characteristics: (a) Thermal magnetic; (b) Full magnetic*

The breaking capacity is a limitation only where the source impedance is very low and highly inductive. Since the minimum source impedance can be readily calculated in the design of many installations for a given location, the suitability of moulded case circuit breakers in such a location can be readily assessed. On the other hand, in improving the power supply which has the effect of reducing the source impedance of a system already fitted with moulded case circuit breakers, due notice should be taken of any limitations of the breaking capacity and provision made in the modified design. Modern current limiting MCCBs afford very low let-through energy and better protection coordination.

Three important applications of MCCBs are: main incoming circuit breaker for secondary side of transformers, distribution feeder protection and motor control. In rural areas where there is a problem of getting the right fuse to replace blown fuses, these breakers can be very useful of providing protection to the distribution transformers.

17.4.4 Miniature Circuit Breakers (MCBs)

These are miniature, moulded case ac air break circuit breakers intended for protecting industrial, commercial and domestic installations as per

IS: 8828–1996. Such circuit breakers are available having voltage rating up to 240/415 V rms and current ratings up to 300 A associated with breaking capacities up to 2400 A at 0.35 power factor. At first sight these breaking capacities might appear small, but as in the case of moulded case circuit breakers, the possible short circuit current is limited by the impedances on the source side of the breaker. If the source consists of direct connection to the low voltage side of a small transformer, the short circuit current is likely to be in this range, though the power factor might be too severe. In addition, the faults in MCB circuits are to a large extent limited by cable resistance. On a single-phase-to-earth fault, 5 metre of 10 mm^2 aluminium cable will limit the short circuit current generally to below 5000 A.

Current limiting technology has resulted in the development of MCBs with fault current handling capacity up to 10 kA, high enough for most of the applications. The fault energy in these MCBs is substantially reduced. Under fault conditions, the contacts are physically pulled apart and electrodynamic forces set up by the fault current help to extinguish arc in less than half a cycle (generally 3–5 ms).

17.4.5 Earth Leakage Circuit Breakers (ELCBs)

These are essentially low voltage circuit breakers designed primarily to protect against the risk of electric shock and fire. They are of two types—voltage operated and current operated—designed to isolate the circuit when dangerous potentials are established on the earth conductors and bonded metalwork. The latter are residual current-operated circuit breakers and are widely used. This device breaks a circuit automatically when residual current (earth leakage) exceeds a predetermined amount. These work on the principle of core balance as shown in Fig. 17.5. The load current is fed through two equal and opposing coils wound to a common transformer magnetic core. The phase and neutral currents in a healthy circuit produce equal and opposing fluxes in the transformer core which induces no voltage in the tripping coil. However, if more current flows in the phase conductor than in the neutral conductor as a result of a fault between the phase and earth, an out-of-balance flux will result in an emf being in the trip coil which will open ELCB and isolate the load [4]. Effective use of ELCB requires good wiring and good earthing (see IS: 12640).

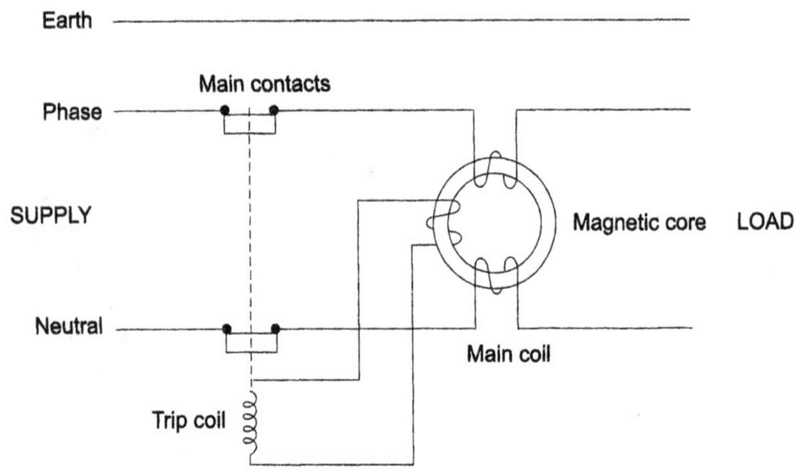

Fig. 17.5 *Principle of earth-leakage circuit breaker*

Leakage currents of 300 mA and above cause insulation failures leading to electrical spark which can spread major fires. Such leakages can be quickly detected by ELCB and human life and property can be saved. Old and substandard building wirings also lead to leakages. These leakages, if allowed to persist and remain undetected and uninterrupted for a long time, lead to a never ending wastage of valuable and scarce electrical energy. Both shock and fire hazards may prove fatal unless proper precautions are taken. Proper earthing is not the only answer. Leakage currents of higher magnitude can cause corrosion through electrolysis and a good earthing can become ineffective. Use of ELCB of 30 mA rated tripping current for shock hazard and 300 mA rating for fire hazard is best solution for safety. The former is normally installed at a socket outlet for the selected appliances and the a latter at the main distribution board for total installation. Total maximum disconnecting time of the circuit is 2 s. For loads 5 kW and above, use of ELCB is mandatory as per IE Rules 61, 71 and 73 prescribe such provisions.

17.4.6 Vacuum Type Circuit Breakers

Vacuum interrupters up to 11, 22 and 33 kV ratings are commonly used. The technique of ac interruption in vacuum consists of separating a pair of current carrying contacts in a high vacuum environment of 10^{-5} torr or more. Unlike other arc interrupters, the current carriers in the arc are mainly metal ions from the contacts, since a negligible amount of gas is present in the arc plasma. The memory of the arc is shortened by the condensation of these metal ions and vapours on to the contacts and shields and the effectiveness of this determines the efficiency of the interruption process at zero current. Vacuum circuit breakers offer the following advantages.

(a) Long life with minimum maintenance. It has 4 times less moving parts than SF_6 circuit breaker
(b) Completely enclosed and sealed construction for indoor/outdoor use
(c) Extremely short and consistent arcing and total break times
(d) Suitability for very fast automatic reclosure
(e) No fire risk
(f) No noise and no emission of gas or air during operation
(g) At the end of its life, a bottle can be quickly removed, discarded and replaced by another spare bottle having an extremely long shelf life

17.4.7 SF_6 Circuit Breakers

The superior arc-quenching ability of SF_6 gas can be attributed to the fact that it is electronegative, which means that its molecules rapidly absorb the free electrons in the arc path between the breaker contacts to form negatively charged ions which are ineffective as current carriers. The electron trapping action results in a rapid build up of the insulation strength after zero current. For effective arc extinction it is, however, necessary to force the gas into the arc, but the properties of SF_6 are such that the gas flow velocity does not need to be high as in the air blast circuit breakers.

Among the advantages claimed for the SF_6 circuit breakers are:

(a) The low gas velocity and pressures employed minimize any tendency towards current chopping, thereby the capacitive currents can be interrupted without restriking.
(b) The closed circuit gas cycle coupled with low gas velocity gives quiet operation as there is no exhaust to atmosphere as in air blast breakers.
(c) The closed gas circuit keeps the interior dry so that there is no moisture problem.

(d) The arc extinguishing properties of SF_6 gas result in very short arcing times so that contact erosion is very little. The contacts can be run at higher temperatures without deterioration.
(e) There are no carbon deposits so that tracking and insulation breakdown is eliminated.
(f) Electrical clearances can be reduced due to the insulating properties of the gas.
(g) As the circuit breaker is totally enclosed and sealed from atmosphere, it is particularly suitable for use in coal mines or in any industry where explosion hazard exists.

SF_6 breakers are used from 1 kV to 800 kV. For transmission and subtransmission voltage 66 kV and above, 'Controlled Pressure Systems' or 'Closed Pressure Systems' are used. Distribution breakers up to 36 kV class are sealed pressure type mostly and are sealed for life of about 20 years. Due to this, SF_6 technology lends itself to diverse circuit switching functions, like load break switches, sectionalizers, reclosers, and capacitor switches.

17.4.8 Air Blast Circuit Breaker

In air blast circuit breakers, compressed air at a pressure of 20–30 kg/cm^2 is employed as an arc quenching medium. Air blast circuit breakers are suitable for operating voltages of 132 kV and above. The *advantages* of air blast circuit breakers over oil circuit breakers are:

- Cheapness and free availability of the interrupting medium, chemical stability and inertness of air.
- High speed operation.
- Elimination of fire hazards.
- Short and consistent arcing time and therefore, less burning of contacts.
- Less maintenance.
- Suitability for frequent operation.
- Facility for high speed reclosure.

The *disadvantages* of an air blast circuit breaker are:

- An air compressor plant has to be installed and maintained.
- Upon arc interruption, the air blast circuit breaker produces a high level of noise when air is discharged to the open atmosphere.
- There is the problem of current chopping.
- There is the problem of restriking voltage.

Switching resistors and equalizing capacitors are generally connected across the interrupters. The switching resistors reduce transient over voltages, and help arc interruption. Capacitors are employed to equalize the voltage across the break. The number of breaks depends upon the system voltage. For example 2 to 4 for 132 kV, 2 to 6 for 220 kV. The breaking capacities are 10,000 MVA at 132 kV, 20,000 MVA at 220 kV.

17.4.9 Fuse Switch Combinations

The accurate current limiting operation and high breaking capacity of fuses has resulted in a number of ingenious combinations of switches and fuse to get an economical unit combining the best performance of each. These combinations find a large number of such applications. Three examples are given below to illustrate their cases.

One example is the limitation by series fuses of short circuit energy in contactors, which are designed only for breaking overload currents. This is a special case of backup protection.

Another example is the use of fuses across the first break to open a load breaking switch. Here the 'limiter' type of fuse finds application, since it is only required to operate on heavy short circuit and does not normally carry current.

A third example is the use of fuses across trip coils to produce rapid, precise and decisive tripping of the breaker at a limiting load current.

IS: 10027–1981 and 13947–Part 1–1993 discuss all the important design considerations for switch fuse units. LT rewirable fuses are used up to maximum fault level of 4 kA.

17.5 Protective Relaying

17.5.1 General Principle

The fuse is a preliminary protective device. As the power capacity and voltage of electrical installations increase and their switching circuits become complicated, fuse protection becomes inadequate. This leads to the development of protective gears based on special, automatic device relays which are called *protective relaying.*

Protective relaying is the basic form of electrical automatic equipment and is indispensable for normal and dependable operation of modern power distribution systems. When a fault occurs, the protection detects and disconnects the faulty section from the system, acting on the circuit breakers for tripping action. When abnormal conditions in current, voltage, frequency and power occur, the protection detects them and depending on the nature of the disturbance, performs the necessary operations to restore the normal conditions or initiates a signal to the operating personnel such as alarm, visual display, etc. or a tripping action to circuit breaker.

17.5.2 Applications

Types

There are two broad categories of protection; *primary* and *backup*. The primary protection is the first line of defence but because 100 per cent reliability not only of the protective scheme, but also of the associated current transformers, voltage transformers and circuit breakers cannot be guaranteed, some form of backup protection must be provided. There are two such forms: *local* and *remote*. Local backup protection is provided at the same location as the primary protection, whereas remote backup protection, as the name implies, is applied at another switching station. An example of remote backup protection is the simple time graded relays as shown in Fig. 17.6. A fault at F_1, would normally be seen first by relay R_1 and isolated by the circuit breaker at R_1. In the event of failure of the relay or associated equipment at R_1, the fault would be isolated by the operation of the relay R_2. Similarly, if a fault is at F_2, in case the relay or allied equipment fails at R_2, the fault would be cleared by relay R_3.

System

Successful application of protective gear involves thorough knowledge of the system to be protected and the method of its operation. The maximum and minimum fault levels for different types of faults occurring at different points of the system must be calculated. The maximum load currents must be known to determine whether the ratio of the minimum fault current to maximum load currents is high enough to enable simple overcurrent operated relays to be used successfully. If the minimum fault current is less than the maximum load current then a

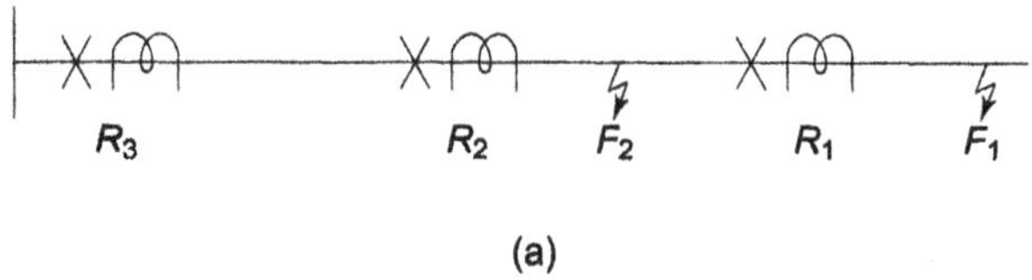

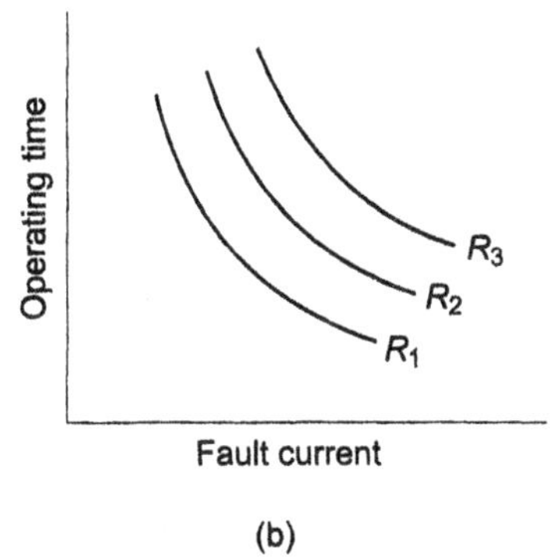

Fig. 17.6 *Time-graded scheme:*
(a) Circuit; (b) Operating characteristics

more complex form of protection involving, e.g., impedance measuring relays or negative phase sequence relays may be required.

Scheme

After the system details have been studied, a suitable protective scheme can be chosen. The choice depends on the following factors.

(a) Can 100 per cent continuity of supply be justified economically?
(b) Do the majority of feeders consist of overhead lines or underground cables?
(c) Are the feeders in the form of ring mains, parallels or radials?
(d) How many switching stations are in series between the supply point and far end of the system?
(e) Are pilot wires readily available?
(f) How is the system earthed?
(g) Are the overhead lines of earthed or nonearthed construction?

The answers to these questions will enable a suitable protective scheme to be selected.

The protective scheme chosen will normally be supplied with samples of the system current and voltage by means of current and voltage transformers. To ensure correct operation of the scheme, it is necessary to choose these instrument transformers so that excessive saturation does not occur which could cause possible loss of discrimination and slowing down of the protection. The effect of the current transformer magnetizing current can also be significant especially with sensitive earth-fault relays as this must be added to the relay operating current when determining the minimum primary current required to operate the relay.

The application of different relays has been explained in IS: 3231 and 3842.

The following are the common protection schemes used.

(a) *Time-grade Overcurrent Protection:* This is based on the time/current principle of protection and is discussed in Section 17.4.5.

(b) *Distance Protection:* In India this protection is normally applied for feeder protection of 66, 110 and 132 kV and above lines. It serves the need for faster clearing times as the fault level increases and also because of the difficulty in grading time/overcurrent relays with the ever increasing number of switching stations creating more stages of protection.

A distance relay measures the distance between the relay location and the point of fault, in terms of impedance (impedance relays) or reactance (reactance relays) or admittance (MHO relays). A single scheme provides both primary and back-up protection.

The effect of arc resistance and power surges plays on important role in the selection of distance relays for a particular distance protective scheme. As the reactance relay remains unaffected by arc resistance, it is preferred for ground fault relaying. As the impedance of a short line is small, the value of arc resistance is comparable to the line impedance and it may cause an appreciable error. Therefore, a relaying unit independent of arc resistance, i.e. a reactance relay, is suitable for the protection of short lines against phase faults. But the reactance relay is more affected by power surges than the MHO and impedance relays. In case of short lines, power surges remain for a shorter period and hence, their effect is unimportant. The predominant factor is therefore, the effect of arc resistance. The effect of power surges stays for a longer period in case of long lines and hence, a relay which is least affected by power surges is preferred for the protection of long lines. The MHO unit is less affected by power surges than the impedance and reactance relays and hence, it is best suited for the protection of long line against phase faults. But it is most affected by arc resistance. As the impedance of a long line is large, the arc resistance will not cause appreciable error and its effect can be neglected. Thus, the predominating factor for the selection of a distance relay for long lines is the effect of power surges. In case of a very long line, elliptical or quadrilateral relays are best suited, as they occupy the least area on the R–X diagram and hence, are least affected by power surges.

An impedance relay is moderately affected, by both power surges as well arc resistance. So it is better suited for medium lines for phase fault relaying. Practical experience also plays an important role in the selection of distance relays for a particular situation.

A distance scheme is capable of providing back-up protection, but it does not provide high-speed protection for the whole length of the line. The circuit breakers do not trip simultaneously at both ends for end-zone faults. The most desirable scheme will be one which includes the best features of both, unit protection and distance protection. This can be achieved by interconnecting the distance relays at both ends of the protected section by carrier signals. Such schemes provide instantaneous tripping for the whole length of the line, as well as back-up protection. There are three types of these schemes: Carrier transfer or intertripping scheme; Carrier blocking scheme; Carrier acceleration scheme.

(c) *Differential Protection:* It consists of pilot wire protection and is quick acting. In India it is generally applied for transformers having a capacity of about 5–10 MVA and above. In some special cases, it is also applied for small transformers up to 0.5 MVA (generator unit auxiliary transformer) for quick clearance of internal faults to limit the damage.

The relay responds to the vector difference between two currents within the zone of protection determined by the location of CTs. It operates on the Merz Price circulating current principle, in which the CTs at each output terminal are balanced against each other. An internal fault creates an unbalance, which results in operation of the differential protection. At the instant of energizing a power transformer, the magnetizing in-rush in the primary winding could cause differential protection to operate. For this, proper discrimination between in-rush current and fault current is provided. The relay is designed to detect high second harmonic content of in-rush current and use this to restrain the relay action. The relay is provided with biased differential element with adjustable setting 15 to 50 per cent by taps and instantaneous high-set unit in differential circuit with pick-up setting of 10 times the rated current.

(d) *Restricted Earth Fault Protection:* Restricted Earth Fault Protection, as shown in Fig. 17.7 is used for winding of the transformer connected in star, where the neutral point is either solidly earthed or earthed through an impedance. The relay used is of high impedance type to make the scheme stable for external faults. For delta connection or ungrounded star winding of the transformer, residual overcurrent relay is used. The REF relay operates only for a ground fault in the transformer. The differential protection of the transformer is supplemented by restricted earth fault protection.

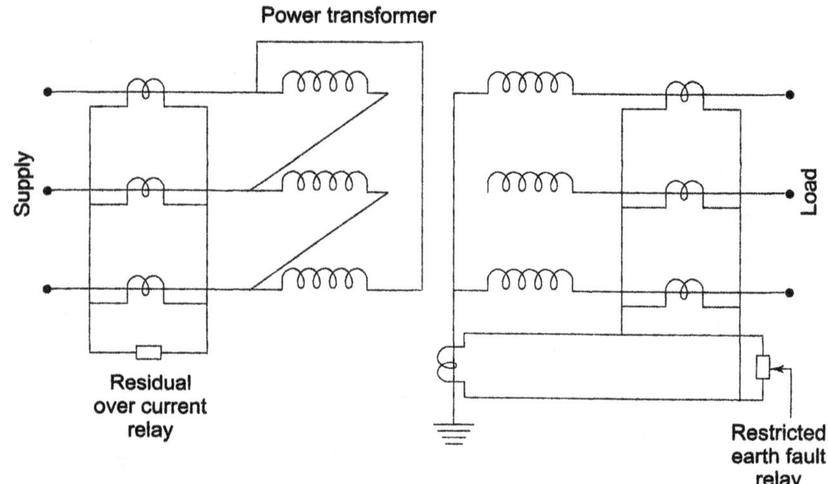

Fig. 17.7 *Restricted earth fault protection of power transformer*

17.5.3 Requirements

Main characteristics of protective relaying equipment are sensitivity, selectivity, speed and reliability. Relaying equipment must be sufficiently sensitive to operate reliably when required under the actual conditions that produce a slight operating tendency. However, it should not operate in a wrong manner. The ability of the protective relay system to operate so as to trip only the minimum number of breakers directly controlling the defective part of the system is called *selectivity of the relaying system.* A protective relay must operate at the required speed and must be reliable. The speed at which relays and circuit breakers operate has a direct bearing on the quality of service to the consumer, stability of the system, and the amount of power that could be transmitted without endangering the life and equipment. Reliability is a matter of design based on long experience. Other things being equal, simplicity and robustness of design contribute largely to reliability. Workmanship, contact pressure, contact materials, prevention of contact contamination and adequate insulation are as important. It is also important to have adequate and reliable sources of current and voltage. Reliability is improved by relay maintenance, proper training of maintenance and operating staff and correct commissioning techniques.

The use of protective relays should be evaluated on the basis of its contribution to the best economy in service to consumers.

17.5.4 Classification of Relays

Digital Relays

Basically, relays can be classified as electromechanical, solid state, thermal and gas-operated relays. With the advent of semi-conductor and micro-processor technology, electro-mechanical relays have been replaced into solid state equivalents as the latter have some inherent advantages as compared to electro-mechanical relays like lower VA burden and smaller size. The solid state relays are now cheaper, versatile and reliable. Static relays contain electronic circuitry and are superior to electromagnetic relays in performance. Microprocessor based digital/numerical protection relays, with their fast response time, better performance characteristics in comparison with electro-mechanical and static relays, and their flexibility of application and increased reliability and availability are best suited for the present-day expanding power system. Different programmes create desired characteristics. Some of the main features/advantages of these relays are:

- Fewer parts and connections
- Continuous monitoring of software/hardware
- Lower burden
- Drastic reduction in size and weight
- Performance of various background tasks like fault location/analysis
- Communication facilities with personal computer system control centre through serial link

All functions can be incorporated into one package. The following factor are considered in the selection of these relays on a cost-benefit basis:

- Improved cost per function
- Disturbance recording
- Decreased maintenance
- Communication capability
- SCADA integration
- Data recording
- Metering
- Age/reliability of existing electromechanical relays

Thermal Relays

Thermal relays are basically used as thermal replicas of the protected apparatus and respond to temperature changes in the same way in which the main apparatus responds. Hence they provide accurate thermal protection against overloads and short circuits to the equipment. Gas-operated relays are actuated by changes in the gas content within them, like Buchholz relays used for transformer protection.

Electromagnetic Relays

In the electro-mechanical range, the relay movements used are induction disc, induction cup, moving coil, attracted armature, moving vane, etc. Induction disc relays are by far the most common type used in time/current graded protection systems where inverse time characteristics are required.

The main principle used in the operation of relays is either electromagnetic attraction or electromagnetic induction. The arrangement consists of electromagnets with an armature or an induction motor carrying contacts, the closing or opening of which controls devices such as trip coils of the circuit breakers. The electromagnets have either current windings or voltage windings or both, the currents passing through the windings producing magnetic fluxes. Torque is developed by the interaction between the fluxes of the same winding or between the fluxes of both windings:

(a) Overcurrent: Torque produced by current winding $= KI^2$

(b) Under/over voltage: Torque produced by voltage winding $= K'V^2$

(c) Directional: Torque produced by interaction of two windings

$$= VI \cos (\phi - \theta)$$

(d) In a relay any of these elements or all the above three elements (a, b, c) may be present and the total torque will be

$$T = KI^2 - K'V^2 + VI\cos(\phi - \theta) - K''$$

where K and K' are the tap settings of windings or constants of I and V, and K'' is the relay mechanical restraint due to a spring or gravity, ϕ is the angle between V and I and θ is the phase angle of relay.

17.5.5 Overcurrent IDMT Type Relays

The overcurrent relay which gives inverse definite minimum time characteristics essentially consists of an ac metre mechanism modified to give the required characteristics. The upper electromagnet (three legged) has two windings. One is connected to the CT in line for the equipment to be protected and is tapped at intervals. The tappings are connected to a plug setting bridge by which the number of turns in use can be adjusted, thus giving the desired current setting. The second winding is energized by induction from the primary and is connected to the winding of the lower electromagnet (U-shaped). By this means the leakage flux from the upper electromagnet and the flux produced by the lower electromagnet are sufficiently displaced in phase to set up a rotational torque on the metal disc suspended between the two magnets. The disc spindle carries a moving contact which bridges two fixed contacts when the disc has rotated through an angle which can be adjusted to give any desired time setting. The definite minimum time characteristics is obtained by saturating the iron in the upper electromagnet so that there is practically no increase in the flux after the current has reached a certain value. This results in the flattening out of the current time characteristics.

Additional directional features can be added to this relay by providing a wattmeter element actuated by both current and voltage. This wattmeter element is prevented from rotating when the power flow is normal but with the reversal of power, rotation can take place. This completes the circuit of the overcurrent element which then operates. IS: 3842 specifies the guidelines for the application of these relays to feeders and transformers. Versatile solid state IDMT relays are available now. Different types of IDMT relays and their uses are discussed below. A comparison of typical characteristics of various types of overcurrent relays is shown in Fig. 17.8.

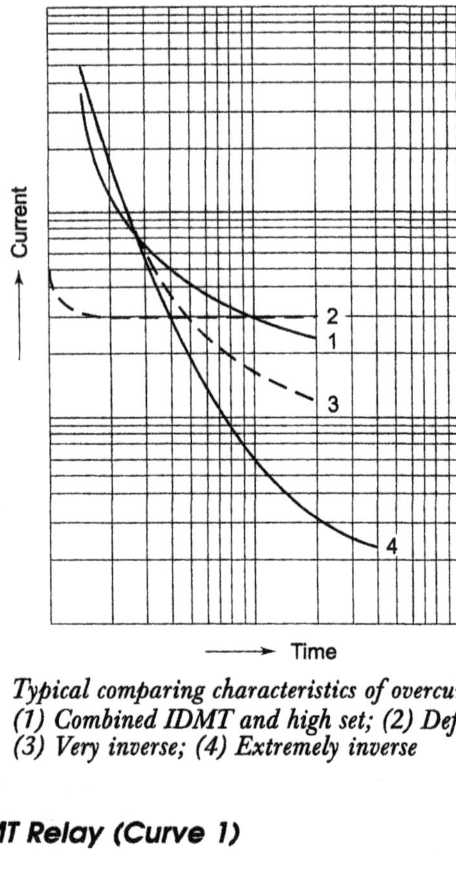

Fig. 17.8 *Typical comparing characteristics of overcurrent relays: (1) Combined IDMT and high set; (2) Definite time; (3) Very inverse; (4) Extremely inverse*

Standard IDMT Relay (Curve 1)

It is used where there is not much reduction in fault currents as the distance from the power source is increased, i.e., it is useful when the line lengths between sections are not very much.

Combined IDMT and High Set Instantaneous Relay (Curve 1)

It is used on long lines or on a transformer feeder where the source impedance is small in comparison with the circuit impedance. High set instantaneous relay reduces the tripping time under maximum short circuit conditions and shortens the time grading space between each breaker. The limitation on the application of the instantaneous relay which is known as transient over-reach of the relay is approximately 3–5 per cent depending on the line angle.

Definite Time Overcurrent Relay (Curve 2)

It provides a ready means of coordinating several relays in series where the system fault current varies very widely due to a change in the source impedance as there is a relatively small change in time with the variation of fault current. Whereas the standard IDMT version is advantageous at higher values of fault current, the definite time relay is preferred for lower current values.

Very Inverse Overcurrent Relay (Curve 3)

It is particularly suitable where substantial reduction in fault level occurs as the distance from the power source increases. The characteristics of this relay are such that its operating time is approximately doubled for a reduction in current from 7 to 4 times the relay plug multiplier setting. This permits use of the same time multiplier for several relays in series.

Extremely Inverse Relay (Curve 4)

The relay time in this is approximately inversely proportional to the square of the current. This makes it suitable for protection of distribution feeder circuits where the feeder is subject to peak currents on switching in, such as the transformer feeder circuit or motor starting circuit. In the case of power circuit supplying refrigerators, water heaters, etc. which remain connected after a prolonged interruption in supply, extremely inverse relay will not only permit the switching-in, but will also provide effective protection against short circuit faults.

These characteristics are similar to those of a HRC fuse and extremely inverse relays can readily coordinate with HRC fuse.

Another application of this relay is in the low voltage distribution circuit in conjunction with auto-reclosures and HRC fuses. The relay is set below the operation of the fuse, but if the fault persists after the reclosure, the auto-reclosure locks itself in a closed position and allows it to blow the fuse. Thus unnecessary blowing of fuses is avoided.

17.5.6 Directional Overcurrent and Earth Fault Relays

Earth fault protection can be given by normal overcurrent relays provided the minimum earth fault current is of sufficient magnitude. It is usual on distribution circuits to provide separate relays either on each phase or on two phases—the latter scheme is sufficient for solidly earthed system. The each fault relay are inverse definite-minimum time or definite-time type and are residually connected as shown in Fig. 17.9. The relay can be set independent of the load current and setting below normal load current (10–40 per cent generally) can be made.

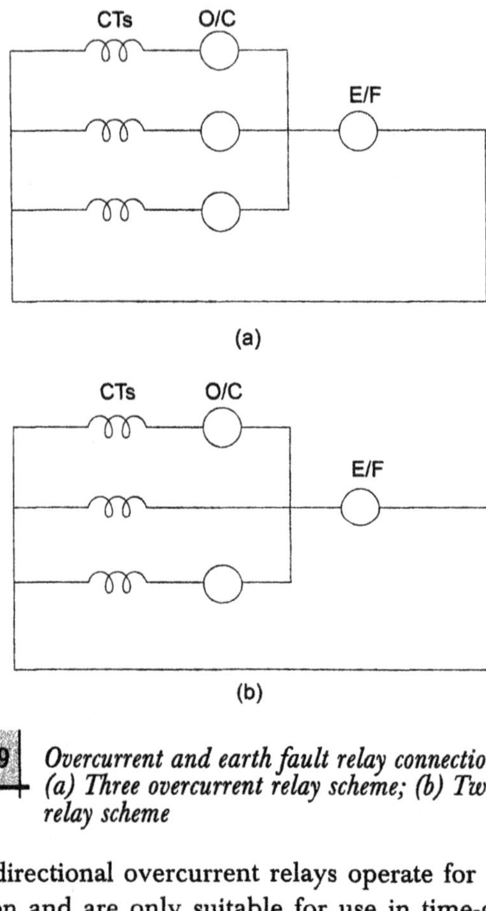

Fig. 17.9 *Overcurrent and earth fault relay connection: (a) Three overcurrent relay scheme; (b) Two overcurrent relay scheme*

The non-directional overcurrent relays operate for current flow in either direction and are only suitable for use in time-graded systems where the power flow is in one direction only. If the fault current can flow in both directions through the relay location, it is necessary to add directional properties to the overcurrent and earth fault relays in order to achieve correct discrimination. Such is the case in the parallel feeders, the directional protection is necessary at the receiving end.

It is essential that the directional units do not allow the overcurrent units to start operating until the fault current flow is in the correct tripping direction. The two quantities applied to the units are the polarizing quantity (voltage) and operating current. For a distribution system connection of 30°, a relay angle with a maximum torque angle of 0° used. The relay angle is defined as the angle between the voltage and current supplied to the relay under balanced three-phase unity-power factor conditions.

Figure 17.10 shows an overcurrent protective scheme for parallel feeders. At the sending end of feeders (at *A* and *B*), non-directional relays are required. At the receiving end of the feeder, directional overcurrent relays placed at *C* and *D* indicate that the relay will operate if current flows in the direction shown by the arrow. If a fault occurs at *F*, the directional relays at *D* trips, as the direction of the current is reversed. The relays at *C* does not trip, as the current flows in the normal direction. The relay at *B* trips for a fault at *F*. Thus, the faulty feeder is isolated and the supply to the healthy feeder is maintained.

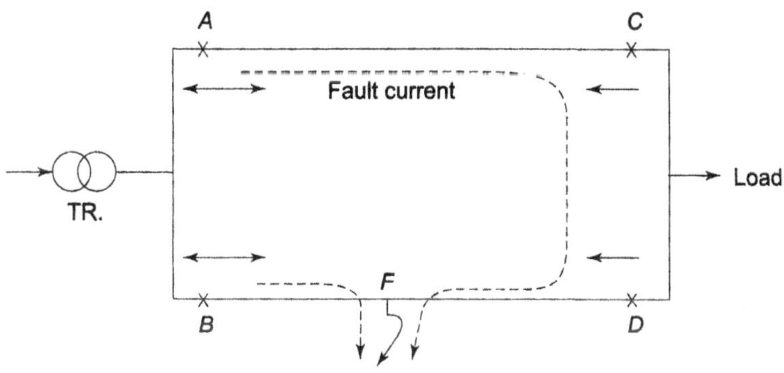

Fig. 17.10 *Typical overcurrent protection scheme for parallel feeders*

If non-directional relays are used at *C* and *D*, both relays placed at *C* and *D* will trip for a fault of *F*. This is not desired as the healthy feeder is also tripped. Due to this very reason, relays at *C* and *D* are directional overcurrent relays. For faults at feeders, the direction of the current at *A* and *B* does not change and hence, relays used at *A* and *B* are non directional.

Figure 17.11 shows an overcurrent scheme for the protection of a ring feeder. It requires two relays; a nondirectional relay is required at one end, and a directional relay at the other end. The operating times for relays are determined by considering the grading, first in one direction and then in the other direction.

If a fault occurs at F_1, the relays at C' and D' will trip to isolate the faulty feeder. The relay at C will not trip as the fault current is not flowing in its tripping direction, though its operating time is the same as that of C'. Similarly, the relays at B and D will not trip as the fault currents are not in their tripping direction, though their operating time is less than the operating time of B' and D' respectively. If a fault occurs at F_2 the relays at A' and D will trip.

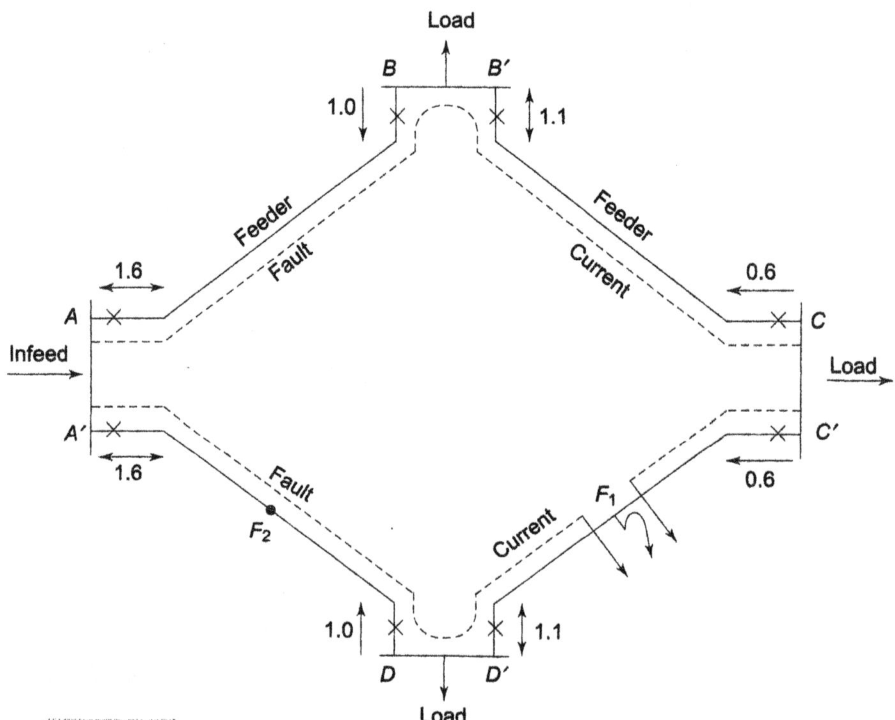

Fig. 17.11 *Typical overcurrent protection scheme for ring main feeders*

17.6 Instrument Transformers

These are current and voltage transformers designed to isolate electrically the high voltage primary circuit from the low voltage secondary circuit and thus provide a safe means of supply for indicating instruments, metres and relays.

17.6.1 Current Transformers

Current transformers are used in power installations for supplying the current circuits of indicating instruments (ammeter, wattmeter, etc.), metres (energy metre, etc.) and protective relays. These transformers are designed to provide a standard secondary current output of 1 or 5 A when the rated current flows through the primary. A fundamental characteristic is its transformation ratio, expressed as the ratio of the rated primary to rated secondary current. Current transformers have two inherent errors: the current ratio and phase errors. These two errors serve as a basis on which current transformers are classified for accuracy. IS: 2705/1992 specifies their classes for metering and protective purposes. The following accuracy classes are used for various purposes:

- 5P, 10P, 15P for general protection
- 5PS, 10PS, 15PS for special protection
- 0.2, 0.5 for general metering
- 0.2S, 0.5S for special metering, having better accuracy at lower value of loads i.e. up to 1 per cent of rated current and is mostly used for electronic meters.

No current transformer should be left with its secondary circuit open. Since no current is able to flow through the secondary winding, when open, there will be no secondary magnetomotive force to act in opposition to the primary one. As a result, the magnetic flux will suddenly rise to a high value and induce a potential of several thousand volts across the secondary terminals. This will not only endanger the winding insulation, but will also pose a serious hazard for the personnel. According to the design of the primary winding, current transformers can be divided into two main groups. Single-turn transformers, such as rod, busbar, core balance and bushing type (ring type) and multi-turn transformer such as wound coil. Epoxy cast CTs are manufactured in range of 33 kV, 22 kV, 11 kV and LT. For 66 kV and above, oil cooled CTs are made.

The measuring CT is required to perform its function over a normal range of load currents. However, the protective CT should be able to supply correct values of secondary currents for fault conditions from full load to twenty times the full load. Knee point voltage and saturation factor are important for these protection CTs. CTs cores should have high knee point/flux density and for metering CTs, knee point is made purposely low so that during fault conditions, the CT may not produce secondary currents heavy enough to damage the metering instruments (ALF is, say, 2.5–4.0). The important characteristics of the protection CTs are.

(a) Rated VA, class of accuracy.
(b) Saturation factor or accuracy limit factor.
(c) Rated short time thermal current.
(d) Rated primary and secondary currents and its ratio.
(e) Knee point voltage.

EXAMPLE: Three 100/5 CTs are used for overcurrent and earth fault as shown in Fig. 17.12a.

The magnetizing characteristics are shown in Fig. 17.12b. Phase fault relays 50–200 per cent range set at 125 per cent. Earth fault relays range 20–80 per cent set at 40 per cent. Relay burden is 3 VA for phase relays and 2.4 VA for earth fault and is 5 amperes relay. Resistance of CT secondary (RCT) is 0.08 ohm. Determine the primary operating current to operate the phase faults and earth fault relays.

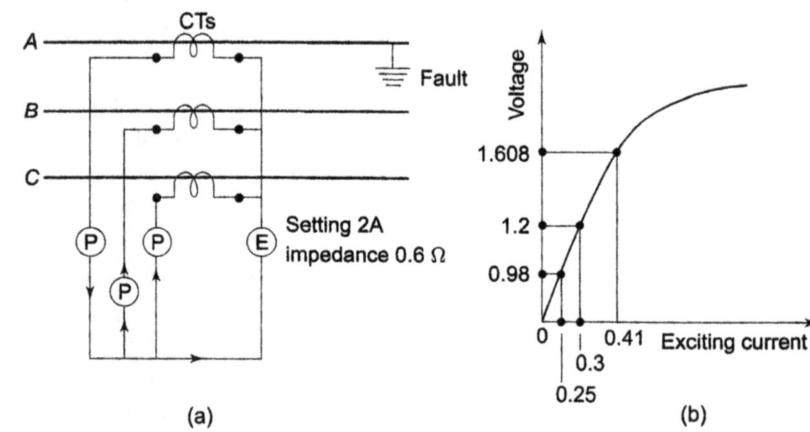

Fig. 17.12 *Current transformer connections: (a) Connection; (b) Magnetizing curve*

Solution: (i) Relay impedance:

3 VA relay impedance = $3.0/(6.25)^2$ = 0.077 ohm at 125 per cent setting

2.4 VA relay impedance = $2.4/(2.0)^2$ = 0.6 ohm at 40 per cent setting

(ii) RCT + R relay = 0.08 + 0.077 = 0.157 ohm

(iii) Primary current for phase faults

At the set value of 6.25 A

Volts from CT = 6.25 × 0.157 = 0.98 volts

From magnetizing curve (refer Fig. 17.12 b) the corresponding magnetizing current is 0.25 A.

Primary current for phase fault relay operation

$= N(I_a + I_{mag})$ where I_{mag} = magnetizing current of CT

$= 20(6.25 + 0.25)$ I_a = relay current

$= 130$ A N = CT ratio

(iv) For earth fault relays

Volts for relay operation $= 2 \times 0.6 = 1.2$ V

From Fig. 17.12b the corresponding magnetizing current is 0.3 A.

Current through A phase relay $= 2.0 + 0.3 + 0.3$

$= 2.6$ A

Volts developed by A phase CT $= 2.6\ (0.077 + 0.08) + 1.2$

$= 1.608$ V

Exciting current for 1.608 V $= 0.41$ A (from Fig. 17.12b)

Total secondary current $= 2.6 + 0.41 = 3.01$ A

Primary current 'A' phase

$$= \frac{100}{5}(3.01)$$

$$= 60.2 \text{ A}$$

Though the earth fault relays are set for operation at 40 per cent, i.e., 40 A of the primary current, they will actually act to operate at 60 A due to the effect of shunting of other phases and exciting current of CTs.

17.6.2 Voltage Transformers

These instrument transformers are used for supplying the voltage circuit of indicative instruments, metres and protective relays. These may be of single-phase or three-phase design and of the dry or oil immersed types. A fundamental rating of the voltage transformer is its transformation ratio and burden, i.e. the total load presented by the instruments connected. All voltage transformers are designed for a standard secondary voltage of 110 or $110/\sqrt{3}$ V. These are divided into classes according to their percentage ratio and phase errors. The errors at which a voltage transformer will operate depend on its burden and the primary voltage.

In practice, we must distinguish between three-limb and five-limb three-phase voltage transformers. The first type has a Y/Y_0–12 connection and an insulated primary neutral. These transformers are only used for measuring/supplying the line voltages to instruments. The second type has a Y_0/Y_0–12 connection. The primary neutral in these transformers is brought out and requires earthing. The five-limb voltage transformers serve for obtaining the line and phase voltages as well as the zero-sequence voltage by means of an auxiliary winding connected as a broken delta. Five-limb VTs are used in capacitor bank protection.

17.7 Overcurrent Schemes

17.7.1 Shortcomings

The inherent shortcomings of overcurrent schemes are:

(a) Inability to distinguish between operating conditions at maximum generation and fault conditions at minimum generation.

(b) Comparatively large fault clearing time involved in clearing the faults.

(c) Increased settings at the generating ends in order to provide suitable discrimination times between sections.

In spite of the above shortcomings, overcurrent relays especially of the inverse definite minimum time type are universally used for distribution lines of LT, 11, 33 and even 66 kV, mainly because of the simplicity and reliability of relays and their maintenance.

However, one has to bear in mind that the coordination of relay settings to a distribution network is an extremely painstaking task and a thorough check of the applicability of the settings is necessary even when a small change is made in the distribution system.

17.7.2 Overcurrent Relays in a Distribution System

The application of overcurrent relays in a system is not simple and requires a thorough checking of the other components for coordination within the system for reliable protection. The following steps are involved for ensuring proper protection:

(a) A single line diagram of the system is drawn out with various elements, such as bus bars, transformers, CTs ratio, etc. marked so that a clear picture is obtained of the system under study.

(b) Information of the relays used, their setting ranges, VA burden, etc. are collected.

(c) Details of CT such as magnetization characteristics, secondary resistance, saturation factor etc. are obtained from the manufacturer's catalogue.

(d) Magnitudes of loads occurring in differential feeders, starting currents due to motor or switching into be ascertained.

(e) Short circuit levels at different points (bus bars) obtained from already available information or calculated from the available data of the feeding stations or assuming infinite capacity of HV buses.

(f) From the data available on breakers, relays (relay error, overshoot time, etc.), the discriminating time which should be allowed is decided.

(g) Current settings are tentatively decided next to allow maximum full load currents continuously.

(h) Taking a log-log graph paper which is more convenient for plotting the characteristics of a number of relays and starting from the farthest point of the distribution system, the current-time characteristics are plotted assuming a tentative time multiplier setting and adjusting whenever necessary to keep a discriminating time decided previously.

(i) The current scale of the characteristics should be chosen for a particular secondary voltage.

17.7.3 Overall Protection Coordination

Coordination is primarily proper selection of protective devices and their settings to develop temporary fault protection zones and limit an outage area to the minimum possible, if a fault is permanent. For this, it is necessary to get the following data:

- Circuit key diagram
- Location of existing protection devices
- Time-Current Characteristics (TCC) curves of protective devices
- Normal load current
- Fault current at every point, where protective devices are located for minimum and maximum generation

After getting this data following steps be taken:

(a) Select initial position at sectionalizing points.

(b) Determine the maximum and minimum values of fault current at each of the selected locations and at the end of a circuit.

(c) Coordinate the protective devices from the sub-station outward, or from the end of the distribution circuit back to the sub-station.
(d) Re-examine the chosen protective devices for current carrying capacity, interrupting capacity and minimum pick-up rating.
(e) Draw a composite TCC curve diagram, showing the coordination of all protective devices employed, with curves drawn for a common base voltage.
(f) Draw a circuit diagram indicating the maximum and minimum fault levels and rating of protective device.

17.8 Unit Protection

Distribution Transformer

Expulsion type fuses on the HV side and HRC fuse or MCCB on the LV side protection is generally provided for 100 kVA transformers and above. Below 100 kVA transformers, the HV side protection is provided by horn gap wire-fuse with air-break switch and the LV side is protected with switch wire-fuse unit or wire-fuse cut-out. As a practice, EDF, France use HRC fuses, ZnO lightning arresters for HV side protection of the distribution transformer.

Distribution Lines

The primary lines may be protected by circuit breakers, auto-reclosers or dropout fuses, cutouts and depending upon the importance of the circuit, the protection is decided. For example, rural feeders may require automatic reclosers which clear almost all the transient faults automatically. Dropout cutouts have been extensively used by various State Electricity Boards. The horn gap fuse with air break switch are increasingly being applied on HT feeders. LT feeders are protected with HRC or wire fuses.

Capacitors

Individual protection for capacitor units is provided by the fuse. In the case of LT capacitor units of rated 5 kVAr and above the protection is generally provided by internal fuses. HT fuses (IS: 9402) of the expulsion type or HRC provided externally are now in common use. Some manufacturers also give the internal 'wire' fuse for individual elements

[9]. According to a study [10], capacitor with internal fuses (refer IS: 12672/1989) have been found to have almost double life as compared to external fuse protected capacitors.

This fuse has a unique function. It must primarily disconnect and isolate a faulted capacitor unit to allow the remaining capacitors to function normally and should also prevent the bursting of the remaining capacitor banks following the flow of too high a value of energy arising from the fault. Ideally speaking, time-current characteristics of the fuse should match the rupturing capacity of the capacitor tank. The selection should match I^2t of a fuse to the I^2t content of the capacitor inrush current. It is desirable to use fuse links having a rated power frequency current of about 1.65 ~ 2 times the rated current of the capacitor and a high energy rating of the order of the same numerical value in kilo Joules along with similar over-rating for cables and switchgear.

PROBLEMS

1. Figure 17.13 below gives the low voltage circuit showing protection with HRC fuses in series. The main circuit is protected by a major fuse and the minor fuse in series protects the sub-circuits. Draw the typical time-current characteristics of the fuses. Also show the co-ordination or discrimination between the fuses when a fault occurs at point 'A' with fault current I_f.

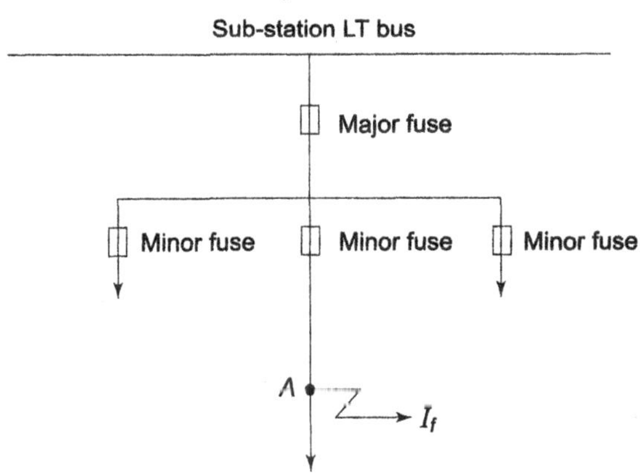

Fig. 17.13 *Fuse protection co-ordination in a distribution system*

2. What is fault impedance and give the fundamental formulae for fault impedances for phase-to-earth, phase-to-phase and three-phase faults?

3. A SF_6, 132 kV system circuit breaker is capable of breaking a symmetrical three-phase current of 11.2 kA. What is its rating?
4. What is the significance of X/R ratio in the protection of the distribution line?
5. Define the fusing factor. Calculate the sizes of fuse you will use for:
   (a) 137.24 kVAr, 6.99 kV shunt capacitors;
   (b) 50 HP, 400 V, star-delta starter motor having a power factor 0.85 lagging and efficiency of 90 per cent; and
   (c) 100 kVA, 11/0.415 kV, distribution transformer (HV and LV both).

BIBLIOGRAPHY

1. Singh, L.P. '*Digital Protection*', New Age International (P) Limited, New Delhi, 1997.
2. 'Protection expanded with electronic sectionalizer', *Chance Tips*, vol. S1, no. 3, (U.S.A.), December, 1990.
3. Jacks, E., *High Rupturing Capacity Fuses—Design & Application for Safety in Electrical System*, E. and F.N. London, 1975.
4. Linsley Trevor '*Basic Electrical Installation Work*', ARNOLD, London 1999, p. 76.
5. Ram Badri and D.N. Vishwakarma, *Power System Protection and Switchgear*, Tata McGraw-Hill, New Delhi, 1995, pp. 260–263.
6. Report of Working Group on Distribution 'Distribution Line Protection Practices—Industry Survey Results, *IEEE Transaction on Power Delivery*, vol. 10, no. 1, January, 1995, pp. 176–86.
7. 'New and Smart Electronic Sectionalizer', *Chance Tips*, June 1993.
8. Bishop, M.T. and et al., 'Overcurrent Protection Alternatives for Underground Distribution Systems, *IEEE Transactions on Power Delivery*, vol. 10, no. 1, January, 1995, pp. 252–57.
9. Held, W. and F.J. Pollmeier, "Internal fuses in modern high voltage capacitors", *Electra*, March, 1974, pp. 32–47.
10. Danemar, Ake and Thomas Lovkvist, "Reliability calculations for capacitor banks," *IEEE Transac.* PAS, vol. 102, September, 1983, p. 3116.
11. *Asian Review of Business and Technology*, Hongkong, May, 1989, p. 49.
12. *Electricity International*, vol. 2, no. 2, U.K., February, 1990, pp. 21–23.

Chapter 18

System Maintenance

18.1 Successful Maintenance

System maintenance includes inspection, preventive maintenance and overhaul. The inspection can be on schedule or off schedule. Well-managed maintenance practices should result in fewer forced outages, win consumer goodwill and lower maintenance costs/kWh of the energy supply. The infrastructure for good maintenance should provide requisite technical training to linemen, line supervisors, and operators; suitable maintenance and operating manuals; tool kits and maintenance materials. At present, most of the electric power distribution system maintenance is carried out as fire fighting or breakdown maintenance. Workmanship is poor and does not conform to codal instructions in most power utilities in the country. Loose joints, loose sags, incorrect fuse sizes, kundi (unauthorised) connection, underground cables lying partly on the surface and public passing over them, no duct for rising cables to the pole top, LT capacitors at distribution transformers often disconnected etc. are the usual violations. Constructing and maintaining the electric power distribution system that delivers energy demands professionals with training and special skills. Qualified technicians fill that role. For example, the following technical skills are important:

- Safety and accident prevention
- Climbing skills
- Pole-framing and construction specifications
- Equipment operations
- Setting and replacing poles
- Transformer installation
- Service and metre installation

- Conductor installation and repair
- Rubber gloving methods
- Underground distribution
- Fusing and system coordination
- Sub-stations and voltage regulation

The successful implementation of maintenance should be based on the following considerations.

(a) The prerequisite for any maintenance programme is that the system is well planned, properly erected with good quality material and well trained and adequately equipped maintenance staff.

(b) Newly erected lines must be inspected thoroughly after rains during the first year of their service.

(c) The minor defects noticed during inspection should be rectified at the time of the inspection itself wherever possible, and the other defects at the earliest possible occasion after chalking out a programme in advance.

(d) In case of the occurrence of any abnormal situation, the equipment should be immediately disconnected from service and the matter reported to higher authorities for further instructions.

(e) Manufacturer's instructions should always be given due consideration while carrying out the maintenance of a particular equipment.

(f) A correct record of all test results and inspections should be maintained. Lines having more trippings, villages having more electrical complaints, areas where distribution transformers are repeated damage, should be thoroughly inspected and emergent maintenance be done.

(g) Condition Monitoring: The distribution system in the country was developed mostly in the 1970s. Nearly 25 per cent of that system has outlived its designed life. In practice, the utility should plan to replace an equipment before there is a significant risk of its failing in service (see Fig. 7.1). However, the equipment may still be capable of giving service. To check the health or to extend the life of ageing equipment, condition-based maintenance is required. Diagnostic testing of equipment in service should be done to check incipient failures. For ageing equipment such as power transformers, capacitors, cables, lightning arresters, etc., incipient fault detection, diagnostic age analysis is necessary to undertake special or emergent repairs/replacement/modernization/renova-tion/recondition programme.

(h) Required safety precautions must be observed while carrying out any maintenance works.

(i) The schedules once adopted by an electric utility should be subjected to periodical review in the light of previous experience to see if improvements are possible, not only to ensure adequate maintenance but also to reduce cost. However, alterations should not be made frequently, otherwise it may not be possible to obtain any correlation between the cost and performance.

(j) Hot-line stick method for maintenance of 11 kV, 33 kV lines/ transformers is desirable in those areas where essentially an uninterrupted supply is required. Workers climb on a pole to a minimum safety clearance from the live conductors and work with a wide selection of epoxycoated fibreglass poles/rods operated tools. Operations such as detailed inspections, changing of pin and tension insulators, jumpers, fuses and changing of decayed poles, etc. can be made with this method. Power System Training Institute, Bangalore (hot-line training centre) has listed twenty hot-line stick maintenance operations on distribution systems. The 11 kV hot-line sticks are now indigenously available in the country. 15 kV insulated tower wagons are being used for hot line working for L.T., street light and 11 kV lines.

(k) Maintenance of transformers and its local distribution system should be carried out together, to ensure a healthy system.

(l) For fast attendance of consumers electricity complaints, good field communication between complaint centres and complaint/ maintenance staff is desirable. Two-way wireless sets (1.47–1.70 MHz) should be provided at consumer service complaint centres and with the mobile staff attending complaints. Interactive Voice Response System (IVRS) at each sub-division should be installed for billing clarifications and for the consumer's supply enquiries.

(m) Infrared Thermal Imaging: Infrared thermal imaging is based on the principle that, all objects emit electromagnetic thermal radiation. Normally invisible to the naked eye, this radiation can be detected using an infrared scanning camera which converts the thermal energy into electronic video signals. By amplifying these signals, it is then possible to view thermal images of the object on a display screen.

The distribution systems can be evaluated for phase load imbalance and poor joints, which rob the equipment of its ability to operate efficiently. It is useful in both aerial and ground surveys of electricity lines and sub-stations to identify the hot spots during peak load conditions [5,21].

18.1.1 Network Diagrams

As stated in Sec. 7.9.3, accurate primary and secondary system diagrams are important for better system maintenance and its extension. The diagrams be prepared manually or with the aid of computer [see Sec. 3.6] with geographical orientation for permanent record with regular updating. The diagram should include:

Overhead System

Primary feeder (11 kV and 22 kV) diagram should include, name of feeder and feeding sub-station; CT ratio; maximum demand; outgoing cable-size, length and its joints position; line route plan with permanent land marks; position of tree growth, etc.; size of conductor and span length, structure type and its mark number; sectionalizing/inter-linking positions; distribution transformers, size and connected load; voltage drop at the tail end.

Similarly, LT network sketch of each distribution transformer should be kept in folio form and must indicate: location and name of distribution transformer along with name plate particulars; type of LT system, type of pole and its mark number; conductor size and span length; line route plan with position of permanent landmarks; total number of consumers categorywise with load; voltage drop at tail end.

Underground Cables

The diagram should show: size, type and rating of cable; route plan with land marks; year/month of laying; position, date of making joint and type of joints; depth of burial and clearance to other cables nearby; size of ducts if any; tail end voltage drop.

Service Record

After releasing the power connection to the consumer, prepare the following service record:

- Service account number
- Name of the consumer

- Address
- Size and type of service
- Service diagram including cable length and position of joints. A typical service is shown in Fig. 18.1.
- Date of connection

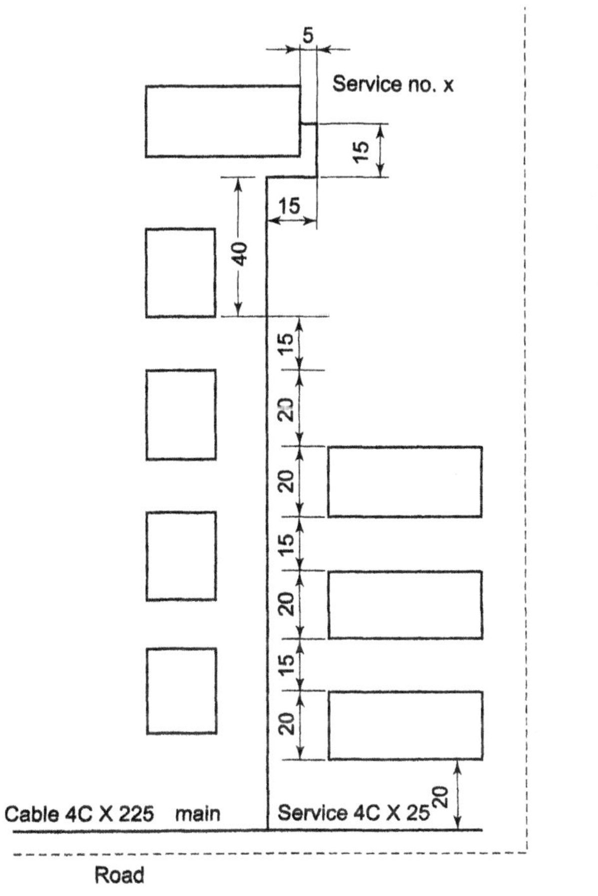

Fig. 18.1 *Typical service line tee-off (All dimensions are in metres)*

18.1.2 Asset Management

Asset Management System (AMS) can provide the accounting equivalent of comprehensive inventory control, property management, logistics management and system maintenance planning. The rational for AMS is that it will answer some basic questions for the utility. Such as:

- What do we own?
- Where is it?
- When did we buy/construct it?
- Is the asset connected to a grid?
- How much is its original O&M cost?
- What is its P/PC balance? Here 'P' stands for delivery of power at acceptable reliability. 'PC' indicates system capability to deliver power over the asset life. P/PC is unity for a good system.

The equipment should be replaced when it reaches its economic life (see Fig. 18.2). Otherwise the equipment maintenance cost will start rising. The economic life is reached when maintenance cost exceeds the annual cost of a replacement of the asset. It is then cheaper to replace the asset with new one.

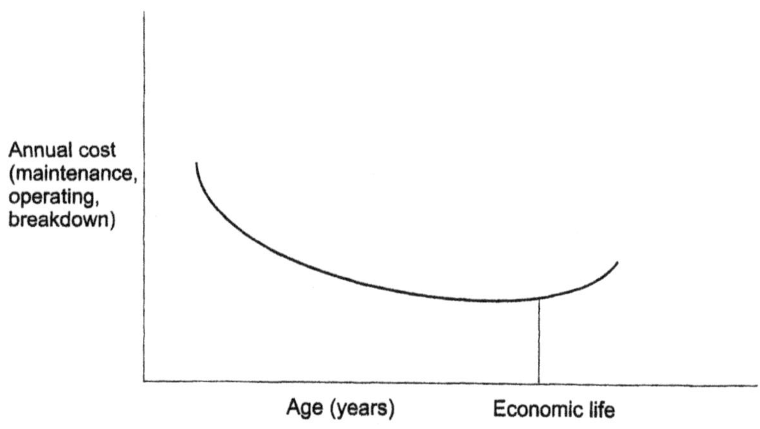

Fig. 18.2 *Economic life of an asset*

18.1.3 Faults Analysis

Most distribution interruptions are initiated by severe weather, with a major contribution being inadequate maintenance. Mostly interruptions are the result of damage from natural hazards, such as lightning, wind storms (hurricanes, tornadoes, etc.), snow storms, birds and animals. Other interruptions are attributable to defective materials, equipment failures and human actions, such as bad workmanship, overloading, vehicles hitting poles, cranes contacting over-head wires, felling of trees, vandalism, excavation equipment damaging buried cable or apparatus. Close monitoring and analysis of faults is necessary to identify and rectify unhealthy pocket of system and improve upon maintenance practices. Consumer electricity complaints be recorded in a register indicating time of receipt, time when attended and nature and location of fault. The Pareto Rule analysis will indicate that roughly 80 per cent of the faults occur on 20 per cent of the distribution network area. In practice, this proportion should move to 60/40 if networks are well-managed. As for example, a case study of monthly analysis of consumer complaints at a consumer complaint centre is given in Table 18.1.

A quick look of the case suggests: (i) immediate attentions of 63 kVA Hem Raj, 63 kVA Teor, 100 kVA Bhajauli transformers and their LD system. (ii) The fault-wise frequency is: HT fuse blown-9; HT conductor broken-1; LT fuse blown-4, LT jumper burnt-6; LT conductor broken 3; Service line burnt-2; Tee-joint loose-28; Metre terminal burnt-1. Loose Tee-joints are very high and proper ferrules/crimping/compounding is desirable for the joints. HT fuse blown are 9 and LT fuse blown are 4, the reasons may be HT fuses are undersized or LT fuses are oversized, as the usual practice is to put AAC/ACSR conductor strands as LT fuses by consumers/staff or close faults due to LT transformer four-core cable damage or birdage at transformer terminals/jumper, which need to be insulated with tape.

18.2 Failures and Maintenance

Failure analysis provides valuable information for preventive maintenance programmes and equipment replacements. According to IEEE Committee report [1], the following information should be included in an equipment failure or outage report.

(i) Type, design, manufacturer and other description for purposes of classification
(ii) Date of installation, location on system, length in the case of line
(iii) Mode of failure (short-circuit, mal-operation, etc.)
(iv) Cause of failure (lightning, tree, etc.)
(v) Times (both out of service and back in service rather than outage duration alone), date, meteorological conditions when the failure occurred
(vi) Type of the outage, forced or scheduled, transient or permanent

Table 18.1

Typical analysis of monthly consumer complaints of a Complaint Centre

Transformer: Fault-Nos.	*Frequency in month*	*Consumers on the transformer*
63 kVA Hem Raj		
Tee joint loose-4	4	10
100 kVA Vill: Teor	7	276
H.T. Fuse blown-1		
Tee joint loose-6		
100 kVA Bhajauli	7	132
Tee joint loose-4		
L.T. Fuse blown-2		
L.T. conductor broken-1		
63 kVA Chandon	5	61
Tee joint loose-3		
H.T. fuse blown-1		
L.T. fuse blown-1		
63 kVA Roorkee	3	10
L.T. jumper burnt-1		
Tee joint loose-1		
H.T. fuse blown-1		
63 kVA Rampur Taprian	4	73
H.T. jumper fuse blown-2		
L.T. fuse blown-1		
L.T. conductor broken-1		
50 kVA Vill: Palheri	6	138
Tee joint loose-2		
L.T. jumper burnt-1		
Metre terminal burnt-1		
H.T. fuse blown-1		
Service line burnt-1		
63 kVA Malikpur Taprian	4	53
L.T. jumper burnt-3		
H.T. fuse blown-1		
63 kVA Vill: Sohali	3	160
L.T. jumper burnt-1		
Tee joint loose-2		
63 kVA Teor	11	14
H.T. fuse blown-2		
Tee joint loose-6		
L.T. conductor broken-1		
Service line burnt-1		
H.T. conductor broken-1		
Total	54	927

18.2.1 Distribution Transformers

The transformer failure rate per annum in different states in India is very high and varies up to 15 per cent. The failure rate of repaired transformer is about twice that of a new transformer, which is mainly due to bad repairs. Besides that, about 3 per cent of new transformers fail during one year warranty period, showing the bad quality of design/manufacture. The failure rate is 0.8 per cent in USA, 3 per cent in Canada and 1 per cent in Australia. IEEE Standards 100/1980 and 500/1984 allows maximum annual failure rate of 3 per cent for distribution transformers.

According to a study, in distribution transformer failures, HT winding account for about 60 per cent failures, about 20 per cent failures pertain to combined LT and HT windings damage and then about 5 per cent account due to LT winding failures and remaining 15 per cent failures are attributed to other miscellaneous causes. Failures are more in the monsoon season as compared to other seasons. Winding faults are mainly due to vibrations and shocks on the winding during line faults, and rough transportation, causing spacers to become loose, ejected and resulting in flashover from one coil to another coil.

Causes of Failures

Failure due to Repeated Faults: When a transformer winding is suddenly subjected to a short circuit current mostly due to either clashing of LT conductors in the loose lines or due to core faults of the LT 4-core cables, a rapid rise in winding temperature can be expected. This temperature rise can cause conductor annealing and insulation decomposition which produces gas. A mechanical weakening of the winding can occur because of thermal aging. Insulation aging as a result of short circuits, however, is of nominal concern. The thermal effect due to short circuiting is proportional to I^2t where I is the rms value of the short circuit/fault current and t is the duration for which the fault current persists. The mechanical effect of short circuit currents is probably the most important aspect of failure mode of transformers due to through faults. The mechanical force on a conductor is proportional to the product of the instantaneous current within the conductor and the electromagnetic flux density magnitude at the conductor. The generated electromagnetic force is proportional to the square of the current. (The offset value of the current is the maximum instantaneous value. For constant reactance, the ratio of the maximum offset wave to the symmetrical current wave is a function of the reactance/resistance ratio of the circuit up to the point of fault as well as the closing angle.) (See Table 17.4.)

Faults on a distribution system are characterized by relatively long duration and repetitive reclosing into the fault. Tests have shown that the continuous operation of transformers causes a gradual relaxation of their strength. Cellulosic insulation undergoes two types of physical change in service. When in compression it is not capable of retaining its original dimension and loading pressure over a long period of time, but it gradually creeps and the fibre assumes a permanent set. This tends to relax the compressive pressure. The second change is the embrittlement of the insulation. These two changes make it unrealistic to anticipate retaining the original short circuit capabilities of transformer.

The cumulative effect of relaxation of compressed insulation is of concern which stems from the evidence that the peak dynamic forces increases substantially as clamping prestress is reduced to very low values.

It would be seen that prestress values within the transformers are generally decreasing, due to short circuits and increased loading and that this increases the compressive stress for each succeeding fault of the same magnitude.

The net effect of all this is a reduction in the mechanical capability of the transformer and consequently, an increase in the probability of transformer failure due to short circuits [11].

Failure due to Bad Protection: Distribution transformers are usually protected by wire fuses though this is not a reliable method. The fuses used sometimes are of higher ratings than the capacity of the transformer and this totally defeats the very purpose of protection. In rural area, fuses of pole-mounted transformers are often changed by the consumer at will as most of the gang-operating switches are without locks.

The transformer, when not protected adequately, may cause failure due to lightning overvoltages. It is desirable to install lightning arresters at or near the transformer terminals.

Failure due to Bad Loading Conditions: Hardly any reference is made to the temperature rise of distribution transformers while they are loaded. Most of the field staff are not well conversant with the loading of transformers in relation to temperature rise. The loads on different phases are not balanced as a result of which normal operation of the transformer in one way or other is impaired. The neutral current should be limited to 2 per cent of the full load current [8].

Failure due to Bad Design and Sub-standard Material: Sometimes transformer failures are due to faulty design and the poor quality of material used by various manufacturers. Poor quality tank welding has resulted in tank bursting. Sealed transformer are not being provided with gas releasing pressure valve and proper cushion of dry air. For example, the varnish for the impregnating of winding insulation varies from manufacturer to manufacturer. As a result, sludge formation in transformer varies. The presence of sludge effects the normal cooling which further aggravates the problem of sludge formation and leading to overheating of the winding.

In service, transformer oil is tested for its dielectric strength and is dehydrated as a corrective measures. Acidity development during the course of operation is very harmful for organic insulation of the windings and thus undermines the useful life of the transformer.

Inspection and Maintenance Check List (17)

The following items of the distribution transformer need to be checked for inspection and maintenance purposes at suitable time intervals.

(a) Oil level and tapping;
(b) Temperature rise of oil/winding;
(c) Unusual noises;
(d) Relief diaphragm for bigger transformers;
(e) External connections: cables, thimble connectors;
(f) Bushing arcing horns (arcing horns gaps may be as per IS: 3716–1978). The recommended gap for 11 kV is 32 mm for double gap arrangement (Table 12.3);
(g) Breather's silica jel: It should be re-conditioned/replaced when necessary;
(h) Cooling system;
(i) Relays and alarm circuit;
(j) Earth resistance (the value should not exceed 5 ohms);
(k) Transformer oil (as per IS: 1866);
(l) Supports;
(m) Surface cleaning;
(n) Lightning arresters;
(o) Fuses: Pole mounting transformers are generally provided with horn gap type fuses with air break switch on the HT side and switch fuse unit on the LT side. For rewirable fuse sizes refer Appendix VIII;

(p) Insulation Resistance [8]: Table 18.2 gives the minimum safe insulation resistances in megaohms at different temperatures (windings to ground). These values hold good for newly commissioned transformers. However, 50 per cent lower values can be accepted for old transformers in service. It may be noted that for every 10°C rise in temperature, the IR values are roughly halved. The measurement is made with 2.5 kV insulation tester;

(q) Noisy transformers are likely to fail in the near future. It signifies loose core/tie bolts.

Table 18.2

Minimum insulation resistance at different temperatures

Voltage of winding	*20°C*	*30°C*	*40°C*	*50°C*	*60°C*
66 kV and above	1200	600	300	150	75
22 and 33 kV	1000	500	250	125	65
11 kV	800	400	200	100	50
below 6.6 kV	400	200	100	50	25

Overhaul

With the passage of time, newly installed transformers in service undergo some shrinkage in the insulation of their windings and consequently, the windings become loose. The transformer oil under the action of heat and in the presence of copper, iron, etc. undergoes certain chemical changes, resulting in the formation of sludge, acid and water. Thus, it becomes essential to overhaul the transformer. The overhauls may be desirable once every ten years but the period depends on the load cycle of the equipment and manufacturer's recommendations.

18.2.2 Circuit Breakers

The circuit breaker may be bulk oil type, minimum oil type, air circuit breaker, vacuum type, and SF_6 type (see IS: 10118).

The inspection, maintenance and overhauls may be done as per manufacturer's instructions. The usual general checks include general cleanliness and lubricating of operating mechanism, oil condition, auxiliary circuits, fuses and alarm operation, relays working, total operating time, battery condition, load setting, contacts condition, insulators and mechanism working, CTs and PTs connections, earthing resistance, arc chutes of air circuit breaker and vacuum bottles of vacuum circuit breakers, arc control pots of oil breakers, sealed units of SF_6 etc. Batteries be checked for any lagging cell, cell voltage and specific gravity.

18.2.3 Lines

All overhead lines should be patrolled periodically at intervals not exceeding three months, from the ground when the line is live. The patrolls should write the inspection notes and pass them on to the maintenance staff for carrying out the necessary repairs. To avoid the close high fault currents, the first 5 km line from the sub-station, must be maintained as zero defect line.

Many breakdowns including slipping of conductor due to loose clamps, loose or missing nuts, cracks/burns on the porcelain of insulator and defects on the suspension fittings can only be discovered or seen by going on top of every pole. This inspection should be carried out by taking a shut down of the line at least once a year and should be done in as little time as possible. Such inspection be done before the start of rainy season. The main repairs should also be carried out, including the replacement of broken and cracked insulator, etc. Other points which cannot be examined from the ground such as defective clamps, sleeves, and connectors, missing bolts, sign of overheating on clamps or connectors, loose binding of conductor with insulators and lightning arresters should be checked and repairs carried out.

When an overhead line trips on fault frequently, it should be inspected to find out the nature of the fault, such as loose sag, loose jumper, birdage, trees or kite-string touching, or other extraneous matter, etc. Find out the amount of repairs involved, with a view to avoid recurrence of such faults in future.

Inspection of maintenance of lines need check on the following items:

(a) Poles: identification number, alignment, top cover of wooden and tubular poles, back filling etc.
(b) Stay: tightness, guy insulators.
(c) X-arms, insulators, clamps: alignment, looseness, breakage etc.
(d) Span, conductors and earthwire: correct span and sag and clearance, loose joints etc.

(e) Jumpers and other line accessories: tightness and orderliness.
(f) GO switch with fuse unit, arcing horns.
(g) Lightning arresters: test these periodically for its operative conditions.
(h) Cable and cable end boxes: IR values be checked every six months in case of feeder cables.
(i) Earthing system: resistance should be less than 15 ohms of each H.T. pole.
(j) LT switch/cut-outs.
(k) Timely tree pruning be done near overhead lines and pole-mounted distribution transformers. Tree trimming is labour-intensive. It should be done regularly to reduce possible outage during severe weather conditions. The resistivity of a tree varies from 1 ohm-metre to 50 ohm-metres depending upon its moisture content [2]. A study indicates that tree-touching faults are predominant in the distribution network. Fault indicators can be installed at strategic points to speed up fault detection and reduce breakdown time (see Section 14.6).

18.2.4 Service Connections

The following items require scheduled attention:
(a) Poles and pole fittings, cut-outs, cable intake pipe or conduit.
(b) Service wire, cables and tee-off points and jumpers. Number of service line tee-offs at a pole should not exceed four to avoid congestion and accidents.

18.3 Porcelain Insulators

Insulator contaminations can be industrial (dust, fog, smoke, cement, fly ash, coal dust, oxides and sulphates of various metals) and marine type (saline deposition). The contaminants due to general dust and dirt are washed away by rains and those due to adhesive dust, i.e., cement, coal petrochemicals, etc. are not washed away by rains. The former can be removed by simple periodical cleaning and the latter by special cleaning methods. The design and maintenance of insulators is generally concerned with the effects of these deposits which can result in partially conducting layers over the surfaces. Flashover does not normally occur until the deposits become wet due to fog, mist or light rain. Heavy rain may, however, assist in washing the deposits away. In the case of greasy deposits such as cement etc. measures of grease coating and periodic washing are necessary. Grease coating comprises hydrocarbons or silicon based grease (Metroark-14 brand, indigenously available) and is applied either manually by spray or by dipping. Petroleum jelly is found to be most satisfactory and economical below 60°C working temperature and above this temperature silicon grease is desirable. The removal and replacement of grease is required every 2–4 years. This is a laborious task and may require plant outage for considerable time. Live washing of insulators is effective which can considerably decrease the outage times. Indian Western Railways are regularly using hot line washing equipment for 25 kV traction line insulators. Tata Electric Co. have maintenance schedules on hot line washing for coastal pollution. Punjab State Electricity Board is using this equipment for hot line washing of 66 kV lines for acid prone pollution of line insulators and accessories. Pollution on the insulators creates nonuniform voltage distribution across the part of the insulator/string etc. and to obtain uniform voltage distribution, application of semiconducting greases can be very helpful. The use of resistance glaze on porcelain instead of conventional glazing permits a leakage current of 1 mA to the flow. The creepage path may be provided of a higher order for industrial pollution. According to IS: 2099–1986, for bushing, the minimum values of the creepage distance (in mm per unit of rated voltage) are as follows:

For normal and lightly polluted atmosphere	16 mm/kV
For medium polluted atmosphere	20 mm/kV
For heavily polluted atmosphere	25 mm/kV
For very heavily polluted atmosphere	31 mm/kV

Fog type insulators may be beneficial but due to special design these may be costly. Polymer insulators are pollution resistant.

18.4 Transformer Oil Maintenance

The causes of deterioration of transformer oil are summarized in Table 18.3.

Tests to judge deterioration of oil are given below.

1. The new oil must withstand 50 kV between two brass sphere electrodes with a gap of 4 mm. After three months of the new oil

in use, it should withstand at least 40 kV and minimum 30 kV for one minute after every year. If the value is lower than this and also does not improve after the removal of moisture by suitable means, the oil has become unserviceable and should be removed from the equipment.

2. Physical observations, such as colour, odour, etc. of oil in service cannot be used to judge its fitness but these may yield important information. Oil darkens with age and, therefore, any sudden change in colour may indicate accelerated deterioration. Acrid and offensive odours are usually associated with a high organic acid or dissolved gas.

Table 18.3

Deterioration of transformer oil

Physical contamination	Chemical contamination	Contamination by gases
(a) Dust, fibres, metallic particles, other solid impurities	Oxidation resulting in acids, sludges and other polar impurities	(a) Dissolved from atmosphere: oxygen, nitrogen, carbon dioxide
(b) Dissolution of varnish		(b) Generated in oil: methane, ethane, acetylene, ethylene, etc.
(c) Free and dissolved water		

3. A lower power factor (loss tangent or tan delta) and a high resistivity value indicate that the oil is in good condition. However, for oils with a high power factor and low resistivity values, further tests are necessary to ascertain the cause of deterioration such as moisture content and dielectric tests.
4. Experience indicates that when the oil has reached a neutralization number of 1.0 mgKOH/g, it should be considered unfit for service. Portable kit to determine the neutralization number is available with Central Power Research Institute (CPRI), Bangalore.
5. The interfacial test is a very good indicator of the condition of the oil. A fall in the interfacial tension below .015 Newtons/m

indicates that sludge formation is imminent and oil should be considered unfit for service. An interfacial tension value of greater than 0.018 N/m ensures freedom from all polar impurities.

Reconditioning and Reclamation: The oil service conditions can be classified as, (i) oils fit for service, (ii) oils with a fair degree of contamination and (iii) heavily contaminated oils.

In the case of (i), no action need to be taken. For (ii), the oil can be made serviceable by simple process of filtration, centrifuging or vacuum dehydration. This is termed as reconditioning. Category (iii) oil needs complex treatment to bring the oil back to the original state. This process is termed *Reclamation* or *regeneration.* The test limits of these classes are given in Table 18.4.

IS: 1866–1983 gives the guidelines for the maintenance of transformer oil in service.

The complex treatment of reclamation is done with the help of korvi-fullers earth as explained in CBI&P Report [9]. This is an indigenous process developed by CPRI, Bangalore. Korvi-fullers earth is available in Karnataka. A similar earth has been developed by the Regional Research Laboratory, Hyderabad and is marketed by D.C.M.

Table 18.4

Tests limits for transformer oil (IS: 1866–1983)

S. No.	*Tests*	*Satisfactory use*	*To be reconditioned*	*To be reclaimed*
1.	Dielectric strength kV (breakdown)	Above 30	Between 20 and 30	Below 20
2.	Neutralization Number mgKOH/gm	Less than 0.5	Above 0.5	Above 1.0
3.	Power factor (tan δ) per cent at 90°C	1.0 or less	Between 1.0 – 1.5	Above 1.5
4.	Resistivity ohm cm at 90°C	0.1×10^{12} or more	Between 0.1×10^{12} and 0.01×10^{12}	Below 0.01×10^{12}
5.	Interfacial tension Newtons/m at 27°C	0.018 or more	Between 0.018 and 0.01	Below 0.01
6.	Flashpoint °C	125 or more	Between 125 and 115	Below 115
7.	Moisture ppm	$\not> 35$	> 35	—

Chemical Works, New Delhi. The earth is renewable during the process. Karnataka Electricity Board has set up this plant at Bangalore. The electricity boards should go in for mobile reclamation unit for field use. Antioxidants (inhibitors) are added to the reclaimed oil to lend stability to the oil for further use. Such an inhibitor (trade name–DBPC) has been tested successfully by CPRI, Bangalore.

Supply undertakings should establish permanent laboratories to reclaim unserviceable transformer oils. The cost of reclaiming is just about 5 per cent of the new oil.

Dissolved Gas Analysis by Gas. Chromatograph [16]: The insulating oil used in transformers is capable of dissolving gases; the amount thus dissolved depends on the kind of gas concerned. Most of the faults that occur in transformers may be attributed to the following causes:

(a) Severe localised overheating (hot spots) due to bad joints or core overheating due to core bolt insulation failure or shaking of core otherwise or overheating of oil by bare conductors.

(b) Arcing or high current discharge or insulation breakdown between adjacent turns, or bad contacts in tap changing arrangement.

(c) Partial discharges or low energy spark discharges.

All the above faults invariably generate gases as a result of the decomposition of oil and other insulating materials, viz. paper, press-board, resin-bonded paper and wood. The decomposed gas of these materials dissolved in oil varies considerably in its constituents. They are hydrogen, methane, ethane, ethylene, acetylene, carbon monoxide and carbon dioxide. The analysis of the above gases dissolved in oil provides a sensitive diagnostic tool for detecting the nature and type of incipient faults. Further, depending upon the type of incipient faults, different gases in different percentages are evolved. The combinations and ratios of different gases indicating the type of fault are well identified. These gases are analysed using the gas chromatograph. IS: 9434 and 10593 detail the interpretation of dissolved gas analysis. Since the sensitivity of the gas chromatograph is very high and the analysis time short, small quantities of oil samples from operating transformers can be drawn periodically without disturbing the power system and the composition of dissolved gases ascertained from time to time. This analysis, if carried out periodically, also clearly indicates the nature of the incipient fault which is likely to occur after a definite period of time depending on the type of gas evolved so that timely remedial measures can be taken to save the life of power transformers.

18.5 Transformer Drying

For smaller transformers of capacity less than 630 kVA, drying out by oil circulation with the dehydration filtration set is sufficient. Short-circuit with hot oil circulation with dehydration set is recommended for medium-size power transformers 630 to 5000 kVA and for transformers above 5 MVA, induction heating is more suitable.

(1) Oil Circulation Method

Oil, core and windings inside the tank are simultaneously dried out with dehydration set. The oil temperature more than 75°C accelerates its ageing properties. The oil dehydration by the conventional streamline filter set is not desirable. Because its heater surface temperature is about 300°C, the oil in contact with the heaters during dehydration process, start carbonizing and degrades. High vacuum pump (759.9 mm Hg) oil dehydration plants now available [20] be used. No degradation of oil during dehydration process occur with these plants due to very small temperature differentials (± 2°C) between oil and the heating porous element surface. The moisture is driven out from the windings into the oil and is removed from the oil by vacuum evaporation and filtering with circulating hot oil temperature not exceeding 70°C.

(2) Short-circuit with Oil-circulation

The above process as stated in (1), is used and the tank sides and top are lagged with some covering like tarpauline. If the transformer radiator is fitted with valves, the oil circulation in the radiator may be prevented by closing the upper radiator valves, leaving the lower valves open. One winding HV or LV should be short-circuited with tap setting, so that entire winding is in circuit. A three-phase supply not exceeding the impedance voltage is applied to the other winding.

The temperature of the winding measured by the resistance or winding temperature indicator and that of top oil by oil temperature gauge should not exceed 80° and 70°C respectively. The temperature should be raised slowly in this final limit, at least after 24 h. With the oil temperature rise, the insulation resistance may fall and will eventually reach a steady value as shown in Fig. 18.3. The temperature should be kept constant until the insulation resistance show a steady and increasing upward trend. When this point is reached, the drying out process is completed and the application of heat may be discontinued. During the cooling down process, the insulation resistance begins to rise and as temperature reaches 60°C, the insulation resistance should be measured and compared for healthiness (see Sec. 16.2.).

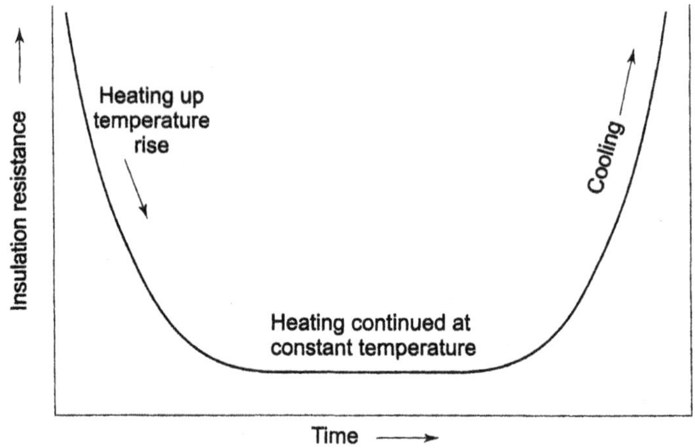

Fig. 18.3 *Variation of insulation resistance with time of drying*

(3) Induction Heating

This consists of generation of heat by eddy currents in the transformer tank, body, winding and core by using induction winding around the tank body. This method is cheap, quick, reliable and is undertaken in the following sequence.

The oil is drained out, dehydrated and tested separately. Radiators and bushings are removed. The transformer core assembly is taken out, inspected and set right for any loose clamping, joints, etc. Then it is put in the tank. All the openings in the transformer tank are sealed with rubber gaskets, so that the tank can withstand an almost full vacuum. The transformer is tested for leak-proofness and a suitable vacuum pump is connected to it through a condenser connection made near the top of the tank, which condenses and collects the moisture from the transformer. A number of thermocouples or thermometers are provided at carefully selected places, to control the temperature of core, winding and the tank. In addition, mercury thermometers are also provided to ascertain the ambient temperature at two suitable places, i.e. at the places which are not affected by the heating of the transformer. The empirical formulae for design of induction winding and bottom strip heaters rating are:

Induction winding power (P), kW

$$= \text{specific power} \times \text{tank height} \times \text{tank perimeter}$$

Induction winding current/phase

$$\text{Amps } I = \frac{P \times 10^3}{3 \times V \times \cos\phi}$$

where, V = supply phase voltage, ($\cos\phi$ is taken as 0.5 to 0.6)

The outer phase turns = $R = B$

$$= \frac{0.75 \times A \times V \times 3^{\frac{1}{2}}}{L}$$

where, A = turn coefficient depending upon induction winding specific power consumption given in Table 18.5. L = Tank perimeter in m

$$\text{The middle phase turns, } Y = \frac{0.3 \times A \times V \times 3^{\frac{1}{2}}}{L}$$

$$\text{Total turns} = R + B + Y$$

Table 18.5

Induction winding specific power (SP) and turns coefficient 'A'

Transformer tank perimeter, m	*SP (kW/m²)*	*Turns coefficient 'A'*
Up to 5	Up to 0.9	Up to 2.12
5–10	0.9 to 1.8	2.12 to 1.59
10 to 12.5	1.8 to 2.4	1.59 to 1.44
12.5 to 15	2.4 to 2.8	1.44 to 1.38

All phase turns are wound in the same direction on suitable wooden X-arms firmer laid in vertical position along the tank body. The turns of phase Y are connected in opposition to phase R and B. The number of turns for phase Y is 40 per cent, that of in phase R or B. This enable to have equitable loading of all phases.

Specific power consumption for bottom-strip heaters is 0.8 kW/m^2 and 0.8–1.4 kW/m^2 for transformer having tank perimeters up to 10 m and 10–15 m respectively. Strip heaters are installed below the bottom of the tank (at a distance of 150 to 220 mm) to provide additional and uniform heating. The transformer tank is heat insulated by suitable heat-insulating material.

It should be noted that, (i) while heating up, the temperature should be built up slowly. (ii) The depressurising and pressurising be done gradually to prevent damage to insulation. (iii) Hot spot temperatures does not exceed 90°C. This can be achieved by switching ON/OFF the supply to the induction winding and to strip heaters if necessary. (iv) After the process is completed, winding dc resistance, no load current, turn ratio, winding insulation resistance and polarization index are measured, compared/checked for healthy condition.

18.6 Maintenance Staff and Tools

For the maintenance of any system, well-trained staff and adequate tools are required.

Exhaustive training may be in four stages:

(a) Training in field work
(b) A series of lectures of a technical and semi-technical nature
(c) Workshop training
(d) Practical and oral examinations.

Field training may consist of the erection of short length line, pole mounting sub-station, erection of supports with proper use of tools, fixing of gangs, etc. at the institute training ground and visits to field works. Lectures may be given on conductor types, jointing, insulators, binding, supports, conductor terminators, staying, earthing, erection, painting, services, patrolling of overhead lines and safety regulations. Workshop training may consist of making different types of joints, binding, jumpers and insulator assembly, soldering, etc.

Maintenance staff may be divided into two categories:

1. Regular maintenance staff whose job would be to perform scheduled maintenance and emergency repairs in case of system faults or major consumer complaints. The staff is organized in the units of gangs under the charge of Junior Engineer or assistant foreman. Each gang should consist of linemen, assistant linemen and skilled labour, their numbers depending upon local conditions.
2. Maintenance staff for consumer complaint centres: This staff attends to consumer complaints. The complaint centres should be easily approachable and strategically located. The complaints may

be attended by different gangs, each consisting of one lineman and one assistant lineman.

18.6.1 Lineman Duties

Lineman is a technical worker authorized to work, inspect and patrol the lines, transformers and connected switches. The general duties of lineman, depending upon the nature of job assigned, i.e. construction works or maintenance works or consumer complaint works are:

- Construction of distribution transformer substations.
- Construction of overhead and underground power lines.
- Construction of local distribution system.
- Laying of service lines to consumers.
- Replacement and dismantlement of works.
- Attending to breakdowns on HT/EHT lines.
- Checking of auto-reclosure/sectionalizer/GO switches at primary feeder and distribution transformer load and ensuring their smooth and correct operation with the help of other staff.
- Patrolling of LT/HT/EHT lines and reporting the defects to his immediate superior through patrol book.
- Checking of the oil level/leakage, fuse sizes in the transformer.
- The lineman will be responsible for ensuring that the transformers are not overloaded by the use of unauthorized running of irrigation pumps and other load and check inadequate protection on HT and LT side; theft of energy will be checked in the specified area.
- The lineman is authorized to obtain the permit-to-work (PTW) in the absence of Junior Engineer or to issue the PTW to himself in the field, by operating the sectionalizing switches, when work on the overhead HT lines is to be attempted for the purpose of maintenance or attending to breakdowns.
- The lineman will report about his findings regarding shortage/ leakage of oil in the distribution transformers, loss of power utility's property, stealing of electric energy, etc., to his immediate superior.
- Checking of general service consumer premises and metre sealing in his area, when assigned by the Assistant Engineer, in which case a sealing plier will be issued to him. He will carry out different

jobs, such as metre change, temporary/permanent disconnection of consumers, etc., as assigned by the Assistant Engineer.

- He will be held responsible for any accidents to human being/ animals due to cause attributed to lapse on his part.
- The lineman posted in the complaint centres will maintain complaint register, attend to consumer complaints and break-downs in LT supply in the areas served by the complaint centres on first priority, and will also be responsible for:
  (i) Checking of the size of the fuse wire on the HT and LT side of distribution transformer in the area of the complaint centre and to replace them if over-sized or blown out.
  (ii) Reporting of the missing seal of the metres, while attending complaints at the consumer premises.
  (iii) Checking and attending to the defects in the LT distribution system.
  (iv) Refixing of the loose supports for the service lines.
  (v) Refixing of loose rag eye bolts for cable mains and submains.
  (vi) Re-sagging of bearer earth wire and the cable services, submains and mains.
  (vii) Cutting of tree branches near the conductors.
  (viii) Patrolling the LT lines at least once in three months and bringing the titled poles in plumb and re-sagging of conductors.
  (ix) Upkeep and maintenance of personal and general T & P in the complaint office.

Besides the above duties, any job specifically assigned by the Junior Engineer/Assistant Engineer in charge is also to be carried out by the lineman.

18.6.2 Tools for Maintenance

A power system distribution division should be a self-contained unit for the proper maintenance of the system under its jurisdiction. Within the division, the following essential tools and materials should be available.

(a) Divisional Level

Cable jointer kit (with cable jointer)	One
Hydraulic compression machine 5 tonne, 10 tonne capacity	One each

Crimping tool kit (2.5–500 mm^2 dies)	One
Power factor metre, earth tester	One each
Infrared camera-therma CAM [21]	One
Insulation tester 2.5 kV	One
Binoculars	One set

(*b*) *Sub-divisional Level*

Each patrol gang (say, one lineman and four assistant linemen) should have:

Clip-on current/voltage tong tester: 0–1000 A, 0–600 V	One
Torch (5 cells)	One
Automatic electric oven for drying silica-gel	One
Bamboo lathi	One
Gum shoes	Two pairs
Patrolling van	One
Insulation tester 1000 V, 500 V	One each
Drilling machine	One

Each maintenance gang (say, two linemen and four assistant linemen) should keep a minimum of the following tools and materials:

Earthing set (HT line)	Two
Lineman tool kit	Two
Sledge hammer 5 kg	One
Crow bar	Four
Pipe Wrench	One
Hand line 13 mm size, 30 m	One
Rope sizes (manila) 16 mm, 25 mm, 30 m long	One each
Pulling and lifting machine 3.5 tonne	One
Crimping tools kit with set of dies (2.5 mm^2 to 95 mm^2)	One
Twisting wrenches for twisting joints	One set
Hacksaw frame with blade	One
Single way pulley	Two
Spanner set	One
Kassi	Two
Augur 30 cm dia	One

For consumer complaints, each gang may be equipped with:

Lineman tool kit	One
Wooden ladder or aluminium ladder 8 m	One

At important complaint centres, the provision of pickup van is desirable.

(c) *Lineman's Tool Kit*

General purpose kit consists of:

Rubber gloves	One pair
Torch (3 cells)	One
Test Pen	One
Test lamp	One
Safety belt	One
Screw wrench 25 mm	One
Insulated side cutting plier	One
Insulated plier 15 cm, 20 cm	One each
Splicing plier 20 cm, 15 cm	One each
Long nose (cutting) plier 15 cm	One
Hand crimping tools 2.5–10 mm^2	One set
Sledge hammer 1/2 kg	One
Screw driver 20 cm, 15 cm, 10 cm	One each
Spanner 16 mm, 18 mm, 30 mm	One each
Measuring ruler	One
Hot chisel	One
Kit bag	One

All tools should be checked periodically for healthy working condition.

18.7 Maintenance Costs

Maintenance costs are annual charges on maintaining the system. It is not easy to estimate the relative magnitudes of maintenance costs of different types of systems, on the basis of an estimate of few good power utilities. The costs can be allowed for as a percentage of capital costs, between 2 per cent and 3 per cent, the higher value being for rural area. It can be argued that a system having higher capital cost, due to the use of automatic reclosers/sectionalizers, etc., will have a lower maintenance cost. The good trained staff could also lower the maintenance costs. Rural Electrification Corporation use 3 per cent of capital cost as annual operation and maintenance charges for sanction of rural electrification schemes. The average annual maintenance costs in most of the power boards are high, up to 7 per cent. According to a study, the average cost of power boards in the northern region are:

Item	Percentage of capital cost
11 kV line	6
LT line (Mains)	5.5
Transformer	5
Consumer submain & service	5
One consumer complaint service (2001)	about Rs 400

This cost is about 15 paise/kWh of energy delivered.

Of the above cost, almost 80 per cent cost is labour and remaining 20 per cent is material and overheads. Overall cost reduction would mean suitable reduction in technical staff and using maintenance free technologies.

PROBLEMS

1. What are the technologies available for carrying out condition-based maintenance? Name the technologies used for capacitors and distribution transformers.
2. What do you understand by low or zero cost maintenance. According to the Pareto rule, nearly 80 per cent of the system faults costs are caused by nearly 20 per cent of the components of the distribution system. Name these components.
3. Explain the causes of large failure of distribution transformers in various power utilities in the country. What are the reasons for displacement of LV and HV coils, and spacers found in the failed transformers?
4. When will you decide to renovate the distribution transformer? Also signify its healthy indication in the field.
5. What are the cost components of system maintenance? How will you optimize the maintenance cost/kWh delivered?

BIBLIOGRAPHY

1. *IEEE Committee Report:* List of transmission and distribution components for use in outage reporting and reliability calculations, IEEE. Transac. Power Apparatus Systems, vol. PAS-95, no. 4, July/August, 1976, pp. 1210–15.
2. Daily W.K. 'Engineering Justification for Tree Trimming, *IEEE Transac. on Power Delivery*, vol. 4, no. 4, October, 1999.
3. A new test for quickly determining the acid content of transformer oil, *Technical Report 0037 PS*, Bangalore, 1967.

4. IS: 6711–1972, Code of practice for maintenance of wooden poles for overhead power and telecommunication lines.
5. 'Infrared Thermography Reduces Maintenance Costs', *Electricity International*, November, 1992, pp. 12–13.
6. IS: 1180–1989 (revised), Distribution transformers outdoor type, three-phase up to and including 100 kVA, 11 kV.
7. Condition Monitoring, IPG International Generation, January, 1993, pp. 33–42.
8. *Proceeding of Seminar on 'Distribution Transformers Failures*, CBI&P, New Delhi, February, 1997, pp. 1–36.
9. "Reclamation of used and unserviceable insulating oils", *Review No. 2*, Central Board of Irrigation and Power, New Delhi, January, 1973.
10. "Studies of Insulator pollution under industrially contaminated conditions", *Technical Report*, CPRI, Bangalore, March, 1979.
11. Johnson, L. and N.B. Tweed, "Feeder faults affect transformers", *Electrical World*, McGraw-Hill, N.J. 15, February, 1977, pp. 40–43.
12. *Annual Research Review*, Research Station, Jabalpur, Madhya Pradesh Electricity Board, Jabalpur, 1978.
13. "Antioxidants to prolong the life of insulating oils", *Technical Report No. 005–Toccheron*, CPRI, Bangalore, 1970.
14. IS: 2615–1989, General requirements for pliers, pincers and nippers.
15. IS: 3650–1981(first revision), Specifications for combination side cutting pliers.
16. Luke, L.E., "Gas detection–A key to transformer health", *Transmission and Distribution*, A Cleworth Publication, Philadelphia, January, 1980, pp. 26–27.
17. IS: 10028, *Code of Practice for Selection, Installation and Maintenance of Transformers.*
18. IS: 5613, *Code of Practice for Design, Installation and Maintenance of Overhead Power Lines.*
19. Pabla, A.S. and P.K. Sharma, "Failure of Distribution Transformers", *Proceeding of 3rd All India Conference on Transformers*, New Delhi, October 21–22, 1983.
20. *Thin Film Vacuum Degassers for Dielective Oil Regeneration,* Brochure, Edwards High Vacuum International, England, June 1987.
21. *Catalog:* Flir Systems AB, world wide thermography centre, Danderyd, Sweden, 2002.

Chapter 19

Electrical Services

19.1 Standards

Every country has an electrical installation code for buildings: for example, in the United Kingdom, IEE regulations; in the United States—National Electrical Code (NEC) and on International level, IEC 364 gives the guidelines on the subject. In India, National Electrical Code (NEC)–1985, covers the provisions of the Indian Electricity Rules–1956 and various relevant Indian Standards.

19.2 Electrical Installations

Proper coordination between architect, buildings engineer and electrical engineer is desirable from planning stage of the installation. The provisions may be required of transformer, switchrooms, emergency stand-by supply arrangement, lift wells and other appurtenant rooms, service cable ducts, rising mains and distribution cables, sub-distribution boards, hooks for fans, openings and chases in floors and walls, etc. Indoor lighting illumination level should be specified in advance (see Appendix VI). The outdoor installations are distinct in nature by virtue of their being exposed to environmental conditions ranging from lighting of public thoroughfares to small outdoor locations around buildings, such as gardens, storage yards, etc. The important considerations for designing any installation are:

- The type of supply and occupancy; envisaged load.
- Ambient temperature and any hazardous atmospheric conditions.

- The importance of continuity of service and future extension.
- The relative capital and maintenance costs of various alternative methods.
- The need for radio and telecommunication interference suppression.
- Safety aspects.
- Energy conservation.

After completion of erection work, the installation is tested as laid down in Part I, section 10 of NEC–1985. The test report with detail of schedule of energy consuming points/apparatus is submitted to power utility for the installation's inspection (as per IE Rule 47, 47A, 48) and to release electric supply service connection. After putting the installation into service, periodic inspection and testing is necessary to maintain the installation in a sound condition.

19.2.1 Hazardous Areas Installations

To ensure an adequately safe level of operation of electrical apparatus in a flammable atmosphere, special electrical provisions may be used in hazardous areas, such as ammunition factories and depots, hydrogen plants etc. As per NEC–1985, the hazardous areas have been classified as zone 0, zone 1, and zone 2 in order of high to low severity. The general guidance is as follows:

- The guidance for selection of electrical equipment for hazardous areas be taken from IS–5571–'Guide for selection of electrical equipment of hazardous areas and from IS: 8240–' Guide for electrical equipment for explosive atmosphere.
- Cables are drawn into screwed, solid drawn or seamless conduit or cables are otherwise suitably protected against mechanical damage.
- Efficient grounding bonding should be provided where protection against stray currents or electrostatic charges is necessary.

19.3 Reception of Electric Supply

The commencement of supply and its control at consumer installation is desirable as per I.E. Rule 50. The typical arrangement is shown in

Fig. 19.1. There should be circuit-breaker or a linked switch with fuse on each live conductor of the supply mains at the point of entry. The wiring throughout the installation be such that there is no break in the neutral wire, except in the form of the switch. Earth terminal (as per I.E. Rule 33) is provided by the utility for the use of the consumer. The service terminal equipment is provided in suitable enclosure as per IS: 8061.

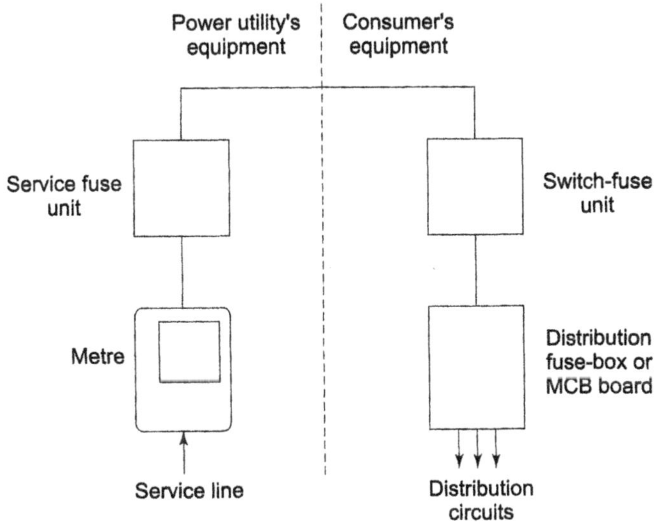

Fig. 19.1 *Typical block diagram of service entry and consumer mains*

19.3.1 Sub-stations

In case of HV supply, a transformer is required. On primary and secondary side of the transformer, a linked switch with fuse or circuit-breaker of adequate rating is required. The secondary side outgoing circuits may be controlled by MCBs or fuses. The whole equipment may be in one pre-fabricated assembly and is called unitized for package sub-stations. Typical single-line layout is shown in Fig. 19.2 The ideal location of a sub-station would be at the load centre (see Sec. 6.7.2). In multistoreyed buildings, sub-station preferably be installed on the lowest floor level with adequate safeguard for any fire, flood, etc., in the locality. The sub-station may be placed adjacent to the air-conditioning plant room. In industrial premises, the sub-station be located near to main plant building.

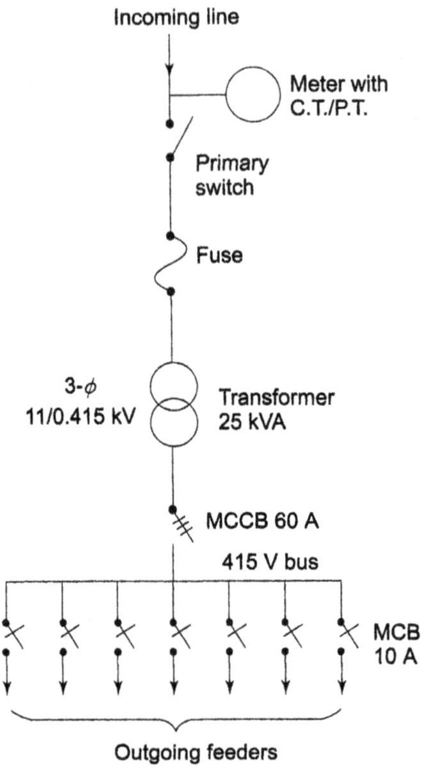

Fig. 19.2 *Typical single-line diagram of package sub-station*

19.3.2 Main Board

The location of the main board should be such that it is easily accessible and as near as practicable to the termination of service line and also have the safeguard against operation by unauthorized persons (see IS: 13032).

19.3.3 Layout and Installation Drawing

The electrical layout drawing should be prepared after all socket outlets, lamps, fans, appliances both fixed and portable, etc., have been selected and best methods of wiring determined. All runs of wiring and exact positions of all points of switch-boxes and other outlets are first marked on the plans of the building and approved by the engineer-in-charge of the owner, before actual commencement of the work. The typical layout is shown in Figs. 19.3 (a, b).

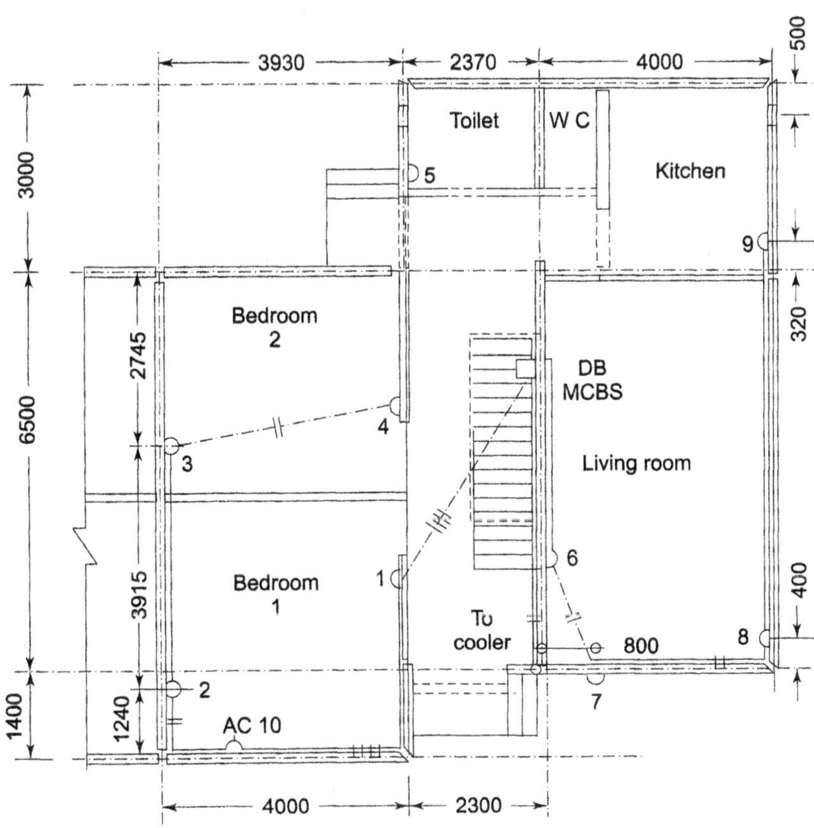

Fig. 19.3 *Typical electrical layouts: (a) Power*

19.4 Supply Voltage

The supply voltage to consumer depends upon the quantum of load (see Sec. 6.6). Generally for load less than 100 kVA, supply voltage may be LT, i.e. 240/415 V, and for load 100 kVA and above, high voltage (say 11 kV or higher) is advantageous. Certain varying load, viz. arc furnace, rolling mills should be given supply even at much higher voltages, such as 33 or 66 kV, so that flicker due to varying loads does not affect other consumers. High voltage supply entails the provision of space, capital cost of providing suitable transformer sub-station, and extra safety precautions. The advantages gained are: cheaper tariff, better earthing, elimination of voltage flicker.

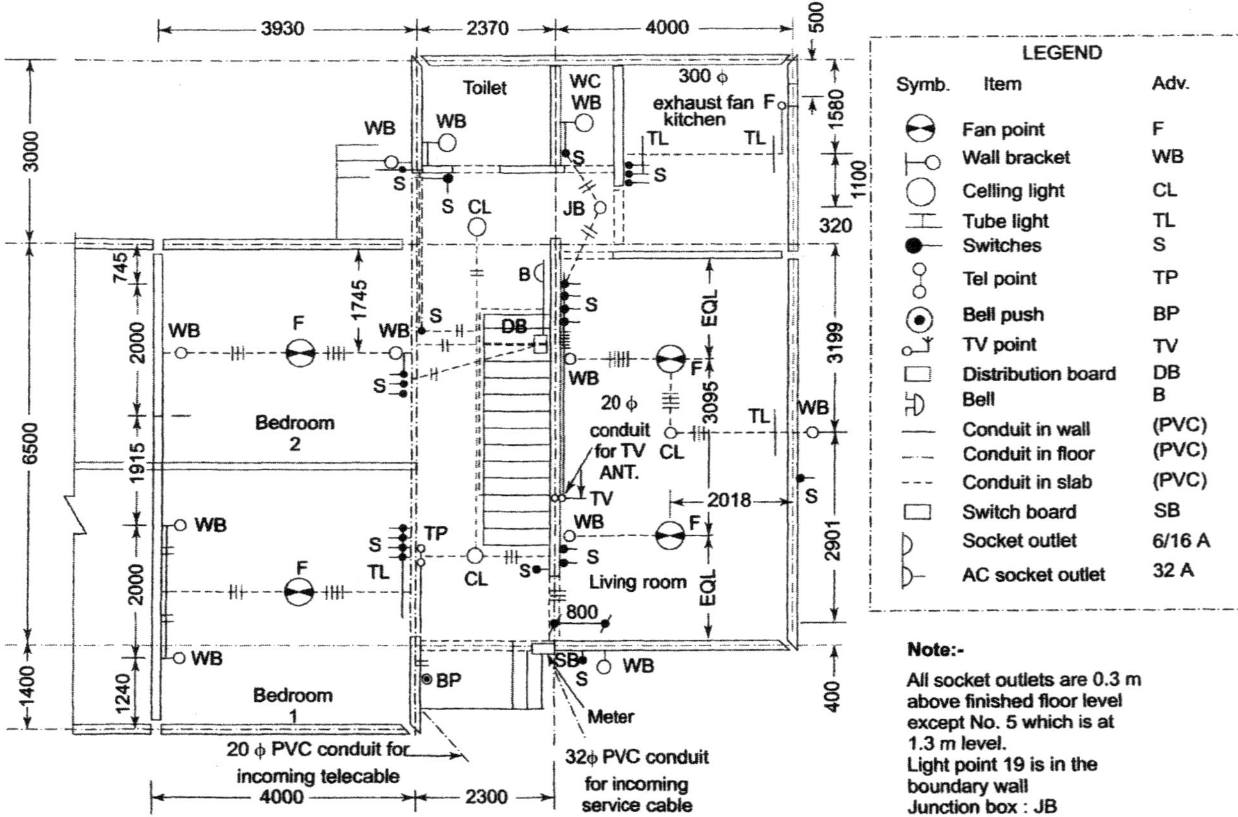

Fig. 19.3 *Typical electrical layouts: (b) Lighting*

19.5 Consumer Supply Arrangement

Supply feeder from main sub-station to consumers may be overhead or underground (see Chapter 11). The latter require high capital cost. The type of cable to be used depends upon the various factors (see Appendix III). For radial laying, the cable size is diminished as distance from the sub-station increases. For example, initial size may be 150 mm^2 and the last portion may be 25 or 10 mm^2 as shown in Fig. 19.4(a). It has been

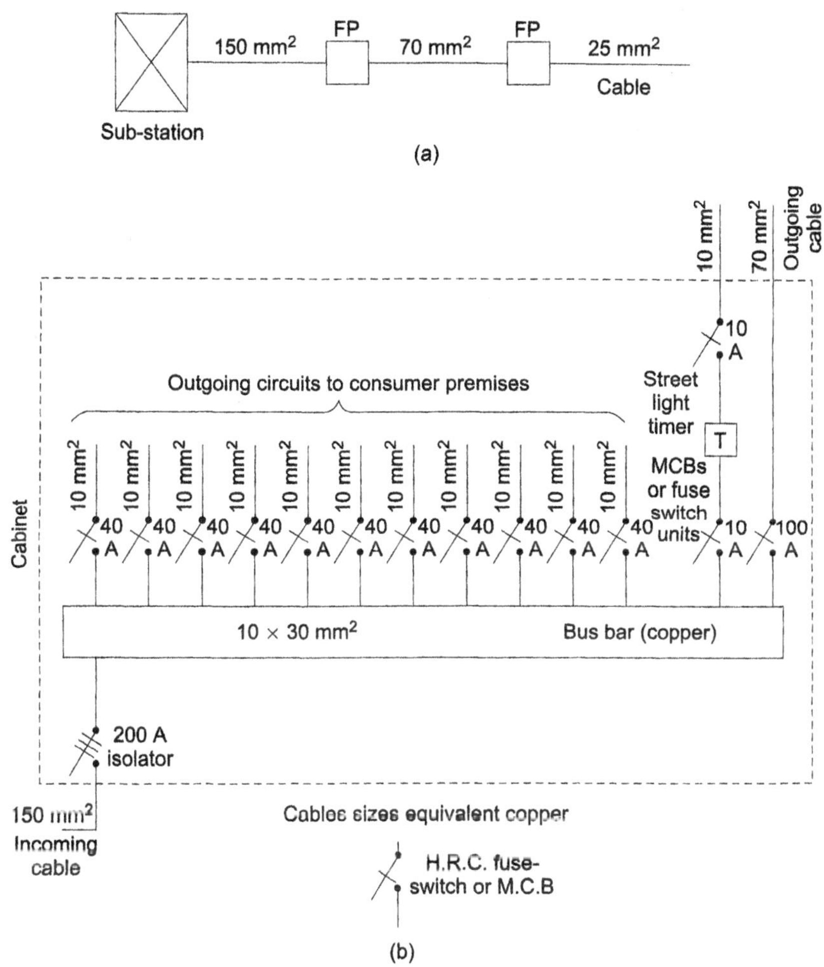

Fig. 19.4 *Typical underground network: (a) radial line (b) single-line diagram of feeder pillar (FP).*

seen that phase-to-earth fault occurring in final portion of the cable can present high resistance as earth path is provided by the armour of cable, the fuse in the sub-station may not blow and burn the cable. To overcome this difficulty, feeder pillars (as per IS–5039) with mains fuses and MCBs/fuses are provided as shown in Fig. 19.4 (b). Whenever cables of lower size are connected, for proper discrimination 2:1 ratio is used for progressive fuses. That is, if the fuse in the sub-station for 150 mm^2 cable was of 200 A and if a 70 mm^2 outgoing cable is to be joined to this, then a 100 A fuse is connected in series at the feeder pillar as shown in Fig. 19.4 (b).

19.5.1 LV Supplies

Generally the distribution system is TN-S for single houses and a protective multiple earth TNCS for multi-storey buildings. A consumer is a PME consumer, when he or she uses the earthing terminal provided by the power utility and the terminal is connected to the neutral at the consumer supply terminals (see Section 19.11).

Small Domestic and Commercial Consumers

The standard supply to consumer having load less than 10 kW is generally single-phase and three-phase supply is given for load above 10 kW. Supply is given via a suitably fused cut-out and with single/three-phase meter.

Medium Industrial or Commercial Consumers

The supplies are given as shown in Fig. 19.6 (a). The consumer apparatus should have the provision of end sealing, which houses Current Transformers (CTs) required for metering, or otherwise separate CTs be installed in metering cabinet.

Rising and Lateral Connections

The installation of rising and lateral connections, is dependent on the physical dimensions of a multi-storey building and its intended use. A separate earthing terminal is provided at each consumer's premises. All consumer meters are installed in a separate room on the ground floor. Alternatively, AMR facility meters may be provided for each apartment in blocks. With wired communication, remote meter-reading can be taken by connecting a handheld computer unit to a magnetic socket connected to the metering bus. The handheld computer unit emits signals to contact each meter on the supply network in turn and access their data (see Fig. 19.5). Two arrangements are used:

Individual Rising Connections As shown in Fig. 19.6 (b), it is often used for small blocks of flats and the standard method of installation is by PVC insulated cables drawn into conduits.

Common Riser Method As shown in Fig. 19.6 (c), it is used for power supplying to large blocks of flats. The incoming cable is fed from a distribution system, the rising neutral conductor and the Protective Conductor (PC) have to be bonded together at both the base and remote end of the riser. When two or more common risers are fed independently, the neutral conductors may be inter-connected and a separate PC is no longer required. The standard method of installation for common risers is normally lightly insulated solid conductor supported on insulators in a preformed chase, with fire barriers at each floor.

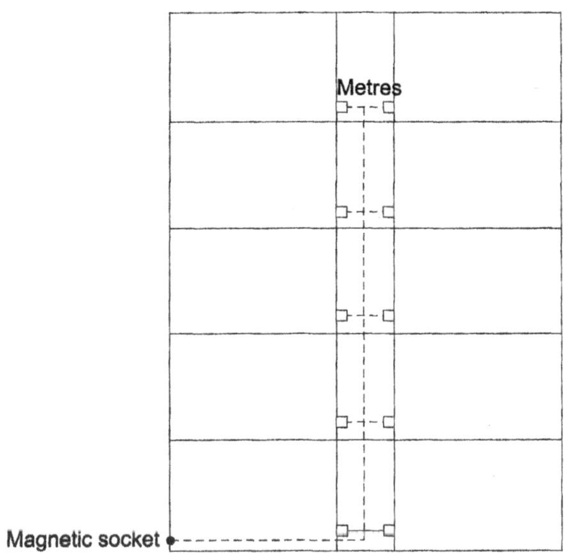

Fig. 19.5 *AMR metering bus in a block of flats with wired communication*

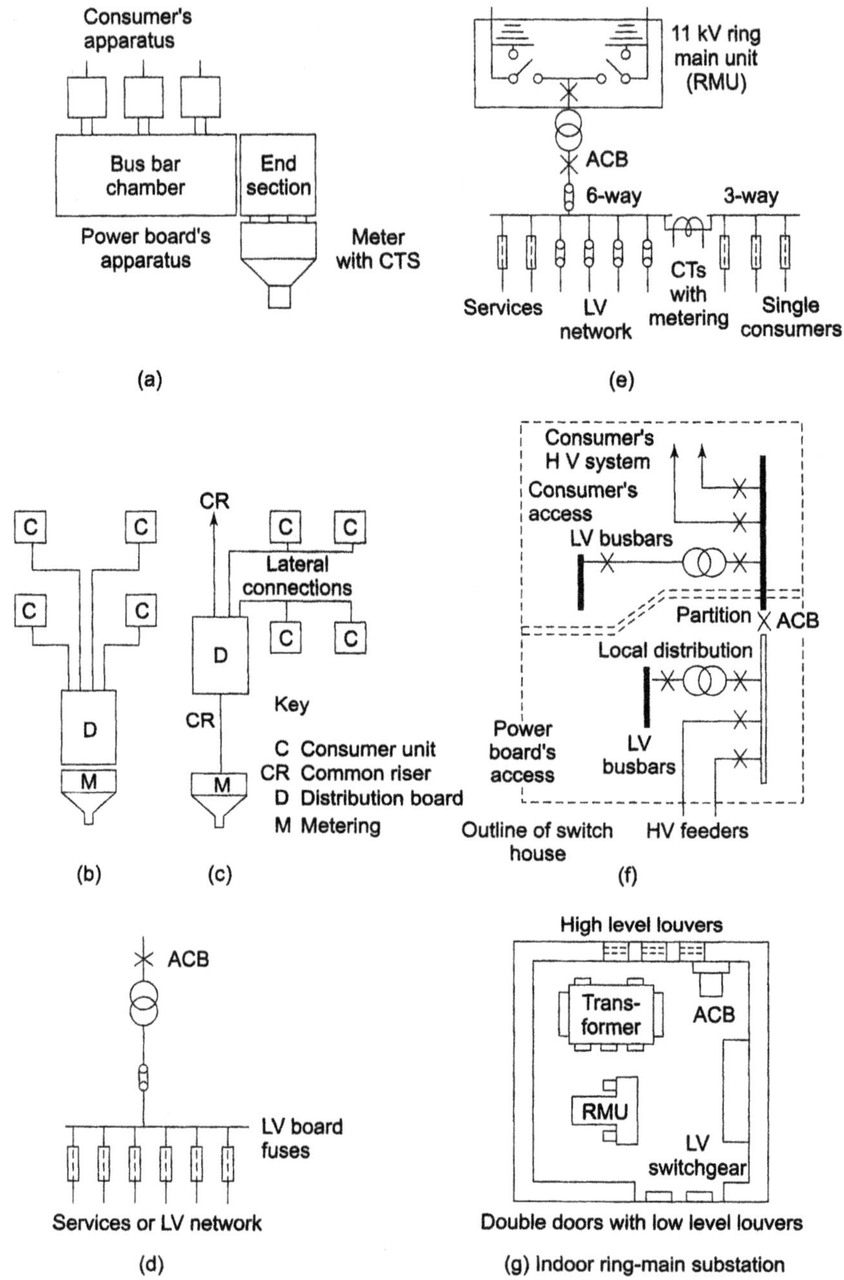

Fig. 19.6 *Typical consumer supplies*

19.5.2 Large LV Supplies

For a radial feeder supply, the incoming supply is normally 11 kV and is taken to the transformer via an HV circuit breaker or switch-fuse. If no interconnection [see Fig. 19.6 (d)] is present, then a circuit-breaker is not normally installed between the transformer and distribution board. However, links are provided to allow electrical isolation of the transformer.

Figure 19.6 (e) gives a typical arrangement used for an interconnected LV distribution system. This system can be extensively used in metropolitan cities, having densely loaded area. The design varies from one area to another, but the main characteristics are similar. There are no fuses on outgoing LV network, except at the boundaries between flat/commercial blocks, where fringe fuses are inserted. A fault on the LV network will almost always burn clear, without loss of supply to the consumers. Fringe fuses interconnect the LV blocks and for a sustained LV fault, it will operate, together with all the circuit-breakers in the faulty block, thereby isolating the fault.

19.5.3 HV Supplies

The HV supply is given to a large supply consumers via one or two HV feeders. The voltage levels vary. The typical 11 kV supply are shown in Fig. 19.6 (f). Single service are taken directly to the consumer switchboard via metering with PT/CTs. Interlocking facilities are normally required to prevent paralleling, in case of ring-main feeders. Separate access is usually required for power utility personnel and so the switchboards are physically separated and operational responsibility is clearly defined. The power utility's typical ring-main indoor sub-station is shown in Fig. 19.6 (g).

19.5.4 Industrial Supply

Industrial complex supply is generally of high voltage, medium voltage and low voltage networks, with diverse nature of loads such as induction motors, induction furnaces, arc furnaces, rolling mills, pumps, fans and compressors, heating power, electronics (static converters, variable speed drives) etc. These loads are continuously evolving as the industry expands, and variable in nature both in terms of active and reactive power requirements, depending upon the process requirements.

Supply system studies should be carried out at the planning stage of industry. It is necessary to investigate the unusual loads or conditions such as:

- Starting requirements of large motors.
- Arc furnace operation.
- Welder operation.
- Continuous process loads i.e., loads that must be kept in operation under all conditions.
- Sensitive loads such as computers, that may be affected by voltage and frequency transients.
- The possibility of demand limiting to reduce power cost.
- Short-circuit level at the point of connection at the consumer's premises.

A typical single-line diagram for industrial supply is given in Fig. 19.7.

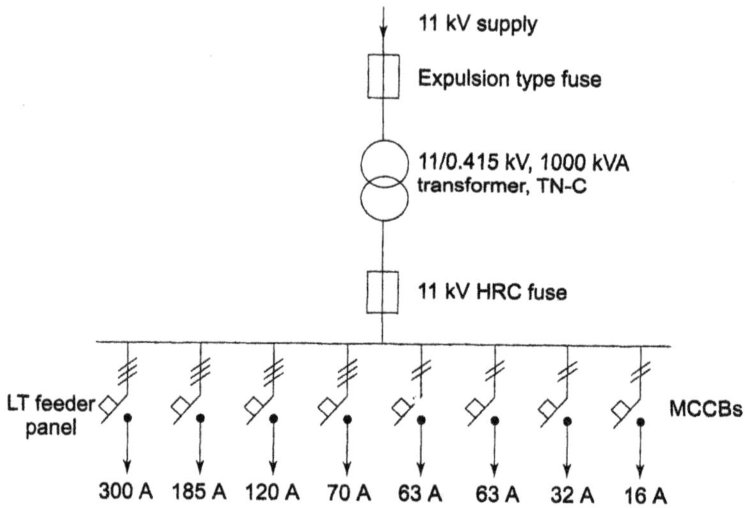

Fig 19.7 *Typical industrial supply distribution board*

19.6 Internal Wiring

Choice of wiring system depends upon safety, cost, durability, appearance, load, supply voltage, weather conditions, type of buildings, future extension. An essential feature is a proper earthing system giving an earth continuity conducting path for protection and earth. Guidelines for wiring installations are given in IS: 732–1989. In conduit wiring, not more than two 90° bends are desirable for any circuit and beyond this limit, a junction box should be used. The different capacity wire sizes are given in Tables 19.1 and 19.2.

19.6.1 Low and Medium Voltage Wiring

(i) Cleat Wiring

The cleats are used to fix and hold wires. These consist of two parts, a base piece and a cap. A special pattern of cleats may be used if necessary where conductors pass round corners, so that there may be no risk of the conductor touching the wall, owing to the sagging or stretching. Cleats are fixed at regular distances not greater than 60 mm apart. This system should not be employed for wiring on damp walls or ceilings.

(ii) Plastic Casing Wiring

In this system of wiring, PVC insulated wiring work is carried in plastic casing enclosure, for low-voltage installation. Plastic casing wiring should not be used in damp places or in ill-ventilated places.

(iii) Tough Rubber Sheathed/PVC-Sheathed Wiring

The wiring are laid on carrier base, which is usually wooden batton or carrier wire and fixed with the help of link clips. PVC sheathed cables are suitable for low and medium voltage installation, and may be exposed to sun and rain or dampness. Tough rubber sheathed system be not used in places exposed to sun, rain and dampness, unless wires are sheathed in protective covering against atmosphere. The links are so arranged that one single link should not hold more than two twin-core TRS or PVC sheathed cables.

(iv) Metal-Sheathed Wiring

Metal sheathed wiring system is suitable for low-voltage installations, and be used in situations where acids and alkalies are likely to be present. Metal sheathed wiring may be used in places exposed to sun, rain, dampness, provided no joint of any description is exposed.

(v) Conduit/Trunk Wiring (see IS: 9537)

Concealed conduit wiring with rigid steel or plastic conduits is suitable for better aesthetic appearance of building, personal safety and

Table 19.1

Single-core PVC insulated copper conductor (unsheathed) cables for flexible wiring in voltage grade 650/1100 V (IS: 694–1990)

Nominal conductor x-section mm^2	Number/nominal dia. (mm) of wire strands	Voltage drop mV/A/m	Amperes capacity		Maximum resistance per km at 20°C Ohms
			in conduit/ or trunking	2 cable unenclosed clipped directly to surface or on cable tray	
1	32/0.2	42	11	12	18.10
1.5	48/0.2	28	13	16	12.10
2.5	80.0.2	17	18	22	7.41
4	56/0.3	11	24	29	4.95
6	84/0.3	7.1	31	37	3.30
10	80/0.4	4.2	42	51	1.91
16	126/0.4	2.7	57	68	1.21
25	196/0.4	1.7	71	86	0.780
35	276/0.4	1.3	91	110	0.554
50	396/0.4	0.97	120	145	0.386

Table 19.2

Four-core PVC insulated copper conductor and sheathed cables for fixed wiring 650/1100 V (IS: 694–1990)

Nominal x-section of conductor mm^2	*Number/nominal dia. (mm) of wires*	*Ampere rating* A	*Voltage drop* mV/A/m	*Maximum resistance per km at 20°C (ohms)*
1	7/0.43	11	37	18.1
1.5	7/0.52	15	24	12.1
2.5	7/0.67	20	15	7.41
4	7/0.85	26	9.2	4.61
6	7/1.04	31	6.2	3.08
10	7/1.35	42	3.7	1.83
16	7/1.70	57	2.3	1.15
25	7/2.14	71	1.5	0.727
35	7/2.50	91	1.1	0.524
50	19/1.78	120	0.84	0.387

protection to wiring from mechanical damage. It is increasingly used in modern domestic and commercial buildings. Surface conduit wiring is used in industrial building and is flexible for carrying out necessary extension/modification whenever desired. Accessories of threaded type have advantage over grip-type or clamp-type, due to better earth continuity for steel conduits and weather-proofness. Trunks are used where wires are very large in numbers or of heavy size. The wire capacities in conduit and trunks with space factor not exceeding 40 per cent and 45 per cent respectively, are given in Tables 19.3 and 19.4 as a general guidance. For aluminium conductors of the same size, the capacities may be reduced by 70 per cent.

Table 19.3

Conduit capacities in number of wires (IS: 694–1990)

Copper wire, mm²	*Conduit diameter (mm)*				
	20	*25*	*32*	*36*	*50*
1.5	7	12	20	—	—
2.5	4	8	12	—	—
4.0	3	6	10	—	—
6.0	3	5	8	—	—
10.0	—	3	6	8	—
16.0	—	—	3	4	5
25.0	—	—	—	4	5
35.0	—	—	—	4	6
50.0	—	—	—	—	4

Table 19.4

Trunk capacities in number of wires (IS: 694–1990)

Copper wire, mm²	*Trunk size*		
	(50 × 50 mm)	*(75 × 75 mm)*	*(100 × 100 mm)*
1.5	45	100	180
2.5	40	90	160
4.0	35	80	140
6.0	29	65	116
10.0	24	54	96

Contd.

Contd.

Copper wire, mm^2	Trunk size		
	(50 × 50 mm)	(75 × 75 mm)	(100 × 100 mm)
16.0	14	35	56
25.0	12	27	48
35.0	10	21	40
50.0	8	18	32
70.0	5	10	20
95.0	4	9	16
120.0	3	5	12
150.0	—	4	7

(vi) Busbar Trunking

Busbar trunking is a means of replacing conventional cabling by a prewired factory built cable management system, which can be installed immediately. It eliminate the need for a large number of distribution boards and long lines. The proliferation of data and communication system in modern office buildings, has necessitated the busbar cable system. Busbar is the most suited to applications where the configuration of cable system is reliable, and requires frequent upgrading or alteration facilities in offices, high-rise buildings, hospitals, industrial buildings, departmental stores etc. It provides power for dedicated service, soft wired options and puts other services to the hard-wired options within the main trunking compartment. To prevent data being wiped out in power surges as data and communication system wiring cables are laid in separate compartments of the trunk system.

(vii) Cable Tray/Ladder

Cable tray/ladder comprises the basic lengths of galvanised steel or aluminium trays, usually in 3 m sections joined together. Cable tray/ladder is an ideal method of supporting cables in workshops, laboratories and power stations. In buildings with false (suspended) ceiling, it is the best method of wiring in the ceiling void.

19.6.2 High Voltage Wiring

For higher loads, HT distribution of voltage level more than 650 V is taken from main sub-station to a number of plant buildings or HT load, such as large motors and furnaces by cables, such as:

- Paper insulated, metal sheathed, bedded and armoured cable having overall serving installed underground, or
- Mineral insulated metal sheathed cable served with PVC laid directly on the ground, or
- PVC insulated, armoured and PVC sheathed cable installed underground or on surface.
- XLPE insulated, armoured cable installed underground or on surface.

The main distribution within plant/building may be done as follows:

- Sheathed armoured cables served with PVC run on cable hangers, or racks fixed to walls or other supports or installed in duct, or
- Sheathed unarmoured cables laid in concrete trench suitably protected by precast concrete covers against any mechanical damage at point of entering and leaving trench, or
- Bare rigid conductors supported on insulators.

19.7 Switchgear

Switchgear choice for any installation depends upon supply voltage, the prospective short-circuit current at the point of supply, the load and nature of installation and the space available. Guidelines for switchgear selection for a particular application are given in IS: 10118 (part-II)–1991 and IS: 13947. Generally an incoming isolator and MCBs (as per IS: 8828–1996) for outgoing circuits are suitable. It will prevent tripping of entire installation, in case of high current fault taking place in any one of sub-outgoing circuits. The fault level in domestic circuits normally do not exceed 1 kA and in case of multistorey buildings or large commercial establishments, the maximum fault level does not exceed 9 kA on LT.

19.7.1 Main Distribution Board

It consists of board/panel assembly in main and sub-mains switch-fuse units or MCBs made as per IS: 2675, 8623, 13032. The rating of the phase bus-bar is 67 per cent of the total rating of the fuse/MCB ways (number of ways multiplied by rating of each fuse way). The neutral bus-bar is of the same rating as of main bus-bar for sizes up to 32 A, and above 32 A sizes, it is of one-half size. All boards are designated by voltage level, number of phases, number of ways/circuits, current rating, etc.

19.7.2 Branch/Local Distribution Boards

It is fuse or MCBs assembly of adequate rating for control and protection of the final sub-circuits. At least one spare way of the same capacity is provided on each distribution board. Overload and short-circuit protection is normally provided by MCBs or fuses either of High Rupturing Capacity (HRC) or rewireable type. Where MCBs are used for protection of sub-main circuit and of the sub-circuits drived therefrom, discrimination in operation may be achieved by adjusting the protective devices of the sub-circuit MCBs to operate at lower current settings and shorter time-lag, than the sub-main MCBs. If fuses are used, the rewireable fuses are restricted to the circuits with short-circuit level up to 4 kA and for higher level cartridge/HRC fuse be used. In multistorey buildings, sub-circuit/sub-main protection be invariably by MCBs to reduce risk of fire due to short-circuit along with ELCB of 300 mA sensitivity.

19.7.3 Switches

A switch is a device designed to make or break the current in one or more electric circuits. The pole of the switch is part of the switch associated with one conductor path of its circuit(s) provided with contacts intended to connect and disconnect the circuit itself.

Bakelite switches (6A, 16A one pole) are used for switching the supply to small electric appliance, light, fan points. Bakelite switches (16A, 20A double-pole) are for bigger appliances in power circuits in domestic and commercial installations. Metal clad switch or switch-fuse combination in standard sizes in the range of 32 to 1200 A are used as isolator for mains sub-main circuits and high power electrical appliance. The switches are usually designated as: (see IS: 3851 and 3854)

SP — Single-pole switch for phase switching.
SPN — Single pole switch for phase switching with neutral link.
DP — Double-pole switch for switching both phase and neutral switching metal-clad or bakelite made.
TP — Triple-pole in three-phase isolating switch and with or without fuse cut-outs.
TPN — TP with additional provision of neutral link.
4 P — Four pole switch for 3 phases and 1 neutral switching.

19.8 Plug and Socket Outlets

It is a safe and convenient means by which free standing or portable apparatus can be connected to an electric supply. The usual ratings are 6A, 16A, 20A for general use and 32A, 63A, 100A, 200A, etc., for industrial use. The plug is a device intended for insertion into a socket outlet. The socket outlet is a fixed device with a contact intended for making a detachable connection. Flexible cords with a terminal plug are used as supply leads from the socket outlet to a portable or some fixed appliances. Socket outlets incorporating overvoltage protection for computer terminals and other sensitive electronic equipment are available in western countries [7] (see IS: 1293–1988).

19.8.1 Miscellaneous Accessories

Other common accessories used are:

- Link clips (IS: 2412)
- Split PVC casing
- Flexible PVC conduit
- Junction boxes
- Boxes for the enclosure of electrical accessories (IS: 5133)
- Ceiling roses (IS: 371)
- Ceiling fans (IS: 374)
- Telephone outlet
- Bell switch
- Change-over switch

19.9 Circuit Loading

Circuits are conductors conveying energy from main board to local distribution board. Circuit may form conductor system or branch of a system such as a ring-main circuit/spurs. Separate circuits from main board are provided for each important service, i.e. automatic fire alarm system; automatic security system; Local Area Network (LAN), etc. The sub-circuits are final conductor path for electric energy to current using points, as shown in Fig. 19.8. For small consumer, such as dwellings, the function of main and local boards are combined in one Distribution Board (DB) (see Tables 19.1 and 19.2 for wire sizes). The voltage drop between the point of supply and any point in the installation should not exceed 3% of the nominal voltage when the circuits are carrying the maximum demand admissible.

The computer draws current in huge gulps and requires high size wires than an equivalent load of heating and lighting. For a computer load, always take a separate circuit from the distribution board from the phase which does not have any heavy equipment connected to it. The voltage between neutral and earth should be less than 5 volts. Surge protection i.e. Metal Oxide Varistors (MOV) and noise filters must be installed on the incoming circuit at the distribution board for computer grade supplies in the commercial building.

EXAMPLE A 1 kW room heater will draw current of 4.34 A at 230 volts i.e. 1000/230 A. Whereas 1 kW computer system with a typical crest factor (peak current) of 3 will draw $3 \times 4.34 = 13$ A of repetitive peak current. The heating effect of this repetitive current is relatively very high as shown in Section 2.3.

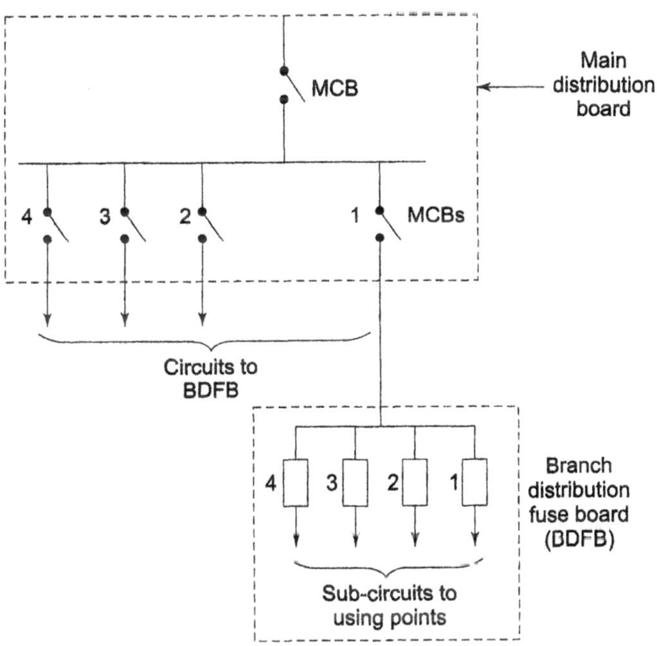

Fig. 19.8 *Circuits and sub-circuits*

19.9.1 Residential and Commercial

In residential and commercial installations, lights, fans and 6A socket-outlets are wired on a common sub-circuit, having not more than a total of ten points. The load of such circuit is generally restricted to 800 W. If separate fans sub-circuit is provided, the number of fans in the sub-circuit should not exceed ten. Power sub-circuits are designed according to the load, not more than two 16A socket outlets on each sub-circuit. The design of wiring length and size should be such that voltage drop in the total circuit from main should not exceed 3 per cent. A typical distribution board with service entry line diagram is shown in Fig. 19.9 and the boards circuit rating are shown in Table 19.5 (see IS: 4648). Power sub-circuit load is restricted to 3000 W.

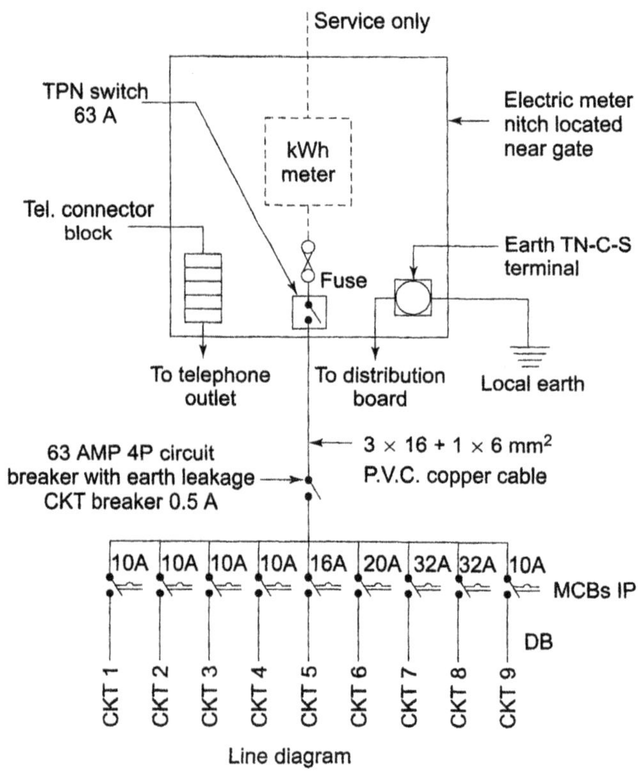

Fig. 19.9 *Typical domestic consumer distribution board with service entry line diagram*

For multistorey buildings, the choice between cables and metal rising main for distribution is made depending upon the load and number of floors to be fed. It is essential to provide separate main circuit for emergency lighting and automatic fire-alarm system. Sub-main wiring to each of the flats/apartments/offices should be independent and metered separately. The meter preferably installed at the ground floor.

Table 19.5

Consumer board circuits

Circuit No.	MCB (SP) A	Conductor size mm^2 (copper)	Points
1	10	1.5	6 lighting points 2 ceiling fans
2	10	1.5	6 lighting points 2 ceiling fans 1 TV socket outlet
3	10	1.5	7 lighting points 1 exhaust fan 1 bell point
4	10	1.5	1 light point first floor and future extension
5	16	2.5	1 air cooler socket-outlet
6	20	2.5	1 air conditioner socket-outlet
7	32	4	Ring circuit for 6, 16A socket-outlets
8	32	4	Ring circuit for 4, 16A socket-outlets
9	10	1.5	Spare

19.9.2 Industrial

In industrial and other similar installations, the use of group control for switching operation is used. Sub-circuits for socket outlets may be kept separate from fans and lights. Normally fans and lights are wired on a common sub-circuit. However, if necessary, separate sub-circuits may be provided for the two. The load on any low voltage sub-circuit should not exceed 3000 W. In case of a new installation, all circuits and sub-circuits are designed by making a provision of 20 per cent increase in load due to any future modification. Power sub-circuits are designed for not more than four outlets on each.

19.9.3 Miscellaneous

At construction sites, stadium, shipyards, open yards of industrial plots, etc., where a large number of high wattage lamps may be required, no restriction of load on any circuit are levied, but circuit conductors should be of adequate size with proper protection.

19.9.4 Radial and Ring-Main Circuit

Where number of socket-outlet points are very large, ring main circuit arrangement (see Fig. 19.10) is economical. A cable is laid round the building in a closed ring. The sockets which lie along the route of the ring are connected directly to it, while those in the outlying positions are connected by spurs. The number of outlets per spur is limited to two. The cable requirement per point is much less in the ring main system and extra outlets can be added without any difficulty. As a matter of fact, a ring feeder can cater to four times the outlet points, as can a radial feeder of the same size at a little extra cost only. With the ring circuits, fuse protection in the individual appliance is provided by a fuse fitted in the plug head or socket. Thus, continuity of power in the circuit is not affected by a faulty appliance. Figure 19.9 shows the provision of two ring circuits No. 7, 8 and circuits 6, 5, for power socket outlets, respectively.

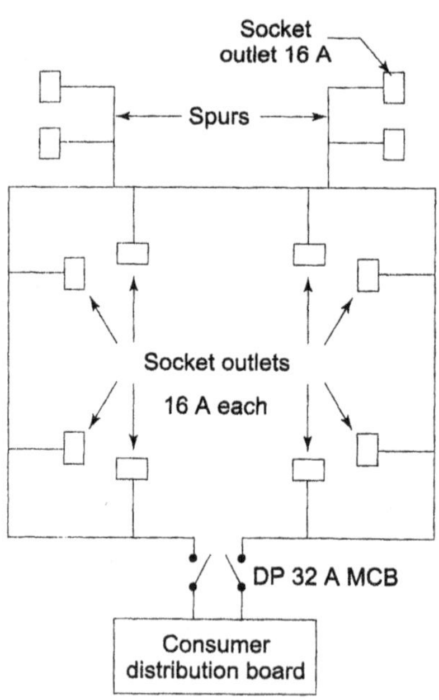

Fig. 19.10 *Typical ring-main circuit*

19.9.5 Efficient Neutral Conductor

The proliferation of switch-mode power supplies and non-linear electronic loads causes high neutral currents dominated by the third harmonic, resulting in a need for increased neutral conductor sizes. Sensitive loads need to be on dedicated branch circuits to prevent interference from other loads that may cause transient disturbances. Local Area Networks (LANs) and other communication systems experience interference due to ground potential differences and noise coupled into the ground system. Efficient neutral conductor requirements are:

- Run a separate neutral conductor for each phase in a three-phase circuit that serves single-phase nonlinear loads.
- When a shared neutral must be used in a three-phase circuit with single-phase non-linear loads, the neutral conductor capacity should be approximately double the phase conductor capacity.
- Delta wye transformers designed for non-linear loads can be used to limit the penetration of high neutral currents. These transformers should be placed as close as possible to the non-linear loads (e.g. in the computer room). The neutral conductors on the secondary of each separately derived system, must be rated on the basis of expected neutral current magnitudes.
- The neutral must have a continuous metallic path between the sub-stations and users services. No disconnecting devices should be installed in the common neutral.
- On single-phase secondary circuits (phase and neutral) the neutral conductor should be large enough to carry almost as much current as the phase conductor. Often the same neutral conductor size is used for both.
- In a 3-phase primary circuit carrying a reasonably balanced non-linear load, the neutral conductor can be considerably small than the phase conductors. This may be 50 per cent in some cases.

19.9.6 Dedicated Supply

Sensitive electronic industrial units need a dedicated power supply having low internal impedance which will increase the fault level at the supply bus. The power supply bus should have a high fault level to reduce the effect of any voltage fluctuations. The power leads (lines) should have low characteristic impedance (Z_c) to reduce any noise. Z_c of 8–20 ohms is considered satisfactory.

19.10 Load Estimation

Circuit design is based on primary load estimates of the installations. The circuit demand is calculated by multiplying the load estimate with suitable demand factor and divided by suitable diversity factor. The average primary load estimates as per NEC are:

19.10.1 Domestic Dwellings

		Wattage
Incandescent lamp		60 W each
Ceiling fan or table fan		60 W each
6A socket		100 W each
Fluorescent tube		
	600 mm	25 W each
	1200 mm	50 W each
	1500 mm	90 W each
Socket outlet 16 A		1000 W each

Diversity factor of 3 is generally taken.

19.10.2 Multistorey Buildings

Commercial complex	*Percentage of total load*	*Diversity factor*
Ventilation heating (Air-conditioning)	45%	1
Power plants (drives)	5%	1.5
Lighting	30%	1.05
Lifts	20%	1.00
Hospitals		
Lighting	20%	1.1
Air-conditioning	15%	1.0
Kitchen	10%	1.6
Sterilizer	10%	1.6
Laundry	10%	1.6
Lifts	15%	1.6
Electromedical installation and other loads	20%	1.6

19.10.3 Recommended Schedule for Points for Dwelling Units

(i) Unit wise

	Area of units m^2				
	33	45	55	85	140
Light points	7	8	10	12	17
Ceiling fans	2	2	3	4	5
6A sockets	2	3	4	5	7
Call bell	—	—	1	1	1

(ii) Room-wise

	Sockets 6A	*Power sockets* 16 A
Bed room	2 to 3	1
Lobby/Lounge room	2 to 3	2
Kitchen	1	2
Drawing room	2	1
Garage	1	1
Bath room	1	1

19.10.4 General Loads by Occupancies

Type of occupancy	*W/*m^2 *of covered area*
Auditoriums/gurudwaras/temples	10
Banks	35
Barber shops and beauty parlours	30
Churches	10
Clubs	20
Court rooms	20
Dwelling units	10
Garages—commercial storage	5
Hospitals	20
Hotels, motels, including apartment house without provisions for cooking by tenants	20
Industrial buildings	20
Office buildings	35
Restaurants	20
Schools	30
Warehouses (storage)	2.5
Assembly halls and auditoriums	10
Corridors	5
Storage spaces	2.5

19.11 Earthing

The purpose of earthing a consumer's installation is to ensure that all exposed conductive parts and extraneous conductive parts associated with the electrical installation are at or near earth potential. The earthing conductor connects a main earthing terminal of the installation to an earth electrode of the installation in case of a three-phase supply or directly to the power utility's earth terminal made available if the connection is single-phase. An earthing conductor (usually green PVC insulated) needs to be installed along with the circuit conductor's wiring. All portable appliances and other plug-in appliances are earthed by the use of 3-pin socket outlets.

Except for equipment provided with double insulation, all the non-current carrying metal parts of electrical installations should be earthed properly. All metal conduit, trunkings, cable sheaths, switchgear, distribution fuse boards, starters, motors and all other parts made of metal, should be bounded together and connected by means of two separate and distinct earthed continuity conductors to efficient earth electrode. The earthing of high frequency equipment should be done separately, as per manufacturers recommendation and generally should not be bonded to other earthing system within the premises, to avoid interference to communication circuits.

19.11.1 Types of System Earthing

The supply system and its earthing can take several forms, which are classified according to the basic method and position of earthing (protective) conductor and its connection to the neutral. Three types of general arrangement of neutral and protective conductors are:

- Having separate neutral and protective conductors throughout the system (TN–S system) [Fig. 19.11(a)]
- In which neutral and protective functions are combined in a single conductor in a part of the system (TN–C–S system) [Fig. 19.11(b)]
- In which neutral and protective function are combined in a single conductor throughout the system (T–T system) [Fig. 19.11(c)]

The choice and effectiveness of earthing arrangement is important to protect a system during faults. In India, TN–S system is applied for consumer installation.

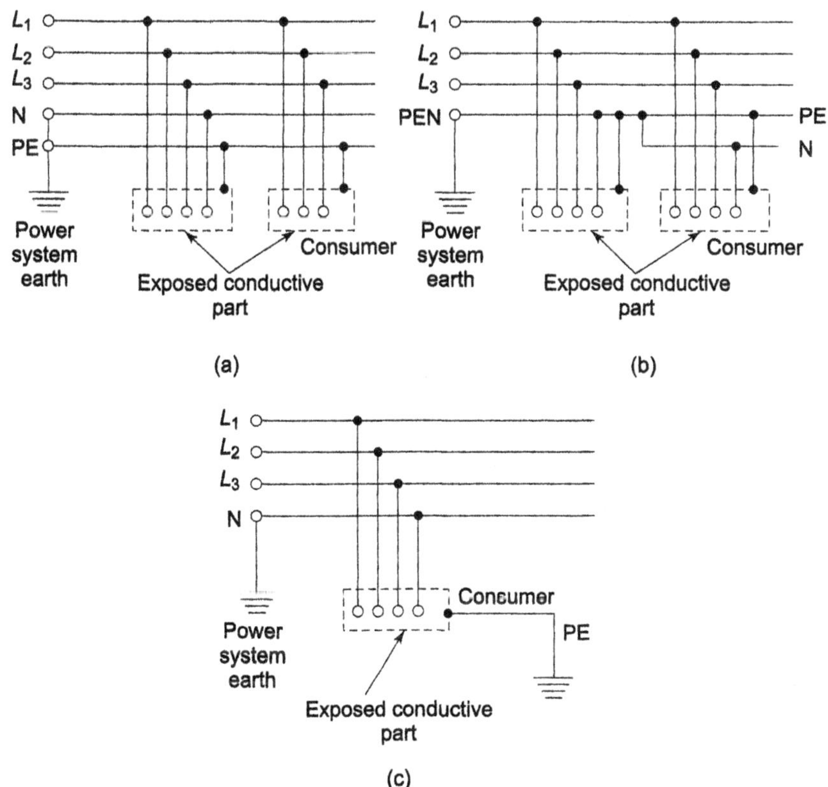

Fig. 19.11 *Basic forms of supply system earthing for three phase supply*

19.11.2 Earth Resistance and Loop Impedance

In direct earthing, a driven or buried electrode is used, the earth resistance as per IS–732 part (II) should:

$$\ngtr \frac{\frac{1}{2} \times \text{Voltage to earth}}{2.5 \times \text{Current rating of largest fuse or OC setting of ciruit breaker}}$$

The multiple earth neutral system is used only where the neutral conductor of supply system is multiple earthed (see Sec. 10.6.5), so that the resistance between the neutral and earth is low enough to reduce the possibility of a dangerous rise of potential in the neutral. Neutral of the secondary side of a medium voltage load-centre transformer should be solidly earthed by not less than two separate and distinct connection, with earth having its own electrode. Earthing of the metal work of electrical installations and apparatus means that it is connected to the general mass of earth in such a manner that leakage from a live conductor to metal enclosure ensures the immediate disconnection of the faulty circuit, without risk of shock or fire. If, as in many electric installations, a fuse or circuit-breaker is intended to open the circuit concerned on earth fault, the resistance (or impedance) of earth fault, path or loop must be low enough to pass sufficient fault current to operate the fuse or circuit-breaker. The impedance of the earth loop at any socket outlet of rating 16A, or less should not exceed 4 Ω. Where a sufficiently low earth-loop impedance cannot be obtained economically, earth-leakage protection must be provided by one or more earth-leakage circuit-breaker (see Sec. 17.3.5) as per IE Rules 61, 71 and 73 sensing 30 mA/300 mA for shock/fire hazards. 300 mA tripping rated ELCB is used at the main distribution board [3]. ELCB 30 mA tripping rating is desirable for protection at the individual socket outlet meant for connecting air-conditioners, electric presses etc.

The value of protective earthing resistance (R_e) should be such that due to dead short-circuit, the current if either blows the fuse or trips the circuit-breaker and that the voltage at any external part of the machine or switchgear connected to earth is less than the safe voltage 32 V as per IS: 3043, i.e.

$$R_e = \frac{32}{I_0}$$

or

$$I_0 = KI_r$$

K is the trip or fusing factor (IS: 3043–1988). I_r is rated current of fuse or short-circuit trip factor of circuit breaker. The fusing factor is 3.5 for HRC fuses and for overload and earth leakage circuit breakers trip factor is 1.35–1.5. It means that the system impedance should be such that on dead short-circuit, the current should be minimum 3.5 times the current rating of the fuse or 1.35–1.5 times the current setting of the circuit-breaker shall flow through these devices.

Each earth system should have earth resistance less than 5 ohms, unless otherwise specified. Also construction of earth system should be such, that it should be possible to test the individual earth electrode. The drawing showing the main earth connection and earth electrodes should be prepared for each installation.

In case of high frequency of more than 10^6 cycles/second discharge currents (lightning discharges), a considerable displacement current in the capacitance of the ground, in parallel with the conductive current in the earth resistance, flows together with the self-inductance of both of these currents through earthing electrode as shown in the equivalent circuit in the Fig. 19.12. Crow feet earth electrode should be used to lower impedance:

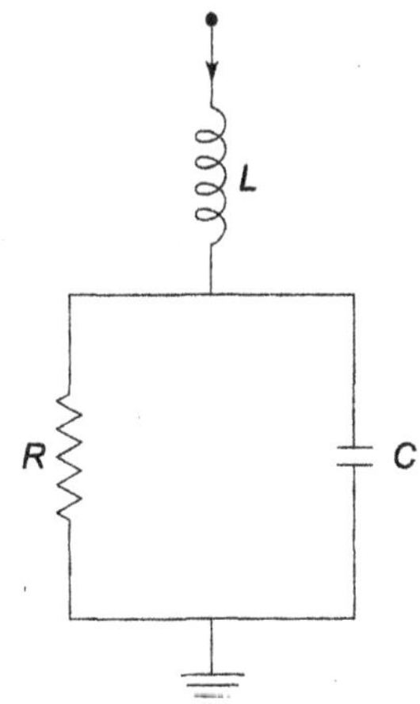

Fig. 19.12 *Equivalent circuit for UHF ground current*

19.12 Lighting

The aim of a lighting scheme is to enhance the peoples ability to see tasks clearly and to provide comfortable working visual environments on economical basis. The principal criteria are the amount of lighting, the balance of lighting, glare and colour rendering. Lighting efficiency is important from energy conservation point-of-view as it consume about 17 per cent of the total power consumption in the country.

(i) Interior Interior illumination of buildings ranging from one or two lighting points in a tiny croft such as those situated in hill areas, to the large installation found in cities or modern block of flats. Public buildings such as theatres, concert halls, hospitals, churches and schools present a wide variation of special demands. Shops and large departmental stores call for both general lighting and shop-window lighting (see IS: 3646 (1)–1992).

(ii) External This may involve the flood-lighting of public buildings, religious places, castles, gardens especially during period of calebrations. Flood-lighting may be demanded for the illumination of sporting arenas, dock-sides, and railway marshalling yards, etc.

(iii) Street Lighting With the increase of high-speed traffic upon our roads, it has become essential, in order to reduce accidents, to provide adequate illumination on all streets, roads, traffic junctions, tunnels, bridges etc.

19.12.1 Lighting Schemes

(i) General Lighting General lighting is obtained by placing a number of lamps, luminaires in a more or less regular arrangement over the whole ceiling or on walls. The result is a horizontal illuminance of a certain average level and with adequate uniformity. General lighting gives uniform lighting conditions, and should be used for rooms in which there are no fixed working places.

(ii) Directional Lighting In this case, light comes predominantly from a preferred direction, usually either by means of a special arrangement of mirrored fluorescent lamp luminaires or by the use of spotlights having wide beam reflectors. This type of lighting is often used for ascent and display or to illuminate surfaces which, in their turn, act as secondary lights sources. Directional lighting may be used in combination with general lighting as a means of overcoming the possibly monotonous effect produced by the latter.

(iii) Localized Lighting In some cases, it may be advantageous to concentrate luminaires in certain areas of the work, in order to produce a sufficiently high illuminance on the main place of interest. This type of lighting is useful for localized working areas in factories, where the production equipment is not likely to be moved.

(iv) Local Lighting Local lighting is produced by placing luminaires close to the visual task, so as to illuminate only a very small area. It is supplemented by one of the other lighting system. Local lighting is recommended when the work involves very critical visual tasks, or the viewing of forms of textures requires that the light come from a particular direction. Local lighting should essentially be used to supplement general lighting.

19.13 Lamps

Lamps fall into three categories:

Incandescent lamps, discharge lamps and fluorescent lamps. The choice depends upon economy, load, colour, suitability to task.

19.13.1 Incandescent Lamps

(a) The incandescent lamps produce light because of the passage of current through its filament, which is heated to incandescence. The higher the temperature, the greater will be portion of radiation falling within the visible region of the spectrum. Tungsten wire is used, because of its relatively high melting point; a temperature of about 3400°K can be achieved, e.g. in certain photographic lamps. The filament of a typical 1000–hour GLS lamps operates at about 2850°K. Efficiency is relatively low, but is increased by use of coiled coil filament and inert gas filling to reduce the rate of evaporation of the filament. (see IS: 418).

(b) Tungsten halogen have bromide or iodide added to the gas filling. This re-cycles tungsten and enables the filament to operate at a higher temperature for a given hours rating, and hence at a higher efficiency.

(c) Reflector lamps with ratings ranging from 40 W to 300 W have internal reflectors of aluminium or diffusing coating. Beam characteristics are determined by filament shape and position relative to the reflector. Some lamps also have a refractive finish to the front glass. Pressed glass lamps have more efficient reflectors and more controlled beams and some of them are suitable for unprotected use outdoors. These lamps contribute to energy conscious lighting, as they direct the light only to where it is required.

19.13.2 Discharge Lamps (12)

(a) Low pressure sodium In these lamps sodium metal is vaporized to a low pressure and produces monochromatic yellow light with very high efficiency and have very long life. These lamps have high visual acuity.

(b) High pressure sodium These lamps operate with sodium vapour at high pressure and produce a golden yellow light with high efficiency and good colour rendering. These have very long life and are used for public lighting, flood lighting and industrial lighting (see IS: 9974).

(c) High pressure mercury These lamps operate with mercury vapour arc at high pressure and have long life. The light from the arc is supplemented by light from phosphors activated by ultra-violet radiation. The lamps have a cool white light. These can be used in factory and public lighting.

(d) Metal halide These lamps with their very high efficiency combined with their excellent colour rendering and long life can be used for industrial lighting and also for flood lighting. These are similar to high pressure mercury lamps, but contain metallic halides, e.g. of thallium, indium and sodium. This enables the lamps designer to tailor the spectral power distribution.

(e) Mercury blended These lamps combine a high pressure mercury discharge tube with a tungsten filament, which acts as a ballast to the discharge tube and also increase emission at the red-end of the spectrum. These lamps have long life and are a direct replacement for incandescent lamps and can be used for industrial and public lighting.

(f) Fluorescent lamps The fluorescent lamps are tube-shaped low-pressure discharge lamp, but are regarded as a separate category of lamps, because the majority of the light is produced by the phosphors activated by the discharge. The phosphors are coated on the inside of the envelope and can be blended to give the lighting properties desired. A choice of colour appearance, e.g. cool or warm is available with a choice of colour rendering. High frequency fluorescent lighting are more energy efficient. Compact SL and PL lamps, (11, 15, 20 W) should replace incandescent lamps. For example, PL 9 W lamp can be compared in light output with a 60 W incandescent lamp (see IS: 2418).

19.14 Luminaires

Luminaires are lamps fittings and are used to redirect light output distribution, to reduce glare from light source to contribute to the decor of a building. Owing to absorption and inter-reflection between adjacent surfaces, no luminaire gives all the light output of the lamp which it surrounds. They are classified according to light source, application or appearance, light distribution (see IS: 10322).

19.15 Lighting Design

19.15.1 Interior

Lighting level is given by average illuminance (see Appendix VI) on a horizontal working plane, considered to be at a height above the floor (normally 0.85 m) and covering the entire floor area. The average illuminance is calculated by the relation:

$$E = \frac{F \times UF \times MF}{A}$$

where E = average illuminance over the working plane (Lux)

F = total luminous flux of the lamps

A = area of working plane (m^2)

UF = utilization factor: It depends upon type of the spacing, the room index and generally ranges from 0.3 to 0.6.

MF = Maintenance factor for the installation generally ranging between 0.8 and 0.9.

In some instances, it may be necessary to apply correction factors for abnormal conditions, e.g. low supply voltage, absorption by all furniture, additional daylight, etc.

Maintenance Factor (MF) It is the ratio of the illuminance halfway through a cleaning cycle, to what the illuminance would be if the installation was clean. Thus, it allows for depreciation of illuminance due to dust and dirt on the lamps, the luminaires and the room surfaces but does not allow for depreciation in the lamp light output. Regular cleaning permits the lighting designer to raise the *MF*, e.g. from 0.8 to 0.9. The cleaning cycle should be planned to fit in with the lamp replacement cycle.

Utilization Factor (UF) It is an indication of the effect of the lighting equipment and the interior combined in producing horizontal illuminance. For example, a *UF* of 0.3 means that the lumens reaching the horizontal plane are 30 per cent of the lumens of the lamps operated bare under standard conditions. The *UF* allows for the direct illuminance and for the indirect illuminance, due to reflections from room surfaces. It is dependent on luminaires light distribution and output ratio; reflections of the ceilings, walls and floor; room index. Generally *UF* is taken for commercial fixtures as 0.5; for industrial fixtures as 0.6, for decorative fixtures from 0.3 to 0.4.

To achieve uniform illumination and remove any dark areas, it is necessary that the area illuminated by a lamp should overlap the area illuminated by an adjacent lamp. For that space-height ratio is important i.e.

$$\text{Space-Height Ratio (SHR)} = \frac{\text{Space between lamps}}{\text{Height of lamps above working plane}}$$

SHR is kept about 1.5 for dispersion type reflectors, fluorescent and direct fitting lamps. It is taken as 1 for lamps with concentrating reflector [2].

EXAMPLE: An office room 6 m × 10 m is illuminated by 15 single lamps luminaires. Each lamp output is 4800 lumens.

$$MF = 0.9, \qquad UF = 0.5$$

Calculate the average illuminance (E):
We know

$$E = \frac{F \times UF \times MF}{A}$$

$$E = \frac{15 \times 4800 \times 0.5 \times 0.9}{6 \times 10}$$

$$= 540 \text{ lux}$$

19.16 Road Lighting

The design of road lighting depends upon choice of lamps, luminaires and their spacing, mounting height, overhang, lateral placement and control schemes (see IS: 1944).

19.16.1 Choice of Lamp

Incandescent lamps are generally used for residential streets when initial cost is to be kept low. These are not usually employed in traffic routes. Mercury Mixed Lamp (MLL) may be employed for modernizing an installation to obtain higher levels without the need for ballasts. High Pressure Mercury Vapour lamp (HPMV) have higher luminous efficiency and longer life than MLL lamp and are suitable for installations where colour rendering is less important and high illumination levels are needed. Mercury halide lamps are improved versions of HPMV lamps and having very much higher efficiencies in the order of 80 Lm/W combined with good colour characteristics. Tubular fluorescent lamps have high luminous efficiency and long life. They are suitable for installations where colour rendering is important. The choice between high pressure mercury vapour lamps and fluorescent tubes in general determined by local considerations of aesthetics and cost of installation. Low pressure sodium vapour lamps are convenient when colour rendering is not important and high luminous efficiency is desired. Their colour is sometimes useful to provide visual guidance and is particularly suitable under foggy conditions. High pressure sodium vapour lamps are improved versions of low pressure sodium vapour, with efficiency of the order of 100 Lm/W with color rendering satisfactory and of dimensions suited to fittings of small size and accurate light control.

19.16.2 Luminaires

IS: 10322 Part 5/Sec. 3–1987 specify the following types of luminaires for public lighting:

(i) Cut-off luminaire This luminaire light distribution is characterized by a rapid reduction of luminous intensity in the region between 80° and the horizontal and should not exceed 10 cd per 1000 lm of flux from the light sources and the intensity at 80° is of 30 cd per 100 lm. The direction of the maximum intensity may vary, but should be below 65°. The principal advantage is the reduction of glare and its use is favoured under the conditions of matt carriageway surfaces, absence of buildings and presence of large trees on road sides, and few intersections and curves.

(ii) Semi-cut-off luminaire Its light distribution is characterized by a less severe reduction in the intensity in the region 80° to 90°. The intensity at the horizontal should not exceed 50 cd per 1000 lm of flux from the light sources and the intensity at 80° is of the order of 1000 cd per 1000 lm. The direction of the maximum intensity may vary, but should be below 75°. The main advantage is a greater flexibility in siting and its use is favoured under the conditions of smooth carriageway surfaces, buildings close to carriageway, and many intersections and curves.

(iii) Non-cut-off luminaire A luminaire whose luminous intensity in directions making an angle equal to or greater than 80° from the downward vertical is not reduced materially and the intensity of which at the horizontal may exceed the values specified for the semi-cut-off distribution, but should not nevertheless exceed 1000 cd. Non-cut-off luminaires are permissible only when a certain amount of glare may be accepted and when the luminaires are of large size and of reduced brightness. In certain cases, they have some advantages in increasing the illumination of facades.

19.16.3 Placement

(i) Spacing It inversly effect the illumination level. A luminaire light distribution and/or the coefficient of utilization are available from manufacturers catalogue. Typical values are shown in Fig. 19.13. If lamp lumen, mounting height, width of road/paved area and spacing between lighting poles are known, the average lux on road surface is given by:

$$\text{Average lux} = \frac{\text{Lamps lumen} \times \text{Coefficient of utilization} \times \text{Mtc. factor}}{\text{Spacing} \times \text{Width of road}}$$

Maintenance factor is taken generally as 0.8

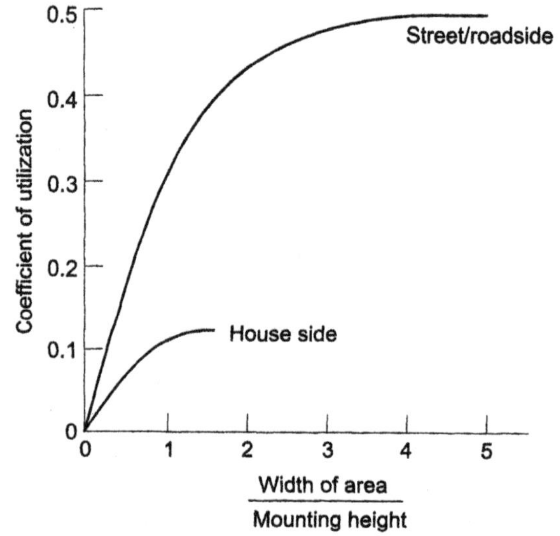

Fig. 19.13 *Typical coefficient of utilization of street light luminaire*

(ii) Mounting height and overhang Glare effect from the street lamps depend on the mounting height and increases with the power of the lamp directed towards the eye and decreases with increases in height range from 7 to 10 m, the higher values being preferred for important metropolitan roads. Highmast lighting is economical at large intersections.

Overhangs on the lighting poles would keep the poles away from the pavement edges, but still allow the lamp to be held above the kerb or towards the pavement. This enables better distribution of light on the pavement and less glare on eyes of road users.

(iii) Lateral placement For free movement of traffic, Indian Road Congress has specified the following horizontal clearances:

For roads with raised kerbs	Minimum 0.3 m and desirable 0.6 m from the edge of raised kerb.
For road without raised kerbs	Minimum 1.5 m from the edge of carriageway, subject to a minimum of 5.0 m from the centre line of the carriageway.

19.16.4 Layout

For straight roads, single side lighting is economical to install; but is suitable only for narrow roads. Due to cost consideration, even in two-lane roads often single side lighting is adopted. For wider roads with three or more lanes, the staggered system or the central lighting system may be adopted.

Special care should be taken while locating the lights on curves. Lights are installed at closer spacing on curves than on straights. The lights are located on the outer side of the curve to provide better visibility. At vertical summit, curve lights should be installed at closer intervals near the summit.

At intersections, due to potential conflicts of vehicular and pedestrian traffic, more illumination is required. For simple intersections in urban area, the illumination should be at least equal to the sum of the illumination values for two roads which form the intersection.

EXAMPLE A street 15 m wide is to be put up with lighting lamps each of 6000 lumens output at the mounting height 7.5 m. Calculate the spacing between light points to produce average 6 lux illumination.

Solution

$$\text{Ratio} = \frac{\text{Street width}}{\text{Mounting height}} = 15/7.5 = 2$$

From Fig. 19.13, coefficient of utilization

$$= 0.44$$

Assuming maintenance factor

$$= 0.8$$

Then required spacing

$$= \frac{6000 \times 0.44 \times 0.8}{6 \times 15} = 23.2 \text{ m}$$

19.16.5 Street Lighting Controls

(i) Automatic control The light switching contactors may be controlled automatically by use of any auto-control devices, such as:

- Photoelectric: achieved by installing suitably mounted photoelectric switches near the control points. The photoelectric switch should be mounted so as to be free from the glare caused by headlights of motor vehicles and protected from the weather.
- Time switches, by which local electrically operated contactors may be controlled through time switches. The time switches may be electrically operated and provided with microprocessor control for seasonal variation adjustment.
- Electronic control switches: These are installed on individual lamp unit and equipped with suitably pre-programmed electronic microprocessor units with yearly seasonal adjustment. These devices are used to switch off-on unwanted lights after peak hour traffic is over.

  (i) Ripple control: This may be achieved by injecting audio frequency impulse through suitable power supply network and installing suitable sensors at control points (see Section 5.4.2).

  (ii) Multiple lamp luminaires: On main city roads, the luminaires adopted could have each two or three lamps, each lamp controlled either by different phase or by separate ripple control signals. In case of necessity, either, one-or two-phase could be switched off or alternately, alternate lamps be switched off by suitable ripple control signals. According to traffic needs, either three or two or one phase are switched on by microprocessor controlled time switch. That is, all the three lamps may be on at peak traffic time, two lamps be in the circuit during medium traffic and then onward time of night, only one lamp will be in the circuit. Also, each phase is rotated on monthly basis, to have uniform burning life of lamps of each phase.

  (iii) Single-lamp luminaires phasing: In this scheme, the street light is given a three-phase, four-wire supply. Every third lamp is put on the same phase. For the first four hours i.e. 6 pm to 10 pm in winter or 7 pm to 11 pm in summer, a three-phase supply is given. This creates high illumination to meet the safety requirements of high traffic during these hours. For the next three hours, the need being essentially for security,

$2/3^{rd}$ of the lights are switched on by switching off one phase. Thereafter, for the rest of the night, another phase is switched off with only $1/3^{rd}$ of the lights on. The phase rotation and switching automation may be carried out as given in (ii) above. With this scheme, nearly 40 per cent energy can be saved over continuous three-phase supply for the whole street light. This scheme has been adopted by the Haryana Urban Development Authority.

19.17 Flood Lighting

Buildings, open spaces like playground, gardens, etc., are flood lighted for aesthetic or functional purpose. Buildings are flood lighted in main direction of view. If the surroundings and background of the building are dark, a relatively small amount of light is needed to make the building lighter than the background. If there are other buildings in the close vicinity, their lighted windows will give a strong impression of brightness. More light will then be needed for the flood-lighting if it is to have any impact. The same is true if, in addition, the background is also bright. Another solution can be found in the creation of a colour contrast instead of brightness contrast. The advantage can be taken of any expense of water in the foreground, such as lake, river or canal. The lighted building will be reflected in the surface of the water, which serves as a black mirror.

Illuminance on a surface is calculated by the point-by-point method. The inverse square law and cosine rule are used to calculate the illuminance at a point on a surface as shown in Fig. 19.14 (a). The illuminance can be expressed as follows:

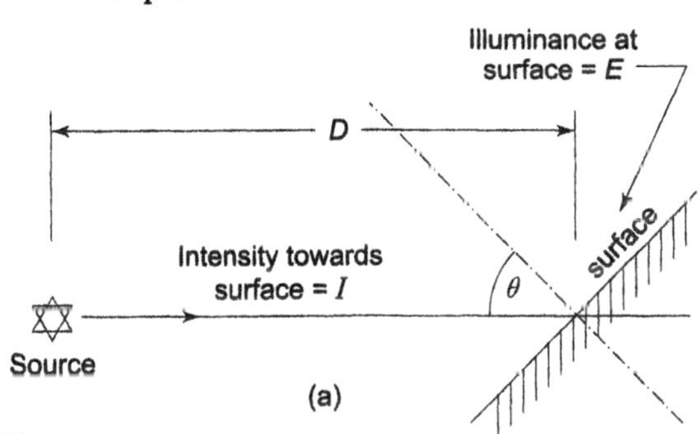

Fig. 19.14 *Flood light illumination:*
(a) At point beneath the source

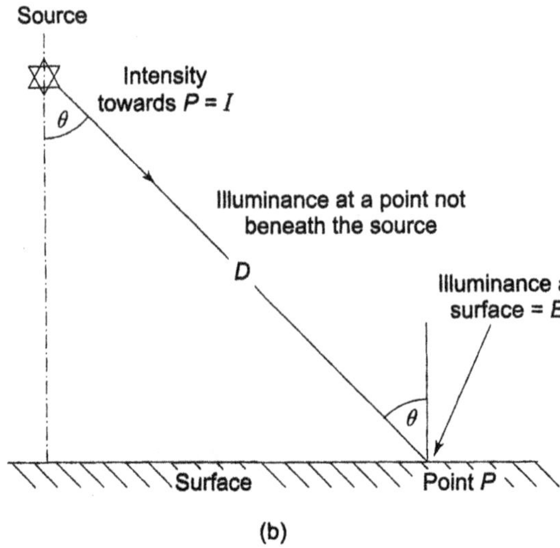

Fig. 19.14 *Flood light illumination:*
(b) At point not beneath the source

$$E = \frac{I \cos \theta}{D^2}$$

where E = illuminance at the point.

I = luminous intensity of the source towards the point (available from luminaire distribution curves)

D = distance from source to point

θ = angle between the line of the incident light and a line at right angle to the surface at the point, i.e. a normal.

When the point on the surface is displaced to one side of a light source [see Fig. 19.14 (b)] the trigonometry involved becomes a little more complex, because of the need to calculate the oblique distance D. For simplicity, it is easier to express D in terms of H and the equation then becomes:

$$E = \frac{I \cos^3 \theta}{H^2}$$

H = mounting height of the source above P.

θ = is equal to the angle between the vertical through the source and the line from the source to P, and can be determined by trigonometry or by drawing to scale and using a protractor. Knowing the angle, the luminous intensity value I may be read from the distribution curve of luminaire.

The lumen method (see Sec. 19.16) can be used to find average illuminance i.e.:

$$E = \frac{F \times MF \times BF}{A \times WLF}$$

where E = average illuminance of surface (Lux)
F = luminous lux of a lamp-fitting
A = area per fitting (m^2)
MF = maintenance factor generally ranging from 0.8 to 0.9
BF = beam factor is used for utilization factor. It is ratio of

$$\frac{\text{Beam luminous lux}}{\text{Lamp luminous lux}}$$

and generally ranges from 0.25 to 0.55
WLF = waste light factor depends upon the dimension of building, etc. It varies from 1.2 for square building to 1.5 for the rectangulars.

19.18 Automatic Fire Alarm System

Automatic fire alarm system is essential to save life and property from growing incidence of fire in high-rise buildings, hospitals, power stations, etc. As per National Building Code–1983, automatic fire alarm system is necessary for apartment building having height beyond 30 m and other buildings higher than 15 m or stories more than five. By this system, the sources of the fire is immediately recognized and precisely located by means of fully automatic operated fire detectors. The detectors are arranged in the ceiling, cable galleries, air-conditioning, ventilation or venting system. These register fire alarms, automatically alert the local fire services in case of fire, close fire doors, open smoke vents and shut down ventilators and air-conditioning plants, computer systems, etc. Thus, it automatically controls fire protection facilities [4], which are essential for rapid and effective fire-fighting. It should be connected to a master alarm unit of the public fire alarm system. The fire brigade can be automatically alerted. Parallel panels can be connected at other locations to signalize the fire, for example, in caretaker's room, or at the fire engine access point. The typical layout is shown in Fig. 19.15. On every detector or in its immediate vicinity should be clearly marked to which line and alarm area, it is assigned. Typical examples are shown in Figs. 19.16 and 19.17.

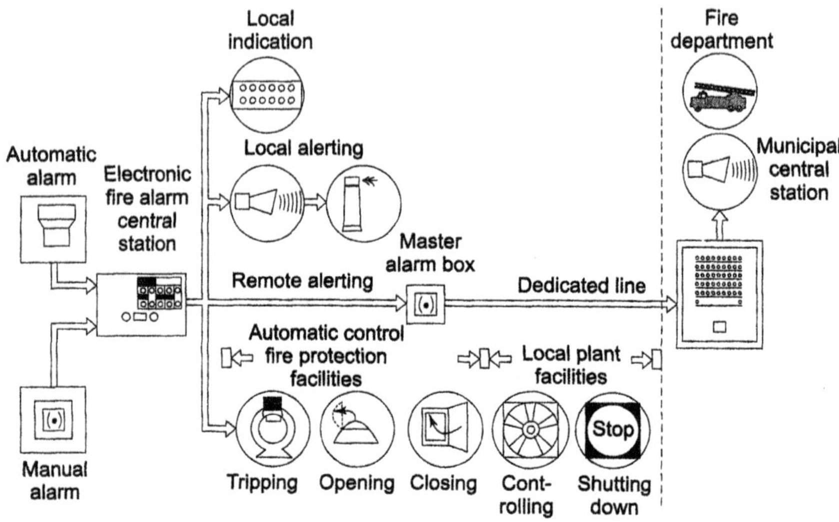

Fig. 19.15 *Layout of a fire alarm system*

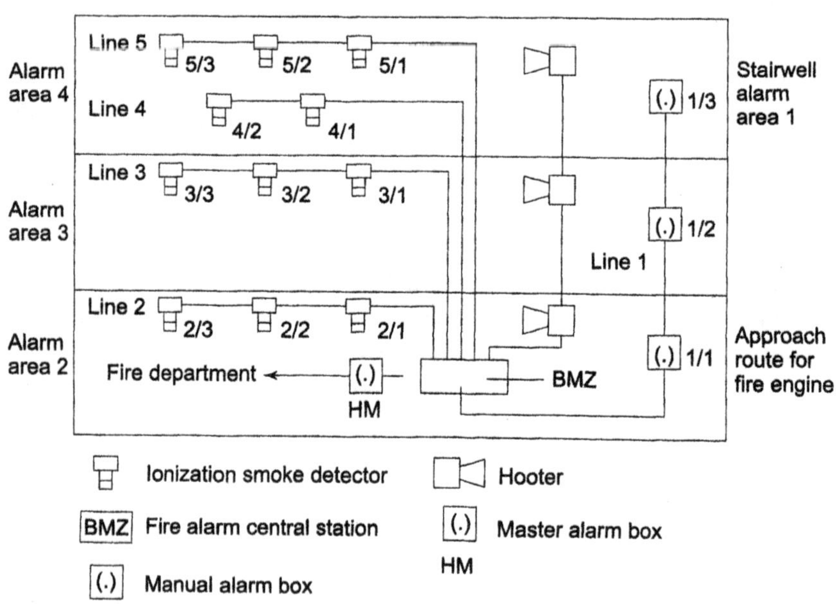

Fig. 19.16 *Subdivision of a protected area into alarm areas and lines*

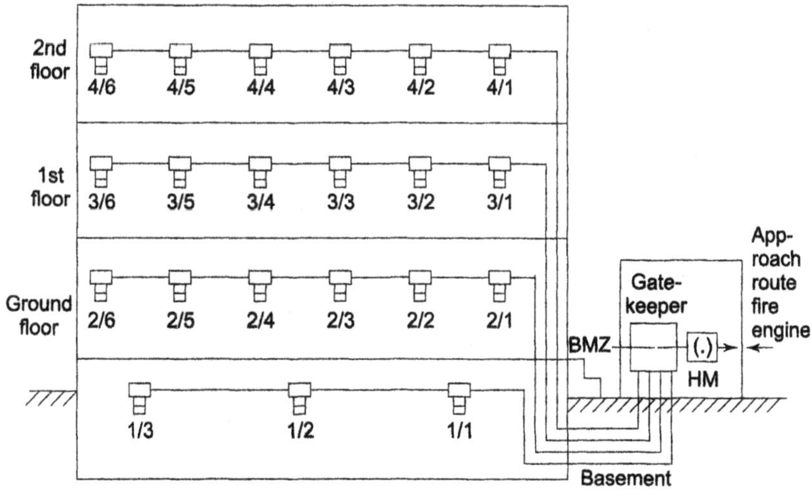

Fig. 19.17 *Typical automatic fire alarm system*
Legend: 1/1: 1st number = line number
2nd number = detector number in line
BMZ: fire alarm central station
HM: master alarm box

19.18.1 Fire Detectors

The detectors are of three types:

(i) Heat detectors (IS: 2175–1988)
- Point or spot type detectors
- Line type detectors

These may be of fixed temperature detector or rate of rise detector.

(ii) Smoke detectors:
- Optical detectors
- Ionization chamber detectors
- Chemical sensitive detectors

(iii) Flame detectors.

19.18.2 Wiring

The equipment and wiring of the fire alarm system should be independent of any other equipment wiring, and should be spaced at least 5 cm away in metallic conduits. No line/zone should have more than 200 fire detectors connected together. Copper conductor, mineral insulated aluminium sheathed wires or PVC fire survival cables are preferred. The sources of supply for the alarm system should be battery with trickle charging and emergency charging (see IS: 1646–1997).

FRLS Cables

PVC, elastromeric and XLPE cables with FRLS (Fire Resistant Low Smoke) characteristics, are flame-retardant and do not propagate fire in addition to low smoke. Apart from FRLS cables, Fire Survival (FS) cables are being manufactured in the country, which guarantee service operation during the fire. These cables will survive fire, even when exposed to a flame temperature of 1,100 degree celsius for three hours, and will operate all the circuits for its rated voltage and thus, proves to very useful for certain essential control and emergency circuits (see IS: 9968 and 6380 and 5831).

19.19 Electrical Call Bell Services

As per IS: 8884–1978, the electric call bell system are:

- Simple Call Bell: It is used for dwelling and small offices as shown in Fig. 19.18 (a).
- Multiple Call Bell: It is meant for hostels, hospitals or similar large buildings, where call points are numerous as given in Fig. 19.18 (b).
- Time Bell: It is used in factories, schools, etc.

19.20 Other Services

19.20.1 Telephone

In offices, colonies, industrial units, hospitals, hotels, banks, apartment complexes, guest houses, there is usual need to lay internal communication system. Generally few direct lines are taken from telephone exchange of Telephone Department and connected to internal system by suitable local exchange. Effective communication with increased efficiency is possible by virtue of Electronic Private Automatic Branch Exchange (EPABXs). These are noiseless and

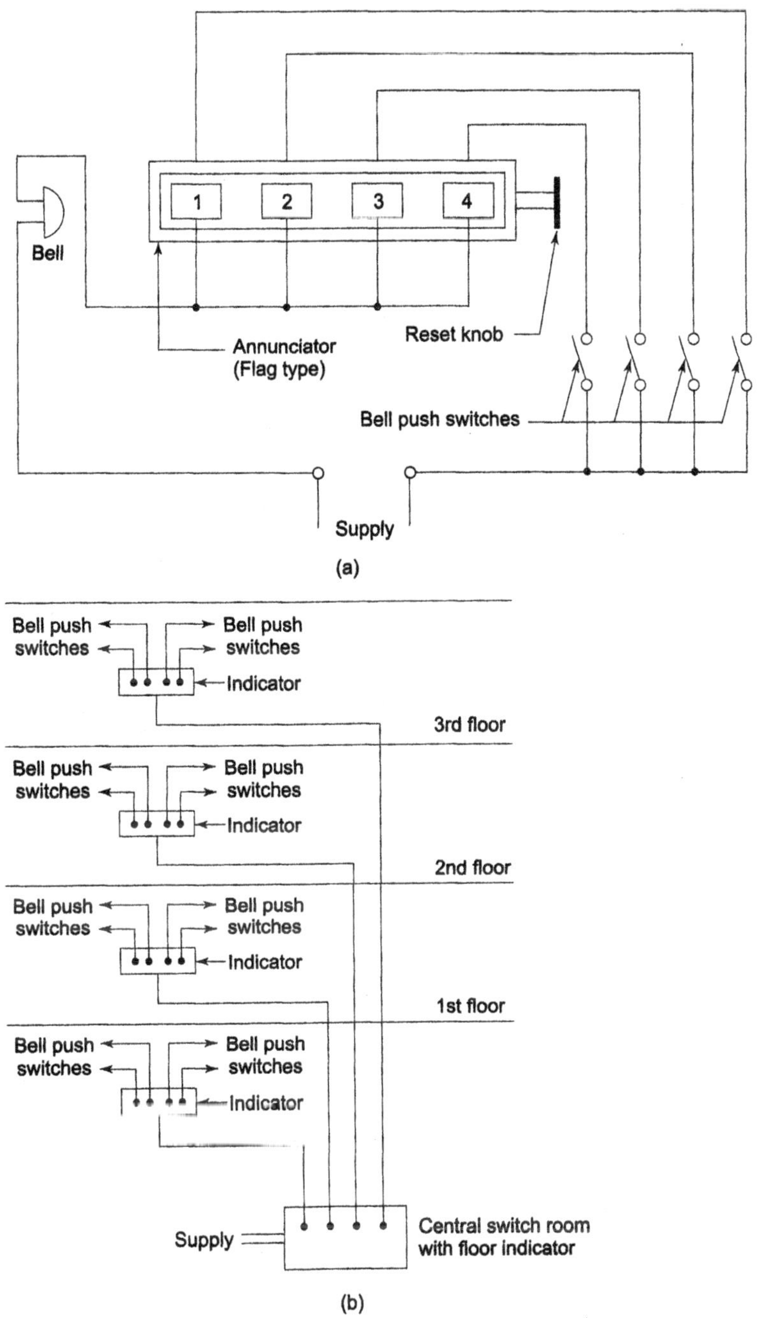

Fig. 19.18 *Call-bell system: (a) Simple call-bell system (b) Multiple call-bell system*

virtually maintenance free. It does not require a separate room for air conditioning. Main features of an EPABX system are:

System features, Extension features, Operator features. Their details are given in Table 19.6.

The internal telephone system may be underground or overhead. For underground system, plastic insulated petroleum jelly filled screened paired cables are used along with the distribution pillars.

Table 19.6

Detailed features of EPABX

System	*Extension*	*Operator*
• Class-of-service restriction	• Extension to extention	• Call queing
• Automatic fault location	• Extension to operator	• Call waiting
• Different tones	• Automatic call back	• Call intercept
• Busy over ride	• Call forwarding	• Emergency reporting
• Line lockout	• Call transfer	• Trunk offering
• Group hunting	• Consult on hold	• Position busy
• Voice/data integration	• Conferencing	• Night answering service
• Privacy of calls	• Paging	
• Direct outward dialling	• Paging access	
• STD barring facility	• Call-pick-up	
	• Fax access	
• Voice over internet protocol		
• Remote dialling		
• Calling Line Identification (CLI) facility		
• Loop dialling		
• Absent message texts		

19.20.2 Television and Address/Music System

Proper conduiting/wiring at construction stage of building can be made for provision of Television Master Antenna System, a cable TV network for flats, building complexes, etc., and address system/music system for mosques, gurdwaras, art houses, concert halls, etc.

19.20.3 Intelligent Buildings

Intelligent buildings [6] are designed for efficiency and productivity. Sensors, monitors and controls optimise energy consumption and are also designed to reduce waiting time for services while allowing convenient control of work environment. Along with usual facilities like lighting, power, lifts, elevators, etc. the other services likely are EPABXs, facsimile, xerox machines, telex, telephone, internet, shared computers, dish antennas, cable TV, security surveillance TV systems, building management equipment, etc. Wiring development in fibre optic cables and associated equipment offer powerful advantages over coaxial or metallic wire links due to their ability to meet almost any need from small point-to-point data links to large local area networks efficiently and at lower cost. One example of the ring or loop LAN is shown in the Fig. 19.19. Generally, intelligent buildings have rooftops made of photovoltaic tiles with separate dc wiring to supply clean dc power for electronic equipment in the building. One of the strengths of intelligent buildings is the ability to sense when the building is occupied or empty and switch on/off the supporting services as per the programming. This approach minimises energy demand, consumption and maintenance cost [1].

Due to growing crime, intelligent fences are becoming important for supermarket stores, banks, production units, dwelling houses, etc. Modern security systems using micro-processors operating round the clock are desirable. These range from a total integrated system with closed circuit television (CCTV) cameras monitoring every angle of a premises to a simple but effective keyless door. Passive infrared detectors (PIR) sensing heat energy emitted by the human body and raising an alarm, through a control box are most suitable for dwelling houses.

19.21 Lightning Protection

The purpose of lightning protection is to divert a lightning discharge, which might otherwise strike a vulnerable part of the structure to be protected.

19.21.1 Principle

During lightning strike, the negatively charged downward stepped leaders of a lightning stroke are attracted to areas of positive charge at

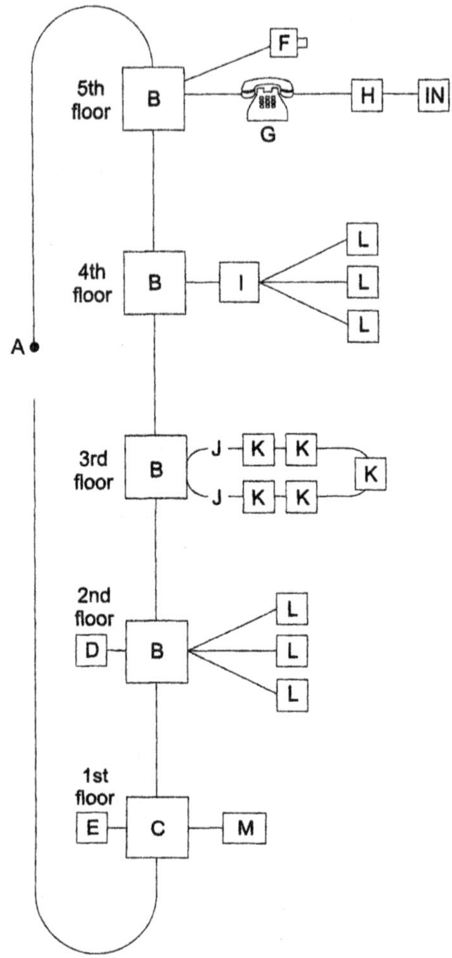

Fig. 19.19 *Fibre optic Loop Local Area Network;*
Legend: A–Fibre optic Loop, B–Slave MUX, C–Master MUX, D–EPBAX, E–LAN Monitor/Controller, F–CCTV, G-Telephone, H–Facsimile, I–PC, J–Sub Loop, K–Personal Computers, L–Computer Terminals, M–Host Computer, IN–Internet

the ground. A correctly installed lightning protection system provides the best possible discharge point of positive ions from the ground to meet the descending stroke. Once the two have met, the system then allows the return stroke to flow with a minimum amount of damage. The process of an upward streamer leaving the lightning protection system to meet a downward stroke is important and gives rise to the zone of protection concept. The length of upward streamer is referred to as the "striking distance". This distance increases with the severity of the stroke and is related to the intensity of the subsequent return stroke. The problems associated with a low intensity stroke and high structures are important, as the striking distance could be short enough not to attract any of the stepped leaders. These can enter under the zone of protection and hit the sides of tall structures.

19.21.2 Practice

Any lightning protection system consists of: (i) air termination, (ii) down conductor, (iii) earth electrodes. The typical systems are given in Table 19.7. Protection guidance should be obtained from IS–2309. "The code of practice for the protection of buildings and allied structures against lightning." The main considerations of protection are: (i) The use to which the structure/building is put, (ii) Type of construction, (iii) Degree of isolation; (iv) Type of country (flat, hilly, mountainous), (v) Frequency of lightning in the area.

Table 19.7

Typical protection systems

Type of building	*Protection*
(i) Building with explosive dust or flammable vapour risk	Internal mounted system with vertical air terminals 1.5 m high and horizontal air terminals spaced 3 to 7.5 m from each other depending on the type of storage and process involved
(ii) Explosive storage buildings and explosive workshops	Integrally mounted system with vertical air terminals 0.3 m high and horizontal air terminals spaced 7.5 m
(iii) Small explosive storage buildings	Vertical pole type air terminal
(iv) Building storing more dangerous type of explosive such as NG	Suspended horizontal air termination at least 2 m higher than the structure and with a spacing of 3 m
(v) Other high rise buildings	Vertical air terminals

19.22 Standby Supplies

The essential loads need standby supplies, i.e. generating set or batteries to enable continuity of supply in the event of failure of mains. Manual or automatic double-throw change–over switches are installed in service room to ensure that two sources of supply are not connected simultaneously. The alternative sources of supply may be provided by battery continuously trickle charged from the electric mains. Self-contained emergency lighting are readily available in market, comprising of a lamp, battery with charger and control equipment.

Uninterruptible Power Supplies (UPS) are important for computer systems, data terminals, traffic control systems, security system, communication equipment, etc. Generally, UPS from 0.3 to 10 kVA rating are single-phase and 10 kVA to 300 kVA rating are three-phase. Metal oxide semiconductor field effect transistors (MOSFETs) are used in the inverter circuit for high efficiency. UPS are classified into three broad categories:

- Offline
- Online
- Line interactive

An offline UPS provides an alternative source of power in case of power failure or high voltage fluctuations. The time taken for change-over to battery power should not take more than 4 milli-seconds so that the system does not detect any change in case of power failure. An on-line UPS [8] offer voltage regulation within ± 3 per cent and frequency variation ± 0.1 per cent and proven well regulated sine-wave output. The general scheme of UPS is shown in Fig. 19.20. The standard UPS sets ranging from 0.3 to 300 kVA are available as per IS: 7204.

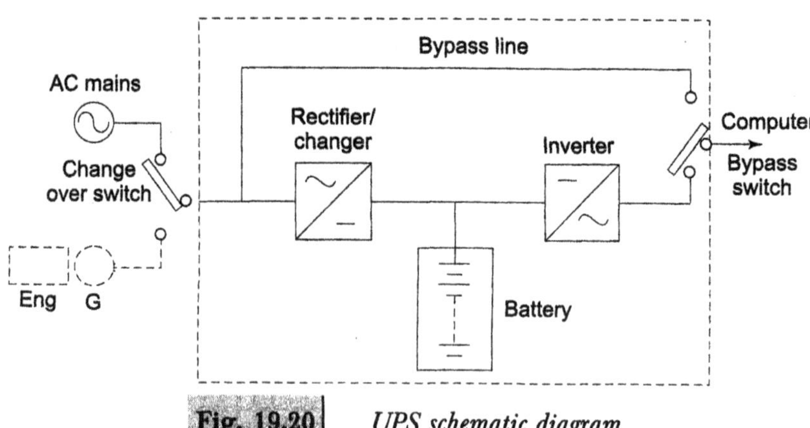

Fig. 19.20 *UPS schematic diagram*

The standby supplies are provided for:

- Multistorey buildings exceeding 24 m height for lighting in common areas, namely corridors staircases, lift lobbies, entrance hall, common toilets, etc.
- Assembly institution buildings.
- Fire lift, fire fighting and control system.
- Security lighting; military establishment.
- Obstruction lights.
- Water supply pump, etc.

19.23 Voltage Stabilizers (5)

19.23.1 Constant Voltage Transformers (Ferroresonant)

The basic constant voltage transformer has a primary winding connected to the mains supply and a loosely coupled secondary winding, which is resonated at mains frequency by a large capacitor. The magnetic core linking the primary and secondary works near to magnetic saturation. Some types will stabilize the rms voltage to ± 1 per cent, but at the same time distort the waveform by up to 20 per cent. Others will stabilize to ± 0.5 per cent with a maximum distortion of 5 per cent. For computer equipment, transient attenuation can be achieved, but this is at the expense of accuracy, which is usually reduced to ± 3 per cent.

19.23.2 Tap Changers

The tap changer operates automatically by selection voltage taps on an auto-transformer, according to the input mains voltage as measured by an electronic monitoring and control unit. The taps are selected either electromechanically or by solid-state devices. Electro-mechanical tap changers give output in steps and switching occurs at a pre-set input voltage. Some electro-mechanical tap changers switch in synchronism with the supply waveform to minimize contact arcing. Relays do not switch over instantaneously and arcing cannot be totally eliminated. In the solid state tap changer, four or more triaces are connected in series with the output taps of the auto-transformer. Because the triaces switch only at the zero (crossover) points of the ac waveform, no radio frequency interference is generated and no waveform distortion introduced.

19.23.3 Electro-mechanical Stabilizers

The electro-mechanical type of stabilizer uses a motor drive variable transformer to provide the automatic control action. A second fixed transformer has its secondary winding connected in series with the line wire, between the incoming ac mains supply and the stabilizer output terminal to the load. The primary winding of this fixed transformer is fed with power from the variable transformer, while the motor drive is controlled by a solid state sensing circuit monitoring the voltage across the load. Any error in the stabilizer output voltage, relative to a set reference voltage, causes the motor to drive the variable transformer in such a direction that produces a correcting voltage in the secondary of the fixed transformer. Being in series with the mains supply, this secondary voltage adds to or subtracts from the mains voltage, according to the correction required.

This type of stabilizer has high efficiency with losses below 2 per cent at full-load, and is capable of sustaining high surge currents. Its accuracy is better than ± 0.5 per cent from zero to full-load and this is not affected by load changes. Waveform distortion is negligible and the action is not dependent on the mains frequency.

19.23.4 Saturable Reactors

In the saturable reactor, or ac transductor stabilizer, the incoming mains voltage is first stepped up by an auto-transformer. Then the auto-transformer output is applied across a potential divider formed by twin windings of a saturable reactor, the output tapped off from this divider becoming the ac output of the stabilizer to the load. A third control winding on the saturable reactor varies the voltage tapped off from the potential divider in accordance with the load voltage error, relative to a reference voltage. The important features of this type of stabilizer are a fast speed of response (3 to 15 mains cycles, depending on type), an accuracy of 0.3 per cent, insensitivity to mains frequency variation, high efficiency (better than 95 per cent) and capability to sustain high surge currents.

19.23.5 Switch-mode Power Supplies

In the rapidly expanding world of micro-electronics, such as computers, televisions, etc., there is growing need for a high-efficiency power supply. This need is met by the switch-mode power supply (SMPS), in which the output power is not regulated continuously, but pulsed at a relatively high frequency. An output filter is included for smoothing the supply voltage. It is essentially a dc-dc converter. The output voltage is measured, compared to a reference, and kept constant by controlling the duty factor of the drive signal applied to the power switches (i.e. transistors). Additional supply voltages are fairly simple to implement in a SMPS with the aid of appropriate auxiliary windings on the secondary of the transformer. In computer equipment, the usual combination is power 5 V supply, and auxiliary 12 V supplies.

Three basic methods of voltage control are:

(i) Direct duty factor control The error signal is amplified, and drives a pulsewidth modulator, which in turn adjusts the duty factor as required.

(ii) Voltage feed forward This is the most commonly used system. Pre-regulation of the duty factor is implemented as a function of the input voltage, enabling the output voltage of the open loop system to be made independent of the input voltage. The control circuit is, therefore, only required for compensating load fluctuations.

(iii) Current mode control A second control circuit (inner loop) inside the voltage control circuit (outer loop) enables switching off the power transistor at a more or less fixed peak value of the current. The effect so obtained is the (quasi) disappearance of the inductor from the output filter. The whole system is then essentially a first order network, with the capacitor in the output filter as the only phase changing element. The stability of the whole supply is excellent.

PROBLEMS

1. (a) What is the permissible voltage drop as per the standard within domestic consumer installation?
   (b) A three-phase, four-wire, 415/240 V system supplies single-phase non-inductive loads of 10 kW, 8 kW and 5 kW connected as shown in Fig. 19.21. Find the line and neutral currents.

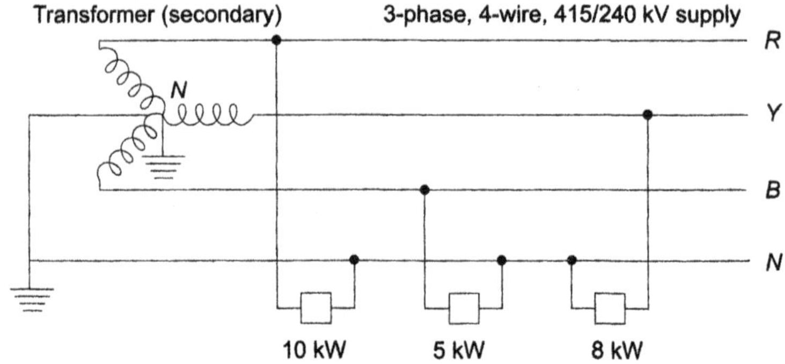

Fig. 19.21 *Supply system with three loads*

2. A consumer installation has an 11/0.415 kV, 500 kVA transformer having primary and secondary winding resistance of 0.42 ohms and 0.0019 ohms respectively. The no-load (core) loss is 2.9 kW and the power factor of the load is 0.8 lagging. What is the load at which the efficiency of the transformer will be the maximum and what will be the maximum efficiency?
3. The phase currents in a delta-connected three-phase load are as follows:
   (a) Between red and yellow lines is 30 A at power factor 0.707 leading;
   (b) Between yellow and blue lines is 20 A at power factor unity; and
   (c) Between blue and red lines is 25 A at power factor 0.866 lagging.
   Calculate the line currents.
4. Which earthing systems are attributed to the following? Draw the diagrams.
   (a) An overhead line supply with no earth,
   (b) A multi-core supply cable with separate neutral and earth,
   (c) A supply cable, in which the functions of earth and neutral are performed by one conductor?
5. What is the characteristic impedance of a cable? Why should its value be low for power supply to sensitive electronic loads?

BIBLIOGRAPHY

1. Park Alan, *Facilities Management*, Macmillan, London, 1994, p. 111.
2. Tripathy, S.C., *Electric Energy Utilisation and Conservation*, Tata McGraw Hill, New Delhi, 1991, pp. 73–105.
3. Steward, W.E. & T A. Stubbs, *Modern Wiring Practice—Design and Installation*, Newness, U.K., 1995.
4. Fire Alarm Systems, *Siemens Handbook*, Siemens Aktiengesellschaft, Munchen, Edition 8/79, pp. 10–61.

5. *Middle East Electronics Journal*, March, 1983, pp. 37–38.
6. *Asian Review of Business and Technology*, October, 1989, p. 40.
7. *Electricity International*, vol. 2, no. 2, U.K., February, 1990, p. 29.
8. Griffith, C. David, *Uninterruptible Power Supplies—Power Conditioners for Critical Equipment*, Marcel Dekker Inc., New York, 1989.
9. Lerch Dr. E. and Dr. R.K. Aggarwal., Importance of System Studies for Industrial Power System Networks, *Siemens Circuits*, October 1995, pp. 12–19.
10. *Asian Electricity*, vol. 16, no. 9, November, 1998, p. 28.
11. *Wiring Rules*, *AS/NZS 3000:2000*, Standards Australia, 2000, pp. 132–143.
12. Fitt Brian and Joe Thornley, *Lighting by Design: A Technical Guide*, Focal Press, Oxford, 1992.

Appendix I

International Electricity Supplies

Alternating current is normally distributed either through 3-phase wye (star) or delta (triangle), 4-wire secondary distribution systems. In the wye or star distribution system, the nominal voltage examples are 208/120, 220/127, 380/220 and 400/230. The higher voltage is 1.732 (the square root of 3) times the lower voltage. In a delta or triangle system, 220/110 and 460/230 are examples of nominal voltages. The higher voltage is always double the lower voltage. The higher voltage is obtained by using 2- or 3-phase wires while the lower voltage is the voltage between the neutral wire (mid-point) and one-phase wire. The higher voltage may used for big loads while the lower voltage is used primarily for lighting and for small appliances. The nominal supply voltages and frequencies for consumers around the world are given in Table A-1.1

Table A-1.1

Supply frequency and voltage

Country	*Frequency (Hz)*	*Supply voltage levels in common use in V otherwise mentioned in kV*
Afghanistan	50, 60	380/220; 220
Algeria	50	10 kV; 5.5 kV; 380/220; 220/127
Angola	50	380/220
Argentina	50	13.2 kV; 6.88 kV; 390/225; 380/220; 220

Contd.

Country	Frequency (Hz)	Supply voltage levels in common use in V otherwise mentioned in kV
Australia	50	22 kV; 11 kV; 19.1 kV and 12.7 kV (SWER); 6.6 kV 440/250; 415/240; 240
Austria	50	20 kV; 10 kV; 5 kV; 400/230
Bahamas	60	240/120; 208/120
Bahrain	50; 60	400/230; 110
Bangladesh	50	11 kV; 400/230; 230
Belgium	50	15 kV; 6 kV; 400/230
Bhutan	50	11 kV; 400/230; 230
Brazil	50, 60	13.8 kV; 11.2 kV; 380/220; 220/127
Bulgaria	50	400/230
Canada	60	27.6/16 kV; 12.5/7.2 kV; 600/347; 240/120; 208/120 600; 480; 240
Chile	50	230
China	50	380/220
CIS	50	20 kV; 10 kV; 6 kV, 3 kV 380/220; 220/127
Colombia	60	220/110; 208/120
Croatia	50	10 kV; 6.6 kV; 380/220; 220
Cuba	60	220/110
Cyprus	50	240
Czech Republic	50	230
Denmark	50	30 kV; 10 kV; 400/230
Dubai	50	6.6 kV; 380/220; 220
Egypt	50	11 kV; 6.6 kV; 380/220; 220
Ethiopia	50	380/220; 220
Fiji	50	11 kV; 415/240; 240
Finland	50	660/380; 400/230
France	50	20 kV; 15 kV; 400/230
Ghana	50	230
Germany	50	20 kV; 10 kV; 6 kV; 400/230; 220, 127

Contd.

Country	Frequency (Hz)	Supply voltage levels in common use in V otherwise mentioned in kV
Greece	50	22 kV; 20 kV, 15 kV; 6.6 kV; 380/220
Guatemala	60	240/120
Hong Kong	50	11 kV; 3.3 kV; 380/220
Hungry	50	20 kV; 10 kV; 380/220; 220
Iceland	50	380/220
India	50	22 kV; 11 kV; 440/250; 415/240; 400/230; 460/230 230
Indonesia	50	380/220; 220/127
Iran	50	20 kV; 11 kV; 400/230; 380/220; 220
Iraq	50	11 kV; 6.6 kV; 3 kV; 380/220; 220; 415/240
Ireland	50	400/230
Israel	50	22 kV; 12.6 kV; 6.3 kV; 400/230; 230
Italy	50	20 kV; 15 kV; 10 kV; 400/230; 220/127; 220
Japan	50, 60, 0	33 kV; 400 V d.c. 22 kV; 6.6 kV; 210/105; 200/100; 100
Jordan	50	380/220; 220
Kenya	50	415/240; 240
Kuwait	50	415/240; 240
Laos	50	400/230
Lebanon	50	380/220; 190/110; 220; 110
Libya	50	400/230; 220/127; 230; 127
Malaysia	50	11 kV; 400/230; 230
Mauritius	50	400/230; 230
Mexico	50	13.8 kV; 13.2 kV; 480/227; 220/127; 220; 120
Mongolia	50	220/127
Myanmar	50	11 kV; 6.6 kV; 400/230; 230
Nepal	50	11 kV; 400/230; 230

Contd.

Country	Frequency (Hz)	Supply voltage levels in common use in V otherwise mentioned in kV
Netherlands	50	10 kV; 3 kV; 380/220; 220
New Zealand	50	11 kV; 415/240; 400/230; 440; 240; 230
Nigeria	50	230
North Korea	60	200/100; 220
Norway	50	20 kV; 10 kV; 5 kV; 380/220; 230
Oman	50	240
Pakistan	50	400/230; 230
Panama	60	240/120; 220/110
Peru	50; 60	220/110
Philippines	60	20 kV; 13.8 kV; 6.2 kV; 4.16 kV; 3.6 kV 240/120; 220/110
Poland	50	400/230
Portugal	50	220
Qatar	50	415/240; 240
Romania	50	20 kV; 10 kV; 6 kV; 400/230
Saudi Arabia	50, 60	380/220; 220/127; 127
Singapore	50	22 kV; 6.6 kV; 400/230; 230
Slovakia	50	22 kV; 15 kV; 6 kV 3 kV; 380/220; 220
Somalia	50	230; 220/110
South Africa	50	11 kV; 400/230
South Korea	60	380/220, 100
Spain	50	15 kV; 11 kV; 380/220; 400/230
Sri Lanka	50	11 kV; 400/230; 230
Sudan	50	230
Sweden	50	20 kV; 10 kV; 6 kV; 400/230
Switzerland	50	16 kV; 11 kV; 6 kV; 400/230
Syria	50	220/115
Taiwan	60	22.8 kV; 11.4 kV; 380/220; 220/110
Tajikistan	50	380/220
Tanzania	50	11 kV; 400/230

Contd.

Country	Frequency (Hz)	Supply voltage levels in common use in V otherwise mentioned in kV
Thailand	50	22 kV; 380/220; 220
Turkey	50	15 kV; 6.3 kV; 380/220; 220
Turkmenistan	50	380/220
UAE	50	11 kV; 6.6 kV; 440/230, 415/240; 380/220
Uganda	50	11 kV; 415/240; 240
Ukraine	50	380/220
United Kingdom	50	22 kV; 11 kV; 6.6 kV; 3.3 kV; 415/240; 400/230
USA	60	20.8 kV; 19.9 kV; 14.4 kV; 13.2 kV; 12.47 kV; 12 kV; 7.2 kV; 4.8 kV; 4.16 kV; 2.4 kV; 480/277; 460/265; 240/120; 208/120
Venezuela	60	220/120
Vietnam	50	15 kV; 380/220; 208/120; 220; 120
Yemen Arab Republic	50	220
Yemen Democratic Republic	50	220
Zaire	50	220
Zambia	50	400/230; 230
Zimbabwe	50	380/220

Appendix II

Conductors and Cable Data

Table A-2.1

All Aluminium (standard) conductor (AAC)
IS: 398 (Part 1): 1996

Nominal aluminium	Stranding and wire diameter	Sectional area	Approximate overall dia	Approximate mass	Calculated resistance at 20°C maximum	Approximate calculated breaking load	Approximate current carrying capacity at 45°C ambient temperature
(1) mm^2	(2) mm	(3) mm^2	(4) mm	(5) kg/km	(6) ohms/km	(7) kN	(8) (A)
25	7/2.21	26.85	6.63	74	1.096	4.52	–
50	7/3.10	52.83	9.30	145	0.5525	8.25	189
100	7/4.39	106.0	13.17	290	0.2752	15.96	295
150	19/3.18	150.9	15.90	415	0.1942	23.28	–
240	19/3.99	237.6	19.95	654	0.1235	35.74	500
300	19/4.65	322.7	32.25	888	0.09107	48.74	608

Note: (i) The resistance (R_T) of the conductor at any other temperature (T) will be:

$R_T = R_{20} [1 + 0.00403 (T - 20)]$

where R_T = resistance at temperature T°C

R_{20} = resistance at temperature 20°C given in the Table.

(ii) Ampere rating is given for conductor working temperature 75°C.

(iii) For all Aluminium Alloy Conductors (AAAC) refer IS: 398 Part IV, and Table A-2.3

Table A-2.2

Aluminium Conductors, Galvanized Steel-reinforced (ACSR) IS: 398 (Part 2): 1996

Nominal aluminium	Stranding and wire diameter Aluminium	Steel	Sectional area of aluminium	Total sectional area	Approximate diameter	Approximate mass	Calculated resistance at 20°C, maximum	Approximate calculated breaking load	Approx. current carrying capacity at 45°C ambient
(1) mm^2	(2) mm	(3) mm	(4) mm^2	(5) mm^2	(6) mm	(7) kg/km	(8) ohm/km	(9) kN	(10) (A)
10	6/1.50	1/1.50	10.60	12.37	4.50	43	2.780	3.97	–
18	6/1.96	1/1.96	18.10	21.12	5.88	73	1.618	6.74	–
20	6/2.11	1/2.11	20.98	24.48	6.33	85	1.394	7.61	107
30	6/2.59	1/2.59	31.61	36.88	7.77	128	0.9289	11.12	139
50	6/3.35	1/3.35	52.88	61.70	10.05	214	0.5524	18.25	193
80	6/4.09	1/4.09	78.83	91.97	12.27	319	0.3712	26.91	250
100	6/4.72	7/1.57	105.0	118.5	14.15	394	0.2792	32.41	300
150	30/2.59	7/2.59	158.1	194.9	18.13	726	0.1871	67.34	398
200	30/3.00	7/3.00	212.1	261.5	21.00	974	0.1390	89.67	482
400	42/3.50	7/1.96	404.1	452.2	26.88	1281	0.07311	88.79	–
420	54/3.18	7/3.18	428.9	484.5	28.62	1621	0.06868	130.32	736
520	54/4.13	7/3.53	528.5	597.0	31.77	1998	0.05595	159.60	835
560	42/4.13	7/2.30	562.7	591.7	31.68	1781	0.05231	120.16	–

Note: (i) The resistance (R_T) of the conductor at any other temperature (T) will be:

$$R_T = R_{20} [1 + 0.00403 (T - 20)]$$

where R_T = resistance at temperature T°C

R_{20} = resistance at temperature 20°C given in the Table.

(ii) Ampere rating is given for conductor working temperature 75°C.

(iii) For Aluminium Conductors Aluminized steel reinforced, refer IS: 398, Part III.

Table A-2.3

All Aluminium alloy conductors (AAAC)
IS: 398 (Part IV): 1994

Actual area	*Stranding and wire dia*	*Approx. overall dia*	*Approx. mass*	*Calculated maximum at 20°C*	*Approx. calculated breaking load*
mm^2	mm	mm	kg/km	Ohms/km	kN
(1)	(2)	(3)	(4)	(5)	(6)
15	3/2.50	5.39	40.15	2.3040	4.33
22	7/2.00	6.00	60.16	1.5410	6.45
34	7/2.50	7.50	94.00	0.990	10.11
55	7/3.15	9.45	149.20	0.6210	16.03
80	7/3.81	11.43	218.26	0.4250	23.41
100	7/4.26	12.78	272.86	0.3390	29.26
125	19/2.89	14.45	342.51	0.2735	36.64
148	19/3.15	15.75	406.91	0.2290	43.50
173	19/3.40	17.00	474.02	0.1969	50.54
200	19/3.66	18.30	549.40	0.1710	58.66
232	19/3.94	19.70	636.67	0.1471	68.05
288	37/3.15	22.05	794.05	0.1182	84.71
346	37/3.45	24.15	952.56	0.0984	101.58
400	37/3.71	25.97	1101.63	0.0829	117.40
465	37/4.00	28.00	1280.50	0.0734	136.38
525	61/3.31	29.79	1448.39	0.0651	146.03
570	61/3.45	31.05	1573.71	0.0598	158.66
604	61/3.55	31.95	1666.00	0.0568	167.99
642	61/3.66	32.94	1771.36	0.0534	178.43
695	61/3.81	34.29	1919.13	0.0492	193.25
767	61/4.00	36.00	2115.54	0.0446	213.01

Note: The resistance (R_T) of the conductor at any temperature (T) will be:

$$R_T = R_{20}\,[1 + 0.0036\,(T - 20)]$$

where R_T = resistance at temperature T°C

R_{20} = resistance at temperature 20°C given in the Table.

Table A-2.4

Voltage regulation—400 V system uniformly distributed load (Figs. 12.4 and 12.5)

Conductor (ACSR) mm^2	km-kVA for line regulation at 0.85 pf (temperature rise 30°C), 70°C conductor temp.			
	5%	6%	7%	8%
20	6.05	7.43	8.89	10.40
30	8.60	10.55	12.60	14.80
80	16.75	20.60	24.60	28.75
100	20.00	24.50	29.20	34.50

Table A-2.5

Voltage regulation—11 kV overhead lines (Fig. 12.6)

Conductor (ACSR) mm^2	km-MVA for line regulation at 0.85 pf (temperature rise 30°C), 70°C conductor					
	5%	6%	7%	8%	9%	10%
20	3.63	4.41	5.21	5.95	6.78	7.63
30	5.17	6.27	7.38	8.52	9.67	10.84
50	7.60	9.24	10.85	12.56	14.30	15.99
80	10.23	12.40	14.60	16.85	19.22	21.56
100	12.43	15.08	17.73	20.50	23.32	26.20

Note: For phase to phase and phase to neutral system, voltage regulation will be respectively 0.5 and 0.28 times the values given in this table. See Section 14.10.3(c).

Table A-2.6

Voltage regulation—33 kV sub-transmission lines (temperature rise 30°C) 70°C conductor temp.

Conductor (ACSR) mm^2	km-MVA for line regulation at 0.85 pf					
	5%	6%	7%	8%	9%	10%
50	67.50	81.61	96.26	111.21	126.46	142.11
80	90.86	110.21	129.89	149.87	168.60	187.33
100	109.10	132.23	155.27	177.46	199.64	221.82
150	141.95	170.66	199.11	227.55	256.00	284.44

Table A-2.7

Conductor (ACSR) mm^2	Voltage regulation km-MVA for 66 kV and 132 kV line at pf of 0.85					
	5%	6%	7%	8%	9%	10%
66 kV lines						
80	418.295	501.954	585.613	669.292	752.931	836.590
100	501.310	601.572	701.834	802.096	902.358	1002.620
150	642.675	771.210	899.745	1028.280	1156.815	1285.350
132 kV lines						
100	1532.035	1838.442	2144.849	2451.663	2757.663	3064.070
150	1920.675	2304.810	2688.945	3073.080	3457.215	3841.350
420	2262.670	2715.204	3167.738	3620.272	4072.806	4525.340

Table A-2.8

Wind load and pressure

Conductor types and sizes	Wind load in kg/metre run of conductor Wind regions of pressure			
	50 kg/m^2	75 kg/m^2	100 kg/m^2	150 kg/m^2
AAC, 7/2.11 mm	0.221	0.331	0.442	—
AAC, 7/3.10 mm	0.310	0.465	0.620	—
ACSR,7/2.11 mm	0.211	0.331	0.422	0.633
ACSR,7/2.59 mm	0.259	0.388	0.518	0.777
ACSR,7/3.35 mm	0.335	0.502	0.670	1.005

Notes: (a) Rural Electrification Corporation (India) has divided the country into three wind pressure zones, viz. 50, 75 and 100 kg/m^2 zones.

(b) For designing structures up to 11 kV system, IS: 5613/1970 specified maximum wind pressure of 150 kg/m^2 for poles and conductors for coastal areas.
(c) In accordance with IE Rule-76, the wind pressure specified by the respective state governments is to be adopted as the basis of designing the structures.
(d) The relation of Pressure (P) and wind velocity (V) as per IS: 5613 is: $P = 0.006\ V^2$.
(e) Refer IS: 802.

Table A-2.9

Line reactances

Line rating	Conductor (ACSR) mm^2	Inductive resistance Ω/km	Line equivalent spacing
132 kV	100	0.464	6000 mm
	150	0.426	
	420	0.414	
33 kV	100	0.362	1200 mm
	150	0.325	
	420	0.312	
11 kV	20	0.3915	910 mm
	30	0.3820	
	50	0.3720	
	80	0.3600	
	100	0.3530	
LT (240/415 V)	20	0.3356	378 mm
	30	0.3278	
	50	0.3185	
	80	0.2906	
	100	0.2837	

Table A-2.10

Nominal average surge impedances (see Sec. 13.7)

Cable: LT & 11 kV	30–40 Ω
Overhead line: LT & 11 kV	~ 400 Ω
Distribution transformer:	~ 3000 Ω
Steel pole:	~ 300 Ω

Note: Surge impedance loading of line = $(V_{line})^2/Z_{surge}$

Table A-2.11

Standard percentage impedances of three-phase transformers

kVA rating range	*Percentage impedance*
1 to 630	4.5
630 to 1250	5
1251 to 3150	6.25
3151 to 6350	7.15
6351 to 12,500	8.33
12501 to 25,000	10

Table A-2.12

Paper insulated, aluminium conductor, armoured, 3-core cables conforming to IS: 692

Nominal cross-sectional area (mm^2)	*Current rating A*		*Conductor a.c. resistance at 65°C ohms per/km*	*Approx. inductive reactance per phase at 50 Hz ohms/km*	*Approx. capacitance per phase micro F/km*	*Voltage drop V/km/A*
	Direct in ground	In air				
(i) 6.35/11 kV belted type:						
35	84	80	1.03	0.107	0.345	1.79
95	155	155	0.378	0.094	0.480	0.67
150	190	200	0.244	0.089	0.565	0.44
185	220	230	0.196	0.087	0.605	0.36
240	250	275	0.150	0.085	0.710	0.30
300	280	310	0.121	0.084	0.750	0.25
400	340	385	0.0956	0.081	0.870	0.21
(ii) 19/33 kV H-type (screened):						
95	155	160	0.378	0.139	0.276	0.70
150	190	210	0.243	0.128	0.339	0.47
185	220	240	0.195	0.124	0.365	0.38
240	245	285	0.149	0.119	0.405	0.32
300	270	320	0.119	0.115	0.443	0.28
400	310	380	0.0937	0.111	0.494	0.24

Table A-2.13

Estimated ac current ratings (A) for cables/wires: (i) PVC 250/440 volts grade single-core aluminium conductor (IS: 694-1990)

Nominal X-section area, mm^2	No. of cables/wires Bunched and enclosed or trunking		No. of cables/wires Clipped direct to a surface or on a cable tray, bunched and unclosed		Defined condition	
	2	4	2	4	Flat or vertical 2/4 cables	Trefoil 3 cables
1.5	12	11	16	14	—	—
2.5	17	15	21	19	—	—
4.0	23	19	27	25	—	—
6	29	24	35	32	—	—
10	40	33	48	43	—	—
16	54	46	65	58	—	—
25	69	58	84	74	—	—
35	83	71	105	90	—	—
50	105	89	125	115	—	—
70	125	110	150	135	165	145
95	—	—	185	165	205	175

Cable Rating Factors:

The cable rating given in the above 'Tables' are based on standard conditions given in Sec. 12.2.2 For actual conditions, suitables multiplying factors be applied for arriving at actual current carrying capacity. These factors are usually given in the manufacturer's catalogue. As a general guide, the rating factors are given in Table A. 2.14.

Table A-2.14

Cable rating multiplying factor

(i) 250 V, 440 V, 650/11 kV Grade PVC cables							
(a) *Temp. °C ambient*	15	20	25	30	35	40	45
For ambient air	—	—	1.25	1.18	1.09	1.0	0.9
For ambient ground	1.17	1.12	1.06	1.0	0.94	0.87	0.79
(b) *Depth of laying (cm)*	75	90	105	120	150	180 or more	
For cable conductor size =< 25 mm^2	1	0.99	0.98	0.97	0.96	0.95	

Contd.

Cable rating multiplying factor

For cable conductor size > 25 mm^2 = < 300 mm^2	1	0.98	0.97	0.96	0.94	0.93
For cable conductor size > 300 mm^2	1	0.97	0.96	0.95	0.92	0.91

(c) *Soil thermal resistivity*

(°C-cm/W)	100	120	150	200	250	300
For conductor						
Size: 1.5 to 6 mm^2	1.1	1.05	1	0.92	0.86	0.81
Size 10 to 50 mm^2	1.15	1.08	1	0.91	0.84	0.77
Size 70 to 150 mm^2	1.17	1.09	1	0.90	0.82	0.76
Size 180 mm^2 and above	1.19	1.10	1	0.89	0.81	0.75

(d) *Rating factors for excess current protection*

Class-Q cartridge fuse	0.714
Class-Q semi-enclosed fuse	0.625
MCB or MCCB	1.0

(ii) *High voltage XLPE cables*

(a) *Depth of laying (cm)*

	Voltage grade	
	Up to 11 kV	*22 and 33 kV*
71 to 90	1.00	1.02
91 to 110	0.98	1.00
111 to 130	0.96	0.97
131 to 150	0.95	0.95

(b) *Ambient temp, °C*	15	20	25	30	35	40	45	50
For ambient air	1.27	1.22	1.17	1.11	1.06	1.00	0.95	0.92
For ambient ground	1.12	1.07	1.04	1.00	0.96	0.91	—	—

(c) *Thermal resistivity of soil (°C cm/W)*	100	120	150	200	250
Rating factor	1.19	1.10	1.00	0.88	0.80

(iii) *High voltage PILC cables*

(a) *Depth of laying (cm)*

	Voltage grade	
	11 kV	*22 kV and 33 kV*
90	1.00	—
105	0.99	1.00
120	0.98	0.99
150	0.96	0.97
180 or more	0.95	0.96

(b) *Ambient temp, °C*	15	20	25	30	35	40	45
For air ambient	—	1.30	1.21	1.10	1.00	0.88	—
For ground ambient	1.20	1.13	1.07	1.0	0.93	0.85	0.76

Contd.

(iv) (a) For voltage drop calculations, the change in resistance with temperature is determined by factor.

$$K_T = 1 + a\,(T_C - T_O),$$

where a = 0.00303 for copper

a = 0.004 for aluminium

T_C = Max. conductor temperature allowed (°C)

T_O = 20°C standard temperature

(v) *Group of cables in close proximity*

No reduction in rating is necessary if

- The horizontal clearance between cables in air or ground is more than twice the overall diameter of individual cables.
- The vertical clearance between cables in air is more than four times the diameter of an individual cable and
- Single core cables are laid in trefoil arrangement with the clearances as above

The above guidelines are thumb rules applicable for laying two or three multi-core/single-core cables. For higher number of cables in a group, there will be significant reduction in the rating of a cable in the group. For example, 10 cables laid direct in the ground in horizontal formation will de-rate the cable in the group by a factor of 0.65. Therefore, the manufacturer's recommendation or catalogue must be consulted to ascertain the rating of a cable in a group.

(vi) The rating in air is based on that the cables are shielded from direct rays of sun and in case of direct sun, the rating be suitably reduced.

(vii) Cables installed in single way ducts, the duct diameter should not be less than twice the diameters of cables. Control and power cables be laid in separate ducts/trenches.

(viii) Minimum clearances desirable: (as per IS: 1255–1983)

Power cable to control cable	200 mm
Power cable to communication cable	300 mm
Power cable to power cable	300 mm
Power cable to water or gas main	300 mm

(ix) The cables can be laid manually or by winch pulling or by motor driven rollers.

Power Cable Laying Merits

Table A-3.1

Method of cable laying (see IS: 1255–1983)

Merit	*Direct in ground*	*In duct*	*On racks in air*	*In tunnel/trench*
Application	General application for all site conditions and types of distribution	Normally for short-runs for crossing railway, road, etc.	In factories, power stations, etc.	In sub-station, stations switch yards, control rooms, etc.
Preferred cable finish	Armoured with serving	Unarmoured	Armoured or unarmoured	Armoured or unarmoured
Conductor current carrying capacity	Very good	Poor	Good if sun-shields are provided	Fair if well ventilated
Initial preparatory work before laying	Nominal	Most significant	Significant	Most significant
Initial cost of laying	Comparatively low	Higher	High	Highest
Cost of repair and maintenance	Highest	High	Fair	Low
Time requirement of locating a fault	Significant	Most significant	Little	Little
Cost of cable replacement for load growth	High	Low	Lower	Lowest

Contd.

Merit	Direct in ground	In duct	On racks in air	In tunnel/trench
Susceptibility to electrolytic/ galvanic corrosion	Considerable	Infrequent	Little	Little
Susceptibility to cable sheath failure by inter-crystalline fatigue	Little	Little	Considerable	Considerable
Susceptibility to mechanical damage	Considerable	Infrequent	Little	Negligible
Damage to adjacent cables following a fault in a cable	Rare	Huge	Considerable	Considerable
Special precautions necessary	Special care may be required in aggressive site conditions: white ants, salts etc.	• Care against abrasion during pulling and damage at duct entry • Proper alignment of duct • Provision for water drainage	• Protection to cable at clamp position • Provision to reduce thrust at joints, if any	• Protection to cable at clamp position • Provision to reduce thrust at joints position • Provision for water drainage, ventilation

Appendix IV

Village Electrification and Pumpset Energization

Table A-4.1

Status of village electrification and pumpset energization in India as on 31-6-2001

States/UTs	*Total no. of villages (as per 1991 census)*	*Percentage villages electrified*	*Number of agriculture pumpsets energized*
Andhra Pradesh	26586	100	1924684
Arunachal Pradesh	3649	60.5	0
Assam	24685	77	3675
Bihar and Jharkhand	67513	71	275062
Chattisgarh	19720	91.7	74065
Goa	360	100	6867
Gujarat	18028	100	699650
Haryana	6759	100.0	421068
Himachal Pradesh	16997	99.3	6241
Jammu & Kashmir	6477	97.5	5621
Karnataka	27066	98.8	1263859
Kerala	1384	100.0	396905
Madhya Pradesh	51806	97.1	1237512
Maharashtra	40412	100	2333961
Manipur	2182	91.7	45
Meghalaya	5484	45.9	65
Mizoram	698	99	0

Contd.

States/UTs	Total no. of villages (as per 1991 census)	Percentage villages electrified	Number of agriculture pumpsets energized
Nagaland	1216	99.7	176
Orissa	46989	75	74625
Punjab	12428	100.0	781976
Rajasthan	37889	94.8	643426
Sikkim	447	100	0
Tamil Nadu	15822	100	1731722
Tripura	855	95.1	2094
Uttaranchal	15681	79.6	17521
Uttar Pradesh	97122	79.3	808534
West Bengal	37910	78.1	111019
UTs	1093	100	38163
Total (all India)	587258	86.53	12858636

Note: A village is treated as electrified, only if electricity is used in inhabitated locality within the revenue boundry of the village for any purpose.

Appendix V

Electrical Accidents

A-1 Accident Incidence

Every year, about 12,000 electrical accidents to human or animal lives are reported in the country, as per Rule 44A of the Indian Electricity Rules, 1956. The situation is more grave as a significant number of accidents are not reported. Major cause of accidents are:

- Non-adherence to specified procedure/practice of construction and maintenance of distribution system
- Lack of adequate supervision
- Non-enforcement of the regulations in respect of use and maintenance of proper safety equipments
- Lack of adequate training in safety precautions for line staff
- Ignorance of the users/common man about the hazards of electricity

(ii) Accident statistics for the year 1984 compiled by the International Brotherhood of Electrical workers show that the causes for purely electrical accidents to workers can be grouped as follows:

Accidents	*Cause*
36%	Lineman did not use personal protective equipment.
29%	Lineman did not provide adequately safety clearance/ insulation/guarding to energized equipment/line.
14%	Lineman did not follow the proper work procedure.
7%	Lineman worked on de-energized lines without adequate temporary protective grounds.

14% Miscellaneous causes, such as tool contacting energized conductor, lineman having poor work concentration due to some mental worries/depressions, etc.

(iii) The proportion of fatal to non-fatal accident in India is 1 to 3, which is very high as compared to advanced countries, (U.S.A., U.K., Australia, Canada) where it is about 1 to 6. This shows the first-aid to accident persons is lacking in the country. That is mainly due to lack of proper training of technical staff.

(iv) On an average, about one-half the number of total human being electrical accidents pertains to animals. The animals are more accident prone, because the effective body resistance is low as compared to a person. First, the animals are affected under extremely adverse conditions, either in damp premises where the floor is conductive or just in the field, where the soil may be damp and, hence, conductive. Second, the value of a step voltage determined by the distance between the forelegs and hindlegs of an animal is greater than the greatest possible distance between the extremities of a human being. Third, there are two routes of current through the body of animals and that is, from leg to leg and from nasal cavity to legs.

A-2 Statutory Provisions

Indian Electricity Rules, 1956 provide the safety precautions regarding (a) Protection: over current and earth fault current and short-circuit, (b) Circuit isolation: switching, caution of double supply and different voltages at the point, proper sectionalizing, redundant lines, (c) Earthing: guarding, use of ELCB as per IS: 12648/1988, (d) Insulations: stay wire insulating, adequate clearance and proper span lengths, (e) Practices: earthing, proper T & P, first-aid, periodical inspection and testing. IS–5216 and IS Handbook SP–31/1986 lay down general safety guidance of life saving techniques, safety instructions and training, working on electrical works in hazardous areas.

A-3 Safety Precautions

Lack of safety procedure may cause fire, accidents such as not using safety belts and falling from overhead pole; coming in contact with snapped overhead conductor; and coming in contact with cables, conductors, joints or apparatus, which were wrongly presumed as dead; leakage of current through poles, stay wire, neutral broken, cut-out, motor, switches and domestic appliances due to bad maintenance; having a person electrocuted and then trying to save the electrocuted person without insulation.

Proper design and maintenance of system could reduce accidents. For example, on the LT lines where low earth resistance is not maintainable, it is advisable to provide neutral line with insulators at supports to avoid accident hazards. Stay wire should always be insulated. Proper clearance of lines as IE rules should be kept.

Staff should be trained to observe safety precautions and procedures including switching out, isolating and earthing of the circuit and issuing of permits-to-work. Before the permits are issued, the dead circuit must be identified and clearly marked. The permit-to-work should not be regarded as means of evading responsibility, but of properly exercising it. In this respect, it is not sufficient to assume that the circuit is dead because it is isolated and earthing must be ensured at both ends; and at the point of working to guard against induced voltage. Hand gloves be used while using earthing rods for temporary earthing.

Public should be educated through public media for safety, such as: (a) Every electrical appliance connected to mains should be treated as live and used carefully; (b) Cattle be not tied with electric poles or stay wire; (c) Wires not be tied to the supports carrying electric lines for drying clothes; (d) Every electricity point and appliances should be suitably earthed. Three-pin plug and sockets be used. The earth should not be connected to water mains or pipes. (e) Whenever electric leakage is detected or shocks felt, the defective wire or appliance should immediately be got checked by a competent person, (f) Fuse is for safety from hazards and therefore only proper capacity fuse may be used, (g) Shock treatment charts should be displayed in the premises by the industrial and agriculture consumers; (h) Hay and wheat stocks should never be made under live electric wire; (i) Any sparking or glowing observed at the poles should be reported to the nearest electricity complaint centre; (j) During rainy season, the metal supports as steel poles, rails, etc., should not be touched.

Appendix VI

Illumination Levels

Illumination Levels: IS-National Electrical Code 1985

Table A-6.1

Installations	Illumination level (lux)
Factory Areas (see IS: 6665)	
Canteens	150
Stores, stairs, corridors, etc.	100
Outdoor stockyards, car parks, internal factory roads	20
Assembly inspection shops:	
Rough work, for example, frame assembly, etc.	150
Medium work, for example, machined parts, etc. Fine work;	300
for example, radio and telephone equipment, etc.	700
Very fine work, for example, assembly of very small precision mechanisms and instruments	1500
Minute works, for example very small instruments	3000
Chemical works: Boiling tanks, driers, crystallizers, filtration plants, bleaching, extractors and electrolytic cells	150
Control rooms:	200–300
Electricity Generating Stations	
Turbine, generator halls	200
Auxiliary equipment, battery rooms, blowers, auxiliary generators, switchgear and transformer chambers	100
Boiler houses (including operation floors), platforms, coil conveyors, pulverizers, feeders, precipitators, soot and slag blowers	70 to 100

Contd.

Installations	*Illumination level (lux)*
Basements	70
Control rooms: Control desks and panels	300
Rear of control panel and switch houses	150
Nuclear reactors and steam plant	150–200
Outdoor coal yards	20
Conveyors	50
Fuel oil delivery headers and oil storage	50
Platforms, boiler and turbine desks	50
Transformers and outdoor switchgear	100
Laboratories and Test Rooms:	
General laboratories and balance rooms	300
Electrical and instrument laboratories	450
Banks:	
Counter, typing and accounting book areas	300
Public areas	150
Garages:	
Parking areas (interior)	70
Washing and polishing, greasing general servicing and pits	150
Repairs	300
Cinemas/Theatres:	
Foyers	150
Auditoria	50
Corridors	70
Stairs	100
Assembly and Concert:	
Foyers and auditoria	100–150
Platforms	450
Corridors	70
Stairs	100
Libraries:	
Shelves (stacks)	70 to 150
Reading rooms (newspapers and magazines)	150 to 300
Reading tables	300 to 700
Cataloguing, sorting and stock rooms	150 to 300
Offices:	
Entrance halls and reception areas	150
Conference rooms and executive and general offices	300
Business machine operation	450

Contd.

Installations	*Illumination level (lux)*
Drawing boards and tracing	450
Corridors and lifts	70
Stairs	100
Telephone Exchanges:	
Manual exchange rooms (on desk)	200
Main distribution frame room	150
School and Colleges:	
Assembly halls:	
General	150
When used for examinations	300
Platforms	300
Museums	150
Art Galleries:	
General	100
Paintings	200
Hospitals: (see IS: 43477)	
Receptions and waiting rooms	150
Operation theatre surgeries:	
General	300
Tables/chairs	Special lighting
Laboratories	300
Corridors	70
Stairs	100
Sight testing (acuity) wall charts and near vision types	450
Wards:	
General	100
Beds	150
Radiology department	100
Casualty and out-patient department	150
Hotels/Restaurants:	
Entrance halls	150
Reception and accounts	300
Dining room	50–100
Bedrooms:	
General	100
Dressing tables, bed heads, etc.	200
Writing tables	300
Corridors	70
Stairs	100

Contd.

Laundries	200
Kitchens	200
Goods and passenger lifts	70
Clock-rooms and toilets	100
Bathrooms	100
Shops/show rooms	150-300
Homes:	
Kitchens	200
Bathrooms	100
Stairs	100
Garage	70
Sewing	700
Homework and sustained reading	300
Bedroom	300
Lounge room	100
Living room	300
Dining room	150
Flood Lighting of Building/Monuments:	
With poor-lit surroundings	25
With bright-lit surroundings	300
Road Lighting-Average Level:	
Group as per IS–1944 (I)	
Group A_1 (Fast traffic roads) (MH: 9–10 m)	30
Group A_2 (Mixed traffic roads) (MH: 9–10 m)	15
Group B_1 (Secondary road for fast traffic) (MH: 7.5–9 m)	8
Group B_2 (Secondary road for mixed traffic) (MH: 7.5–9 m)	4
Group C (Secondary road for medium traffic)	6
Group D	
Junctions	30
Foot bridge	6
Group E	
(MH: 10–15 m)	20
Car parks	10

Note: MH = mounting height; The ratio of minimum to average illuminating is 0.40: The illumination on main city roads can be varied according to traffic prevailing at different times of night.

Appendix VII

Optimum Specific Energy Consumption

Better house-keeping, modern competitive technology, relevant co-generation/captive generation can reduce specific electric energy consumption per unit of industrial production. The figures given in Table A-7.1, are signals to reduce specific energy consumption further. This data is also important for strategic energy forecasting and energy audit.

Table A-7.1

Optimum Specific Electric Energy Consumption

Industry	*Specific energy consumption*
Steel rolling mill	30–40 kWh/tonne of steel
Arc furnace	500–530 kWh/tonne of steel
Induction furnace	470–500 kWh/tonne of steel
Aluminium	1300 kWh/tonne of aluminium produced
Paper mill	700–800 kWh/tonne of paper
Textile mill	1300–1400 kWh/tonne of cotton/polyster cloth
Woollen mill	1500–1800 kWh/tonne of woollen cloth
Roller flour mill	50–65 kWh/tonne of wheat handled
Vanaspati mill	240–270 kWh/tonne of ghee
Oil mill	50–60 kWh/tonne of oil seeds
Sugar mill	60–70 kWh/tonne of sugarcane (5000–3000 TCD plants)
Rich sheller	20–25 kWh/tonne of rice

Contd.

Industry	*Specific energy consumption*
Fertilizer plant	15–20 kWh/tonne of fertilizer
Ginning mill	80–90 kWh/tonne of raw cotton
Milk plant	50–55 kWh/tonne of milk handled
Breweries	8–11 kWh/hectoliter ready for sale of beer
Agriculture tubewell	1000–1200 kWh/H.P./year
Cement	(a) 95–120 kWh/tonne of cement for dry and wet process respectively (IS: 125) (b) 87–110 kWh/tonne of cement for dry and wet process respectively (International Standard)
Urea	125 kWh/tonne of urea

Appendix VIII

Tinned Copper Fuse Wire

(1) LT Fuses up to 650 V: Table A–8.1 depictes the fuse wire sizes and their rating as per Indian Standard: 9926–1981 for fuse wires. The current at which the fuse blows (fusing current) will depend upon the construction of fuse carriers and bases (see IS: 2086–1993).

Table A-8.1

Rating, size and allowable resistance

Rated current A	Nominal diameter mm	Tolerance ± t mm	Permissible resistance at 20°C ohms/m	
			Maximum	Minimum
6	0.20	0.003	0.5644	0.5250
10	0.35	0.004	0.1834	0.1730
16	0.50	0.005	0.0898	0.0848
20	0.63	0.006	0.0566	0.0535
25	0.75	0.008	0.0400	0.0376
32	0.85	0.009	0.0311	0.0293
40	1.25	0.011	0.0143	0.0136
63	1.50	0.015	0.0099	0.0094
80	1.80	0.018	0.0069	0.0065
100	2.00	0.020	0.0056	0.0053

(2) H T Fuses

Table A-8.2

Horn gap fuses for H T side of distribution transformers

kVA capacity	22 kV SWG (dia. mm)	11 kV SWG (dia. mm)
25	40(0.122)	38(0.152)
50	38(0.152)	33(0.254)
63	38(0.152)	33(0.254)
100	35(0.213)	32(0.274)
200	32(0.274)	26(0.457)

Appendix IX

Power Factors

Table A-9.1

Natural power factors

Equipment/Installation	*Average natural power factor*
Neon sign	0.5 to 0.55
Window type air conditioners	0.62 to 0.85
Hair dryers	0.7 to 0.8
Liquidiser	0.8
Mixer	0.8
Coffee grinder	0.75
Refrigerator	0.65
Freezer	0.7
Shaver	0.6
Table fan	0.5 to 0.6
Ceiling fan	0.5 to 0.7
Cabin fan	0.5 to 0.6
Exhaust fan	0.6 to 0.7
Sewing machine	0.7 to 0.8
Washing machine	0.6 to 0.7
Radio	0.9
Vacuum cleaner	0.7
Tube light	0.5 to 0.9
Clock	0.9
Electronic equipment	0.4 to 0.95
Mercury vapour lamp	0.4–0.6
Industrial induction motor:	
— No load	0.18
— 25% full load	0.56

Contd.

Equipment/Installation	Average natural power factor
— 50% full load	0.74
— 75% full load	0.81
— 100% full load	0.85
— 125% full load	0.86
Cold storage	0.76 to 0.80
Cinemas	0.78 to 0.80
Metal pressing	0.57 to 0.72
Plastic moulding	0.57 to 0.73
Film studies	0.65 to 0.74
Heavy engineering work	0.48 to 0.75
Rubber extrusion and moulding	0.48
Pharmaceuticals	0.75 to 0.86
Oil and paint manufacturing	0.51 to 0.69
Biscuit factory	0.60
Printing press	0.65 to 0.75
Food processing	0.63
Laundries	0.92
Floor mills	0.61
Gas works	0.87
Oil mills	0.51 to 0.59
Woollen mills	0.70
Potteries	0.61
Cigarette manufacturing	0.80
Foundries	0.59
Structural engineering	0.53 to 0.68
Chemicals	0.72 to 0.87
Municipal pumping stations	0.65 to 0.75
Oil terminals	0.64 to 0.83
Rolling mills	0.60 to 0.72
Irrigation pumps	0.62 to 0.80
Repair shop, automatic lathe, workshop, spinning mill weaving mill	0.6
Welding shop	0.5-0.6
Heat treatment shop, steel works, rolling mills	0.65-0.8
Textile plant	0.65-0.75
Cement plant	0.8-0.85
Office building	0.8-0.85

Note: IS: 7752 (Part–1)1975 specifies average values of the power factor for different types of electrical installations.

Ambient Temperature and Average Altitude

Table A-10.1

Ambient temperature and average altitude of states/territories

State/Territory	*Max. ambient temp. (°C) (IS: 802–1973)*	*Average altitude above M.S.L. (approx. m)*
Andhra Pradesh	45	200
Assam	40	—
Bihar	45	100
Delhi	48	220
Goa	40	—
Gujarat	50	70
Haryana	48	220
Himachal Pradesh	43	2200
Jammu	48	400
Kashmir	—	1600
Karnataka	40	650
Kerala	37	250
Madhya Pradesh	46	370
Maharashtra	45–47	280
Manipur	40	1500
Meghalaya	40	1500
Nagaland	43	1500
Orissa	45–49	100

Contd.

State/Territory	*Max. ambient temp. (°C) (IS: 802–1973)*	*Average altitude above M.S.L. (approx. m)*
Punjab	48	220
Rajasthan	50	340
Sikkim	37	—
Tamil Nadu	43	—
Tripura	40	—
Uttar Pradesh	48	120
West Bengal	43	60
DVC areas	45	

Source: Irrigation and Power, Central Board of Irrigation and Power, New Delhi, April 1979, p. 164. (Also see IS: 5613–Part–1, Section–1–1985 and IS: 802, Part–1, Section–1–1995.)

Note: Reference ambient temperature °C as per IS: 9676/1980 = 1/3 (Max. ambient temperature + max. daily average ambient temperature + max. yearly average ambient temperature + 10)°C.

Appendix XI

Transmission and Distribution Losses in States

Table A-11.1

Transmission and distribution losses during 2000–01

Electricity Board	*T & D Losses (%)*
Andhra Pradesh	31.0
Assam	35.0
Bihar and Jharkhand	22.0
Delhi	45
Gujarat	19.4
Haryana	25.0
Himachal Pradesh	17.3
Jammu & Kashmir	46.5
Karnataka	30.0
Kerala	20.8
Madhya Pradesh and Chhatisgarh	20.5
Maharashtra	16.7
Meghalaya	20.3
Orissa	40
Punjab	16.5
Rajasthan	22
Tamil Nadu	16.5
Uttar Pradesh	25
West Bengal	28

Electricity Board	T & D Losses (%)
Pondichery	13
Arunachal Pradesh	18
Goa	24
Manipur	30
Mizoram	40
Nagaland	29.2
Sikkim	20
Tripura	28

Source: Annual Report on the Working of State Electricity Boards and Electricity Departments, Power and Energy Division, Planning Commission, Govt. of India. June 2001

Index

www.ingramcontent.com/pod-product-compliance
Lightning Source LLC
Chambersburg PA
CBHW071252180926
51613CB00001B/19
9780071447836